Instructor's Solutions Manual

to accompany

Chemistry

Eighth Edition

Raymond Chang
Williams College

Brandon Cruickshank
Northern Arizona University

McGraw Hill **Higher Education**

Boston   Burr Ridge, IL   Dubuque, IA   Madison, WI   New York   San Francisco   St. Louis
Bangkok   Bogotá   Caracas   Kuala Lumpur   Lisbon   London   Madrid   Mexico City
Milan   Montreal   New Delhi   Santiago   Seoul   Singapore   Sydney   Taipei   Toronto

The McGraw·Hill Companies

Instructor's Solutions Manual to accompany
CHEMISTRY, EIGHTH EDITION
RAYMOND CHANG AND BRANDON CRUICKSHANK

Published by McGraw-Hill Higher Education, an imprint of The McGraw-Hill Companies, Inc.,
1221 Avenue of the Americas, New York, NY 10020. Copyright © 2005, 2002, 1998 by
The McGraw-Hill Companies, Inc. All rights reserved.

1 2 3 4 5 6 7 8 9 0 DCD DCD 0 9 8 7 6 5 4

ISBN 0-07-287283-7

www.mhhe.com

CONTENTS

CHAPTER 1
CHEMISTRY: THE STUDY OF CHANGE

1.3   **(a)**   Quantitative. This statement clearly involves a measurable distance.

       (b)   Qualitative. This is a value judgment. There is no numerical scale of measurement for artistic excellence.

       (c)   Qualitative. If the numerical values for the densities of ice and water were given, it would be a quantitative statement.

       (d)   Qualitative. Another value judgment.

       (e)   Qualitative. Even though numbers are involved, they are not the result of measurement.

1.4   **(a)**   hypothesis   **(b)**   law   **(c)**   theory

1.11   **(a)**   Chemical property. Oxygen gas is consumed in a combustion reaction; its composition and identity are changed.

       (b)   Chemical property. The fertilizer is consumed by the growing plants; it is turned into vegetable matter (different composition).

       (c)   Physical property. The measurement of the boiling point of water does not change its identity or composition.

       (d)   Physical property. The measurement of the densities of lead and aluminum does not change their composition.

       (e)   Chemical property. When uranium undergoes nuclear decay, the products are chemically different substances.

1.12   **(a)**   Physical change. The helium isn't changed in any way by leaking out of the balloon.

       (b)   Chemical change in the battery.

       (c)   Physical change. The orange juice concentrate can be regenerated by evaporation of the water.

       (d)   Chemical change. Photosynthesis changes water, carbon dioxide, etc., into complex organic matter.

       (e)   Physical change. The salt can be recovered unchanged by evaporation.

1.13   Li, lithium; F, fluorine; P, phosphorus; Cu, copper; As, arsenic; Zn, zinc; Cl, chlorine; Pt, platinum; Mg, magnesium; U, uranium; Al, aluminum; Si, silicon; Ne, neon.

1.14   **(a)**   K   **(b)**   Sn   **(c)**   Cr   **(d)**   B   **(e)**   Ba
       (f)   Pu   **(g)**   S   **(h)**   Ar   **(i)**   Hg

1.15   **(a)**   element   **(b)**   compound   **(c)**   element   **(d)**   compound

1.16   **(a)**   homogeneous mixture   **(b)**   element   **(c)**   compound
       (d)   homogeneous mixture   **(e)**   heterogeneous mixture   **(f)**   homogeneous mixture
       (g)   heterogeneous mixture

1.21 $\text{density} = \dfrac{\text{mass}}{\text{volume}} = \dfrac{586 \text{ g}}{188 \text{ mL}} = \textbf{3.12 g/mL}$

1.22   **Strategy:** We are given the density and volume of a liquid and asked to calculate the mass of the liquid. Rearrange the density equation, Equation (1.1) of the text, to solve for mass.

$$\text{density} = \dfrac{\text{mass}}{\text{volume}}$$

Solution:

$$\textbf{mass} = \text{density} \times \text{volume}$$

$$\textbf{mass of ethanol} = \dfrac{0.798 \text{ g}}{1 \text{ mL}} \times 17.4 \text{ mL} = \textbf{13.9 g}$$

1.23 $? \, °\text{C} = (°\text{F} - 32°\text{F}) \times \dfrac{5°\text{C}}{9°\text{F}}$

(a) $? \, °\text{C} = (95°\text{F} - 32°\text{F}) \times \dfrac{5°\text{C}}{9°\text{F}} = \textbf{35°C}$

(b) $? \, °\text{C} = (12°\text{F} - 32°\text{F}) \times \dfrac{5°\text{C}}{9°\text{F}} = \textbf{−11°C}$

(c) $? \, °\text{C} = (102°\text{F} - 32°\text{F}) \times \dfrac{5°\text{C}}{9°\text{F}} = \textbf{39°C}$

(d) $? \, °\text{C} = (1852°\text{F} - 32°\text{F}) \times \dfrac{5°\text{C}}{9°\text{F}} = \textbf{1011°C}$

(e) $? \, °\text{F} = \left(°\text{C} \times \dfrac{9°\text{F}}{5°\text{C}} \right) + 32°\text{F}$

$? \, °\text{F} = \left(-273.15°\text{C} \times \dfrac{9°\text{F}}{5°\text{C}} \right) + 32°\text{F} = \textbf{−459.67°F}$

1.24   **Strategy:** Find the appropriate equations for converting between Fahrenheit and Celsius and between Celsius and Fahrenheit given in Section 1.7 of the text. Substitute the temperature values given in the problem into the appropriate equation.

(a)   Conversion from Fahrenheit to Celsius.

$$? \, °\text{C} = (°\text{F} - 32°\text{F}) \times \dfrac{5°\text{C}}{9°\text{F}}$$

$$? \, °\text{C} = (105°\text{F} - 32°\text{F}) \times \dfrac{5°\text{C}}{9°\text{F}} = \textbf{41°C}$$

(b)   Conversion from Celsius to Fahrenheit.

$$? \, °\text{F} = \left(°\text{C} \times \dfrac{9°\text{F}}{5°\text{C}} \right) + 32°\text{F}$$

$$? \, °\text{F} = \left(-11.5°\text{C} \times \dfrac{9°\text{F}}{5°\text{C}} \right) + 32°\text{F} = \textbf{11.3°F}$$

(c)   Conversion from Celsius to Fahrenheit.

$$? \,°F = \left(°C \times \frac{9°F}{5°C} \right) + 32°F$$

$$? \,°F = \left(6.3 \times 10^3 \,°C \times \frac{9°F}{5°C} \right) + 32°F = \mathbf{1.1 \times 10^4 \,°F}$$

(d)   Conversion from Fahrenheit to Celsius.

$$? \,°C = (°F - 32°F) \times \frac{5°C}{9°F}$$

$$? \,°C = (451°F - 32°F) \times \frac{5°C}{9°F} = \mathbf{233°C}$$

1.25 $K = (°C + 273°C) \dfrac{1 \,K}{1°C}$

(a) $K = 113°C + 273°C = \mathbf{386 \,K}$

(b) $K = 37°C + 273°C = \mathbf{3.10 \times 10^2 \,K}$

(c) $K = 357°C + 273°C = \mathbf{6.30 \times 10^2 \,K}$

1.26   (a) $K = (°C + 273°C) \dfrac{1 \,K}{1°C}$

              $°C = K - 273 = 77 \,K - 273 = \mathbf{-196°C}$

(b) $°C = 4.2 \,K - 273 = \mathbf{-269°C}$

(c) $°C = 601 \,K - 273 = \mathbf{328°C}$

1.29   (a) 2.7×10^{-8} (b) 3.56×10^2 (c) 4.7764×10^4 (d) 9.6×10^{-2}

1.30   (a) 10^{-2} indicates that the decimal point must be moved two places to the left.

$$1.52 \times 10^{-2} = \mathbf{0.0152}$$

(b) 10^{-8} indicates that the decimal point must be moved 8 places to the left.

$$7.78 \times 10^{-8} = \mathbf{0.0000000778}$$

1.31   (a) $145.75 + (2.3 \times 10^{-1}) = 145.75 + 0.23 = \mathbf{1.4598 \times 10^2}$

(b) $\dfrac{79500}{2.5 \times 10^2} = \dfrac{7.95 \times 10^4}{2.5 \times 10^2} = \mathbf{3.2 \times 10^2}$

(c) $(7.0 \times 10^{-3}) - (8.0 \times 10^{-4}) = (7.0 \times 10^{-3}) - (0.80 \times 10^{-3}) = \mathbf{6.2 \times 10^{-3}}$

(d) $(1.0 \times 10^4) \times (9.9 \times 10^6) = \mathbf{9.9 \times 10^{10}}$

1.32   **(a)**   Addition using scientific notation.

Strategy: Let's express scientific notation as $N \times 10^n$. When adding numbers using scientific notation, we must write each quantity with the same exponent, n. We can then add the N parts of the numbers, keeping the exponent, n, the same.

Solution: Write each quantity with the same exponent, n.

Let's write 0.0095 in such a way that $n = -3$. We have decreased 10^n by 10^3, so we must increase N by 10^3. Move the decimal point 3 places to the right.

$$0.0095 = 9.5 \times 10^{-3}$$

Add the N parts of the numbers, keeping the exponent, n, the same.

$$\begin{array}{r} 9.5 \times 10^{-3} \\ +\ \ 8.5 \times 10^{-3} \\ \hline \mathbf{18.0 \times 10^{-3}} \end{array}$$

The usual practice is to express N as a number between 1 and 10. Since we must *decrease* N by a factor of 10 to express N between 1 and 10 (1.8), we must *increase* 10^n by a factor of 10. The exponent, n, is increased by 1 from -3 to -2.

$$18.0 \times 10^{-3} = \mathbf{1.8 \times 10^{-2}}$$

(b)   Division using scientific notation.

Strategy: Let's express scientific notation as $N \times 10^n$. When dividing numbers using scientific notation, divide the N parts of the numbers in the usual way. To come up with the correct exponent, n, we *subtract* the exponents.

Solution: Make sure that all numbers are expressed in scientific notation.

$$653 = 6.53 \times 10^2$$

Divide the N parts of the numbers in the usual way.

$$6.53 \div 5.75 = 1.14$$

Subtract the exponents, n.

$$1.14 \times 10^{+2 - (-8)} = 1.14 \times 10^{+2 + 8} = \mathbf{1.14 \times 10^{10}}$$

(c)   Subtraction using scientific notation.

Strategy: Let's express scientific notation as $N \times 10^n$. When subtracting numbers using scientific notation, we must write each quantity with the same exponent, n. We can then subtract the N parts of the numbers, keeping the exponent, n, the same.

Solution: Write each quantity with the same exponent, n.

Let's write 850,000 in such a way that $n = 5$. This means to move the decimal point five places to the left.

$$850,000 = 8.5 \times 10^5$$

Subtract the N parts of the numbers, keeping the exponent, n, the same.

$$\begin{array}{r} 8.5 \times 10^5 \\ -\ 9.0 \times 10^5 \\ \hline -0.5 \times 10^5 \end{array}$$

The usual practice is to express N as a number between 1 and 10. Since we must *increase* N by a factor of 10 to express N between 1 and 10 (5), we must *decrease* 10^n by a factor of 10. The exponent, n, is decreased by 1 from 5 to 4.

$$-0.5 \times 10^5 = -5 \times 10^4$$

(d)   Multiplication using scientific notation.

Strategy: Let's express scientific notation as $N \times 10^n$. When multiplying numbers using scientific notation, multiply the N parts of the numbers in the usual way. To come up with the correct exponent, n, we *add* the exponents.

Solution: Multiply the N parts of the numbers in the usual way.

$$3.6 \times 3.6 = 13$$

Add the exponents, n.

$$13 \times 10^{-4 + (+6)} = 13 \times 10^2$$

The usual practice is to express N as a number between 1 and 10. Since we must *decrease* N by a factor of 10 to express N between 1 and 10 (1.3), we must *increase* 10^n by a factor of 10. The exponent, n, is increased by 1 from 2 to 3.

$$13 \times 10^2 = 1.3 \times 10^3$$

1.33 **(a)** four   **(b)** two   **(c)** five   **(d)** two, three, or four
         (e) three   **(f)** one   **(g)** one   **(h)** two

1.34 **(a)** one   **(b)** three   **(c)** three   **(d)** four
         (e) two or three   **(f)** one   **(g)** one or two

1.35 **(a)** 10.6 m   **(b)** 0.79 g   **(c)** 16.5 cm^2

1.36 **(a)** Division

Strategy: The number of significant figures in the answer is determined by the original number having the smallest number of significant figures.

Solution:

$$\frac{7.310 \text{ km}}{5.70 \text{ km}} = 1.28\mathbf{3}$$

The 3 (bolded) is a nonsignificant digit because the original number 5.70 only has three significant digits. Therefore, the answer has only three significant digits.

The correct answer rounded off to the correct number of significant figures is:

$$\textbf{1.28}\quad \text{(Why are there no units?)}$$

(b)   Subtraction

Strategy: The number of significant figures to the right of the decimal point in the answer is determined by the lowest number of digits to the right of the decimal point in any of the original numbers.

Solution: Writing both numbers in decimal notation, we have

$$
\begin{array}{r}
0.00326 \text{ mg} \\
-\ 0.0000788 \text{ mg} \\
\hline
0.0031\mathbf{812} \text{ mg}
\end{array}
$$

The bolded numbers are nonsignificant digits because the number 0.00326 has five digits to the right of the decimal point. Therefore, we carry five digits to the right of the decimal point in our answer.

The correct answer rounded off to the correct number of significant figures is:

$$0.00318 \text{ mg} = 3.18 \times 10^{-3} \text{ mg}$$

(c)   Addition

Strategy: The number of significant figures to the right of the decimal point in the answer is determined by the lowest number of digits to the right of the decimal point in any of the original numbers.

Solution: Writing both numbers with exponents = +7, we have

$$(0.402 \times 10^7 \text{ dm}) + (7.74 \times 10^7 \text{ dm}) = \mathbf{8.14 \times 10^7 \text{ dm}}$$

Since 7.74×10^7 has only two digits to the right of the decimal point, two digits are carried to the right of the decimal point in the final answer.

1.37   **(a)** $? \text{ dm} = 22.6 \text{ m} \times \dfrac{1 \text{ dm}}{0.1 \text{ m}} = \mathbf{226 \text{ dm}}$

(b) $? \text{ kg} = 25.4 \text{ mg} \times \dfrac{0.001 \text{ g}}{1 \text{ mg}} \times \dfrac{1 \text{ kg}}{1000 \text{ g}} = \mathbf{2.54 \times 10^{-5} \text{ kg}}$

(c) $? \text{ L} = 556 \text{ mL} \times \dfrac{1 \times 10^{-3} \text{ L}}{1 \text{ mL}} = \mathbf{0.556 \text{ L}}$

(d) $? \dfrac{\text{g}}{\text{cm}^3} = \dfrac{10.6 \text{ kg}}{1 \text{ m}^3} \times \dfrac{1000 \text{ g}}{1 \text{ kg}} \times \left(\dfrac{1 \times 10^{-2} \text{ m}}{1 \text{ cm}} \right)^3 = \mathbf{0.0106 \text{ g/cm}^3}$

1.38   **(a)**

Strategy: The problem may be stated as

$$? \text{ mg} = 242 \text{ lb}$$

A relationship between pounds and grams is given on the end sheet of your text (1 lb = 453.6 g). This relationship will allow conversion from pounds to grams. A metric conversion is then needed to convert grams to milligrams (1 mg = 1×10^{-3} g). Arrange the appropriate conversion factors so that pounds and grams cancel, and the unit milligrams is obtained in your answer.

Solution: The sequence of conversions is

$$lb \rightarrow grams \rightarrow mg$$

Using the following conversion factors,

$$\frac{453.6 \text{ g}}{1 \text{ lb}} \qquad \frac{1 \text{ mg}}{1 \times 10^{-3} \text{ g}}$$

we obtain the answer in one step:

$$? \text{ mg} = 242 \text{ lb} \times \frac{453.6 \text{ g}}{1 \text{ lb}} \times \frac{1 \text{ mg}}{1 \times 10^{-3} \text{ g}} = 1.10 \times 10^{8} \text{ mg}$$

Check: Does your answer seem reasonable? Should 242 lb be equivalent to 110 million mg? How many mg are in 1 lb? There are 453,600 mg in 1 lb.

(b)
Strategy: The problem may be stated as

$$? \text{ m}^3 = 68.3 \text{ cm}^3$$

Recall that $1 \text{ cm} = 1 \times 10^{-2}$ m. We need to set up a conversion factor to convert from cm^3 to m^3.

Solution: We need the following conversion factor so that centimeters cancel and we end up with meters.

$$\frac{1 \times 10^{-2} \text{ m}}{1 \text{ cm}}$$

Since this conversion factor deals with length and we want volume, it must therefore be cubed to give

$$\frac{1 \times 10^{-2} \text{ m}}{1 \text{ cm}} \times \frac{1 \times 10^{-2} \text{ m}}{1 \text{ cm}} \times \frac{1 \times 10^{-2} \text{ m}}{1 \text{ cm}} = \left(\frac{1 \times 10^{-2} \text{ m}}{1 \text{ cm}} \right)^3$$

We can write

$$? \text{ m}^3 = 68.3 \text{ cm}^3 \times \left(\frac{1 \times 10^{-2} \text{ m}}{1 \text{ cm}} \right)^3 = 6.83 \times 10^{-5} \text{ m}^3$$

Check: We know that $1 \text{ cm}^3 = 1 \times 10^{-6} \text{ m}^3$. We started with $6.83 \times 10^{1} \text{ cm}^3$. Multiplying this quantity by 1×10^{-6} gives 6.83×10^{-5}.

(c)
Strategy: The problem may be stated as

$$? \text{ L} = 7.2 \text{ m}^3$$

In Chapter 1 of the text, a conversion is given between liters and cm^3 ($1 \text{ L} = 1000 \text{ cm}^3$). If we can convert m^3 to cm^3, we can then convert to liters. Recall that $1 \text{ cm} = 1 \times 10^{-2}$ m. We need to set up two conversion factors to convert from m^3 to L. Arrange the appropriate conversion factors so that m^3 and cm^3 cancel, and the unit liters is obtained in your answer.

Solution: The sequence of conversions is

$$m^3 \rightarrow cm^3 \rightarrow L$$

Using the following conversion factors,

$$\left(\frac{1\ cm}{1 \times 10^{-2}\ m}\right)^3 \qquad \frac{1\ L}{1000\ cm^3}$$

the answer is obtained in one step:

$$?\ L = 7.2\ m^3 \times \left(\frac{1\ cm}{1 \times 10^{-2}\ m}\right)^3 \times \frac{1\ L}{1000\ cm^3} = 7.2 \times 10^3\ L$$

Check: From the above conversion factors you can show that $1\ m^3 = 1 \times 10^3$ L. Therefore, $7\ m^3$ would equal 7×10^3 L, which is close to the answer.

(d)
Strategy: The problem may be stated as

$$?\ lb = 28.3\ \mu g$$

A relationship between pounds and grams is given on the end sheet of your text (1 lb = 453.6 g). This relationship will allow conversion from grams to pounds. If we can convert from μg to grams, we can then convert from grams to pounds. Recall that $1\ \mu g = 1 \times 10^{-6}$ g. Arrange the appropriate conversion factors so that μg and grams cancel, and the unit pounds is obtained in your answer.

Solution: The sequence of conversions is

$$\mu g \rightarrow g \rightarrow lb$$

Using the following conversion factors,

$$\frac{1 \times 10^{-6}\ g}{1\ \mu g} \qquad \frac{1\ lb}{453.6\ g}$$

we can write

$$?\ lb = 28.3\ \mu g \times \frac{1 \times 10^{-6}\ g}{1\ \mu g} \times \frac{1\ lb}{453.6\ g} = 6.24 \times 10^{-8}\ lb$$

Check: Does the answer seem reasonable? What number does the prefix μ represent? Should 28.3 μg be a very small mass?

1.39 $\dfrac{1255\ m}{1\ s} \times \dfrac{1\ mi}{1609\ m} \times \dfrac{3600\ s}{1\ h} = \mathbf{2808\ mi/h}$

1.40 **Strategy:** The problem may be stated as

$$?\ s = 365.24\ days$$

You should know conversion factors that will allow you to convert between days and hours, between hours and minutes, and between minutes and seconds. Make sure to arrange the conversion factors so that days, hours, and minutes cancel, leaving units of seconds for the answer.

Solution: The sequence of conversions is

$$\text{days} \rightarrow \text{hours} \rightarrow \text{minutes} \rightarrow \text{seconds}$$

Using the following conversion factors,

$$\frac{24\ h}{1\ day} \qquad \frac{60\ min}{1\ h} \qquad \frac{60\ s}{1\ min}$$

we can write

$$?\ s\ =\ 365.24\ day \times \frac{24\ h}{1\ day} \times \frac{60\ min}{1\ h} \times \frac{60\ s}{1\ min}\ =\ 3.1557 \times 10^7\ s$$

Check: Does your answer seem reasonable? Should there be a very large number of seconds in 1 year?

1.41 $(93 \times 10^6\ mi) \times \dfrac{1.609\ km}{1\ mi} \times \dfrac{1000\ m}{1\ km} \times \dfrac{1\ s}{3.00 \times 10^8\ m} \times \dfrac{1\ min}{60\ s}\ =\ \mathbf{8.3\ min}$

1.42   **(a)** $?\ in/s\ =\ \dfrac{1\ mi}{13\ min} \times \dfrac{5280\ ft}{1\ mi} \times \dfrac{12\ in}{1\ ft} \times \dfrac{1\ min}{60\ s}\ =\ \mathbf{81\ in/s}$

(b) $?\ m/min\ =\ \dfrac{1\ mi}{13\ min} \times \dfrac{1609\ m}{1\ mi}\ =\ \mathbf{1.2 \times 10^2\ m/min}$

(c) $?\ km/h\ =\ \dfrac{1\ mi}{13\ min} \times \dfrac{1609\ m}{1\ mi} \times \dfrac{1\ km}{1000\ m} \times \dfrac{60\ min}{1\ h}\ =\ \mathbf{7.4\ km/h}$

1.43 $6.0\ ft \times \dfrac{1\ m}{3.28\ ft}\ =\ \mathbf{1.8\ m}$

$168\ lb \times \dfrac{453.6\ g}{1\ lb} \times \dfrac{1\ kg}{1000\ g}\ =\ \mathbf{76.2\ kg}$

1.44 $?\ km/h\ =\ \dfrac{55\ mi}{1\ h} \times \dfrac{1.609\ km}{1\ mi}\ =\ \mathbf{88\ km/h}$

1.45 $\dfrac{62\ m}{1\ s} \times \dfrac{1\ mi}{1609\ m} \times \dfrac{3600\ s}{1\ h}\ =\ \mathbf{1.4 \times 10^2\ mph}$

1.46 $0.62\ ppm\ Pb\ =\ \dfrac{0.62\ g\ Pb}{1 \times 10^6\ g\ blood}$

$6.0 \times 10^3\ g\ of\ blood \times \dfrac{0.62\ g\ Pb}{1 \times 10^6\ g\ blood}\ =\ \mathbf{3.7 \times 10^{-3}\ g\ Pb}$

1.47   **(a)** $1.42\ yr \times \dfrac{365\ day}{1\ yr} \times \dfrac{24\ h}{1\ day} \times \dfrac{3600\ s}{1\ h} \times \dfrac{3.00 \times 10^8\ m}{1\ s} \times \dfrac{1\ mi}{1609\ m}\ =\ \mathbf{8.35 \times 10^{12}\ mi}$

(b) $32.4 \text{ yd} \times \dfrac{36 \text{ in}}{1 \text{ yd}} \times \dfrac{2.54 \text{ cm}}{1 \text{ in}} = \mathbf{2.96 \times 10^3 \text{ cm}}$

(c) $\dfrac{3.0 \times 10^{10} \text{ cm}}{1 \text{ s}} \times \dfrac{1 \text{ in}}{2.54 \text{ cm}} \times \dfrac{1 \text{ ft}}{12 \text{ in}} = \mathbf{9.8 \times 10^8 \text{ ft/s}}$

1.48   **(a)** $? \text{ m} = 185 \text{ nm} \times \dfrac{1 \times 10^{-9} \text{ m}}{1 \text{ nm}} = \mathbf{1.85 \times 10^{-7} \text{ m}}$

(b) $? \text{ s} = (4.5 \times 10^9 \text{ yr}) \times \dfrac{365 \text{ day}}{1 \text{ yr}} \times \dfrac{24 \text{ h}}{1 \text{ day}} \times \dfrac{3600 \text{ s}}{1 \text{ h}} = \mathbf{1.4 \times 10^{17} \text{ s}}$

(c) $? \text{ m}^3 = 71.2 \text{ cm}^3 \times \left(\dfrac{0.01 \text{ m}}{1 \text{ cm}}\right)^3 = \mathbf{7.12 \times 10^{-5} \text{ m}^3}$

(d) $? \text{ L} = 88.6 \text{ m}^3 \times \left(\dfrac{1 \text{ cm}}{1 \times 10^{-2} \text{ m}}\right)^3 \times \dfrac{1 \text{ L}}{1000 \text{ cm}^3} = \mathbf{8.86 \times 10^4 \text{ L}}$

1.49   density $= \dfrac{2.70 \text{ g}}{1 \text{ cm}^3} \times \dfrac{1 \text{ kg}}{1000 \text{ g}} \times \left(\dfrac{1 \text{ cm}}{0.01 \text{ m}}\right)^3 = \mathbf{2.70 \times 10^3 \text{ kg/m}^3}$

1.50   density $= \dfrac{0.625 \text{ g}}{1 \text{ L}} \times \dfrac{1 \text{ L}}{1000 \text{ mL}} \times \dfrac{1 \text{ mL}}{1 \text{ cm}^3} = \mathbf{6.25 \times 10^{-4} \text{ g/cm}^3}$

1.51

	Substance	Qualitative Statement	Quantitative Statement
(a)	water	colorless liquid	freezes at 0°C
(b)	carbon	black solid (graphite)	density = 2.26 g/cm^3
(c)	iron	rusts easily	density = 7.86 g/cm^3
(d)	hydrogen gas	colorless gas	melts at −255.3°C
(e)	sucrose	tastes sweet	at 0°C, 179 g of sucrose dissolves in 100 g of H_2O
(f)	table salt	tastes salty	melts at 801°C
(g)	mercury	liquid at room temperature	boils at 357°C
(h)	gold	a precious metal	density = 19.3 g/cm^3
(i)	air	a mixture of gases	contains 20% oxygen by volume

1.52   See Section 1.6 of your text for a discussion of these terms.

(a)   <u>Chemical property</u>. Iron has changed its composition and identity by chemically combining with oxygen and water.

(b)   <u>Chemical property</u>. The water reacts with chemicals in the air (such as sulfur dioxide) to produce acids, thus changing the composition and identity of the water.

(c)   <u>Physical property</u>. The color of the hemoglobin can be observed and measured without changing its composition or identity.

(d)   <u>Physical property</u>. The evaporation of water does not change its chemical properties. Evaporation is a change in matter from the liquid state to the gaseous state.

(e) <u>Chemical property</u>. The carbon dioxide is chemically converted into other molecules.

1.53 (95.0 × 10^9 lb of sulfuric acid) × $\dfrac{1 \text{ ton}}{2.0 \times 10^3 \text{ lb}}$ = **4.75 × 10^7 tons of sulfuric acid**

1.54 Volume of rectangular bar = length × width × height

density = $\dfrac{m}{V}$ = $\dfrac{52.7064 \text{ g}}{(8.53 \text{ cm})(2.4 \text{ cm})(1.0 \text{ cm})}$ = **2.6 g/cm^3**

1.55 mass = density × volume

(a) **mass** = (19.3 g/cm^3) × [$\frac{4}{3}$ π(10.0 cm)^3] = **8.08 × 10^4 g**

(b) **mass** = (21.4 g/cm^3) × $\left(0.040 \text{ mm} \times \dfrac{1 \text{ cm}}{10 \text{ mm}} \right)^3$ = **1.4 × 10^{-6} g**

(c) **mass** = (0.798 g/mL)(50.0 mL) = **39.9 g**

1.56 You are asked to solve for the inner diameter of the tube. If you can calculate the volume that the mercury occupies, you can calculate the radius of the cylinder, $V_{\text{cylinder}} = \pi r^2 h$ (r is the inner radius of the cylinder, and h is the height of the cylinder). The cylinder diameter is $2r$.

$$\text{volume of Hg filling cylinder} = \frac{\text{mass of Hg}}{\text{density of Hg}}$$

$$\text{volume of Hg filling cylinder} = \frac{105.5 \text{ g}}{13.6 \text{ g/cm}^3} = 7.76 \text{ cm}^3$$

Next, solve for the radius of the cylinder.

$$\text{Volume of cylinder} = \pi r^2 h$$

$$r = \sqrt{\frac{\text{volume}}{\pi \times h}}$$

$$r = \sqrt{\frac{7.76 \text{ cm}^3}{\pi \times 12.7 \text{ cm}}} = 0.441 \text{ cm}$$

The cylinder diameter equals $2r$.

Cylinder diameter = $2r$ = 2(0.441 cm) = **0.882 cm**

1.57 From the mass of the water and its density, we can calculate the volume that the water occupies. The volume that the water occupies is equal to the volume of the flask.

$$\text{volume} = \frac{\text{mass}}{\text{density}}$$

Mass of water = 87.39 g − 56.12 g = 31.27 g

$$\textbf{Volume of the flask} = \frac{\text{mass}}{\text{density}} = \frac{31.27 \text{ g}}{0.9976 \text{ g/cm}^3} = \textbf{31.35 cm}^3$$

1.58 $\dfrac{343 \text{ m}}{1 \text{ s}} \times \dfrac{1 \text{ mi}}{1609 \text{ m}} \times \dfrac{3600 \text{ s}}{1 \text{ h}} = \textbf{767 mph}$

1.59   The volume of silver is equal to the volume of water it displaces.

$$\text{Volume of silver} = 260.5 \text{ mL} - 242.0 \text{ mL} = 18.5 \text{ mL} = 18.5 \text{ cm}^3$$

$$\text{density} = \frac{194.3 \text{ g}}{18.5 \text{ cm}^3} = \textbf{10.5 g/cm}^3$$

1.60   In order to work this problem, you need to understand the physical principles involved in the experiment in Problem 1.59. The volume of the water displaced must equal the volume of the piece of silver. If the silver did not sink, would you have been able to determine the volume of the piece of silver?

The liquid must be *less dense* than the ice in order for the ice to sink. The temperature of the experiment must be maintained at or below *0°C* to prevent the ice from melting.

1.61 $\text{density} = \dfrac{\text{mass}}{\text{volume}} = \dfrac{1.20 \times 10^4 \text{ g}}{1.05 \times 10^3 \text{ cm}^3} = \textbf{11.4 g/cm}^3$

1.62 $\text{Volume} = \dfrac{\text{mass}}{\text{density}}$

$$\textbf{Volume occupied by Li} = \frac{1.20 \times 10^3 \text{ g}}{0.53 \text{ g}/\text{cm}^3} = \textbf{2.3} \times \textbf{10}^3 \textbf{ cm}^3$$

1.63   For the Fahrenheit thermometer, we must convert the possible error of 0.1°F to °C.

$$? \text{ °C} = 0.1\text{°F} \times \frac{5\text{°C}}{9\text{°F}} = 0.056\text{°C}$$

The percent error is the amount of uncertainty in a measurement divided by the value of the measurement, converted to percent by multiplication by 100.

$$\text{Percent error} = \frac{\text{known error in a measurement}}{\text{value of the measurement}} \times 100\%$$

For the Fahrenheit thermometer, $\textbf{percent error} = \dfrac{0.056\text{°C}}{38.9\text{°C}} \times 100\% = \textbf{0.14\%}$

For the Celsius thermometer, $\textbf{percent error} = \dfrac{0.1\text{°C}}{38.9\text{°C}} \times 100\% = \textbf{0.26\%}$

Which thermometer is more accurate?

1.64   To work this problem, we need to convert from cubic feet to L. Some tables will have a conversion factor of $28.3 \text{ L} = 1 \text{ ft}^3$, but we can also calculate it using the factor-label method described in Section 1.9 of the text.

First, converting from cubic feet to liters:

$$(5.0 \times 10^7 \text{ ft}^3) \times \left(\frac{12 \text{ in}}{1 \text{ ft}}\right)^3 \times \left(\frac{2.54 \text{ cm}}{1 \text{ in}}\right)^3 \times \frac{1 \text{ mL}}{1 \text{ cm}^3} \times \frac{1 \times 10^{-3} \text{ L}}{1 \text{ mL}} = 1.4 \times 10^9 \text{ L}$$

The mass of vanillin (in g) is:

$$\frac{2.0 \times 10^{-11} \text{ g vanillin}}{1 \text{ L}} \times (1.4 \times 10^9 \text{ L}) = 2.8 \times 10^{-2} \text{ g vanillin}$$

The cost is:

$$(2.8 \times 10^{-2} \text{ g vanillin}) \times \frac{\$112}{50 \text{ g vanillin}} = \mathbf{\$0.063} = \mathbf{6.3¢}$$

1.65 $? \, °F = \left(°C \times \dfrac{9°F}{5°C}\right) + 32°F$

Let temperature $= t$

$t = \dfrac{9}{5}t + 32°F$

$t - \dfrac{9}{5}t = 32°F$

$-\dfrac{4}{5}t = 32°F$

$\mathbf{\mathit{t} = -40°F = -40°C}$

1.66   There are $78.3 + 117.3 = 195.6$ Celsius degrees between $0°S$ and $100°S$. We can write this as a unit factor.

$$\left(\frac{195.6°C}{100°S}\right)$$

Set up the equation like a Celsius to Fahrenheit conversion. We need to subtract $117.3°C$, because the zero point on the new scale is $117.3°C$ lower than the zero point on the Celsius scale.

$$? \, °C = \left(\frac{195.6°C}{100°S}\right)(? \, °S) - 117.3°C$$

Solving for $? \, °S$ gives: $? \, °S = (? \, °C + 117.3°C)\left(\dfrac{100°S}{195.6°C}\right)$

For $25°C$ we have: $\mathbf{? \, °S} = (25°C + 117.3°C)\left(\dfrac{100°S}{195.6°C}\right) = \mathbf{73°S}$

1.67   The key to solving this problem is to realize that all the oxygen needed must come from the 4% difference (20% − 16%) between inhaled and exhaled air.

The 240 mL of pure oxygen/min requirement comes from the 4% of inhaled air that is oxygen.

240 mL of pure oxygen/min $=$ (0.04)(volume of inhaled air/min)

$$\text{Volume of inhaled air/min} = \frac{240 \text{ mL of oxygen/min}}{0.04} = 6000 \text{ mL of inhaled air/min}$$

Since there are 12 breaths per min,

$$\textbf{volume of air/breath} = \frac{6000 \text{ mL of inhaled air}}{1 \text{ min}} \times \frac{1 \text{ min}}{12 \text{ breaths}} = \textbf{5} \times \textbf{10}^2 \textbf{ mL/breath}$$

1.68   **(a)** $\dfrac{6000 \text{ mL of inhaled air}}{1 \text{ min}} \times \dfrac{0.001 \text{ L}}{1 \text{ mL}} \times \dfrac{60 \text{ min}}{1 \text{ h}} \times \dfrac{24 \text{ h}}{1 \text{ day}} = \textbf{8.6} \times \textbf{10}^3 \textbf{ L of air/day}$

(b) $\dfrac{8.6 \times 10^3 \text{ L of air}}{1 \text{ day}} \times \dfrac{2.1 \times 10^{-6} \text{ L CO}}{1 \text{ L of air}} = \textbf{0.018 L CO/day}$

1.69   The mass of the seawater is:

$$(1.5 \times 10^{21} \text{ L}) \times \frac{1 \text{ mL}}{0.001 \text{ L}} \times \frac{1.03 \text{ g}}{1 \text{ mL}} = 1.5 \times 10^{24} \text{ g} = 1.5 \times 10^{21} \text{ kg seawater}$$

Seawater is 3.1% NaCl by mass. The total mass of NaCl in kilograms is:

$$\textbf{mass NaCl (kg)} = (1.5 \times 10^{21} \text{ kg seawater}) \times \frac{3.1\% \text{ NaCl}}{100\% \text{ seawater}} = \textbf{4.7} \times \textbf{10}^{19} \textbf{ kg NaCl}$$

$$\textbf{mass NaCl (tons)} = (4.7 \times 10^{19} \text{ kg}) \times \frac{2.205 \text{ lb}}{1 \text{ kg}} \times \frac{1 \text{ ton}}{2000 \text{ lb}} = \textbf{5.2} \times \textbf{10}^{16} \textbf{ tons NaCl}$$

1.70   First, calculate the volume of 1 kg of seawater from the density and the mass. We chose 1 kg of seawater, because the problem gives the amount of Mg in every kg of seawater. The density of seawater is given in Problem 1.69.

$$\text{volume} = \frac{\text{mass}}{\text{density}}$$

$$\text{volume of 1 kg of seawater} = \frac{1000 \text{ g}}{1.03 \text{ g/mL}} = 971 \text{ mL} = 0.971 \text{ L}$$

In other words, there are 1.3 g of Mg in every 0.971 L of seawater.

Next, let's convert tons of Mg to grams of Mg.

$$(8.0 \times 10^4 \text{ tons Mg}) \times \frac{2000 \text{ lb}}{1 \text{ ton}} \times \frac{453.6 \text{ g}}{1 \text{ lb}} = 7.3 \times 10^{10} \text{ g Mg}$$

Volume of seawater needed to extract 8.0×10^4 ton Mg $=$

$$(7.3 \times 10^{10} \text{ g Mg}) \times \frac{0.971 \text{ L seawater}}{1.3 \text{ g Mg}} = \textbf{5.5} \times \textbf{10}^{10} \textbf{ L of seawater}$$

1.71   Assume that the crucible is platinum. Let's calculate the volume of the crucible and then compare that to the volume of water that the crucible displaces.

$$\text{volume} = \frac{\text{mass}}{\text{density}}$$

$$\text{Volume of crucible} = \frac{860.2 \text{ g}}{21.45 \text{ g/cm}^3} = \textbf{40.10 cm}^3$$

$$\text{Volume of water displaced} = \frac{(860.2 - 820.2)\text{g}}{0.9986 \text{ g/cm}^3} = \textbf{40.1 cm}^3$$

The volumes are the same (within experimental error), so the crucible is made of platinum.

1.72   Volume = surface area × depth

Recall that $1 \text{ L} = 1 \text{ dm}^3$. Let's convert the surface area to units of dm^2 and the depth to units of dm.

$$\text{surface area} = (1.8 \times 10^8 \text{ km}^2) \times \left(\frac{1000 \text{ m}}{1 \text{ km}}\right)^2 \times \left(\frac{1 \text{ dm}}{0.1 \text{ m}}\right)^2 = 1.8 \times 10^{16} \text{ dm}^2$$

$$\text{depth} = (3.9 \times 10^3 \text{ m}) \times \frac{1 \text{ dm}}{0.1 \text{ m}} = 3.9 \times 10^4 \text{ dm}$$

Volume = surface area × depth = $(1.8 \times 10^{16} \text{ dm}^2)(3.9 \times 10^4 \text{ dm}) = 7.0 \times 10^{20} \text{ dm}^3 = \textbf{7.0} \times \textbf{10}^{20} \textbf{ L}$

1.73   **(a)** $2.41 \text{ troy oz Au} \times \dfrac{31.103 \text{ g Au}}{1 \text{ troy oz Au}} = \textbf{75.0 g Au}$

(b)   1 troy oz = **31.103 g**

$? \text{ g in 1 oz} = 1 \text{ oz} \times \dfrac{1 \text{ lb}}{16 \text{ oz}} \times \dfrac{453.6 \text{ g}}{1 \text{ lb}} = \textbf{28.35 g}$

A troy ounce is heavier than an ounce.

1.74   Volume of sphere $= \dfrac{4}{3}\pi r^3$

$$\text{Volume} = \frac{4}{3}\pi \left(\frac{15 \text{ cm}}{2}\right)^3 = 1.8 \times 10^3 \text{ cm}^3$$

$$\text{mass} = \text{volume} \times \text{density} = (1.8 \times 10^3 \text{ cm}^3) \times \frac{22.57 \text{ g Os}}{1 \text{ cm}^3} \times \frac{1 \text{ kg}}{1000 \text{ g}} = \textbf{41 kg Os}$$

$$41 \text{ kg Os} \times \frac{2.205 \text{ lb}}{1 \text{ kg}} = \textbf{9.0} \times \textbf{10}^1 \textbf{ lb Os}$$

1.75   **(a)** $\dfrac{|0.798 \text{ g/mL} - 0.802 \text{ g/mL}|}{0.798 \text{ g/mL}} \times 100\% = \textbf{0.5\%}$

(b) $\dfrac{|0.864 \text{ g} - 0.837 \text{ g}|}{0.864 \text{ g}} \times 100\% = \textbf{3.1\%}$

1.76 $62 \text{ kg} = 6.2 \times 10^4 \text{ g}$

O: $(6.2 \times 10^4 \text{ g})(0.65) = \textbf{4.0} \times \textbf{10}^{\textbf{4}} \textbf{ g O}$ N: $(6.2 \times 10^4 \text{ g})(0.03) = \textbf{2} \times \textbf{10}^{\textbf{3}} \textbf{ g N}$

C: $(6.2 \times 10^4 \text{ g})(0.18) = \textbf{1.1} \times \textbf{10}^{\textbf{4}} \textbf{ g C}$ Ca: $(6.2 \times 10^4 \text{ g})(0.016) = \textbf{9.9} \times \textbf{10}^{\textbf{2}} \textbf{ g Ca}$

H: $(6.2 \times 10^4 \text{ g})(0.10) = \textbf{6.2} \times \textbf{10}^{\textbf{3}} \textbf{ g H}$ P: $(6.2 \times 10^4 \text{ g})(0.012) = \textbf{7.4} \times \textbf{10}^{\textbf{2}} \textbf{ g P}$

1.77   3 minutes 44.39 seconds $= 224.39$ seconds

Time to run 1500 meters is:

$$1500 \text{ m} \times \frac{1 \text{ mi}}{1609 \text{ m}} \times \frac{224.39 \text{ s}}{1 \text{ mi}} = \textbf{209.19 s} = \textbf{3 min 29.19 s}$$

1.78 $? \,^{\circ}\text{C} = (7.3 \times 10^2 - 273) \text{ K} = \textbf{4.6} \times \textbf{10}^{\textbf{2}} \,^{\circ}\textbf{C}$

$$? \,^{\circ}\text{F} = \left((4.6 \times 10^2 \,^{\circ}\text{C}) \times \frac{9 \,^{\circ}\text{F}}{5 \,^{\circ}\text{C}} \right) + 32\,^{\circ}\text{F} = \textbf{8.6} \times \textbf{10}^{\textbf{2}} \,^{\circ}\textbf{F}$$

1.79 $? \text{ g Cu} = (5.11 \times 10^3 \text{ kg ore}) \times \dfrac{34.63\% \text{ Cu}}{100\% \text{ ore}} \times \dfrac{1000 \text{ g}}{1 \text{ kg}} = \textbf{1.77} \times \textbf{10}^{\textbf{6}} \textbf{ g Cu}$

1.80 $(8.0 \times 10^4 \text{ tons Au}) \times \dfrac{2000 \text{ lb Au}}{1 \text{ ton Au}} \times \dfrac{16 \text{ oz Au}}{1 \text{ lb Au}} \times \dfrac{\$350}{1 \text{ oz Au}} = \textbf{\$9.0} \times \textbf{10}^{\textbf{11}} \textbf{ or 900 billion dollars}$

1.81 $? \textbf{ g Au} = \dfrac{4.0 \times 10^{-12} \text{ g Au}}{1 \text{ mL seawater}} \times \dfrac{1 \text{ mL}}{0.001 \text{ L}} \times (1.5 \times 10^{21} \text{ L seawater}) = \textbf{6.0} \times \textbf{10}^{\textbf{12}} \textbf{ g Au}$

value of gold $= (6.0 \times 10^{12} \text{ g Au}) \times \dfrac{1 \text{ lb}}{453.6 \text{ g}} \times \dfrac{16 \text{ oz}}{1 \text{ lb}} \times \dfrac{\$350}{1 \text{ oz}} = \textbf{\$7.4} \times \textbf{10}^{\textbf{13}}$

No one has become rich mining gold from the ocean, because the cost of recovering the gold would outweigh the price of the gold.

1.82 $? \textbf{ Fe atoms} = 4.9 \text{ g Fe} \times \dfrac{1.1 \times 10^{22} \text{ Fe atoms}}{1.0 \text{ g Fe}} = \textbf{5.4} \times \textbf{10}^{\textbf{22}} \textbf{ Fe atoms}$

1.83   mass of Earth's crust $= (5.9 \times 10^{21} \text{ tons}) \times \dfrac{0.50\% \text{ crust}}{100\% \text{ Earth}} = 3.0 \times 10^{19} \text{ tons}$

mass of silicon in crust $= (3.0 \times 10^{19} \text{ tons crust}) \times \dfrac{27.2\% \text{ Si}}{100\% \text{ crust}} \times \dfrac{2000 \text{ lb}}{1 \text{ ton}} \times \dfrac{1 \text{ kg}}{2.205 \text{ lb}} = \textbf{7.4} \times \textbf{10}^{\textbf{21}} \textbf{ kg Si}$

1.84 $10 \text{ cm} = 0.1 \text{ m}$. We need to find the number of times the 0.1 m wire must be cut in half until the piece left is 1.3×10^{-10} m long. Let n be the number of times we can cut the Cu wire in half. We can write:

$$\left(\frac{1}{2}\right)^n \times 0.1 \text{ m} = 1.3 \times 10^{-10} \text{ m}$$

$$\left(\frac{1}{2}\right)^n = 1.3 \times 10^{-9} \text{ m}$$

Taking the log of both sides of the equation:

$$n \log\left(\frac{1}{2}\right) = \log(1.3 \times 10^{-9})$$

$$n = 30 \text{ times}$$

1.85 $(40 \times 10^6 \text{ cars}) \times \dfrac{5000 \text{ mi}}{1 \text{ car}} \times \dfrac{1 \text{ gal gas}}{20 \text{ mi}} \times \dfrac{9.5 \text{ kg } CO_2}{1 \text{ gal gas}} = \mathbf{9.5 \times 10^{10} \text{ kg } CO_2}$

1.86   Volume $=$ area $\times$ thickness.

From the density, we can calculate the volume of the Al foil.

$$\text{Volume} = \frac{\text{mass}}{\text{density}} = \frac{3.636 \text{ g}}{2.699 \text{ g}/\text{cm}^3} = 1.347 \text{ cm}^3$$

Convert the unit of area from ft^2 to cm^2.

$$1.000 \text{ ft}^2 \times \left(\frac{12 \text{ in}}{1 \text{ ft}}\right)^2 \times \left(\frac{2.54 \text{ cm}}{1 \text{ in}}\right)^2 = 929.0 \text{ cm}^2$$

$$\textbf{thickness} = \frac{\text{volume}}{\text{area}} = \frac{1.347 \text{ cm}^3}{929.0 \text{ cm}^2} = 1.450 \times 10^{-3} \text{ cm} = \mathbf{1.450 \times 10^{-2} \text{ mm}}$$

1.87   **(a)**   homogeneous
(b)   heterogeneous. The air will contain particulate matter, clouds, etc. This mixture is not homogeneous.

1.88   First, let's calculate the mass (in g) of water in the pool. We perform this conversion because we know there is 1 g of chlorine needed per million grams of water.

$$(2.0 \times 10^4 \text{ gallons } H_2O) \times \frac{3.79 \text{ L}}{1 \text{ gallon}} \times \frac{1 \text{ mL}}{0.001 \text{ L}} \times \frac{1 \text{ g}}{1 \text{ mL}} = 7.6 \times 10^7 \text{ g } H_2O$$

Next, let's calculate the mass of chlorine that needs to be added to the pool.

$$(7.6 \times 10^7 \text{ g } H_2O) \times \frac{1 \text{ g chlorine}}{1 \times 10^6 \text{ g } H_2O} = 76 \text{ g chlorine}$$

The chlorine solution is only 6 percent chlorine by mass. We can now calculate the volume of chlorine solution that must be added to the pool.

$$76 \text{ g chlorine} \times \frac{100\% \text{ soln}}{6\% \text{ chlorine}} \times \frac{1 \text{ mL soln}}{1 \text{ g soln}} = 1.3 \times 10^3 \text{ mL of chlorine solution}$$

1.89 $(2.0 \times 10^{22} \text{ J}) \times \dfrac{1 \text{ yr}}{1.8 \times 10^{20} \text{ J}} = 1.1 \times 10^2 \text{ yr}$

1.90   We assume that the thickness of the oil layer is equivalent to the length of one oil molecule. We can calculate the thickness of the oil layer from the volume and surface area.

$$40 \text{ m}^2 \times \left(\frac{1 \text{ cm}}{0.01 \text{ m}}\right)^2 = 4.0 \times 10^5 \text{ cm}^2$$

$0.10 \text{ mL} = 0.10 \text{ cm}^3$

$$\text{Volume} = \text{surface area} \times \text{thickness}$$

$$\text{thickness} = \frac{\text{volume}}{\text{surface area}} = \frac{0.10 \text{ cm}^3}{4.0 \times 10^5 \text{ cm}^2} = 2.5 \times 10^{-7} \text{ cm}$$

Converting to nm:

$$(2.5 \times 10^{-7} \text{ cm}) \times \frac{0.01 \text{ m}}{1 \text{ cm}} \times \frac{1 \text{ nm}}{1 \times 10^{-9} \text{ m}} = 2.5 \text{ nm}$$

1.91   The mass of water used by 50,000 people in 1 year is:

$$50,000 \text{ people} \times \frac{150 \text{ gal water}}{1 \text{ person each day}} \times \frac{3.79 \text{ L}}{1 \text{ gal.}} \times \frac{1000 \text{ mL}}{1 \text{ L}} \times \frac{1 \text{ g H}_2\text{O}}{1 \text{ mL H}_2\text{O}} \times \frac{365 \text{ days}}{1 \text{ yr}} = 1.04 \times 10^{13} \text{ g H}_2\text{O/yr}$$

A concentration of 1 ppm of fluorine is needed. In other words, 1 g of fluorine is needed per million grams of water. NaF is 45.0% fluorine by mass. The amount of NaF needed per year in kg is:

$$(1.04 \times 10^{13} \text{ g H}_2\text{O}) \times \frac{1 \text{ g F}}{10^6 \text{ g H}_2\text{O}} \times \frac{100\% \text{ NaF}}{45\% \text{ F}} \times \frac{1 \text{ kg}}{1000 \text{ g}} = 2.3 \times 10^4 \text{ kg NaF}$$

An average person uses 150 gallons of water per day. This is equal to 569 L of water. If only 6 L of water is used for drinking and cooking, 563 L of water is used for purposes in which NaF is not necessary. Therefore the amount of NaF wasted is:

$$\frac{563 \text{ L}}{569 \text{ L}} \times 100\% = 99\%$$

1.92   **(a)** $\dfrac{\$1.30}{15.0 \text{ ft}^3} \times \left(\dfrac{1 \text{ ft}}{12 \text{ in}}\right)^3 \times \left(\dfrac{1 \text{ in}}{2.54 \text{ cm}}\right)^3 \times \dfrac{1 \text{ cm}^3}{1 \text{ mL}} \times \dfrac{1 \text{ mL}}{0.001 \text{ L}} = \$3.06 \times 10^{-3}/\text{L}$

(b) $2.1 \text{ L water} \times \dfrac{0.304 \text{ ft}^3 \text{ gas}}{1 \text{ L water}} \times \dfrac{\$1.30}{15.0 \text{ ft}^3} = \$0.055 = 5.5\cancel{\text{c}}$

1.93 To calculate the density of the pheromone, you need the mass of the pheromone, and the volume that it occupies. The mass is given in the problem. First, let's calculate the volume of the cylinder. Converting the radius and height to cm gives:

$$0.50 \text{ mi} \times \frac{1609 \text{ m}}{1 \text{ mi}} \times \frac{1 \text{ cm}}{0.01 \text{ m}} = 8.0 \times 10^4 \text{ cm}$$

$$40 \text{ ft} \times \frac{12 \text{ in}}{1 \text{ ft}} \times \frac{2.54 \text{ cm}}{1 \text{ in}} = 1.2 \times 10^3 \text{ cm}$$

$$\text{volume of a cylinder} = \text{area} \times \text{height} = \pi r^2 \times h$$

$$\text{volume} = \pi (8.0 \times 10^4 \text{ cm})^2 \times (1.2 \times 10^3 \text{ cm}) = 2.4 \times 10^{13} \text{ cm}^3$$

Density of gases is usually expressed in g/L. Let's convert the volume to liters.

$$(2.4 \times 10^{13} \text{ cm}^3) \times \frac{1 \text{ mL}}{1 \text{ cm}^3} \times \frac{1 \text{ L}}{1000 \text{ mL}} = 2.4 \times 10^{10} \text{ L}$$

$$\textbf{density} = \frac{\text{mass}}{\text{volume}} = \frac{1.0 \times 10^{-8} \text{ g}}{2.4 \times 10^{10} \text{ L}} = \textbf{4.2} \times \textbf{10}^{-19} \textbf{ g/L}$$

CHAPTER 2
ATOMS, MOLECULES, AND IONS

2.7 First, convert 1 cm to picometers.

$$1 \text{ cm} \times \frac{0.01 \text{ m}}{1 \text{ cm}} \times \frac{1 \text{ pm}}{1 \times 10^{-12} \text{ m}} = 1 \times 10^{10} \text{ pm}$$

$$? \text{ He atoms } = (1 \times 10^{10} \text{ pm}) \times \frac{1 \text{ He atom}}{1 \times 10^{2} \text{ pm}} = 1 \times 10^{8} \text{ He atoms}$$

2.8 Note that you are given information to set up the unit factor relating meters and miles.

$$r_{\text{atom}} = 10^{4} \, r_{\text{nucleus}} = 10^{4} \times 2.0 \text{ cm} \times \frac{1 \text{ m}}{100 \text{ cm}} \times \frac{1 \text{ mi}}{1609 \text{ m}} = 0.12 \text{ mi}$$

2.13 For iron, the atomic number Z is 26. Therefore the mass number A is:

$$A = 26 + 28 = 54$$

2.14 **Strategy:** The 239 in Pu-239 is the mass number. The **mass number (A)** is the total number of neutrons and protons present in the nucleus of an atom of an element. You can look up the atomic number (number of protons) on the periodic table.

Solution:

mass number = number of protons + number of neutrons

number of neutrons = mass number − number of protons = 239 − 94 = **145**

2.15

Isotope	$^{3}_{2}\text{He}$	$^{4}_{2}\text{He}$	$^{24}_{12}\text{Mg}$	$^{25}_{12}\text{Mg}$	$^{48}_{22}\text{Ti}$	$^{79}_{35}\text{Br}$	$^{195}_{78}\text{Pt}$
No. Protons	2	2	12	12	22	35	78
No. Neutrons	1	2	12	13	26	44	117

2.16

Isotope	$^{15}_{7}\text{N}$	$^{33}_{16}\text{S}$	$^{63}_{29}\text{Cu}$	$^{84}_{38}\text{Sr}$	$^{130}_{56}\text{Ba}$	$^{186}_{74}\text{W}$	$^{202}_{80}\text{Hg}$
No. Protons	7	16	29	38	56	74	80
No. Neutrons	8	17	34	46	74	112	122
No. Electrons	7	16	29	38	56	74	80

2.17 **(a)** $^{23}_{11}\text{Na}$ **(b)** $^{64}_{28}\text{Ni}$

2.18 The accepted way to denote the atomic number and mass number of an element X is as follows:

$$^{A}_{Z}\text{X}$$

where,

A = mass number
Z = atomic number

(a) $^{186}_{74}W$ (b) $^{201}_{80}Hg$

2.23   Helium and Selenium are nonmetals whose name ends with *ium*. (Tellerium is a metalloid whose name ends in *ium*.)

2.24   (a)   Metallic character increases as you progress down a group of the periodic table. For example, moving down Group 4A, the nonmetal carbon is at the top and the metal lead is at the bottom of the group.

   (b)   Metallic character decreases from the left side of the table (where the metals are located) to the right side of the table (where the nonmetals are located).

2.25   The following data were measured at 20°C.

   (a)   Li (0.53 g/cm^3)   K (0.86 g/cm^3) H_2O (0.98 g/cm^3)

   (b)   Au (19.3 g/cm^3)   Pt (21.4 g/cm^3)   Hg (13.6 g/cm^3)

   (c)   Os (22.6 g/cm^3)

   (d)   Te (6.24 g/cm^3)

2.26   F and Cl are Group 7A elements; they should have similar chemical properties. Na and K are both Group 1A elements; they should have similar chemical properties. P and N are both Group 5A elements; they should have similar chemical properties.

2.31   (a)   This is a polyatomic molecule that is an elemental form of the substance. It is not a compound.
   (b)   This is a polyatomic molecule that is a compound.
   (c)   This is a diatomic molecule that is a compound.

2.32   (a)   This is a diatomic molecule that is a compound.
   (b)   This is a polyatomic molecule that is a compound.
   (c)   This is a polyatomic molecule that is the elemental form of the substance. It is not a compound.

2.33   **Elements:** N_2, S_8, H_2
   Compounds: NH_3, NO, CO, CO_2, SO_2

2.34   There are more than two correct answers for each part of the problem.

   (a) H_2 and F_2 (b)   HCl and CO   (c) S_8 and P_4
   (d) H_2O and $C_{12}H_{22}O_{11}$ (sucrose)

2.35

Ion	Na^+	Ca^{2+}	Al^{3+}	Fe^{2+}	I^-	F^-	S^{2-}	O^{2-}	N^{3-}
No. protons	11	20	13	26	53	9	16	8	7
No. electrons	10	18	10	24	54	10	18	10	10

2.36   The **atomic number (Z)** is the number of protons in the nucleus of each atom of an element. You can find this on a periodic table. The number of **electrons** in an *ion* is equal to the number of protons minus the charge on the ion.

       number of electrons (ion) = number of protons − charge on the ion

Ion	K^+	Mg^{2+}	Fe^{3+}	Br^-	Mn^{2+}	C^{4-}	Cu^{2+}
No. protons	19	12	26	35	25	6	29
No. electrons	18	10	23	36	23	10	27

2.43   **(a)** CN   **(b)** CH   **(c)** C_9H_{20} **(d)** P_2O_5 **(e)** BH_3

2.44   **Strategy:** An *empirical formula* tells us which elements are present and the *simplest* whole-number ratio of their atoms. Can you divide the subscripts in the formula by some factor to end up with smaller whole-number subscripts?

Solution:

(a)   Dividing both subscripts by 2, the simplest whole number ratio of the atoms in Al_2Br_6 is **$AlBr_3$**.

(b)   Dividing all subscripts by 2, the simplest whole number ratio of the atoms in $Na_2S_2O_4$ is **$NaSO_2$**.

(c)   The molecular formula as written, **N_2O_5**, contains the simplest whole number ratio of the atoms present. In this case, the molecular formula and the empirical formula are the same.

(d)   The molecular formula as written, **$K_2Cr_2O_7$**, contains the simplest whole number ratio of the atoms present. In this case, the molecular formula and the empirical formula are the same.

2.45   The molecular formula of glycine is **$C_2H_5NO_2$**.

2.46   The molecular formula of ethanol is **C_2H_6O**.

2.47   Compounds of metals with nonmetals are usually ionic. Nonmetal-nonmetal compounds are usually molecular.

Ionic: LiF, $BaCl_2$, KCl

Molecular: $SiCl_4$, B_2H_6, C_2H_4

2.48   Compounds of metals with nonmetals are usually ionic. Nonmetal-nonmetal compounds are usually molecular.

Ionic: NaBr, BaF_2, CsCl.

Molecular: CH_4, CCl_4, ICl, NF_3

2.55   **(a)** potassium dihydrogen phosphate   **(h)** iodic acid
(b) potassium hydrogen phosphate   **(i)** phosphorus pentafluoride
(c) hydrogen bromide (molecular compound)   **(j)** tetraphosphorus hexoxide
(d) hydrobromic acid   **(k)** cadmium iodide
(e) lithium carbonate   **(l)** strontium sulfate
(f) potassium dichromate   **(m)** aluminum hydroxide
(g) ammonium nitrite

2.56   **Strategy:** When naming ionic compounds, our reference for the names of cations and anions is Table 2.3 of the text. Keep in mind that if a metal can form cations of different charges, we need to use the Stock system. In the Stock system, Roman numerals are used to specify the charge of the cation. The metals that have only one charge in ionic compounds are the alkali metals (+1), the alkaline earth metals (+2), Ag^+, Zn^{2+}, Cd^{2+}, and Al^{3+}.

When naming acids, binary acids are named differently than oxoacids. For binary acids, the name is based on the nonmetal. For oxoacids, the name is based on the polyatomic anion. For more detail, see Section 2.7 of the text.

Solution:

(a) This is an ionic compound in which the metal cation (K^+) has only one charge. The correct name is **potassium hypochlorite**. Hypochlorite is a polyatomic ion with one less O atom than the chlorite ion, ClO_2^-

(b) **silver carbonate**

(c) This is an oxoacid that contains the nitrite ion, NO_2^-. The "-ite" suffix is changed to "-ous". The correct name is **nitrous acid.**

(d) **potassium permanganate**   (e) **cesium chlorate**   (f) **potassium ammonium sulfate**

(g) This is an ionic compound in which the metal can form more than one cation. Use a Roman numeral to specify the charge of the Fe ion. Since the oxide ion has a −2 charge, the Fe ion has a +2 charge. The correct name is **iron(II) oxide.**

(h) **iron(III) oxide**

(i) This is an ionic compound in which the metal can form more than one cation. Use a Roman numeral to specify the charge of the Ti ion. Since each of the four chloride ions has a −1 charge (total of −4), the Ti ion has a +4 charge. The correct name is **titanium(IV) chloride.**

(j) **sodium hydride**   (k) **lithium nitride**   (l) **sodium oxide**

(m) This is an ionic compound in which the metal cation (Na^+) has only one charge. The O_2^{2-} ion is called the peroxide ion. Each oxygen has a −1 charge. You can determine that each oxygen only has a −1 charge, because each of the two Na ions has a +1 charge. Compare this to sodium oxide in part (l). The correct name is **sodium peroxide.**

2.57 (a) $RbNO_2$ (b) K_2S (c) NaHS   (d) $Mg_3(PO_4)_2$ (e) $CaHPO_4$

(f) KH_2PO_4 (g) IF_7 (h) $(NH_4)_2SO_4$ (i) $AgClO_4$ (j) BCl_3

2.58 **Strategy:** When writing formulas of molecular compounds, the prefixes specify the number of each type of atom in the compound.

When writing formulas of ionic compounds, the subscript of the cation is numerically equal to the charge of the anion, and the subscript of the anion is numerically equal to the charge on the cation. If the charges of the cation and anion are numerically equal, then no subscripts are necessary. Charges of common cations and anions are listed in Table 2.3 of the text. Keep in mind that Roman numerals specify the charge of the cation, *not* the number of metal atoms. Remember that a Roman numeral is not needed for some metal cations, because the charge is known. These metals are the alkali metals (+1), the alkaline earth metals (+2), Ag^+, Zn^{2+}, Cd^{2+}, and Al^{3+}.

When writing formulas of oxoacids, you must know the names and formulas of polyatomic anions (see Table 2.3 of the text).

Solution:

(a) The Roman numeral I tells you that the Cu cation has a +1 charge. Cyanide has a −1 charge. Since, the charges are numerically equal, no subscripts are necessary in the formula. The correct formula is **CuCN.**

(b)   Strontium is an alkaline earth metal. It only forms a +2 cation. The polyatomic ion chlorite, ClO_2^-, has a −1 charge. Since the charges on the cation and anion are numerically different, the subscript of the cation is numerically equal to the charge on the anion, and the subscript of the anion is numerically equal to the charge on the cation. The correct formula is **$Sr(ClO_2)_2$**.

(c)   Perbromic tells you that the anion of this oxoacid is perbromate, BrO_4^-. The correct formula is **$HBrO_4(aq)$**. Remember that (*aq*) means that the substance is dissolved in water.

(d)   Hydroiodic tells you that the anion of this binary acid is iodide, I^-. The correct formula is **$HI(aq)$**.

(e)   Na is an alkali metal. It only forms a +1 cation. The polyatomic ion ammonium, NH_4^+, has a +1 charge and the polyatomic ion phosphate, PO_4^{3-}, has a −3 charge. To balance the charge, you need 2 Na^+ cations. The correct formula is **$Na_2(NH_4)PO_4$**.

(f)   The Roman numeral II tells you that the Pb cation has a +2 charge. The polyatomic ion carbonate, CO_3^{2-}, has a −2 charge. Since, the charges are numerically equal, no subscripts are necessary in the formula. The correct formula is **$PbCO_3$**.

(g)   The Roman numeral II tells you that the Sn cation has a +2 charge. Fluoride has a −1 charge. Since the charges on the cation and anion are numerically different, the subscript of the cation is numerically equal to the charge on the anion, and the subscript of the anion is numerically equal to the charge on the cation. The correct formula is **SnF_2**.

(h)   This is a molecular compound. The Greek prefixes tell you the number of each type of atom in the molecule. The correct formula is **P_4S_{10}**.

(i)   The Roman numeral II tells you that the Hg cation has a +2 charge. Oxide has a −2 charge. Since, the charges are numerically equal, no subscripts are necessary in the formula. The correct formula is **HgO**.

(j)   The Roman numeral I tells you that the Hg cation has a +1 charge. However, this cation exists as Hg_2^{2+}. Iodide has a −1 charge. You need two iodide ion to balance the +2 charge of Hg_2^{2+}. The correct formula is **Hg_2I_2**.

(k)   This is a molecular compound. The Greek prefixes tell you the number of each type of atom in the molecule. The correct formula is **SeF_6**.

2.59   Uranium is radioactive. It loses mass because it constantly emits alpha (α) particles.

2.60   Changing the electrical charge of an atom usually has a major effect on its chemical properties. The two electrically neutral carbon isotopes should have nearly identical chemical properties.

2.61   The number of protons = 65 − 35 = 30. The element that contains 30 protons is zinc, Zn. There are two fewer electrons than protons, so the charge of the cation is +2. The symbol for this cation is **Zn^{2+}**.

2.62   Atomic number = 127 − 74 = 53. This anion has 53 protons, so it is an iodide ion. Since there is one more electron than protons, the ion has a −1 charge. The correct symbol is **I^-**.

2.63   **(a)**   Species with the same number of protons and electrons will be neutral. **A, F, G.**
          (b)   Species with more electrons than protons will have a negative charge. **B, E.**
          (c)   Species with more protons than electrons will have a positive charge. **C, D.**
          (d)   A: $^{10}_{5}B$ B: $^{14}_{7}N^{3-}$ C: $^{39}_{19}K^+$ D: $^{66}_{30}Zn^{2+}$ E: $^{81}_{35}Br^-$ F: $^{11}_{5}B$ G: $^{19}_{9}F$

2.64   NaCl is an ionic compound; it doesn't form molecules.

2.65   **Yes.** The law of multiple proportions requires that the masses of sulfur combining with phosphorus must be in the ratios of small whole numbers. For the three compounds shown, four phosphorus atoms combine with three, seven, and ten sulfur atoms, respectively. If the atom ratios are in small whole number ratios, then the mass ratios must also be in small whole number ratios.

2.66   The species and their identification are as follows:

(a)	SO_2	molecule and compound	**(g)**	O_3	element and molecule
(b)	S_8	element and molecule	**(h)**	CH_4	molecule and compound
(c)	Cs	element	**(i)**	KBr	compound
(d)	N_2O_5	molecule and compound	**(j)**	S	element
(e)	O	element	**(k)**	P_4	element and molecule
(f)	O_2	element and molecule	**(l)**	LiF	compound

2.67   **(a)**   This is an ionic compound. Prefixes are *not* used. The correct name is barium chloride.

       (b)   Iron has a +3 charge in this compound. The correct name is iron(III) oxide.

       (c) NO_2^- is the nitrite ion. The correct name is cesium nitrite.

       (d)   Magnesium is an alkaline earth metal, which always has a +2 charge in ionic compounds. The roman numeral is not necessary. The correct name is magnesium bicarbonate.

2.68   **(a)**   Ammonium is NH_4^+, not NH_3^+. The formula should be **$(NH_4)_2CO_3$**.

       (b)   Calcium has a +2 charge and hydroxide has a −1 charge. The formula should be **$Ca(OH)_2$**.

       (c)   Sulfide is S^{2-}, not SO_3^{2-}. The correct formula is **CdS**.

       (d)   Dichromate is $Cr_2O_7^{2-}$, not $Cr_2O_4^{2-}$. The correct formula is **$ZnCr_2O_7$**.

2.69

Symbol	$^{11}_{5}B$	$^{54}_{26}Fe^{2+}$	$^{31}_{15}P^{3-}$	$^{196}_{79}Au$	$^{222}_{86}Rn$
Protons	5	26	15	79	86
Neutrons	6	28	16	117	136
Electrons	5	24	18	79	86
Net Charge	0	+2	−3	0	0

2.70   **(a)**   Ionic compounds are typically formed between metallic and nonmetallic elements.

       (b)   In general the transition metals, the actinides and lanthanides have variable charges.

2.71   **(a)** Li^+, alkali metals always have a +1 charge in ionic compounds

       (b) S^{2-}

       (c) I^-, halogens have a −1 charge in ionic compounds

       (d) N^{3-}

       (e) Al^{3+}, aluminum always has a +3 charge in ionic compounds

       (f) Cs^+, alkali metals always have a +1 charge in ionic compounds

       (g) Mg^{2+}, alkaline earth metals always have a +2 charge in ionic compounds.

2.72   The symbol ^{23}Na provides more information than $_{11}Na$. The mass number plus the chemical symbol identifies a specific isotope of Na (sodium) while combining the atomic number with the chemical symbol tells you nothing new. Can other isotopes of sodium have different atomic numbers?

2.73   The binary Group 7A element acids are: HF, hydrofluoric acid; HCl, hydrochloric acid; HBr, hydrobromic acid; HI, hydroiodic acid. Oxoacids containing Group 7A elements (using the specific examples for chlorine) are: $HClO_4$, perchloric acid; $HClO_3$, chloric acid; $HClO_2$, chlorous acid: HClO, hypochlorous acid.

Examples of oxoacids containing other Group A-block elements are: H_3BO_3, boric acid (Group 3A); H_2CO_3, carbonic acid (Group 4A); HNO_3, nitric acid and H_3PO_4, phosphoric acid (Group 5A); and H_2SO_4, sulfuric acid (Group 6A). Hydrosulfuric acid, H_2S, is an example of a binary Group 6A acid while HCN, hydrocyanic acid, contains both a Group 4A and 5A element.

2.74   Mercury (Hg) and bromine (Br_2)

2.75   **(a)**

Isotope	$_2^4He$	$_{10}^{20}Ne$	$_{18}^{40}Ar$	$_{36}^{84}Kr$	$_{54}^{132}Xe$
No. Protons	2	10	18	36	54
No. Neutrons	2	10	22	48	78

(b)

neutron/proton ratio	1.00	1.00	1.22	1.33	1.44

The neutron/proton ratio increases with increasing atomic number.

2.76 H_2, N_2, O_2, F_2, Cl_2, He, Ne, Ar, Kr, Xe, Rn

2.77   Cu, Ag, and Au are fairly chemically unreactive. This makes them specially suitable for making coins and jewelry, that you want to last a very long time.

2.78   They do not have a strong tendency to form compounds. Helium, neon, and argon are chemically inert.

2.79   Magnesium and strontium are also alkaline earth metals. You should expect the charge of the metal to be the same (+2). **MgO** and **SrO**.

2.80   All isotopes of radium are radioactive. It is a radioactive decay product of uranium-238. Radium itself does *not* occur naturally on Earth.

2.81   **(a)**   Berkelium (Berkeley, CA); Europium (Europe); Francium (France); Scandium (Scandinavia); Ytterbium (Ytterby, Sweden); Yttrium (Ytterby, Sweden).
    (b)   Einsteinium (Albert Einstein); Fermium (Enrico Fermi); Curium (Marie and Pierre Curie); Mendelevium (Dmitri Mendeleev); Lawrencium (Ernest Lawrence).
    (c)   Arsenic, Cesium, Chlorine, Chromium, Iodine.

2.82   Argentina is named after silver (argentum, Ag).

2.83   The mass of fluorine reacting with hydrogen and deuterium would be the same. The ratio of F atom to hydrogen (or deuterium) is 1:1 in both compounds. This does not violate the law of definite proportions. When the law of definite proportions was formulated, scientists did not know of the existence of isotopes.

2.84   **(a)**   NaH, sodium hydride   **(b)** B_2O_3, diboron trioxide   **(c)** Na_2S, sodium sulfide
    (d) AlF_3, aluminum fluoride   **(e)** OF_2, oxygen difluoride   **(f)** $SrCl_2$, strontium chloride

2.85   **(a)**   Br   **(b)**   Rn   **(c)**   Se   **(d)**   Rb   **(e)**   Pb

2.86   All of these are molecular compounds. We use prefixes to express the number of each atom in the molecule. The names are nitrogen trifluoride (NF_3), phosphorus pentabromide (PBr_5), and sulfur dichloride (SCl_2).

2.87

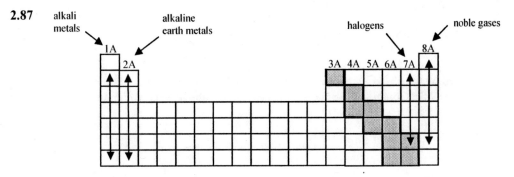

The metalloids are shown in gray.

2.88

Cation	Anion	Formula	Name
Mg^{2+}	HCO_3^-	$Mg(HCO_3)_2$	Magnesium bicarbonate
Sr^{2+}	Cl^-	$SrCl_2$	**Strontium chloride**
Fe^{3+}	NO_2^-	$Fe(NO_2)_3$	**Iron(III) nitrite**
Mn^{2+}	ClO_3^-	$Mn(ClO_3)_2$	Manganese(II) chlorate
Sn^{4+}	Br^-	$SnBr_4$	**Tin(IV) bromide**
Co^{2+}	PO_4^{3-}	$Co_3(PO_4)_2$	**Cobalt(II) phosphate**
Hg_2^{2+}	I^-	Hg_2I_2	**Mercury(I) iodide**
Cu^+	CO_3^{2-}	Cu_2CO_3	**Copper(I) carbonate**
Li^+	N^{3-}	Li_3N	Lithium nitride
Al^{3+}	S^{2-}	Al_2S_3	**Aluminum sulfide**

2.89   **(a)** $CO_2(s)$, solid carbon dioxide   **(f)** $Ca(OH)_2$, calcium hydroxide

       (b) NaCl, sodium chloride   **(g)** $NaHCO_3$, sodium bicarbonate

       (c) N_2O, nitrous oxide   **(h)** $Na_2CO_3 \cdot 10H_2O$, sodium carbonate decahydrate

       (d) $CaCO_3$, calcium carbonate   **(i)** $CaSO_4 \cdot 2H_2O$, calcium sulfate dihydrate

       (e) CaO, calcium oxide   **(j)** $Mg(OH)_2$, magnesium hydroxide

2.90   **(a)** Rutherford's experiment is described in detail in Section 2.2 of the text. From the average magnitude of scattering, Rutherford estimated the number of protons (based on electrostatic interactions) in the nucleus.

       (b) Assuming that the nucleus is spherical, the volume of the nucleus is:

$$V = \frac{4}{3}\pi r^3 = \frac{4}{3}\pi(3.04 \times 10^{-13}\text{ cm})^3 = 1.18 \times 10^{-37}\text{ cm}^3$$

       The density of the nucleus can now be calculated.

$$d = \frac{m}{V} = \frac{3.82 \times 10^{-23}\text{ g}}{1.18 \times 10^{-37}\text{ cm}^3} = \mathbf{3.24 \times 10^{14}\text{ g/cm}^3}$$

To calculate the density of the space occupied by the electrons, we need both the mass of 11 electrons, and the volume occupied by these electrons.

The mass of 11 electrons is:

$$11 \text{ electrons} \times \frac{9.1095 \times 10^{-28} \text{ g}}{1 \text{ electron}} = 1.0020 \times 10^{-26} \text{ g}$$

The volume occupied by the electrons will be the difference between the volume of the atom and the volume of the nucleus. The volume of the nucleus was calculated above. The volume of the atom is calculated as follows:

$$186 \text{ pm} \times \frac{1 \times 10^{-12} \text{ m}}{1 \text{ pm}} \times \frac{1 \text{ cm}}{1 \times 10^{-2} \text{ m}} = 1.86 \times 10^{-8} \text{ cm}$$

$$V_{atom} = \frac{4}{3}\pi r^3 = \frac{4}{3}\pi(1.86 \times 10^{-8} \text{ cm})^3 = 2.70 \times 10^{-23} \text{ cm}^3$$

$$V_{electrons} = V_{atom} - V_{nucleus} = (2.70 \times 10^{-23} \text{ cm}^3) - (1.18 \times 10^{-37} \text{ cm}^3) = 2.70 \times 10^{-23} \text{ cm}^3$$

As you can see, the volume occupied by the nucleus is insignificant compared to the space occupied by the electrons.

The density of the space occupied by the electrons can now be calculated.

$$d = \frac{m}{V} = \frac{1.0020 \times 10^{-26} \text{ g}}{2.70 \times 10^{-23} \text{ cm}^3} = 3.71 \times 10^{-4} \text{ g/cm}^3$$

The above results do support Rutherford's model. Comparing the space occupied by the electrons to the volume of the nucleus, it is clear that most of the atom is empty space. Rutherford also proposed that the nucleus was a *dense* central core with most of the mass of the atom concentrated in it. Comparing the density of the nucleus with the density of the space occupied by the electrons also supports Rutherford's model.

2.91

S	N
B	I

CHAPTER 3
MASS RELATIONSHIPS IN CHEMICAL REACTIONS

3.5 $(34.968 \text{ amu})(0.7553) + (36.956 \text{ amu})(0.2447) = \textbf{35.45 amu}$

3.6   **Strategy:** Each isotope contributes to the average atomic mass based on its relative abundance. Multiplying the mass of an isotope by its fractional abundance (not percent) will give the contribution to the average atomic mass of that particular isotope.

It would seem that there are two unknowns in this problem, the fractional abundance of ^{6}Li and the fractional abundance of ^{7}Li. However, these two quantities are not independent of each other; they are related by the fact that they must sum to 1. Start by letting x be the fractional abundance of ^{6}Li. Since the sum of the two abundance's must be 1, we can write

$$\text{Abundance } ^7\text{Li} = (1 - x)$$

Solution:

$$\textbf{Average atomic mass of Li} = 6.941 \text{ amu} = x(6.0151 \text{ amu}) + (1 - x)(7.0160 \text{ amu})$$
$$6.941 = 1.0009x + 7.0160$$
$$1.0009x = 0.075$$
$$\boldsymbol{x = 0.075}$$

$x = 0.075$ corresponds to a natural abundance of ^{6}Li of **7.5 percent**. The natural abundance of ^{7}Li is $(1 - x) = 0.925$ or **92.5 percent**.

3.7   The unit factor required is $\left(\dfrac{6.022 \times 10^{23} \text{ amu}}{1 \text{ g}} \right)$

$$\textbf{? g} = 13.2 \text{ amu} \times \frac{1 \text{ g}}{6.022 \times 10^{23} \text{ amu}} = \textbf{2.19} \times \textbf{10}^{-23} \textbf{ g}$$

3.8   The unit factor required is $\left(\dfrac{6.022 \times 10^{23} \text{ amu}}{1 \text{ g}} \right)$

$$\textbf{? amu} = 8.4 \text{ g} \times \frac{6.022 \times 10^{23} \text{ amu}}{1 \text{ g}} = \textbf{5.1} \times \textbf{10}^{24} \textbf{ amu}$$

3.11   In one year:

$$(6.5 \times 10^9 \text{ people}) \times \frac{365 \text{ days}}{1 \text{ yr}} \times \frac{24 \text{ hr}}{1 \text{ day}} \times \frac{3600 \text{ s}}{1 \text{ hr}} \times \frac{2 \text{ particles}}{1 \text{ person}} = 4.1 \times 10^{17} \text{ particles/yr}$$

$$\textbf{Total time} = \frac{6.022 \times 10^{23} \text{ particles}}{4.1 \times 10^{17} \text{ particles/yr}} = \textbf{1.5} \times \textbf{10}^6 \textbf{ yr}$$

3.12   The thickness of the book in miles would be:

$$\frac{0.0036 \text{ in}}{1 \text{ page}} \times \frac{1 \text{ ft}}{12 \text{ in}} \times \frac{1 \text{ mi}}{5280 \text{ ft}} \times (6.022 \times 10^{23} \text{ pages}) = 3.4 \times 10^{16} \text{ mi}$$

The distance, in miles, traveled by light in one year is:

$$1.00 \text{ yr} \times \frac{365 \text{ day}}{1 \text{ yr}} \times \frac{24 \text{ h}}{1 \text{ day}} \times \frac{3600 \text{ s}}{1 \text{ h}} \times \frac{3.00 \times 10^8 \text{ m}}{1 \text{ s}} \times \frac{1 \text{ mi}}{1609 \text{ m}} = 5.88 \times 10^{12} \text{ mi}$$

The thickness of the book in light-years is:

$$(3.4 \times 10^{16} \text{ mi}) \times \frac{1 \text{ light-yr}}{5.88 \times 10^{12} \text{ mi}} = \mathbf{5.8 \times 10^3 \text{ light-yr}}$$

It will take light 5.8×10^3 years to travel from the first page to the last one!

3.13 $5.10 \text{ mol S} \times \dfrac{6.022 \times 10^{23} \text{ S atoms}}{1 \text{ mol S}} = \mathbf{3.07 \times 10^{24} \text{ S atoms}}$

3.14 $(6.00 \times 10^9 \text{ Co atoms}) \times \dfrac{1 \text{ mol Co}}{6.022 \times 10^{23} \text{ Co atoms}} = \mathbf{9.96 \times 10^{-15} \text{ mol Co}}$

3.15 $77.4 \text{ g of Ca} \times \dfrac{1 \text{ mol Ca}}{40.08 \text{ g Ca}} = \mathbf{1.93 \text{ mol Ca}}$

3.16   **Strategy:** We are given moles of gold and asked to solve for grams of gold. What conversion factor do we need to convert between moles and grams? Arrange the appropriate conversion factor so moles cancel, and the unit grams is obtained for the answer.

Solution: The conversion factor needed to covert between moles and grams is the molar mass. In the periodic table (see inside front cover of the text), we see that the molar mass of Au is 197.0 g. This can be expressed as

$$1 \text{ mol Au} = 197.0 \text{ g Au}$$

From this equality, we can write two conversion factors.

$$\frac{1 \text{ mol Au}}{197.0 \text{ g Au}} \quad \text{and} \quad \frac{197.0 \text{ g Au}}{1 \text{ mol Au}}$$

The conversion factor on the right is the correct one. Moles will cancel, leaving the unit grams for the answer.

We write

$$\mathbf{?\ g\ Au} = 15.3 \text{ mol Au} \times \frac{197.0 \text{ g Au}}{1 \text{ mol Au}} = \mathbf{3.01 \times 10^3 \text{ g Au}}$$

Check: Does a mass of 3010 g for 15.3 moles of Au seem reasonable? What is the mass of 1 mole of Au?

3.17 **(a)** $\dfrac{200.6 \text{ g Hg}}{1 \text{ mol Hg}} \times \dfrac{1 \text{ mol Hg}}{6.022 \times 10^{23} \text{ Hg atoms}} = \mathbf{3.331 \times 10^{-22} \text{ g/Hg atom}}$

       (b) $\dfrac{20.18 \text{ g Ne}}{1 \text{ mol Ne}} \times \dfrac{1 \text{ mol Ne}}{6.022 \times 10^{23} \text{ Ne atoms}} = \mathbf{3.351 \times 10^{-23} \text{ g/Ne atom}}$

3.18 **(a)**

Strategy: We can look up the molar mass of arsenic (As) on the periodic table (74.92 g/mol). We want to find the mass of a single atom of arsenic (unit of g/atom). Therefore, we need to convert from the unit mole in the denominator to the unit atom in the denominator. What conversion factor is needed to convert between moles and atoms? Arrange the appropriate conversion factor so mole in the denominator cancels, and the unit atom is obtained in the denominator.

Solution: The conversion factor needed is Avogadro's number. We have

$$1 \text{ mol} = 6.022 \times 10^{23} \text{ particles (atoms)}$$

From this equality, we can write two conversion factors.

$$\frac{1 \text{ mol As}}{6.022 \times 10^{23} \text{ As atoms}} \quad \text{and} \quad \frac{6.022 \times 10^{23} \text{ As atoms}}{1 \text{ mol As}}$$

The conversion factor on the left is the correct one. Moles will cancel, leaving the unit atoms in the denominator of the answer.

We write

$$? \text{ g/As atom} = \frac{74.92 \text{ g As}}{1 \text{ mol As}} \times \frac{1 \text{ mol As}}{6.022 \times 10^{23} \text{ As atoms}} = \mathbf{1.244 \times 10^{-22} \text{ g/As atom}}$$

(b) Follow same method as part (a).

$$? \text{ g/Ni atom} = \frac{58.69 \text{ g Ni}}{1 \text{ mol Ni}} \times \frac{1 \text{ mol Ni}}{6.022 \times 10^{23} \text{ Ni atoms}} = \mathbf{9.746 \times 10^{-23} \text{ g/Ni atom}}$$

Check: Should the mass of a single atom of As or Ni be a very small mass?

3.19 $1.00 \times 10^{12} \text{ Pb atoms} \times \dfrac{1 \text{ mol Pb}}{6.022 \times 10^{23} \text{ Pb atoms}} \times \dfrac{207.2 \text{ g Pb}}{1 \text{ mol Pb}} = \mathbf{3.44 \times 10^{-10} \text{ g Pb}}$

3.20 **Strategy:** The question asks for atoms of Cu. We cannot convert directly from grams to atoms of copper. What unit do we need to convert grams of Cu to in order to convert to atoms? What does Avogadro's number represent?

Solution: To calculate the number of Cu atoms, we first must convert grams of Cu to moles of Cu. We use the molar mass of copper as a conversion factor. Once moles of Cu are obtained, we can use Avogadro's number to convert from moles of copper to atoms of copper.

$$1 \text{ mol Cu} = 63.55 \text{ g Cu}$$

The conversion factor needed is

$$\frac{1 \text{ mol Cu}}{63.55 \text{ g Cu}}$$

Avogadro's number is the key to the second conversion. We have

$$1 \text{ mol} = 6.022 \times 10^{23} \text{ particles (atoms)}$$

From this equality, we can write two conversion factors.

$$\frac{1 \text{ mol Cu}}{6.022 \times 10^{23} \text{ Cu atoms}} \quad \text{and} \quad \frac{6.022 \times 10^{23} \text{ Cu atoms}}{1 \text{ mol Cu}}$$

The conversion factor on the right is the one we need because it has number of Cu atoms in the numerator, which is the unit we want for the answer.

Let's complete the two conversions in one step.

$$\text{grams of Cu} \rightarrow \text{moles of Cu} \rightarrow \text{number of Cu atoms}$$

$$\textbf{? atoms of Cu} = 3.14 \text{ g Cu} \times \frac{1 \text{ mol Cu}}{63.55 \text{ g Cu}} \times \frac{6.022 \times 10^{23} \text{ Cu atoms}}{1 \text{ mol Cu}} = \textbf{2.98} \times \textbf{10}^{\textbf{22}} \textbf{ Cu atoms}$$

Check: Should 3.14 g of Cu contain fewer than Avogadro's number of atoms? What mass of Cu would contain Avogadro's number of atoms?

3.21   For hydrogen: $1.10 \text{ g H} \times \dfrac{1 \text{ mol H}}{1.008 \text{ g H}} \times \dfrac{6.022 \times 10^{23} \text{ H atoms}}{1 \text{ mol H}} = \textbf{6.57} \times \textbf{10}^{\textbf{23}} \textbf{ H atoms}$

For chromium: $14.7 \text{ g Cr} \times \dfrac{1 \text{ mol Cr}}{52.00 \text{ g Cr}} \times \dfrac{6.022 \times 10^{23} \text{ Cr atoms}}{1 \text{ mol Cr}} = \textbf{1.70} \times \textbf{10}^{\textbf{23}} \textbf{ Cr atoms}$

There are more hydrogen atoms than chromium atoms.

3.22 $2 \text{ Pb atoms} \times \dfrac{1 \text{ mol Pb}}{6.022 \times 10^{23} \text{ Pb atoms}} \times \dfrac{207.2 \text{ g Pb}}{1 \text{ mol Pb}} = 6.881 \times 10^{-22} \text{ g Pb}$

$(5.1 \times 10^{-23} \text{ mol He}) \times \dfrac{4.003 \text{ g He}}{1 \text{ mol He}} = 2.0 \times 10^{-22} \text{ g He}$

2 atoms of lead have a greater mass than 5.1×10^{-23} mol of helium.

3.23   Using the appropriate atomic masses,

(a) CH_4 12.01 amu + 4(1.008 amu) = **16.04 amu**
(b) NO_2 14.01 amu + 2(16.00 amu) = **46.01 amu**
(c) SO_3 32.07 amu + 3(16.00 amu) = **80.07 amu**
(d) C_6H_6 6(12.01 amu) + 6(1.008 amu) = **78.11 amu**
(e)   NaI   22.99 amu + 126.9 amu = **149.9 amu**
(f) K_2SO_4 2(39.10 amu) + 32.07 amu + 4(16.00 amu) = **174.27 amu**
(g) $Ca_3(PO_4)_2$ 3(40.08 amu) + 2(30.97 amu) + 8(16.00 amu) = **310.18 amu**

3.24   **Strategy:** How do molar masses of different elements combine to give the molar mass of a compound?

Solution: To calculate the molar mass of a compound, we need to sum all the molar masses of the elements in the molecule. For each element, we multiply its molar mass by the number of moles of that element in one mole of the compound. We find molar masses for the elements in the periodic table (inside front cover of the text).

(a)   **molar mass Li_2CO_3** $= 2(6.941 \text{ g}) + 12.01 \text{ g} + 3(16.00 \text{ g}) = $ **73.89 g**

(b)   **molar mass CS_2** $= 12.01 \text{ g} + 2(32.07 \text{ g}) = $ **76.15 g**

(c)   **molar mass $CHCl_3$** $= 12.01 \text{ g} + 1.008 \text{ g} + 3(35.45 \text{ g}) = $ **119.37 g**

(d)   **molar mass $C_6H_8O_6$** $= 6(12.01 \text{ g}) + 8(1.008 \text{ g}) + 6(16.00 \text{ g}) = $ **176.12 g**

(e)   **molar mass KNO_3** $= 39.10 \text{ g} + 14.01 \text{ g} + 3(16.00 \text{ g}) = $ **101.11 g**

(f)   **molar mass Mg_3N_2** $= 3(24.31 \text{ g}) + 2(14.01 \text{ g}) = $ **100.95 g**

3.25   To find the molar mass (g/mol), we simply divide the mass (in g) by the number of moles.

$$\frac{152 \text{ g}}{0.372 \text{ mol}} = \textbf{409 g/mol}$$

3.26   **Strategy:** We are given grams of ethane and asked to solve for molecules of ethane. We cannot convert directly from grams ethane to molecules of ethane. What unit do we need to obtain first before we can convert to molecules? How should Avogadro's number be used here?

Solution: To calculate number of ethane molecules, we first must convert grams of ethane to moles of ethane. We use the molar mass of ethane as a conversion factor. Once moles of ethane are obtained, we can use Avogadro's number to convert from moles of ethane to molecules of ethane.

molar mass of $C_2H_6 = 2(12.01 \text{ g}) + 6(1.008 \text{ g}) = 30.07 \text{ g}$

The conversion factor needed is

$$\frac{1 \text{ mol } C_2H_6}{30.07 \text{ g } C_2H_6}$$

Avogadro's number is the key to the second conversion. We have

$$1 \text{ mol} = 6.022 \times 10^{23} \text{ particles (molecules)}$$

From this equality, we can write the conversion factor:

$$\frac{6.022 \times 10^{23} \text{ ethane molecules}}{1 \text{ mol ethane}}$$

Let's complete the two conversions in one step.

grams of ethane $\rightarrow$ moles of ethane $\rightarrow$ number of ethane molecules

$$\textbf{? molecules of } \mathbf{C_2H_6} = 0.334 \text{ g } C_2H_6 \times \frac{1 \text{ mol } C_2H_6}{30.07 \text{ g } C_2H_6} \times \frac{6.022 \times 10^{23} \text{ } C_2H_6 \text{ molecules}}{1 \text{ mol } C_2H_6}$$

$$= \textbf{6.69} \times \textbf{10}^{\textbf{21}} \textbf{ } \mathbf{C_2H_6} \textbf{ molecules}$$

Check: Should 0.334 g of ethane contain fewer than Avogadro's number of molecules? What mass of ethane would contain Avogadro's number of molecules?

3.27 $1.50 \text{ g glucose} \times \dfrac{1 \text{ mol glucose}}{180.2 \text{ g glucose}} \times \dfrac{6.022 \times 10^{23} \text{ molecules glucose}}{1 \text{ mol glucose}} \times \dfrac{6 \text{ C atoms}}{1 \text{ molecule glucose}}$

$= \mathbf{3.01 \times 10^{22} \ C \ atoms}$

The ratio of O atoms to C atoms in glucose is 1:1. Therefore, there are the same number of O atoms in glucose as C atoms, so the number of O atoms = $\mathbf{3.01 \times 10^{22} \ O \ atoms}$.

The ratio of H atoms to C atoms in glucose is 2:1. Therefore, there are twice as many H atoms in glucose as C atoms, so the number of H atoms = $2(3.01 \times 10^{22} \text{ atoms}) = \mathbf{6.02 \times 10^{22} \ H \ atoms}$.

3.28   **Strategy:** We are asked to solve for the number of N, C, O, and H atoms in 1.68×10^4 g of urea. We cannot convert directly from grams urea to atoms. What unit do we need to obtain first before we can convert to atoms? How should Avogadro's number be used here? How many atoms of N, C, O, or H are in 1 molecule of urea?

Solution: Let's first calculate the number of N atoms in 1.68×10^4 g of urea. First, we must convert grams of urea to number of molecules of urea. This calculation is similar to Problem 3.26. The molecular formula of urea shows there are two N atoms in one urea molecule, which will allow us to convert to atoms of N. We need to perform three conversions:

grams of urea $\rightarrow$ moles of urea $\rightarrow$ molecules of urea $\rightarrow$ atoms of N

The conversion factors needed for each step are: 1) the molar mass of urea, 2) Avogadro's number, and 3) the number of N atoms in 1 molecule of urea.

We complete the three conversions in one calculation.

$? \text{ atoms of N} = (1.68 \times 10^4 \text{ g urea}) \times \dfrac{1 \text{ mol urea}}{60.06 \text{ g urea}} \times \dfrac{6.022 \times 10^{23} \text{ urea molecules}}{1 \text{ mol urea}} \times \dfrac{2 \text{ N atoms}}{1 \text{ molecule urea}}$

$= \mathbf{3.37 \times 10^{26} \ N \ atoms}$

The above method utilizes the ratio of molecules (urea) to atoms (nitrogen). We can also solve the problem by reading the formula as the ratio of moles of urea to moles of nitrogen by using the following conversions:

grams of urea $\rightarrow$ moles of urea $\rightarrow$ moles of N $\rightarrow$ atoms of N

Try it.

Check: Does the answer seem reasonable? We have 1.68×10^4 g urea. How many atoms of N would 60.06 g of urea contain?

We could calculate the number of atoms of the remaining elements in the same manner, or we can use the atom ratios from the molecular formula. The carbon atom to nitrogen atom ratio in a urea molecule is 1:2, the oxygen atom to nitrogen atom ratio is 1:2, and the hydrogen atom to nitrogen atom ration is 4:2.

$? \text{ atoms of C} = (3.37 \times 10^{26} \text{ N atoms}) \times \dfrac{1 \text{ C atom}}{2 \text{ N atoms}} = \mathbf{1.69 \times 10^{26} \ C \ atoms}$

$$? \text{ atoms of O } = (3.37 \times 10^{26} \text{ N atoms}) \times \frac{1 \text{ O atom}}{2 \text{ N atoms}} = 1.69 \times 10^{26} \text{ O atoms}$$

$$? \text{ atoms of H } = (3.37 \times 10^{26} \text{ N atoms}) \times \frac{4 \text{ H atoms}}{2 \text{ N atoms}} = 6.74 \times 10^{26} \text{ H atoms}$$

3.29   The molar mass of $C_{19}H_{38}O$ is 282.5 g.

$$1.0 \times 10^{-12} \text{ g} \times \frac{1 \text{ mol}}{282.5 \text{ g}} \times \frac{6.022 \times 10^{23} \text{ molecules}}{1 \text{ mol}} = 2.1 \times 10^9 \text{ molecules}$$

Notice that even though 1.0×10^{-12} g is an extremely small mass, it still is comprised of over a billion pheromone molecules!

3.30 $\text{Mass of water } = 2.56 \text{ mL} \times \frac{1.00 \text{ g}}{1.00 \text{ mL}} = 2.56 \text{ g}$

Molar mass of $H_2O = (16.00 \text{ g}) + 2(1.008 \text{ g}) = 18.02 \text{ g/mol}$

$$? \; H_2O \text{ molecules } = 2.56 \text{ g } H_2O \times \frac{1 \text{ mol } H_2O}{18.02 \text{ g } H_2O} \times \frac{6.022 \times 10^{23} \text{ molecules } H_2O}{1 \text{ mol } H_2O}$$

$$= 8.56 \times 10^{22} \text{ molecules}$$

3.33   Since there are only two isotopes of carbon, there are only two possibilities for $CF_4{}^+$.

$${}^{12}_{6}C \, {}^{19}_{9}F_4^+ \text{ (molecular mass 88 amu) and } {}^{13}_{6}C \, {}^{19}_{9}F_4^+ \text{ (molecular mass 89 amu)}$$

There would be two peaks in the mass spectrum.

3.34   Since there are two hydrogen isotopes, they can be paired in three ways: ${}^1H-{}^1H$, ${}^1H-{}^2H$, and ${}^2H-{}^2H$. There will then be three choices for each sulfur isotope. We can make a table showing all the possibilities (masses in amu):

	${}^{32}S$	${}^{33}S$	${}^{34}S$	${}^{36}S$
${}^1H_2$	34	35	36	38
${}^1H{}^2H$	35	36	37	39
${}^2H_2$	36	37	38	40

There will be **seven peaks** of the following mass numbers: 34, 35, 36, 37, 38, 39, and 40.

Very accurate (and expensive!) mass spectrometers can detect the mass difference between two ${}^1H$ and one ${}^2H$. How many peaks would be detected in such a "high resolution" mass spectrum?

3.39   Molar mass of $SnO_2 = (118.7 \text{ g}) + 2(16.00 \text{ g}) = 150.7 \text{ g}$

$$\%Sn = \frac{118.7 \text{ g/mol}}{150.7 \text{ g/mol}} \times 100\% = 78.77\%$$

$$\%O = \frac{(2)(16.00 \text{ g/mol})}{150.7 \text{ g/mol}} \times 100\% = 21.23\%$$

3.40   **Strategy:** Recall the procedure for calculating a percentage. Assume that we have 1 mole of $CHCl_3$. The percent by mass of each element (C, H, and Cl) is given by the mass of that element in 1 mole of $CHCl_3$ divided by the molar mass of $CHCl_3$, then multiplied by 100 to convert from a fractional number to a percentage.

Solution: The molar mass of $CHCl_3$ = 12.01 g/mol + 1.008 g/mol + 3(35.45 g/mol) = 119.4 g/mol. The percent by mass of each of the elements in $CHCl_3$ is calculated as follows:

$$\%C = \frac{12.01 \text{ g/mol}}{119.4 \text{ g/mol}} \times 100\% = \mathbf{10.06\%}$$

$$\%H = \frac{1.008 \text{ g/mol}}{119.4 \text{ g/mol}} \times 100\% = \mathbf{0.8442\%}$$

$$\%Cl = \frac{3(35.45) \text{ g/mol}}{119.4 \text{ g/mol}} \times 100\% = \mathbf{89.07\%}$$

Check: Do the percentages add to 100%? The sum of the percentages is (10.06% + 0.8442% + 89.07%) = 99.97%. The small discrepancy from 100% is due to the way we rounded off.

3.41   The molar mass of cinnamic alcohol is 134.17 g/mol.

(a) $$\%C = \frac{(9)(12.01 \text{ g/mol})}{134.17 \text{ g/mol}} \times 100\% = \mathbf{80.56\%}$$

$$\%H = \frac{(10)(1.008 \text{ g/mol})}{134.17 \text{ g/mol}} \times 100\% = \mathbf{7.51\%}$$

$$\%O = \frac{16.00 \text{ g/mol}}{134.17 \text{ g/mol}} \times 100\% = \mathbf{11.93\%}$$

(b) $$0.469 \text{ g } C_9H_{10}O \times \frac{1 \text{ mol } C_9H_{10}O}{134.17 \text{ g } C_9H_{10}O} \times \frac{6.022 \times 10^{23} \text{ molecules } C_9H_{10}O}{1 \text{ mol } C_9H_{10}O}$$

$$= \mathbf{2.11 \times 10^{21} \text{ molecules } C_9H_{10}O}$$

3.42

Compound	Molar mass (g)	N% by mass
(a) $(NH_2)_2CO$	60.06	$\dfrac{2(14.01 \text{ g})}{60.06 \text{ g}} \times 100\% = 46.65\%$
(b) NH_4NO_3	80.05	$\dfrac{2(14.01 \text{ g})}{80.05 \text{ g}} \times 100\% = 35.00\%$
(c) $HNC(NH_2)_2$	59.08	$\dfrac{3(14.01 \text{ g})}{59.08 \text{ g}} \times 100\% = 71.14\%$
(d) NH_3	17.03	$\dfrac{14.01 \text{ g}}{17.03 \text{ g}} \times 100\% = 82.27\%$

Ammonia, $\mathbf{NH_3}$, is the richest source of nitrogen on a mass percentage basis.

3.43   Assume you have exactly 100 g of substance.

$$n_C = 44.4 \text{ g C} \times \frac{1 \text{ mol C}}{12.01 \text{ g C}} = 3.70 \text{ mol C}$$

$$n_H = 6.21 \text{ g H} \times \frac{1 \text{ mol H}}{1.008 \text{ g H}} = 6.16 \text{ mol H}$$

$$n_S = 39.5 \text{ g S} \times \frac{1 \text{ mol S}}{32.07 \text{ g S}} = 1.23 \text{ mol S}$$

$$n_O = 9.86 \text{ g O} \times \frac{1 \text{ mol O}}{16.00 \text{ g O}} = 0.616 \text{ mol O}$$

Thus, we arrive at the formula $C_{3.70}H_{6.16}S_{1.23}O_{0.616}$. Dividing by the smallest number of moles (0.616 mole) gives the empirical formula, $C_6H_{10}S_2O$.

To determine the molecular formula, divide the molar mass by the empirical mass.

$$\frac{\text{molar mass}}{\text{empirical molar mass}} = \frac{162 \text{ g}}{162.3 \text{ g}} \approx 1$$

Hence, the molecular formula and the empirical formula are the same, $C_6H_{10}S_2O$.

3.44   **METHOD 1:**

Step 1: Assume you have exactly 100 g of substance. 100 g is a convenient amount, because all the percentages sum to 100%. The percentage of oxygen is found by difference:

$$100\% - (19.8\% + 2.50\% + 11.6\%) = 66.1\%$$

In 100 g of PAN there will be 19.8 g C, 2.50 g H, 11.6 g N, and 66.1 g O.

Step 2: Calculate the number of moles of each element in the compound. Remember, an *empirical formula* tells us which elements are present and the simplest whole-number ratio of their atoms. This ratio is also a mole ratio. Use the molar masses of these elements as conversion factors to convert to moles.

$$n_C = 19.8 \text{ g C} \times \frac{1 \text{ mol C}}{12.01 \text{ g C}} = 1.65 \text{ mol C}$$

$$n_H = 2.50 \text{ g H} \times \frac{1 \text{ mol H}}{1.008 \text{ g H}} = 2.48 \text{ mol H}$$

$$n_N = 11.6 \text{ g N} \times \frac{1 \text{ mol N}}{14.01 \text{ g N}} = 0.828 \text{ mol N}$$

$$n_O = 66.1 \text{ g O} \times \frac{1 \text{ mol O}}{16.00 \text{ g O}} = 4.13 \text{ mol O}$$

Step 3: Try to convert to whole numbers by dividing all the subscripts by the smallest subscript. The formula is $C_{1.65}H_{2.48}N_{0.828}O_{4.13}$. Dividing the subscripts by 0.828 gives the empirical formula, $C_2H_3NO_5$.

To determine the molecular formula, remember that the molar mass/empirical mass will be an integer greater than or equal to one.

$$\frac{\text{molar mass}}{\text{empirical molar mass}} \geq 1 \text{ (integer values)}$$

In this case,

$$\frac{\text{molar mass}}{\text{empirical molar mass}} = \frac{120 \text{ g}}{121.05 \text{ g}} \approx 1$$

Hence, the molecular formula and the empirical formula are the same, $C_2H_3NO_5$.

METHOD 2:

Step 1: Multiply the mass % (converted to a decimal) of each element by the molar mass to convert to grams of each element. Then, use the molar mass to convert to moles of each element.

$$n_C = (0.198) \times (120 \text{ g}) \times \frac{1 \text{ mol C}}{12.01 \text{ g C}} = 1.98 \text{ mol C} \approx \textbf{2 mol C}$$

$$n_H = (0.0250) \times (120 \text{ g}) \times \frac{1 \text{ mol H}}{1.008 \text{ g H}} = 2.98 \text{ mol H} \approx \textbf{3 mol H}$$

$$n_N = (0.116) \times (120 \text{ g}) \times \frac{1 \text{ mol N}}{14.01 \text{ g N}} = 0.994 \text{ mol N} \approx \textbf{1 mol N}$$

$$n_O = (0.661) \times (120 \text{ g}) \times \frac{1 \text{ mol O}}{16.00 \text{ g O}} = 4.96 \text{ mol O} \approx \textbf{5 mol O}$$

Step 2: Since we used the molar mass to calculate the moles of each element present in the compound, this method directly gives the molecular formula. The formula is $C_2H_3NO_5$.

Step 3: Try to reduce the molecular formula to a simpler whole number ratio to determine the empirical formula. The formula is already in its simplest whole number ratio. The molecular and empirical formulas are the same. The empirical formula is $C_2H_3NO_5$.

3.45 $24.6 \text{ g Fe}_2O_3 \times \dfrac{1 \text{ mol Fe}_2O_3}{159.7 \text{ g Fe}_2O_3} \times \dfrac{2 \text{ mol Fe}}{1 \text{ mol Fe}_2O_3} = \textbf{0.308 mol Fe}$

3.46   Using unit factors we convert:

g of Hg $\rightarrow$ mol Hg $\rightarrow$ mol S $\rightarrow$ g S

$$\textbf{? g S} = 246 \text{ g Hg} \times \frac{1 \text{ mol Hg}}{200.6 \text{ g Hg}} \times \frac{1 \text{ mol S}}{1 \text{ mol Hg}} \times \frac{32.07 \text{ g S}}{1 \text{ mol S}} = \textbf{39.3 g S}$$

3.47   The balanced equation is: $2Al(s) + 3I_2(s) \longrightarrow 2AlI_3(s)$

Using unit factors, we convert: g of Al $\rightarrow$ mol of Al $\rightarrow$ mol of I_2 $\rightarrow$ g of I_2

$$20.4 \text{ g Al} \times \frac{1 \text{ mol Al}}{26.98 \text{ g Al}} \times \frac{3 \text{ mol I}_2}{2 \text{ mol Al}} \times \frac{253.8 \text{ g I}_2}{1 \text{ mol I}_2} = \textbf{288 g I}_2$$

3.48   **Strategy:** Tin(II) fluoride is composed of Sn and F. The mass due to F is based on its percentage by mass in the compound. How do we calculate mass percent of an element?

Solution: First, we must find the mass % of fluorine in SnF_2. Then, we convert this percentage to a fraction and multiply by the mass of the compound (24.6 g), to find the mass of fluorine in 24.6 g of SnF_2.

The percent by mass of fluorine in tin(II) fluoride, is calculated as follows:

$$\text{mass \% F} = \frac{\text{mass of F in 1 mol } SnF_2}{\text{molar mass of } SnF_2} \times 100\%$$

$$= \frac{2(19.00 \text{ g})}{156.7 \text{ g}} \times 100\% = 24.25\% \text{ F}$$

Converting this percentage to a fraction, we obtain $24.25/100 = 0.2425$.

Next, multiply the fraction by the total mass of the compound.

? g F in 24.6 g SnF_2 = (0.2425)(24.6 g) = 5.97 g F

Check: As a ball-park estimate, note that the mass percent of F is roughly 25 percent, so that a quarter of the mass should be F. One quarter of approximately 24 g is 6 g, which is close to the answer.

> **Note:** This problem could have been worked in a manner similar to Problem 3.46. You could complete the following conversions:
> $$\text{g of } SnF_2 \rightarrow \text{mol of } SnF_2 \rightarrow \text{mol of F} \rightarrow \text{g of F}$$

3.49   In each case, assume 100 g of compound.

(a) $2.1 \text{ g H} \times \dfrac{1 \text{ mol H}}{1.008 \text{ g H}} = 2.1 \text{ mol H}$

$65.3 \text{ g O} \times \dfrac{1 \text{ mol O}}{16.00 \text{ g O}} = 4.08 \text{ mol O}$

$32.6 \text{ g S} \times \dfrac{1 \text{ mol S}}{32.07 \text{ g S}} = 1.02 \text{ mol S}$

This gives the formula $H_{2.1}S_{1.02}O_{4.08}$. Dividing by 1.02 gives the empirical formula, **H_2SO_4**.

(b) $20.2 \text{ g Al} \times \dfrac{1 \text{ mol Al}}{26.98 \text{ g Al}} = 0.749 \text{ mol Al}$

$79.8 \text{ g Cl} \times \dfrac{1 \text{ mol Cl}}{35.45 \text{ g Cl}} = 2.25 \text{ mol Cl}$

This gives the formula, $Al_{0.749}Cl_{2.25}$. Dividing by 0.749 gives the empirical formula, **$AlCl_3$**.

3.50   (a)
Strategy: In a chemical formula, the subscripts represent the ratio of the number of moles of each element that combine to form the compound. Therefore, we need to convert from mass percent to moles in order to determine the empirical formula. If we assume an exactly 100 g sample of the compound, do we know the mass of each element in the compound? How do we then convert from grams to moles?

Solution: If we have 100 g of the compound, then each percentage can be converted directly to grams. In this sample, there will be 40.1 g of C, 6.6 g of H, and 53.3 g of O. Because the subscripts in the formula represent a mole ratio, we need to convert the grams of each element to moles. The conversion factor needed is the molar mass of each element. Let *n* represent the number of moles of each element so that

$$n_C = 40.1 \text{ g C} \times \frac{1 \text{ mol C}}{12.01 \text{ g C}} = 3.34 \text{ mol C}$$

$$n_H = 6.6 \text{ g H} \times \frac{1 \text{ mol H}}{1.008 \text{ g H}} = 6.5 \text{ mol H}$$

$$n_O = 53.3 \text{ g O} \times \frac{1 \text{ mol O}}{16.00 \text{ g O}} = 3.33 \text{ mol O}$$

Thus, we arrive at the formula $C_{3.34}H_{6.5}O_{3.33}$, which gives the identity and the mole ratios of atoms present. However, chemical formulas are written with whole numbers. Try to convert to whole numbers by dividing all the subscripts by the smallest subscript (3.33).

$$\textbf{C:} \; \frac{3.34}{3.33} \approx 1 \qquad \textbf{H:} \; \frac{6.5}{3.33} \approx 2 \qquad \textbf{O:} \; \frac{3.33}{3.33} = 1$$

This gives the empirical formula, **CH_2O**.

Check: Are the subscripts in CH_2O reduced to the smallest whole numbers?

(b) Following the same procedure as part (a), we find:

$$n_C = 18.4 \text{ g C} \times \frac{1 \text{ mol C}}{12.01 \text{ g C}} = 1.53 \text{ mol C}$$

$$n_N = 21.5 \text{ g N} \times \frac{1 \text{ mol N}}{14.01 \text{ g N}} = 1.53 \text{ mol N}$$

$$n_K = 60.1 \text{ g K} \times \frac{1 \text{ mol K}}{39.10 \text{ g K}} = 1.54 \text{ mol K}$$

Dividing by the smallest number of moles (1.53 mol) gives the empirical formula, **KCN**.

3.51   The molar mass of $CaSiO_3$ is 116.17 g/mol.

$$\%Ca = \frac{40.08 \text{ g}}{116.17 \text{ g}} = \textbf{34.50\%}$$

$$\%Si = \frac{28.09 \text{ g}}{116.17 \text{ g}} = \textbf{24.18\%}$$

$$\%O = \frac{(3)(16.00 \text{ g})}{116.17 \text{ g}} = \textbf{41.32\%}$$

Check to see that the percentages sum to 100%. $(34.50\% + 24.18\% + 41.32\%) = 100.00\%$

3.52   The empirical molar mass of CH is approximately 13.02 g. Let's compare this to the molar mass to determine the molecular formula.

Recall that the molar mass divided by the empirical mass will be an integer greater than or equal to one.

$$\frac{\text{molar mass}}{\text{empirical molar mass}} \geq 1 \text{ (integer values)}$$

In this case,

$$\frac{\text{molar mass}}{\text{empirical molar mass}} = \frac{78 \text{ g}}{13.02 \text{ g}} \approx 6$$

Thus, there are six CH units in each molecule of the compound, so the molecular formula is $(CH)_6$, or C_6H_6.

3.53 Find the molar mass corresponding to each formula.

For $C_4H_5N_2O$: $4(12.01 \text{ g}) + 5(1.008 \text{ g}) + 2(14.01 \text{ g}) + (16.00 \text{ g}) = 97.10 \text{ g}$

For $C_8H_{10}N_4O_2$: $8(12.01 \text{ g}) + 10(1.008 \text{ g}) + 4(14.01 \text{ g}) + 2(16.00 \text{ g}) = 194.20 \text{ g}$

The molecular formula is $C_8H_{10}N_4O_2$.

3.54 **METHOD 1:**

Step 1: Assume you have exactly 100 g of substance. 100 g is a convenient amount, because all the percentages sum to 100%. In 100 g of MSG there will be 35.51 g C, 4.77 g H, 37.85 g O, 8.29 g N, and 13.60 g Na.

Step 2: Calculate the number of moles of each element in the compound. Remember, an *empirical formula* tells us which elements are present and the simplest whole-number ratio of their atoms. This ratio is also a mole ratio. Let n_C, n_H, n_O, n_N, and n_{Na} be the number of moles of elements present. Use the molar masses of these elements as conversion factors to convert to moles.

$$n_C = 35.51 \text{ g C} \times \frac{1 \text{ mol C}}{12.01 \text{ g C}} = 2.957 \text{ mol C}$$

$$n_H = 4.77 \text{ g H} \times \frac{1 \text{ mol H}}{1.008 \text{ g H}} = 4.73 \text{ mol H}$$

$$n_O = 37.85 \text{ g O} \times \frac{1 \text{ mol O}}{16.00 \text{ g O}} = 2.366 \text{ mol O}$$

$$n_N = 8.29 \text{ g N} \times \frac{1 \text{ mol N}}{14.01 \text{ g N}} = 0.592 \text{ mol N}$$

$$n_{Na} = 13.60 \text{ g Na} \times \frac{1 \text{ mol Na}}{22.99 \text{ g Na}} = 0.5916 \text{ mol Na}$$

Thus, we arrive at the formula $C_{2.957}H_{4.73}O_{2.366}N_{0.592}Na_{0.5916}$, which gives the identity and the ratios of atoms present. However, chemical formulas are written with whole numbers.

Step 3: Try to convert to whole numbers by dividing all the subscripts by the smallest subscript.

$$C: \frac{2.957}{0.5916} = 4.998 \approx 5 \qquad H: \frac{4.73}{0.5916} = 8.00 \qquad O: \frac{2.366}{0.5916} = 3.999 \approx 4$$

$$N: \frac{0.592}{0.5916} = 1.00 \qquad Na: \frac{0.5916}{0.5916} = 1$$

This gives us the empirical formula for MSG, $C_5H_8O_4NNa$.

To determine the molecular formula, remember that the molar mass/empirical mass will be an integer greater than or equal to one.

$$\frac{\text{molar mass}}{\text{empirical molar mass}} \geq 1 \ (\text{integer values})$$

In this case,

$$\frac{\text{molar mass}}{\text{empirical molar mass}} = \frac{169 \text{ g}}{169.11 \text{ g}} \approx 1$$

Hence, the molecular formula and the empirical formula are the same, $C_5H_8O_4NNa$. It should come as no surprise that the empirical and molecular formulas are the same since MSG stands for *monosodium*glutamate.

METHOD 2:

Step 1: Multiply the mass % (converted to a decimal) of each element by the molar mass to convert to grams of each element. Then, use the molar mass to convert to moles of each element.

$$n_C = (0.3551) \times (169 \text{ g}) \times \frac{1 \text{ mol C}}{12.01 \text{ g C}} = 5.00 \text{ mol C}$$

$$n_H = (0.0477) \times (169 \text{ g}) \times \frac{1 \text{ mol H}}{1.008 \text{ g H}} = 8.00 \text{ mol H}$$

$$n_O = (0.3785) \times (169 \text{ g}) \times \frac{1 \text{ mol O}}{16.00 \text{ g O}} = 4.00 \text{ mol O}$$

$$n_N = (0.0829) \times (169 \text{ g}) \times \frac{1 \text{ mol N}}{14.01 \text{ g N}} = 1.00 \text{ mol N}$$

$$n_{Na} = (0.1360) \times (169 \text{ g}) \times \frac{1 \text{ mol Na}}{22.99 \text{ g Na}} = 1.00 \text{ mol Na}$$

Step 2: Since we used the molar mass to calculate the moles of each element present in the compound, this method directly gives the molecular formula. The formula is $C_5H_8O_4NNa$.

3.59   The balanced equations are as follows:

(a) $2C + O_2 \rightarrow 2CO$

(b) $2CO + O_2 \rightarrow 2CO_2$

(c) $H_2 + Br_2 \rightarrow 2HBr$

(d) $2K + 2H_2O \rightarrow 2KOH + H_2$

(e) $2Mg + O_2 \rightarrow 2MgO$

(f) $2O_3 \rightarrow 3O_2$

(g) $2H_2O_2 \rightarrow 2H_2O + O_2$

(h) $N_2 + 3H_2 \rightarrow 2NH_3$

(i) $Zn + 2AgCl \rightarrow ZnCl_2 + 2Ag$

(j) $S_8 + 8O_2 \rightarrow 8SO_2$

(k) $2NaOH + H_2SO_4 \rightarrow Na_2SO_4 + 2H_2O$

(l) $Cl_2 + 2NaI \rightarrow 2NaCl + I_2$

(m) $3KOH + H_3PO_4 \rightarrow K_3PO_4 + 3H_2O$

(n) $CH_4 + 4Br_2 \rightarrow CBr_4 + 4HBr$

3.60   The balanced equations are as follows:

(a) $2N_2O_5 \rightarrow 2N_2O_4 + O_2$

(b) $2KNO_3 \rightarrow 2KNO_2 + O_2$

(c) $NH_4NO_3 \rightarrow N_2O + 2H_2O$

(d) $NH_4NO_2 \rightarrow N_2 + 2H_2O$

(e) $2NaHCO_3 \rightarrow Na_2CO_3 + H_2O + CO_2$

(f) $P_4O_{10} + 6H_2O \rightarrow 4H_3PO_4$

(g) $2HCl + CaCO_3 \rightarrow CaCl_2 + H_2O + CO_2$

(h) $2Al + 3H_2SO_4 \rightarrow Al_2(SO_4)_3 + 3H_2$

(i) $CO_2 + 2KOH \rightarrow K_2CO_3 + H_2O$

(j) $CH_4 + 2O_2 \rightarrow CO_2 + 2H_2O$

(k) $Be_2C + 4H_2O \rightarrow 2Be(OH)_2 + CH_4$

(l) $3Cu + 8HNO_3 \rightarrow 3Cu(NO_3)_2 + 2NO + 4H_2O$

(m) $S + 6HNO_3 \rightarrow H_2SO_4 + 6NO_2 + 2H_2O$

(n) $2NH_3 + 3CuO \rightarrow 3Cu + N_2 + 3H_2O$

3.63   On the reactants side there are 8 A atoms and 4 B atoms. On the products side, there are 4 C atoms and 4 D atoms. Writing an equation,

$$8A + 4B \rightarrow 4C + 4D$$

Chemical equations are typically written with the smallest set of whole number coefficients. Dividing the equation by four gives,

$$2A + B \rightarrow C + D$$

The correct answer is choice **(c)**.

3.64   On the reactants side there are 6 A atoms and 4 B atoms. On the products side, there are 4 C atoms and 2 D atoms. Writing an equation,

$$6A + 4B \rightarrow 4C + 2D$$

Chemical equations are typically written with the smallest set of whole number coefficients. Dividing the equation by two gives,

$$3A + 2B \rightarrow 2C + D$$

The correct answer is choice **(d)**.

3.65   The mole ratio from the balanced equation is 2 moles CO_2 : 2 moles CO.

$$3.60 \text{ mol CO} \times \frac{2 \text{ mol CO}_2}{2 \text{ mol CO}} = \textbf{3.60 mol CO}_2$$

3.66 $Si(s) + 2Cl_2(g) \longrightarrow SiCl_4(l)$

Strategy: Looking at the balanced equation, how do we compare the amounts of Cl_2 and $SiCl_4$? We can compare them based on the mole ratio from the balanced equation.

Solution: Because the balanced equation is given in the problem, the mole ratio between Cl_2 and $SiCl_4$ is known: 2 moles $Cl_2 \simeq 1$ mole $SiCl_4$. From this relationship, we have two conversion factors.

$$\frac{2 \text{ mol Cl}_2}{1 \text{ mol SiCl}_4} \quad \text{and} \quad \frac{1 \text{ mol SiCl}_4}{2 \text{ mol Cl}_2}$$

Which conversion factor is needed to convert from moles of $SiCl_4$ to moles of Cl_2? The conversion factor on the left is the correct one. Moles of $SiCl_4$ will cancel, leaving units of "mol Cl_2" for the answer. We calculate moles of Cl_2 reacted as follows:

$$\textbf{? mol Cl}_2 \textbf{ reacted} = 0.507 \text{ mol SiCl}_4 \times \frac{2 \text{ mol Cl}_2}{1 \text{ mol SiCl}_4} = \textbf{1.01 mol Cl}_2$$

Check: Does the answer seem reasonable? Should the moles of Cl_2 reacted be *double* the moles of $SiCl_4$ produced?

3.67   Starting with the amount of ammonia produced (6.0 moles), we can use the mole ratio from the balanced equation to calculate the moles of H_2 and N_2 that reacted to produce 6.0 moles of NH_3.

$$3H_2(g) + N_2(g) \rightarrow 2NH_3(g)$$

$$? \text{ mol } H_2 = 6.0 \text{ mol } NH_3 \times \frac{3 \text{ mol } H_2}{2 \text{ mol } NH_3} = \textbf{9.0 mol } H_2$$

$$? \text{ mol } N_2 = 6.0 \text{ mol } NH_3 \times \frac{1 \text{ mol } N_2}{2 \text{ mol } NH_3} = \textbf{3.0 mol } N_2$$

3.68   Starting with the 5.0 moles of C_4H_{10}, we can use the mole ratio from the balanced equation to calculate the moles of CO_2 formed.

$$2C_4H_{10}(g) + 13O_2(g) \rightarrow 8CO_2(g) + 10H_2O(l)$$

$$? \text{ mol } CO_2 = 5.0 \text{ mol } C_4H_{10} \times \frac{8 \text{ mol } CO_2}{2 \text{ mol } C_4H_{10}} = 20 \text{ mol } CO_2 = \textbf{2.0} \times \textbf{10}^{\textbf{1}} \textbf{ mol } CO_2$$

3.69   It is convenient to use the unit ton-mol in this problem. We normally use a g-mol. 1 g-mol SO_2 has a mass of 64.07 g. In a similar manner, 1 ton-mol of SO_2 has a mass of 64.07 tons. We need to complete the following conversions: tons $SO_2 \rightarrow$ ton-mol $SO_2 \rightarrow$ ton-mol $S \rightarrow$ ton S.

$$(2.6 \times 10^7 \text{ tons } SO_2) \times \frac{1 \text{ ton-mol } SO_2}{64.07 \text{ ton } SO_2} \times \frac{1 \text{ ton-mol } S}{1 \text{ ton-mol } SO_2} \times \frac{32.07 \text{ ton } S}{1 \text{ ton-mol } S} = \textbf{1.3} \times \textbf{10}^7 \textbf{ tons } S$$

3.70   **(a)** $2NaHCO_3 \longrightarrow Na_2CO_3 + H_2O + CO_2$

(b)   Molar mass $NaHCO_3 = 22.99 \text{ g} + 1.008 \text{ g} + 12.01 \text{ g} + 3(16.00 \text{ g}) = 84.01 \text{ g}$
Molar mass $CO_2 = 12.01 \text{ g} + 2(16.00 \text{ g}) = 44.01 \text{ g}$

The balanced equation shows one mole of CO_2 formed from two moles of $NaHCO_3$.

$$\textbf{mass } NaHCO_3 = 20.5 \text{ g } CO_2 \times \frac{1 \text{ mol } CO_2}{44.01 \text{ g } CO_2} \times \frac{2 \text{ mol } NaHCO_3}{1 \text{ mol } CO_2} \times \frac{84.01 \text{ g } NaHCO_3}{1 \text{ mol } NaHCO_3}$$

$$= \textbf{78.3 g } NaHCO_3$$

3.71   The balanced equation shows a mole ratio of 1 mole HCN : 1 mole KCN.

$$0.140 \text{ g KCN} \times \frac{1 \text{ mol KCN}}{65.12 \text{ g KCN}} \times \frac{1 \text{ mol HCN}}{1 \text{ mol KCN}} \times \frac{27.03 \text{ g HCN}}{1 \text{ mol HCN}} = \textbf{0.0581 g HCN}$$

3.72 $C_6H_{12}O_6 \longrightarrow 2C_2H_5OH + 2CO_2$
glucose   ethanol

Strategy: We compare glucose and ethanol based on the *mole ratio* in the balanced equation. Before we can determine moles of ethanol produced, we need to convert to moles of glucose. What conversion factor is needed to convert from grams of glucose to moles of glucose? Once moles of ethanol are obtained, another conversion factor is needed to convert from moles of ethanol to grams of ethanol.

Solution: The molar mass of glucose will allow us to convert from grams of glucose to moles of glucose. The molar mass of glucose = 6(12.01 g) + 12(1.008 g) + 6(16.00 g) = 180.16 g. The balanced equation is given, so the mole ratio between glucose and ethanol is known; that is 1 mole glucose $\simeq$ 2 moles ethanol. Finally, the molar mass of ethanol will convert moles of ethanol to grams of ethanol. This sequence of three conversions is summarized as follows:

grams of glucose $\rightarrow$ moles of glucose $\rightarrow$ moles of ethanol $\rightarrow$ grams of ethanol

$$? \text{ g C}_2\text{H}_5\text{OH} = 500.4 \text{ g C}_6\text{H}_{12}\text{O}_6 \times \frac{1 \text{ mol C}_6\text{H}_{12}\text{O}_6}{180.16 \text{ g C}_6\text{H}_{12}\text{O}_6} \times \frac{2 \text{ mol C}_2\text{H}_5\text{OH}}{1 \text{ mol C}_6\text{H}_{12}\text{O}_6} \times \frac{46.07 \text{ g C}_2\text{H}_5\text{OH}}{1 \text{ mol C}_2\text{H}_5\text{OH}}$$

$$= \textbf{255.9 g C}_2\textbf{H}_5\textbf{OH}$$

Check: Does the answer seem reasonable? Should the mass of ethanol produced be approximately half the mass of glucose reacted? Twice as many moles of ethanol are produced compared to the moles of glucose reacted, but the molar mass of ethanol is about one-fourth that of glucose.

The liters of ethanol can be calculated from the density and the mass of ethanol.

$$\text{volume} = \frac{\text{mass}}{\text{density}}$$

$$\textbf{Volume of ethanol obtained} = \frac{255.9 \text{ g}}{0.789 \text{ g/mL}} = 324 \text{ mL} = \textbf{0.324 L}$$

3.73   The mass of water lost is just the difference between the initial and final masses.

$$\text{Mass H}_2\text{O lost} = 15.01 \text{ g} - 9.60 \text{ g} = 5.41 \text{ g}$$

$$\text{moles of H}_2\text{O} = 5.41 \text{ g H}_2\text{O} \times \frac{1 \text{ mol H}_2\text{O}}{18.02 \text{ g H}_2\text{O}} = \textbf{0.300 mol H}_2\textbf{O}$$

3.74   The balanced equation shows that eight moles of KCN are needed to combine with four moles of Au.

$$\textbf{? mol KCN} = 29.0 \text{ g Au} \times \frac{1 \text{ mol Au}}{197.0 \text{ g Au}} \times \frac{8 \text{ mol KCN}}{4 \text{ mol Au}} = \textbf{0.294 mol KCN}$$

3.75   The balanced equation is: $CaCO_3(s) \longrightarrow CaO(s) + CO_2(g)$

$$1.0 \text{ kg CaCO}_3 \times \frac{1000 \text{ g}}{1 \text{ kg}} \times \frac{1 \text{ mol CaCO}_3}{100.1 \text{ g CaCO}_3} \times \frac{1 \text{ mol CaO}}{1 \text{ mol CaCO}_3} \times \frac{56.08 \text{ g CaO}}{1 \text{ mol CaO}} = \textbf{5.6} \times \textbf{10}^2 \textbf{ g CaO}$$

3.76 **(a)** $NH_4NO_3(s) \longrightarrow N_2O(g) + 2H_2O(g)$

(b) Starting with moles of NH_4NO_3, we can use the mole ratio from the balanced equation to find moles of N_2O. Once we have moles of N_2O, we can use the molar mass of N_2O to convert to grams of N_2O. Combining the two conversions into one calculation, we have:

$$\text{mol } NH_4NO_3 \rightarrow \text{mol } N_2O \rightarrow \text{g } N_2O$$

$$\textbf{? g } N_2O = 0.46 \text{ mol } NH_4NO_3 \times \frac{1 \text{ mol } N_2O}{1 \text{ mol } NH_4NO_3} \times \frac{44.02 \text{ g } N_2O}{1 \text{ mol } N_2O} = \textbf{2.0} \times \textbf{10}^1 \textbf{ g } N_2O$$

3.77 The quantity of ammonia needed is:

$$1.00 \times 10^8 \text{ g } (NH_4)_2SO_4 \times \frac{1 \text{ mol } (NH_4)_2SO_4}{132.2 \text{ g } (NH_4)_2SO_4} \times \frac{2 \text{ mol } NH_3}{1 \text{ mol } (NH_4)_2SO_4} \times \frac{17.03 \text{ g } NH_3}{1 \text{ mol } NH_3} \times \frac{1 \text{ kg}}{1000 \text{ g}}$$

$$= \textbf{2.58} \times \textbf{10}^4 \textbf{ kg } NH_3$$

3.78 The balanced equation for the decomposition is :

$$2KClO_3(s) \longrightarrow 2KCl(s) + 3O_2(g)$$

$$\textbf{? g } O_2 = 46.0 \text{ g } KClO_3 \times \frac{1 \text{ mol } KClO_3}{122.6 \text{ g } KClO_3} \times \frac{3 \text{ mol } O_2}{2 \text{ mol } KClO_3} \times \frac{32.00 \text{ g } O_2}{1 \text{ mol } O_2} = \textbf{18.0 g } O_2$$

3.81 $2A + B \rightarrow C$

(a) The number of B atoms shown in the diagram is 5. The balanced equation shows 2 moles A $\simeq$ 1 mole B. Therefore, we need 10 atoms of A to react completely with 5 atoms of B. There are only 8 atoms of A present in the diagram. There are not enough atoms of A to react completely with B.

A is the limiting reagent.

(b) There are 8 atoms of A. Since the mole ratio between A and B is 2:1, 4 atoms of B will react with 8 atoms of A, leaving 1 atom of B in excess. The mole ratio between A and C is also 2:1. When 8 atoms of A react, 4 molecules of C will be produced.

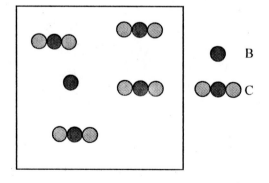

3.82 $N_2 + 3H_2 \rightarrow 2NH_3$

    (a)   The number of N_2 molecules shown in the diagram is 3. The balanced equation shows
        3 moles $H_2 \simeq 1$ mole N_2. Therefore, we need 9 molecules of H_2 to react completely with 3 molecules
        of N_2. There are 10 molecules of H_2 present in the diagram. H_2 is in excess.

    N_2 is the limiting reagent.

    (b)   9 molecules of H_2 will react with 3 molecules of N_2, leaving 1 molecule of H_2 in excess. The mole
        ratio between N_2 and NH_3 is 1:2. When 3 molecules of N_2 react, 6 molecules of NH_3 will be produced.

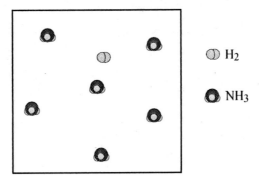

3.83   This is a limiting reagent problem. Let's calculate the moles of NO_2 produced assuming complete reaction
    for each reactant.

    $2NO(g) + O_2(g) \rightarrow 2NO_2(g)$

$$0.886 \text{ mol NO} \times \frac{2 \text{ mol NO}_2}{2 \text{ mol NO}} = 0.886 \text{ mol NO}_2$$

$$0.503 \text{ mol O}_2 \times \frac{2 \text{ mol NO}_2}{1 \text{ mol O}_2} = 1.01 \text{ mol NO}_2$$

    NO is the **limiting reagent**; it limits the amount of product produced. The amount of product produced is
    0.886 mole NO_2.

3.84   **Strategy:** Note that this reaction gives the amounts of both reactants, so it is likely to be a limiting reagent
    problem. The reactant that produces fewer moles of product is the limiting reagent because it limits the
    amount of product that can be produced. How do we convert from the amount of reactant to amount of
    product? Perform this calculation for each reactant, then compare the moles of product, NO_2, formed by the
    given amounts of O_3 and NO to determine which reactant is the limiting reagent.

    Solution: We carry out two separate calculations. First, starting with 0.740 g O_3, we calculate the number
    of moles of NO_2 that could be produced if all the O_3 reacted. We complete the following conversions.

        grams of $O_3 \rightarrow$ moles of $O_3 \rightarrow$ moles of NO_2

    Combining these two conversions into one calculation, we write

$$? \text{ mol NO}_2 = 0.740 \text{ g O}_3 \times \frac{1 \text{ mol O}_3}{48.00 \text{ g O}_3} \times \frac{1 \text{ mol NO}_2}{1 \text{ mol O}_3} = 0.0154 \text{ mol NO}_2$$

Second, starting with 0.670 g of NO, we complete similar conversions.

grams of NO $\rightarrow$ moles of NO $\rightarrow$ moles of NO_2

Combining these two conversions into one calculation, we write

$$? \text{ mol NO}_2 = 0.670 \text{ g NO} \times \frac{1 \text{ mol NO}}{30.01 \text{ g NO}} \times \frac{1 \text{ mol NO}_2}{1 \text{ mol NO}} = 0.0223 \text{ mol NO}_2$$

The initial amount of **O_3** limits the amount of product that can be formed; therefore, it is the **limiting reagent**.

The problem asks for grams of NO_2 produced. We already know the moles of NO_2 produced, 0.0154 mole. Use the molar mass of NO_2 as a conversion factor to convert to grams (Molar mass NO_2 = 46.01 g).

$$\mathbf{?\ g\ NO_2} = 0.0154 \text{ mol NO}_2 \times \frac{46.01 \text{ g NO}_2}{1 \text{ mol NO}_2} = \mathbf{0.709\ g\ NO_2}$$

Check: Does your answer seem reasonable? 0.0154 mole of product is formed. What is the mass of 1 mole of NO_2?

Strategy: Working backwards, we can determine the amount of NO that reacted to produce 0.0154 mole of NO_2. The amount of NO left over is the difference between the initial amount and the amount reacted.

Solution: Starting with 0.0154 mole of NO_2, we can determine the moles of NO that reacted using the mole ratio from the balanced equation. We can calculate the initial moles of NO starting with 0.670 g and using molar mass of NO as a conversion factor.

$$\text{mol NO reacted} = 0.0154 \text{ mol NO}_2 \times \frac{1 \text{ mol NO}}{1 \text{ mol NO}_2} = 0.0154 \text{ mol NO}$$

$$\text{mol NO initial} = 0.670 \text{ g NO} \times \frac{1 \text{ mol NO}}{30.01 \text{ g NO}} = 0.0223 \text{ mol NO}$$

mol NO remaining = mol NO initial – mol NO reacted.

mol NO remaining = 0.0223 mol NO – 0.0154 mol NO = **0.0069 mol NO**

3.85   (a)   The balanced equation is: $C_3H_8(g) + 5O_2(g) \longrightarrow 3CO_2(g) + 4H_2O(l)$

(b)   The balanced equation shows a mole ratio of 3 moles CO_2 : 1 mole C_3H_8. The mass of CO_2 produced is:

$$3.65 \text{ mol C}_3\text{H}_8 \times \frac{3 \text{ mol CO}_2}{1 \text{ mol C}_3\text{H}_8} \times \frac{44.01 \text{ g CO}_2}{1 \text{ mol CO}_2} = \mathbf{482\ g\ CO_2}$$

3.86   This is a limiting reagent problem. Let's calculate the moles of Cl_2 produced assuming complete reaction for each reactant.

$$0.86 \text{ mol MnO}_2 \times \frac{1 \text{ mol Cl}_2}{1 \text{ mol MnO}_2} = 0.86 \text{ mol Cl}_2$$

$$48.2 \text{ g HCl} \times \frac{1 \text{ mol HCl}}{36.46 \text{ g HCl}} \times \frac{1 \text{ mol Cl}_2}{4 \text{ mol HCl}} = 0.330 \text{ mol Cl}_2$$

HCl is the limiting reagent; it limits the amount of product produced. It will be used up first. The amount of product produced is 0.330 mole Cl_2. Let's convert this to grams.

$$? \text{ g Cl}_2 = 0.330 \text{ mol Cl}_2 \times \frac{70.90 \text{ g Cl}_2}{1 \text{ mol Cl}_2} = \textbf{23.4 g Cl}_2$$

3.89   The balanced equation is given: $CaF_2 + H_2SO_4 \longrightarrow CaSO_4 + 2HF$

The balanced equation shows a mole ratio of 2 moles HF : 1 mole CaF_2. The theoretical yield of HF is:

$$(6.00 \times 10^3 \text{ g CaF}_2) \times \frac{1 \text{ mol CaF}_2}{78.08 \text{ g CaF}_2} \times \frac{2 \text{ mol HF}}{1 \text{ mol CaF}_2} \times \frac{20.01 \text{ g HF}}{1 \text{ mol HF}} \times \frac{1 \text{ kg}}{1000 \text{ g}} = 3.08 \text{ kg HF}$$

The actual yield is given in the problem (2.86 kg HF).

$$\% \text{ yield } = \frac{\text{actual yield}}{\text{theoretical yield}} \times 100\%$$

$$\textbf{\% yield } = \frac{2.86 \text{ kg}}{3.08 \text{ kg}} \times 100\% = \textbf{92.9\%}$$

3.90   **(a)**   Start with a balanced chemical equation. It's given in the problem. We use NG as an abbreviation for nitroglycerin. The molar mass of NG = 227.1 g/mol.

$$4C_3H_5N_3O_9 \longrightarrow 6N_2 + 12CO_2 + 10H_2O + O_2$$

Map out the following strategy to solve this problem.

$$\text{g NG} \rightarrow \text{ mol NG} \rightarrow \text{ mol O}_2 \rightarrow \text{ g O}_2$$

Calculate the grams of O_2 using the strategy above.

$$\textbf{? g O}_2 = 2.00 \times 10^2 \text{ g NG} \times \frac{1 \text{ mol NG}}{227.1 \text{ g NG}} \times \frac{1 \text{ mol O}_2}{4 \text{ mol NG}} \times \frac{32.00 \text{ g O}_2}{1 \text{ mol O}_2} = \textbf{7.05 g O}_2$$

(b)   The theoretical yield was calculated in part (a), and the actual yield is given in the problem (6.55 g). The percent yield is:

$$\% \text{ yield } = \frac{\text{actual yield}}{\text{theoretical yield}} \times 100\%$$

$$\textbf{\% yield } = \frac{6.55 \text{ g O}_2}{7.05 \text{ g O}_2} \times 100\% = \textbf{92.9\%}$$

3.91   The balanced equation shows a mole ratio of 1 mole TiO_2 : 1 mole $FeTiO_3$. The molar mass of $FeTiO_3$ is 151.7 g/mol, and the molar mass of TiO_2 is 79.88 g/mol. The theoretical yield of TiO_2 is:

$$8.00 \times 10^6 \text{ g FeTiO}_3 \times \frac{1 \text{ mol FeTiO}_3}{151.7 \text{ g FeTiO}_3} \times \frac{1 \text{ mol TiO}_2}{1 \text{ mol FeTiO}_3} \times \frac{79.88 \text{ g TiO}_2}{1 \text{ mol TiO}_2} \times \frac{1 \text{ kg}}{1000 \text{ g}}$$

$$= \textbf{4.21} \times \textbf{10}^3 \textbf{ kg TiO}_2$$

The actual yield is given in the problem (3.67×10^3 kg TiO_2).

$$\% \text{ yield} = \frac{\text{actual yield}}{\text{theoretical yield}} \times 100\% = \frac{3.67 \times 10^3 \text{ kg}}{4.21 \times 10^3 \text{ kg}} \times 100\% = \textbf{87.2\%}$$

3.92   The actual yield of ethylene is 481 g. Let's calculate the yield of ethylene if the reaction is 100 percent efficient. We can calculate this from the definition of percent yield. We can then calculate the mass of hexane that must be reacted.

$$\% \text{ yield} = \frac{\text{actual yield}}{\text{theoretical yield}} \times 100\%$$

$$42.5\% \text{ yield} = \frac{481 \text{ g } C_2H_4}{\text{theoretical yield}} \times 100\%$$

$$\text{theoretical yield } C_2H_4 = 1.13 \times 10^3 \text{ g } C_2H_4$$

The mass of hexane that must be reacted is:

$$(1.13 \times 10^3 \text{ g } C_2H_4) \times \frac{1 \text{ mol } C_2H_4}{28.05 \text{ g } C_2H_4} \times \frac{1 \text{ mol } C_6H_{14}}{1 \text{ mol } C_2H_4} \times \frac{86.15 \text{ g } C_6H_{14}}{1 \text{ mol } C_6H_{14}} = \textbf{3.47} \times \textbf{10}^\textbf{3} \textbf{ g } \textbf{C}_\textbf{6}\textbf{H}_\textbf{14}$$

3.93   This is a limiting reagent problem. Let's calculate the moles of Li_3N produced assuming complete reaction for each reactant.

$$6Li(s) + N_2(g) \rightarrow 2Li_3N(s)$$

$$12.3 \text{ g Li} \times \frac{1 \text{ mol Li}}{6.941 \text{ g Li}} \times \frac{2 \text{ mol } Li_3N}{6 \text{ mol Li}} = 0.591 \text{ mol } Li_3N$$

$$33.6 \text{ g } N_2 \times \frac{1 \text{ mol } N_2}{28.02 \text{ g } N_2} \times \frac{2 \text{ mol } Li_3N}{1 \text{ mol } N_2} = 2.40 \text{ mol } Li_3N$$

Li is the limiting reagent; it limits the amount of product produced. The amount of product produced is 0.591 mole Li_3N. Let's convert this to grams.

$$? \text{ g } Li_3N = 0.591 \text{ mol } Li_3N \times \frac{34.83 \text{ g } Li_3N}{1 \text{ mol } Li_3N} = \textbf{20.6 g } \textbf{Li}_\textbf{3}\textbf{N}$$

This is the theoretical yield of Li_3N. The actual yield is given in the problem (5.89 g). The percent yield is:

$$\% \text{ yield} = \frac{\text{actual yield}}{\text{theoretical yield}} \times 100\% = \frac{5.89 \text{ g}}{20.6 \text{ g}} \times 100\% = \textbf{28.6\%}$$

3.94   This is a limiting reagent problem. Let's calculate the moles of S_2Cl_2 produced assuming complete reaction for each reactant.

$$S_8(l) + 4Cl_2(g) \rightarrow 4S_2Cl_2(l)$$

$$4.06 \text{ g } S_8 \times \frac{1 \text{ mol } S_8}{256.6 \text{ g } S_8} \times \frac{4 \text{ mol } S_2Cl_2}{1 \text{ mol } S_8} = 0.0633 \text{ mol } S_2Cl_2$$

$$6.24 \text{ g Cl}_2 \times \frac{1 \text{ mol Cl}_2}{70.90 \text{ g Cl}_2} \times \frac{4 \text{ mol S}_2\text{Cl}_2}{4 \text{ mol Cl}_2} = 0.0880 \text{ mol S}_2\text{Cl}_2$$

S_8 is the limiting reagent; it limits the amount of product produced. The amount of product produced is 0.0633 mole S_2Cl_2. Let's convert this to grams.

$$? \text{ g S}_2\text{Cl}_2 = 0.0633 \text{ mol S}_2\text{Cl}_2 \times \frac{135.04 \text{ g S}_2\text{Cl}_2}{1 \text{ mol S}_2\text{Cl}_2} = \mathbf{8.55 \text{ g S}_2\text{Cl}_2}$$

This is the theoretical yield of S_2Cl_2. The actual yield is given in the problem (6.55 g). The percent yield is:

$$\% \text{ yield} = \frac{\text{actual yield}}{\text{theoretical yield}} \times 100\% = \frac{6.55 \text{ g}}{8.55 \text{ g}} \times 100\% = \mathbf{76.6\%}$$

3.95 All the carbon from the hydrocarbon reactant ends up in CO_2, and all the hydrogen from the hydrocarbon reactant ends up in water. In the diagram, we find 4 CO_2 molecules and 6 H_2O molecules. This gives a ratio between carbon and hydrogen of 4:12. We write the formula C_4H_{12}, which reduces to the empirical formula CH_3. The empirical molar mass equals approximately 15 g, which is half the molar mass of the hydrocarbon. Thus, the molecular formula is double the empirical formula or C_2H_6. Since this is a combustion reaction, the other reactant is O_2. We write:

$$C_2H_6 + O_2 \rightarrow CO_2 + H_2O$$

Balancing the equation,

$$\mathbf{2C_2H_6 + 7O_2 \rightarrow 4CO_2 + 6H_2O}$$

3.96 $2H_2(g) + O_2(g) \rightarrow 2H_2O(g)$

We start with 8 molecules of H_2 and 3 molecules of O_2. The balanced equation shows 2 moles $H_2 \simeq$ 1 mole O_2. If 3 molecules of O_2 react, 6 molecules of H_2 will react, leaving 2 molecules of H_2 in excess. The balanced equation also shows 1 mole $O_2 \simeq$ 2 moles H_2O. If 3 molecules of O_2 react, 6 molecules of H_2O will be produced.

After complete reaction, there will be **2 molecules of H_2** and **6 molecules of H_2O**. The correct diagram is choice **(b)**.

3.97 First, let's convert to moles of HNO_3 produced.

$$1.00 \text{ ton HNO}_3 \times \frac{2000 \text{ lb}}{1 \text{ ton}} \times \frac{453.6 \text{ g}}{1 \text{ lb}} \times \frac{1 \text{ mol HNO}_3}{63.02 \text{ g HNO}_3} = 1.44 \times 10^4 \text{ mol HNO}_3$$

Now, we will work in the reverse direction to calculate the amount of reactant needed to produce 1.44×10^3 mol of HNO_3. Realize that since the problem says to assume an 80% yield for each step, the amount of reactant needed in each step will be *larger* by a factor of $\frac{100\%}{80\%}$, compared to a standard stoichiometry calculation where a 100% yield is assumed.

Referring to the balanced equation in the *last step*, we calculate the moles of NO_2.

$$(1.44 \times 10^4 \text{ mol HNO}_3) \times \frac{2 \text{ mol NO}_2}{1 \text{ mol HNO}_3} \times \frac{100\%}{80\%} = 3.60 \times 10^4 \text{ mol NO}_2$$

Now, let's calculate the amount of NO needed to produce 3.60×10^4 mol NO_2. Following the same procedure as above, and referring to the balanced equation in the *middle step*, we calculate the moles of NO.

$$(3.60 \times 10^4 \text{ mol NO}_2) \times \frac{1 \text{ mol NO}}{1 \text{ mol NO}_2} \times \frac{100\%}{80\%} = 4.50 \times 10^4 \text{ mol NO}$$

Now, let's calculate the amount of NH_3 needed to produce 4.5×10^4 mol NO. Referring to the balanced equation in the *first step*, the moles of NH_3 is:

$$(4.50 \times 10^4 \text{ mol NO}) \times \frac{4 \text{ mol NH}_3}{4 \text{ mol NO}} \times \frac{100\%}{80\%} = 5.63 \times 10^4 \text{ mol NH}_3$$

Finally, converting to grams of NH_3:

$$5.63 \times 10^4 \text{ mol NH}_3 \times \frac{17.03 \text{ g NH}_3}{1 \text{ mol NH}_3} = \mathbf{9.59 \times 10^5 \text{ g NH}_3}$$

3.98   We assume that all the Cl in the compound ends up as HCl and all the O ends up as H_2O. Therefore, we need to find the number of moles of Cl in HCl and the number of moles of O in H_2O.

$$\text{mol Cl} = 0.233 \text{ g HCl} \times \frac{1 \text{ mol HCl}}{36.46 \text{ g HCl}} \times \frac{1 \text{ mol Cl}}{1 \text{ mol HCl}} = 0.00639 \text{ mol Cl}$$

$$\text{mol O} = 0.403 \text{ g H}_2\text{O} \times \frac{1 \text{ mol H}_2\text{O}}{18.02 \text{ g H}_2\text{O}} \times \frac{1 \text{ mol O}}{1 \text{ mol H}_2\text{O}} = 0.0224 \text{ mol O}$$

Dividing by the smallest number of moles (0.00639 mole) gives the formula, $ClO_{3.5}$. Multiplying both subscripts by two gives the empirical formula, $\mathbf{Cl_2O_7}$.

3.99   The number of moles of Y in 84.10 g of Y is:

$$27.22 \text{ g X} \times \frac{1 \text{ mol X}}{33.42 \text{ g X}} \times \frac{1 \text{ mol Y}}{1 \text{ mol X}} = 0.8145 \text{ mol Y}$$

The molar mass of Y is:

$$\text{molar mass Y} = \frac{84.10 \text{ g Y}}{0.8145 \text{ mol Y}} = 103.3 \text{ g/mol}$$

The atomic mass of Y is **103.3 amu**.

3.100   The symbol "O" refers to moles of oxygen atoms, not oxygen molecule (O_2). Look at the molecular formulas given in parts (a) and (b). What do they tell you about the relative amounts of carbon and oxygen?

(a) $0.212 \text{ mol C} \times \frac{1 \text{ mol O}}{1 \text{ mol C}} = \mathbf{0.212 \text{ mol O}}$

(b) $0.212 \text{ mol C} \times \dfrac{2 \text{ mol O}}{1 \text{ mol C}} = \textbf{0.424 mol O}$

3.101   The observations mean either that the amount of the more abundant isotope was increasing or the amount of the less abundant isotope was decreasing. One possible explanation is that the less abundant isotope was undergoing radioactive decay, and thus its mass would decrease with time.

3.102   This is a calculation involving percent composition. Remember,

$$\text{percent by mass of each element} = \frac{\text{mass of element in 1 mol of compound}}{\text{molar mass of compound}} \times 100\%$$

The molar masses are: Al, 26.98 g/mol; $Al_2(SO_4)_3$, 342.2 g/mol; H_2O, 18.02 g/mol. Thus, using x as the number of H_2O molecules,

$$\text{mass \% Al} = \left(\frac{2(\text{molar mass of Al})}{\text{molar mass of } Al_2(SO_4)_3 + x(\text{molar mass of } H_2O)} \right) \times 100\%$$

$$8.20\% = \left(\frac{2(26.98 \text{ g})}{342.2 \text{ g} + x(18.02 \text{ g})} \right) \times 100\%$$

$$x = \textbf{17.53}$$

Rounding off to a whole number of water molecules, $x = 18$. Therefore, the formula is $\textbf{Al}_2(\textbf{SO}_4)_3 \cdot \textbf{18 H}_2\textbf{O}$.

3.103   Molar mass of $C_4H_8Cl_2S = 4(12.01 \text{ g}) + 8(1.008 \text{ g}) + 2(35.45 \text{ g}) + 32.07 \text{ g} = 159.1 \text{ g}$

$$\textbf{\%C} = \frac{4(12.01 \text{ g/mol})}{159.1 \text{ g/mol}} \times 100\% = \textbf{30.19\%}$$

$$\textbf{\%H} = \frac{8(1.008 \text{ g/mol})}{159.1 \text{ g/mol}} \times 100\% = \textbf{5.069\%}$$

$$\textbf{\%Cl} = \frac{2(35.45 \text{ g/mol})}{159.1 \text{ g/mol}} \times 100\% = \textbf{44.56\%}$$

$$\textbf{\%S} = \frac{32.07 \text{ g/mol}}{159.1 \text{ g/mol}} \times 100\% = \textbf{20.16\%}$$

3.104   The number of carbon atoms in a 24-carat diamond is:

$$24 \text{ carat} \times \frac{200 \text{ mg C}}{1 \text{ carat}} \times \frac{0.001 \text{ g C}}{1 \text{ mg C}} \times \frac{1 \text{ mol C}}{12.01 \text{ g C}} \times \frac{6.022 \times 10^{23} \text{ atoms C}}{1 \text{ mol C}} = \textbf{2.4} \times \textbf{10}^{23} \textbf{ atoms C}$$

3.105   The amount of Fe that reacted is: $\dfrac{1}{8} \times 664 \text{ g} = 83.0 \text{ g reacted}$

The amount of Fe remaining is: $664 \text{ g} - 83.0 \text{ g} = \textbf{581 g remaining}$

Thus, 83.0 g of Fe reacts to form the compound Fe_2O_3, which has two moles of Fe atoms per 1 mole of compound. The mass of Fe_2O_3 produced is:

$$83.0 \text{ g Fe} \times \frac{1 \text{ mol Fe}}{55.85 \text{ g Fe}} \times \frac{1 \text{ mol Fe}_2O_3}{2 \text{ mol Fe}} \times \frac{159.7 \text{ g Fe}_2O_3}{1 \text{ mol Fe}_2O_3} = \textbf{119 g Fe}_2\textbf{O}_3$$

The final mass of the iron bar and rust is: 581 g Fe + 119 g Fe$_2$O$_3$ = **700 g**

3.106   The mass of oxygen in MO is 39.46 g − 31.70 g = 7.76 g O. Therefore, for every 31.70 g of M, there is 7.76 g of O in the compound MO. The molecular formula shows a mole ratio of 1 mole M : 1 mole O. First, calculate moles of M that react with 7.76 g O.

$$\text{mol M} = 7.76 \text{ g O} \times \frac{1 \text{ mol O}}{16.00 \text{ g O}} \times \frac{1 \text{ mol M}}{1 \text{ mol O}} = 0.485 \text{ mol M}$$

$$\text{molar mass M} = \frac{31.70 \text{ g M}}{0.485 \text{ mol M}} = 65.4 \text{ g/mol}$$

Thus, the atomic mass of M is **65.4 amu**. The metal is most likely **Zn**.

3.107   **(a)** $\text{Zn}(s) + \text{H}_2\text{SO}_4(aq) \longrightarrow \text{ZnSO}_4(aq) + \text{H}_2(g)$

(b)   We assume that a pure sample would produce the theoretical yield of H$_2$. The balanced equation shows a mole ratio of 1 mole H$_2$: 1 mole Zn. The theoretical yield of H$_2$ is:

$$3.86 \text{ g Zn} \times \frac{1 \text{ mol Zn}}{65.39 \text{ g Zn}} \times \frac{1 \text{ mol H}_2}{1 \text{ mol Zn}} \times \frac{2.016 \text{ g H}_2}{1 \text{ mol H}_2} = 0.119 \text{ g H}_2$$

$$\textbf{percent purity} = \frac{0.0764 \text{ g H}_2}{0.119 \text{ g H}_2} \times 100\% = \textbf{64.2\%}$$

(c)   We assume that the impurities are inert and do not react with the sulfuric acid to produce hydrogen.

3.108   The wording of the problem suggests that the actual yield is less than the theoretical yield. The percent yield will be equal to the percent purity of the iron(III) oxide. We find the theoretical yield :

$$(2.62 \times 10^3 \text{ kg Fe}_2O_3) \times \frac{1000 \text{ g Fe}_2O_3}{1 \text{ kg Fe}_2O_3} \times \frac{1 \text{ mol Fe}_2O_3}{159.7 \text{ g Fe}_2O_3} \times \frac{2 \text{ mol Fe}}{1 \text{ mol Fe}_2O_3} \times \frac{55.85 \text{ g Fe}}{1 \text{ mol Fe}} \times \frac{1 \text{ kg Fe}}{1000 \text{ g Fe}}$$

$$= 1.83 \times 10^3 \text{ kg Fe.}$$

$$\text{percent yield} = \frac{\text{actual yield}}{\text{theoretical yield}} \times 100\%$$

$$\textbf{percent yield} = \frac{1.64 \times 10^3 \text{ kg Fe}}{1.83 \times 10^3 \text{ kg Fe}} \times 100\% = \textbf{89.6\%} = \textbf{purity of Fe}_2\textbf{O}_3$$

3.109   The balanced equation is: $\text{C}_6\text{H}_{12}\text{O}_6 + 6\text{O}_2 \longrightarrow 6\text{CO}_2 + 6\text{H}_2\text{O}$

$$\frac{5.0 \times 10^2 \text{ g glucose}}{1 \text{ day}} \times \frac{1 \text{ mol glucose}}{180.2 \text{ g glucose}} \times \frac{6 \text{ mol CO}_2}{1 \text{ mol glucose}} \times \frac{44.01 \text{ g CO}_2}{1 \text{ mol CO}_2} \times \frac{365 \text{ days}}{1 \text{ yr}} \times (6.5 \times 10^9 \text{ people})$$

$$= \textbf{1.7} \times \textbf{10}^{15} \textbf{ g CO}_2\textbf{/yr}$$

3.110   The carbohydrate contains 40 percent carbon; therefore, the remaining 60 percent is hydrogen and oxygen. The problem states that the hydrogen to oxygen ratio is 2:1. We can write this 2:1 ratio as H_2O.

Assume 100 g of compound.

$$40.0 \text{ g C} \times \frac{1 \text{ mol C}}{12.01 \text{ g C}} = 3.33 \text{ mol C}$$

$$60.0 \text{ g H}_2\text{O} \times \frac{1 \text{ mol H}_2\text{O}}{18.02 \text{ g H}_2\text{O}} = 3.33 \text{ mol H}_2\text{O}$$

Dividing by 3.33 gives **CH_2O** for the empirical formula.

To find the molecular formula, divide the molar mass by the empirical mass.

$$\frac{\text{molar mass}}{\text{empirical mass}} = \frac{178 \text{ g}}{30.03 \text{ g}} \approx 6$$

Thus, there are six CH_2O units in each molecule of the compound, so the molecular formula is $(CH_2O)_6$, or **$C_6H_{12}O_6$**.

3.111   The molar mass of chlorophyll is 893.5 g/mol. Finding the mass of a 0.0011-mol sample:

$$0.0011 \text{ mol chlorophyll} \times \frac{893.5 \text{ g chlorophyll}}{1 \text{ mol chlorophyll}} = 0.98 \text{ g chlorophyll}$$

The chlorophyll sample has the greater mass.

3.112   If we assume 100 g of compound, the masses of Cl and X are 67.2 g and 32.8 g, respectively. We can calculate the moles of Cl.

$$67.2 \text{ g Cl} \times \frac{1 \text{ mol Cl}}{35.45 \text{ g Cl}} = 1.90 \text{ mol Cl}$$

Then, using the mole ratio from the chemical formula (XCl_3), we can calculate the moles of X contained in 32.8 g.

$$1.90 \text{ mol Cl} \times \frac{1 \text{ mol X}}{3 \text{ mol Cl}} = 0.633 \text{ mol X}$$

0.633 mole of X has a mass of 32.8 g. Calculating the molar mass of X:

$$\frac{32.8 \text{ g X}}{0.633 \text{ mol X}} = \textbf{51.8 g/mol}$$

The element is most likely **chromium** (molar mass = 52.00 g/mol).

3.113   **(a)**   The molar mass of hemoglobin is:

$$2952(12.01 \text{ g}) + 4664(1.008 \text{ g}) + 812(14.01 \text{ g}) + 832(16.00 \text{ g}) + 8(32.07 \text{ g}) + 4(55.85 \text{ g})$$

$$= \textbf{6.532} \times \textbf{10}^{\textbf{4}} \textbf{ g}$$

(b)   To solve this problem, the following conversions need to be completed:

L → mL → red blood cells → hemoglobin molecules → mol hemoglobin → mass hemoglobin

We will use the following abbreviations: RBC = red blood cells, HG = hemoglobin

$$5.00 \text{ L} \times \frac{1 \text{ mL}}{0.001 \text{ L}} \times \frac{5.0 \times 10^9 \text{ RBC}}{1 \text{ mL}} \times \frac{2.8 \times 10^8 \text{ HG molecules}}{1 \text{ RBC}} \times \frac{1 \text{ mol HG}}{6.022 \times 10^{23} \text{ molecules HG}} \times \frac{6.532 \times 10^4 \text{ g HG}}{1 \text{ mol HG}}$$

$$= \mathbf{7.6 \times 10^2 \text{ g HG}}$$

3.114   A 100 g sample of myoglobin contains 0.34 g of iron (0.34% Fe). The number of moles of Fe is:

$$0.34 \text{ g Fe} \times \frac{1 \text{ mol Fe}}{55.85 \text{ g Fe}} = 6.1 \times 10^{-3} \text{ mol Fe}$$

Since there is one Fe atom in a molecule of myoglobin, the moles of myoglobin also equal 6.1×10^{-3} mole. The molar mass of myoglobin can be calculated.

$$\textbf{molar mass myoglobin} = \frac{100 \text{ g myoglobin}}{6.1 \times 10^{-3} \text{ mol myoglobin}} = \mathbf{1.6 \times 10^4 \text{ g/mol}}$$

3.115   **(a)** $8.38 \text{ g KBr} \times \dfrac{1 \text{ mol KBr}}{119.0 \text{ g KBr}} \times \dfrac{6.022 \times 10^{23} \text{ KBr}}{1 \text{ mol KBr}} \times \dfrac{1 \text{ K}^+ \text{ ion}}{1 \text{ KBr}} = \mathbf{4.24 \times 10^{22} \text{ K}^+ \text{ ions}}$

Since there is one Br^- for every one K^+, the number of Br^- ions = $\mathbf{4.24 \times 10^{22} \text{ Br}^- \text{ ions}}$

(b) $5.40 \text{ g Na}_2\text{SO}_4 \times \dfrac{1 \text{ mol Na}_2\text{SO}_4}{142.05 \text{ g Na}_2\text{SO}_4} \times \dfrac{6.022 \times 10^{23} \text{ Na}_2\text{SO}_4}{1 \text{ mol Na}_2\text{SO}_4} \times \dfrac{2 \text{ Na}^+ \text{ ions}}{1 \text{ Na}_2\text{SO}_4} = \mathbf{4.58 \times 10^{22} \text{ Na}^+ \text{ ions}}$

Since there are two Na^+ for every one SO_4^{2-}, the number of SO_4^{2-} ions = $\mathbf{2.29 \times 10^{22} \text{ SO}_4^{2-} \text{ ions}}$

(c) $7.45 \text{ g Ca}_3(\text{PO}_4)_2 \times \dfrac{1 \text{ mol Ca}_3(\text{PO}_4)_2}{310.2 \text{ g Ca}_3(\text{PO}_4)_2} \times \dfrac{6.022 \times 10^{23} \text{ Ca}_3(\text{PO}_4)_2}{1 \text{ mol Ca}_3(\text{PO}_4)_2} \times \dfrac{3 \text{ Ca}^{2+} \text{ ions}}{1 \text{ Ca}_3(\text{PO}_4)_2}$

$$= \mathbf{4.34 \times 10^{22} \text{ Ca}^{2+} \text{ ions}}$$

Since there are three Ca^{2+} for every two PO_4^{3-}, the number of PO_4^{3-} ions is:

$$4.34 \times 10^{22} \text{ Ca}^{2+} \text{ ions} \times \frac{2 \text{ PO}_4^{3-} \text{ ions}}{3 \text{ Ca}^{2+} \text{ ions}} = \mathbf{2.89 \times 10^{22} \text{ PO}_4^{3-} \text{ ions}}$$

3.116   Assume 100 g of sample. Then,

$$\text{mol Na} = 32.08 \text{ g Na} \times \frac{1 \text{ mol Na}}{22.99 \text{ g Na}} = 1.395 \text{ mol Na}$$

$$\text{mol O} = 36.01 \text{ g O} \times \frac{1 \text{ mol O}}{16.00 \text{ g O}} = 2.251 \text{ mol O}$$

$$\text{mol Cl} = 19.51 \text{ g Cl} \times \frac{1 \text{ mol Cl}}{35.45 \text{ g Cl}} = 0.5504 \text{ mol Cl}$$

Since Cl is only contained in NaCl, the moles of Cl equals the moles of Na contained in NaCl.

$$\text{mol Na (in NaCl)} = 0.5504 \text{ mol}$$

The number of moles of Na in the remaining two compounds is: 1.395 mol − 0.5504 mol = 0.8446 mol Na.

To solve for moles of the remaining two compounds, let

$$x = \text{moles of } Na_2SO_4$$
$$y = \text{moles of } NaNO_3$$

Then, from the mole ratio of Na and O in each compound, we can write

$$2x + y = \text{mol Na} = 0.8446 \text{ mol}$$
$$4x + 3y = \text{mol O} = 2.251 \text{ mol}$$

Solving two equations with two unknowns gives

$$x = 0.1414 = \text{mol } Na_2SO_4 \quad \text{and} \quad y = 0.5618 = \text{mol } NaNO_3$$

Finally, we convert to mass of each compound to calculate the mass percent of each compound in the sample. Remember, the sample size is 100 g.

$$\textbf{mass\% NaCl} = 0.5504 \text{ mol NaCl} \times \frac{58.44 \text{ g NaCl}}{1 \text{ mol NaCl}} \times \frac{1}{100 \text{ g sample}} \times 100\% = \textbf{32.17\% NaCl}$$

$$\textbf{mass\% Na}_2\textbf{SO}_4 = 0.1414 \text{ mol } Na_2SO_4 \times \frac{142.1 \text{ g } Na_2SO_4}{1 \text{ mol } Na_2SO_4} \times \frac{1}{100 \text{ g sample}} \times 100\% = \textbf{20.09\% Na}_2\textbf{SO}_4$$

$$\textbf{mass\% NaNO}_3 = 0.5618 \text{ mol } NaNO_3 \times \frac{85.00 \text{ g } NaNO_3}{1 \text{ mol } NaNO_3} \times \frac{1}{100 \text{ g sample}} \times 100\% = \textbf{47.75\% NaNO}_3$$

3.117 **(a)** 16 amu, CH_4 17 amu, NH_3 18 amu, H_2O 64 amu, SO_2

(b) The formula C_3H_8 can also be written as $CH_3CH_2CH_3$. A CH_3 fragment could break off from this molecule giving a peak at 15 amu. No fragment of CO_2 can have a mass of 15 amu. Therefore, the substance responsible for the mass spectrum is most likely C_3H_8.

(c) First, let's calculate the masses of CO_2 and C_3H_8.

molecular mass CO_2 = 12.00000 amu + 2(15.99491 amu) = 43.98982 amu

molecular mass C_3H_8 = 3(12.00000 amu) + 8(1.00797 amu) = 44.06376 amu

These masses differ by only 0.07394 amu. The measurements must be precise to **±0.030 amu**.

43.98982 + 0.030 amu = 44.02 amu

44.06376 − 0.030 amu = 44.03 amu

3.118 The mass percent of an element in a compound can be calculated as follows:

$$\text{percent by mass of each element} = \frac{\text{mass of element in 1 mol of compound}}{\text{molar mass of compound}} \times 100\%$$

The molar mass of $Ca_3(PO_4)_2$ = 310.18 g/mol

$$\textbf{\% Ca} = \frac{(3)(40.08 \text{ g})}{310.18 \text{ g}} \times 100\% = \textbf{38.76\% Ca}$$

$$\% \; P \; = \; \frac{(2)(30.97 \; g)}{310.18 \; g} \times 100\% \; = \; \mathbf{19.97\% \; P}$$

$$\% \; O \; = \; \frac{(8)(16.00 \; g)}{310.18 \; g} \times 100\% \; = \; \mathbf{41.27\% \; O}$$

3.119   (a)   First, calculate the mass of C in CO_2, the mass of H in H_2O, and the mass of N in NH_3. For now, we will carry more than 3 significant figures and then round to the correct number at the end of the problem.

$$? \; g \; C \; = \; 3.94 \; g \; CO_2 \times \frac{1 \; mol \; CO_2}{44.01 \; g \; CO_2} \times \frac{1 \; mol \; C}{1 \; mol \; CO_2} \times \frac{12.01 \; g \; C}{1 \; mol \; C} \; = \; 1.075 \; g \; C$$

$$? \; g \; H \; = \; 1.89 \; g \; H_2O \times \frac{1 \; mol \; H_2O}{18.02 \; g \; H_2O} \times \frac{2 \; mol \; H}{1 \; mol \; H_2O} \times \frac{1.008 \; g \; H}{1 \; mol \; H} \; = \; 0.2114 \; g \; H$$

$$? \; g \; N \; = \; 0.436 \; g \; NH_3 \times \frac{1 \; mol \; NH_3}{17.03 \; g \; NH_3} \times \frac{1 \; mol \; N}{1 \; mol \; NH_3} \times \frac{14.01 \; g \; N}{1 \; mol \; N} \; = \; 0.3587 \; g \; N$$

Next, we can calculate the %C, %H, and the %N in each sample, then we can calculate the %O by difference.

$$\%C \; = \; \frac{1.075 \; g \; C}{2.175 \; g \; sample} \times 100\% \; = \; 49.43\% \; C$$

$$\%H \; = \; \frac{0.2114 \; g \; H}{2.175 \; g \; sample} \times 100\% \; = \; 9.720\% \; H$$

$$\%N \; = \; \frac{0.3587 \; g \; N}{1.873 \; g \; sample} \times 100\% \; = \; 19.15\% \; N$$

The % O $= 100\% - (49.43\% + 9.720\% + 19.15\%) = 21.70\% \; O$

Assuming 100 g of compound, calculate the moles of each element.

$$? \; mol \; C \; = \; 49.43 \; g \; C \times \frac{1 \; mol \; C}{12.01 \; g \; C} \; = \; 4.116 \; mol \; C$$

$$? \; mol \; H \; = \; 9.720 \; g \; H \times \frac{1 \; mol \; H}{1.008 \; g \; H} \; = \; 9.643 \; mol \; H$$

$$? \; mol \; N \; = \; 19.15 \; g \; N \times \frac{1 \; mol \; N}{14.01 \; g \; N} \; = \; 1.367 \; mol \; N$$

$$? \; mol \; O \; = \; 21.70 \; g \; O \times \frac{1 \; mol \; O}{16.00 \; g \; O} \; = \; 1.356 \; mol \; O$$

Thus, we arrive at the formula $C_{4.116}H_{9.643}N_{1.367}O_{1.356}$. Dividing by 1.356 gives the empirical formula, $\mathbf{C_3H_7NO}$.

(b)   The empirical molar mass is 73.10 g. Since the approximate molar mass of lysine is 150 g, we have:

$$\frac{150 \; g}{73.10 \; g} \; \approx \; 2$$

Therefore, the molecular formula is $(C_3H_7NO)_2$ or $\mathbf{C_6H_{14}N_2O_2}$.

3.120   **Yes**. The number of hydrogen atoms in one gram of hydrogen molecules is the same as the number in one gram of hydrogen atoms. There is no difference in mass, only in the way that the particles are arranged.

Would the mass of 100 dimes be the same if they were stuck together in pairs instead of separated?

3.121   The mass of one fluorine atom is 19.00 amu. The mass of one mole of fluorine atoms is 19.00 g. Multiplying the mass of one atom by Avogadro's number gives the mass of one mole of atoms. We can write:

$$\frac{19.00 \text{ amu}}{1 \text{ F atom}} \times (6.022 \times 10^{23} \text{ F atoms}) = 19.00 \text{ g F}$$

or,

$$\mathbf{6.022 \times 10^{23} \text{ amu} = 1 \text{ g}}$$

This is why Avogadro's numbers has sometimes been described as a conversion factor between amu and grams.

3.122   Since we assume that water exists as either H_2O or D_2O, the natural abundances are 99.985 percent and 0.015 percent, respectively. If we convert to molecules of water (both H_2O or D_2O), we can calculate the molecules that are D_2O from the natural abundance (0.015%).

The necessary conversions are:

mL water $\rightarrow$ g water $\rightarrow$ mol water $\rightarrow$ molecules water $\rightarrow$ molecules D_2O

$$400 \text{ mL water} \times \frac{1 \text{ g water}}{1 \text{ mL water}} \times \frac{1 \text{ mol water}}{18.02 \text{ g water}} \times \frac{6.022 \times 10^{23} \text{ molecules}}{1 \text{ mol water}} \times \frac{0.015\% \text{ molecules } D_2O}{100\% \text{ molecules water}}$$

$$= \mathbf{2.01 \times 10^{21} \text{ molecules } D_2O}$$

3.123   There can only be one chlorine per molecule, since two chlorines have a combined mass in excess of 70 amu. Since the ^{35}Cl isotope is more abundant, let's subtract 35 amu from the mass corresponding to the more intense peak.

$$50 \text{ amu} - 35 \text{ amu} = 15 \text{ amu}$$

15 amu equals the mass of one ^{12}C and three ^1H. To explain the two peaks, we have:

molecular mass $^{12}C^1H_3^{35}Cl$ = 12 amu + 3(1 amu) + 35 amu = 50 amu
molecular mass $^{12}C^1H_3^{37}Cl$ = 12 amu + 3(1 amu) + 37 amu = 52 amu

^{35}Cl is three times more abundant than ^{37}Cl; therefore, the 50 amu peak will be three times more intense than the 52 amu peak.

3.124   First, we can calculate the moles of oxygen.

$$2.445 \text{ g C} \times \frac{1 \text{ mol C}}{12.01 \text{ g C}} \times \frac{1 \text{ mol O}}{1 \text{ mol C}} = 0.2036 \text{ mol O}$$

Next, we can calculate the molar mass of oxygen.

$$\text{molar mass O} = \frac{3.257 \text{ g O}}{0.2036 \text{ mol O}} = 16.00 \text{ g/mol}$$

If 1 mole of oxygen atoms has a mass of 16.00 g, then 1 atom of oxygen has an **atomic mass of 16.00 amu.**

3.125 The molecular formula for Cl_2O_7 means that there are 2 Cl atoms for every 7 O atoms or 2 moles of Cl atoms for every 7 moles of O atoms. We can write:

$$\text{mole ratio} = \frac{2 \text{ mol Cl}}{7 \text{ mol O}} = \frac{1 \text{ mol Cl}_2}{3.5 \text{ mol O}_2}$$

3.126 **(a)** The mass of chlorine is **5.0 g**.

(b) From the percent by mass of Cl, we can calculate the mass of chlorine in 60.0 g of $NaClO_3$.

$$\text{mass \% Cl} = \frac{35.45 \text{ g Cl}}{106.44 \text{ g compound}} \times 100\% = 33.31\% \text{ Cl}$$

$$\text{mass Cl} = 60.0 \text{ g} \times 0.3331 = \textbf{20.0 g Cl}$$

(c) 0.10 mol of KCl contains 0.10 mol of Cl.

$$0.10 \text{ mol Cl} \times \frac{35.45 \text{ g Cl}}{1 \text{ mol Cl}} = \textbf{3.5 g Cl}$$

(d) From the percent by mass of Cl, we can calculate the mass of chlorine in 30.0 g of $MgCl_2$.

$$\text{mass \% Cl} = \frac{(2)(35.45 \text{ g Cl})}{95.21 \text{ g compound}} \times 100\% = 74.47\% \text{ Cll}$$

$$\text{mass Cl} = 30.0 \text{ g} \times 0.7447 = \textbf{22.3 g Cl}$$

(e) The mass of Cl can be calculated from the molar mass of Cl_2.

$$0.50 \text{ mol Cl}_2 \times \frac{70.90 \text{ g Cl}}{1 \text{ mol Cl}_2} = \textbf{35.45 g Cl}$$

Thus, **(e) 0.50 mol Cl$_2$** contains the greatest mass of chlorine.

3.127 **(a)** We need to compare the mass % of K in both KCl and K_2SO_4.

$$\%K \text{ in KCl} = \frac{39.10 \text{ g}}{74.55 \text{ g}} \times 100\% = 52.45\% \text{ K}$$

$$\%K \text{ in K}_2SO_4 = \frac{2(39.10 \text{ g})}{174.27 \text{ g}} \times 100\% = 44.87\% \text{ K}$$

The price is dependent on the %K.

$$\frac{\text{Price of K}_2SO_4}{\text{Price of KCl}} = \frac{\%K \text{ in K}_2SO_4}{\%K \text{ in KCl}}$$

$$\text{Price of K}_2SO_4 = \text{Price of KCl} \times \frac{\%K \text{ in K}_2SO_4}{\%K \text{ in KCl}}$$

$$\textbf{Price of K}_2\textbf{SO}_4 = \frac{\$0.55}{\text{kg}} \times \frac{44.87\%}{52.45\%} = \textbf{\$0.47/kg}$$

(b)   First, calculate the number of moles of K in 1.00 kg of KCl.

$$(1.00 \times 10^3 \text{ g KCl}) \times \frac{1 \text{ mol KCl}}{74.55 \text{ g KCl}} \times \frac{1 \text{ mol K}}{1 \text{ mol KCl}} = 13.4 \text{ mol K}$$

Next, calculate the amount of K_2O needed to supply 13.4 mol K.

$$13.4 \text{ mol K} \times \frac{1 \text{ mol K}_2\text{O}}{2 \text{ mol K}} \times \frac{94.20 \text{ g K}_2\text{O}}{1 \text{ mol K}_2\text{O}} \times \frac{1 \text{ kg}}{1000 \text{ g}} = \textbf{0.631 kg K}_2\textbf{O}$$

3.128   Both compounds contain only Pt and Cl. The percent by mass of Pt can be calculated by subtracting the percent Cl from 100 percent.

Compound A: Assume 100 g of compound.

$$26.7 \text{ g Cl} \times \frac{1 \text{ mol Cl}}{35.45 \text{ g Cl}} = 0.753 \text{ mol Cl}$$

$$73.3 \text{ g Pt} \times \frac{1 \text{ mol Pt}}{195.1 \text{ g Pt}} = 0.376 \text{ mol Pt}$$

Dividing by the smallest number of moles (0.376 mole) gives the empirical formula, **PtCl₂**.

Compound B: Assume 100 g of compound.

$$42.1 \text{ g Cl} \times \frac{1 \text{ mol Cl}}{35.45 \text{ g Cl}} = 1.19 \text{ mol Cl}$$

$$57.9 \text{ g Pt} \times \frac{1 \text{ mol Pt}}{195.1 \text{ g Pt}} = 0.297 \text{ mol Pt}$$

Dividing by the smallest number of moles (0.297 mole) gives the empirical formula, **PtCl₄**.

3.129   The mass of the metal (X) in the metal oxide is 1.68 g. The mass of oxygen in the metal oxide is 2.40 g − 1.68 g = 0.72 g oxygen. Next, find the number of moles of the metal and of the oxygen.

$$\text{moles X} = 1.68 \text{ g} \times \frac{1 \text{ mol X}}{55.9 \text{ g X}} = 0.0301 \text{ mol X}$$

$$\text{moles O} = 0.72 \text{ g} \times \frac{1 \text{ mol O}}{16.00 \text{ g O}} = 0.045 \text{ mol O}$$

This gives the formula $X_{0.0301}O_{0.045}$. Dividing by the smallest number of moles (0.0301 moles) gives the formula $X_{1.00}O_{1.5}$. Multiplying by two gives the empirical formula, **X₂O₃**.

The balanced equation is: $X_2O_3(s) + 3CO(g) \longrightarrow 2X(s) + 3CO_2(g)$

3.130   Both compounds contain only Mn and O. When the first compound is heated, oxygen gas is evolved. Let's calculate the empirical formulas for the two compounds, then we can write a balanced equation.

(a) Compound X: Assume 100 g of compound.

$$63.3 \text{ g Mn} \times \frac{1 \text{ mol Mn}}{54.94 \text{ g Mn}} = 1.15 \text{ mol Mn}$$

$$36.7 \text{ g O} \times \frac{1 \text{ mol O}}{16.00 \text{ g O}} = 2.29 \text{ mol O}$$

Dividing by the smallest number of moles (1.15 moles) gives the empirical formula, MnO_2.

Compound Y: Assume 100 g of compound.

$$72.0 \text{ g Mn} \times \frac{1 \text{ mol Mn}}{54.94 \text{ g Mn}} = 1.31 \text{ mol Mn}$$

$$28.0 \text{ g O} \times \frac{1 \text{ mol O}}{16.00 \text{ g O}} = 1.75 \text{ mol O}$$

Dividing by the smallest number of moles gives $MnO_{1.33}$. Recall that an empirical formula must have whole number coefficients. Multiplying by a factor of 3 gives the empirical formula Mn_3O_4.

(b) The unbalanced equation is: $MnO_2 \longrightarrow Mn_3O_4 + O_2$

Balancing by inspection gives: $3MnO_2 \longrightarrow Mn_3O_4 + O_2$

3.131   The mass of the water is the difference between 1.936 g of the hydrate and the mass of *water-free* (anhydrous) $BaCl_2$. First, we need to start with a balanced equation for the reaction. Upon treatment with sulfuric acid, $BaCl_2$ dissolves, losing its waters of hydration.

$$BaCl_2(aq) + H_2SO_4(aq) \longrightarrow BaSO_4(s) + 2HCl(aq)$$

Next, calculate the mass of anhydrous $BaCl_2$ based on the amount of $BaSO_4$ produced.

$$1.864 \text{ g BaSO}_4 \times \frac{1 \text{ mol BaSO}_4}{233.4 \text{ g BaSO}_4} \times \frac{1 \text{ mol BaCl}_2}{1 \text{ mol BaSO}_4} \times \frac{208.2 \text{ g BaCl}_2}{1 \text{ mol BaCl}_2} = 1.663 \text{ g BaCl}_2$$

The mass of water is (1.936 g − 1.663 g) = 0.273 g H_2O. Next, we convert the mass of H_2O and the mass of $BaCl_2$ to moles to determine the formula of the hydrate.

$$0.273 \text{ g H}_2\text{O} \times \frac{1 \text{ mol H}_2\text{O}}{18.02 \text{ g H}_2\text{O}} = 0.0151 \text{ mol H}_2\text{O}$$

$$1.663 \text{ g BaCl}_2 \times \frac{1 \text{ mol BaCl}_2}{208.2 \text{ g BaCl}_2} = 0.00799 \text{ mol BaCl}_2$$

The ratio of the number of moles of H_2O to the number of moles of $BaCl_2$ is 0.0151/0.00799 = **1.89**. We round this number to **2**, which is the value of x. The formula of the hydrate is $BaCl_2 \cdot 2H_2O$.

3.132 SO_2 is converted to H_2SO_4 by reaction with water. The mole ratio between SO_2 and H_2SO_4 is 1:1.

This is a unit conversion problem. You should come up with the following strategy to solve the problem.

tons $SO_2 \rightarrow$ ton-mol $SO_2 \rightarrow$ ton-mol $H_2SO_4 \rightarrow$ tons H_2SO_4

$$? \textbf{ tons H}_2\textbf{SO}_4 = (4.0 \times 10^5 \text{ tons SO}_2) \times \frac{1 \text{ ton-mol SO}_2}{64.07 \text{ tons SO}_2} \times \frac{1 \text{ ton-mol H}_2\text{SO}_4}{1 \text{ ton-mol SO}_2} \times \frac{98.09 \text{ tons H}_2\text{SO}_4}{1 \text{ ton-mol H}_2\text{SO}_4}$$

$$= \textbf{6.1} \times \textbf{10}^5 \textbf{ tons H}_2\textbf{SO}_4$$

> **Tip:** You probably won't come across a ton-mol that often in chemistry. However, it was convenient to use in this problem. We normally use a g-mol. 1 g-mol SO_2 has a mass of 64.07 g. In a similar manner, 1 ton-mol of SO_2 has a mass of 64.07 tons.

3.133   The mass of water lost upon heating the mixture is (5.020 g – 2.988 g) = 2.032 g water. Next, if we let x = mass of $CuSO_4 \cdot 5H_2O$, then the mass of $MgSO_4 \cdot 7H_2O$ is $(5.020 - x)$g. We can calculate the amount of water lost by each salt based on the mass % of water in each hydrate. We can write:

(mass $CuSO_4 \cdot 5H_2O$)(% H_2O) + (mass $MgSO_4 \cdot 7H_2O$)(% H_2O) = total mass H_2O = 2.032 g H_2O

Calculate the % H_2O in each hydrate.

$$\% \ H_2O \ (CuSO_4 \cdot 5H_2O) = \frac{(5)(18.02 \ g)}{249.7 \ g} \times 100\% = 36.08\% \ H_2O$$

$$\% \ H_2O \ (MgSO_4 \cdot 7H_2O) = \frac{(7)(18.02 \ g)}{246.5 \ g} \times 100\% = 51.17\% \ H_2O$$

Substituting into the equation above gives:

$(x)(0.3608) + (5.020 - x)(0.5117) = 2.032$ g

$0.1509x = 0.5367$

$x = 3.557$ g = mass of $CuSO_4 \cdot 5H_2O$

Finally, the percent by mass of $CuSO_4 \cdot 5H_2O$ in the mixture is:

$$\frac{3.557 \ g}{5.020 \ g} \times 100\% = \textbf{70.86\%}$$

3.134   We assume that the increase in mass results from the element nitrogen. The mass of nitrogen is:

0.378 g – 0.273 g = 0.105 g N

The empirical formula can now be calculated. Convert to moles of each element.

$$0.273 \ g \ Mg \times \frac{1 \ mol \ Mg}{24.31 \ g \ Mg} = 0.0112 \ mol \ Mg$$

$$0.105 \ g \ N \times \frac{1 \ mol \ N}{14.01 \ g \ N} = 0.00749 \ mol \ N$$

Dividing by the smallest number of moles gives $Mg_{1.5}N$. Recall that an empirical formula must have whole number coefficients. Multiplying by a factor of 2 gives the empirical formula **Mg_3N_2**. The name of this compound is **magnesium nitride**.

3.135   The balanced equations are:

$$CH_4 + 2O_2 \longrightarrow CO_2 + 2H_2O \qquad\qquad 2C_2H_6 + 7O_2 \longrightarrow 4CO_2 + 6H_2O$$

If we let x = mass of CH_4, then the mass of C_2H_6 is $(13.43 - x)$ g.

Next, we need to calculate the mass of CO_2 and the mass of H_2O produced by both CH_4 and C_2H_6. The sum of the masses of CO_2 and H_2O will add up to 64.84 g.

$$? \text{ g } CO_2 \text{ (from } CH_4) = x \text{ g } CH_4 \times \frac{1 \text{ mol } CH_4}{16.04 \text{ g } CH_4} \times \frac{1 \text{ mol } CO_2}{1 \text{ mol } CH_4} \times \frac{44.01 \text{ g } CO_2}{1 \text{ mol } CO_2} = 2.744x \text{ g } CO_2$$

$$? \text{ g } H_2O \text{ (from } CH_4) = x \text{ g } CH_4 \times \frac{1 \text{ mol } CH_4}{16.04 \text{ g } CH_4} \times \frac{2 \text{ mol } H_2O}{1 \text{ mol } CH_4} \times \frac{18.02 \text{ g } H_2O}{1 \text{ mol } H_2O} = 2.247x \text{ g } H_2O$$

$$? \text{ g } CO_2 \text{ (from } C_2H_6) = (13.43 - x) \text{ g } C_2H_6 \times \frac{1 \text{ mol } C_2H_6}{30.07 \text{ g } C_2H_6} \times \frac{4 \text{ mol } CO_2}{2 \text{ mol } C_2H_6} \times \frac{44.01 \text{ g } CO_2}{1 \text{ mol } CO_2}$$

$$= 2.927(13.43 - x) \text{ g } CO_2$$

$$? \text{ g } H_2O \text{ (from } C_2H_6) = (13.43 - x) \text{ g } C_2H_6 \times \frac{1 \text{ mol } C_2H_6}{30.07 \text{ g } C_2H_6} \times \frac{6 \text{ mol } H_2O}{2 \text{ mol } C_2H_6} \times \frac{18.02 \text{ g } H_2O}{1 \text{ mol } H_2O}$$

$$= 1.798(13.43 - x) \text{ g } H_2O$$

Summing the masses of CO_2 and H_2O:

$$2.744x \text{ g} + 2.247x \text{ g} + 2.927(13.43 - x) \text{ g} + 1.798(13.43 - x) \text{ g} = 64.84 \text{ g}$$

$$0.266x = 1.383$$

$$x = 5.20 \text{ g}$$

The fraction of CH_4 in the mixture is $\dfrac{5.20 \text{ g}}{13.43 \text{ g}} = \mathbf{0.387}$

3.136   *Step 1:* Calculate the mass of C in 55.90 g CO_2, and the mass of H in 28.61 g H_2O. This is a factor-label problem. To calculate the mass of each component, you need the molar masses and the correct mole ratio.

You should come up with the following strategy:

$$\text{g } CO_2 \rightarrow \text{mol } CO_2 \rightarrow \text{mol C} \rightarrow \text{g C}$$

Step 2: $? \text{ g C} = 55.90 \text{ g } CO_2 \times \dfrac{1 \text{ mol } CO_2}{44.01 \text{ g } CO_2} \times \dfrac{1 \text{ mol C}}{1 \text{ mol } CO_2} \times \dfrac{12.01 \text{ g C}}{1 \text{ mol C}} = 15.25 \text{ g C}$

Similarly,

$$? \text{ g H} = 28.61 \text{ g } H_2O \times \frac{1 \text{ mol } H_2O}{18.02 \text{ g } H_2O} \times \frac{2 \text{ mol H}}{1 \text{ mol } H_2O} \times \frac{1.008 \text{ g H}}{1 \text{ mol H}} = 3.201 \text{ g H}$$

Since the compound contains C, H, and Pb, we can calculate the mass of Pb by difference.

$$51.36 \text{ g} = \text{mass C} + \text{mass H} + \text{mass Pb}$$

$$51.36 \text{ g} = 15.25 \text{ g} + 3.201 \text{ g} + \text{mass Pb}$$

$$\text{mass Pb} = 32.91 \text{ g Pb}$$

Step 3: Calculate the number of moles of each element present in the sample. Use molar mass as a conversion factor.

$$? \text{ mol C} = 15.25 \text{ g C} \times \frac{1 \text{ mol C}}{12.01 \text{ g C}} = 1.270 \text{ mol C}$$

Similarly,

$$? \text{ mol H} = 3.201 \text{ g H} \times \frac{1 \text{ mol H}}{1.008 \text{ g H}} = 3.176 \text{ mol H}$$

$$? \text{ mol Pb} = 32.91 \text{ g Pb} \times \frac{1 \text{ mol Pb}}{207.2 \text{ g Pb}} = 0.1588 \text{ mol Pb}$$

Thus, we arrive at the formula $Pb_{0.1588}C_{1.270}H_{3.176}$, which gives the identity and the ratios of atoms present. However, chemical formulas are written with whole numbers.

Step 4: Try to convert to whole numbers by dividing all the subscripts by the smallest subscript.

$$Pb: \frac{0.1588}{0.1588} = 1.00 \qquad C: \frac{1.270}{0.1588} \approx 8 \qquad H: \frac{3.176}{0.1588} \approx 20$$

This gives the empirical formula, **PbC_8H_{20}**.

3.137   First, calculate the mass of C in CO_2 and the mass of H in H_2O.

$$? \text{ g C} = 30.2 \text{ g CO}_2 \times \frac{1 \text{ mol CO}_2}{44.01 \text{ g CO}_2} \times \frac{1 \text{ mol C}}{1 \text{ mol CO}_2} \times \frac{12.01 \text{ g C}}{1 \text{ mol C}} = 8.24 \text{ g C}$$

$$? \text{ g H} = 14.8 \text{ g H}_2\text{O} \times \frac{1 \text{ mol H}_2\text{O}}{18.02 \text{ g H}_2\text{O}} \times \frac{2 \text{ mol H}}{1 \text{ mol H}_2\text{O}} \times \frac{1.008 \text{ g H}}{1 \text{ mol H}} = 1.66 \text{ g H}$$

Since the compound contains C, H, and O, we can calculate the mass of O by difference.

$$12.1 \text{ g} = \text{mass C} + \text{mass H} + \text{mass O}$$

$$12.1 \text{ g} = 8.24 \text{ g} + 1.66 \text{ g} + \text{mass O}$$

$$\text{mass O} = 2.2 \text{ g O}$$

Next, calculate the moles of each element.

$$? \text{ mol C} = 8.24 \text{ g C} \times \frac{1 \text{ mol C}}{12.01 \text{ g C}} = 0.686 \text{ mol C}$$

$$? \text{ mol H} = 1.66 \text{ g H} \times \frac{1 \text{ mol H}}{1.008 \text{ g H}} = 1.65 \text{ mol H}$$

$$? \text{ mol O} = 2.2 \text{ g O} \times \frac{1 \text{ mol O}}{16.00 \text{ g O}} = 0.14 \text{ mol O}$$

Thus, we arrive at the formula $C_{0.686}H_{1.65}O_{0.14}$. Dividing by 0.14 gives the empirical formula, **$C_5H_{12}O$**.

3.138   **(a)**   The following strategy can be used to convert from the volume of the Mg cube to the number of Mg atoms.

$$\text{cm}^3 \rightarrow \text{grams} \rightarrow \text{moles} \rightarrow \text{atoms}$$

$$1.0 \text{ cm}^3 \times \frac{1.74 \text{ g Mg}}{1 \text{ cm}^3} \times \frac{1 \text{ mol Mg}}{24.31 \text{ g Mg}} \times \frac{6.022 \times 10^{23} \text{ Mg atoms}}{1 \text{ mol Mg}} = \textbf{4.3} \times \textbf{10}^{22} \textbf{ Mg atoms}$$

(b)   Since 74 percent of the available space is taken up by Mg atoms, 4.3×10^{22} atoms occupy the following volume:

$$0.74 \times 1.0 \text{ cm}^3 = 0.74 \text{ cm}^3$$

We are trying to calculate the radius of a single Mg atom, so we need the volume occupied by a single Mg atom.

$$\text{volume Mg atom} = \frac{0.74 \text{ cm}^3}{4.3 \times 10^{22} \text{ Mg atoms}} = 1.7 \times 10^{-23} \text{ cm}^3/\text{Mg atom}$$

The volume of a sphere is $\frac{4}{3}\pi r^3$. Solving for the radius:

$$V = 1.7 \times 10^{-23} \text{ cm}^3 = \frac{4}{3}\pi r^3$$

$$r^3 = 4.1 \times 10^{-24} \text{ cm}^3$$

$$r = 1.6 \times 10^{-8} \text{ cm}$$

Converting to picometers:

$$\textbf{radius Mg atom} = (1.6 \times 10^{-8} \text{ cm}) \times \frac{0.01 \text{ m}}{1 \text{ cm}} \times \frac{1 \text{ pm}}{1 \times 10^{-12} \text{ m}} = \textbf{1.6} \times \textbf{10}^2 \textbf{ pm}$$

3.139   The balanced equations for the combustion of sulfur and the reaction of SO_2 with CaO are:

$$S(s) + O_2(g) \longrightarrow SO_2(g) \qquad\qquad SO_2(g) + CaO(s) \longrightarrow CaSO_3(s)$$

First, find the amount of sulfur present in the daily coal consumption.

$$(6.60 \times 10^6 \text{ kg coal}) \times \frac{1.6\% \text{ S}}{100\%} = 1.06 \times 10^5 \text{ kg S} = 1.06 \times 10^8 \text{ g S}$$

The daily amount of CaO needed is:

$$(1.06 \times 10^8 \text{ g S}) \times \frac{1 \text{ mol S}}{32.07 \text{ g S}} \times \frac{1 \text{ mol SO}_2}{1 \text{ mol S}} \times \frac{1 \text{ mol CaO}}{1 \text{ mol SO}_2} \times \frac{56.08 \text{ g CaO}}{1 \text{ mol CaO}} \times \frac{1 \text{ kg}}{1000 \text{ g}} = \textbf{1.85} \times \textbf{10}^5 \textbf{ kg CaO}$$

3.140   The molar mass of air can be calculated by multiplying the mass of each component by its abundance and adding them together. Recall that nitrogen gas and oxygen gas are diatomic.

molar mass air $= (0.7808)(28.02 \text{ g/mol}) + (0.2095)(32.00 \text{ g/mol}) + (0.0097)(39.95 \text{ g/mol}) = \textbf{28.97 g/mol}$

3.141   (a)   Assuming the die pack with no empty space between die, the volume of one mole of die is:

$$(1.5 \text{ cm})^3 \times (6.022 \times 10^{23}) = \textbf{2.0} \times \textbf{10}^{24} \textbf{ cm}^3$$

(b) $6371 \text{ km} = 6.371 \times 10^8 \text{ cm}$

Volume $= \text{area} \times \text{height } (h)$

$$h = \frac{\text{volume}}{\text{area}} = \frac{2.0 \times 10^{24} \text{ cm}^3}{4\pi(6.371 \times 10^8 \text{ cm})^2} = 3.9 \times 10^5 \text{ cm} = \mathbf{3.9 \times 10^3 \text{ m}}$$

3.142   The surface area of the water can be calculated assuming that the dish is circular.

surface area of water $= \pi r^2 = \pi(10 \text{ cm})^2 = 3.1 \times 10^2 \text{ cm}^2$

The cross-sectional area of one stearic acid molecule in cm^2 is:

$$0.21 \text{ nm}^2 \times \left(\frac{1 \times 10^{-9} \text{ m}}{1 \text{ nm}}\right)^2 \times \left(\frac{1 \text{ cm}}{0.01 \text{ m}}\right)^2 = 2.1 \times 10^{-15} \text{ cm}^2/\text{molecule}$$

Assuming that there is no empty space between molecules, we can calculate the number of stearic acid molecules that will fit in an area of $3.1 \times 10^2 \text{ cm}^2$.

$$(3.1 \times 10^2 \text{ cm}^2) \times \frac{1 \text{ molecule}}{2.1 \times 10^{-15} \text{ cm}^2} = 1.5 \times 10^{17} \text{ molecules}$$

Next, we can calculate the moles of stearic acid in the 1.4×10^{-4} g sample. Then, we can calculate Avogadro's number (the number of molecules per mole).

$$1.4 \times 10^{-4} \text{ g stearic acid} \times \frac{1 \text{ mol stearic acid}}{284.5 \text{ g stearic acid}} = 4.9 \times 10^{-7} \text{ mol stearic acid}$$

$$\textbf{Avogadro's number } (N_A) = \frac{1.5 \times 10^{17} \text{ molecules}}{4.9 \times 10^{-7} \text{ mol}} = \mathbf{3.1 \times 10^{23} \text{ molecules/mol}}$$

3.143   The balanced equations for the combustion of octane are:

$$2C_8H_{18} + 25O_2 \longrightarrow 16CO_2 + 18H_2O$$

$$2C_8H_{18} + 17O_2 \longrightarrow 16CO + 18H_2O$$

The quantity of octane burned is 2650 g (1 gallon with a density of 2.650 kg/gallon). Let x be the mass of octane converted to CO_2; therefore, $(2650 - x)$ g is the mass of octane converted to CO.

The amounts of CO_2 and H_2O produced by x g of octane are:

$$x \text{ g } C_8H_{18} \times \frac{1 \text{ mol } C_8H_{18}}{114.2 \text{ g } C_8H_{18}} \times \frac{16 \text{ mol } CO_2}{2 \text{ mol } C_8H_{18}} \times \frac{44.01 \text{ g } CO_2}{1 \text{ mol } CO_2} = 3.083x \text{ g } CO_2$$

$$x \text{ g } C_8H_{18} \times \frac{1 \text{ mol } C_8H_{18}}{114.2 \text{ g } C_8H_{18}} \times \frac{18 \text{ mol } H_2O}{2 \text{ mol } C_8H_{18}} \times \frac{18.02 \text{ g } H_2O}{1 \text{ mol } H_2O} = 1.420x \text{ g } H_2O$$

The amounts of CO and H_2O produced by $(2650 - x)$ g of octane are:

$$(2650 - x) \text{ g } C_8H_{18} \times \frac{1 \text{ mol } C_8H_{18}}{114.2 \text{ g } C_8H_{18}} \times \frac{16 \text{ mol } CO}{2 \text{ mol } C_8H_{18}} \times \frac{28.01 \text{ g } CO}{1 \text{ mol } CO} = (5200 - 1.962x) \text{ g } CO$$

$$(2650 - x) \text{ g } C_8H_{18} \times \frac{1 \text{ mol } C_8H_{18}}{114.2 \text{ g } C_8H_{18}} \times \frac{18 \text{ mol } H_2O}{2 \text{ mol } C_8H_{18}} \times \frac{18.02 \text{ g } H_2O}{1 \text{ mol } H_2O} = (3763 - 1.420x) \text{ g } H_2O$$

The total mass of $CO_2 + CO + H_2O$ produced is 11530 g. We can write:

$$11530 \text{ g} = 3.083x + 1.420x + 5200 - 1.962x + 3763 - 1.420x$$

$$x = 2290 \text{ g}$$

Since x is the amount of octane converted to CO_2, we can now calculate the efficiency of the process.

$$\textbf{efficiency} = \frac{\text{g octane converted}}{\text{g octane total}} \times 100\% = \frac{2290 \text{ g}}{2650 \text{ g}} \times 100\% = \textbf{86.49\%}$$

3.144   **(a)**   The balanced chemical equation is:

$$C_3H_8(g) + 3H_2O(g) \longrightarrow 3CO(g) + 7H_2(g)$$

(b)   You should come up with the following strategy to solve this problem. In this problem, we use kg-mol to save a couple of steps.

$$\text{kg } C_3H_8 \rightarrow \text{mol } C_3H_8 \rightarrow \text{mol } H_2 \rightarrow \text{kg } H_2$$

$$\textbf{? kg } \textbf{H}_2 = (2.84 \times 10^3 \text{ kg } C_3H_8) \times \frac{1 \text{ kg-mol } C_3H_8}{44.09 \text{ kg } C_3H_8} \times \frac{7 \text{ kg-mol } H_2}{1 \text{ kg-mol } C_3H_8} \times \frac{2.016 \text{ kg } H_2}{1 \text{ kg-mol } H_2}$$

$$= \textbf{9.09} \times \textbf{10}^2 \textbf{ kg } \textbf{H}_2$$

3.145   For the first step of the synthesis, the yield is 90% or 0.9. For the second step, the yield will be 90% of 0.9 or $(0.9 \times 0.9) = 0.81$. For the third step, the yield will be 90% of 0.81 or $(0.9 \times 0.9 \times 0.9) = 0.73$. We see that the yield will be:

$$\text{Yield} = (0.9)^n$$

where n = number of steps in the reaction. For 30 steps,

$$\text{Yield} = (0.9)^{30} = 0.04 = \textbf{4\%}$$

CHAPTER 4
REACTIONS IN AQUEOUS SOLUTIONS

4.7   (a) is a strong electrolyte. The compound dissociates completely into ions in solution.
(b) is a nonelectrolyte. The compound dissolves in water, but the molecules remain intact.
(c) is a weak electrolyte. A small amount of the compound dissociates into ions in water.

4.8   When NaCl dissolves in water it dissociates into Na^+ and Cl^- ions. When the ions are hydrated, the water molecules will be oriented so that the negative end of the water dipole interacts with the positive sodium ion, and the positive end of the water dipole interacts with the negative chloride ion. The negative end of the water dipole is near the oxygen atom, and the positive end of the water dipole is near the hydrogen atoms. The diagram that best represents the hydration of NaCl when dissolved in water is choice **(c)**.

4.9   Ionic compounds, strong acids, and strong bases (metal hydroxides) are strong electrolytes (completely broken up into ions of the compound). Weak acids and weak bases are weak electrolytes. Molecular substances other than acids or bases are nonelectrolytes.

   (a) very weak electrolyte   **(b)**   strong electrolyte (ionic compound)

   (c) strong electrolyte (strong acid)   **(d)**   weak electrolyte (weak acid)

   (e) nonelectrolyte (molecular compound - neither acid nor base)

4.10   Ionic compounds, strong acids, and strong bases (metal hydroxides) are strong electrolytes (completely broken up into ions of the compound). Weak acids and weak bases are weak electrolytes. Molecular substances other than acids or bases are nonelectrolytes.

   (a) strong electrolyte (ionic)   **(b)**   nonelectrolyte

   (c) weak electrolyte (weak base)   **(d)**   strong electrolyte (strong base)

4.11   Since solutions must be electrically neutral, any flow of positive species (cations) must be balanced by the flow of negative species (anions). Therefore, the correct answer is **(d)**.

4.12   **(a)** Solid NaCl does not conduct. The ions are locked in a rigid lattice structure.

   (b) Molten NaCl conducts. The ions can move around in the liquid state.

   (c) Aqueous NaCl conducts. NaCl dissociates completely to $Na^+(aq)$ and $Cl^-(aq)$ in water.

4.13   Measure the conductance to see if the solution carries an electrical current. If the solution is conducting, then you can determine whether the solution is a strong or weak electrolyte by comparing its conductance with that of a known strong electrolyte.

4.14   Since HCl dissolved in water conducts electricity, then HCl(aq) must actually exists as $H^+(aq)$ cations and $Cl^-(aq)$ anions. Since HCl dissolved in benzene solvent does not conduct electricity, then we must assume that the HCl molecules in benzene solvent do not ionize, but rather exist as un-ionized molecules.

4.17   Refer to Table 4.2 of the text to solve this problem. AgCl is insoluble in water. It will precipitate from solution. $NaNO_3$ is soluble in water and will remain as Na^+ and NO_3^- ions in solution. Diagram **(c)** best represents the mixture.

4.18   Refer to Table 4.2 of the text to solve this problem. $Mg(OH)_2$ is insoluble in water. It will precipitate from solution. KCl is soluble in water and will remain as K^+ and Cl^- ions in solution. Diagram **(b)** best represents the mixture.

4.19   Refer to Table 4.2 of the text to solve this problem.

   (a) $Ca_3(PO_4)_2$ is *insoluble*.
   (b) $Mn(OH)_2$ is *insoluble*.
   (c) $AgClO_3$ is *soluble*.
   (d) K_2S is *soluble*.

4.20   **Strategy:** Although it is not necessary to memorize the solubilities of compounds, you should keep in mind the following useful rules: all ionic compounds containing alkali metal cations, the ammonium ion, and the nitrate, bicarbonate, and chlorate ions are soluble. For other compounds, refer to Table 4.2 of the text.

   Solution:
   (a) $CaCO_3$ is **insoluble**. Most carbonate compounds are insoluble.
   (b) $ZnSO_4$ is **soluble**. Most sulfate compounds are soluble.
   (c) $Hg(NO_3)_2$ is **soluble**. All nitrate compounds are soluble.
   (d) $HgSO_4$ is **insoluble**. Most sulfate compounds are soluble, but those containing Ag^+, Ca^{2+}, Ba^{2+}, Hg^{2+}, and Pb^{2+} are insoluble.
   (e) NH_4ClO_4 is **soluble**. All ammonium compounds are soluble.

4.21   **(a)**   Ionic: $2Ag^+(aq) + 2NO_3^-(aq) + 2Na^+(aq) + SO_4^{2-}(aq) \longrightarrow Ag_2SO_4(s) + 2Na^+(aq) + 2NO_3^-(aq)$

   Net ionic: $2Ag^+(aq) + SO_4^{2-}(aq) \longrightarrow Ag_2SO_4(s)$

   (b)   Ionic: $Ba^{2+}(aq) + 2Cl^-(aq) + Zn^{2+}(aq) + SO_4^{2-}(aq) \longrightarrow BaSO_4(s) + Zn^{2+}(aq) + 2Cl^-(aq)$

   Net ionic: $Ba^{2+}(aq) + SO_4^{2-}(aq) \longrightarrow BaSO_4(s)$

   (c)   Ionic: $2NH_4^+(aq) + CO_3^{2-}(aq) + Ca^{2+}(aq) + 2Cl^-(aq) \longrightarrow CaCO_3(s) + 2NH_4^+(aq) + 2Cl^-(aq)$

   Net ionic: $Ca^{2+}(aq) + CO_3^{2-}(aq) \longrightarrow CaCO_3(s)$

4.22   **(a)**
   Strategy: Recall that an *ionic equation* shows dissolved ionic compounds in terms of their free ions. A *net ionic equation* shows only the species that actually take part in the reaction. What happens when ionic compounds dissolve in water? What ions are formed from the dissociation of Na_2S and $ZnCl_2$? What happens when the cations encounter the anions in solution?

Solution: In solution, Na_2S dissociates into Na^+ and S^{2-} ions and $ZnCl_2$ dissociates into Zn^{2+} and Cl^- ions. According to Table 4.2 of the text, zinc ions (Zn^{2+}) and sulfide ions (S^{2-}) will form an insoluble compound, zinc sulfide (ZnS), while the other product, $NaCl$, is soluble and remains in solution. This is a precipitation reaction. The balanced molecular equation is:

$$Na_2S(aq) + ZnCl_2(aq) \longrightarrow ZnS(s) + 2NaCl(aq)$$

The ionic and net ionic equations are:

$$\textit{Ionic: } 2Na^+(aq) + S^{2-}(aq) + Zn^{2+}(aq) + 2Cl^-(aq) \longrightarrow ZnS(s) + 2Na^+(aq) + 2Cl^-(aq)$$

$$\textit{Net ionic: } Zn^{2+}(aq) + S^{2-}(aq) \longrightarrow ZnS(s)$$

Check: Note that because we balanced the molecular equation first, the net ionic equation is balanced as to the number of atoms on each side, and the number of positive and negative charges on the left-hand side of the equation is the same.

(b)
Strategy: What happens when ionic compounds dissolve in water? What ions are formed from the dissociation of K_3PO_4 and $Sr(NO_3)_2$? What happens when the cations encounter the anions in solution?

Solution: In solution, K_3PO_4 dissociates into K^+ and PO_4^{3-} ions and $Sr(NO_3)_2$ dissociates into Sr^{2+} and NO_3^- ions. According to Table 4.2 of the text, strontium ions (Sr^{2+}) and phosphate ions (PO_4^{3-}) will form an insoluble compound, strontium phosphate [$Sr_3(PO_4)_2$], while the other product, KNO_3, is soluble and remains in solution. This is a precipitation reaction. The balanced molecular equation is:

$$2K_3PO_4(aq) + 3Sr(NO_3)_2(aq) \longrightarrow Sr_3(PO_4)_2(s) + 6KNO_3(aq)$$

The ionic and net ionic equations are:

$$\textit{Ionic: } 6K^+(aq) + 2PO_4^{3-}(aq) + 3Sr^{2+}(aq) + 6NO_3^-(aq) \longrightarrow Sr_3(PO_4)_2(s) + 6K^+(aq) + 6NO_3^-(aq)$$

$$\textit{Net ionic: } 3Sr^{2+}(aq) + 2PO_4^{3-}(aq) \longrightarrow Sr_3(PO_4)_2(s)$$

Check: Note that because we balanced the molecular equation first, the net ionic equation is balanced as to the number of atoms on each side, and the number of positive and negative charges on the left-hand side of the equation is the same.

(c)
Strategy: What happens when ionic compounds dissolve in water? What ions are formed from the dissociation of $Mg(NO_3)_2$ and $NaOH$? What happens when the cations encounter the anions in solution?

Solution: In solution, $Mg(NO_3)_2$ dissociates into Mg^{2+} and NO_3^- ions and $NaOH$ dissociates into Na^+ and OH^- ions. According to Table 4.2 of the text, magnesium ions (Mg^{2+}) and hydroxide ions (OH^-) will form an insoluble compound, magnesium hydroxide [$Mg(OH)_2$], while the other product, $NaNO_3$, is soluble and remains in solution. This is a precipitation reaction. The balanced molecular equation is:

$$Mg(NO_3)_2(aq) + 2NaOH(aq) \longrightarrow Mg(OH)_2(s) + 2NaNO_3(aq)$$

The ionic and net ionic equations are:

$$\textit{Ionic: } Mg^{2+}(aq) + 2NO_3^-(aq) + 2Na^+(aq) + 2OH^-(aq) \longrightarrow Mg(OH)_2(s) + 2Na^+(aq) + 2NO_3^-(aq)$$

$$\textit{Net ionic: } Mg^{2+}(aq) + 2OH^-(aq) \longrightarrow Mg(OH)_2(s)$$

Check: Note that because we balanced the molecular equation first, the net ionic equation is balanced as to the number of atoms on each side, and the number of positive and negative charges on the left-hand side of the equation is the same.

4.23   **(a)**   Both reactants are soluble ionic compounds. The other possible ion combinations, Na_2SO_4 and $Cu(NO_3)_2$, are also soluble.

(b)   Both reactants are soluble. Of the other two possible ion combinations, KCl is soluble, but $BaSO_4$ is insoluble and will precipitate.

$$Ba^{2+}(aq) + SO_4^{2-}(aq) \rightarrow BaSO_4(s)$$

4.24   **(a)**   Add chloride ions. KCl is soluble, but AgCl is not.

(b)   Add hydroxide ions. $Ba(OH)_2$ is soluble, but $Pb(OH)_2$ is insoluble.

(c)   Add carbonate ions. $(NH_4)_2CO_3$ is soluble, but $CaCO_3$ is insoluble.

(d)   Add sulfate ions. $CuSO_4$ is soluble, but $BaSO_4$ is insoluble.

4.31   **(a)**   HI dissolves in water to produce H^+ and I^-, so HI is a **Brønsted acid**.

(b) CH_3COO^- can accept a proton to become acetic acid CH_3COOH, so it is a **Brønsted base**.

(c) $H_2PO_4^-$ can either accept a proton, H^+, to become H_3PO_4 and thus behaves as a **Brønsted base**, or can donate a proton in water to yield H^+ and HPO_4^{2-}, thus behaving as a **Brønsted acid**.

(d) HSO_4^- can either accept a proton, H^+, to become H_2SO_4 and thus behaves as a **Brønsted base**, or can donate a proton in water to yield H^+ and SO_4^{2-}, thus behaving as a **Brønsted acid**.

4.32   **Strategy:** What are the characteristics of a Brønsted acid? Does it contain at least an H atom? With the exception of ammonia, most Brønsted bases that you will encounter at this stage are anions.

Solution:
(a) PO_4^{3-} in water can accept a proton to become HPO_4^{2-}, and is thus a **Brønsted base**.

(b) ClO_2^- in water can accept a proton to become $HClO_2$, and is thus a **Brønsted base**.

(c) NH_4^+ dissolved in water can donate a proton H^+, thus behaving as a **Brønsted acid**.

(d) HCO_3^- can either accept a proton to become H_2CO_3, thus behaving as a **Brønsted base**. Or, HCO_3^- can donate a proton to yield H^+ and CO_3^{2-}, thus behaving as a **Brønsted acid**.

Comment: The HCO_3^- species is said to be *amphoteric* because it possesses both acidic and basic properties.

4.33   Recall that strong acids and strong bases are strong electrolytes. They are completely ionized in solution. An <u>ionic equation</u> will show strong acids and strong bases in terms of their free ions. A <u>net ionic equation</u> shows only the species that actually take part in the reaction.

(a)   <u>Ionic</u>: $H^+(aq) + Br^-(aq) + NH_3(aq) \longrightarrow NH_4^+(aq) + Br^-(aq)$

<u>Net ionic</u>: $H^+(aq) + NH_3(aq) \longrightarrow NH_4^+(aq)$

(b)   Ionic: $3Ba^{2+}(aq) + 6OH^-(aq) + 2H_3PO_4(aq) \longrightarrow Ba_3(PO_4)_2(s) + 6H_2O(l)$

Net ionic: $3Ba^{2+}(aq) + 6OH^-(aq) + 2H_3PO_4(aq) \longrightarrow Ba_3(PO_4)_2(s) + 6H_2O(l)$

(c)   Ionic: $2H^+(aq) + 2ClO_4^-(aq) + Mg^{2+}(aq) + 2OH^-(aq) \longrightarrow Mg^{2+}(aq) + 2ClO_4^-(aq) + 2H_2O(l)$

Net ionic: $2H^+(aq) + 2OH^-(aq) \longrightarrow 2H_2O(l)$ or $H^+(aq) + OH^-(aq) \longrightarrow H_2O(l)$

4.34   **Strategy:** Recall that strong acids and strong bases are strong electrolytes. They are completely ionized in solution. An *ionic equation* will show strong acids and strong bases in terms of their free ions. Weak acids and weak bases are weak electrolytes. They only ionize to a small extent in solution. Weak acids and weak bases are shown as molecules in ionic and net ionic equations. A *net ionic equation* shows only the species that actually take part in the reaction.

(a)

Solution: CH_3COOH is a weak acid. It will be shown as a molecule in the ionic equation. KOH is a strong base. It completely ionizes to K^+ and OH^- ions. Since CH_3COOH is an acid, it donates an H^+ to the base, OH^-, producing water. The other product is the salt, CH_3COOK, which is soluble and remains in solution. The balanced molecular equation is:

$$CH_3COOH(aq) + KOH(aq) \longrightarrow CH_3COOK(aq) + H_2O(l)$$

The ionic and net ionic equations are:

Ionic: $CH_3COOH(aq) + K^+(aq) + OH^-(aq) \longrightarrow CH_3COO^-(aq) + K^+(aq) + H_2O(l)$

Net ionic: $CH_3COOH(aq) + OH^-(aq) \longrightarrow CH_3COO^-(aq) + H_2O(l)$

(b)

Solution: H_2CO_3 is a weak acid. It will be shown as a molecule in the ionic equation. NaOH is a strong base. It completely ionizes to Na^+ and OH^- ions. Since H_2CO_3 is an acid, it donates an H^+ to the base, OH^-, producing water. The other product is the salt, Na_2CO_3, which is soluble and remains in solution. The balanced molecular equation is:

$$H_2CO_3(aq) + 2NaOH(aq) \longrightarrow Na_2CO_3(aq) + 2H_2O(l)$$

The ionic and net ionic equations are:

Ionic: $H_2CO_3(aq) + 2Na^+(aq) + 2OH^-(aq) \longrightarrow 2Na^+(aq) + CO_3^{2-}(aq) + 2H_2O(l)$

Net ionic: $H_2CO_3(aq) + 2OH^-(aq) \longrightarrow CO_3^{2-}(aq) + 2H_2O(l)$

(c)

Solution: HNO_3 is a strong acid. It completely ionizes to H^+ and NO_3^- ions. $Ba(OH)_2$ is a strong base. It completely ionizes to Ba^{2+} and OH^- ions. Since HNO_3 is an acid, it donates an H^+ to the base, OH^-, producing water. The other product is the salt, $Ba(NO_3)_2$, which is soluble and remains in solution. The balanced molecular equation is:

$$2HNO_3(aq) + Ba(OH)_2(aq) \longrightarrow Ba(NO_3)_2(aq) + 2H_2O(l)$$

The ionic and net ionic equations are:

Ionic: $2H^+(aq) + 2NO_3^-(aq) + Ba^{2+}(aq) + 2OH^-(aq) \longrightarrow Ba^{2+}(aq) + 2NO_3^-(aq) + 2H_2O(l)$

Net ionic: $2H^+(aq) + 2OH^-(aq) \longrightarrow 2H_2O(l)$ or $H^+(aq) + OH^-(aq) \longrightarrow H_2O(l)$

4.43   Even though the problem doesn't ask you to assign oxidation numbers, you need to be able to do so in order to determine what is being oxidized or reduced.

	(i) Half Reactions	(ii) Oxidizing Agent	(iii) Reducing Agent
(a)	$Sr \rightarrow Sr^{2+} + 2e^-$ $O_2 + 4e^- \rightarrow 2O^{2-}$	O_2	Sr
(b)	$Li \rightarrow Li^+ + e^-$ $H_2 + 2e^- \rightarrow 2H^-$	H_2	Li
(c)	$Cs \rightarrow Cs^+ + e^-$ $Br_2 + 2e^- \rightarrow 2Br^-$	Br_2	Cs
(d)	$Mg \rightarrow Mg^{2+} + 2e^-$ $N_2 + 6e^- \rightarrow 2N^{3-}$	N_2	Mg

4.44   **Strategy:** In order to break a redox reaction down into an oxidation half-reaction and a reduction half-reaction, you should first assign oxidation numbers to all the atoms in the reaction. In this way, you can determine which element is oxidized (loses electrons) and which element is reduced (gains electrons).

Solution: In each part, the reducing agent is the reactant in the first half-reaction and the oxidizing agent is the reactant in the second half-reaction. The coefficients in each half-reaction have been reduced to smallest whole numbers.

(a)   The product is an ionic compound whose ions are Fe^{3+} and O^{2-}.

$$Fe \longrightarrow Fe^{3+} + 3e^-$$
$$O_2 + 4e^- \longrightarrow 2O^{2-}$$

O_2 is the oxidizing agent; Fe is the reducing agent.

(b) Na^+ does not change in this reaction. It is a "spectator ion."

$$2Br^- \longrightarrow Br_2 + 2e^-$$
$$Cl_2 + 2e^- \longrightarrow 2Cl^-$$

Cl_2 is the oxidizing agent; Br^- is the reducing agent.

(c)   Assume SiF_4 is made up of Si^{4+} and F^-.

$$Si \longrightarrow Si^{4+} + 4e^-$$
$$F_2 + 2e^- \longrightarrow 2F^-$$

F_2 is the oxidizing agent; Si is the reducing agent.

(d)   Assume HCl is made up of H^+ and Cl^-.

$$H_2 \longrightarrow 2H^+ + 2e^-$$
$$Cl_2 + 2e^- \longrightarrow 2Cl^-$$

Cl_2 is the oxidizing agent; H_2 is the reducing agent.

4.45   The oxidation number for hydrogen is +1 (rule 4), and for oxygen is –2 (rule 3). The oxidation number for sulfur in S_8 is zero (rule 1). Remember that in a neutral molecule, the sum of the oxidation numbers of all the atoms must be zero, and in an ion the sum of oxidation numbers of all elements in the ion must equal the net charge of the ion (rule 6).

$$H_2S\ (-2),\ S^{2-}\ (-2),\ HS^-\ (-2)\ <\ S_8\ (0)\ <\ SO_2\ (+4)\ <\ SO_3\ (+6),\ H_2SO_4\ (+6)$$

The number in parentheses denotes the oxidation number of sulfur.

4.46   **Strategy:** In general, we follow the rules listed in Section 4.4 of the text for assigning oxidation numbers. Remember that all alkali metals have an oxidation number of +1 in ionic compounds, and in most cases hydrogen has an oxidation number of +1 and oxygen has an oxidation number of –2 in their compounds.

Solution: All the compounds listed are neutral compounds, so the oxidation numbers must sum to zero (Rule 6, Section 4.4 of the text).

Let the oxidation number of P = x.

(a) $x + 1 + (3)(-2) = 0,\ x = +5$ **(d)** $x + (3)(+1) + (4)(-2) = 0,\ x = +5$
(b) $x + (3)(+1) + (2)(-2) = 0,\ x = +1$ **(e)** $2x + (4)(+1) + (7)(-2) = 0,\ 2x = 10,\ x = +5$
(c) $x + (3)(+1) + (3)(-2) = 0,\ x = +3$ **(f)** $3x + (5)(+1) + (10)(-2) = 0,\ 3x = 15,\ x = +5$

The molecules in part (a), (e), and (f) can be made by strongly heating the compound in part (d). Are these oxidation-reduction reactions?

Check: In each case, does the sum of the oxidation numbers of all the atoms equal the net charge on the species, in this case zero?

4.47   See Section 4.4 of the text.

(a)   C<u>l</u>F: F –1 (rule 5), **Cl +1 (rule 6)**   **(b)**   I<u>F</u>$_7$: F –1 (rule 5), **I +7 (rules 5 and 6)**

(c)   <u>C</u>H$_4$: H +1 (rule 4), **C –4 (rule 6)**   **(d)**   <u>C</u>$_2$H$_2$: H +1 (rule 4), **C –1 (rule 6)**

(e)   <u>C</u>$_2$H$_4$: H +1 (rule 4), **C –2 (rule 6),**   **(f)**   K$_2$<u>Cr</u>O$_4$: K +1 (rule 2), O –2 (rule 3), **Cr +6 (rule 6)**

(g)   K$_2$<u>Cr</u>$_2$O$_7$: K +1 (rule 2), O –2 (rule 3), **Cr +6 (rule 6)**

(h)   K<u>Mn</u>O$_4$: K +1 (rule 2), O –2 (rule 3), **Mn +7 (rule 6)**

(i)   NaH<u>C</u>O$_3$: Na +1 (rule 2), H +1 (rule 4), O –2 (rule 3), **C +4 (rule 6)**

(j)   <u>Li</u>$_2$: **Li 0 (rule 1)**   **(k)**   Na<u>I</u>O$_3$: Na +1 (rule 2), O –2 (rule 3), **I +5 (rule 6)**

(l)   K<u>O</u>$_2$: K +1 (rule 2), **O –1/2 (rule 6)**   **(m)**   <u>P</u>F$_6^-$: F –1 (rule 5), **P +5 (rule 6)**

(n)   K<u>Au</u>Cl$_4$: K +1 (rule 2), Cl –1 (rule 5), **Au +3 (rule 6)**

4.48   All are free elements, so all have an oxidation number of **zero**.

4.49   **(a)** <u>Cs</u>$_2$O, **+1**   **(b)** Ca<u>I</u>$_2$, **–1**   **(c)** <u>Al</u>$_2$O$_3$, **+3**   **(d)** H$_3$<u>As</u>O$_3$, **+3**   **(e)** <u>Ti</u>O$_2$, **+4**
(f) <u>Mo</u>O$_4^{2-}$, **+6**   **(g)** <u>Pt</u>Cl$_4^{2-}$, **+2**   **(h)** <u>Pt</u>Cl$_6^{2-}$, **+4**   **(i)** <u>Sn</u>F$_2$, **+2**   **(j)** <u>Cl</u>F$_3$, **+3**
(k) <u>Sb</u>F$_6^-$, **+5**

4.50 **(a)** N: −3   **(b)** O: −1/2   **(c)** C: −1   **(d)** C: +4

(e) C: +3   **(f)** O: −2   **(g)** B: +3   **(h)** W: +6

4.51 If nitric acid is a strong oxidizing agent and zinc is a strong reducing agent, then zinc metal will probably reduce nitric acid when the two react; that is, N will gain electrons and the oxidation number of N must decrease. Since the oxidation number of nitrogen in nitric acid is +5 (verify!), then the nitrogen-containing product must have a smaller oxidation number for nitrogen. The only compound in the list that doesn't have a nitrogen oxidation number less than +5 is N_2O_5, (what is the oxidation number of N in N_2O_5?). This is never a product of the reduction of nitric acid.

4.52 **Strategy:** *Hydrogen displacement*: Any metal above hydrogen in the activity series will displace it from water or from an acid. Metals below hydrogen will *not* react with either water or an acid.

Solution: Only **(b)** Li and **(d)** Ca are above hydrogen in the activity series, so they are the only metals in this problem that will react with water.

4.53 In order to work this problem, you need to assign the oxidation numbers to all the elements in the compounds. In each case oxygen has an oxidation number of −2 (rule 3). These oxidation numbers should then be compared to the range of possible oxidation numbers that each element can have. Molecular oxygen is a powerful oxidizing agent. In SO_3 alone, the oxidation number of the element bound to oxygen (S) is at its maximum value (+6); the sulfur cannot be oxidized further. The other elements bound to oxygen in this problem have less than their maximum oxidation number and can undergo further oxidation.

4.54 **(a)** $Cu(s) + HCl(aq) \rightarrow$ no reaction, since $Cu(s)$ is less reactive than the hydrogen from acids.

(b) $I_2(s) + NaBr(aq) \rightarrow$ no reaction, since $I_2(s)$ is less reactive than $Br_2(l)$.

(c) $Mg(s) + CuSO_4(aq) \rightarrow MgSO_4(aq) + Cu(s)$, since $Mg(s)$ is more reactive than Cu(s).

Net ionic equation: $Mg(s) + Cu^{2+}(aq) \rightarrow Mg^{2+}(aq) + Cu(s)$

(d) $Cl_2(g) + 2KBr(aq) \rightarrow Br_2(l) + 2KCl(aq)$, since $Cl_2(g)$ is more reactive than $Br_2(l)$

Net ionic equation: $Cl_2(g) + 2Br^-(aq) \rightarrow 2Cl^-(aq) + Br_2(l)$

4.55 **(a)** Disproportionation reaction   **(b)** Displacement reaction
        (c) Decomposition reaction   **(d)** Combination reaction

4.56 **(a)** Combination reaction
        (b) Decomposition reaction
        (c) Displacement reaction
        (d) Disproportionation reaction

4.59 First, calculate the moles of KI needed to prepare the solution.

$$\text{mol KI} = \frac{2.80 \text{ mol KI}}{1000 \text{ mL soln}} \times (5.00 \times 10^2 \text{ mL soln}) = 1.40 \text{ mol KI}$$

Converting to grams of KI:

$$1.40 \text{ mol KI} \times \frac{166.0 \text{ g KI}}{1 \text{ mol KI}} = \textbf{232 g KI}$$

4.60   **Strategy:** How many moles of $NaNO_3$ does 250 mL of a 0.707 *M* solution contain? How would you convert moles to grams?

Solution: From the molarity (0.707 *M*), we can calculate the moles of $NaNO_3$ needed to prepare 250 mL of solution.

$$\text{Moles } NaNO_3 = \frac{0.707 \text{ mol } NaNO_3}{1000 \text{ mL soln}} \times 250 \text{ mL soln} = 0.177 \text{ mol}$$

Next, we use the molar mass of $NaNO_3$ as a conversion factor to convert from moles to grams.

$\mathcal{M}(NaNO_3) = 85.00$ g/mol.

$$0.177 \text{ mol } NaNO_3 \times \frac{85.00 \text{ g } NaNO_3}{1 \text{ mol } NaNO_3} = 15.0 \text{ g } NaNO_3$$

To make the solution, **dissolve 15.0 g of $NaNO_3$ in enough water to make 250 mL of solution.**

Check: As a ball-park estimate, the mass should be given by [molarity (mol/L) × volume (L) = moles × molar mass (g/mol) = grams]. Let's round the molarity to 1 *M* and the molar mass to 80 g, because we are simply making an estimate. This gives: [1 mol/L × (1/4)L × 80 g = 20 g]. This is close to our answer of 15.0 g.

4.61 $\text{mol} = M \times L$

60.0 mL = 0.0600 L

$$\textbf{mol MgCl}_2 = \frac{0.100 \text{ mol } MgCl_2}{1 \text{ L soln}} \times 0.0600 \text{ L soln} = \textbf{6.00} \times \textbf{10}^{-3} \textbf{ mol MgCl}_2$$

4.62   Since the problem asks for grams of solute (KOH), you should be thinking that you can calculate moles of solute from the molarity and volume of solution. Then, you can convert moles of solute to grams of solute.

$$? \text{ moles KOH solute} = \frac{5.50 \text{ moles solute}}{1000 \text{ mL solution}} \times 35.0 \text{ mL solution} = 0.193 \text{ mol KOH}$$

The molar mass of KOH is 56.11 g/mol. Use this conversion factor to calculate grams of KOH.

$$\textbf{? grams KOH} = 0.193 \text{ mol KOH} \times \frac{56.11 \text{ g KOH}}{1 \text{ mol KOH}} = \textbf{10.8 g KOH}$$

4.63   Molar mass of $C_2H_5OH = 46.07$ g/mol; molar mass of $C_{12}H_{22}O_{11} = 342.3$ g/mol; molar mass of $NaCl = 58.44$ g/mol.

(a) $? \text{ mol } C_2H_5OH = 29.0 \text{ g } C_2H_5OH \times \dfrac{1 \text{ mol } C_2H_5OH}{46.07 \text{ g } C_2H_5OH} = 0.629 \text{ mol } C_2H_5OH$

$$\textbf{Molarity} = \frac{\text{mol solute}}{\text{L of soln}} = \frac{0.629 \text{ mol } C_2H_5OH}{0.545 \text{ L soln}} = \textbf{1.15 } \textbf{\textit{M}}$$

(b) $? \text{ mol } C_{12}H_{22}O_{11} = 15.4 \text{ g } C_{12}H_{22}O_{11} \times \dfrac{1 \text{ mol } C_{12}H_{22}O_{11}}{342.3 \text{ g } C_{12}H_{22}O_{11}} = 0.0450 \text{ mol } C_{12}H_{22}O_{11}$

$$\text{Molarity} = \frac{\text{mol solute}}{\text{L of soln}} = \frac{0.0450 \text{ mol } C_{12}H_{22}O_{11}}{74.0 \times 10^{-3} \text{ L soln}} = \textbf{0.608 } \textbf{\textit{M}}$$

(c) $? \text{ mol NaCl} = 9.00 \text{ g NaCl} \times \dfrac{1 \text{ mol NaCl}}{58.44 \text{ g NaCl}} = 0.154 \text{ mol NaCl}$

$$\text{Molarity} = \frac{\text{mol solute}}{\text{L of soln}} = \frac{0.154 \text{ mol NaCl}}{86.4 \times 10^{-3} \text{ L soln}} = \textbf{1.78 } \textbf{\textit{M}}$$

4.64   (a) $? \text{ mol CH}_3\text{OH} = 6.57 \text{ g CH}_3\text{OH} \times \dfrac{1 \text{ mol CH}_3\text{OH}}{32.04 \text{ g CH}_3\text{OH}} = 0.205 \text{ mol CH}_3\text{OH}$

$$M = \frac{0.205 \text{ mol CH}_3\text{OH}}{0.150 \text{ L}} = \textbf{1.37 } \textbf{\textit{M}}$$

(b) $? \text{ mol CaCl}_2 = 10.4 \text{ g CaCl}_2 \times \dfrac{1 \text{ mol CaCl}_2}{111.0 \text{ g CaCl}_2} = 0.0937 \text{ mol CaCl}_2$

$$M = \frac{0.0937 \text{ mol CaCl}_2}{0.220 \text{ L}} = \textbf{0.426 } \textbf{\textit{M}}$$

(c) $? \text{ mol C}_{10}\text{H}_8 = 7.82 \text{ g C}_{10}\text{H}_8 \times \dfrac{1 \text{ mol C}_{10}\text{H}_8}{128.2 \text{ g C}_{10}\text{H}_8} = 0.0610 \text{ mol C}_{10}\text{H}_8$

$$M = \frac{0.0610 \text{ mol C}_{10}\text{H}_8}{0.0852 \text{ L}} = \textbf{0.716 } \textbf{\textit{M}}$$

4.65   First, calculate the moles of each solute. Then, you can calculate the volume (in L) from the molarity and the number of moles of solute.

(a) $? \text{ mol NaCl} = 2.14 \text{ g NaCl} \times \dfrac{1 \text{ mol NaCl}}{58.44 \text{ g NaCl}} = 0.0366 \text{ mol NaCl}$

$$\text{L soln} = \frac{\text{mol solute}}{\text{Molarity}} = \frac{0.0366 \text{ mol NaCl}}{0.270 \text{ mol/L}} = 0.136 \text{ L} = \textbf{136 mL soln}$$

(b) $? \text{ mol C}_2\text{H}_5\text{OH} = 4.30 \text{ g C}_2\text{H}_5\text{OH} \times \dfrac{1 \text{ mol C}_2\text{H}_5\text{OH}}{46.07 \text{ g C}_2\text{H}_5\text{OH}} = 0.0933 \text{ mol C}_2\text{H}_5\text{OH}$

$$\text{L soln} = \frac{\text{mol solute}}{\text{Molarity}} = \frac{0.0933 \text{ mol C}_2\text{H}_5\text{OH}}{1.50 \text{ mol/L}} = 0.0622 \text{ L} = \textbf{62.2 mL soln}$$

(c) $? \text{ mol CH}_3\text{COOH} = 0.85 \text{ g CH}_3\text{COOH} \times \dfrac{1 \text{ mol CH}_3\text{COOH}}{60.05 \text{ g CH}_3\text{COOH}} = 0.014 \text{ mol CH}_3\text{COOH}$

$$\text{L soln} = \frac{\text{mol solute}}{\text{Molarity}} = \frac{0.014 \text{ mol CH}_3\text{COOH}}{0.30 \text{ mol/L}} = 0.047 \text{ L} = \textbf{47 mL soln}$$

4.66   A 250 mL sample of 0.100 M solution contains 0.0250 mol of solute (mol = $M \times$ L). The computation in each case is the same:

(a) $0.0250 \text{ mol CsI} \times \dfrac{259.8 \text{ g CsI}}{1 \text{ mol CsI}}$ = **6.50 g CsI**

(b) $0.0250 \text{ mol H}_2\text{SO}_4 \times \dfrac{98.09 \text{ g H}_2\text{SO}_4}{1 \text{ mol H}_2\text{SO}_4}$ = **2.45 g H$_2$SO$_4$**

(c) $0.0250 \text{ mol Na}_2\text{CO}_3 \times \dfrac{106.0 \text{ g Na}_2\text{CO}_3}{1 \text{ mol Na}_2\text{CO}_3}$ = **2.65 g Na$_2$CO$_3$**

(d) $0.0250 \text{ mol K}_2\text{Cr}_2\text{O}_7 \times \dfrac{294.2 \text{ g K}_2\text{Cr}_2\text{O}_7}{1 \text{ mol K}_2\text{Cr}_2\text{O}_7}$ = **7.36 g K$_2$Cr$_2$O$_7$**

(e) $0.0250 \text{ mol KMnO}_4 \times \dfrac{158.0 \text{ g KMnO}_4}{1 \text{ mol KMnO}_4}$ = **3.95 g KMnO$_4$**

4.69 $M_{\text{initial}} V_{\text{initial}} = M_{\text{final}} V_{\text{final}}$

You can solve the equation algebraically for V_{initial}. Then substitute in the given quantities to solve for the volume of 2.00 M HCl needed to prepare 1.00 L of a 0.646 M HCl solution.

$$V_{\text{initial}} = \frac{M_{\text{final}} \times V_{\text{final}}}{M_{\text{initial}}} = \frac{0.646\ M \times 1.00\ \text{L}}{2.00\ M} = \textbf{0.323 L} = \textbf{323 mL}$$

To prepare the 0.646 M solution, you would dilute 323 mL of the 2.00 M HCl solution to a final volume of 1.00 L.

4.70   **Strategy:** Because the volume of the final solution is greater than the original solution, this is a dilution process. Keep in mind that in a dilution, the concentration of the solution decreases, but the number of moles of the solute remains the same.

Solution: We prepare for the calculation by tabulating our data.

M_i = 0.866 M M_f = ?

V_i = 25.0 mL V_f = 500 mL

We substitute the data into Equation (4.2) of the text.

$M_i V_i = M_f V_f$

$(0.866\ M)(25.0\ \text{mL}) = M_f(500\ \text{mL})$

$$M_f = \frac{(0.866\ M)(25.0\ \text{mL})}{500\ \text{mL}} = \textbf{0.0433}\ M$$

4.71 $M_{\text{initial}} V_{\text{initial}} = M_{\text{final}} V_{\text{final}}$

You can solve the equation algebraically for V_{initial}. Then substitute in the given quantities to solve the for the volume of 4.00 M HNO$_3$ needed to prepare 60.0 mL of a 0.200 M HNO$_3$ solution.

$$V_{\text{initial}} = \frac{M_{\text{final}} \times V_{\text{final}}}{M_{\text{initial}}} = \frac{0.200\ M \times 60.00\ \text{mL}}{4.00\ M} = \textbf{3.00 mL}$$

To prepare the 0.200 M solution, you would dilute 3.00 mL of the 4.00 M HNO₃ solution to a final volume of 60.0 mL.

4.72   You need to calculate the final volume of the dilute solution. Then, you can subtract 505 mL from this volume to calculate the amount of water that should be added.

$$V_{\text{final}} = \frac{M_{\text{initial}} V_{\text{initial}}}{M_{\text{final}}} = \frac{(0.125\ M)(505\ \text{mL})}{(0.100\ M)} = 631\ \text{mL}$$

$$(631 - 505)\ \text{mL} = \textbf{126 mL of water}$$

4.73   Moles of KMnO₄ in the first solution:

$$\frac{1.66\ \text{mol}}{1000\ \text{mL soln}} \times 35.2\ \text{mL} = 0.0584\ \text{mol KMnO}_4$$

Moles of KMnO₄ in the second solution:

$$\frac{0.892\ \text{mol}}{1000\ \text{mL soln}} \times 16.7\ \text{mL} = 0.0149\ \text{mol KMnO}_4$$

The total volume is 35.2 mL + 16.7 mL = 51.9 mL. The concentration of the final solution is:

$$M = \frac{(0.0584 + 0.0149)\,\text{mol}}{51.9 \times 10^{-3}\ \text{L}} = \textbf{1.41}\ \boldsymbol{M}$$

4.74   Moles of calcium nitrate in the first solution:

$$\frac{0.568\ \text{mol}}{1000\ \text{mL soln}} \times 46.2\ \text{mL soln} = 0.0262\ \text{mol Ca(NO}_3)_2$$

Moles of calcium nitrate in the second solution:

$$\frac{1.396\ \text{mol}}{1000\ \text{mL soln}} \times 80.5\ \text{mL soln} = 0.112\ \text{mol Ca(NO}_3)_2$$

The volume of the combined solutions = 46.2 mL + 80.5 mL = 126.7 mL. The concentration of the final solution is:

$$M = \frac{(0.0262 + 0.112)\,\text{mol}}{0.1267\ \text{L}} = \textbf{1.09}\ \boldsymbol{M}$$

4.77   The balanced equation is: $CaCl_2(aq) + 2AgNO_3(aq) \longrightarrow Ca(NO_3)_2(aq) + 2AgCl(s)$

We need to determine the limiting reagent. Ag^+ and Cl^- combine in a 1:1 mole ratio to produce AgCl. Let's calculate the amount of Ag^+ and Cl^- in solution.

$$\text{mol Ag}^+ = \frac{0.100\ \text{mol Ag}^+}{1000\ \text{mL soln}} \times 15.0\ \text{mL soln} = 1.50 \times 10^{-3}\ \text{mol Ag}^+$$

$$\text{mol Cl}^- = \frac{0.150 \text{ mol CaCl}_2}{1000 \text{ mL soln}} \times \frac{2 \text{ mol Cl}^-}{1 \text{ mol CaCl}_2} \times 30.0 \text{ mL soln} = 9.00 \times 10^{-3} \text{ mol Cl}^-$$

Since Ag^+ and Cl^- combine in a 1:1 mole ratio, $AgNO_3$ is the limiting reagent. Only 1.50×10^{-3} mole of AgCl can form. Converting to grams of AgCl:

$$1.50 \times 10^{-3} \text{ mol AgCl} \times \frac{143.4 \text{ g AgCl}}{1 \text{ mol AgCl}} = \textbf{0.215 g AgCl}$$

4.78 **Strategy:** We want to calculate the mass % of Ba in the original compound. Let's start with the definition of mass %.

want to calculate   need to find

$$\text{mass \% Ba} = \frac{\text{mass Ba}}{\text{mass of sample}} \times 100\%$$

given

The mass of the sample is given in the problem (0.6760 g). Therefore we need to find the mass of Ba in the original sample. We assume the precipitation is quantitative, that is, that all of the barium in the sample has been precipitated as barium sulfate. From the mass of $BaSO_4$ produced, we can calculate the mass of Ba. There is 1 mole of Ba in 1 mole of $BaSO_4$.

Solution: First, we calculate the mass of Ba in 0.4105 g of the $BaSO_4$ precipitate. The molar mass of $BaSO_4$ is 233.4 g/mol.

$$? \text{ mass of Ba} = 0.4105 \text{ g BaSO}_4 \times \frac{1 \text{ mol BaSO}_4}{233.4 \text{ g BaSO}_4} \times \frac{1 \text{ mol Ba}}{1 \text{ mol BaSO}_4} \times \frac{137.3 \text{ g Ba}}{1 \text{ mol Ba}}$$

$$= 0.2415 \text{ g Ba}$$

Next, we calculate the mass percent of Ba in the unknown compound.

$$\textbf{\% Ba by mass} = \frac{0.2415 \text{ g}}{0.6760 \text{ g}} \times 100\% = \textbf{35.72\%}$$

4.79 The net ionic equation is: $Ag^+(aq) + Cl^-(aq) \longrightarrow AgCl(s)$

One mole of Cl^- is required per mole of Ag^+. First, find the number of moles of Ag^+.

$$\text{mol Ag}^+ = \frac{0.0113 \text{ mol Ag}^+}{1000 \text{ mL soln}} \times (2.50 \times 10^2 \text{ mL soln}) = 2.83 \times 10^{-3} \text{ mol Ag}^+$$

Now, calculate the mass of NaCl using the mole ratio from the balanced equation.

$$(2.83 \times 10^{-3} \text{ mol Ag}^+) \times \frac{1 \text{ mol Cl}^-}{1 \text{ mol Ag}^+} \times \frac{1 \text{ mol NaCl}}{1 \text{ mol Cl}^-} \times \frac{58.44 \text{ g NaCl}}{1 \text{ mol NaCl}} = \textbf{0.165 g NaCl}$$

4.80   The net ionic equation is: $Cu^{2+}(aq) + S^{2-}(aq) \longrightarrow CuS(s)$

The answer sought is the molar concentration of Cu^{2+}, that is, moles of Cu^{2+} ions per liter of solution. The factor-label method is used to convert, in order:

$$g \text{ of CuS} \rightarrow \text{moles CuS} \rightarrow \text{moles } Cu^{2+} \rightarrow \text{moles } Cu^{2+} \text{ per liter soln}$$

$$[Cu^{2+}] = 0.0177 \text{ g CuS} \times \frac{1 \text{ mol CuS}}{95.62 \text{ g CuS}} \times \frac{1 \text{ mol } Cu^{2+}}{1 \text{ mol CuS}} \times \frac{1}{0.800 \text{ L}} = \mathbf{2.31 \times 10^{-4}} \textit{ M}$$

4.85   The reaction between KHP ($KHC_8H_4O_4$) and KOH is:

$$KHC_8H_4O_4(aq) + KOH(aq) \rightarrow H_2O(l) + K_2C_8H_4O_4(aq)$$

We know the volume of the KOH solution, and we want to calculate the molarity of the KOH solution.

$$\begin{array}{ccc} \text{want to calculate} & & \text{need to find} \\ \searrow & & \swarrow \\ \textit{M} \text{ of KOH} = & \dfrac{\text{mol KOH}}{\text{L of KOH soln}} & \\ & \nwarrow & \\ & \text{given} & \end{array}$$

If we can determine the moles of KOH in the solution, we can then calculate the molarity of the solution. From the mass of KHP and its molar mass, we can calculate moles of KHP. Then, using the mole ratio from the balanced equation, we can calculate moles of KOH.

$$? \text{ mol KOH} = 0.4218 \text{ g KHP} \times \frac{1 \text{ mol KHP}}{204.2 \text{ g KHP}} \times \frac{1 \text{ mol KOH}}{1 \text{ mol KHP}} = 2.066 \times 10^{-3} \text{ mol KOH}$$

From the moles and volume of KOH, we calculate the molarity of the KOH solution.

$$\textbf{M of KOH} = \frac{\text{mol KOH}}{\text{L of KOH soln}} = \frac{2.066 \times 10^{-3} \text{ mol KOH}}{18.68 \times 10^{-3} \text{ L soln}} = \mathbf{0.1106} \textit{ M}$$

4.86   The reaction between HCl and NaOH is:

$$HCl(aq) + NaOH(aq) \rightarrow H_2O(l) + NaCl(aq)$$

We know the volume of the NaOH solution, and we want to calculate the molarity of the NaOH solution.

$$\begin{array}{ccc} \text{want to calculate} & & \text{need to find} \\ \searrow & & \swarrow \\ \textit{M} \text{ of NaOH} = & \dfrac{\text{mol NaOH}}{\text{L of NaOH soln}} & \\ & \nwarrow & \\ & \text{given} & \end{array}$$

If we can determine the moles of NaOH in the solution, we can then calculate the molarity of the solution. From the volume and molarity of HCl, we can calculate moles of HCl. Then, using the mole ratio from the balanced equation, we can calculate moles of NaOH.

$$? \text{ mol NaOH} = 17.4 \text{ mL HCl} \times \frac{0.312 \text{ mol HCl}}{1000 \text{ mL soln}} \times \frac{1 \text{ mol NaOH}}{1 \text{ mol HCl}} = 5.43 \times 10^{-3} \text{ mol NaOH}$$

From the moles and volume of NaOH, we calculate the molarity of the NaOH solution.

$$\textbf{\textit{M} of NaOH} = \frac{\text{mol NaOH}}{\text{L of NaOH soln}} = \frac{5.43 \times 10^{-3} \text{ mol NaOH}}{25.0 \times 10^{-3} \text{ L soln}} = \textbf{0.217 \textit{M}}$$

4.87   **(a)**   In order to have the correct mole ratio to solve the problem, you must start with a balanced chemical equation.

$$\text{HCl}(aq) + \text{NaOH}(aq) \longrightarrow \text{NaCl}(aq) + \text{H}_2\text{O}(l)$$

From the molarity and volume of the HCl solution, you can calculate moles of HCl. Then, using the mole ratio from the balanced equation above, you can calculate moles of NaOH.

$$? \text{ mol NaOH} = 25.00 \text{ mL} \times \frac{2.430 \text{ mol HCl}}{1000 \text{ mL soln}} \times \frac{1 \text{ mol NaOH}}{1 \text{ mol HCl}} = 6.075 \times 10^{-2} \text{ mol NaOH}$$

Solving for the volume of NaOH:

$$\text{liters of solution} = \frac{\text{moles of solute}}{M}$$

$$\textbf{volume of NaOH} = \frac{6.075 \times 10^{-2} \text{ mol NaOH}}{1.420 \text{ mol/L}} = 4.278 \times 10^{-2} \text{ L} = \textbf{42.78 mL}$$

(b)   This problem is similar to part (a). The difference is that the mole ratio between base and acid is 2:1.

$$\text{H}_2\text{SO}_4(aq) + 2\text{NaOH}(aq) \longrightarrow \text{Na}_2\text{SO}_4(aq) + \text{H}_2\text{O}(l)$$

$$? \text{ mol NaOH} = 25.00 \text{ mL} \times \frac{4.500 \text{ mol H}_2\text{SO}_4}{1000 \text{ mL soln}} \times \frac{2 \text{ mol NaOH}}{1 \text{ mol H}_2\text{SO}_4} = 0.2250 \text{ mol NaOH}$$

$$\textbf{volume of NaOH} = \frac{0.2250 \text{ mol NaOH}}{1.420 \text{ mol/L}} = 0.1585 \text{ L} = \textbf{158.5 mL}$$

(c)   This problem is similar to parts (a) and (b). The difference is that the mole ratio between base and acid is 3:1.

$$\text{H}_3\text{PO}_4(aq) + 3\text{NaOH}(aq) \longrightarrow \text{Na}_3\text{PO}_4(aq) + 3\text{H}_2\text{O}(l)$$

$$? \text{ mol NaOH} = 25.00 \text{ mL} \times \frac{1.500 \text{ mol H}_3\text{PO}_4}{1000 \text{ mL soln}} \times \frac{3 \text{ mol NaOH}}{1 \text{ mol H}_3\text{PO}_4} = 0.1125 \text{ mol NaOH}$$

$$\textbf{volume of NaOH} = \frac{0.1125 \text{ mol NaOH}}{1.420 \text{ mol/L}} = 0.07923 \text{ L} = \textbf{79.23 mL}$$

4.88   **Strategy:** We know the molarity of the HCl solution, and we want to calculate the volume of the HCl solution.

$$M \text{ of HCl} = \frac{\overset{\text{given}}{\text{mol HCl}}}{\underset{\text{want to calculate}}{\overset{\text{need to find}}{\text{L of HCl soln}}}}$$

If we can determine the moles of HCl, we can then use the definition of molarity to calculate the volume of HCl needed. From the volume and molarity of NaOH or $Ba(OH)_2$, we can calculate moles of NaOH or $Ba(OH)_2$. Then, using the mole ratio from the balanced equation, we can calculate moles of HCl.

Solution:

(a)   In order to have the correct mole ratio to solve the problem, you must start with a balanced chemical equation.

$$HCl(aq) + NaOH(aq) \longrightarrow NaCl(aq) + H_2O(l)$$

$$? \text{ mol HCl} = 10.0 \text{ mL} \times \frac{0.300 \text{ mol NaOH}}{1000 \text{ mL of solution}} \times \frac{1 \text{ mol HCl}}{1 \text{ mol NaOH}} = 3.00 \times 10^{-3} \text{ mol HCl}$$

From the molarity and moles of HCl, we calculate volume of HCl required to neutralize the NaOH.

$$\text{liters of solution} = \frac{\text{moles of solute}}{M}$$

$$\textbf{volume of HCl} = \frac{3.00 \times 10^{-3} \text{ mol HCl}}{0.500 \text{ mol/L}} = \textbf{6.00} \times \textbf{10}^{-3} \textbf{ L} = \textbf{6.00 mL}$$

(b)   This problem is similar to part (a). The difference is that the mole ratio between acid and base is 2:1.

$$2HCl(aq) + Ba(OH)_2(aq) \longrightarrow BaCl_2(aq) + 2H_2O(l)$$

$$? \text{ mol HCl} = 10.0 \text{ mL} \times \frac{0.200 \text{ mol Ba(OH)}_2}{1000 \text{ mL of solution}} \times \frac{2 \text{ mol HCl}}{1 \text{ mol Ba(OH)}_2} = 4.00 \times 10^{-3} \text{ mol HCl}$$

$$\textbf{volume of HCl} = \frac{4.00 \times 10^{-3} \text{ mol HCl}}{0.500 \text{ mol/L}} = \textbf{8.00} \times \textbf{10}^{-3} \textbf{ L} = \textbf{8.00 mL}$$

4.91   The balanced equation is given in the problem. The mole ratio between Fe^{2+} and $Cr_2O_7^{2-}$ is 6:1.

First, calculate the moles of Fe^{2+} that react with $Cr_2O_7^{2-}$.

$$26.00 \text{ mL soln} \times \frac{0.0250 \text{ mol Cr}_2O_7^{2-}}{1000 \text{ mL soln}} \times \frac{6 \text{ mol Fe}^{2+}}{1 \text{ mol Cr}_2O_7^{2-}} = 3.90 \times 10^{-3} \text{ mol Fe}^{2+}$$

The molar concentration of Fe^{2+} is:

$$M = \frac{3.90 \times 10^{-3} \text{ mol } Fe^{2+}}{25.0 \times 10^{-3} \text{ L soln}} = \textbf{0.156 } \boldsymbol{M}$$

4.92   **Strategy:** We want to calculate the grams of SO_2 in the sample of air. From the molarity and volume of $KMnO_4$, we can calculate moles of $KMnO_4$. Then, using the mole ratio from the balanced equation, we can calculate moles of SO_2. How do we convert from moles of SO_2 to grams of SO_2?

Solution: The balanced equation is given in the problem.

$$5SO_2 + 2MnO_4^- + 2H_2O \longrightarrow 5SO_4^{2-} + 2Mn^{2+} + 4H^+$$

The moles of $KMnO_4$ required for the titration are:

$$\frac{0.00800 \text{ mol } KMnO_4}{1000 \text{ mL soln}} \times 7.37 \text{ mL} = 5.90 \times 10^{-5} \text{ mol } KMnO_4$$

We use the mole ratio from the balanced equation and the molar mass of SO_2 as conversion factors to convert to grams of SO_2.

$$(5.90 \times 10^{-5} \text{ mol } KMnO_4) \times \frac{5 \text{ mol } SO_2}{2 \text{ mol } KMnO_4} \times \frac{64.07 \text{ g } SO_2}{1 \text{ mol } SO_2} = \textbf{9.45} \times \textbf{10}^{-3} \textbf{ g } \boldsymbol{SO_2}$$

4.93   The balanced equation is given in problem 4.91. The mole ratio between Fe^{2+} and $Cr_2O_7^{2-}$ is 6:1. First, calculate the moles of $Cr_2O_7^{2-}$ that reacted.

$$23.30 \text{ mL soln} \times \frac{0.0194 \text{ mol } Cr_2O_7^{2-}}{1000 \text{ mL soln}} = 4.52 \times 10^{-4} \text{ mol } Cr_2O_7^{2-}$$

Use the mole ratio from the balanced equation to calculate the mass of iron that reacted.

$$(4.52 \times 10^{-4} \text{ mol } Cr_2O_7^{2-}) \times \frac{6 \text{ mol } Fe^{2+}}{1 \text{ mol } Cr_2O_7^{2-}} \times \frac{55.85 \text{ g } Fe^{2+}}{1 \text{ mol } Fe^{2+}} = 0.151 \text{ g } Fe^{2+}$$

The percent by mass of iron in the ore is:

$$\frac{0.151 \text{ g}}{0.2792 \text{ g}} \times 100\% = \textbf{54.1\%}$$

4.94   The balanced equation is given in the problem.

$$2MnO_4^- + 5H_2O_2 + 6H^+ \longrightarrow 5O_2 + 2Mn^{2+} + 8H_2O$$

First, calculate the moles of potassium permanganate in 36.44 mL of solution.

$$\frac{0.01652 \text{ mol } KMnO_4}{1000 \text{ mL soln}} \times 36.44 \text{ mL} = 6.020 \times 10^{-4} \text{ mol } KMnO_4$$

Next, calculate the moles of hydrogen peroxide using the mole ratio from the balanced equation.

$$(6.020 \times 10^{-4} \text{ mol KMnO}_4) \times \frac{5 \text{ mol H}_2\text{O}_2}{2 \text{ mol KMnO}_4} = 1.505 \times 10^{-3} \text{ mol H}_2\text{O}_2$$

Finally, calculate the molarity of the H_2O_2 solution. The volume of the solution is 0.02500 L.

$$\textbf{Molarity of H}_2\textbf{O}_2 = \frac{1.505 \times 10^{-3} \text{ mol H}_2\text{O}_2}{0.02500 \text{ L}} = \textbf{0.06020 } \textit{M}$$

4.95   The balanced equation shows that 2 moles of electrons are lost for each mole of SO_3^{2-} that reacts. The electrons are gained by IO_3^-. We need to find the moles of electrons gained for each mole of IO_3^- that reacts. Then, we can calculate the final oxidation state of iodine.

The number of moles of electrons lost by SO_3^{2-} is:

$$32.5 \text{ mL} \times \frac{0.500 \text{ mol SO}_3^{2-}}{1000 \text{ mL soln}} \times \frac{2 \text{ mol e}^-}{1 \text{ mol SO}_3^{2-}} = 0.0325 \text{ mol e}^- \text{ lost}$$

The number of moles of iodate, IO_3^-, that react is:

$$1.390 \text{ g KIO}_3 \times \frac{1 \text{ mol KIO}_3}{214.0 \text{ g KIO}_3} \times \frac{1 \text{ mol IO}_3^-}{1 \text{ mol KIO}_3} = 6.495 \times 10^{-3} \text{ mol IO}_3^-$$

6.495×10^{-3} mole of IO_3^- gain 0.0325 mole of electrons. The number of moles of electrons gained per mole of IO_3^- is:

$$\frac{0.0325 \text{ mol e}^-}{6.495 \times 10^{-3} \text{ mol IO}_3^-} = 5.00 \text{ mol e}^-/\text{mol IO}_3^-$$

The oxidation number of iodine in IO_3^- is +5. Since 5 moles of electrons are gained per mole of IO_3^-, the final oxidation state of iodine is +5 − 5 = **0**. The iodine containing product of the reaction is most likely elemental iodine, I_2.

4.96   First, calculate the moles of $KMnO_4$ in 24.0 mL of solution.

$$\frac{0.0100 \text{ mol KMnO}_4}{1000 \text{ mL soln}} \times 24.0 \text{ mL} = 2.40 \times 10^{-4} \text{ mol KMnO}_4$$

Next, calculate the mass of oxalic acid needed to react with 2.40×10^{-4} mol $KMnO_4$. Use the mole ratio from the balanced equation.

$$(2.40 \times 10^{-4} \text{ mol KMnO}_4) \times \frac{5 \text{ mol H}_2\text{C}_2\text{O}_4}{2 \text{ mol KMnO}_4} \times \frac{90.04 \text{ g H}_2\text{C}_2\text{O}_4}{1 \text{ mol H}_2\text{C}_2\text{O}_4} = 0.0540 \text{ g H}_2\text{C}_2\text{O}_4$$

The original sample had a mass of 1.00 g. The mass percent of $H_2C_2O_4$ in the sample is:

$$\textbf{mass \%} = \frac{0.0540 \text{ g}}{1.00 \text{ g}} \times 100\% = \textbf{5.40\% H}_2\textbf{C}_2\textbf{O}_4$$

4.97   The first titration oxidizes Fe^{2+} to Fe^{3+}. This titration gives the amount of Fe^{2+} in solution. Zn metal is added to reduce all Fe^{3+} back to Fe^{2+}. The second titration oxidizes all the Fe^{2+} back to Fe^{3+}. We can find the amount of Fe^{3+} in the original solution by difference.

Titration #1: The mole ratio between Fe^{2+} and MnO_4^- is 5:1.

$$23.0 \text{ mL soln} \times \frac{0.0200 \text{ mol MnO}_4^-}{1000 \text{ mL soln}} \times \frac{5 \text{ mol Fe}^{2+}}{1 \text{ mol MnO}_4^-} = 2.30 \times 10^{-3} \text{ mol Fe}^{2+}$$

$$[Fe^{2+}] = \frac{\text{mol solute}}{\text{L of soln}} = \frac{2.30 \times 10^{-3} \text{ mol Fe}^{2+}}{25.0 \times 10^{-3} \text{ L soln}} = \textbf{0.0920 } \textbf{\textit{M}}$$

Titration #2: The mole ratio between Fe^{2+} and MnO_4^- is 5:1.

$$40.0 \text{ mL soln} \times \frac{0.0200 \text{ mol MnO}_4^-}{1000 \text{ mL soln}} \times \frac{5 \text{ mol Fe}^{2+}}{1 \text{ mol MnO}_4^-} = 4.00 \times 10^{-3} \text{ mol Fe}^{2+}$$

In this second titration, there are more moles of Fe^{2+} in solution. This is due to Fe^{3+} in the original solution being reduced by Zn to Fe^{2+}. The number of moles of Fe^{3+} in solution is:

$$(4.00 \times 10^{-3} \text{ mol}) - (2.30 \times 10^{-3} \text{ mol}) = 1.70 \times 10^{-3} \text{ mol Fe}^{3+}$$

$$[Fe^{3+}] = \frac{\text{mol solute}}{\text{L of soln}} = \frac{1.70 \times 10^{-3} \text{ mol Fe}^{3+}}{25.0 \times 10^{-3} \text{ L soln}} = \textbf{0.0680 } \textbf{\textit{M}}$$

4.98   The balanced equation is:

$$2MnO_4^- + 16H^+ + 5C_2O_4^{2-} \longrightarrow 2Mn^{2+} + 10CO_2 + 8H_2O$$

$$\text{mol MnO}_4^- = \frac{9.56 \times 10^{-4} \text{ mol MnO}_4^-}{1000 \text{ mL of soln}} \times 24.2 \text{ mL} = 2.31 \times 10^{-5} \text{ mol MnO}_4^-$$

Using the mole ratio from the balanced equation, we can calculate the mass of Ca^{2+} in the 10.0 mL sample of blood.

$$(2.31 \times 10^{-5} \text{ mol MnO}_4^-) \times \frac{5 \text{ mol C}_2\text{O}_4^{2-}}{2 \text{ mol MnO}_4^-} \times \frac{1 \text{ mol Ca}^{2+}}{1 \text{ mol C}_2\text{O}_4^{2-}} \times \frac{40.08 \text{ g Ca}^{2+}}{1 \text{ mol Ca}^{2+}} = 2.31 \times 10^{-3} \text{ g Ca}^{2+}$$

Converting to mg/mL:

$$\frac{2.31 \times 10^{-3} \text{ g Ca}^{2+}}{10.0 \text{ mL of blood}} \times \frac{1 \text{ mg}}{0.001 \text{ g}} = \textbf{0.231 mg Ca}^{2+}\textbf{/mL of blood}$$

4.99   In redox reactions the oxidation numbers of elements change. To test whether an equation represents a redox process, assign the oxidation numbers to each of the elements in the reactants and products. If oxidation numbers change, it is a redox reaction.

(a)   On the left the oxidation number of chlorine in Cl_2 is zero (rule 1). On the right it is -1 in Cl^- (rule 2) and $+1$ in OCl^- (rules 3 and 5). Since chlorine is both oxidized and reduced, this is a disproportionation redox reaction.

(b) The oxidation numbers of calcium and carbon do not change. This is not a redox reaction; it is a precipitation reaction.

(c) The oxidation numbers of nitrogen and hydrogen do not change. This is not a redox reaction; it is an acid-base reaction.

(d) The oxidation numbers of carbon, chlorine, chromium, and oxygen do not change. This is not a redox reaction; it doesn't fit easily into any category, but could be considered as a type of combination reaction.

(e) The oxidation number of calcium changes from 0 to +2, and the oxidation number of fluorine changes from 0 to −1. This is a combination redox reaction.

The remaining parts (f) through (j) can be worked the same way.

(f) Redox   **(g)** Precipitation   **(h)** Redox   **(i)** Redox   **(j)** Redox

4.100 First, the gases could be tested to see if they supported combustion. O_2 would support combustion, CO_2 would not. Second, if CO_2 is bubbled through a solution of calcium hydroxide [$Ca(OH)_2$], a white precipitate of $CaCO_3$ forms. No reaction occurs when O_2 is bubbled through a calcium hydroxide solution.

4.101 Choice **(d)**, 0.20 M $Mg(NO_3)_2$, should be the best conductor of electricity; the total ion concentration in this solution is 0.60 M. The total ion concentrations for solutions (a) and (c) are 0.40 M and 0.50 M, respectively. We can rule out choice (b), because acetic acid is a weak electrolyte.

4.102 Starting with a balanced chemical equation:

$$Mg(s) + 2HCl(aq) \longrightarrow MgCl_2(aq) + H_2(g)$$

From the mass of Mg, you can calculate moles of Mg. Then, using the mole ratio from the balanced equation above, you can calculate moles of HCl reacted.

$$4.47 \text{ g Mg} \times \frac{1 \text{ mol Mg}}{24.31 \text{ g Mg}} \times \frac{2 \text{ mol HCl}}{1 \text{ mol Mg}} = 0.368 \text{ mol HCl reacted}$$

Next we can calculate the number of moles of HCl in the original solution.

$$\frac{2.00 \text{ mol HCl}}{1000 \text{ mL soln}} \times (5.00 \times 10^2 \text{ mL}) = 1.00 \text{ mol HCl}$$

Moles HCl remaining $= 1.00 \text{ mol} - 0.368 \text{ mol} = 0.632 \text{ mol HCl}$

$$\text{\textbf{conc. of HCl after reaction}} = \frac{\text{mol HCl}}{\text{L soln}} = \frac{0.632 \text{ mol HCl}}{0.500 \text{ L}} = 1.26 \text{ mol/L} = \textbf{1.26 } \textit{M}$$

4.103 The balanced equation for the displacement reaction is:

$$Zn(s) + CuSO_4(aq) \longrightarrow ZnSO_4(aq) + Cu(s)$$

The moles of $CuSO_4$ that react with 7.89 g of zinc are:

$$7.89 \text{ g Zn} \times \frac{1 \text{ mol Zn}}{65.39 \text{ g Zn}} \times \frac{1 \text{ mol CuSO}_4}{1 \text{ mol Zn}} = 0.121 \text{ mol CuSO}_4$$

The volume of the 0.156 M $CuSO_4$ solution needed to react with 7.89 g Zn is:

$$\text{L of soln} = \frac{\text{mole solute}}{M} = \frac{0.121 \text{ mol } CuSO_4}{0.156 \text{ mol/L}} = \textbf{0.776 L} = \textbf{776 mL}$$

Would you expect Zn to displace Cu^{2+} from solution, as shown in the equation?

4.104   The balanced equation is:

$$2HCl(aq) + Na_2CO_3(s) \longrightarrow CO_2(g) + H_2O(l) + 2NaCl(aq)$$

The mole ratio from the balanced equation is 2 moles HCl : 1 mole Na_2CO_3. The moles of HCl needed to react with 0.256 g of Na_2CO_3 are:

$$0.256 \text{ g } Na_2CO_3 \times \frac{1 \text{ mol } Na_2CO_3}{106.0 \text{ g } Na_2CO_3} \times \frac{2 \text{ mol HCl}}{1 \text{ mol } Na_2CO_3} = 4.83 \times 10^{-3} \text{ mol HCl}$$

$$\textbf{Molarity HCl} = \frac{\text{moles HCl}}{\text{L soln}} = \frac{4.83 \times 10^{-3} \text{ mol HCl}}{0.0283 \text{ L soln}} = 0.171 \text{ mol/L} = \textbf{0.171 } \textbf{\textit{M}}$$

4.105   The neutralization reaction is: $HA(aq) + NaOH(aq) \longrightarrow NaA(aq) + H_2O(l)$

The mole ratio between the acid and NaOH is 1:1. The moles of HA that react with NaOH are:

$$20.27 \text{ mL soln} \times \frac{0.1578 \text{ mol NaOH}}{1000 \text{ mL soln}} \times \frac{1 \text{ mol HA}}{1 \text{ mol NaOH}} = 3.199 \times 10^{-3} \text{ mol HA}$$

3.664 g of the acid reacted with the base. The molar mass of the acid is:

$$\textbf{Molar mass} = \frac{3.664 \text{ g HA}}{3.199 \times 10^{-3} \text{ mol HA}} = \textbf{1145 g/mol}$$

4.106   Starting with a balanced chemical equation:

$$CH_3COOH(aq) + NaOH(aq) \longrightarrow CH_3COONa(aq) + H_2O(l)$$

From the molarity and volume of the NaOH solution, you can calculate moles of NaOH. Then, using the mole ratio from the balanced equation above, you can calculate moles of CH_3COOH.

$$5.75 \text{ mL solution} \times \frac{1.00 \text{ mol NaOH}}{1000 \text{ mL of solution}} \times \frac{1 \text{ mol } CH_3COOH}{1 \text{ mol NaOH}} = 5.75 \times 10^{-3} \text{ mol } CH_3COOH$$

$$\textbf{Molarity CH}_3\textbf{COOH} = \frac{5.75 \times 10^{-3} \text{ mol } CH_3COOH}{0.0500 \text{ L}} = \textbf{0.115 } \textbf{\textit{M}}$$

4.107   Let's call the original solution, soln 1; the first dilution, soln 2; and the second dilution, soln 3. Start with the concentration of soln 3, 0.00383 M. From the concentration and volume of soln 3, we can find the concentration of soln 2. Then, from the concentration and volume of soln 2, we can find the concentration of soln 1, the original solution.

$$M_2V_2 = M_3V_3$$

$$M_2 = \frac{M_3V_3}{V_2} = \frac{(0.00383\ M)(1.000 \times 10^3\ \text{mL})}{25.00\ \text{mL}} = 0.153\ M$$

$$M_1V_1 = M_2V_2$$

$$M_1 = \frac{M_2V_2}{V_1} = \frac{(0.153\ M)(125.0\ \text{mL})}{15.00\ \text{mL}} = \mathbf{1.28\ M}$$

4.108 The balanced equation is:

$$Zn(s) + 2AgNO_3(aq) \longrightarrow Zn(NO_3)_2(aq) + 2Ag(s)$$

Let x = mass of Ag produced. We can find the mass of Zn reacted in terms of the amount of Ag produced.

$$x\ \text{g Ag} \times \frac{1\ \text{mol Ag}}{107.9\ \text{g Ag}} \times \frac{1\ \text{mol Zn}}{2\ \text{mol Ag}} \times \frac{65.39\ \text{g Zn}}{1\ \text{mol Zn}} = 0.303x\ \text{g Zn reacted}$$

The mass of Zn remaining will be:

$$2.50\ \text{g} - \text{amount of Zn reacted} = 2.50\ \text{g Zn} - 0.303x\ \text{g Zn}$$

The final mass of the strip, 3.37 g, equals the mass of Ag produced + the mass of Zn remaining.

$$3.37\ \text{g} = x\ \text{g Ag} + (2.50\ \text{g Zn} - 0.303\ x\ \text{g Zn})$$

$$x = \mathbf{1.25\ g} = \textbf{mass of Ag produced}$$

$$\textbf{mass of Zn remaining} = 3.37\ \text{g} - 1.25\ \text{g} = \mathbf{2.12\ g\ Zn}$$

or

$$\textbf{mass of Zn remaining} = 2.50\ \text{g Zn} - 0.303x\ \text{g Zn} = 2.50\ \text{g} - (0.303)(1.25\ \text{g}) = \mathbf{2.12\ g\ Zn}$$

4.109 The balanced equation is: $Ba(OH)_2(aq) + Na_2SO_4(aq) \longrightarrow BaSO_4(s) + 2NaOH(aq)$

moles $Ba(OH)_2$: (2.27 L)(0.0820 mol/L) = 0.186 mol $Ba(OH)_2$
moles Na_2SO_4: (3.06 L)(0.0664 mol/L) = 0.203 mol Na_2SO_4

Since the mole ratio between $Ba(OH)_2$ and Na_2SO_4 is 1:1, $Ba(OH)_2$ is the limiting reagent. The mass of $BaSO_4$ formed is:

$$0.186\ \text{mol Ba(OH)}_2 \times \frac{1\ \text{mol BaSO}_4}{1\ \text{mol Ba(OH)}_2} \times \frac{233.4\ \text{g BaSO}_4}{\text{mol BaSO}_4} = \mathbf{43.4\ g\ BaSO_4}$$

4.110 The balanced equation is: $HNO_3(aq) + NaOH(aq) \longrightarrow NaNO_3(aq) + H_2O(l)$

$$\text{mol HNO}_3 = \frac{0.211\ \text{mol HNO}_3}{1000\ \text{mL soln}} \times 10.7\ \text{mL soln} = 2.26 \times 10^{-3}\ \text{mol HNO}_3$$

$$\text{mol NaOH} = \frac{0.258\ \text{mol NaOH}}{1000\ \text{mL soln}} \times 16.3\ \text{mL soln} = 4.21 \times 10^{-3}\ \text{mol NaOH}$$

Since the mole ratio from the balanced equation is 1 mole NaOH : 1 mole HNO_3, then 2.26×10^{-3} mol HNO_3 will react with 2.26×10^{-3} mol NaOH.

mol NaOH remaining $= (4.21 \times 10^{-3} \text{ mol}) - (2.26 \times 10^{-3} \text{ mol}) = 1.95 \times 10^{-3}$ mol NaOH

$10.7 \text{ mL} + 16.3 \text{ mL} = 27.0 \text{ mL} = 0.027 \text{ L}$

$$\textbf{molarity NaOH} = \frac{1.95 \times 10^{-3} \text{ mol NaOH}}{0.027 \text{ L}} = \textbf{0.0722 } \textit{M}$$

4.111 **(a)** Magnesium hydroxide is insoluble in water. It can be prepared by mixing a solution containing Mg^{2+} ions such as $MgCl_2(aq)$ or $Mg(NO_3)_2(aq)$ with a solution containing hydroxide ions such as $NaOH(aq)$. $Mg(OH)_2$ will precipitate, which can then be collected by filtration. The net ionic reaction is:

$$Mg^{2+}(aq) + 2OH^-(aq) \rightarrow Mg(OH)_2(s)$$

(b) The balanced equation is: $2HCl + Mg(OH)_2 \longrightarrow MgCl_2 + 2H_2O$

The moles of $Mg(OH)_2$ in 10 mL of milk of magnesia are:

$$10 \text{ mL soln} \times \frac{0.080 \text{ g Mg(OH)}_2}{1 \text{ mL soln}} \times \frac{1 \text{ mol Mg(OH)}_2}{58.33 \text{ g Mg(OH)}_2} = 0.014 \text{ mol Mg(OH)}_2$$

$$\text{Moles of HCl reacted} = 0.014 \text{ mol Mg(OH)}_2 \times \frac{2 \text{ mol HCl}}{1 \text{ mol Mg(OH)}_2} = 0.028 \text{ mol HCl}$$

$$\textbf{Volume of HCl} = \frac{\text{mol solute}}{M} = \frac{0.028 \text{ mol HCl}}{0.035 \text{ mol/L}} = \textbf{0.80 L}$$

4.112 The balanced equations for the two reactions are:

$$X(s) + H_2SO_4(aq) \longrightarrow XSO_4(aq) + H_2(g)$$

$$H_2SO_4(aq) + 2NaOH(aq) \longrightarrow Na_2SO_4(aq) + 2H_2O(l)$$

First, let's find the number of moles of excess acid from the reaction with NaOH.

$$0.0334 \text{ L} \times \frac{0.500 \text{ mol NaOH}}{1 \text{ L soln}} \times \frac{1 \text{ mol H}_2SO_4}{2 \text{ mol NaOH}} = 8.35 \times 10^{-3} \text{ mol H}_2SO_4$$

The original number of moles of acid was:

$$0.100 \text{ L} \times \frac{0.500 \text{ mol H}_2SO_4}{1 \text{ L soln}} = 0.0500 \text{ mol H}_2SO_4$$

The amount of sulfuric acid that reacted with the metal, X, is

$$(0.0500 \text{ mol H}_2SO_4) - (8.35 \times 10^{-3} \text{ mol H}_2SO_4) = 0.0417 \text{ mol H}_2SO_4.$$

Since the mole ratio from the balanced equation is 1 mole X : 1 mole H_2SO_4, then the amount of X that reacted is 0.0417 mol X.

$$\textbf{molar mass X} = \frac{1.00 \text{ g X}}{0.0417 \text{ mol X}} = \textbf{24.0 g/mol}$$

The element is **magnesium**.

4.113 Add a known quantity of compound in a given quantity of water. Filter and recover the undissolved compound, then dry and weigh it. The difference in mass between the original quantity and the recovered quantity is the amount that dissolved in the water.

4.114 First, calculate the number of moles of glucose present.

$$\frac{0.513 \text{ mol glucose}}{1000 \text{ mL soln}} \times 60.0 \text{ mL} = 0.0308 \text{ mol glucose}$$

$$\frac{2.33 \text{ mol glucose}}{1000 \text{ mL soln}} \times 120.0 \text{ mL} = 0.280 \text{ mol glucose}$$

Add the moles of glucose, then divide by the total volume of the combined solutions to calculate the molarity.

60.0 mL + 120.0 mL = 180.0 mL = 0.180 L

$$\textbf{Molarity of final solution} = \frac{(0.0308 + 0.280)\text{ mol glucose}}{0.180 \text{ L}} = 1.73 \text{ mol/L} = \textbf{1.73 } \boldsymbol{M}$$

4.115 First, you would accurately measure the electrical conductance of pure water. The conductance of a solution of the slightly soluble ionic compound X should be greater than that of pure water. The increased conductance would indicate that some of the compound X had dissolved.

4.116 Iron(II) compounds can be oxidized to iron(III) compounds. The sample could be tested with a small amount of a strongly colored oxidizing agent like a $KMnO_4$ solution, which is a deep purple color. A loss of color would imply the presence of an oxidizable substance like an iron(II) salt.

4.117 The three chemical tests might include:

(1) Electrolysis to ascertain if hydrogen and oxygen were produced,

(2) The reaction with an alkali metal to see if a base and hydrogen gas were produced, and

(3) The dissolution of a metal oxide to see if a base was produced (or a nonmetal oxide to see if an acid was produced).

4.118 Since both of the original solutions were strong electrolytes, you would expect a mixture of the two solutions to also be a strong electrolyte. However, since the light dims, the mixture must contain fewer ions than the original solution. Indeed, H^+ from the sulfuric acid reacts with the OH^- from the barium hydroxide to form water. The barium cations react with the sulfate anions to form insoluble barium sulfate.

$$2H^+(aq) + SO_4^{2-}(aq) + Ba^{2+}(aq) + 2OH^-(aq) \longrightarrow 2H_2O(l) + BaSO_4(s)$$

Thus, the reaction depletes the solution of ions and the conductivity decreases.

4.119 **(a)** Check with litmus paper, react with carbonate or bicarbonate to see if CO_2 gas is produced, react with a base and check with an indicator.

(b) Titrate a known quantity of acid with a standard NaOH solution. Since it is a monoprotic acid, the moles of NaOH reacted equals the moles of the acid. Dividing the mass of acid by the number of moles gives the molar mass of the acid.

(c) Visually compare the conductivity of the acid with a standard NaCl solution of the same molar concentration. A strong acid will have a similar conductivity to the NaCl solution. The conductivity of a weak acid will be considerably less than the NaCl solution.

4.120   You could test the conductivity of the solutions. Sugar is a nonelectrolyte and an aqueous sugar solution will not conduct electricity; whereas, NaCl is a strong electrolyte when dissolved in water. Silver nitrate could be added to the solutions to see if silver chloride precipitated. In this particular case, the solutions could also be tasted.

4.121   **(a)** $Pb(NO_3)_2(aq) + Na_2SO_4(aq) \longrightarrow PbSO_4(s) + 2NaNO_3(aq)$

$Pb^{2+}(aq) + SO_4^{2-}(aq) \longrightarrow PbSO_4(s)$

(b)   First, calculate the moles of Pb^{2+} in the polluted water.

$$0.00450 \text{ g Na}_2SO_4 \times \frac{1 \text{ mol Na}_2SO_4}{142.1 \text{ g Na}_2SO_4} \times \frac{1 \text{ mol Pb(NO}_3)_2}{1 \text{ mol Na}_2SO_4} \times \frac{1 \text{ mol Pb}^{2+}}{1 \text{ mol Pb(NO}_3)_2} = 3.17 \times 10^{-5} \text{ mol Pb}^{2+}$$

The volume of the polluted water sample is 500 mL (0.500 L). The molar concentration of Pb^{2+} is:

$$[Pb^{2+}] = \frac{\text{mol Pb}^{2+}}{\text{L of soln}} = \frac{3.17 \times 10^{-5} \text{ mol Pb}^{2+}}{0.500 \text{ L soln}} = 6.34 \times 10^{-5} \, M$$

4.122   In a redox reaction, the oxidizing agent gains one or more electrons. In doing so, the oxidation number of the element gaining the electrons must become more negative. In the case of chlorine, the −1 oxidation number is already the most negative state possible. The chloride ion *cannot* accept any more electrons; therefore, hydrochloric acid is *not* an oxidizing agent.

4.123   **(a)**   An acid and a base react to form water and a salt. Potassium iodide is a salt; therefore, the acid and base are chosen to produce this salt.

$KOH(aq) + HI(aq) \longrightarrow KI(aq) + H_2O(l)$

The water could be evaporated to isolate the KI.

(b)   Acids react with carbonates to form carbon dioxide gas. Again, chose the acid and carbonate salt so that KI is produced.

$2HI(aq) + K_2CO_3(aq) \longrightarrow 2KI(aq) + CO_2(g) + H_2O(l)$

4.124   The reaction is too violent. This could cause the hydrogen gas produced to ignite, and an explosion could result.

4.125   All three products are water insoluble. Use this information in formulating your answer.

(a) $MgCl_2(aq) + 2NaOH(aq) \longrightarrow \textbf{Mg(OH)}_2\textbf{(s)} + 2NaCl(aq)$

(b) $AgNO_3(aq) + NaI(aq) \longrightarrow \textbf{AgI(s)} + NaNO_3(aq)$

(c) $3Ba(OH)_2(aq) + 2H_3PO_4(aq) \longrightarrow \textbf{Ba}_3\textbf{(PO}_4)_2\textbf{(s)} + 6H_2O(l)$

4.126   The solid sodium bicarbonate would be the better choice. The hydrogen carbonate ion, HCO_3^-, behaves as a Brønsted base to accept a proton from the acid.

$HCO_3^-(aq) + H^+(aq) \longrightarrow H_2CO_3(aq) \longrightarrow H_2O(l) + CO_2(g)$

The heat generated during the reaction of hydrogen carbonate with the acid causes the carbonic acid, H_2CO_3, that was formed to decompose to water and carbon dioxide.

The reaction of the spilled sulfuric acid with sodium hydroxide would produce sodium sulfate, Na_2SO_4, and water. There is a possibility that the Na_2SO_4 could precipitate. Also, the sulfate ion, SO_4^{2-} is a weak base; therefore, the "neutralized" solution would actually be *basic*.

$$H_2SO_4(aq) + 2NaOH(aq) \longrightarrow Na_2SO_4(aq) + 2H_2O(l)$$

Also, NaOH is a caustic substance and therefore is not safe to use in this manner.

4.127 **(a)** A soluble sulfate salt such as sodium sulfate or sulfuric acid could be added. Barium sulfate would precipitate leaving sodium ions in solution.

(b) Potassium carbonate, phosphate, or sulfide could be added which would precipitate the magnesium cations, leaving potassium cations in solution.

(c) Add a soluble silver salt such as silver nitrate. AgBr would precipitate, leaving nitrate ions in solution.

(d) Add a solution containing a cation other than ammonium or a Group 1A cation to precipitate the phosphate ions; the nitrate ions will remain in solution.

(e) Add a solution containing a cation other than ammonium or a Group 1A cation to precipitate the carbonate ions; the nitrate ions will remain in solution.

4.128 **(a)** Table salt, NaCl, is very soluble in water and is a strong electrolyte. Addition of $AgNO_3$ will precipitate AgCl.

(b) Table sugar or sucrose, $C_{12}H_{22}O_{11}$, is soluble in water and is a nonelectrolyte.

(c) Aqueous acetic acid, CH_3COOH, the primary ingredient of vinegar, is a weak electrolyte. It exhibits all of the properties of acids (Section 4.3).

(d) Baking soda, $NaHCO_3$, is a water-soluble strong electrolyte. It reacts with acid to release CO_2 gas. Addition of $Ca(OH)_2$ results in the precipitation of $CaCO_3$.

(e) Washing soda, $Na_2CO_3 \cdot 10H_2O$, is a water-soluble strong electrolyte. It reacts with acids to release CO_2 gas. Addition of a soluble alkaline-earth salt will precipitate the alkaline-earth carbonate. Aqueous washing soda is also slightly basic (Section 4.3).

(f) Boric acid, H_3BO_3, is weak electrolyte and a weak acid.

(g) Epsom salt, $MgSO_4 \cdot 7H_2O$, is a water-soluble strong electrolyte. Addition of $Ba(NO_3)_2$ results in the precipitation of $BaSO_4$. Addition of hydroxide precipitates $Mg(OH)_2$.

(h) Sodium hydroxide, NaOH, is a strong electrolyte and a strong base. Addition of $Ca(NO_3)_2$ results in the precipitation of $Ca(OH)_2$.

(i) Ammonia, NH_3, is a sharp-odored gas that when dissolved in water is a weak electrolyte and a weak base. NH_3 in the gas phase reacts with HCl gas to produce solid NH_4Cl.

(j) Milk of magnesia, $Mg(OH)_2$, is an insoluble, strong base that reacts with acids. The resulting magnesium salt may be soluble or insoluble.

(k) $CaCO_3$ is an insoluble salt that reacts with acid to release CO_2 gas. $CaCO_3$ is discussed in the Chemistry in Action essays entitled, "An Undesirable Precipitation Reaction" and "Metal from the Sea" in Chapter 4.

With the exception of NH_3 and vinegar, all the compounds in this problem are white solids.

4.129 Reaction 1: $SO_3^{2-}(aq) + H_2O_2(aq) \longrightarrow SO_4^{2-}(aq) + H_2O(l)$

Reaction 2: $SO_4^{2-}(aq) + BaCl_2(aq) \longrightarrow BaSO_4(s) + 2Cl^-(aq)$

4.130   The balanced equation for the reaction is:

$$XCl(aq) + AgNO_3(aq) \longrightarrow AgCl(s) + XNO_3(aq) \qquad \text{where } X = \text{Na, or K}$$

From the amount of AgCl produced, we can calculate the moles of XCl reacted (X = Na, or K).

$$1.913 \text{ g AgCl} \times \frac{1 \text{ mol AgCl}}{143.4 \text{ g AgCl}} \times \frac{1 \text{ mol XCl}}{1 \text{ mol AgCl}} = 0.01334 \text{ mol XCl}$$

Let x = number of moles NaCl. Then, the number of moles of KCl = 0.01334 mol − x. The sum of the NaCl and KCl masses must equal the mass of the mixture, 0.8870 g. We can write:

mass NaCl + mass KCl = 0.8870 g

$$\left[x \text{ mol NaCl} \times \frac{58.44 \text{ g NaCl}}{1 \text{ mol NaCl}} \right] + \left[(0.01334 - x) \text{ mol KCl} \times \frac{74.55 \text{ g KCl}}{1 \text{ mol KCl}} \right] = 0.8870 \text{ g}$$

$$x = 6.673 \times 10^{-3} = \text{moles NaCl}$$

$$\text{mol KCl} = 0.01334 - x = 0.01334 \text{ mol} - (6.673 \times 10^{-3} \text{ mol}) = 6.667 \times 10^{-3} \text{ mol KCl}$$

Converting moles to grams:

$$\text{mass NaCl} = (6.673 \times 10^{-3} \text{ mol NaCl}) \times \frac{58.44 \text{ g NaCl}}{1 \text{ mol NaCl}} = 0.3900 \text{ g NaCl}$$

$$\text{mass KCl} = (6.667 \times 10^{-3} \text{ mol KCl}) \times \frac{74.55 \text{ g KCl}}{1 \text{ mol KCl}} = 0.4970 \text{ g KCl}$$

The percentages by mass for each compound are:

$$\textbf{\% NaCl} = \frac{0.3900 \text{ g}}{0.8870 \text{ g}} \times 100\% = \textbf{43.97\% NaCl}$$

$$\textbf{\% KCl} = \frac{0.4970 \text{ g}}{0.8870 \text{ g}} \times 100\% = \textbf{56.03\% KCl}$$

4.131 Cl_2O (Cl = +1) Cl_2O_3 (Cl = +3) ClO_2 (Cl = +4) Cl_2O_6 (Cl = +6)
Cl_2O_7 (Cl = +7)

4.132   The number of moles of oxalic acid in 5.00×10^2 mL is:

$$\frac{0.100 \text{ mol H}_2\text{C}_2\text{O}_4}{1000 \text{ mL soln}} \times (5.00 \times 10^2 \text{ mL}) = 0.0500 \text{ mol H}_2\text{C}_2\text{O}_4$$

The balanced equation shows a mole ratio of 1 mol Fe_2O_3 : 6 mol $H_2C_2O_4$. The mass of rust that can be removed is:

$$0.0500 \text{ mol H}_2\text{C}_2\text{O}_4 \times \frac{1 \text{ mol Fe}_2\text{O}_3}{6 \text{ mol H}_2\text{C}_2\text{O}_4} \times \frac{159.7 \text{ g Fe}_2\text{O}_3}{1 \text{ mol Fe}_2\text{O}_3} = \textbf{1.33 g Fe}_2\textbf{O}_3$$

4.133   Since aspirin is a monoprotic acid, it will react with NaOH in a 1:1 mole ratio.

First, calculate the moles of aspirin in the tablet.

$$12.25 \text{ mL soln} \times \frac{0.1466 \text{ mol NaOH}}{1000 \text{ mL soln}} \times \frac{1 \text{ mol aspirin}}{1 \text{ mol NaOH}} = 1.796 \times 10^{-3} \text{ mol aspirin}$$

Next, convert from moles of aspirin to grains of aspirin.

$$1.796 \times 10^{-3} \text{ mol aspirin} \times \frac{180.2 \text{ g aspirin}}{1 \text{ mol aspirin}} \times \frac{1 \text{ grain}}{0.0648 \text{ g}} = \textbf{4.99 grains aspirin in one tablet}$$

4.134   The precipitation reaction is: $Ag^+(aq) + Br^-(aq) \longrightarrow AgBr(s)$

In this problem, the relative amounts of NaBr and $CaBr_2$ are not known. However, the total amount of Br^- in the mixture can be determined from the amount of AgBr produced. Let's find the number of moles of Br^-.

$$1.6930 \text{ g AgBr} \times \frac{1 \text{ mol AgBr}}{187.8 \text{ g AgBr}} \times \frac{1 \text{ mol Br}^-}{1 \text{ mol AgBr}} = 9.015 \times 10^{-3} \text{ mol Br}^-$$

The amount of Br^- comes from both NaBr and $CaBr_2$. Let x = number of moles NaBr. Then, the number of moles of $CaBr_2 = \dfrac{9.015 \times 10^{-3} \text{ mol} - x}{2}$. The moles of $CaBr_2$ are divided by 2, because 1 mol of $CaBr_2$ produces 2 moles of Br^-. The sum of the NaBr and $CaBr_2$ masses must equal the mass of the mixture, 0.9157 g. We can write:

mass NaBr + mass $CaBr_2$ = 0.9157 g

$$\left[x \text{ mol NaBr} \times \frac{102.9 \text{ g NaBr}}{1 \text{ mol NaBr}} \right] + \left[\left(\frac{9.015 \times 10^{-3} - x}{2} \right) \text{mol CaBr}_2 \times \frac{199.9 \text{ g CaBr}_2}{1 \text{ mol CaBr}_2} \right] = 0.9157 \text{ g}$$

$$2.95x = 0.01465$$

$$x = 4.966 \times 10^{-3} = \text{moles NaBr}$$

Converting moles to grams:

$$\text{mass NaBr} = (4.966 \times 10^{-3} \text{ mol NaBr}) \times \frac{102.9 \text{ g NaBr}}{1 \text{ mol NaBr}} = 0.5110 \text{ g NaBr}$$

The percentage by mass of NaBr in the mixture is:

$$\textbf{\% NaBr} = \frac{0.5110 \text{ g}}{0.9157 \text{ g}} \times 100\% = \textbf{55.80\% NaBr}$$

4.135   **(a)**   The balanced equations are:

1) $Cu(s) + 4HNO_3(aq) \longrightarrow Cu(NO_3)_2(aq) + 2NO_2(g) + 2H_2O(l)$ Redox

2) $Cu(NO_3)_2(aq) + 2NaOH(aq) \longrightarrow Cu(OH)_2(s) + 2NaNO_3(aq)$ Precipitation

3) $Cu(OH)_2(s) \xrightarrow{\text{heat}} CuO(s) + H_2O(g)$ Decomposition

4) $CuO(s) + H_2SO_4(aq) \longrightarrow CuSO_4(aq) + H_2O(l)$ Acid-Base

5) $CuSO_4(aq) + Zn(s) \longrightarrow Cu(s) + ZnSO_4(aq)$ Redox

6) $Zn(s) + 2HCl(aq) \longrightarrow ZnCl_2(aq) + H_2(g)$ Redox

(b) We start with 65.6 g Cu, which is $65.6 \text{ g Cu} \times \dfrac{1 \text{ mol Cu}}{63.55 \text{ g Cu}} = 1.03 \text{ mol Cu}$. The mole ratio between

product and reactant in each reaction is 1:1. Therefore, the theoretical yield in each reaction is 1.03 moles.

1) $1.03 \text{ mol} \times \dfrac{187.6 \text{ g Cu(NO}_3)_2}{1 \text{ mol Cu(NO}_3)_2} = \textbf{193 g Cu(NO}_3)_2$

2) $1.03 \text{ mol} \times \dfrac{97.57 \text{ g Cu(OH)}_2}{1 \text{ mol Cu(OH)}_2} = \textbf{1.00} \times \textbf{10}^2 \textbf{ g Cu(OH)}_2$

3) $1.03 \text{ mol} \times \dfrac{79.55 \text{ g CuO}}{1 \text{ mol CuO}} = \textbf{81.9 g CuO}$

4) $1.03 \text{ mol} \times \dfrac{159.6 \text{ g CuSO}_4}{1 \text{ mol CuSO}_4} = \textbf{164 g CuSO}_4$

5) $1.03 \text{ mol} \times \dfrac{63.55 \text{ g Cu}}{1 \text{ mol Cu}} = \textbf{65.5 g Cu}$

(c) All of the reaction steps are clean and almost quantitative; therefore, the recovery yield should be high.

4.136 There are two moles of Cl^- per one mole of $CaCl_2$.

(a) $25.3 \text{ g CaCl}_2 \times \dfrac{1 \text{ mol CaCl}_2}{111.0 \text{ g CaCl}_2} \times \dfrac{2 \text{ mol Cl}^-}{1 \text{ mol CaCl}_2} = 0.456 \text{ mol Cl}^-$

$\textbf{Molarity Cl}^- = \dfrac{0.456 \text{ mol Cl}^-}{0.325 \text{ L soln}} = 1.40 \text{ mol/L} = \textbf{1.40 } \textbf{\textit{M}}$

(b) We need to convert from mol/L to grams in 0.100 L.

$\dfrac{1.40 \text{ mol Cl}^-}{1 \text{ L soln}} \times \dfrac{35.45 \text{ g Cl}}{1 \text{ mol Cl}^-} \times 0.100 \text{ L soln} = \textbf{4.96 g Cl}^-$

4.137 **(a)** $CaF_2(s) + H_2SO_4(aq) \longrightarrow 2HF(g) + CaSO_4(s)$

$2NaCl(s) + H_2SO_4(aq) \longrightarrow 2HCl(aq) + Na_2SO_4(aq)$

(b) HBr and HI **cannot** be prepared similarly, because Br^- and I^- would be oxidized to the element, Br_2 and I_2, respectively.

$2NaBr(s) + 2H_2SO_4(aq) \longrightarrow Br_2(l) + SO_2(g) + Na_2SO_4(aq) + 2H_2O(l)$

(c) $PBr_3(l) + 3H_2O(l) \longrightarrow 3HBr(g) + H_3PO_3(aq)$

4.138 **(a)** The precipitation reaction is:

$$Mg^{2+}(aq) + 2OH^-(aq) \longrightarrow Mg(OH)_2(s)$$

The acid-base reaction is:

$$Mg(OH)_2(s) + 2HCl(aq) \longrightarrow MgCl_2(aq) + 2H_2O(l)$$

The redox reactions are:

$$Mg^{2+} + 2e^- \longrightarrow Mg$$
$$\underline{2Cl^- \longrightarrow Cl_2 + 2e^-}$$
$$MgCl_2 \longrightarrow Mg + Cl_2$$

(b) NaOH is much more expensive than CaO.

(c) Dolomite has the advantage of being an additional source of magnesium that can also be recovered.

4.139 **Electric furnace method:**

$$P_4(s) + 5O_2(g) \longrightarrow P_4O_{10}(s) \qquad \text{redox}$$
$$P_4O_{10}(s) + 6H_2O(l) \longrightarrow 4H_3PO_4(aq) \qquad \text{acid-base}$$

Wet process:

$$Ca_5(PO_4)_3F(s) + 5H_2SO_4(aq) \longrightarrow 3H_3PO_4(aq) + HF(aq) + 5CaSO_4(s)$$

This is a precipitation and an acid-base reaction.

4.140 **(a)** $NH_4^+(aq) + OH^-(aq) \longrightarrow NH_3(aq) + H_2O(l)$

(b) From the amount of NaOH needed to neutralize the 0.2041 g sample, we can find the amount of the 0.2041 g sample that is NH_4NO_3.

First, calculate the moles of NaOH.

$$\frac{0.1023 \text{ mol NaOH}}{1000 \text{ mL of soln}} \times 24.42 \text{ mL soln} = 2.498 \times 10^{-3} \text{ mol NaOH}$$

Using the mole ratio from the balanced equation, we can calculate the amount of NH_4NO_3 that reacted.

$$(2.498 \times 10^{-3} \text{ mol NaOH}) \times \frac{1 \text{ mol NH}_4NO_3}{1 \text{ mol NaOH}} \times \frac{80.05 \text{ g NH}_4NO_3}{1 \text{ mol NH}_4NO_3} = 0.2000 \text{ g NH}_4NO_3$$

The purity of the NH_4NO_3 sample is:

$$\textbf{\% purity} = \frac{0.2000 \text{ g}}{0.2041 \text{ g}} \times 100\% = \textbf{97.99\%}$$

4.141 In a redox reaction, electrons must be transferred between reacting species. In other words, oxidation numbers must change in a redox reation. In both O_2 (molecular oxygen) and O_3 (ozone), the oxidation number of oxygen is zero. This is *not* a redox reaction.

4.142 Using the rules for assigning oxidation numbers given in Section 4.4, H is +1, F is −1, so the oxidation number of O must be **zero**.

4.143 (a)

$$OH^- + H_3O^+ \longrightarrow H_2O + H_2O$$

(b)

$$NH_4^+ + NH_2^- \longrightarrow NH_3 + NH_3$$

4.144 The balanced equation is:

$$3CH_3CH_2OH + 2K_2Cr_2O_7 + 8H_2SO_4 \longrightarrow 3CH_3COOH + 2Cr_2(SO_4)_3 + 2K_2SO_4 + 11H_2O$$

From the amount of $K_2Cr_2O_7$ required to react with the blood sample, we can calculate the mass of ethanol (CH_3CH_2OH) in the 10.0 g sample of blood.

First, calculate the moles of $K_2Cr_2O_7$ reacted.

$$\frac{0.07654 \text{ mol } K_2Cr_2O_7}{1000 \text{ mL soln}} \times 4.23 \text{ mL} = 3.24 \times 10^{-4} \text{ mol } K_2Cr_2O_7$$

Next, using the mole ratio from the balanced equation, we can calculate the mass of ethanol that reacted.

$$3.24 \times 10^{-4} \text{ mol } K_2Cr_2O_7 \times \frac{3 \text{ mol ethanol}}{2 \text{ mol } K_2Cr_2O_7} \times \frac{46.07 \text{ g ethanol}}{1 \text{ mol ethanol}} = 0.0224 \text{ g ethanol}$$

The percent ethanol by mass is:

$$\textbf{\% by mass ethanol} = \frac{0.0224 \text{ g}}{10.0 \text{ g}} \times 100\% = \textbf{0.224\%}$$

This is well above the legal limit of 0.1 percent by mass ethanol in the blood. The individual should be prosecuted for drunk driving.

4.145 Notice that nitrogen is in its highest possible oxidation state (+5) in nitric acid. It is reduced as it decomposes to NO_2.

$$4HNO_3 \longrightarrow 4NO_2 + O_2 + 2H_2O$$

The yellow color of "old" nitric acid is caused by the production of small amounts of NO_2 which is a brown gas. This process is accelerated by light.

4.146 (a) $Zn(s) + H_2SO_4(aq) \longrightarrow ZnSO_4(aq) + H_2(g)$

(b) $2KClO_3(s) \longrightarrow 2KCl(s) + 3O_2(g)$

(c) $Na_2CO_3(s) + 2HCl(aq) \longrightarrow 2NaCl(aq) + CO_2(g) + H_2O(l)$

(d) $NH_4NO_2(s) \xrightarrow{\text{heat}} N_2(g) + 2H_2O(g)$

4.147 **(a)** The precipitate $CaSO_4$ formed over Ca preventing the Ca from reacting with the sulfuric acid.

(b) Aluminum is protected by a tenacious oxide layer with the composition Al_2O_3.

(c) These metals react more readily with water.

$$2Na(s) + 2H_2O(l) \longrightarrow 2NaOH(aq) + H_2(g)$$

(d) The metal should be placed below Fe and above H.

(e) Any metal above Al in the activity series will react with Al^{3+}. Metals from Mg to Li will work.

4.148 Because the volume of the solution changes (increases or decreases) when the solid dissolves.

4.149 Place the following metals in the correct positions on the periodic table framework provided in the problem.

(a) Li, Na   **(b)** Mg, Fe   **(c)** Zn, Cd

Two metals that do not react with water or acid are Ag and Au.

4.150 NH_4Cl exists as NH_4^+ and Cl^-. To form NH_3 and HCl, a proton (H^+) is transferred from NH_4^+ to Cl^-. Therefore, this is a Brønsted acid-base reaction.

4.151 **(a)**
First Solution:

$$0.8214 \text{ g KMnO}_4 \times \frac{1 \text{ mol KMnO}_4}{158.0 \text{ g KMnO}_4} = 5.199 \times 10^{-3} \text{ mol KMnO}_4$$

$$M = \frac{\text{mol solute}}{\text{L of soln}} = \frac{5.199 \times 10^{-3} \text{ mol KMnO}_4}{0.5000 \text{ L}} = 1.040 \times 10^{-2} M$$

Second Solution:

$$M_1 V_1 = M_2 V_2$$
$$(1.040 \times 10^{-2} M)(2.000 \text{ mL}) = M_2(1000 \text{ mL})$$
$$M_2 = 2.080 \times 10^{-5} M$$

Third Solution:

$$M_1 V_1 = M_2 V_2$$
$$(2.080 \times 10^{-5} M)(10.00 \text{ mL}) = M_2(250.0 \text{ mL})$$
$$\mathbf{M_2 = 8.320 \times 10^{-7} M}$$

(b) From the molarity and volume of the final solution, we can calculate the moles of $KMnO_4$. Then, the mass can be calculated from the moles of $KMnO_4$.

$$\frac{8.320 \times 10^{-7} \text{ mol KMnO}_4}{1000 \text{ mL of soln}} \times 250 \text{ mL} = 2.080 \times 10^{-7} \text{ mol KMnO}_4$$

$$2.080 \times 10^{-7} \text{ mol KMnO}_4 \times \frac{158.0 \text{ g KMnO}_4}{1 \text{ mol KMnO}_4} = \mathbf{3.286 \times 10^{-5} \text{ g KMnO}_4}$$

This mass is too small to directly weigh accurately.

CHAPTER 5
GASES

5.13 $562 \text{ mmHg} \times \dfrac{1 \text{ atm}}{760 \text{ mmHg}} = \textbf{0.739 atm}$

5.14   **Strategy:** Because 1 atm = 760 mmHg, the following conversion factor is needed to obtain the pressure in atmospheres.

$$\dfrac{1 \text{ atm}}{760 \text{ mmHg}}$$

For the second conversion, 1 atm = 101.325 kPa.

Solution:

$$? \textbf{ atm } = 606 \text{ mmHg} \times \dfrac{1 \text{ atm}}{760 \text{ mmHg}} = \textbf{0.797 atm}$$

$$? \textbf{ kPa } = 0.797 \text{ atm} \times \dfrac{101.325 \text{ kPa}}{1 \text{ atm}} = \textbf{80.8 kPa}$$

5.17   **(a)**   If the final temperature of the sample is above the boiling point, it would still be in the gas phase. The diagram that best represents this is choice **(d)**.

   (b)   If the final temperature of the sample is below its boiling point, it will condense to a liquid. The liquid will have a vapor pressure, so some of the sample will remain in the gas phase. The diagram that best represents this is choice **(b)**.

5.18   **(1)**   Recall that $V \propto \dfrac{1}{P}$. As the pressure is tripled, the volume will decrease to $\frac{1}{3}$ of its original volume, assuming constant n and T. The correct choice is **(b)**.

   (2)   Recall that $V \propto T$. As the temperature is doubled, the volume will also double, assuming constant n and P. The correct choice is **(a)**. The depth of color indicates the density of the gas. As the volume increases at constant moles of gas, the density of the gas will decrease. This decrease in gas density is indicated by the lighter shading.

   (3)   Recall that $V \propto n$. Starting with n moles of gas, adding another n moles of gas ($2n$ total) will double the volume. The correct choice is **(c)**. The density of the gas will remain the same as moles are doubled and volume is doubled.

   (4)   Recall that $V \propto T$ and $V \propto \dfrac{1}{P}$. Halving the temperature would decrease the volume to $\frac{1}{2}$ its original volume. However, reducing the pressure to $\frac{1}{4}$ its original value would increase the volume by a factor of 4. Combining the two changes, we have

$$\dfrac{1}{2} \times 4 = 2$$

The volume will double. The correct choice is **(a)**.

5.19 $P_1 = 0.970$ atm $P_2 = 0.541$ atm

$V_1 = 725$ mL $V_2 = ?$

$$P_1V_1 = P_2V_2$$

$$V_2 = \frac{P_1V_1}{P_2} = \frac{(0.970 \text{ atm})(725 \text{ mL})}{0.541 \text{ atm}} = \mathbf{1.30 \times 10^3 \text{ mL}}$$

5.20   Temperature and amount of gas do not change in this problem ($T_1 = T_2$ and $n_1 = n_2$). Pressure and volume change; it is a Boyle's law problem.

$$\frac{P_1V_1}{n_1T_1} = \frac{P_2V_2}{n_2T_2}$$

$$P_1V_1 = P_2V_2$$

$V_2 = 0.10 \, V_1$

$$P_2 = \frac{P_1V_1}{V_2}$$

$$P_2 = \frac{(5.3 \text{ atm})V_1}{0.10V_1} = \mathbf{53 \text{ atm}}$$

5.21 $P_1 = 1.00$ atm $= 760$ mmHg $P_2 = ?$

$V_1 = 5.80$ L $V_2 = 9.65$ L

$$P_1V_1 = P_2V_2$$

$$P_2 = \frac{P_1V_1}{V_2} = \frac{(760 \text{ mmHg})(5.80 \text{ L})}{9.65 \text{ L}} = \mathbf{457 \text{ mmHg}}$$

5.22   **(a)**

Strategy: The amount of gas and its temperature remain constant, but both the pressure and the volume change. What equation would you use to solve for the final volume?

Solution: We start with Equation (5.9) of the text.

$$\frac{P_1V_1}{n_1T_1} = \frac{P_2V_2}{n_2T_2}$$

Because $n_1 = n_2$ and $T_1 = T_2$,

$$P_1V_1 = P_2V_2$$

which is Boyle's Law. The given information is tabulated below.

Initial conditions	Final Conditions
$P_1 = 1.2$ atm	$P_2 = 6.6$ atm
$V_1 = 3.8$ L	$V_2 = ?$

The final volume is given by:

$$V_2 = \frac{P_1 V_1}{P_2}$$

$$V_2 = \frac{(1.2 \text{ atm})(3.8 \text{ L})}{(6.6 \text{ atm})} = \textbf{0.69 L}$$

Check: When the pressure applied to the sample of air is increased from 1.2 atm to 6.6 atm, the volume occupied by the sample will decrease. Pressure and volume are inversely proportional. The final volume calculated is less than the initial volume, so the answer seems reasonable.

(b)
Strategy: The amount of gas and its temperature remain constant, but both the pressure and the volume change. What equation would you use to solve for the final pressure?

Solution: You should also come up with the equation $P_1 V_1 = P_2 V_2$ for this problem. The given information is tabulated below.

Initial conditions	Final Conditions
$P_1 = 1.2$ atm	$P_2 = ?$
$V_1 = 3.8$ L	$V_2 = 0.075$ L

The final pressure is given by:

$$P_2 = \frac{P_1 V_1}{V_2}$$

$$P_2 = \frac{(1.2 \text{ atm})(3.8 \text{ L})}{(0.075 \text{ L})} = \textbf{61 atm}$$

Check: To decrease the volume of the gas fairly dramatically from 3.8 L to 0.075 L, the pressure must be increased substantially. A final pressure of 61 atm seems reasonable.

5.23 $T_1 = 25° + 273° = 298$ K $T_2 = 88° + 273° = 361$ K
    $V_1 = 36.4$ L $V_2 = ?$

$$\frac{V_1}{T_1} = \frac{V_2}{T_2}$$

$$V_2 = \frac{V_1 T_2}{T_1} = \frac{(36.4 \text{ L})(361 \text{ K})}{298 \text{ K}} = \textbf{44.1 L}$$

5.24   **Strategy:** The amount of gas and its pressure remain constant, but both the temperature and the volume change. What equation would you use to solve for the final temperature? What temperature unit should we use?

Solution: We start with Equation (5.9) of the text.

$$\frac{P_1 V_1}{n_1 T_1} = \frac{P_2 V_2}{n_2 T_2}$$

Because $n_1 = n_2$ and $P_1 = P_2$,

$$\frac{V_1}{T_1} = \frac{V_2}{T_2}$$

which is Charles' Law. The given information is tabulated below.

Initial conditions	Final Conditions
$T_1 = (88 + 273)K = 361$ K	$T_2 = ?$
$V_1 = 9.6$ L	$V_2 = 3.4$ L

The final temperature is given by:

$$T_2 = \frac{T_1 V_2}{V_1}$$

$$T_2 = \frac{(361 \text{ K})(3.4 \text{ L})}{(9.6 \text{ L})} = \mathbf{1.3 \times 10^2 \text{ K}}$$

5.25   The balanced equation is: $4NH_3(g) + 5O_2(g) \longrightarrow 4NO(g) + 6H_2O(g)$

Recall that Avogadro's Law states that the volume of a gas is directly proportional to the number of moles of gas at constant temperature and pressure. The ammonia and nitric oxide coefficients in the balanced equation are the same, so **one volume** of nitric oxide must be obtained from **one volume** of ammonia.

Could you have reached the same conclusion if you had noticed that nitric oxide is the only nitrogen-containing product and that ammonia is the only nitrogen-containing reactant?

5.26   This is a gas stoichiometry problem that requires knowledge of Avogadro's law to solve. Avogadro's law states that the volume of a gas is directly proportional to the number of moles of gas at constant temperature and pressure.

The volume ratio, 1 vol. Cl_2 : 3 vol. F_2 : 2 vol. product, can be written as a mole ratio, 1 mol Cl_2 : 3 mol F_2 : 2 mol product.

Attempt to write a balanced chemical equation. The subscript of F in the product will be three times the Cl subscript, because there are three times as many F atoms reacted as Cl atoms.

$$1Cl_2(g) + 3F_2(g) \longrightarrow 2Cl_xF_{3x}(g)$$

Balance the equation. The x must equal one so that there are two Cl atoms on each side of the equation. If $x = 1$, the subscript on F is 3.

$$Cl_2(g) + 3F_2(g) \longrightarrow 2ClF_3(g)$$

The formula of the product is **ClF₃**.

5.31 $n = \dfrac{PV}{RT} = \dfrac{(4.7 \text{ atm})(2.3 \text{ L})}{\left(0.0821 \dfrac{\text{L} \cdot \text{atm}}{\text{mol} \cdot \text{K}}\right)(273 + 32)\text{K}} = \mathbf{0.43 \text{ mol}}$

5.32 **Strategy:** This problem gives the amount, volume, and temperature of CO gas. Is the gas undergoing a change in any of its properties? What equation should we use to solve for the pressure? What temperature unit should be used?

Solution: Because no changes in gas properties occur, we can use the ideal gas equation to calculate the pressure. Rearranging Equation (5.8) of the text, we write:

$$P = \frac{nRT}{V}$$

$$P = \frac{(6.9 \text{ mol})\left(0.0821\dfrac{\text{L} \cdot \text{atm}}{\text{mol} \cdot \text{K}}\right)(62 + 273)\,\text{K}}{30.4 \text{ L}} = \textbf{6.2 atm}$$

5.33 We solve the ideal gas equation for V.

$$V = \frac{nRT}{P} = \frac{(5.6 \text{ mol})\left(0.0821\dfrac{\text{L} \cdot \text{atm}}{\text{mol} \cdot \text{K}}\right)(128 + 273)\text{K}}{9.4 \text{ atm}} = \textbf{2.0} \times \textbf{10}^{\textbf{1}}\textbf{ L}$$

5.34 In this problem, the moles of gas and the volume the gas occupies are constant ($V_1 = V_2$ and $n_1 = n_2$). Temperature and pressure change.

$$\frac{P_1 V_1}{n_1 T_1} = \frac{P_2 V_2}{n_2 T_2}$$

$$\frac{P_1}{T_1} = \frac{P_2}{T_2}$$

The given information is tabulated below.

Initial conditions	Final Conditions
$T_1 = (25 + 273)\text{K} = 298\text{ K}$	$T_2 = ?$
$P_1 = 0.800$ atm	$P_2 = 2.00$ atm

The final temperature is given by:

$$T_2 = \frac{T_1 P_2}{P_1}$$

$$T_2 = \frac{(298 \text{ K})(2.00 \text{ atm})}{(0.800 \text{ atm})} = \textbf{745 K} = \textbf{472°C}$$

5.35

Initial Conditions	Final Conditions
$P_1 = 1.2$ atm	$P_2 = 3.00 \times 10^{-3}$ atm
$V_1 = 2.50$ L	$V_2 = ?$
$T_1 = (25 + 273)\text{K} = 298\text{ K}$	$T_2 = (-23 + 273)\text{K} = 250\text{ K}$

$$\frac{P_1 V_1}{T_1} = \frac{P_2 V_2}{T_2}$$

$$V_2 = \frac{P_1 V_1 T_2}{T_1 P_2} = \frac{(1.2 \text{ atm})(2.50 \text{ L})(250 \text{ K})}{(298 \text{ K})(3.00 \times 10^{-3} \text{ atm})} = 8.4 \times 10^2 \text{ L}$$

5.36   In this problem, the moles of gas and the volume the gas occupies are constant ($V_1 = V_2$ and $n_1 = n_2$). Temperature and pressure change.

$$\frac{P_1 V_1}{n_1 T_1} = \frac{P_2 V_2}{n_2 T_2}$$

$$\frac{P_1}{T_1} = \frac{P_2}{T_2}$$

The given information is tabulated below.

Initial conditions	Final Conditions
$T_1 = 273$ K	$T_2 = (250 + 273)$K $= 523$ K
$P_1 = 1.0$ atm	$P_2 = ?$

The final pressure is given by:

$$P_2 = \frac{P_1 T_2}{T_1}$$

$$P_2 = \frac{(1.0 \text{ atm})(523 \text{ K})}{273 \text{ K}} = 1.9 \text{ atm}$$

5.37   Note that the statement "...its absolute temperature is decreased by one-half" implies that $\frac{T_2}{T_1} = 0.50$.

Similarly, the statement "...pressure is decreased to one-third of its original pressure" indicates $\frac{P_2}{P_1} = \frac{1}{3}$.

$$\frac{P_1 V_1}{T_1} = \frac{P_2 V_2}{T_2}$$

$$V_2 = V_1 \left(\frac{P_1}{P_2}\right)\left(\frac{T_2}{T_1}\right) = 6.0 \text{ L} \times 3.0 \times 0.50 = 9.0 \text{ L}$$

5.38   In this problem, the moles of gas and the pressure on the gas are constant ($n_1 = n_2$ and $P_1 = P_2$). Temperature and volume are changing.

$$\frac{P_1 V_1}{n_1 T_1} = \frac{P_2 V_2}{n_2 T_2}$$

$$\frac{V_1}{T_1} = \frac{V_2}{T_2}$$

The given information is tabulated below.

Initial conditions	Final Conditions
$T_1 = (20.1 + 273)$ K $= 293.1$ K	$T_2 = (36.5 + 273)$K $= 309.5$ K
$V_1 = 0.78$ L	$V_2 = ?$

The final volume is given by:

$$V_2 = \frac{V_1 T_2}{T_1}$$

$$V_2 = \frac{(0.78 \text{ L})(309.5 \text{ K})}{(293.1 \text{ K})} = \textbf{0.82 L}$$

5.39 $\dfrac{P_1 V_1}{T_1} = \dfrac{P_2 V_2}{T_2}$

$$V_1 = \frac{P_2 V_2 T_1}{P_1 T_2} = \frac{(0.60 \text{ atm})(94 \text{ mL})(66 + 273)\text{K}}{(0.85 \text{ atm})(45 + 273)\text{K}} = \textbf{71 mL}$$

5.40 In the problem, temperature and pressure are given. If we can determine the moles of CO_2, we can calculate the volume it occupies using the ideal gas equation.

$$? \text{ mol } CO_2 = 88.4 \text{ g } CO_2 \times \frac{1 \text{ mol } CO_2}{44.01 \text{ g } CO_2} = 2.01 \text{ mol } CO_2$$

We now substitute into the ideal gas equation to calculate volume of CO_2.

$$V_{CO_2} = \frac{nRT}{P} = \frac{(2.01 \text{ mol})\left(0.0821\dfrac{\text{L} \cdot \text{atm}}{\text{mol} \cdot \text{K}}\right)(273 \text{ K})}{(1 \text{ atm})} = \textbf{45.1 L}$$

Alternatively, we could use the fact that 1 mole of an ideal gas occupies a volume of 22.41 L at STP. After calculating the moles of CO_2, we can use this fact as a conversion factor to convert to volume of CO_2.

$$? \text{ L } CO_2 = 2.01 \text{ mol } CO_2 \times \frac{22.41 \text{ L}}{1 \text{ mol}} = \textbf{45.0 L } CO_2$$

The slight difference in the results of our two calculations is due to rounding the volume occupied by 1 mole of an ideal gas to 22.41 L.

5.41 $\dfrac{P_1 V_1}{T_1} = \dfrac{P_2 V_2}{T_2}$

$$V_2 = \frac{P_1 V_1 T_2}{P_2 T_1} = \frac{(772 \text{ mmHg})(6.85 \text{ L})(273 \text{ K})}{(760 \text{ mmHg})(35 + 273)\text{K}} = \textbf{6.17 L}$$

5.42 The molar mass of $CO_2 = 44.01$ g/mol. Since $PV = nRT$, we write:

$$P = \frac{nRT}{V}$$

$$P = \frac{\left(0.050 \text{ g} \times \dfrac{1 \text{ mol}}{44.01 \text{ g}}\right)\left(0.0821\dfrac{\text{L} \cdot \text{atm}}{\text{mol} \cdot \text{K}}\right)(30 + 273)\text{K}}{4.6 \text{ L}} = \textbf{6.1} \times \textbf{10}^{-3} \textbf{ atm}$$

5.43   Solve for the number of moles of gas using the ideal gas equation.

$$n = \frac{PV}{RT} = \frac{(1.00 \text{ atm})(0.280 \text{ L})}{\left(0.0821 \dfrac{\text{L} \cdot \text{atm}}{\text{mol} \cdot \text{K}}\right)(273 \text{ K})} = 0.0125 \text{ mol}$$

Solving for the molar mass:

$$\mathcal{M} = \frac{\text{mass (in g)}}{\text{mol}} = \frac{0.400 \text{ g}}{0.0125 \text{ mol}} = \textbf{32.0 g/mol}$$

5.44   **Strategy:** We can calculate the molar mass of a gas if we know its density, temperature, and pressure. What temperature and pressure units should we use?

Solution: We need to use Equation (5.11) of the text to calculate the molar mass of the gas.

$$\mathcal{M} = \frac{dRT}{P}$$

Before substituting into the above equation, we need to calculate the density and check that the other known quantities (P and T) have the appropriate units.

$$d = \frac{7.10 \text{ g}}{5.40 \text{ L}} = 1.31 \text{ g/L}$$

$$T = 44° + 273° = 317 \text{ K}$$

$$P = 741 \text{ torr} \times \frac{1 \text{ atm}}{760 \text{ torr}} = 0.975 \text{ atm}$$

Calculate the molar mass by substituting in the known quantities.

$$\mathcal{M} = \frac{\left(1.31 \dfrac{\text{g}}{\text{L}}\right)\left(0.0821 \dfrac{\text{L} \cdot \text{atm}}{\text{mol} \cdot \text{K}}\right)(317 \text{ K})}{0.975 \text{ atm}} = \textbf{35.0 g/mol}$$

Alternatively, we can solve for the molar mass by writing:

$$\text{molar mass of compound} = \frac{\text{mass of compound}}{\text{moles of compound}}$$

Mass of compound is given in the problem (7.10 g), so we need to solve for moles of compound in order to calculate the molar mass.

$$n = \frac{PV}{RT}$$

$$n = \frac{(0.975 \text{ atm})(5.40 \text{ L})}{\left(0.0821 \dfrac{\text{L} \cdot \text{atm}}{\text{mol} \cdot \text{K}}\right)(317 \text{ K})} = 0.202 \text{ mol}$$

Now, we can calculate the molar mass of the gas.

$$\text{molar mass of compound} = \frac{\text{mass of compound}}{\text{moles of compound}} = \frac{7.10 \text{ g}}{0.202 \text{ mol}} = \textbf{35.1 g/mol}$$

5.45 First calculate the moles of ozone (O_3) using the ideal gas equation.

$$n = \frac{PV}{RT} = \frac{(1.0 \times 10^{-3} \text{ atm})(1.0 \text{ L})}{\left(0.0821 \dfrac{\text{L} \cdot \text{atm}}{\text{mol} \cdot \text{K}}\right)(250 \text{ K})} = 4.9 \times 10^{-5} \text{ mol } O_3$$

Use Avogadro's number as a conversion factor to convert to molecules of O_3.

$$\text{molecules } \mathbf{O_3} = (4.9 \times 10^{-5} \text{ mol } O_3) \times \frac{6.022 \times 10^{23} \text{ } O_3 \text{ molecules}}{1 \text{ mol } O_3} = \mathbf{3.0 \times 10^{19} \text{ } O_3 \text{ molecules}}$$

5.46 The number of particles in 1 L of gas at STP is:

$$\text{Number of particles} = 1.0 \text{ L} \times \frac{1 \text{ mol}}{22.414 \text{ L}} \times \frac{6.022 \times 10^{23} \text{ particles}}{1 \text{ mol}} = 2.7 \times 10^{22} \text{ particles}$$

$$\text{Number of } \mathbf{N_2} \text{ molecules} = \left(\frac{78\%}{100\%}\right)(2.7 \times 10^{22} \text{ particles}) = \mathbf{2.1 \times 10^{22} \text{ } N_2 \text{ molecules}}$$

$$\text{Number of } \mathbf{O_2} \text{ molecules} = \left(\frac{21\%}{100\%}\right)(2.7 \times 10^{22} \text{ particles}) = \mathbf{5.7 \times 10^{21} \text{ } O_2 \text{ molecules}}$$

$$\text{Number of } \mathbf{Ar \ atoms} = \left(\frac{1\%}{100\%}\right)(2.7 \times 10^{22} \text{ particles}) = \mathbf{3 \times 10^{20} \text{ Ar atoms}}$$

5.47 The density is given by:

$$\text{density} = \frac{\text{mass}}{\text{volume}} = \frac{4.65 \text{ g}}{2.10 \text{ L}} = \textbf{2.21 g/L}$$

Solving for the molar mass:

$$\text{molar mass} = \frac{dRT}{P} = \frac{(2.21 \text{ g/L})\left(0.0821 \dfrac{\text{L} \cdot \text{atm}}{\text{mol} \cdot \text{K}}\right)(27 + 273)\text{K}}{(1.00 \text{ atm})} = \textbf{54.4 g/mol}$$

5.48 The density can be calculated from the ideal gas equation.

$$d = \frac{P\mathcal{M}}{RT}$$

$\mathcal{M} = 1.008 \text{ g/mol} + 79.90 \text{ g/mol} = 80.91 \text{ g/mol}$

$T = 46° + 273° = 319 \text{ K}$

$$P = 733 \text{ mmHg} \times \frac{1 \text{ atm}}{760 \text{ mmHg}} = 0.964 \text{ atm}$$

$$d = \frac{(0.964 \text{ atm})\left(\dfrac{80.91 \text{ g}}{1 \text{ mol}}\right)}{319 \text{ K}} \times \frac{\text{mol} \cdot \text{K}}{0.0821 \text{ L} \cdot \text{atm}} = 2.98 \text{ g/L}$$

Alternatively, we can solve for the density by writing:

$$\text{density} = \frac{\text{mass}}{\text{volume}}$$

Assuming that we have 1 mole of HBr, the mass is 80.91 g. The volume of the gas can be calculated using the ideal gas equation.

$$V = \frac{nRT}{P}$$

$$V = \frac{(1 \text{ mol})\left(0.0821 \dfrac{\text{L} \cdot \text{atm}}{\text{mol} \cdot \text{K}}\right)(319 \text{ K})}{0.964 \text{ atm}} = 27.2 \text{ L}$$

Now, we can calculate the density of HBr gas.

$$\text{density} = \frac{\text{mass}}{\text{volume}} = \frac{80.91 \text{ g}}{27.2 \text{ L}} = 2.97 \text{ g/L}$$

5.49   METHOD 1:

The empirical formula can be calculated from mass percent data. The molar mass can be calculated using the ideal gas equation. The molecular formula can then be determined.

To calculate the empirical formula, assume 100 g of substance.

$$64.9 \text{ g C} \times \frac{1 \text{ mol C}}{12.01 \text{ g C}} = 5.40 \text{ mol C}$$

$$13.5 \text{ g H} \times \frac{1 \text{ mol H}}{1.008 \text{ g H}} = 13.4 \text{ mol H}$$

$$21.6 \text{ g O} \times \frac{1 \text{ mol O}}{16.00 \text{ g O}} = 1.35 \text{ mol O}$$

This gives the formula $C_{5.40}H_{13.4}O_{1.35}$. Dividing by 1.35 gives the empirical formula, $C_4H_{10}O$.

To calculate the molar mass, first calculate the number of moles of gas using the ideal gas equation.

$$n = \frac{PV}{RT} = \frac{\left(750 \text{ mmHg} \times \dfrac{1 \text{ atm}}{760 \text{ mmHg}}\right)(1.00 \text{ L})}{\left(0.0821 \dfrac{\text{L} \cdot \text{atm}}{\text{mol} \cdot \text{K}}\right)(120 + 273)\text{K}} = 0.0306 \text{ mol}$$

Solving for the molar mass:

$$\mathcal{M} = \frac{\text{mass (in g)}}{\text{mol}} = \frac{2.30 \text{ g}}{0.0306 \text{ mol}} = 75.2 \text{ g/mol}$$

The empirical mass is 74.0 g/mol which is essentially the same as the molar mass. In this case, the molecular formula is the same as the empirical formula, $C_4H_{10}O$.

METHOD 2:

First calculate the molar mass using the ideal gas equation.

$$n = \frac{PV}{RT} = \frac{\left(750 \text{ mmHg} \times \dfrac{1 \text{ atm}}{760 \text{ mmHg}}\right)(1.00 \text{ L})}{\left(0.0821 \dfrac{\text{L} \cdot \text{atm}}{\text{mol} \cdot \text{K}}\right)(120 + 273)\text{K}} = 0.0306 \text{ mol}$$

Solving for the molar mass:

$$\mathcal{M} = \frac{\text{mass (in g)}}{\text{mol}} = \frac{2.30 \text{ g}}{0.0306 \text{ mol}} = \textbf{75.2 g/mol}$$

Next, multiply the mass % (converted to a decimal) of each element by the molar mass to convert to grams of each element. Then, use the molar mass to convert to moles of each element.

$$n_C = (0.649) \times (75.2 \text{ g}) \times \frac{1 \text{ mol C}}{12.01 \text{ g C}} = \textbf{4.06 mol C}$$

$$n_H = (0.135) \times (75.2 \text{ g}) \times \frac{1 \text{ mol H}}{1.008 \text{ g H}} = \textbf{10.07 mol H}$$

$$n_O = (0.216) \times (75.2 \text{ g}) \times \frac{1 \text{ mol O}}{16.00 \text{ g O}} = \textbf{1.02 mol O}$$

Since we used the molar mass to calculate the moles of each element present in the compound, this method directly gives the molecular formula. The formula is $C_4H_{10}O$.

5.50   This is an extension of an ideal gas law calculation involving molar mass. If you determine the molar mass of the gas, you will be able to determine the molecular formula from the empirical formula.

$$\mathcal{M} = \frac{dRT}{P}$$

Calculate the density, then substitute its value into the equation above.

$$d = \frac{0.100 \text{ g}}{22.1 \text{ mL}} \times \frac{1000 \text{ mL}}{1 \text{ L}} = 4.52 \text{ g/L}$$

$T(\text{K}) = 20° + 273° = 293 \text{ K}$

$$\mathcal{M} = \frac{\left(4.52 \dfrac{\text{g}}{\text{L}}\right)\left(0.0821 \dfrac{\text{L} \cdot \text{atm}}{\text{mol} \cdot \text{K}}\right)(293 \text{ K})}{1.02 \text{ atm}} = 107 \text{ g/mol}$$

Compare the empirical mass to the molar mass.

empirical mass = 32.07 g/mol + 4(19.00 g/mol) = 108.07 g/mol

Remember, the molar mass will be a whole number multiple of the empirical mass. In this case, the $\dfrac{\text{molar mass}}{\text{empirical mass}} \approx 1$. Therefore, the molecular formula is the same as the empirical formula, **SF_4**.

5.51   In addition to a mole ratio, the coefficients from a balanced equation can represent the volume ratio in which the gases in the equation react and are produced. Recall that Avogadro's Law states that $V \propto n$. See Figure 5.10 of the text. We can use this volume ratio to convert from liters of NO to liters of NO_2.

$$9.0 \text{ L NO} \times \frac{2 \text{ volumes } NO_2}{2 \text{ volumes NO}} = \textbf{9.0 L } \textbf{NO}_2$$

5.52   **Strategy:** From the moles of CH_4 reacted, we can calculate the moles of CO_2 produced. From the balanced equation, we see that 1 mol $CH_4 \simeq$ 1 mol CO_2. Once moles of CO_2 are determined, we can use the ideal gas equation to calculate the volume of CO_2.

Solution: First let's calculate moles of CO_2 produced.

$$? \text{ mol } CO_2 = 15.0 \text{ mol } CH_4 \times \frac{1 \text{ mol } CO_2}{1 \text{ mol } CH_4} = 15.0 \text{ mol } CO_2$$

Now, we can substitute moles, temperature, and pressure into the ideal gas equation to solve for volume of CO_2.

$$V = \frac{nRT}{P}$$

$$V_{CO_2} = \frac{(15.0 \text{ mol})\left(0.0821\dfrac{\text{L} \cdot \text{atm}}{\text{mol} \cdot \text{K}}\right)(23 + 273)\text{K}}{0.985 \text{ atm}} = \textbf{3.70} \times \textbf{10}^2 \textbf{ L}$$

5.53   If we can calculate the moles of S, we can use the mole ratio from the balanced equation to calculate the moles of SO_2. Once we know the moles of SO_2, we can determine the volume of SO_2 using the ideal gas equation.

$$(2.54 \times 10^3 \text{ g S}) \times \frac{1 \text{ mol S}}{32.07 \text{ g S}} \times \frac{1 \text{ mol } SO_2}{1 \text{ mol S}} = 79.2 \text{ mol } SO_2$$

$$V = \frac{nRT}{P} = \frac{(79.2 \text{ mol})\left(0.0821\dfrac{\text{L} \cdot \text{atm}}{\text{mol} \cdot \text{K}}\right)(303.5 \text{ K})}{1.12 \text{ atm}} = 1.76 \times 10^3 \text{ L} = \textbf{1.76} \times \textbf{10}^6 \textbf{ mL } \textbf{SO}_2$$

5.54   From the amount of glucose reacted (5.97 g), we can calculate the theoretical yield of CO_2. We can then compare the theoretical yield to the actual yield given in the problem (1.44 L) to determine the percent yield.

First, let's determine the moles of CO_2 that can be produced theoretically. Then, we can use the ideal gas equation to determine the volume of CO_2.

$$? \text{ mol } CO_2 = 5.97 \text{ g glucose} \times \frac{1 \text{ mol glucose}}{180.2 \text{ g glucose}} \times \frac{2 \text{ mol } CO_2}{1 \text{ mol glucose}} = 0.0663 \text{ mol } CO_2$$

Now, substitute moles, pressure, and temperature into the ideal gas equation to calculate the volume of CO_2.

$$V = \frac{nRT}{P}$$

$$V_{CO_2} = \frac{(0.0663 \text{ mol})\left(0.0821\dfrac{\text{L}\cdot\text{atm}}{\text{mol}\cdot\text{K}}\right)(293 \text{ K})}{0.984 \text{ atm}} = 1.62 \text{ L}$$

This is the theoretical yield of CO_2. The actual yield, which is given in the problem, is 1.44 L. We can now calculate the percent yield.

$$\text{percent yield} = \frac{\text{actual yield}}{\text{theoretical yield}} \times 100\%$$

$$\textbf{percent yield} = \frac{1.44 \text{ L}}{1.62 \text{ L}} \times 100\% = \textbf{88.9\%}$$

5.55 If you determine the molar mass of the gas, you will be able to determine the molecular formula from the empirical formula. First, let's calculate the molar mass of the compound.

$$n = \frac{PV}{RT} = \frac{\left(97.3 \text{ mmHg} \times \dfrac{1 \text{ atm}}{760 \text{ mmHg}}\right)(0.378 \text{ L})}{\left(0.0821 \dfrac{\text{L}\cdot\text{atm}}{\text{mol}\cdot\text{K}}\right)(77 + 273)\text{K}} = 0.00168 \text{ mol}$$

Solving for the molar mass:

$$\mathcal{M} = \frac{\text{mass (in g)}}{\text{mol}} = \frac{0.2324 \text{ g}}{0.00168 \text{ mol}} = \textbf{138 g/mol}$$

To calculate the empirical formula, first we need to find the mass of F in 0.2631 g of CaF_2.

$$0.2631 \text{ g CaF}_2 \times \frac{1 \text{ mol CaF}_2}{78.08 \text{ g CaF}_2} \times \frac{2 \text{ mol F}}{1 \text{ mol CaF}_2} \times \frac{19.00 \text{ g F}}{1 \text{ mol F}} = 0.1280 \text{ g F}$$

Since the compound only contains P and F, the mass of P in the 0.2324 g sample is:

$$0.2324 \text{ g} - 0.1280 \text{ g} = 0.1044 \text{ g P}$$

Now, we can convert masses of P and F to moles of each substance.

$$? \text{ mol P} = 0.1044 \text{ g P} \times \frac{1 \text{ mol P}}{30.97 \text{ g P}} = 0.003371 \text{ mol P}$$

$$? \text{ mol F} = 0.1280 \text{ g F} \times \frac{1 \text{ mol F}}{19.00 \text{ g F}} = 0.006737 \text{ mol F}$$

Thus, we arrive at the formula $P_{0.003371}F_{0.006737}$. Dividing by the smallest number of moles (0.003371 mole) gives the empirical formula PF_2.

To determine the molecular formula, divide the molar mass by the empirical mass.

$$\frac{\text{molar mass}}{\text{empirical molar mass}} = \frac{138 \text{ g}}{68.97 \text{ g}} \approx 2$$

Hence, the molecular formula is $(PF_2)_2$ or P_2F_4.

5.56   **Strategy:** We can calculate the moles of M reacted, and the moles of H_2 gas produced. By comparing the number of moles of M reacted to the number of moles H_2 produced, we can determine the mole ratio in the balanced equation.

Solution: First let's calculate the moles of the metal (M) reacted.

$$\text{mol M} = 0.225 \text{ g M} \times \frac{1 \text{ mol M}}{27.0 \text{ g M}} = 8.33 \times 10^{-3} \text{ mol M}$$

Solve the ideal gas equation algebraically for n_{H_2}. Then, calculate the moles of H_2 by substituting the known quantities into the equation.

$$P = 741 \text{ mmHg} \times \frac{1 \text{ atm}}{760 \text{ mmHg}} = 0.975 \text{ atm}$$

$$T = 17° + 273° = 290 \text{ K}$$

$$n_{H_2} = \frac{PV_{H_2}}{RT}$$

$$n_{H_2} = \frac{(0.975 \text{ atm})(0.303 \text{ L})}{\left(0.0821 \dfrac{\text{L} \cdot \text{atm}}{\text{mol} \cdot \text{K}}\right)(290 \text{ K})} = 1.24 \times 10^{-2} \text{ mol H}_2$$

Compare the number moles of H_2 produced to the number of moles of M reacted.

$$\frac{1.24 \times 10^{-2} \text{ mol H}_2}{8.33 \times 10^{-3} \text{ mol M}} \approx 1.5$$

This means that the mole ratio of H_2 to M is 1.5 : 1.

We can now write the balanced equation since we know the mole ratio between H_2 and M.

The unbalanced equation is:

$$M(s) + HCl(aq) \longrightarrow 1.5H_2(g) + M_xCl_y(aq)$$

We have 3 atoms of H on the products side of the reaction, so a 3 must be placed in front of HCl. The ratio of M to Cl on the reactants side is now 1 : 3. Therefore the formula of the metal chloride must be MCl_3. The balanced equation is:

$$M(s) + 3HCl(aq) \longrightarrow 1.5H_2(g) + MCl_3(aq)$$

From the formula of the metal chloride, we determine that the charge of the metal is +3. Therefore, the formula of the metal oxide and the metal sulfate are M_2O_3 and $M_2(SO_4)_3$, respectively.

5.57   The balanced equation for the reaction is: $NH_3(g) + HCl(g) \longrightarrow NH_4Cl(s)$

First, we must determine which of the two reactants is the limiting reagent. We find the number of moles of each reactant.

$$? \text{ mol } NH_3 = 73.0 \text{ g } NH_3 \times \frac{1 \text{ mol } NH_3}{17.03 \text{ g } NH_3} = 4.29 \text{ mol } NH_3$$

$$? \text{ mol } HCl = 73.0 \text{ g } HCl \times \frac{1 \text{ mol } HCl}{36.46 \text{ g } HCl} = 2.00 \text{ mol } HCl$$

Since NH_3 and HCl react in a 1:1 mole ratio, HCl is the limiting reagent. The mass of NH_4Cl formed is:

$$? \textbf{ g } \mathbf{NH_4Cl} = 2.00 \text{ mol } HCl \times \frac{1 \text{ mol } NH_4Cl}{1 \text{ mol } HCl} \times \frac{53.49 \text{ g } NH_4Cl}{1 \text{ mol } NH_4Cl} = \textbf{107 g } \mathbf{NH_4Cl}$$

The gas remaining is ammonia, NH_3. The number of moles of NH_3 remaining is $(4.29 - 2.00)$ mol $= 2.29$ mol NH_3. The volume of NH_3 gas is:

$$V_{NH_3} = \frac{n_{NH_3} RT}{P} = \frac{(2.29 \text{ mol})\left(0.0821 \dfrac{\text{L} \cdot \text{atm}}{\text{mol} \cdot \text{K}}\right)(14 + 273)\text{K}}{\left(752 \text{ mmHg} \times \dfrac{1 \text{ atm}}{760 \text{ mmHg}}\right)} = \textbf{54.5 L } \mathbf{NH_3}$$

5.58   From the moles of CO_2 produced, we can calculate the amount of calcium carbonate that must have reacted. We can then determine the percent by mass of $CaCO_3$ in the 3.00 g sample.

The balanced equation is:

$$CaCO_3(s) + 2HCl(aq) \longrightarrow CO_2(g) + CaCl_2(aq) + H_2O(l)$$

The moles of CO_2 produced can be calculated using the ideal gas equation.

$$n_{CO_2} = \frac{PV_{CO_2}}{RT}$$

$$n_{CO_2} = \frac{\left(792 \text{ mmHg} \times \dfrac{1 \text{ atm}}{760 \text{ mmHg}}\right)(0.656 \text{ L})}{\left(0.0821 \dfrac{\text{L} \cdot \text{atm}}{\text{mol} \cdot \text{K}}\right)(20 + 273 \text{ K})} = \textbf{2.84} \times \textbf{10}^{-2} \textbf{ mol } \mathbf{CO_2}$$

The balanced equation shows a 1:1 mole ratio between CO_2 and $CaCO_3$. Therefore, 2.84×10^{-2} mole of $CaCO_3$ must have reacted.

$$? \text{ g } CaCO_3 \text{ reacted} = (2.84 \times 10^{-2} \text{ mol } CaCO_3) \times \frac{100.1 \text{ g } CaCO_3}{1 \text{ mol } CaCO_3} = 2.84 \text{ g } CaCO_3$$

The percent by mass of the $CaCO_3$ sample is:

$$\% \text{ CaCO}_3 = \frac{2.84 \text{ g}}{3.00 \text{ g}} \times 100\% = \textbf{94.7\%}$$

Assumption: The impurity (or impurities) must not react with HCl to produce CO_2 gas.

5.59   The balanced equation is: $H_2(g) + Cl_2(g) \longrightarrow 2HCl(g)$

At STP, 1 mole of an ideal gas occupies a volume of 22.41 L. We can use this as a conversion factor to find the moles of H_2 reacted. Then, we can calculate the mass of HCl produced.

$$? \text{ mol } H_2 \text{ reacted } = 5.6 \text{ L } H_2 \times \frac{1 \text{ mol } H_2}{22.41 \text{ L } H_2} = 0.25 \text{ mol } H_2$$

The mass of HCl produced is:

$$? \text{ g HCl } = 0.25 \text{ mol } H_2 \times \frac{2 \text{ mol HCl}}{1 \text{ mol } H_2} \times \frac{36.46 \text{ g HCl}}{1 \text{ mol HCl}} = \textbf{18 g HCl}$$

5.60   The balanced equation is:

$$C_2H_5OH(l) + 3O_2(g) \longrightarrow 2CO_2(g) + 3H_2O(l)$$

The moles of O_2 needed to react with 227 g ethanol are:

$$227 \text{ g } C_2H_5OH \times \frac{1 \text{ mol } C_2H_5OH}{46.07 \text{ g } C_2H_5OH} \times \frac{3 \text{ mol } O_2}{1 \text{ mol } C_2H_5OH} = 14.8 \text{ mol } O_2$$

14.8 moles of O_2 correspond to a volume of:

$$V_{O_2} = \frac{n_{O_2}RT}{P} = \frac{(14.8 \text{ mol } O_2)\left(0.0821\dfrac{L \cdot atm}{mol \cdot K}\right)(35 + 273 \text{ K})}{\left(790 \text{ mmHg} \times \dfrac{1 \text{ atm}}{760 \text{ mmHg}}\right)} = 3.60 \times 10^2 \text{ L } O_2$$

Since air is 21.0 percent O_2 by volume, we can write:

$$V_{air} = V_{O_2}\left(\frac{100\% \text{ air}}{21\% \ O_2}\right) = (3.60 \times 10^2 \text{ L } O_2)\left(\frac{100\% \text{ air}}{21\% \ O_2}\right) = \textbf{1.71} \times \textbf{10}^\textbf{3} \textbf{ L air}$$

5.63   First, we calculate the mole fraction of each component of the mixture. Then, we can calculate the partial pressure of each component using the equation, $P_i = X_i P_T$.

The number of moles of the combined gases is:

$$n = n_{CH_4} + n_{C_2H_6} + n_{C_3H_8} = 0.31 \text{ mol} + 0.25 \text{ mol} + 0.29 \text{ mol} = 0.85 \text{ mol}$$

$$X_{CH_4} = \frac{0.31 \text{ mol}}{0.85 \text{ mol}} = 0.36 \qquad X_{C_2H_6} = \frac{0.25 \text{ mol}}{0.85 \text{ mol}} = 0.29 \qquad X_{C_3H_8} = \frac{0.29 \text{ mol}}{0.85 \text{ mol}} = 0.34$$

The partial pressures are:

$$P_{CH_4} = X_{CH_4} \times P_{total} = 0.36 \times 1.50 \text{ atm} = \textbf{0.54 atm}$$

$$P_{C_2H_6} = X_{C_2H_6} \times P_{total} = 0.29 \times 1.50 \text{ atm} = \textbf{0.44 atm}$$

$$P_{C_3H_8} = X_{C_3H_8} \times P_{total} = 0.34 \times 1.50 \text{ atm} = \textbf{0.51 atm}$$

5.64   Dalton's law states that the total pressure of the mixture is the sum of the partial pressures.

  (a) $P_{\text{total}} = 0.32 \text{ atm} + 0.15 \text{ atm} + 0.42 \text{ atm} = \textbf{0.89 atm}$

  (b)   We know:

Initial conditions	Final Conditions
$P_1 = (0.15 + 0.42)\text{atm} = 0.57 \text{ atm}$	$P_2 = 1.0 \text{ atm}$
$T_1 = (15 + 273)\text{K} = 288 \text{ K}$	$T_2 = 273 \text{ K}$
$V_1 = 2.5 \text{ L}$	$V_2 = ?$

$$\frac{P_1 V_1}{n_1 T_1} = \frac{P_2 V_2}{n_2 T_2}$$

Because $n_1 = n_2$, we can write:

$$V_2 = \frac{P_1 V_1 T_2}{P_2 T_1}$$

$$V_2 = \frac{(0.57 \text{ atm})(2.5 \text{ L})(273 \text{ K})}{(1.0 \text{ atm})(288 \text{ K})} = \textbf{1.4 L at STP}$$

5.65   Since volume is proportional to the number of moles of gas present, we can directly convert the volume percents to mole fractions.

$$X_{N_2} = 0.7808 \qquad X_{O_2} = 0.2094 \qquad X_{Ar} = 0.0093 \qquad X_{CO_2} = 0.0005$$

  (a)   For each gas, $P_i = X_i P_T = X_i(1.00 \text{ atm})$.

$$P_{N_2} = \textbf{0.781 atm}, \qquad P_{O_2} = \textbf{0.209 atm}, \qquad P_{Ar} = \textbf{9.3} \times \textbf{10}^{-3} \textbf{ atm}, \qquad P_{CO_2} = \textbf{5} \times \textbf{10}^{-4} \textbf{ atm}$$

  (b)   Concentration (mol/L) is $c = \dfrac{n}{V} = \dfrac{P}{RT}$. Therefore, we have:

$$c_{N_2} = \frac{0.781 \text{ atm}}{\left(0.0821 \dfrac{\text{L} \cdot \text{atm}}{\text{mol} \cdot \text{K}}\right)(273 \text{ K})} = \textbf{3.48} \times \textbf{10}^{-2} \textbf{ M}$$

Similarly, $c_{O_2} = \textbf{9.32} \times \textbf{10}^{-3} \textbf{ M}$, $c_{Ar} = \textbf{4.1} \times \textbf{10}^{-4} \textbf{ M}$, $c_{CO_2} = \textbf{2} \times \textbf{10}^{-5} \textbf{ M}$

5.66 $P_{\text{Total}} = P_1 + P_2 + P_3 + \ldots + P_n$

In this case,

$$P_{\text{Total}} = P_{Ne} + P_{He} + P_{H_2O}$$

$$P_{Ne} = P_{\text{Total}} - P_{He} - P_{H_2O}$$

$$P_{Ne} = 745 \text{ mm Hg} - 368 \text{ mmHg} - 28.3 \text{ mmHg} = \textbf{349 mmHg}$$

5.67   If we can calculate the moles of H_2 gas collected, we can determine the amount of Na that must have reacted. We can calculate the moles of H_2 gas using the ideal gas equation.

$$P_{H_2} = P_{Total} - P_{H_2O} = 1.00 \text{ atm} - 0.0313 \text{ atm} = 0.97 \text{ atm}$$

The number of moles of hydrogen gas collected is:

$$n_{H_2} = \frac{P_{H_2}V}{RT} = \frac{(0.97 \text{ atm})(0.246 \text{ L})}{\left(0.0821 \dfrac{\text{L} \cdot \text{atm}}{\text{mol} \cdot \text{K}}\right)(25 + 273)\text{K}} = 0.0098 \text{ mol } H_2$$

The balanced equation shows a 2:1 mole ratio between Na and H_2. The mass of Na consumed in the reaction is:

$$? \textbf{ g Na} = 0.0098 \text{ mol } H_2 \times \frac{2 \text{ mol Na}}{1 \text{ mol } H_2} \times \frac{22.99 \text{ g Na}}{1 \text{ mol Na}} = \textbf{0.45 g Na}$$

5.68   **Strategy:** To solve for moles of H_2 generated, we must first calculate the partial pressure of H_2 in the mixture. What gas law do we need? How do we convert from moles of H_2 to amount of Zn reacted?

Solution: Dalton's law of partial pressure states that

$$P_{Total} = P_1 + P_2 + P_3 + \ldots + P_n$$

In this case,

$$P_{Total} = P_{H_2} + P_{H_2O}$$

$$P_{H_2} = P_{Total} - P_{H_2O}$$

$$P_{H_2} = 0.980 \text{ atm} - (23.8 \text{ mmHg})\left(\frac{1 \text{ atm}}{760 \text{ mmHg}}\right) = 0.949 \text{ atm}$$

Now that we know the pressure of H_2 gas, we can calculate the moles of H_2. Then, using the mole ratio from the balanced equation, we can calculate moles of Zn.

$$n_{H_2} = \frac{P_{H_2}V}{RT}$$

$$n_{H_2} = \frac{(0.949 \text{ atm})(7.80 \text{ L})}{(25 + 273)\text{ K}} \times \frac{\text{mol} \cdot \text{K}}{0.0821 \text{ L} \cdot \text{atm}} = 0.303 \text{ mol } H_2$$

Using the mole ratio from the balanced equation and the molar mass of zinc, we can now calculate the grams of zinc consumed in the reaction.

$$? \textbf{ g Zn} = 0.303 \text{ mol } H_2 \times \frac{1 \text{ mol Zn}}{1 \text{ mol } H_2} \times \frac{65.39 \text{ g Zn}}{1 \text{ mol Zn}} = \textbf{19.8 g Zn}$$

5.69 In the mixture, the temperature and volume occupied are the same for the two gases, so the pressure should be proportional to the number of moles. Recall that $P_i = X_i P_T$. The mole fraction of oxygen is:

$$X_{O_2} = \frac{P_{O_2}}{P_{\text{total}}} = \frac{0.20 \text{ atm}}{4.2 \text{ atm}} = 0.048$$

In other words 4.8% of the gas particles are oxygen molecules, which occupy **4.8%** of the volume.

5.70 $P_i = X_i P_T$

We need to determine the mole fractions of each component in order to determine their partial pressures. To calculate mole fraction, write the balanced chemical equation to determine the correct mole ratio.

$$2NH_3(g) \longrightarrow N_2(g) + 3H_2(g)$$

The mole fractions of H_2 and N_2 are:

$$X_{H_2} = \frac{3 \text{ mol}}{3 \text{ mol} + 1 \text{ mol}} = 0.750$$

$$X_{N_2} = \frac{1 \text{ mol}}{3 \text{ mol} + 1 \text{ mol}} = 0.250$$

The partial pressures of H_2 and N_2 are:

$$P_{H_2} = X_{H_2} P_T = (0.750)(866 \text{ mmHg}) = \textbf{650 mmHg}$$

$$P_{N_2} = X_{N_2} P_T = (0.250)(866 \text{ mmHg}) = \textbf{217 mmHg}$$

5.77 $u_{\text{rms}} = \sqrt{\dfrac{3RT}{\mathcal{M}}}$

O_2: $u_{\textbf{rms}} = \sqrt{\dfrac{3(8.314 \text{ J/K} \cdot \text{mol})(65 + 273)\text{K}}{32.00 \times 10^{-3} \text{ kg/mol}}} = \textbf{513 m/s}$

UF_6: $u_{\textbf{rms}} = \sqrt{\dfrac{3(8.314 \text{ J/K} \cdot \text{mol})(65 + 273)\text{K}}{352.00 \times 10^{-3} \text{ kg/mol}}} = \textbf{155 m/s}$

As should be the case, the heavier gas, UF_6, has a smaller average velocity than the lighter gas, O_2.

5.78 **Strategy:** To calculate the root-mean-square speed, we use Equation (5.14) of the text. What units should we use for R and $\mathcal{M}$ so the u_{rms} will be expressed in units of m/s.

Solution: To calculate u_{rms}, the units of R should be 8.314 J/mol·K, and because 1 J = 1 kg·m^2/s^2, the units of molar mass must be kg/mol.

First, let's calculate the molar masses ($\mathcal{M}$) of N_2, O_2, and O_3. Remember, $\mathcal{M}$ must be in units of kg/mol.

$$\mathcal{M}_{N_2} = 2(14.01 \text{ g/mol}) = 28.02 \frac{\text{g}}{\text{mol}} \times \frac{1 \text{ kg}}{1000 \text{ g}} = 0.02802 \text{ kg/mol}$$

$$\mathcal{M}_{O_2} = 2(16.00 \text{ g/mol}) = 32.00 \frac{\text{g}}{\text{mol}} \times \frac{1 \text{ kg}}{1000 \text{ g}} = 0.03200 \text{ kg/mol}$$

$$\mathcal{M}_{O_3} = 3(16.00 \text{ g/mol}) = 48.00 \frac{\text{g}}{\text{mol}} \times \frac{1 \text{ kg}}{1000 \text{ g}} = 0.04800 \text{ kg/mol}$$

Now, we can substitute into Equation (5.14) of the text.

$$u_{rms} = \sqrt{\frac{3RT}{\mathcal{M}}}$$

$$u_{rms}(N_2) = \sqrt{\frac{(3)\left(8.314 \dfrac{\text{J}}{\text{mol}\cdot\text{K}}\right)(-23 + 273)\text{K}}{\left(0.02802 \dfrac{\text{kg}}{\text{mol}}\right)}}$$

$$u_{rms}(N_2) = 472 \text{ m/s}$$

Similarly,

$$u_{rms}(O_2) = 441 \text{ m/s} \qquad u_{rms}(O_3) = 360 \text{ m/s}$$

Check: Since the molar masses of the gases increase in the order: $N_2 < O_2 < O_3$, we expect the lightest gas (N_2) to move the fastest on average and the heaviest gas (O_3) to move the slowest on average. This is confirmed in the above calculation.

5.79   **(a)**   Inversely proportional to density
       (b)   Independent of temperature
       (c)   Decreases with increasing pressure
       (d)   Increases with increasing volume
       (e)   Inversely proportional to size

5.80   **RMS speed** $= \sqrt{\dfrac{\left(2.0^2 + 2.2^2 + 2.6^2 + 2.7^2 + 3.3^2 + 3.5^2\right)(\text{m/s})^2}{6}} = $ **2.8 m/s**

Average speed $= \dfrac{(2.0 + 2.2 + 2.6 + 2.7 + 3.3 + 3.5)\text{m/s}}{6} = $ **2.7 m/s**

The root-mean-square value is always greater than the average value, because squaring favors the larger values compared to just taking the average value.

5.85   In this problem, we are comparing the pressure as determined by the van der waals' equation with that determined by the ideal gas equation.

van der waals' equation:

We find the pressure by first solving algebraically for P.

$$P = \frac{nRT}{(V - nb)} - \frac{an^2}{V^2}$$

where $n = 2.50$ mol, $V = 5.00$ L, $T = 450$ K, $a = 3.59$ atm·L^2/mol^2, and $b = 0.0427$ L/mol

$$P = \frac{(2.50 \text{ mol})\left(0.0821\dfrac{\text{L}\cdot\text{atm}}{\text{mol}\cdot\text{K}}\right)(450 \text{ K})}{[(5.00 \text{ L}) - (2.50 \text{ mol} \times 0.0427 \text{ L/mol})]} - \frac{\left(3.59\dfrac{\text{atm}\cdot\text{L}^2}{\text{mol}^2}\right)(2.50 \text{ mol})^2}{(5.00 \text{ L})^2} = \textbf{18.0 atm}$$

ideal gas equation:

$$P = \frac{nRT}{V} = \frac{(2.50 \text{ mol})\left(0.0821\dfrac{\text{L}\cdot\text{atm}}{\text{mol}\cdot\text{K}}\right)(450 \text{ K})}{(5.00 \text{ L})} = \textbf{18.5 atm}$$

Since the pressure calculated using van der waals' equation is comparable to the pressure calculated using the ideal gas equation, we conclude that CO_2 behaves fairly ideally under these conditions.

5.86   **Strategy:** In this problem we can determine if the gas deviates from ideal behavior, by comparing the ideal pressure with the actual pressure. We can calculate the ideal gas pressure using the ideal gas equation, and then compare it to the actual pressure given in the problem. What temperature unit should we use in the calculation?

Solution: We convert the temperature to units of Kelvin, then substitute the given quantities into the ideal gas equation.

$$T(\text{K}) = 27°\text{C} + 273° = 300 \text{ K}$$

$$P = \frac{nRT}{V} = \frac{(10.0 \text{ mol})\left(0.0821\dfrac{\text{L}\cdot\text{atm}}{\text{mol}\cdot\text{K}}\right)(300 \text{ K})}{1.50 \text{ L}} = 164 \text{ atm}$$

Now, we can compare the ideal pressure to the actual pressure by calculating the percent error.

$$\% \text{ error} = \frac{164 \text{ atm} - 130 \text{ atm}}{130 \text{ atm}} \times 100\% = 26.2\%$$

Based on the large percent error, we conclude that under this condition of high pressure, the gas behaves in a **non-ideal** manner.

5.87   (a)   Neither the amount of gas in the tire nor its volume change appreciably. The pressure is proportional to the temperature. Therefore, as the temperature rises, the pressure increases.

(b)   As the paper bag is hit, its volume decreases so that its pressure increases. The popping sound occurs when the bag is broken.

(c)   As the balloon rises, the pressure outside decreases steadily, and the balloon expands.

(d)   The pressure inside the bulb is greater than 1 atm.

5.88   When a and b are zero, the van der Waals equation simply becomes the ideal gas equation. In other words, an ideal gas has zero for the a and b values of the van der Waals equation. It therefore stands to reason that the gas with the smallest values of a and b will behave most like an ideal gas under a specific set of pressure and temperature conditions. Of the choices given in the problem, the gas with the smallest a and b values is **Ne** (see Table 5.4).

5.89   You can map out the following strategy to solve for the total volume of gas.

grams nitroglycerin $\rightarrow$ moles nitroglycerin $\rightarrow$ moles products $\rightarrow$ volume of products

$$? \text{ mol products} = 2.6 \times 10^2 \text{ g nitroglycerin} \times \frac{1 \text{ mol nitroglycerin}}{227.09 \text{ g nitroglycerin}} \times \frac{29 \text{ mol product}}{4 \text{ mol nitroglycerin}} = 8.3 \text{ mol}$$

Calculating the volume of products:

$$V_{\text{product}} = \frac{n_{\text{product}} RT}{P} = \frac{(8.3 \text{ mol})\left(0.0821 \dfrac{\text{L} \cdot \text{atm}}{\text{mol} \cdot \text{K}}\right)(298 \text{ K})}{(1.2 \text{ atm})} = 1.7 \times 10^2 \text{ L}$$

The relationship between partial pressure and P_{total} is:

$$P_i = X_i P_T$$

Calculate the mole fraction of each gaseous product, then calculate its partial pressure using the equation above.

$$X_{\text{component}} = \frac{\text{moles component}}{\text{total moles all components}}$$

$$X_{CO_2} = \frac{12 \text{ mol } CO_2}{29 \text{ mol product}} = 0.41$$

Similarly, $X_{H_2O} = 0.34$, $X_{N_2} = 0.21$, and $X_{O_2} = 0.034$

$$P_{CO_2} = X_{CO_2} P_T$$

$$P_{CO_2} = (0.41)(1.2 \text{ atm}) = \textbf{0.49 atm}$$

Similarly, $P_{H_2O} = \textbf{0.41 atm}$, $P_{N_2} = \textbf{0.25 atm}$, and $P_{O_2} = \textbf{0.041 atm}$.

5.90 We need to determine the molar mass of the gas. Comparing the molar mass to the empirical mass will allow us to determine the molecular formula.

$$n = \frac{PV}{RT} = \frac{(0.74 \text{ atm})\left(97.2 \text{ mL} \times \dfrac{0.001 \text{ L}}{1 \text{ mL}}\right)}{\left(0.0821 \dfrac{\text{L} \cdot \text{atm}}{\text{mol} \cdot \text{K}}\right)(200 + 273) \text{K}} = 1.85 \times 10^{-3} \text{ mol}$$

$$\text{molar mass} = \frac{0.145 \text{ g}}{1.85 \times 10^{-3} \text{ mol}} = 78.4 \text{ g/mol}$$

The empirical mass of CH = 13.02 g/mol

$$\text{Since } \frac{78.4 \text{ g/mol}}{13.02 \text{ g/mol}} = 6.02 \approx 6 \text{, the molecular formula is } (CH)_6 \text{ or } \textbf{C}_6\textbf{H}_6.$$

5.91 **(a)** $NH_4NO_2(s) \longrightarrow N_2(g) + 2H_2O(l)$

(b) Map out the following strategy to solve the problem.

volume $N_2 \rightarrow$ moles $N_2 \rightarrow$ moles $NH_4NO_2 \rightarrow$ grams NH_4NO_2

First, calculate the moles of N_2 using the ideal gas equation.

$$T(K) = 22° + 273° = 295 \text{ K}$$

$$V = 86.2 \text{ mL} \times \frac{1 \text{ L}}{1000 \text{ mL}} = 0.0862 \text{ L}$$

$$n_{N_2} = \frac{P_{N_2} V}{RT}$$

$$n_{N_2} = \frac{(1.20 \text{ atm})(0.0862 \text{ L})}{\left(0.0821 \dfrac{\text{L} \cdot \text{atm}}{\text{mol} \cdot \text{K}}\right)(295 \text{ K})} = 4.27 \times 10^{-3} \text{ mol}$$

Next, calculate the mass of NH_4NO_2 needed to produce 4.27×10^{-3} mole of N_2.

$$? \text{ g } NH_4NO_2 = (4.27 \times 10^{-3} \text{ mol } N_2) \times \frac{1 \text{ mol } NH_4NO_2}{1 \text{ mol } N_2} \times \frac{64.05 \text{ g } NH_4NO_2}{1 \text{ mol } NH_4NO_2} = 0.273 \text{ g}$$

5.92   The reaction is: $HCO_3^-(aq) + H^+(aq) \longrightarrow H_2O(l) + CO_2(g)$

The mass of HCO_3^- reacted is:

$$3.29 \text{ g tablet} \times \frac{32.5\% \text{ } HCO_3^-}{100\% \text{ tablet}} = 1.07 \text{ g } HCO_3^-$$

$$\text{mol } CO_2 \text{ produced} = 1.07 \text{ g } HCO_3^- \times \frac{1 \text{ mol } HCO_3^-}{61.02 \text{ g } HCO_3^-} \times \frac{1 \text{ mol } CO_2}{1 \text{ mol } HCO_3^-} = 0.0175 \text{ mol } CO_2$$

$$V_{CO_2} = \frac{n_{CO_2} RT}{P} = \frac{(0.0175 \text{ mol } CO_2)\left(0.0821 \dfrac{\text{L} \cdot \text{atm}}{\text{mol} \cdot \text{K}}\right)(37 + 273)\text{K}}{(1.00 \text{ atm})} = 0.445 \text{ L} = \mathbf{445 \text{ mL}}$$

5.93   No, because an ideal gas cannot be liquefied, since the assumption is that there are no intermolecular forces in an ideal gas.

5.94   **(a)**   The number of moles of $Ni(CO)_4$ formed is:

$$86.4 \text{ g Ni} \times \frac{1 \text{ mol Ni}}{58.69 \text{ g Ni}} \times \frac{1 \text{ mol } Ni(CO)_4}{1 \text{ mol Ni}} = 1.47 \text{ mol } Ni(CO)_4$$

The pressure of $Ni(CO)_4$ is:

$$P = \frac{nRT}{V} = \frac{(1.47 \text{ mol})\left(0.0821 \dfrac{\text{L} \cdot \text{atm}}{\text{mol} \cdot \text{K}}\right)(43 + 273)\text{K}}{4.00 \text{ L}} = \mathbf{9.53 \text{ atm}}$$

(b) $Ni(CO)_4$ decomposes to produce more moles of gas (CO), which increases the pressure.

$$Ni(CO)_4(g) \longrightarrow Ni(s) + 4CO(g)$$

5.95   The partial pressure of carbon dioxide is higher in the winter because carbon dioxide is utilized less by photosynthesis in plants.

5.96   Using the ideal gas equation, we can calculate the moles of gas.

$$n = \frac{PV}{RT} = \frac{(1.1 \text{ atm})\left(5.0 \times 10^2 \text{ mL} \times \dfrac{0.001 \text{ L}}{1 \text{ mL}}\right)}{\left(0.0821\dfrac{\text{L} \cdot \text{atm}}{\text{mol} \cdot \text{K}}\right)(37 + 273)\text{K}} = 0.0216 \text{ mol gas}$$

Next, use Avogadro's number to convert to molecules of gas.

$$0.0216 \text{ mol gas} \times \frac{6.022 \times 10^{23} \text{ molecules}}{1 \text{ mol gas}} = \mathbf{1.30 \times 10^{22} \text{ molecules of gas}}$$

The most common gases present in exhaled air are: **CO_2, O_2, N_2, and H_2O.**

5.97   **(a)**   Write a balanced chemical equation.

$$2NaHCO_3(s) \longrightarrow Na_2CO_3(s) + CO_2(g) + H_2O(g)$$

First, calculate the moles of CO_2 produced.

$$? \text{ mol } CO_2 = 5.0 \text{ g NaHCO}_3 \times \frac{1 \text{ mol NaHCO}_3}{84.01 \text{ g NaHCO}_3} \times \frac{1 \text{ mol } CO_2}{2 \text{ mol NaHCO}_3} = 0.030 \text{ mol}$$

Next, calculate the volume of CO_2 produced using the ideal gas equation.

$$T(\text{K}) = 180° + 273° = 453 \text{ K}$$

$$V_{CO_2} = \frac{n_{CO_2} RT}{P}$$

$$V_{CO_2} = \frac{(0.030 \text{ mol})\left(0.0821\dfrac{\text{L} \cdot \text{atm}}{\text{mol} \cdot \text{K}}\right)(453 \text{ K})}{(1.3 \text{ atm})} = \mathbf{0.86 \text{ L}}$$

(b)   The balanced chemical equation for the decomposition of NH_4HCO_3 is

$$NH_4HCO_3(s) \longrightarrow NH_3(g) + CO_2(g) + H_2O(g)$$

The advantage in using the ammonium salt is that more gas is produced per gram of reactant. The disadvantage is that one of the gases is ammonia. The strong odor of ammonia would ***not*** make the ammonium salt a good choice for baking.

5.98   Mass of the Earth's atmosphere = (surface area of the earth in cm^2) × (mass per 1 cm^2 column)

Mass of a single column of air with a surface area of 1 cm^2 area is:

$$76.0 \text{ cm} \times 13.6 \text{ g/cm}^3 = 1.03 \times 10^3 \text{ g/cm}^2$$

The surface area of the Earth in cm^2 is:

$$4\pi r^2 = 4\pi(6.371 \times 10^8 \text{ cm})^2 = 5.10 \times 10^{18} \text{ cm}^2$$

$$\textbf{Mass of atmosphere} = (5.10 \times 10^{18} \text{ cm}^2)(1.03 \times 10^3 \text{ g/cm}^2) = 5.25 \times 10^{21} \text{ g} = \textbf{5.25} \times \textbf{10}^{18} \textbf{ kg}$$

5.99   First, calculate the moles of H_2 formed.

$$? \text{ mol } H_2 = 3.12 \text{ g Al} \times \frac{1 \text{ mol Al}}{26.98 \text{ g}} \times \frac{3 \text{ mol } H_2}{2 \text{ mol Al}} = 0.173 \text{ mol}$$

Next, calculate the volume of H_2 produced using the ideal gas equation.

$$V_{H_2} = \frac{n_{H_2} RT}{P} = \frac{(0.173 \text{ mol})\left(0.0821 \dfrac{\text{L} \cdot \text{atm}}{\text{mol} \cdot \text{K}}\right)(296 \text{ K})}{(1.00 \text{ atm})} = \textbf{4.20 L}$$

5.100   To calculate the molarity of NaOH, we need moles of NaOH and volume of the NaOH solution. The volume is given in the problem; therefore, we need to calculate the moles of NaOH. The moles of NaOH can be calculated from the reaction of NaOH with HCl. The balanced equation is:

$$NaOH(aq) + HCl(aq) \longrightarrow H_2O(l) + NaCl(aq)$$

The number of moles of HCl gas is found from the ideal gas equation. $V = 0.189$ L, $T = (25 + 273)\text{K} = 298$ K, and $P = 108 \text{ mmHg} \times \dfrac{1 \text{ atm}}{760 \text{ mmHg}} = 0.142 \text{ atm}$.

$$n_{HCl} = \frac{PV_{HCl}}{RT} = \frac{(0.142 \text{ atm})(0.189 \text{ L})}{\left(0.0821 \dfrac{\text{L} \cdot \text{atm}}{\text{mol} \cdot \text{K}}\right)(298 \text{ K})} = 1.10 \times 10^{-3} \text{ mol HCl}$$

The moles of NaOH can be calculated using the mole ratio from the balanced equation.

$$(1.10 \times 10^{-3} \text{ mol HCl}) \times \frac{1 \text{ mol NaOH}}{1 \text{ mol HCl}} = 1.10 \times 10^{-3} \text{ mol NaOH}$$

The molarity of the NaOH solution is:

$$M = \frac{\text{mol NaOH}}{\text{L of soln}} = \frac{1.10 \times 10^{-3} \text{ mol NaOH}}{0.0157 \text{ L soln}} = 0.0701 \text{ mol/L} = \textbf{0.0701 } \textbf{\textit{M}}$$

5.101   **(a)** $C_3H_8(g) + 5O_2(g) \longrightarrow 3CO_2(g) + 4H_2O(g)$

(b)   From the balanced equation, we see that there is a 1:3 mole ratio between C_3H_8 and CO_2.

$$? \text{ L } CO_2 = 7.45 \text{ g } C_3H_8 \times \frac{1 \text{ mol } C_3H_8}{44.09 \text{ g } C_3H_8} \times \frac{3 \text{ mol } CO_2}{1 \text{ mol } C_3H_8} \times \frac{22.414 \text{ L } CO_2}{1 \text{ mol } CO_2} = \textbf{11.4 L } CO_2$$

5.102   To calculate the partial pressures of He and Ne, the total pressure of the mixture is needed. To calculate the total pressure of the mixture, we need the total number of moles of gas in the mixture (mol He + mol Ne).

$$n_{He} = \frac{PV}{RT} = \frac{(0.63 \text{ atm})(1.2 \text{ L})}{\left(0.0821\dfrac{\text{L} \cdot \text{atm}}{\text{mol} \cdot \text{K}}\right)(16 + 273)\text{K}} = 0.032 \text{ mol He}$$

$$n_{Ne} = \frac{PV}{RT} = \frac{(2.8 \text{ atm})(3.4 \text{ L})}{\left(0.0821\dfrac{\text{L} \cdot \text{atm}}{\text{mol} \cdot \text{K}}\right)(16 + 273)\text{K}} = 0.40 \text{ mol Ne}$$

The total pressure is:

$$P_{Total} = \frac{(n_{He} + n_{Ne})RT}{V_{Total}} = \frac{(0.032 + 0.40)\text{mol}\left(0.0821\dfrac{\text{L} \cdot \text{atm}}{\text{mol} \cdot \text{K}}\right)(16 + 273)\text{K}}{(1.2 + 3.4)\text{L}} = 2.2 \text{ atm}$$

$P_i = X_i P_T$. The partial pressures of He and Ne are:

$$P_{He} = \frac{0.032 \text{ mol}}{(0.032 + 0.40)\text{mol}} \times 2.2 \text{ atm} = \mathbf{0.16 \text{ atm}}$$

$$P_{Ne} = \frac{0.40 \text{ mol}}{(0.032 + 0.40)\text{mol}} \times 2.2 \text{ atm} = \mathbf{2.0 \text{ atm}}$$

5.103   Calculate the initial number of moles of NO and O_2 using the ideal gas equation.

$$n_{NO} = \frac{P_{NO}V}{RT} = \frac{(0.500 \text{ atm})(4.00 \text{ L})}{\left(0.0821\dfrac{\text{L} \cdot \text{atm}}{\text{mol} \cdot \text{K}}\right)(298 \text{ K})} = 0.0817 \text{ mol NO}$$

$$n_{O_2} = \frac{P_{O_2}V}{RT} = \frac{(1.00 \text{ atm})(2.00 \text{ L})}{\left(0.0821\dfrac{\text{L} \cdot \text{atm}}{\text{mol} \cdot \text{K}}\right)(298 \text{ K})} = 0.0817 \text{ mol O}_2$$

Determine which reactant is the limiting reagent. The number of moles of NO and O_2 calculated above are equal; however, the balanced equation shows that twice as many moles of NO are needed compared to O_2. Thus, NO is the limiting reagent.

Determine the molar amounts of NO, O_2, and NO_2 after complete reaction.

mol NO = 0 mol (All NO is consumed during reaction)

$$\textbf{mol NO}_2 = 0.0817 \text{ mol NO} \times \frac{2 \text{ mol NO}_2}{2 \text{ mol NO}} = \mathbf{0.0817 \text{ mol NO}_2}$$

$$\text{mol O}_2 \text{ consumed} = 0.0817 \text{ mol NO} \times \frac{1 \text{ mol O}_2}{2 \text{ mol NO}} = 0.0409 \text{ mol O}_2 \text{ consumed}$$

$$\textbf{mol O}_2 \textbf{ remaining} = 0.0817 \text{ mol O}_2 \text{ initial} - 0.0409 \text{ mol O}_2 \text{ consumed} = \mathbf{0.0408 \text{ mol O}_2}$$

Calculate the partial pressures of O_2 and NO_2 using the ideal gas equation.

Volume of entire apparatus $= 2.00 \text{ L} + 4.00 \text{ L} = 6.00 \text{ L}$

$T(\text{K}) = 25° + 273° = 298 \text{ K}$

$$P_{O_2} = \frac{n_{O_2}RT}{V} = \frac{(0.0408 \text{ mol})\left(0.0821\frac{\text{L}\cdot\text{atm}}{\text{mol}\cdot\text{K}}\right)(298 \text{ K})}{(6.00 \text{ L})} = \mathbf{0.166 \text{ atm}}$$

$$P_{NO_2} = \frac{n_{NO_2}RT}{V} = \frac{(0.0817 \text{ mol})\left(0.0821\frac{\text{L}\cdot\text{atm}}{\text{mol}\cdot\text{K}}\right)(298 \text{ K})}{(6.00 \text{ L})} = \mathbf{0.333 \text{ atm}}$$

5.104 When the water enters the flask from the dropper, some hydrogen chloride dissolves, creating a partial vacuum. Pressure from the atmosphere forces more water up the vertical tube.

5.105 **(a)** First, the total pressure (P_{Total}) of the mixture of carbon dioxide and hydrogen must be determined at a given temperature in a container of known volume. Next, the carbon dioxide can be removed by reaction with sodium hydroxide.

$$CO_2(g) + 2NaOH(aq) \longrightarrow Na_2CO_3(aq) + H_2O(l)$$

The pressure of the hydrogen gas that remains can now be measured under the same conditions of temperature and volume. Finally, the partial pressure of CO_2 can be calculated.

$$P_{CO_2} = P_{Total} - P_{H_2}$$

(b) The most direct way to measure the partial pressures would be to use a mass spectrometer to measure the mole fractions of the gases. The partial pressures could then be calculated from the mole fractions and the total pressure. Another way to measure the partial pressures would be to realize that helium has a much lower boiling point than nitrogen. Therefore, nitrogen gas can be removed by lowering the temperature until nitrogen liquefies. Helium will remain as a gas. As in part (a), the total pressure is measured first. Then, the pressure of helium can be measured after the nitrogen is removed. Finally, the pressure of nitrogen is simply the difference between the total pressure and the pressure of helium.

5.106 Use the ideal gas equation to calculate the moles of water produced.

$$n_{H_2O} = \frac{PV}{RT} = \frac{(24.8 \text{ atm})(2.00 \text{ L})}{\left(0.0821\frac{\text{L}\cdot\text{atm}}{\text{mol}\cdot\text{K}}\right)(120 + 273)\text{K}} = 1.54 \text{ mol } H_2O$$

Next, we can determine the mass of H_2O in the 54.2 g sample. Subtracting the mass of H_2O from 54.2 g will give the mass of $MgSO_4$ in the sample.

$$1.54 \text{ mol } H_2O \times \frac{18.02 \text{ g } H_2O}{1 \text{ mol } H_2O} = 27.8 \text{ g } H_2O$$

$$\text{Mass MgSO}_4 = 54.2 \text{ g sample} - 27.8 \text{ g } H_2O = 26.4 \text{ g MgSO}_4$$

Finally, we can calculate the moles of $MgSO_4$ in the sample. Comparing moles of $MgSO_4$ to moles of H_2O will allow us to determine the correct mole ratio in the formula.

$$26.4 \text{ g MgSO}_4 \times \frac{1 \text{ mol MgSO}_4}{120.4 \text{ g MgSO}_4} = 0.219 \text{ mol MgSO}_4$$

$$\frac{\text{mol H}_2\text{O}}{\text{mol MgSO}_4} = \frac{1.54 \text{ mol}}{0.219 \text{ mol}} = 7.03$$

Therefore, the mole ratio between H_2O and $MgSO_4$ in the compound is 7 : 1. Thus, the value of $x = 7$, and the formula is **$MgSO_4 \cdot 7H_2O$**.

5.107   The reactions are: $Na_2CO_3(s) + 2HCl(aq) \longrightarrow 2NaCl(aq) + H_2O(l) + CO_2(g)$

$MgCO_3(s) + 2HCl(aq) \longrightarrow MgCl_2(aq) + H_2O(l) + CO_2(g)$

First, let's calculate the moles of CO_2 produced using the ideal gas equation.

$$n_{CO_2} = \frac{PV}{RT} = \frac{(1.24 \text{ atm})(1.67 \text{ L})}{\left(0.0821 \dfrac{\text{L} \cdot \text{atm}}{\text{mol} \cdot \text{K}}\right)(26 + 273)\text{K}} = 0.0844 \text{ mol CO}_2$$

Since there is a 1:1 mole ratio between CO_2 and both reactants (Na_2CO_3 and $MgCO_3$), then 0.0844 mole of the mixture must have reacted.

$$\text{mol Na}_2\text{CO}_3 + \text{mol MgCO}_3 = 0.0844 \text{ mol}$$

Let x be the mass of Na_2CO_3 in the mixture, then $(7.63 - x)$ is the mass of $MgCO_3$ in the mixture.

$$\left[x \text{ g Na}_2\text{CO}_3 \times \frac{1 \text{ mol Na}_2\text{CO}_3}{106.0 \text{ g Na}_2\text{CO}_3} \right] + \left[(7.63 - x)\text{g MgCO}_3 \times \frac{1 \text{ mol MgCO}_3}{84.32 \text{ g MgCO}_3} \right] = 0.0844 \text{ mol}$$

$$0.009434x - 0.0119x + 0.0905 = 0.0844$$

$$x = 2.47 \text{ g} = \text{mass of Na}_2\text{CO}_3 \text{ in the mixture}$$

The percent composition by mass of Na_2CO_3 in the mixture is:

$$\textbf{mass \% Na}_2\textbf{CO}_3 = \frac{\text{mass Na}_2\text{CO}_3}{\text{mass of mixture}} \times 100\% = \frac{2.47 \text{ g}}{7.63 \text{ g}} \times 100\% = \textbf{32.4\% Na}_2\textbf{CO}_3$$

5.108   The circumference of the cylinder is $= 2\pi r = 2\pi\left(\dfrac{15.0 \text{ cm}}{2}\right) = 47.1 \text{ cm}$

(a)   The speed at which the target is moving equals:

$$\text{speed of target} = \text{circumference} \times \text{revolutions/sec}$$

$$\textbf{speed of target} = \frac{47.1 \text{ cm}}{1 \text{ revolution}} \times \frac{130 \text{ revolutions}}{1 \text{ s}} \times \frac{0.01 \text{ m}}{1 \text{ cm}} = \textbf{61.2 m/s}$$

(b) $2.80 \text{ cm} \times \dfrac{0.01 \text{ m}}{1 \text{ cm}} \times \dfrac{1 \text{ s}}{61.2 \text{ m}} = \textbf{4.58} \times \textbf{10}^{-4} \textbf{ s}$

(c) The Bi atoms must travel across the cylinder to hit the target. This distance is the diameter of the cylinder, which is 15.0 cm. The Bi atoms travel this distance in 4.58×10^{-4} s.

$$\frac{15.0 \text{ cm}}{4.58 \times 10^{-4} \text{ s}} \times \frac{0.01 \text{ m}}{1 \text{ cm}} = \textbf{328 m/s}$$

$$u_{rms} = \sqrt{\frac{3RT}{\mathcal{M}}} = \sqrt{\frac{3(8.314 \text{ J/K} \cdot \text{mol})(850 + 273)\text{K}}{209.0 \times 10^{-3} \text{ kg/mol}}} = \textbf{366 m/s}$$

The magnitudes of the speeds are comparable, but not identical. This is not surprising since 328 m/s is the velocity of a particular Bi atom, and u_{rms} is an average value.

5.109 Using the ideal gas equation, we can calculate the moles of water that would be vaporized. We can then convert to mass of water vaporized.

$$n = \frac{PV}{RT} = \frac{\left(187.5 \text{ mmHg} \times \frac{1 \text{ atm}}{760 \text{ mmHg}}\right)(2.500 \text{ L})}{\left(0.0821\frac{\text{L} \cdot \text{atm}}{\text{mol} \cdot \text{K}}\right)(65 + 273)\text{K}} = 0.0222 \text{ mol H}_2\text{O}$$

$$\textbf{? g H}_2\textbf{O vaporized} = 0.0222 \text{ mol H}_2\text{O} \times \frac{18.02 \text{ g H}_2\text{O}}{1 \text{ mol H}_2\text{O}} = \textbf{0.400 g H}_2\textbf{O}$$

5.110 The moles of O_2 can be calculated from the ideal gas equation. The mass of O_2 can then be calculated using the molar mass as a conversion factor.

$$n_{O_2} = \frac{PV}{RT} = \frac{(132 \text{ atm})(120 \text{ L})}{\left(0.0821\frac{\text{L} \cdot \text{atm}}{\text{mol} \cdot \text{K}}\right)(22 + 273)\text{K}} = 654 \text{ mol O}_2$$

$$\textbf{? g O}_2 = 654 \text{ mol O}_2 \times \frac{32.00 \text{ g O}_2}{1 \text{ mol O}_2} = \textbf{2.09} \times \textbf{10}^4 \textbf{ g O}_2$$

The volume of O_2 gas under conditions of 1.00 atm pressure and a temperature of 22°C can be calculated using the ideal gas equation. The moles of $O_2 = 654$ moles.

$$V_{O_2} = \frac{n_{O_2}RT}{P} = \frac{(654 \text{ mol})\left(0.0821\frac{\text{L} \cdot \text{atm}}{\text{mol} \cdot \text{K}}\right)(22 + 273)\text{K}}{1.00 \text{ atm}} = \textbf{1.58} \times \textbf{10}^4 \textbf{ L O}_2$$

5.111 The air inside the egg expands with increasing temperature. The increased pressure can cause the egg to crack.

5.112 The fruit ripens more rapidly because the quantity (partial pressure) of ethylene gas inside the bag increases.

5.113 The balanced equation is: $CO_2(g) + 2NH_3(g) \longrightarrow (NH_2)_2CO(s) + H_2O(g)$

First, we can calculate the moles of NH_3 needed to produce 1.0 ton of urea. Then, we can use the ideal gas equation to calculate the volume of NH_3.

$$? \text{ mol NH}_3 = 1.0 \text{ ton urea} \times \frac{2000 \text{ lb}}{1 \text{ ton}} \times \frac{453.6 \text{ g}}{1 \text{ lb}} \times \frac{1 \text{ mol urea}}{60.06 \text{ g urea}} \times \frac{2 \text{ mol NH}_3}{1 \text{ mol urea}} = 3.0 \times 10^4 \text{ mol NH}_3$$

$$V_{\text{NH}_3} = \frac{nRT}{P} = \frac{(3.0 \times 10^4 \text{ mol})\left(0.0821\dfrac{\text{L} \cdot \text{atm}}{\text{mol} \cdot \text{K}}\right)(200 + 273)\text{K}}{150 \text{ atm}} = 7.8 \times 10^3 \text{ L NH}_3$$

5.114 As the pen is used the amount of ink decreases, increasing the volume inside the pen. As the volume increases, the pressure inside the pen decreases. The hole is needed to equalize the pressure as the volume inside the pen increases.

5.115 **(a)** This is a Boyle's Law problem, pressure and volume vary. Assume that the pressure at the water surface is 1 atm. The pressure that the diver experiences 36 ft below water is:

$$1 \text{ atm} + \left(36 \text{ ft} \times \frac{1 \text{ atm}}{33 \text{ ft}}\right) = 2.1 \text{ atm}$$

$$P_1 V_1 = P_2 V_2 \qquad \text{or} \qquad \frac{V_1}{V_2} = \frac{P_2}{P_1}$$

$$\frac{V_1}{V_2} = \frac{2.1 \text{ atm}}{1 \text{ atm}} = \mathbf{2.1}$$

The diver's lungs would increase in volume **2.1 times** by the time he reaches the surface.

(b) $P_{\text{O}_2} = X_{\text{O}_2} P_{\text{T}}$

$$P_{\text{O}_2} = \frac{n_{\text{O}_2}}{n_{\text{O}_2} + n_{\text{N}_2}} P_{\text{T}}$$

At constant temperature and pressure, the volume of a gas is directly proportional to the number of moles of gas. We can write:

$$P_{\text{O}_2} = \frac{V_{\text{O}_2}}{V_{\text{O}_2} + V_{\text{N}_2}} P_{\text{T}}$$

We know the partial pressure of O_2 in air, and we know the total pressure exerted on the diver. Plugging these values into the above equation gives:

$$0.20 \text{ atm} = \frac{V_{\text{O}_2}}{V_{\text{O}_2} + V_{\text{N}_2}}(4.00 \text{ atm})$$

$$\frac{V_{\text{O}_2}}{V_{\text{O}_2} + V_{\text{N}_2}} = \frac{0.20 \text{ atm}}{4.00 \text{ atm}} = \mathbf{0.050}$$

In other words, the air that the diver breathes must have an oxygen content of **5% by volume**.

5.116 **(a)** $NH_4NO_3(s) \longrightarrow N_2O(g) + 2H_2O(l)$

(b) $R = \dfrac{PV}{nT} = \dfrac{\left(718 \text{ mmHg} \times \dfrac{1 \text{ atm}}{760 \text{ mmHg}}\right)(0.340 \text{ L})}{\left(0.580 \text{ g } N_2O \times \dfrac{1 \text{ mol } N_2O}{44.02 \text{ g } N_2O}\right)(24 + 273)K} = 0.0821 \dfrac{L \cdot atm}{mol \cdot K}$

5.117 Since Ne and NH_3 are at the same temperature, they have the same average kinetic energy.

$$KE = \frac{1}{2}m\overline{u^2}$$

or

$$\overline{u^2} = \frac{2KE}{m}$$

Recall that mass must have units of kg because kinetic energy has units of Joules. $\left(1 \text{ J} = \dfrac{1 \text{ kg} \cdot m^2}{s^2}\right)$

We need to calculate the mass of one Ne atom in kg.

$$\frac{20.18 \text{ g Ne}}{1 \text{ mol Ne}} \times \frac{1 \text{ mol Ne}}{6.022 \times 10^{23} \text{ Ne atoms}} \times \frac{1 \text{ kg}}{1000 \text{ g}} = 3.351 \times 10^{-26} \text{ kg/Ne atom}$$

Solving for $\overline{u^2}$:

$$\overline{u^2} = \frac{2KE}{m} = \frac{2(7.1 \times 10^{-21} \text{ J/atom})}{3.351 \times 10^{-26} \text{ kg/atom}} = 4.2 \times 10^5 \text{ m}^2/\text{s}^2$$

5.118 The value of a indicates how strongly molecules of a given type of gas attract one anther. **C_6H_6** has the greatest intermolecular attractions due to its larger size compared to the other choices. Therefore, it has the largest a value.

5.119 Using the ideal gas equation, we can calculate the moles of gas in the syringe. Then, knowing the mass of the gas, we can calculate the molar mass of the gas. Finally, comparing the molar mass to the empirical mass will allow us to determine the molecular formula of the gas.

$$n = \frac{PV}{RT} = \frac{(1 \text{ atm})(5.58 \times 10^{-3} \text{ L})}{\left(0.0821 \dfrac{L \cdot atm}{mol \cdot K}\right)(45 + 273)K} = 2.14 \times 10^{-4} \text{ mol}$$

$$\textbf{molar mass} = \frac{\text{mass (in g)}}{\text{mol}} = \frac{0.0184 \text{ g}}{2.14 \times 10^{-4} \text{ mol}} = \textbf{86.0 g/mol}$$

The empirical molar mass of CH_2 is 14.0 g/mol. Dividing the molar mass by the empirical molar mass gives:

$$\frac{86.0 \text{ g/mol}}{14.0 \text{ g/mol}} \approx 6$$

Therefore, the molecular formula is $(CH_2)_6$ or **C_6H_{12}**.

5.120   The gases inside the mine were a mixture of carbon dioxide, carbon monoxide, methane, and other harmful compounds. The low atmospheric pressure caused the gases to flow out of the mine (the gases in the mine were at a higher pressure), and the man suffocated.

5.121   **(a)** $CaO(s) + CO_2(g) \longrightarrow CaCO_3(s)$

$BaO(s) + CO_2(g) \longrightarrow BaCO_3(s)$

(b)   First, we need to find the number of moles of CO_2 consumed in the reaction. We can do this by calculating the initial moles of CO_2 in the flask and then comparing it to the CO_2 remaining after the reaction.

Initially: $n_{CO_2} = \dfrac{PV}{RT} = \dfrac{\left(746 \text{ mmHg} \times \dfrac{1 \text{ atm}}{760 \text{ mmHg}}\right)(1.46 \text{ L})}{\left(0.0821 \dfrac{\text{L} \cdot \text{atm}}{\text{mol} \cdot \text{K}}\right)(35 + 273)\text{K}} = 0.0567 \text{ mol } CO_2$

Remaining: $n_{CO_2} = \dfrac{PV}{RT} = \dfrac{\left(252 \text{ mmHg} \times \dfrac{1 \text{ atm}}{760 \text{ mmHg}}\right)(1.46 \text{ L})}{\left(0.0821 \dfrac{\text{L} \cdot \text{atm}}{\text{mol} \cdot \text{K}}\right)(35 + 273)\text{K}} = 0.0191 \text{ mol } CO_2$

Thus, the amount of CO_2 consumed in the reaction is: $(0.0567 \text{ mol} - 0.0191 \text{ mol}) = 0.0376 \text{ mol } CO_2$.

Since the mole ratio between CO_2 and both reactants (CaO and BaO) is 1:1, 0.0376 mole of the mixture must have reacted. We can write:

mol CaO + mol BaO = 0.0376 mol

Let x = mass of CaO in the mixture, then $(4.88 - x)$ = mass of BaO in the mixture. We can write:

$$\left[x \text{ g CaO} \times \dfrac{1 \text{ mol CaO}}{56.08 \text{ g CaO}}\right] + \left[(4.88 - x)\text{g BaO} \times \dfrac{1 \text{ mol BaO}}{153.3 \text{ g BaO}}\right] = 0.0376 \text{ mol}$$

$0.01783x - 0.006523x + 0.0318 = 0.0376$

$x = 0.513 \text{ g}$ = mass of CaO in the mixture

mass of BaO in the mixture = $4.88 - x = 4.37 \text{ g}$

The percent compositions by mass in the mixture are:

CaO: $\dfrac{0.513 \text{ g}}{4.88 \text{ g}} \times 100\% = \mathbf{10.5\%}$ **BaO**: $\dfrac{4.37 \text{ g}}{4.88 \text{ g}} \times 100\% = \mathbf{89.5\%}$

5.122   **(a)**   This is a Boyle's law problem.

$$P_{tire}V_{tire} = P_{air}V_{air}$$

$$(5.0 \text{ atm})(0.98 \text{ L}) = (1.0 \text{ atm})V_{air}$$

$$V_{air} = \mathbf{4.90 \text{ L}}$$

(b) Pressure in the tire – atmospheric pressure = gauge pressure

$$\text{Pressure in the tire} - 1.0 \text{ atm} = 5.0 \text{ atm}$$

Pressure in the tire = 6.0 atm

(c) Again, this is a Boyle's law problem.

$$P_{\text{pump}}V_{\text{pump}} = P_{\text{gauge}}V_{\text{gauge}}$$

$$(1 \text{ atm})(0.33V_{\text{tire}}) = P_{\text{gauge}}V_{\text{gauge}}$$

$$P_{\text{gauge}} = 0.33 \text{ atm}$$

This is the gauge pressure after one pump stroke. After three strokes, the gauge pressure will be $(3 \times 0.33 \text{ atm})$, or approximately **1 atm**. This is assuming that the initial gauge pressure was zero.

5.123 **(a)** $\dfrac{188 \text{ g CO}}{1 \text{ hr}} \times \dfrac{1 \text{ mol CO}}{28.01 \text{ g CO}} \times \dfrac{1 \text{ hr}}{60 \text{ min}} = \textbf{0.112 mol CO/min}$

(b) 1000 ppm means that there are 1000 particles of gas per 1,000,000 particles of air. The pressure of a gas is directly proportional to the number of particles of gas. We can calculate the partial pressure of CO in atmospheres, assuming that atmospheric pressure is 1 atm.

$$\frac{1000 \text{ particles}}{1,000,000 \text{ particles}} \times 1 \text{ atm} = 1.0 \times 10^{-3} \text{ atm}$$

A partial pressure of 1.0×10^{-3} atm CO is lethal.

The volume of the garage (in L) is:

$$(6.0 \text{ m} \times 4.0 \text{ m} \times 2.2 \text{ m}) \times \left(\frac{1 \text{ cm}}{0.01 \text{ m}}\right)^3 \times \frac{1 \text{ L}}{1000 \text{ cm}^3} = 5.3 \times 10^4 \text{ L}$$

From part (a), we know the rate of CO production per minute. In one minute the partial pressure of CO will be:

$$P_{\text{CO}} = \frac{nRT}{V} = \frac{(0.112 \text{ mol})\left(0.0821\dfrac{\text{L} \cdot \text{atm}}{\text{mol} \cdot \text{K}}\right)(20 + 273)\text{K}}{5.3 \times 10^4 \text{ L}} = 5.1 \times 10^{-5} \text{ atm CO/min}$$

How many minutes will it take for the partial pressure of CO to reach the lethal level, 1.0×10^{-3} atm?

$$\textbf{? min} = (1.0 \times 10^{-3} \text{ atm CO}) \times \frac{1 \text{ min}}{5.1 \times 10^{-5} \text{ atm CO}} = \textbf{2.0} \times \textbf{10}^{\textbf{1}} \textbf{ min}$$

5.124 **(a)** First, let's convert the concentration of hydrogen from atoms/cm^3 to mol/L. The concentration in mol/L can be substituted into the ideal gas equation to calculate the pressure of hydrogen.

$$\frac{1 \text{ H atom}}{1 \text{ cm}^3} \times \frac{1 \text{ mol H}}{6.022 \times 10^{23} \text{ H atoms}} \times \frac{1 \text{ cm}^3}{1 \text{ mL}} \times \frac{1 \text{ mL}}{0.001 \text{ L}} = \frac{2 \times 10^{-21} \text{ mol H}}{\text{L}}$$

The pressure of H is:

$$P = \left(\frac{n}{V}\right)RT = \left(\frac{2 \times 10^{-21} \text{ mol}}{1 \text{ L}}\right)\left(0.0821\frac{\text{L}\cdot\text{atm}}{\text{mol}\cdot\text{K}}\right)(3 \text{ K}) = \mathbf{5 \times 10^{-22} \text{ atm}}$$

(b)   From part (a), we know that 1 L contains 2×10^{-21} mole of H atoms. We convert to the volume that contains 1.0 g of H atoms.

$$\frac{1 \text{ L}}{2 \times 10^{-21} \text{ mol H}} \times \frac{1 \text{ mol H}}{1.008 \text{ g H}} = \mathbf{5 \times 10^{20} \text{ L/g of H}}$$

Note: This volume is about that of all the water on Earth!

5.125   **(a)**   First, convert density to units of g/L.

$$\frac{0.426 \text{ kg}}{1 \text{ m}^3} \times \frac{1000 \text{ g}}{1 \text{ kg}} \times \left(\frac{0.01 \text{ m}}{1 \text{ cm}}\right)^3 \times \frac{1000 \text{ cm}^3}{1 \text{ L}} = 0.426 \text{ g/L}$$

Let's assume a volume of 1.00 L of air. This air sample will have a mass of 0.426 g. Converting to moles of air:

$$0.426 \text{ g air} \times \frac{1 \text{ mol air}}{29.0 \text{ g air}} = 0.0147 \text{ mol air}$$

Now, we can substitute into the ideal gas equation to calculate the air temperature.

$$T = \frac{PV}{nR} = \frac{\left(210 \text{ mmHg} \times \dfrac{1 \text{ atm}}{760 \text{ mmHg}}\right)(1.00 \text{ L})}{(0.0147 \text{ mol})\left(0.0821\dfrac{\text{L}\cdot\text{atm}}{\text{mol}\cdot\text{K}}\right)} = \mathbf{229 \text{ K}} = \mathbf{-44°C}$$

(b)   To determine the percent decrease in oxygen gas, let's compare moles of O_2 at the top of Mt. Everest to the moles of O_2 at sea level.

$$\frac{n_{O_2} \text{(Mt. Everest)}}{n_{O_2} \text{(sea level)}} = \frac{\dfrac{P_{O_2} \text{(Mt. Everest)}V}{RT}}{\dfrac{P_{O_2} \text{(sea level)}V}{RT}}$$

$$\frac{n_{O_2} \text{(Mt. Everest)}}{n_{O_2} \text{(sea level)}} = \frac{P_{O_2} \text{(Mt. Everest)}}{P_{O_2} \text{(sea level)}} = \frac{210 \text{ mmHg}}{760 \text{ mmHg}} = 0.276$$

This calculation indicates that there is only 27.6% as much oxygen at the top of Mt. Everest compared to sea level. Therefore, the percent decrease in oxygen gas from sea level to the top of Mt. Everest is **72.4%**.

5.126   From Table 5.3, the equilibrium vapor pressure at 30°C is 31.82 mmHg.

Converting 3.9×10^3 Pa to units of mmHg:

$$(3.9 \times 10^3 \text{ Pa}) \times \frac{760 \text{ mmHg}}{1.01325 \times 10^5 \text{ Pa}} = 29 \text{ mmHg}$$

$$\text{Relative Humidity} = \frac{\text{partial pressure of water vapor}}{\text{equilibrium vapor pressure}} \times 100\% = \frac{29 \text{ mmHg}}{31.82 \text{ mmHg}} \times 100\% = \mathbf{91\%}$$

5.127 At the same temperature and pressure, the same volume contains the same number of moles of gases. Since water has a lower molar mass (18.02 g/mol) than air (about 29 g/mol), moisture laden air weighs less than dry air.

Under conditions of constant temperature and volume, moist air exerts a lower pressure than dry air. Hence, a low-pressure front indicates that the air is moist.

5.128 The volume of one alveoli is:

$$V = \frac{4}{3}\pi r^3 = \frac{4}{3}\pi(0.0050 \text{ cm})^3 = (5.2 \times 10^{-7} \text{ cm}^3) \times \frac{1 \text{ mL}}{1 \text{ cm}^3} \times \frac{0.001 \text{ L}}{1 \text{ mL}} = 5.2 \times 10^{-10} \text{ L}$$

The number of moles of air in one alveoli can be calculated using the ideal gas equation.

$$n = \frac{PV}{RT} = \frac{(1.0 \text{ atm})(5.2 \times 10^{-10} \text{ L})}{\left(0.0821\dfrac{\text{L} \cdot \text{atm}}{\text{mol} \cdot \text{K}}\right)(37 + 273)\text{K}} = 2.0 \times 10^{-11} \text{ mol of air}$$

Since the air inside the alveoli is 14 percent oxygen, the moles of oxygen in one alveoli equals:

$$(2.0 \times 10^{-11} \text{ mol of air}) \times \frac{14\% \text{ oxygen}}{100\% \text{ air}} = 2.8 \times 10^{-12} \text{ mol } O_2$$

Converting to O_2 molecules:

$$(2.8 \times 10^{-12} \text{ mol } O_2) \times \frac{6.022 \times 10^{23} \, O_2 \text{ molecules}}{1 \text{ mol } O_2} = \mathbf{1.7 \times 10^{12} \, O_2 \text{ molecules}}$$

5.129 **(a)** We can calculate the moles of mercury vapor using the ideal gas equation, but first we need to know the volume of the room in liters.

$$V_{\text{room}} = (15.2 \text{ m})(6.6 \text{ m})(2.4 \text{ m}) \times \left(\frac{1 \text{ cm}}{0.01 \text{ m}}\right)^3 \times \frac{1 \text{ L}}{1000 \text{ cm}^3} = 2.4 \times 10^5 \text{ L}$$

$$n_{\text{Hg}} = \frac{PV}{RT} = \frac{(1.7 \times 10^{-6} \text{ atm})(2.4 \times 10^5 \text{ L})}{\left(0.0821\dfrac{\text{L} \cdot \text{atm}}{\text{mol} \cdot \text{K}}\right)(20 + 273)\text{K}} = 0.017 \text{ mol Hg}$$

Converting to mass:

$$? \text{ g Hg} = 0.017 \text{ mol Hg} \times \frac{200.6 \text{ g Hg}}{1 \text{ mol Hg}} = \mathbf{3.4 \text{ g Hg}}$$

(b) The concentration of Hg vapor in the room is:

$$\frac{3.4 \text{ g Hg}}{(15.2 \text{ m})(6.6 \text{ m})(2.4 \text{ m})} = 0.014 \text{ g Hg/m}^3 = \mathbf{14 \text{ mg Hg/m}^3}$$

Yes, this far exceeds the air quality regulation of 0.050 mg Hg/m^3 of air.

(c)   Physical: The sulfur powder covers the Hg surface, thus retarding the rate of evaporation.

Chemical: Sulfur reacts slowly with Hg to form HgS. HgS has no measurable vapor pressure.

5.130   The molar mass of a gas can be calculated using Equation (5.12) of the text.

$$\mathcal{M} = \frac{dRT}{P} = \frac{\left(1.33\ \frac{g}{L}\right)\left(0.0821\frac{L\cdot atm}{mol\cdot K}\right)(150 + 273)K}{\left(764\ mmHg \times \frac{1\ atm}{760\ mmHg}\right)} = 45.9\ g/mol$$

Some nitrogen oxides and their molar masses are:

NO   30 g/mol   N$_2$O   44 g/mol   NO$_2$ 46 g/mol

The nitrogen oxide is most likely **NO$_2$**, although N$_2$O cannot be completely ruled out.

5.131   Assuming a volume of 1.00 L, we can determine the average molar mass of the mixture.

$$n_{mixture} = \frac{PV}{RT} = \frac{(0.98\ atm)(1.00\ L)}{\left(0.0821\frac{L\cdot atm}{mol\cdot K}\right)(25 + 273)K} = 0.0401\ mol$$

The average molar mass of the mixture is:

$$\overline{\mathcal{M}} = \frac{2.7\ g}{0.0401\ mol} = \textbf{67 g/mol}$$

Now, we can calculate the mole fraction of each component of the mixture. Once we determine the mole fractions, we can calculate the partial pressure of each gas.

$$X_{NO_2}\mathcal{M}_{NO_2} + X_{N_2O_4}\mathcal{M}_{N_2O_4} = 67\ g/mol$$

$$X_{NO_2}(46.01\ g/mol) + (1 - X_{NO_2})(92.02\ g/mol) = 67\ g/mol$$

$$46.01X_{NO_2} - 92.02X_{NO_2} + 92.02 = 67$$

$$X_{NO_2} = 0.54 \qquad and \qquad X_{N_2O_4} = 1 - 0.54 = 0.46$$

Finally, the partial pressures are:

$$P_{NO_2} = X_{NO_2}P_T \qquad\qquad\qquad P_{N_2O_4} = X_{N_2O_4}P_T$$

$$\textbf{\textit{P}}_{\textbf{NO}_2} = (0.54)(0.98\ atm) = \textbf{0.53 atm} \qquad \textbf{\textit{P}}_{\textbf{N}_2\textbf{O}_4} = (0.46)(0.98\ atm) = \textbf{0.45 atm}$$

5.132   When calculating root-mean-square speed, remember that the molar mass must be in units of kg/mol.

$$u_{\textbf{rms}} = \sqrt{\frac{3RT}{\mathcal{M}}} = \sqrt{\frac{3(8.314\ J/mol\cdot K)(1.7 \times 10^{-7}\ K)}{85.47 \times 10^{-3}\ kg/mol}} = \textbf{7.0} \times \textbf{10}^{-3}\ \textbf{m/s}$$

The mass of one Rb atom in kg is:

$$\frac{85.47 \text{ g Rb}}{1 \text{ mol Rb}} \times \frac{1 \text{ mol Rb}}{6.022 \times 10^{23} \text{ Rb atoms}} \times \frac{1 \text{ kg}}{1000 \text{ g}} = 1.419 \times 10^{-25} \text{ kg/Rb atom}$$

$$\overline{KE} = \frac{1}{2} m \overline{u^2} = \frac{1}{2}(1.419 \times 10^{-25} \text{ kg})(7.0 \times 10^{-3} \text{ m/s})^2 = \mathbf{3.5 \times 10^{-30} \text{ J}}$$

5.133 First, calculate the moles of hydrogen gas needed to fill a 4.1 L life belt.

$$n_{H_2} = \frac{PV}{RT} = \frac{(0.97 \text{ atm})(4.1 \text{ L})}{\left(0.0821 \dfrac{\text{L} \cdot \text{atm}}{\text{mol} \cdot \text{K}}\right)(12 + 273)\text{K}} = 0.17 \text{ mol H}_2$$

The balanced equation shows a mole ratio between H_2 and LiH of 1:1. Therefore, 0.17 mole of LiH is needed. Converting to mass in grams:

$$? \text{ g LiH} = 0.17 \text{ mol LiH} \times \frac{7.949 \text{ g LiH}}{1 \text{ mol LiH}} = \mathbf{1.4 \text{ g LiH}}$$

5.134 The molar volume is the volume of 1 mole of gas under the specified conditions.

$$V = \frac{nRT}{P} = \frac{(1 \text{ mol})\left(0.0821 \dfrac{\text{L} \cdot \text{atm}}{\text{mol} \cdot \text{K}}\right)(220 \text{ K})}{\left(6.0 \text{ mmHg} \times \dfrac{1 \text{ atm}}{760 \text{ mmHg}}\right)} = \mathbf{2.3 \times 10^3 \text{ L}}$$

5.135 $P_{CO_2} = (0.965) \times (9.0 \times 10^6 \text{ Pa}) = \mathbf{8.7 \times 10^6 \text{ Pa}}$

$P_{N_2} = (0.035) \times (9.0 \times 10^6 \text{ Pa}) = \mathbf{3.2 \times 10^5 \text{ Pa}}$

$P_{SO_2} = (1.5 \times 10^{-4}) \times (9.0 \times 10^6 \text{ Pa}) = \mathbf{1.4 \times 10^3 \text{ Pa}}$

5.136 The volume of the bulb can be calculated using the ideal gas equation. Pressure and temperature are given in the problem. Moles of air must be calculated before the volume can be determined.

Mass of air $= 91.6843 \text{ g} - 91.4715 \text{ g} = 0.2128 \text{ g air}$

Molar mass of air $= (0.78 \times 28.02 \text{ g/mol}) + (0.21 \times 32.00 \text{ g/mol}) + (0.01 \times 39.95 \text{ g/mol}) = 29 \text{ g/mol}$

moles air $= 0.2128 \text{ g air} \times \dfrac{1 \text{ mol air}}{29 \text{ g air}} = 7.3 \times 10^{-3} \text{ mol air}$

Now, we can calculate the volume of the bulb.

$$V_{bulb} = \frac{nRT}{P} = \frac{(7.3 \times 10^{-3} \text{ mol})\left(0.0821 \dfrac{\text{L} \cdot \text{atm}}{\text{mol} \cdot \text{K}}\right)(23 + 273)\text{K}}{\left(744 \text{ mmHg} \times \dfrac{1 \text{ atm}}{760 \text{ mmHg}}\right)} = 0.18 \text{ L} = \mathbf{1.8 \times 10^2 \text{ mL}}$$

5.137 **(a)** **(i)** Since the two He samples are at the same temperature, their rms speeds and the average kinetic energies are the same.

(ii) The He atoms in V_1 (smaller volume) collide with the walls more frequently. Since the average kinetic energies are the same, the force exerted in the collisions is the same in both flasks.

(b) **(i)** The rms speed is greater at the higher temperature, T_2.

(ii) The He atoms at the higher temperature, T_2, collide with the walls with greater frequency and with greater force.

(c) **(i)** False. The rms speed is greater for the lighter gas, He.

(ii) True. The gases are at the same temperature.

(iii) True. $u_{rms} = \sqrt{\dfrac{(3)\left(8.314\dfrac{J}{mol \cdot K}\right)(74 + 273)K}{4.003 \times 10^{-3} \, kg/mol}} = \mathbf{1.47 \times 10^3 \ m/s}$

5.138 In Problem 5.98, the mass of the Earth's atmosphere was determined to be 5.25×10^{18} kg. Assuming that the molar mass of air is 29.0 g/mol, we can calculate the number of molecules in the atmosphere.

(a) $(5.25 \times 10^{18} \text{ kg air}) \times \dfrac{1000 \text{ g}}{1 \text{ kg}} \times \dfrac{1 \text{ mol air}}{29.0 \text{ g air}} \times \dfrac{6.022 \times 10^{23} \text{ molecules air}}{1 \text{ mol air}} = \mathbf{1.09 \times 10^{44} \text{ molecules}}$

(b) First, calculate the moles of air exhaled in every breath. (500 mL = 0.500 L)

$n = \dfrac{PV}{RT} = \dfrac{(1 \text{ atm})(0.500 \text{ L})}{\left(0.0821\dfrac{L \cdot atm}{mol \cdot K}\right)(37 + 273)K} = 1.96 \times 10^{-2} \text{ mol air/breath}$

Next, convert to molecules of air per breath.

$1.96 \times 10^{-2} \text{ mol air/breath} \times \dfrac{6.022 \times 10^{23} \text{ molecules air}}{1 \text{ mol air}} = \mathbf{1.18 \times 10^{22} \text{ molecules/breath}}$

(c) $\dfrac{1.18 \times 10^{22} \text{ molecules}}{1 \text{ breath}} \times \dfrac{12 \text{ breaths}}{1 \text{ min}} \times \dfrac{60 \text{ min}}{1 \text{ h}} \times \dfrac{24 \text{ h}}{1 \text{ day}} \times \dfrac{365 \text{ days}}{1 \text{ yr}} \times 35 \text{ yr} = \mathbf{2.60 \times 10^{30} \text{ molecules}}$

(d) Fraction of molecules in the atmosphere exhaled by Mozart is:

$\dfrac{2.60 \times 10^{30} \text{ molecules}}{1.09 \times 10^{44} \text{ molecules}} = \mathbf{2.39 \times 10^{-14}}$

Or,

$\dfrac{1}{2.39 \times 10^{-14}} = 4.18 \times 10^{13}$

Thus, about 1 molecule of air in every 4×10^{13} molecules was exhaled by Mozart.

In a single breath containing 1.18×10^{22} molecules, we would breathe in on average:

$$(1.18 \times 10^{22} \text{ molecules}) \times \frac{1 \text{ Mozart air molecule}}{4 \times 10^{13} \text{ air molecules}} = \textbf{3} \times \textbf{10}^8 \textbf{ molecules that Mozart exhaled}$$

(e)   We made the following assumptions:

1.   Complete mixing of air in the atmosphere.
2.   That no molecules escaped to the outer atmosphere.
3.   That no molecules were used up during metabolism, nitrogen fixation, and so on.

5.139   (a)   The plots dip due to intermolecular attractions between gas particles. Consider the approach of a particular molecule toward the wall of a container. The intermolecular attractions exerted by its neighbors tend to soften the impact made by this molecule against the wall. The overall effect is a lower gas pressure than we would expect for an ideal gas. Thus, PV/RT decreases. The plots rise because at higher pressures (smaller volumes), the molecules are close together and repulsive forces among them become dominant. Repulsive forces increase the force of impact of the gas molecules with the walls of the container. The overall effect is a greater gas pressure than we would expect for an ideal gas. Hence, $PV/RT > 1$ and the curves rise above the horizontal line.

(b)   For 1 mole of an ideal gas, $PV/RT = 1$, no matter what the pressure of the gas. At very low pressures, all gases behave ideally; therefore, PV/RT converges to 1 as the pressure approaches zero. As the pressure of a gas approaches zero, a gas behaves more and more like an ideal gas.

(c)   The intercept on the ideal gas line means that $PV/RT = 1$. However, this does ***not*** mean that the gas behaves ideally. It just means that at this particular pressure molecular attraction is equal to molecular repulsion so the net interaction is zero.

5.140   The ideal gas law can be used to calculate the moles of water vapor per liter.

$$\frac{n}{V} = \frac{P}{RT} = \frac{1.0 \text{ atm}}{(0.0821 \frac{\text{L} \cdot \text{atm}}{\text{mol} \cdot \text{K}})(100 + 273)\text{K}} = 0.033 \frac{\text{mol}}{\text{L}}$$

We eventually want to find the distance between molecules. Therefore, let's convert moles to molecules, and convert liters to a volume unit that will allow us to get to distance (m^3).

$$\left(\frac{0.033 \text{ mol}}{1 \text{ L}}\right)\left(\frac{6.022 \times 10^{23} \text{ molecules}}{1 \text{ mol}}\right)\left(\frac{1000 \text{ L}}{1 \text{ m}^3}\right) = 2.0 \times 10^{25} \frac{\text{molecules}}{\text{m}^3}$$

This is the number of ideal gas molecules in a cube that is 1 meter on each side. Assuming an equal distribution of molecules along the three mutually perpendicular directions defined by the cube, a linear density in one direction may be found:

$$\left(\frac{2.0 \times 10^{25} \text{ molecules}}{1 \text{ m}^3}\right)^{\frac{1}{3}} = 2.7 \times 10^8 \frac{\text{molecules}}{\text{m}}$$

This is the number of molecules on a line *one* meter in length. The distance between each molecule is given by:

$$\frac{1 \text{ m}}{2.70 \times 10^8} = 3.7 \times 10^{-9} \text{ m} = \textbf{3.7 nm}$$

Assuming a water molecule to be a sphere with a diameter of 0.3 nm, the water molecules are separated by over 12 times their diameter: $\dfrac{3.7 \text{ nm}}{0.3 \text{ nm}} \approx 12$ times.

A similar calculation is done for liquid water. Starting with density, we convert to molecules per cubic meter.

$$\frac{0.96 \text{ g}}{1 \text{ cm}^3} \times \frac{1 \text{ mol H}_2\text{O}}{18.02 \text{ g H}_2\text{O}} \times \frac{6.022 \times 10^{23} \text{ molecules}}{1 \text{ mol H}_2\text{O}} \times \left(\frac{100 \text{ cm}}{1 \text{ m}}\right)^3 = 3.2 \times 10^{28} \frac{\text{molecules}}{\text{m}^3}$$

This is the number of liquid water molecules in *one* cubic meter. From this point, the calculation is the same as that for water vapor, and the space between molecules is found using the same assumptions.

$$\left(\frac{3.2 \times 10^{28} \text{ molecules}}{1 \text{ m}^3}\right)^{\frac{1}{3}} = 3.2 \times 10^9 \frac{\text{molecules}}{\text{m}}$$

$$\frac{1 \text{ m}}{3.2 \times 10^9} = 3.1 \times 10^{-10} \text{ m} = \mathbf{0.31 \text{ nm}}$$

Assuming a water molecule to be a sphere with a diameter of 0.3 nm, to one significant figure, the water molecules are touching each other in the liquid phase.

5.141   **Radon**, because it is radioactive so that its mass is constantly changing (decreasing). The number of radon atoms is not constant.

5.142   Since the $R = 8.314$ J/mol·K and $1 \text{ J} = 1 \dfrac{\text{kg} \cdot \text{m}^2}{\text{s}^2}$, then the mass substituted into the equation must have units of kg and the height must have units of meters.

29 g/mol = 0.029 kg/mol
5.0 km = 5.0×10^3 m

Substituting the given quantities into the equation, we find the atmospheric pressure at 5.0 km to be:

$$P = P_0 e^{-\frac{g\mathcal{M}h}{RT}}$$

$$P = (1.0 \text{ atm})e^{-\left(\frac{(9.8 \text{ m/s}^2)(0.029 \text{ kg/mol})(5.0 \times 10^3 \text{ m})}{(8.314 \text{ J/mol·K})(278 \text{ K})}\right)}$$

$$\mathbf{P = 0.54 \text{ atm}}$$

5.143   We need to find the total moles of gas present in the flask after the reaction. Then, we can use the ideal gas equation to calculate the total pressure inside the flask. Since we are given the amounts of both reactants, we need to find which reactant is used up first. This is a limiting reagent problem. Let's calculate the moles of each reactant present.

$$5.72 \text{ g C} \times \frac{1 \text{ mol C}}{12.01 \text{ g C}} = 0.476 \text{ mol C}$$

$$68.4 \text{ g } O_2 \times \frac{1 \text{ mol } O_2}{32.00 \text{ g } O_2} = 2.14 \text{ mol } O_2$$

The mole ratio between C and O_2 in the balanced equation is 1:1. Therefore, C is the limiting reagent. The amount of C remaining after complete reaction is 0 moles. Since the mole ratio between C and O_2 is 1:1, the amount of O_2 that reacts is 0.476 mole. The amount of O_2 remaining after reaction is:

moles O_2 remaining = moles O_2 initial − moles O_2 reacted = 2.14 mol − 0.476 mol = 1.66 mol O_2

The amount of CO_2 produced in the reaction is :

$$0.476 \text{ mol C} \times \frac{1 \text{ mol } CO_2}{1 \text{ mol C}} = 0.476 \text{ mol } CO_2$$

The total moles of gas present after reaction are:

total mol of gas = mol CO_2 + mol O_2 = 0.476 mol + 1.66 mol = 2.14 mol

Using the ideal gas equation, we can now calculate the total pressure inside the flask.

$$P = \frac{nRT}{V} = \frac{(2.14 \text{ mol})(0.0821 \text{ L} \cdot \text{atm} / \text{mol} \cdot \text{K})(182 + 273)\text{K}}{8.00 \text{ L}} = \textbf{9.99 atm}$$

5.144   The reaction between Zn and HCl is: $Zn(s) + 2HCl(aq) \rightarrow H_2(g) + ZnCl_2(aq)$

From the amount of $H_2(g)$ produced, we can determine the amount of Zn reacted. Then, using the original mass of the sample, we can calculate the mass % of Zn in the sample.

$$n_{H_2} = \frac{PV_{H_2}}{RT}$$

$$n_{H_2} = \frac{\left(728 \text{ mmHg} \times \dfrac{1 \text{ atm}}{760 \text{ mmHg}}\right)(1.26 \text{ L})}{\left(0.0821 \dfrac{\text{L} \cdot \text{atm}}{\text{mol} \cdot \text{K}}\right)(22 + 273)\text{K}} = 0.0498 \text{ mol } H_2$$

Since the mole ratio between H_2 and Zn is 1:1, the amount of Zn reacted is also 0.0498 mole. Converting to grams of Zn, we find:

$$0.0498 \text{ mol Zn} \times \frac{65.39 \text{ g Zn}}{1 \text{ mol Zn}} = 3.26 \text{ g Zn}$$

The mass percent of Zn in the 6.11 g sample is:

$$\textbf{mass \% Zn} = \frac{\text{mass Zn}}{\text{mass sample}} \times 100\% = \frac{3.26 \text{ g}}{6.11 \text{ g}} \times 100\% = \textbf{53.4\%}$$

5.145   We know that the root-mean-square speed (u_{rms}) of a gas can be calculated as follows:

$$u_{rms} = \sqrt{\frac{3RT}{\mathcal{M}}}$$

The rate of diffusion (r) will be directly proportional to the root-mean-square speed. Gases moving at greater speeds will diffuse faster. For two different gases we can write the rates of diffusion as follows:

$$r_1 = \sqrt{\frac{3RT}{\mathcal{M}_1}} \qquad r_2 = \sqrt{\frac{3RT}{\mathcal{M}_2}}$$

Dividing r_1 by r_2 gives:

$$\frac{r_1}{r_2} = \frac{\sqrt{\dfrac{3RT}{\mathcal{M}_1}}}{\sqrt{\dfrac{3RT}{\mathcal{M}_2}}}$$

Canceling $3RT$ from the equation gives:

$$\frac{r_1}{r_2} = \sqrt{\frac{\dfrac{1}{\mathcal{M}_1}}{\dfrac{1}{\mathcal{M}_2}}} = \sqrt{\frac{\mathcal{M}_2}{\mathcal{M}_1}}$$

Does the derived equation make sense? Assume that gas₁ is a lighter gas (has a smaller molar mass) than gas₂. Dividing a larger molar mass ($\mathcal{M}_2$) by a smaller molar mass ($\mathcal{M}_1$) will give a number larger than 1. This indicates that the lighter gas will diffuse at a faster rate compared to the heavier gas.

5.146 **(a)** The equation to calculate the root-mean-square speed is Equation (5.16) of the text. Let's calculate u_{mp} and u_{rms} at 25°C (298 K). Recall that the molar mass of N_2 must be in units of kg/mol, because the units of R are J/mol·K and 1 J = 1 kg·m²/s².

$$u_{mp} = \sqrt{\frac{2RT}{\mathcal{M}}} \qquad\qquad u_{rms} = \sqrt{\frac{3RT}{\mathcal{M}}}$$

$$u_{mp} = \sqrt{\frac{2(8.314\ \text{J/mol}\cdot\text{K})(298\ \text{K})}{0.02802\ \text{kg/mol}}} \qquad\qquad u_{rms} = \sqrt{\frac{3(8.314\ \text{J/mol}\cdot\text{K})(298\ \text{K})}{0.02802\ \text{kg/mol}}}$$

$$u_{mp} = \textbf{421 m/s} \qquad\qquad u_{rms} = \textbf{515 m/s}$$

The most probable speed (u_{mp}) will always be slower than the root-mean-square speed. We can derive a general relation between the two speeds.

$$\frac{u_{mp}}{u_{rms}} = \frac{\sqrt{\dfrac{2RT}{\mathcal{M}}}}{\sqrt{\dfrac{3RT}{\mathcal{M}}}} = \sqrt{\frac{\dfrac{2RT}{\mathcal{M}}}{\dfrac{3RT}{\mathcal{M}}}}$$

$$\frac{u_{mp}}{u_{rms}} = \sqrt{\frac{2}{3}} = 0.816$$

This relation indicates that the most probable speed (u_{mp}) will be 81.6% of the root-mean-square speed (u_{rms}) at a given temperature.

(b)   We can derive a relationship between the most probable speeds at T_1 and T_2.

$$\frac{u_{mp}(1)}{u_{mp}(2)} = \frac{\sqrt{\dfrac{2RT_1}{\mathcal{M}}}}{\sqrt{\dfrac{2RT_2}{\mathcal{M}}}} = \sqrt{\frac{\dfrac{2RT_1}{\mathcal{M}}}{\dfrac{2RT_2}{\mathcal{M}}}}$$

$$\frac{u_{mp}(1)}{u_{mp}(2)} = \sqrt{\frac{T_1}{T_2}}$$

Looking at the diagram, let's assume that the most probable speed at $T_1 = 300$ K is 500 m/s, and the most probable speed at T_2 is 1000 m/s. Substitute into the above equation to solve for T_2.

$$\frac{500 \text{ m/s}}{1000 \text{ m/s}} = \sqrt{\frac{300 \text{ K}}{T_2}}$$

$$(0.5)^2 = \frac{300}{T_2}$$

$$T_2 = \textbf{1200 K}$$

5.147   When the drum is dented, the volume decreases and therefore we expect the pressure to increase. However, the pressure due to acetone vapor (400 mmHg) will not change as long as the temperature stays at 18°C (vapor pressure is constant at a given temperature). As the pressure increases, more acetone vapor will condense to liquid. Assuming that air does not dissolve in acetone, the pressure inside the drum will increase due an increase in the pressure due to air. Initially, the total pressure is 750 mmHg. The pressure due to air initially is:

$$P_T = P_{air} + P_{acetone}$$

$$P_{air} = P_T - P_{acetone} = 750 \text{ mmHg} - 400 \text{ mmHg} = 350 \text{ mmHg}$$

The initial volume of vapor in the drum is:

$$V_1 = 25.0 \text{ gal} - 15.4 \text{ gal} = 9.6 \text{ gal}$$

When the drum is dented, the volume the vapor occupies decreases to:

$$V_2 = 20.4 \text{ gal} - 15.4 \text{ gal} = 5.0 \text{ gal}$$

The same number of air molecules now occupies a smaller volume. The pressure increases according to Boyle's Law.

$$P_1 V_1 = P_2 V_2$$
$$(350 \text{ mmHg})(9.6 \text{ gal}) = P_2 (5.0 \text{ gal})$$
$$P_2 = 672 \text{ mmHg}$$

This is the pressure due to air. The pressure due to acetone vapor is still 400 mmHg. The total pressure inside the drum after the accident is:

$$P_T = P_{air} + P_{acetone}$$

$$P_T = 672 \text{ mmHg} + 400 \text{ mmHg} = \textbf{1072 mmHg}$$

CHAPTER 6
THERMOCHEMISTRY

6.15 Recall that the work in gas expansion is equal to the product of the external, opposing pressure and the change in volume.

(a) $w = -P\Delta V$

$w = -(0)(5.4 - 1.6)L = 0$

(b) $w = -P\Delta V$

$w = -(0.80 \text{ atm})(5.4 - 1.6)L = -3.0 \text{ L·atm}$

To convert the answer to joules, we write

$$w = -3.0 \text{ L·atm} \times \frac{101.3 \text{ J}}{1 \text{ L·atm}} = -3.0 \times 10^2 \text{ J}$$

(c) $w = -P\Delta V$

$w = -(3.7 \text{ atm})(5.4 - 1.6)L = -14 \text{ L·atm}$

To convert the answer to joules, we write

$$w = -14 \text{ L·atm} \times \frac{101.3 \text{ J}}{1 \text{ L·atm}} = -1.4 \times 10^3 \text{ J}$$

6.16 **(a)** Because the external pressure is zero, no work is done in the expansion.

$$w = -P\Delta V = -(0)(89.3 - 26.7)\text{mL}$$

$$w = 0$$

(b) The external, opposing pressure is 1.5 atm, so

$$w = -P\Delta V = -(1.5 \text{ atm})(89.3 - 26.7)\text{mL}$$

$$w = -94 \text{ mL·atm} \times \frac{0.001 \text{ L}}{1 \text{ mL}} = -0.094 \text{ L·atm}$$

To convert the answer to joules, we write:

$$w = -0.094 \text{ L·atm} \times \frac{101.3 \text{ J}}{1 \text{ L·atm}} = -9.5 \text{ J}$$

(c) The external, opposing pressure is 2.8 atm, so

$$w = -P\Delta V = -(2.8 \text{ atm})(89.3 - 26.7)\text{mL}$$

$$w = (-1.8 \times 10^2 \text{ mL·atm}) \times \frac{0.001 \text{ L}}{1 \text{ mL}} = -0.18 \text{ L·atm}$$

To convert the answer to joules, we write:

$$w = -0.18 \text{ L} \cdot \text{atm} \times \frac{101.3 \text{ J}}{1 \text{ L} \cdot \text{atm}} = -18 \text{ J}$$

6.17 An expansion implies an increase in volume, therefore w must be -325 J (see the defining equation for pressure-volume work.) If the system absorbs heat, q must be $+127$ J. The change in energy (internal energy) is:

$$\Delta E = q + w = 127 \text{ J} - 325 \text{ J} = -198 \text{ J}$$

6.18 **Strategy:** Compression is work done on the gas, so what is the sign for w? Heat is released by the gas to the surroundings. Is this an endothermic or exothermic process? What is the sign for q?

Solution: To calculate the energy change of the gas (ΔE), we need Equation (6.1) of the text. Work of compression is positive and because heat is given off by the gas, q is negative. Therefore, we have:

$$\Delta E = q + w = -26 \text{ kJ} + 74 \text{ kJ} = 48 \text{ kJ}$$

As a result, the energy of the gas increases by 48 kJ.

6.19 We first find the number of moles of hydrogen gas formed in the reaction:

$$50.0 \text{ g Sn} \times \frac{1 \text{ mol Sn}}{118.7 \text{ g Sn}} \times \frac{1 \text{ mol H}_2}{1 \text{ mol Sn}} = 0.421 \text{ mol H}_2$$

The next step is to find the volume occupied by the hydrogen gas under the given conditions. This is the change in volume.

$$V = \frac{nRT}{P} = \frac{(0.421 \text{ mol})(0.0821 \text{ L} \cdot \text{atm} / \text{K} \cdot \text{mol})(298 \text{ K})}{1.00 \text{ atm}} = 10.3 \text{ L H}_2$$

The pressure-volume work done is then:

$$w = -P\Delta V = -(1.00 \text{ atm})(10.3 \text{ L}) = -10.3 \text{ L} \cdot \text{atm} \times \frac{101.3 \text{ J}}{1 \text{ L} \cdot \text{atm}} = -1.04 \times 10^3 \text{ J}$$

6.20 **Strategy:** The work done in gas expansion is equal to the product of the external, opposing pressure and the change in volume.

$$w = -P\Delta V$$

We assume that the volume of liquid water is zero compared to that of steam. How do we calculate the volume of the steam? What is the conversion factor between L·atm and J?

Solution: First, we need to calculate the volume that the water vapor will occupy (V_f).

Using the ideal gas equation:

$$V_{H_2O} = \frac{n_{H_2O} RT}{P} = \frac{(1 \text{ mol})\left(0.0821 \dfrac{\text{L} \cdot \text{atm}}{\text{mol} \cdot \text{K}}\right)(373 \text{ K})}{(1.0 \text{ atm})} = 31 \text{ L}$$

It is given that the volume occupied by liquid water is negligible. Therefore,

$$\Delta V = V_f - V_i = 31\,L - 0\,L = 31\,L$$

Now, we substitute P and ΔV into Equation (6.3) of the text to solve for w.

$$w = -P\Delta V = -(1.0\,\text{atm})(31\,L) = 31\,\text{L·atm}$$

The problems asks for the work done in units of joules. The following conversion factor can be obtained from Appendix 2 of the text.

$$1\,\text{L·atm} = 101.3\,J$$

Thus, we can write:

$$w = -31\,\text{L·atm} \times \frac{101.3\,J}{1\,\text{L·atm}} = -3.1 \times 10^3\,\textbf{J}$$

Check: Because this is gas expansion (work is done by the system on the surroundings), the work done has a negative sign.

6.25   The equation as written shows that 879 kJ of heat is released when two moles of ZnS react. We want to calculate the amount of heat released when 1 g of ZnS reacts.

Let $\Delta H°$ be the heat change per gram of ZnS roasted. We write:

$$\Delta H° = \frac{-879\,kJ}{2\,\text{mol ZnS}} \times \frac{1\,\text{mol ZnS}}{97.46\,\text{g ZnS}} = -4.51\,\textbf{kJ/g ZnS}$$

6.26   **Strategy:** The thermochemical equation shows that for every 2 moles of NO_2 produced, 114.6 kJ of heat are given off (note the negative sign). We can write a conversion factor from this information.

$$\frac{-114.6\,kJ}{2\,\text{mol NO}_2}$$

How many moles of NO_2 are in 1.26×10^4 g of NO_2? What conversion factor is needed to convert between grams and moles?

Solution: We need to first calculate the number of moles of NO_2 in 1.26×10^4 g of the compound. Then, we can convert to the number of kilojoules produced from the exothermic reaction. The sequence of conversions is:

$$\text{grams of NO}_2 \rightarrow \text{moles of NO}_2 \rightarrow \text{kilojoules of heat generated}$$

Therefore, the heat given off is:

$$(1.26 \times 10^4\,\text{g NO}_2) \times \frac{1\,\text{mol NO}_2}{46.01\,\text{g NO}_2} \times \frac{-114.6\,kJ}{2\,\text{mol NO}_2} = -1.57 \times 10^4\,\textbf{kJ}$$

6.27   We can calculate ΔE using Equation (6.10) of the text.

$$\Delta E = \Delta H - RT\Delta n$$

We initially have 2.0 moles of gas. Since our products are 2.0 moles of H_2 and 1.0 mole of O_2, there is a net gain of 1 mole of gas (2 reactant $\rightarrow$ 3 product). Thus, $\Delta n = +1$. Looking at the equation given in the problem, it requires 483.6 kJ to decompose 2.0 moles of water ($\Delta H = 483.6$ kJ). Substituting into the above equation:

$$\Delta E = 483.6 \times 10^3 \text{ J} - (8.314 \text{ J/mol·K})(398 \text{ K})(+1 \text{ mol})$$

$$\Delta E = \mathbf{4.80 \times 10^5 \text{ J}} = \mathbf{4.80 \times 10^2 \text{ kJ}}$$

6.28   We initially have 6 moles of gas (3 moles of chlorine and 3 moles of hydrogen). Since our product is 6 moles of hydrogen chloride, there is no change in the number of moles of gas. Therefore there is no volume change; $\Delta V = 0$.

$$w = -P\Delta V = -(1 \text{ atm})(0 \text{ L}) = 0$$

$-P\Delta V = 0$, so

$$\Delta E^\circ = \Delta H^\circ - P\Delta V$$

$$\Delta E = \Delta H$$

$$\Delta H = 3\Delta H^\circ_{\text{rxn}} = 3(-184.6 \text{ kJ/mol}) = -553.8 \text{ kJ/mol}$$

We need to multiply $\Delta H^\circ_{\text{rxn}}$ by three, because the question involves the formation of 6 moles of HCl; whereas, the equation as written only produces 2 moles of HCl.

$$\Delta E^\circ = \Delta H^\circ = \mathbf{-553.8 \text{ kJ/mol}}$$

6.33 $\textbf{Specific heat} = \dfrac{C}{m} = \dfrac{85.7 \text{ J/°C}}{362 \text{ g}} = \mathbf{0.237 \text{ J/g·°C}}$

6.34 $q = m_{\text{Cu}}s_{\text{Cu}}\Delta t = (6.22 \times 10^3 \text{ g})(0.385 \text{ J/g·°C})(324.3°C - 20.5°C) = 7.28 \times 10^5 \text{ J} = \mathbf{728 \text{ kJ}}$

6.35   See Table 6.1 of the text for the specific heat of Hg.

$$q = ms\Delta t = (366 \text{ g})(0.139 \text{ J/g·°C})(12.0°C - 77.0°C) = \mathbf{-3.31 \times 10^3 \text{ J}} = \mathbf{-3.31 \text{ kJ}}$$

6.36   **Strategy:** We know the masses of gold and iron as well as the initial temperatures of each. We can look up the specific heats of gold and iron in Table 6.2 of the text. Assuming no heat is lost to the surroundings, we can equate the heat lost by the iron sheet to the heat gained by the gold sheet. With this information, we can solve for the final temperature of the combined metals.

Solution: Treating the calorimeter as an isolated system (no heat lost to the surroundings), we can write:

$$q_{\text{Au}} + q_{\text{Fe}} = 0$$

or

$$q_{\text{Au}} = -q_{\text{Fe}}$$

The heat gained by the gold sheet is given by:

$$q_{\text{Au}} = m_{\text{Au}}s_{\text{Au}}\Delta t = (10.0 \text{ g})(0.129 \text{ J/g·°C})(t_f - 18.0)°C$$

where m and s are the mass and specific heat, and $\Delta t = t_{\text{final}} - t_{\text{initial}}$.

The heat lost by the iron sheet is given by:

$$q_{\text{Fe}} = m_{\text{Fe}}s_{\text{Fe}}\Delta t = (20.0 \text{ g})(0.444 \text{ J/g·°C})(t_f - 55.6)°C$$

Substituting into the equation derived above, we can solve for t_f.

$$q_{Au} = -q_{Fe}$$

$$(10.0 \text{ g})(0.129 \text{ J/g·°C})(t_f - 18.0)°C = -(20.0 \text{ g})(0.444 \text{ J/g·°C})(t_f - 55.6)°C$$

$$1.29 \, t_f - 23.2 = -8.88 \, t_f + 494$$

$$10.2 \, t_f = 517$$

$$t_f = \mathbf{50.7°C}$$

Check: Must the final temperature be between the two starting values?

6.37 The heat gained by the calorimeter is:

$$q = C_p \Delta t$$
$$q = (3024 \text{ J/°C})(1.126°C) = 3.405 \times 10^3 \text{ J}$$

The amount of heat given off by burning Mg in kJ/g is:

$$(3.405 \times 10^3 \text{ J}) \times \frac{1 \text{ kJ}}{1000 \text{ J}} \times \frac{1}{0.1375 \text{ g Mg}} = \mathbf{24.76 \text{ kJ/g Mg}}$$

The amount of heat given off by burning Mg in kJ/mol is:

$$\frac{24.76 \text{ kJ}}{1 \text{ g Mg}} \times \frac{24.31 \text{ g Mg}}{1 \text{ mol Mg}} = \mathbf{601.9 \text{ kJ/mol Mg}}$$

If the reaction were endothermic, what would happen to the temperature of the calorimeter and the water?

6.38 **Strategy:** The neutralization reaction is exothermic. 56.2 kJ of heat are released when 1 mole of H^+ reacts with 1 mole of OH^-. Assuming no heat is lost to the surroundings, we can equate the heat lost by the reaction to the heat gained by the combined solution. How do we calculate the heat released during the reaction? Are we reacting 1 mole of H^+ with 1 mole of OH^-? How do we calculate the heat absorbed by the combined solution?

Solution: Assuming no heat is lost to the surroundings, we can write:

$$q_{soln} + q_{rxn} = 0$$

or

$$q_{soln} = -q_{rxn}$$

First, let's set up how we would calculate the heat gained by the solution,

$$q_{soln} = m_{soln} s_{soln} \Delta t$$

where m and s are the mass and specific heat of the solution and $\Delta t = t_f - t_i$.

We assume that the specific heat of the solution is the same as the specific heat of water, and we assume that the density of the solution is the same as the density of water (1.00 g/mL). Since the density is 1.00 g/mL, the mass of 400 mL of solution (200 mL + 200 mL) is 400 g.

Substituting into the equation above, the heat gained by the solution can be represented as:

$$q_{soln} = (4.00 \times 10^2 \text{ g})(4.184 \text{ J/g·°C})(t_f - 20.48°C)$$

Next, let's calculate q_{rxn}, the heat released when 200 mL of 0.862 M HCl are mixed with 200 mL of 0.431 M Ba(OH)$_2$. The equation for the neutralization is:

$$2HCl(aq) + Ba(OH)_2(aq) \longrightarrow 2H_2O(l) + BaCl_2(aq)$$

There is exactly enough Ba(OH)$_2$ to neutralize all the HCl. Note that 2 mole HCl $\triangleq$ 1 mole Ba(OH)$_2$, and that the concentration of HCl is double the concentration of Ba(OH)$_2$. The number of moles of HCl is:

$$(2.00 \times 10^2 \text{ mL}) \times \frac{0.862 \text{ mol HCl}}{1000 \text{ mL}} = 0.172 \text{ mol HCl}$$

The amount of heat released when 1 mole of H$^+$ is reacted is given in the problem (-56.2 kJ/mol). The amount of heat liberated when 0.172 mole of H$^+$ is reacted is:

$$q_{rxn} = 0.172 \text{ mol} \times \frac{-56.2 \times 10^3 \text{ J}}{1 \text{ mol}} = -9.67 \times 10^3 \text{ J}$$

Finally, knowing that the heat lost by the reaction equals the heat gained by the solution, we can solve for the final temperature of the mixed solution.

$$q_{soln} = -q_{rxn}$$
$$(4.00 \times 10^2 \text{ g})(4.184 \text{ J/g}\cdot°\text{C})(t_f - 20.48°\text{C}) = -(-9.67 \times 10^3 \text{ J})$$
$$(1.67 \times 10^3)t_f - (3.43 \times 10^4) = 9.67 \times 10^3 \text{ J}$$
$$t_f = \textbf{26.3°C}$$

6.45   **CH$_4$(g) and H(g).** All the other choices are elements in their most stable form ($\Delta H_f° = 0$). The most stable form of hydrogen is H$_2$(g).

6.46   The standard enthalpy of formation of any element in its most stable form is zero. Therefore, since $\Delta H_f°(O_2) = 0$, **O$_2$** is the more stable form of the element oxygen at this temperature.

6.47   H$_2$O(l) $\rightarrow$ H$_2$O(g)   Endothermic

$$\Delta H_{rxn}° = \Delta H_f°[H_2O(g)] - \Delta H_f°[H_2O(l)] > 0$$

$\Delta H_f°[H_2O(l)]$ is more negative since $\Delta H_{rxn}° > 0$.

You could also solve the problem by realizing that H$_2$O(l) is the stable form of water at 25°C, and therefore will have the more negative $\Delta H_f°$ value.

6.48   **(a)**   Br$_2$(l) is the most stable form of bromine at 25°C; therefore, $\Delta H_f°[Br_2(l)] = 0$. Since Br$_2$(g) is less stable than Br$_2$(l), $\Delta H_f°[Br_2(g)] > 0$.

(b) $I_2(s)$ is the most stable form of iodine at 25°C; therefore, $\Delta H_f^\circ[I_2(s)] = 0$. Since $I_2(g)$ is less stable than $I_2(s)$, $\Delta H_f^\circ[I_2(g)] > 0$.

6.49 $2H_2O_2(l) \rightarrow 2H_2O(l) + O_2(g)$

Because $H_2O(l)$ has a more negative ΔH_f° than $H_2O_2(l)$.

6.50   **Strategy:** What is the reaction for the formation of Ag_2O from its elements? What is the ΔH_f° value for an element in its standard state?

Solution: The balanced equation showing the formation of $Ag_2O(s)$ from its elements is:

$$2Ag(s) + \tfrac{1}{2}O_2(g) \longrightarrow Ag_2O(s)$$

Knowing that the standard enthalpy of formation of any element in its most stable form is zero, and using Equation (6.18) of the text, we write:

$$\Delta H_{rxn}^\circ = \Sigma n \Delta H_f^\circ(\text{products}) - \Sigma m \Delta H_f^\circ(\text{reactants})$$

$$\Delta H_{rxn}^\circ = [\Delta H_f^\circ(Ag_2O)] - [2\Delta H_f^\circ(Ag) + \tfrac{1}{2}\Delta H_f^\circ(O_2)]$$

$$\Delta H_{rxn}^\circ = [\Delta H_f^\circ(Ag_2O)] - [0 + 0]$$

$$\mathbf{\Delta H_f^\circ(Ag_2O) = \Delta H_{rxn}^\circ}$$

In a similar manner, you should be able to show that $\mathbf{\Delta H_f^\circ(CaCl_2) = \Delta H_{rxn}^\circ}$ for the reaction

$$Ca(s) + Cl_2(g) \longrightarrow CaCl_2(s)$$

6.51 $\Delta H^\circ = [\Delta H_f^\circ(CaO) + \Delta H_f^\circ(CO_2)] - \Delta H_f^\circ(CaCO_3)$

$\mathbf{\Delta H^\circ} = [(1)(-635.6 \text{ kJ/mol}) + (1)(-393.5 \text{ kJ/mol})] - (1)(-1206.9 \text{ kJ/mol}) = \mathbf{177.8 \text{ kJ/mol}}$

6.52   **Strategy:** The enthalpy of a reaction is the difference between the sum of the enthalpies of the products and the sum of the enthalpies of the reactants. The enthalpy of each species (reactant or product) is given by the product of the stoichiometric coefficient and the standard enthalpy of formation, ΔH_f°, of the species.

Solution: We use the ΔH_f° values in Appendix 3 and Equation (6.18) of the text.

$$\Delta H_{rxn}^\circ = \Sigma n \Delta H_f^\circ(\text{products}) - \Sigma m \Delta H_f^\circ(\text{reactants})$$

(a) $HCl(g) \rightarrow H^+(aq) + Cl^-(aq)$

$$\Delta H_{rxn}^\circ = \Delta H_f^\circ(H^+) + \Delta H_f^\circ(Cl^-) - \Delta H_f^\circ(HCl)$$

$$-74.9 \text{ kJ/mol} = 0 + \Delta H_f^\circ(Cl^-) - (1)(-92.3 \text{ kJ/mol})$$

$$\mathbf{\Delta H_f^\circ(Cl^-) = -167.2 \text{ kJ/mol}}$$

(b) The neutralization reaction is:

$$H^+(aq) + OH^-(aq) \rightarrow H_2O(l)$$

and,

$$\Delta H_{rxn}^\circ = \Delta H_f^\circ[H_2O(l)] - [\Delta H_f^\circ(H^+) + \Delta H_f^\circ(OH^-)]$$

$$\Delta H_f^\circ[H_2O(l)] = -285.8 \text{ kJ/mol} \text{ (See Appendix 3 of the text.)}$$

$$\Delta H_{rxn}^\circ = (1)(-285.8 \text{ kJ/mol}) - [(1)(0 \text{ kJ/mol}) + (1)(-229.6 \text{ kJ/mol})] = \textbf{–56.2 kJ/mol}$$

6.53 **(a)** $\Delta H^\circ = 2\Delta H_f^\circ(H_2O) - 2\Delta H_f^\circ(H_2) - \Delta H_f^\circ(O_2)$

   $\Delta H^\circ = (2)(-285.8 \text{ kJ/mol}) - (2)(0) - (1)(0) = \textbf{–571.6 kJ/mol}$

   (b) $\Delta H^\circ = 4\Delta H_f^\circ(CO_2) + 2\Delta H_f^\circ(H_2O) - 2\Delta H_f^\circ(C_2H_2) - 5\Delta H_f^\circ(O_2)$

   $\Delta H^\circ = (4)(-393.5 \text{ kJ/mol}) + (2)(-285.8 \text{ kJ/mol}) - (2)(226.6 \text{ kJ/mol}) - (5)(0) = \textbf{–2599 kJ/mol}$

6.54 **(a)** $\Delta H^\circ = [2\Delta H_f^\circ(CO_2) + 2\Delta H_f^\circ(H_2O)] - [\Delta H_f^\circ(C_2H_4) + 3\Delta H_f^\circ(O_2)]]$

   $\Delta H^\circ = [(2)(-393.5 \text{ kJ/mol}) + (2)(-285.8 \text{ kJ/mol})] - [(1)(52.3 \text{ kJ/mol}) + (3)(0)]$

   $\Delta H^\circ = \textbf{–1411 kJ/mol}$

   (b) $\Delta H^\circ = [2\Delta H_f^\circ(H_2O) + 2\Delta H_f^\circ(SO_2)] - [2\Delta H_f^\circ(H_2S) + 3\Delta H_f^\circ(O_2)]$

   $\Delta H^\circ = [(2)(-285.8 \text{ kJ/mol}) + (2)(-296.1 \text{ kJ/mol})] - [(2)(-20.15 \text{ kJ/mol}) + (3)(0)]$

   $\Delta H^\circ = \textbf{–1124 kJ/mol}$

6.55 The given enthalpies are in units of kJ/g. We must convert them to units of kJ/mol.

   (a) $\dfrac{-22.6 \text{ kJ}}{1 \text{ g}} \times \dfrac{32.04 \text{ g}}{1 \text{ mol}} = \textbf{–724 kJ/mol}$

   (b) $\dfrac{-29.7 \text{ kJ}}{1 \text{ g}} \times \dfrac{46.07 \text{ g}}{1 \text{ mol}} = \textbf{–1.37} \times \textbf{10}^\textbf{3} \textbf{ kJ/mol}$

   (c) $\dfrac{-33.4 \text{ kJ}}{1 \text{ g}} \times \dfrac{60.09 \text{ g}}{1 \text{ mol}} = \textbf{–2.01} \times \textbf{10}^\textbf{3} \textbf{ kJ/mol}$

6.56 $\Delta H_{rxn}^\circ = \sum n\Delta H_f^\circ(\text{products}) - \sum m\Delta H_f^\circ(\text{reactants})$

   The reaction is:

$$H_2(g) \longrightarrow H(g) + H(g)$$

and,

$$\Delta H_{rxn}^\circ = [\Delta H_f^\circ(H) + \Delta H_f^\circ(H)] - \Delta H_f^\circ(H_2)$$

$$\Delta H_f^\circ(H_2) = 0$$

$$\Delta H^{\circ}_{rxn} = 436.4 \text{ kJ/mol} = 2\Delta H^{\circ}_{f}(\text{H}) - (1)(0)$$

$$\Delta H^{\circ}_{f}(\text{H}) = \frac{436.4 \text{ kJ/mol}}{2} = \textbf{218.2 kJ/mol}$$

6.57 $\Delta H^{\circ} = 6\Delta H^{\circ}_{f}(\text{CO}_2) + 6\Delta H^{\circ}_{f}(\text{H}_2\text{O}) - [\Delta H^{\circ}_{f}(\text{C}_6\text{H}_{12}) + 9\Delta H^{\circ}_{f}(\text{O}_2)]$

$\boldsymbol{\Delta H^{\circ}} = (6)(-393.5 \text{ kJ/mol}) + (6)(-285.8 \text{ kJ/mol}) - (1)(-151.9 \text{ kJ/mol}) - (1)(0)$

$= \textbf{-3924 kJ/mol}$

Why is the standard heat of formation of oxygen zero?

6.58   The equation as written shows that 879 kJ of heat is released when two moles of ZnS react. We want to calculate the amount of heat released when 1 g of ZnS reacts.

Let ΔH° be the heat change per gram of ZnS roasted. We write:

$$\boldsymbol{\Delta H^{\circ}} = \frac{-879 \text{ kJ}}{2 \text{ mol ZnS}} \times \frac{1 \text{ mol ZnS}}{97.46 \text{ g ZnS}} = \textbf{-4.51 kJ/g ZnS}$$

6.59   The amount of heat given off is:

$$(1.26 \times 10^4 \text{ g NH}_3) \times \frac{1 \text{ mol NH}_3}{17.03 \text{ g NH}_3} \times \frac{-92.6 \text{ kJ}}{2 \text{ mol NH}_3} = \textbf{-3.43} \times \textbf{10}^{\textbf{4}} \textbf{ kJ}$$

6.60 $\Delta H^{\circ}_{rxn} = \sum n\Delta H^{\circ}_{f}(\text{products}) - \sum m\Delta H^{\circ}_{f}(\text{reactants})$

The balanced equation for the reaction is:

$$\text{CaCO}_3(s) \longrightarrow \text{CaO}(s) + \text{CO}_2(g)$$

$$\Delta H^{\circ}_{rxn} = [\Delta H^{\circ}_{f}(\text{CaO}) + \Delta H^{\circ}_{f}(\text{CO}_2)] - \Delta H^{\circ}_{f}(\text{CaCO}_3)$$

$$\Delta H^{\circ}_{rxn} = [(1)(-635.6 \text{ kJ/mol}) + (1)(-393.5 \text{ kJ/mol})] - (1)(-1206.9 \text{ kJ/mol}) = 177.8 \text{ kJ/mol}$$

The enthalpy change calculated above is the enthalpy change if 1 mole of CO_2 is produced. The problem asks for the enthalpy change if 66.8 g of CO_2 are produced. We need to use the molar mass of CO_2 as a conversion factor.

$$\boldsymbol{\Delta H^{\circ}} = 66.8 \text{ g CO}_2 \times \frac{1 \text{ mol CO}_2}{44.01 \text{ g CO}_2} \times \frac{177.8 \text{ kJ}}{1 \text{ mol CO}_2} = \textbf{2.70} \times \textbf{10}^{\textbf{2}} \textbf{ kJ}$$

6.61

Reaction	ΔH° (kJ/mol)
$\text{S(rhombic)} + \text{O}_2(g) \rightarrow \text{SO}_2(g)$	-296.06
$\text{SO}_2(g) \rightarrow \text{S(monoclinic)} + \text{O}_2(g)$	296.36
$\text{S(rhombic)} \rightarrow \text{S(monoclinic)}$	$\Delta H^{\circ}_{rxn} = \textbf{0.30 kJ/mol}$

Which is the more stable allotropic form of sulfur?

6.62 **Strategy:** Our goal is to calculate the enthalpy change for the formation of C_2H_6 from is elements C and H_2. This reaction does not occur directly, however, so we must use an indirect route using the information given in the three equations, which we will call equations (a), (b), and (c).

Solution: Here is the equation for the formation of C_2H_6 from its elements.

$$2C(graphite) + 3H_2(g) \longrightarrow C_2H_6(g) \qquad \Delta H^{\circ}_{rxn} = ?$$

Looking at this reaction, we need two moles of graphite as a reactant. So, we multiply Equation (a) by two to obtain:

(d) $2C(graphite) + 2O_2(g) \longrightarrow 2CO_2(g)$ $\qquad \Delta H^{\circ}_{rxn} = 2(-393.5 \text{ kJ/mol}) = -787.0 \text{ kJ/mol}$

Next, we need three moles of H_2 as a reactant. So, we multiply Equation (b) by three to obtain:

(e) $3H_2(g) + \frac{3}{2}O_2(g) \longrightarrow 3H_2O(l)$ $\qquad \Delta H^{\circ}_{rxn} = 3(-285.8 \text{ kJ/mol}) = -857.4 \text{ kJ/mol}$

Last, we need one mole of C_2H_6 as a product. Equation (c) has two moles of C_2H_6 as a reactant, so we need to reverse the equation and divide it by 2.

(f) $2CO_2(g) + 3H_2O(l) \longrightarrow C_2H_6(g) + \frac{7}{2}O_2(g)$ $\quad \Delta H^{\circ}_{rxn} = \frac{1}{2}(3119.6 \text{ kJ/mol}) = 1559.8 \text{ kJ/mol}$

Adding Equations (d), (e), and (f) together, we have:

Reaction	ΔH° (kJ/mol)
(d) $2C(graphite) + 2O_2(g) \longrightarrow 2CO_2(g)$	−787.0
(e) $3H_2(g) + \frac{3}{2}O_2(g) \longrightarrow 3H_2O(l)$	−857.4
(f) $2CO_2(g) + 3H_2O(l) \longrightarrow C_2H_6(g) + \frac{7}{2}O_2(g)$	1559.8
$2C(graphite) + 3H_2(g) \longrightarrow C_2H_6(g)$	**$\Delta H^{\circ} = -84.6$ kJ/mol**

6.63

Reaction	ΔH° (kJ/mol)
$CO_2(g) + 2H_2O(l) \rightarrow CH_3OH(l) + \frac{3}{2}O_2(g)$	726.4
$C(graphite) + O_2(g) \rightarrow CO_2(g)$	−393.5
$2H_2(g) + O_2(g) \rightarrow 2H_2O(l)$	2(−285.8)
$C(graphite) + 2H_2(g) + \frac{1}{2}O_2(g) \rightarrow CH_3OH(l)$ $\quad \Delta H^{\circ}_{rxn} = -238.7$ kJ/mol	

We have just calculated an enthalpy at standard conditions, which we abbreviate ΔH°_{rxn}. In this case, the reaction in question was for the formation of *one* mole of CH_3OH *from its elements* in their standard state. Therefore, the ΔH°_{rxn} that we calculated is also, by definition, the standard heat of formation ΔH°_f of CH_3OH (**−238.7 kJ/mol**).

6.64 The second and third equations can be combined to give the first equation.

$$2Al(s) + \frac{3}{2}O_2(g) \longrightarrow Al_2O_3(s) \qquad \Delta H^{\circ} = -1601 \text{ kJ/mol}$$

$$Fe_2O_3(s) \longrightarrow 2Fe(s) + \frac{3}{2}O_2(g) \qquad \Delta H^{\circ} = 821 \text{ kJ/mol}$$

$$2Al(s) + Fe_2O_3(s) \longrightarrow 2Fe(s) + Al_2O_3(s) \qquad \Delta H^{\circ} = -780 \text{ kJ/mol}$$

6.71   In a chemical reaction the same elements and the same numbers of atoms are always on both sides of the equation. This provides a consistent reference which allows the energy change in the reaction to be interpreted in terms of the chemical or physical changes that have occurred. In a nuclear reaction the same elements are not always on both sides of the equation and no common reference point exists.

6.72   Rearrange the equations as necessary so they can be added to yield the desired equation.

$$
\begin{array}{lll}
2B \longrightarrow A & -\Delta H_1 \\
\underline{A \longrightarrow C} & \underline{\Delta H_2} \\
2B \longrightarrow C & \boldsymbol{\Delta H^{\circ} = \Delta H_2 - \Delta H_1}
\end{array}
$$

6.73   The reaction corresponding to standard enthalpy of formation, ΔH_f°, of $AgNO_2(s)$ is:

$$Ag(s) + \tfrac{1}{2} N_2(g) + O_2(g) \rightarrow AgNO_2(s)$$

Rather than measuring the enthalpy directly, we can use the enthalpy of formation of $AgNO_3(s)$ and the ΔH_{rxn}° provided.

$$AgNO_3(s) \rightarrow AgNO_2(s) + \tfrac{1}{2} O_2(g)$$

$$\Delta H_{rxn}^{\circ} = \Delta H_f^{\circ}(AgNO_2) + \tfrac{1}{2}\Delta H_f^{\circ}(O_2) - \Delta H_f^{\circ}(AgNO_3)$$

$$78.67 \text{ kJ/mol} = \Delta H_f^{\circ}(AgNO_2) + 0 - (-123.02 \text{ kJ/mol})$$

$$\boldsymbol{\Delta H_f^{\circ}(AgNO_2) = -44.35 \text{ kJ/mol}}$$

6.74   **(a)** $\Delta H_{rxn}^{\circ} = \sum n \Delta H_f^{\circ}(\text{products}) - \sum m \Delta H_f^{\circ}(\text{reactants})$

$$\Delta H_{rxn}^{\circ} = [4\Delta H_f^{\circ}(NH_3) + \Delta H_f^{\circ}(N_2)] - 3\Delta H_f^{\circ}(N_2H_4)$$

$$\boldsymbol{\Delta H_{rxn}^{\circ}} = [(4)(-46.3 \text{ kJ/mol}) + (0)] - (3)(50.42 \text{ kJ/mol}) = \boldsymbol{-336.5 \text{ kJ/mol}}$$

(b)   The balanced equations are:

(1) $N_2H_4(l) + O_2(g) \longrightarrow N_2(g) + 2H_2O(l)$

(2) $4NH_3(g) + 3O_2(g) \longrightarrow 2N_2(g) + 6H_2O(l)$

The standard enthalpy change for equation (1) is:

$$\Delta H_{rxn}^{\circ} = \Delta H_f^{\circ}(N_2) + 2\Delta H_f^{\circ}[H_2O(l)] - \{\Delta H_f^{\circ}[N_2H_4(l)] + \Delta H_f^{\circ}(O_2)\}$$

$$\Delta H_{rxn}^{\circ} = [(1)(0) + (2)(-285.8 \text{ kJ/mol})] - [(1)(50.42 \text{ kJ/mol}) + (1)(0)] = -622.0 \text{ kJ/mol}$$

The standard enthalpy change for equation (2) is:

$$\Delta H_{rxn}^{\circ} = [2\Delta H_f^{\circ}(N_2) + 6\Delta H_f^{\circ}(H_2O)] - [4\Delta H_f^{\circ}(NH_3) + 3\Delta H_f^{\circ}(O_2)]$$

$$\Delta H_{rxn}^{\circ} = [(2)(0) + (6)(-285.8 \text{ kJ/mol})] - [(4)(-46.3 \text{ kJ/mol}) + (3)(0)] = -1529.6 \text{ kJ/mol}$$

We can now calculate the enthalpy change per kilogram of each substance. ΔH°_{rxn} above is in units of kJ/mol. We need to convert to kJ/kg.

$$N_2H_4(l): \quad \Delta H^{\circ}_{rxn} = \frac{-622.0 \text{ kJ}}{1 \text{ mol } N_2H_4} \times \frac{1 \text{ mol } N_2H_4}{32.05 \text{ g } N_2H_4} \times \frac{1000 \text{ g}}{1 \text{ kg}} = -1.941 \times 10^4 \text{ kJ/kg } N_2H_4$$

$$NH_3(g): \quad \Delta H^{\circ}_{rxn} = \frac{-1529.6 \text{ kJ}}{4 \text{ mol } NH_3} \times \frac{1 \text{ mol } NH_3}{17.03 \text{ g } NH_3} \times \frac{1000 \text{ g}}{1 \text{ kg}} = -2.245 \times 10^4 \text{ kJ/kg } NH_3$$

Since **ammonia, NH₃**, releases more energy per kilogram of substance, it would be a better fuel.

6.75 We initially have 8 moles of gas (2 of nitrogen and 6 of hydrogen). Since our product is 4 moles of ammonia, there is a net loss of 4 moles of gas (8 reactant → 4 product). The corresponding volume loss is

$$V = \frac{nRT}{P} = \frac{(4.0 \text{ mol})(0.0821 \text{ L} \cdot \text{atm} / \text{K} \cdot \text{mol})(298 \text{ K})}{1 \text{ atm}} = 98 \text{ L}$$

$$w = -P\Delta V = -(1 \text{ atm})(-98 \text{ L}) = 98 \text{ L} \cdot \text{atm} \times \frac{101.3 \text{ J}}{1 \text{ L} \cdot \text{atm}} = 9.9 \times 10^3 \text{ J} = 9.9 \text{ kJ}$$

$$\Delta H = \Delta E + P\Delta V \quad \text{or} \quad \Delta E = \Delta H - P\Delta V$$

Using ΔH as -185.2 kJ = (2 × -92.6 kJ), (because the question involves the formation of 4 moles of ammonia, not 2 moles of ammonia for which the standard enthalpy is given in the question), and $-P\Delta V$ as 9.9 kJ (for which we just solved):

$$\Delta E = -185.2 \text{ kJ} + 9.9 \text{ kJ} = -175.3 \text{ kJ}$$

6.76 The reaction is, $2Na(s) + Cl_2(g) \rightarrow 2NaCl(s)$. First, let's calculate $\Delta H°$ for this reaction using ΔH°_f values in Appendix 3.

$$\Delta H^{\circ}_{rxn} = 2\Delta H^{\circ}_f(NaCl) - [2\Delta H^{\circ}_f(Na) + \Delta H^{\circ}_f(Cl_2)]$$

$$\Delta H^{\circ}_{rxn} = 2(-411.0 \text{ kJ/mol}) - [2(0) + 0] = -822.0 \text{ kJ/mol}$$

This is the amount of heat released when 1 mole of Cl_2 reacts (see balanced equation). We are not reacting 1 mole of Cl_2, however. From the volume and density of Cl_2, we can calculate grams of Cl_2. Then, using the molar mass of Cl_2 as a conversion factor, we can calculate moles of Cl_2. Combining these two calculations into one step, we find moles of Cl_2 to be:

$$2.00 \text{ L } Cl_2 \times \frac{1.88 \text{ g } Cl_2}{1 \text{ L } Cl_2} \times \frac{1 \text{ mol } Cl_2}{70.90 \text{ g } Cl_2} = 0.0530 \text{ mol } Cl_2$$

Finally, we can use the ΔH°_{rxn} calculated above to find the amount of heat released when 0.0530 mole of Cl_2 reacts.

$$0.0530 \text{ mol } Cl_2 \times \frac{-822.0 \text{ kJ}}{1 \text{ mol } Cl_2} = -43.6 \text{ kJ}$$

6.77 (a) Although we cannot measure ΔH°_{rxn} for this reaction, the reverse process, is the combustion of glucose. We could easily measure ΔH°_{rxn} for this combustion in a bomb calorimeter.

$$C_6H_{12}O_6(s) + 6O_2(g) \longrightarrow 6CO_2(g) + 6H_2O(l)$$

(b)   We can calculate ΔH°_{rxn} using standard enthalpies of formation.

$$\Delta H^\circ_{rxn} = \Delta H^\circ_f[C_6H_{12}O_6(s)] + 6\Delta H^\circ_f[O_2(g)] - \{6\Delta H^\circ_f[CO_2(g)] + 6\Delta H^\circ_f[H_2O(l)]\}$$

$$\Delta H^\circ_{rxn} = [(1)(-1274.5 \text{ kJ/mol}) + 0] - [(6)(-393.5 \text{ kJ/mol}) + (6)(-285.8 \text{ kJ/mol})] = 2801.3 \text{ kJ/mol}$$

ΔH°_{rxn} has units of kJ/1 mol glucose. We want the ΔH° change for 7.0×10^{14} kg glucose. We need to calculate how many moles of glucose are in 7.0×10^{14} kg glucose. You should come up with the following strategy to solve the problem.

kg glucose $\rightarrow$ g glucose $\rightarrow$ mol glucose $\rightarrow$ kJ (ΔH°)

$$\Delta H^\circ = (7.0 \times 10^{14} \text{ kg}) \times \frac{1000 \text{ g}}{1 \text{ kg}} \times \frac{1 \text{ mol } C_6H_{12}O_6}{180.2 \text{ g } C_6H_{12}O_6} \times \frac{2801.3 \text{ kJ}}{1 \text{ mol } C_6H_{12}O_6} = \mathbf{1.1 \times 10^{19} \text{ kJ}}$$

6.78   The initial and final states of this system are identical. Since enthalpy is a state function, its value depends only upon the state of the system. The enthalpy change is **zero**.

6.79   From the balanced equation we see that there is a 1:2 mole ratio between hydrogen and sodium. The number of moles of hydrogen produced is:

$$0.34 \text{ g Na} \times \frac{1 \text{ mol Na}}{22.99 \text{ g Na}} \times \frac{1 \text{ mol } H_2}{2 \text{ mol Na}} = 7.4 \times 10^{-3} \text{ mol } H_2$$

Using the ideal gas equation, we write:

$$V = \frac{nRT}{P} = \frac{(7.4 \times 10^{-3} \text{ mol})(0.0821 \text{ L} \cdot \text{atm} / \text{K} \cdot \text{mol})(273 \text{ K})}{(1 \text{ atm})} = 0.17 \text{ L } H_2$$

$$\Delta V = 0.17 \text{ L}$$

$$w = -P\Delta V = -(1.0 \text{ atm})(0.17 \text{ L}) = -0.17 \text{ L} \cdot \text{atm} \times \frac{101.3 \text{ J}}{1 \text{ L} \cdot \text{atm}} = \mathbf{-17 \text{ J}}$$

6.80 $H(g) + Br(g) \longrightarrow HBr(g) \qquad \Delta H^\circ_{rxn} = ?$

Rearrange the equations as necessary so they can be added to yield the desired equation.

$$H(g) \longrightarrow \tfrac{1}{2} H_2(g) \qquad\qquad\qquad \Delta H^\circ_{rxn} = \tfrac{1}{2}(-436.4 \text{ kJ/mol}) = -218.2 \text{ kJ/mol}$$

$$Br(g) \longrightarrow \tfrac{1}{2} Br_2(g) \qquad\qquad\qquad \Delta H^\circ_{rxn} = \tfrac{1}{2}(-192.5 \text{ kJ/mol}) = -96.25 \text{ kJ/mol}$$

$$\tfrac{1}{2} H_2(g) + \tfrac{1}{2} Br_2(g) \longrightarrow HBr(g) \qquad \Delta H^\circ_{rxn} = \tfrac{1}{2}(-72.4 \text{ kJ/mol}) = -36.2 \text{ kJ/mol}$$

$$H(g) + Br(g) \longrightarrow 2HBr(g) \qquad \mathbf{\Delta H^\circ = -350.7 \text{ kJ/mol}}$$

6.81   Using the balanced equation, we can write:

$$\Delta H^\circ_{rxn} = [2\Delta H^\circ_f(CO_2) + 4\Delta H^\circ_f(H_2O)] - [2\Delta H^\circ_f(CH_3OH) + 3\Delta H^\circ_f(O_2)]$$

$$-1452.8 \text{ kJ/mol} = (2)(-393.5 \text{ kJ/mol}) + (4)(-285.8 \text{ kJ/mol}) - (2)\Delta H^\circ_f(CH_3OH) - (3)(0 \text{ kJ/mol})$$

$$477.4 \text{ kJ/mol} = -(2) \, \Delta H_f^\circ \, (CH_3OH)$$

$$\Delta H_f^\circ \, (CH_3OH) = \mathbf{-238.7 \text{ kJ/mol}}$$

6.82 $q_{system} = 0 = q_{metal} + q_{water} + q_{calorimeter}$

$q_{metal} + q_{water} + q_{calorimeter} = 0$

$m_{metal}s_{metal}(t_{final} - t_{initial}) + m_{water}s_{water}(t_{final} - t_{initial}) + C_{calorimeter}(t_{final} - t_{initial}) = 0$

All the needed values are given in the problem. All you need to do is plug in the values and solve for s_{metal}.

$(44.0 \text{ g})(s_{metal})(28.4 - 99.0)°C + (80.0 \text{ g})(4.184 \text{ J/g·°C})(28.4 - 24.0)°C + (12.4 \text{ J/°C})(28.4 - 24.0)°C = 0$

$(-3.11 \times 10^3)s_{metal} \, (g \cdot °C) = -1.53 \times 10^3 \text{ J}$

$s_{metal} = \mathbf{0.492 \text{ J/g·°C}}$

6.83 The original volume of ammonia is:

$$V = \frac{nRT}{P} = \frac{(1.00 \text{ mol})(0.0821 \text{ L·atm / K·mol})(298 \text{ K})}{14.0 \text{ atm}} = 1.75 \text{ L NH}_3$$

(a) $T_2 = \dfrac{P_2 V_2 T_1}{P_1 V_1} = \dfrac{(1 \text{ atm})(23.5 \text{ L})(298 \text{ K})}{(14.0 \text{ atm})(1.75 \text{ L})} = 286 \text{ K}$

(b) $\Delta t = (286 - 298)°C = -12°C$

$q = ms\Delta t = (17.03 \text{ g})(0.0258 \text{ J/g·°C})(-12°C) = \mathbf{-5.27 \text{ J}}$

$w = -P\Delta V = -(1 \text{ atm})(23.5 - 1.75)\text{L} = -21.75 \text{ L·atm} \times \dfrac{101.3 \text{ J}}{1 \text{ L·atm}} = \mathbf{-2.20 \times 10^3 \text{ J}}$

$\Delta E = q + w = -5.27 \text{ J} - (2.20 \times 10^3 \text{ J}) = \mathbf{-2.21 \times 10^3 \text{ J} = -2.21 \text{ kJ}}$

6.84 A good starting point would be to calculate the standard enthalpy for both reactions.

Calculate the standard enthalpy for the reaction: $C(s) + \frac{1}{2} O_2(g) \longrightarrow CO(g)$

This reaction corresponds to the standard enthalpy of formation of CO, so we use the value of -110.5 kJ/mol (see Appendix 3 of the text).

Calculate the standard enthalpy for the reaction: $C(s) + H_2O(g) \longrightarrow CO(g) + H_2(g)$

$\Delta H_{rxn}^\circ = [\Delta H_f^\circ(CO) + \Delta H_f^\circ(H_2)] - [\Delta H_f^\circ(C) + \Delta H_f^\circ(H_2O)]$

$\Delta H_{rxn}^\circ = [(1)(-110.5 \text{ kJ/mol}) + (1)(0)] - [(1)(0) + (1)(-241.8 \text{ kJ/mol})] = 131.3 \text{ kJ/mol}$

The first reaction, which is exothermic, can be used to promote the second reaction, which is endothermic. Thus, the two gases are produced alternately.

6.85   Let's start with the combustion of methane: $CH_4(g) + O_2(g) \rightarrow CO_2(g) + 2H_2O(l)$

$$\Delta H^\circ_{rxn} = \Delta H^\circ_f(CO_2) + 2\Delta H^\circ_f(H_2O) - \Delta H^\circ_f(CH_4)$$

$$\Delta H^\circ_{rxn} = (1)(-393.5 \text{ kJ/mol}) + (2)(-285.8 \text{ kJ/mol}) - (1)(-74.85 \text{ kJ/mol}) = -890.3 \text{ kJ/mol}$$

Now, let's calculate the heat produced by the combustion of water gas. We will consider the combustion of H_2 and CO separately. We can look up the ΔH°_f of $H_2O(l)$ in Appendix 3.

$$H_2(g) + \tfrac{1}{2}O_2(g) \rightarrow H_2O(l) \qquad \Delta H^\circ_{rxn} = -285.8 \text{ kJ/mol}$$

For the combustion of CO(g), we use ΔH°_f values from Appendix 3 to calculate the ΔH°_{rxn}.

$$CO(g) + \tfrac{1}{2}O_2(g) \rightarrow CO_2(g) \qquad \Delta H^\circ_{rxn} = ?$$

$$\Delta H^\circ_{rxn} = \Delta H^\circ_f(CO_2) - \Delta H^\circ_f(CO) - \tfrac{1}{2}\Delta H^\circ_f(O_2)$$

$$\Delta H^\circ_{rxn} = (1)(-393.5 \text{ kJ/mol}) - (1)(-110.5 \text{ kJ/mol}) = -283.0 \text{ kJ/mol}$$

The ΔH°_{rxn} values calculated above are for the combustion of 1 mole of H_2 and 1 mole of CO, which equals 2 moles of water gas. The total heat produced during the combustion of *1 mole* of water gas is:

$$\frac{-(285.8 + 283.0)\text{kJ/mol}}{2} = -284.4 \text{ kJ/mol of water gas}$$

which is less heat than that produced by the combustion of 1 mole of methane.

Additionally, CO is very toxic. Natural gas (methane) is easier to obtain compared to carrying out the high temperature process of producing water gas.

6.86   First, calculate the energy produced by 1 mole of octane, C_8H_{18}.

$$C_8H_{18}(l) + \tfrac{25}{2}O_2(g) \longrightarrow 8CO_2(g) + 9H_2O(l)$$

$$\Delta H^\circ_{rxn} = 8\Delta H^\circ_f(CO_2) + 9\Delta H^\circ_f[H_2O(l)] - [\Delta H^\circ_f(C_8H_{18}) + \tfrac{25}{2}\Delta H^\circ_f(O_2)]$$

$$\Delta H^\circ_{rxn} = [(8)(-393.5 \text{ kJ/mol}) + (9)(-285.8 \text{ kJ/mol})] - [(1)(-249.9 \text{ kJ/mol}) + (\tfrac{25}{2})(0)]$$

$$= -5470 \text{ kJ/mol}$$

The problem asks for the energy produced by the combustion of 1 gallon of octane. ΔH°_{rxn} above has units of kJ/mol octane. We need to convert from kJ/mol octane to kJ/gallon octane. The heat of combustion for 1 gallon of octane is:

$$\Delta H^\circ = \frac{-5470 \text{ kJ}}{1 \text{ mol octane}} \times \frac{1 \text{ mol octane}}{114.2 \text{ g octane}} \times \frac{2660 \text{ g}}{1 \text{ gal}} = -1.274 \times 10^5 \text{ kJ/gal}$$

The combustion of hydrogen corresponds to the standard heat of formation of water:

$$H_2(g) + \tfrac{1}{2}O_2(g) \longrightarrow H_2O(l)$$

Thus, ΔH°_{rxn} is the same as ΔH°_f for $H_2O(l)$, which has a value of -285.8 kJ/mol. The number of moles of hydrogen required to produce 1.274×10^5 kJ of heat is:

$$n_{H_2} = (1.274 \times 10^5 \text{ kJ}) \times \frac{1 \text{ mol } H_2}{285.8 \text{ kJ}} = 445.8 \text{ mol } H_2$$

Finally, use the ideal gas law to calculate the volume of gas corresponding to 445.8 moles of H_2 at 25°C and 1 atm.

$$V_{H_2} = \frac{n_{H_2} RT}{P} = \frac{(445.8 \text{ mol})\left(0.0821\dfrac{L \cdot atm}{mol \cdot K}\right)(298 \text{ K})}{(1 \text{ atm})} = \mathbf{1.09 \times 10^4 \text{ L}}$$

That is, the volume of hydrogen that is energy-equivalent to 1 gallon of gasoline is over **10,000 liters** at 1 atm and 25°C!

6.87   The reaction for the combustion of octane is:

$$C_8H_{18}(l) + \tfrac{25}{2} O_2(g) \rightarrow 8CO_2(g) + 9H_2O(l)$$

ΔH°_{rxn} for this reaction was calculated in problem 6.86 from standard enthalpies of formation.

$$\Delta H^\circ_{rxn} = -5470 \text{ kJ/mol}$$

$$\text{Heat/gal of octane} = \frac{5470 \text{ kJ}}{1 \text{ mol } C_8H_{18}} \times \frac{1 \text{ mol } C_8H_{18}}{114.2 \text{ g}} \times \frac{0.7025 \text{ g}}{1 \text{ mL}} \times \frac{3785 \text{ mL}}{1 \text{ gal}}$$

$$\text{Heat/gal of octane} = 1.27 \times 10^5 \text{ kJ/gal gasoline}$$

The reaction for the combustion of ethanol is:

$$C_2H_5OH(l) + 3O_2(g) \rightarrow 2CO_2(g) + 3H_2O(l)$$

$$\Delta H^\circ_{rxn} = 2\Delta H^\circ_f(CO_2) + 3\Delta H^\circ_f(H_2O) - \Delta H^\circ_f(C_2H_5OH) - 3\Delta H^\circ_f(O_2)$$

$$\Delta H^\circ_{rxn} = (2)(-393.5 \text{ kJ/mol}) + (3)(-285.8 \text{ kJ/mol}) - (1)(-277.0 \text{ kJ/mol}) = -1367 \text{ kJ/mol}$$

$$\text{Heat/gal of ethanol} = \frac{1367 \text{ kJ}}{1 \text{ mol } C_2H_5OH} \times \frac{1 \text{ mol } C_2H_5OH}{46.07 \text{ g}} \times \frac{0.7894 \text{ g}}{1 \text{ mL}} \times \frac{3785 \text{ mL}}{1 \text{ gal}}$$

$$\text{Heat/gal of ethanol} = 8.87 \times 10^4 \text{ kJ/gal ethanol}$$

For ethanol, what would the cost have to be to supply the same amount of heat per dollar of gasoline? For gasoline, it cost $1.20 to provide 1.27×10^5 kJ of heat.

$$\frac{\$1.20}{1.27 \times 10^5 \text{ kJ}} \times \frac{8.87 \times 10^4 \text{ kJ}}{1 \text{ gal ethanol}} = \mathbf{\$0.84/gal \ ethanol}$$

6.88   The combustion reaction is: $C_2H_6(l) + \tfrac{7}{2} O_2(g) \longrightarrow 2CO_2(g) + 3H_2O(l)$

The heat released during the combustion of 1 mole of ethane is:

$$\Delta H^\circ_{rxn} = [2\Delta H^\circ_f(CO_2) + 3\Delta H^\circ_f(H_2O)] - [\Delta H^\circ_f(C_2H_6) + \tfrac{7}{2}\Delta H^\circ_f(O_2)]$$

$$\Delta H^\circ_{rxn} = [(2)(-393.5 \text{ kJ/mol}) + (3)(-285.8 \text{ kJ/mol})] - [(1)(-84.7 \text{ kJ/mol}) + (\tfrac{7}{2})(0)] = -1560 \text{ kJ/mol}$$

The heat required to raise the temperature of the water to 98°C is:

$$q = m_{H_2O}s_{H_2O}\Delta t = (855 \text{ g})(4.184 \text{ J/g·°C})(98.0 - 25.0)°C = 2.61 \times 10^5 \text{ J} = 261 \text{ kJ}$$

The combustion of 1 mole of ethane produces 1560 kJ; the number of moles required to produce 261 kJ is:

$$261 \text{ kJ} \times \frac{1 \text{ mol ethane}}{1560 \text{ kJ}} = 0.167 \text{ mol ethane}$$

The volume of ethane is:

$$V_{ethane} = \frac{nRT}{P} = \frac{(0.167 \text{ mol})\left(0.0821\dfrac{\text{L·atm}}{\text{mol·K}}\right)(296 \text{ K})}{\left(752 \text{ mmHg} \times \dfrac{1 \text{ atm}}{760 \text{ mmHg}}\right)} = \textbf{4.10 L}$$

6.89   As energy consumers, we are interested in the availability of ***usable*** energy.

6.90   The heat gained by the liquid nitrogen must be equal to the heat lost by the water.

$$q_{N_2} = -q_{H_2O}$$

If we can calculate the heat lost by the water, we can calculate the heat gained by 60.0 g of the nitrogen.

$$\text{Heat lost by the water} = q_{H_2O} = m_{H_2O}s_{H_2O}\Delta t$$

$$q_{H_2O} = (2.00 \times 10^2 \text{ g})(4.184 \text{ J/g·°C})(41.0 - 55.3)°C = -1.20 \times 10^4 \text{ J}$$

The heat gained by 60.0 g nitrogen is the opposite sign of the heat lost by the water.

$$q_{N_2} = -q_{H_2O}$$

$$q_{N_2} = 1.20 \times 10^4 \text{ J}$$

The problem asks for the molar heat of vaporization of liquid nitrogen. Above, we calculated the amount of heat necessary to vaporize 60.0 g of liquid nitrogen. We need to convert from J/60.0 g N_2 to J/mol N_2.

$$\Delta H_{vap} = \frac{1.20 \times 10^4 \text{ J}}{60.0 \text{ g } N_2} \times \frac{28.02 \text{ g } N_2}{1 \text{ mol } N_2} = \textbf{5.60} \times \textbf{10}^3 \textbf{ J/mol} = \textbf{5.60 kJ/mol}$$

6.91   The evaporation of ethanol is an endothermic process with a fairly high $\Delta H°$ value. When the liquid evaporates, it absorbs heat from your body, hence the cooling effect.

6.92   Recall that the standard enthalpy of formation (ΔH_f°) is defined as the heat change that results when 1 mole of a compound is formed from its elements at a pressure of 1 atm. Only in choice **(a)** does $\Delta H_{rxn}^\circ = \Delta H_f^\circ$. In choice (b), C(diamond) is *not* the most stable form of elemental carbon under standard conditions; C(graphite) is the most stable form.

6.93 $w = -P\Delta V = -(1.0 \text{ atm})(0.0196 - 0.0180)\text{L} = -1.6 \times 10^{-3} \text{ L·atm}$

Using the conversion factor 1 L·atm = 101.3 J:

$$w = (-1.6 \times 10^{-3} \text{ L} \cdot \text{atm}) \times \frac{101.3 \text{ J}}{\text{L} \cdot \text{atm}} = -0.16 \text{ J}$$

0.16 J of work are done by water as it expands on freezing.

6.94　**(a)**　No work is done by a gas expanding in a vacuum, because the pressure exerted on the gas is zero.

(b)　$w = -P\Delta V$

$w = -(0.20 \text{ atm})(0.50 - 0.050)\text{L} = -0.090 \text{ L} \cdot \text{atm}$

Converting to units of joules:

$$w = -0.090 \text{ L} \cdot \text{atm} \times \frac{101.3 \text{ J}}{\text{L} \cdot \text{atm}} = -9.1 \text{ J}$$

(c)　The gas will expand until the pressure is the same as the applied pressure of 0.20 atm. We can calculate its final volume using the ideal gas equation.

$$V = \frac{nRT}{P} = \frac{(0.020 \text{ mol})\left(0.0821\dfrac{\text{L} \cdot \text{atm}}{\text{mol} \cdot \text{K}}\right)(273 + 20)\text{K}}{0.20 \text{ atm}} = 2.4 \text{ L}$$

The amount of work done is:

$w = -P\Delta V = (0.20 \text{ atm})(2.4 - 0.050)\text{L} = -0.47 \text{ L} \cdot \text{atm}$

Converting to units of joules:

$$w = -0.47 \text{ L} \cdot \text{atm} \times \frac{101.3 \text{ J}}{\text{L} \cdot \text{atm}} = -48 \text{ J}$$

6.95　The equation corresponding to the standard enthalpy of formation of diamond is:

$$\text{C(graphite)} \longrightarrow \text{C(diamond)}$$

Adding the equations:

$\text{C(graphite)} + \text{O}_2(g) \longrightarrow \text{CO}_2(g)$	$\Delta H° = -393.5 \text{ kJ/mol}$
$\text{CO}_2(g) \longrightarrow \text{C(diamond)} + \text{O}_2(g)$	$\Delta H° = 395.4 \text{ kJ/mol}$
$\text{C(graphite)} \longrightarrow \text{C(diamond)}$	$\Delta H° = 1.9 \text{ kJ/mol}$

Since the reverse reaction of changing diamond to graphite is exothermic, need you worry about any diamonds that you might have changing to graphite?

6.96　**(a)**　The more closely packed, the greater the mass of food. Heat capacity depends on both the mass and specific heat.

$$C = ms$$

The heat capacity of the food is greater than the heat capacity of air; hence, the cold in the freezer will be retained longer.

(b)　Tea and coffee are mostly water; whereas, soup might contain vegetables and meat. Water has a higher heat capacity than the other ingredients in soup; therefore, coffee and tea retain heat longer than soup.

6.97   The balanced equation is:

$$C_6H_{12}O_6(s) \longrightarrow 2C_2H_5OH(l) + 2CO_2(g)$$

$$\Delta H^{\circ}_{rxn} = [2\Delta H^{\circ}_f(C_2H_5OH) + 2\Delta H^{\circ}_f(CO_2)] - \Delta H^{\circ}_f(C_6H_{12}O_6)$$

$$\Delta H^{\circ}_{rxn} = [(2)(-276.98 \text{ kJ/mol}) + (2)(-393.5 \text{ kJ/mol})] - (1)(-1274.5 \text{ kJ/mol}) = \mathbf{-66.5 \text{ kJ/mol}}$$

6.98 $4Fe(s) + 3O_2(g) \rightarrow 2Fe_2O_3(s)$. This equation represents twice the standard enthalpy of formation of Fe_2O_3. From Appendix 3, the standard enthalpy of formation of $Fe_2O_3 = -822.2$ kJ/mol. So, ΔH° for the given reaction is:

$$\Delta H^{\circ}_{rxn} = (2)(-822.2 \text{ kJ/mol}) = -1644 \text{ kJ/mol}$$

Looking at the balanced equation, this is the amount of heat released when four moles of Fe react. But, we are reacting 250 g of Fe, not 4 moles. We can convert from grams of Fe to moles of Fe, then use ΔH° as a conversion factor to convert to kJ.

$$250 \text{ g Fe} \times \frac{1 \text{ mol Fe}}{55.85 \text{ g Fe}} \times \frac{-1644 \text{ kJ}}{4 \text{ mol Fe}} = \mathbf{-1.84 \times 10^3 \text{ kJ}}$$

6.99   One conversion factor needed to solve this problem is the molar mass of water. The other conversion factor is given in the problem. It takes 44.0 kJ of energy to vaporize 1 mole of water.

$$\frac{1 \text{ mol } H_2O}{44.0 \text{ kJ}}$$

You should come up with the following strategy to solve the problem.

$$4000 \text{ kJ} \rightarrow \text{mol } H_2O \rightarrow \text{g } H_2O$$

$$\mathbf{? \text{ g } H_2O} = 4000 \text{ kJ} \times \frac{1 \text{ mol } H_2O}{44.0 \text{ kJ}} \times \frac{18.02 \text{ g } H_2O}{1 \text{ mol } H_2O} = \mathbf{1.64 \times 10^3 \text{ g } H_2O}$$

6.100   The heat required to raise the temperature of 1 liter of water by 1°C is:

$$4.184 \frac{J}{g \cdot ^{\circ}C} \times \frac{1 \text{ g}}{1 \text{ mL}} \times \frac{1000 \text{ mL}}{1 \text{ L}} \times 1^{\circ}C = 4184 \text{ J/L}$$

Next, convert the volume of the Pacific Ocean to liters.

$$(7.2 \times 10^8 \text{ km}^3) \times \left(\frac{1000 \text{ m}}{1 \text{ km}}\right)^3 \times \left(\frac{100 \text{ cm}}{1 \text{ m}}\right)^3 \times \frac{1 \text{ L}}{1000 \text{ cm}^3} = 7.2 \times 10^{20} \text{ L}$$

The amount of heat needed to raise the temperature of 7.2×10^{20} L of water is:

$$(7.2 \times 10^{20} \text{ L}) \times \frac{4184 \text{ J}}{1 \text{ L}} = 3.0 \times 10^{24} \text{ J}$$

Finally, we can calculate the number of atomic bombs needed to produce this much heat.

$$(3.0 \times 10^{24} \text{ J}) \times \frac{1 \text{ atomic bomb}}{1.0 \times 10^{15} \text{ J}} = \mathbf{3.0 \times 10^9 \text{ atomic bombs}} = \mathbf{3.0 \text{ billion atomic bombs}}$$

6.101   First calculate the final volume of CO_2 gas:

$$V = \frac{nRT}{P} = \frac{\left(19.2 \text{ g} \times \dfrac{1 \text{ mol}}{44.01 \text{ g}}\right)(0.0821 \text{ L}\cdot\text{atm}/\text{K}\cdot\text{mol})(295 \text{ K})}{0.995 \text{ atm}} = 10.6 \text{ L}$$

$$w = -P\Delta V = -(0.995 \text{ atm})(10.6 \text{ L}) = -10.5 \text{ L}\cdot\text{atm}$$

$$w = -10.5 \text{ L}\cdot\text{atm} \times \frac{101.3 \text{ J}}{1 \text{ L}\cdot\text{atm}} = -1.06 \times 10^3 \text{ J} = -1.06 \text{ kJ}$$

The expansion work done is 1.06 kJ.

6.102   **Strategy:** The heat released during the reaction is absorbed by both the water and the calorimeter. How do we calculate the heat absorbed by the water? How do we calculate the heat absorbed by the calorimeter? How much heat is released when 1.9862 g of benzoic acid are reacted? The problem gives the amount of heat that is released when 1 mole of benzoic acid is reacted (−3226.7 kJ/mol).

Solution: The heat of the reaction (combustion) is absorbed by both the water and the calorimeter.

$$q_{rxn} = -(q_{water} + q_{cal})$$

If we can calculate both q_{water} and q_{rxn}, then we can calculate q_{cal}. First, let's calculate the heat absorbed by the water.

$$q_{water} = m_{water}s_{water}\Delta t$$
$$q_{water} = (2000 \text{ g})(4.184 \text{ J/g}\cdot\text{°C})(25.67 - 21.84)\text{°C} = 3.20 \times 10^4 \text{ J} = 32.0 \text{ kJ}$$

Next, let's calculate the heat released (q_{rxn}) when 1.9862 g of benzoic acid are burned. ΔH_{rxn} is given in units of kJ/mol. Let's convert to q_{rxn} in kJ.

$$q_{rxn} = 1.9862 \text{ g benzoic acid} \times \frac{1 \text{ mol benzoic acid}}{122.1 \text{ g benzoic acid}} \times \frac{-3226.7 \text{ kJ}}{1 \text{ mol benzoic acid}} = -52.49 \text{ kJ}$$

And,

$$q_{cal} = -q_{rxn} - q_{water}$$
$$q_{cal} = 52.49 \text{ kJ} - 32.0 \text{ kJ} = 20.5 \text{ kJ}$$

To calculate the heat capacity of the bomb calorimeter, we can use the following equation:

$$q_{cal} = C_{cal}\Delta t$$

$$C_{cal} = \frac{q_{cal}}{\Delta t} = \frac{20.5 \text{ kJ}}{(25.67 - 21.84)\text{°C}} = \mathbf{5.35 \text{ kJ/°C}}$$

6.103   **(a)**   The number of moles of water present in 500 g of water is:

$$\text{moles of } H_2O = 500 \text{ g } H_2O \times \frac{1 \text{ mol } H_2O}{18.02 \text{ g } H_2O} = 27.7 \text{ mol } H_2O$$

From the equation for the production of $Ca(OH)_2$, we have 1 mol $H_2O \simeq$ 1 mol CaO $\simeq$ 1 mol $Ca(OH)_2$. Therefore, the heat generated by the reaction is:

$$27.7 \text{ mol Ca(OH)}_2 \times \frac{-65.2 \text{ kJ}}{1 \text{ mol Ca(OH)}_2} = -1.81 \times 10^3 \text{ kJ}$$

Knowing the specific heat and the number of moles of $Ca(OH)_2$ produced, we can calculate the temperature rise using Equation (6.12) of the text. First, we need to find the mass of $Ca(OH)_2$ in 27.7 moles.

$$27.7 \text{ mol Ca(OH)}_2 \times \frac{74.10 \text{ g Ca(OH)}_2}{1 \text{ mol Ca(OH)}_2} = 2.05 \times 10^3 \text{ g Ca(OH)}_2$$

From Equation (6.12) of the text, we write:

$$q = ms\Delta t$$

Rearranging, we get

$$\Delta t = \frac{q}{ms}$$

$$\Delta t = \frac{1.81 \times 10^6 \text{ J}}{(2.05 \times 10^3 \text{ g})(1.20 \text{ J/g} \cdot {}^\circ\text{C})} = 736{}^\circ\text{C}$$

and the final temperature is

$$\Delta t = t_{\text{final}} - t_{\text{initial}}$$

$$t_{\text{final}} = 736{}^\circ\text{C} + 25{}^\circ\text{C} = \mathbf{761{}^\circ C}$$

A temperature of 761°C is high enough to ignite wood.

(b) The reaction is:

$$CaO(s) + H_2O(l) \rightarrow Ca(OH)_2(s)$$

$$\Delta H^\circ_{\text{rxn}} = \Delta H^\circ_{\text{f}}[Ca(OH)_2] - [\Delta H^\circ_{\text{f}}(CaO) + \Delta H^\circ_{\text{f}}(H_2O)]$$

$\Delta H^\circ_{\text{rxn}}$ is given in the problem (−65.2 kJ/mol). Also, the $\Delta H^\circ_{\text{f}}$ values of CaO and H_2O are given. Thus, we can solve for $\Delta H^\circ_{\text{f}}$ of $Ca(OH)_2$.

$$-65.2 \text{ kJ/mol} = \Delta H^\circ_{\text{f}}[Ca(OH)_2] - [(1)(-635.6 \text{ kJ/mol} + (1)(-285.8 \text{ kJ/mol})]$$

$$\Delta H^\circ_{\text{f}}[Ca(OH)_2] = \mathbf{-986.6 \ kJ/mol}$$

6.104   First, let's calculate the standard enthalpy of reaction.

$$\Delta H^\circ_{\text{rxn}} = 2\Delta H^\circ_{\text{f}}(CaSO_4) - [2\Delta H^\circ_{\text{f}}(CaO) + 2\Delta H^\circ_{\text{f}}(SO_2) + \Delta H^\circ_{\text{f}}(O_2)]$$

$$= (2)(-1432.7 \text{ kJ/mol}) - [(2)(-635.6 \text{ kJ/mol}) + (2)(-296.1 \text{ kJ/mol}) + 0]$$

$$= -1002 \text{ kJ/mol}$$

This is the enthalpy change for every 2 moles of SO_2 that are removed. The problem asks to calculate the enthalpy change for this process if 6.6×10^5 g of SO_2 are removed.

$$(6.6 \times 10^5 \text{ g } SO_2) \times \frac{1 \text{ mol } SO_2}{64.07 \text{ g } SO_2} \times \frac{-1002 \text{ kJ}}{2 \text{ mol } SO_2} = \mathbf{-5.2 \times 10^6 \text{ kJ}}$$

6.105 Volume of room $= (2.80 \text{ m} \times 10.6 \text{ m} \times 17.2 \text{ m}) \times \frac{1000 \text{ L}}{1 \text{ m}^3} = 5.10 \times 10^5 \text{ L}$

$PV = nRT$

$$n_{air} = \frac{PV}{RT} = \frac{(1.0 \text{ atm})(5.10 \times 10^5 \text{ L})}{(0.0821 \text{ L atm} / \text{K} \cdot \text{mol})(32 + 273 \text{ K})} = 2.04 \times 10^4 \text{ mol air}$$

$$\text{mass air} = (2.04 \times 10^4 \text{ mol air}) \times \frac{29.0 \text{ g air}}{1 \text{ mol air}} = 5.9 \times 10^5 \text{ g air}$$

Heat to be removed from air:

$$q = m_{air}s_{air}\Delta t$$
$$q = (5.9 \times 10^5 \text{ g})(1.2 \text{ J/g} \cdot {}^\circ\text{C})(-8.2{}^\circ\text{C})$$
$$q = -5.8 \times 10^6 \text{ J} = -5.8 \times 10^3 \text{ kJ}$$

Conservation of energy:

$$q_{air} + q_{salt} = 0$$

$$q_{air} + n\Delta H_{fus} = 0$$

$$q_{air} + \frac{m_{salt}}{\mathcal{M}_{salt}}\Delta H_{fus} = 0$$

$$m_{salt} = \frac{-q_{air}\mathcal{M}_{salt}}{\Delta H_{fus}}$$

$$m_{salt} = \frac{-(-5.8 \times 10^3 \text{ kJ})(322.3 \text{ g/mol})}{74.4 \text{ kJ/mol}} = 2.5 \times 10^4 \text{ g}$$

$$m_{salt} = \mathbf{25 \text{ kg}}$$

6.106 First, we need to calculate the volume of the balloon.

$$V = \frac{4}{3}\pi r^3 = \frac{4}{3}\pi(8 \text{ m})^3 = (2.1 \times 10^3 \text{ m}^3) \times \frac{1000 \text{ L}}{1 \text{ m}^3} = 2.1 \times 10^6 \text{ L}$$

(a) We can calculate the mass of He in the balloon using the ideal gas equation.

$$n_{He} = \frac{PV}{RT} = \frac{\left(98.7 \text{ kPa} \times \dfrac{1 \text{ atm}}{1.01325 \times 10^2 \text{ kPa}}\right)(2.1 \times 10^6 \text{ L})}{\left(0.0821\dfrac{\text{L} \cdot \text{atm}}{\text{mol} \cdot \text{K}}\right)(273 + 18)\text{K}} = 8.6 \times 10^4 \text{ mol He}$$

$$\mathbf{mass\ He} = (8.6 \times 10^4 \text{ mol He}) \times \frac{4.003 \text{ g He}}{1 \text{ mol He}} = \mathbf{3.4 \times 10^5 \text{ g He}}$$

(b) Work done $= -P\Delta V$

$$= -\left(98.7 \text{ kPa} \times \frac{1 \text{ atm}}{1.01325 \times 10^2 \text{ kPa}}\right)(2.1 \times 10^6 \text{ L})$$

$$= (-2.0 \times 10^6 \text{ L} \cdot \text{atm}) \times \frac{101.3 \text{ J}}{1 \text{ L} \cdot \text{atm}}$$

$$\mathbf{Work\ done} = \mathbf{-2.0 \times 10^8 \text{ J}}$$

6.107   The heat produced by the reaction heats the solution and the calorimeter: $q_{rxn} = -(q_{soln} + q_{cal})$

$$q_{soln} = ms\Delta t = (50.0 \text{ g})(4.184 \text{ J/g} \cdot °\text{C})(22.17°\text{C} - 19.25°\text{C}) = 611 \text{ J}$$

$$q_{cal} = C\Delta t = (98.6 \text{ J/°C})(22.17°\text{C} - 19.25°\text{C}) = 288 \text{ J}$$

$$-q_{rxn} = (q_{soln} + q_{cal}) = (611 + 288)\text{J} = 899 \text{ J}$$

The 899 J produced was for 50.0 mL of a 0.100 M $AgNO_3$ solution.

$$50.0 \text{ mL} \times \frac{0.100 \text{ mol Ag}^+}{1000 \text{ mL soln}} = 5.00 \times 10^{-3} \text{ mol Ag}^+$$

On a molar basis the heat produced was:

$$\frac{899 \text{ J}}{5.00 \times 10^{-3} \text{ mol Ag}^+} = 1.80 \times 10^5 \text{ J/mol Ag}^+ = 180 \text{ kJ/mol Ag}^+$$

The balanced equation involves 2 moles of Ag^+, so the heat produced is 2 mol $\times$ 180 kJ/mol = 360 kJ

Since the reaction produces heat (or by noting the sign convention above), then:

$$q_{rxn} = \mathbf{-360 \text{ kJ/mol Zn}} \text{ (or } -360 \text{ kJ/2 mol Ag}^+\text{)}$$

6.108   **(a)**   The heat needed to raise the temperature of the water from 3°C to 37°C can be calculated using the equation:

$$q = ms\Delta t$$

First, we need to calculate the mass of the water.

$$4 \text{ glasses of water} \times \frac{2.5 \times 10^2 \text{ mL}}{1 \text{ glass}} \times \frac{1 \text{ g water}}{1 \text{ mL water}} = 1.0 \times 10^3 \text{ g water}$$

The heat needed to raise the temperature of 1.0×10^3 g of water is:

$$q = ms\Delta t = (1.0 \times 10^3 \text{ g})(4.184 \text{ J/g} \cdot °\text{C})(37 - 3)°\text{C} = 1.4 \times 10^5 \text{ J} = \mathbf{1.4 \times 10^2 \text{ kJ}}$$

(b)   We need to calculate both the heat needed to melt the snow and also the heat needed to heat liquid water form 0°C to 37°C (normal body temperature).

The heat needed to melt the snow is:

$$(8.0 \times 10^2 \text{ g}) \times \frac{1 \text{ mol}}{18.02 \text{ g}} \times \frac{6.01 \text{ kJ}}{1 \text{ mol}} = 2.7 \times 10^2 \text{ kJ}$$

The heat needed to raise the temperature of the water from 0°C to 37°C is:

$$q = ms\Delta t = (8.0 \times 10^2 \text{ g})(4.184 \text{ J/g·°C})(37 - 0)°C = 1.2 \times 10^5 \text{ J} = 1.2 \times 10^2 \text{ kJ}$$

The total heat lost by your body is:

$$(2.7 \times 10^2 \text{ kJ}) + (1.2 \times 10^2 \text{ kJ}) = \mathbf{3.9 \times 10^2 \text{ kJ}}$$

6.109 We assume that when the car is stopped, its kinetic energy is completely converted into heat (friction of the brakes and friction between the tires and the road). Thus,

$$q = \frac{1}{2}mu^2$$

Thus the amount of heat generated must be proportional to the braking distance, d:

$$d \propto q$$
$$d \propto u^2$$

Therefore, as u increases to $2u$, d increases to $(2u)^2 = 4u^2$ which is proportional to $4d$.

6.110 **(a)** $\Delta H° = \Delta H_f°(F^-) + \Delta H_f°(H_2O) - [\Delta H_f°(HF) + \Delta H_f°(OH^-)]$

$\Delta H° = [(1)(-329.1 \text{ kJ/mol}) + (1)(-285.8 \text{ kJ/mol})] - [(1)(-320.1 \text{ kJ/mol}) + (1)(-229.6 \text{ kJ/mol})]$

$\mathbf{\Delta H° = -65.2 \text{ kJ/mol}}$

(b) We can add the equation given in part (a) to that given in part (b) to end up with the equation we are interested in.

$HF(aq) + OH^-(aq) \longrightarrow F^-(aq) + H_2O(l)$	$\Delta H° = -65.2 \text{ kJ/mol}$
$H_2O(l) \longrightarrow H^+(aq) + OH^-(aq)$	$\Delta H° = +56.2 \text{ kJ/mol}$
$\mathbf{HF(aq) \longrightarrow H^+(aq) + F^-(aq)}$	$\mathbf{\Delta H° = -9.0 \text{ kJ/mol}}$

6.111 Water has a larger specific heat than air. Thus cold, damp air can extract more heat from the body than cold, dry air. By the same token, hot, humid air can deliver more heat to the body.

6.112 The equation we are interested in is the formation of CO from its elements.

$$C(\text{graphite}) + \tfrac{1}{2}O_2(g) \longrightarrow CO(g) \qquad \Delta H° = ?$$

Try to add the given equations together to end up with the equation above.

$C(\text{graphite}) + O_2(g) \longrightarrow CO_2(g)$	$\Delta H° = -393.5 \text{ kJ/mol}$
$CO_2(g) \longrightarrow CO(g) + \tfrac{1}{2}O_2(g)$	$\Delta H° = +283.0 \text{ kJ/mol}$
$\mathbf{C(\text{graphite}) + \tfrac{1}{2}O_2(g) \longrightarrow CO(g)}$	$\mathbf{\Delta H° = -110.5 \text{ kJ/mol}}$

We cannot obtain ΔH_f° for CO directly, because burning graphite in oxygen will form both CO and CO_2.

6.113   Energy intake for mechanical work:

$$0.17 \times 500 \text{ g} \times \frac{3000 \text{ J}}{1 \text{ g}} = 2.6 \times 10^5 \text{ J}$$

$$2.6 \times 10^5 \text{ J} = mgh$$

$$1 \text{ J} = 1 \text{ kg·m}^2\text{s}^{-2}$$

$$2.6 \times 10^5 \frac{\text{kg·m}^2}{\text{s}^2} = (46 \text{ kg})(9.8 \text{ m/s}^2)h$$

$$h = 5.8 \times 10^2 \text{ m}$$

6.114   **(a)**   mass $= 0.0010$ kg

Potential energy $= mgh$

$$= (0.0010 \text{ kg})(9.8 \text{ m/s}^2)(51 \text{ m})$$

Potential energy $= 0.50$ J

(b)   Kinetic energy $= \dfrac{1}{2}mu^2 = 0.50$ J

$$\frac{1}{2}(0.0010 \text{ kg})u^2 = 0.50 \text{ J}$$

$$u^2 = 1.0 \times 10^3 \text{ m}^2/\text{s}^2$$

$u = 32$ m/s

(c) $q = ms\Delta t$

$$0.50 \text{ J} = (1.0 \text{ g})(4.184 \text{ J/g°C})\Delta t$$

$\Delta t = 0.12$°C

6.115   For Al: $(0.900 \text{ J/g·°C})(26.98 \text{ g}) = 24.3$ J/°C

This law does not hold for Hg because it is a liquid.

6.116   The reaction we are interested in is the formation of ethanol from its elements.

$$2\text{C(graphite)} + \tfrac{1}{2}\text{O}_2(g) + 3\text{H}_2(g) \longrightarrow \text{C}_2\text{H}_5\text{OH}(l)$$

Along with the reaction for the combustion of ethanol, we can add other reactions together to end up with the above reaction.

Reversing the reaction representing the combustion of ethanol gives:

$$2\text{CO}_2(g) + 3\text{H}_2\text{O}(l) \longrightarrow \text{C}_2\text{H}_5\text{OH}(l) + 3\text{O}_2\,(g) \qquad\qquad \Delta H^{\circ} = +1367.4 \text{ kJ/mol}$$

We need to add equations to add C (graphite) and remove H_2O from the reactants side of the equation. We write:

$$2CO_2(g) + 3H_2O(l) \longrightarrow C_2H_5OH(l) + 3O_2(g) \qquad \Delta H° = +1367.4 \text{ kJ/mol}$$

$$2C(graphite) + 2O_2(g) \longrightarrow 2CO_2(g) \qquad \Delta H° = 2(-393.5 \text{ kJ/mol})$$

$$3H_2(g) + \tfrac{3}{2}O_2(g) \longrightarrow 3H_2O(l) \qquad \Delta H° = 3(-285.8 \text{ kJ/mol})$$

$$\mathbf{2C(graphite) + \tfrac{1}{2}O_2(g) + 3H_2(g) \longrightarrow C_2H_5OH(l)} \qquad \mathbf{\Delta H_f° = -277.0 \text{ kJ/mol}}$$

6.117   (a) $C_6H_6(l) + \tfrac{15}{2}O_2(g) \rightarrow 6CO_2(g) + 3H_2O(l) \qquad \Delta H° = -3267.4 \text{ kJ/mol}$

   (b) $C_2H_2(g) + \tfrac{5}{2}O_2(g) \rightarrow 2CO_2(g) + H_2O(l) \qquad \Delta H° = -1299.4 \text{ kJ/mol}$

   (c) $C(graphite) + O_2 \rightarrow CO_2(g) \qquad \Delta H° = -393.5 \text{ kJ/mol}$

   (d) $H_2(g) + \tfrac{1}{2}O_2(g) \rightarrow H_2O(l) \qquad \Delta H° = -285.8 \text{ kJ/mol}$

Using Hess's Law, we can add the equations in the following manner to calculate the standard enthalpies of formation of C_2H_2 and C_6H_6.

C_2H_2: $-$(b) + 2(c) + (d)

$$2C(graphite) + H_2(g) \rightarrow C_2H_2(g) \qquad \Delta H° = +226.6 \text{ kJ/mol}$$

Therefore, $\Delta H_f°(C_2H_2) = \mathbf{226.6 \text{ kJ/mol}}$

C_6H_6: $-$(a) + 6(c) + 3(d)

$$6C(graphite) + 3H_2(g) \rightarrow C_6H_6(l) \qquad \Delta H° = 49.0 \text{ kJ/mol}$$

Therefore, $\Delta H_f°(C_6H_6) = \mathbf{49.0 \text{ kJ/mol}}$

Finally:

$$3C_2H_2(g) \rightarrow C_6H_6(l)$$

$$\mathbf{\Delta H_{rxn} = (1)(49.0 \text{ kJ/mol}) - (3)(226.6 \text{ kJ/mol}) = -630.8 \text{ kJ/mol}}$$

6.118   Heat gained by ice = Heat lost by the soft drink

$m_{ice} \times 334 \text{ J/g} = -m_{sd}s_{sd}\Delta t$

$m_{ice} \times 334 \text{ J/g} = -(361 \text{ g})(4.184 \text{ J/g·°C})(0 - 23)°C$

$\mathbf{m_{ice} = 104 \text{ g}}$

6.119   The heat required to heat 200 g of water (assume $d = 1$ g/mL) from 20°C to 100°C is:

$$q = ms\Delta t$$

$$q = (200 \text{ g})(4.184 \text{ J/g·°C})(100 - 20)°C = 6.7 \times 10^4 \text{ J}$$

Since 50% of the heat from the combustion of methane is lost to the surroundings, twice the amount of heat needed must be produced during the combustion: $2(6.7 \times 10^4 \text{ J}) = 1.3 \times 10^5 \text{ J} = 1.3 \times 10^2 \text{ kJ}$.

Use standard enthalpies of formation (see Appendix 3) to calculate the heat of combustion of methane.

$$CH_4(g) + 2O_2(g) \rightarrow CO_2(g) + 2H_2O(l) \qquad \Delta H° = -890.3 \text{ kJ/mol}$$

The number of moles of methane needed to produce 1.3×10^2 kJ of heat is:

$$(1.3 \times 10^2 \text{ kJ}) \times \frac{1 \text{ mol CH}_4}{890.3 \text{ kJ}} = 0.15 \text{ mol CH}_4$$

The volume of 0.15 mole CH_4 at 1 atm and 20°C is:

$$V = \frac{nRT}{P} = \frac{(0.15 \text{ mol})(0.0821 \text{ L atm}/\text{K} \cdot \text{mol})(293 \text{ K})}{1.0 \text{ atm}} = 3.6 \text{ L}$$

Since we have the volume of methane needed in units of liters, let's convert the cost of natural gas per 15 ft^3 to the cost per liter.

$$\frac{\$1.30}{15 \text{ ft}^3} \times \left(\frac{1 \text{ ft}}{12 \text{ in}}\right)^3 \times \left(\frac{1 \text{ in}}{2.54 \text{ cm}}\right)^3 \times \frac{1000 \text{ cm}^3}{1 \text{ L}} = \frac{\$3.1 \times 10^{-3}}{1 \text{ L CH}_4}$$

The cost for 3.6 L of methane is:

$$3.6 \text{ L CH}_4 \times \frac{\$3.1 \times 10^{-3}}{1 \text{ L CH}_4} = \textbf{\$0.011 or about 1.1¢}$$

6.120 From Chapter 5, we saw that the kinetic energy (or internal energy) of 1 mole of a gas is $\frac{3}{2} RT$. For 1 mole of an ideal gas, $PV = RT$. We can write:

$$\text{internal energy} = \frac{3}{2} RT = \frac{3}{2} PV$$

$$= \frac{3}{2} (1.2 \times 10^5 \text{ Pa})(5.5 \times 10^3 \text{ m}^3)$$

$$= 9.9 \times 10^8 \text{ Pa} \cdot \text{m}^3$$

$$1 \text{ Pa} \cdot \text{m}^3 = 1 \frac{\text{N}}{\text{m}^2} \text{m}^3 = 1 \text{ N} \cdot \text{m} = 1 \text{ J}$$

Therefore, the internal energy is $\mathbf{9.9 \times 10^8}$ **J**.

The final temperature of the copper metal can be calculated. (10 tons = 9.07×10^6 g)

$$q = m_{Cu}s_{Cu}\Delta t$$

$$9.9 \times 10^8 \text{ J} = (9.07 \times 10^6 \text{ g})(0.385 \text{ J/g°C})(t_f - 21°C)$$

$$(3.49 \times 10^6)t_f = 1.06 \times 10^9$$

$$t_f = \textbf{304°C}$$

6.121 Energy must be supplied to break a chemical bond. By the same token, energy is released when a bond is formed.

6.122 **(a)** $CaC_2(s) + 2H_2O(l) \longrightarrow Ca(OH)_2(s) + C_2H_2(g)$

(b) The reaction for the combustion of acetylene is:

$$2C_2H_2(g) + 5O_2(g) \longrightarrow 4CO_2(g) + 2H_2O(l)$$

We can calculate the enthalpy change for this reaction from standard enthalpy of formation values given in Appendix 3 of the text.

$$\Delta H_{rxn}^\circ = [4\Delta H_f^\circ(CO_2) + 2\Delta H_f^\circ(H_2O)] - [2\Delta H_f^\circ(C_2H_2) + 5\Delta H_f^\circ(O_2)]$$

$$\Delta H_{rxn}^\circ = [(4)(-393.5 \text{ kJ/mol}) + (2)(-285.8 \text{ kJ/mol})] - [(2)(226.6 \text{ kJ/mol}) + (5)(0)]$$

$$\Delta H_{rxn}^\circ = -2599 \text{ kJ/mol}$$

Looking at the balanced equation, this is the amount of heat released when two moles of C_2H_2 are reacted. The problem asks for the amount of heat that can be obtained starting with 74.6 g of CaC_2. From this amount of CaC_2, we can calculate the moles of C_2H_2 produced.

$$74.6 \text{ g } CaC_2 \times \frac{1 \text{ mol } CaC_2}{64.10 \text{ g } CaC_2} \times \frac{1 \text{ mol } C_2H_2}{1 \text{ mol } CaC_2} = 1.16 \text{ mol } C_2H_2$$

Now, we can use the ΔH_{rxn}° calculated above as a conversion factor to determine the amount of heat obtained when 1.16 moles of C_2H_2 are burned.

$$1.16 \text{ mol } C_2H_2 \times \frac{2599 \text{ kJ}}{2 \text{ mol } C_2H_2} = \mathbf{1.51 \times 10^3 \text{ kJ}}$$

6.123 Since the humidity is very low in deserts, there is little water vapor in the air to trap and hold the heat radiated back from the ground during the day. Once the sun goes down, the temperature drops dramatically. 40°F temperature drops between day and night are common in desert climates. Coastal regions have much higher humidity levels compared to deserts. The water vapor in the air retains heat, which keeps the temperature at a more constant level during the night. In addition, sand and rocks in the desert have small specific heats compared with water in the ocean. The water absorbs much more heat during the day compared to sand and rocks, which keeps the temperature warmer at night.

6.124 When 1.034 g of naphthalene are burned, 41.56 kJ of heat are evolved. Let's convert this to the amount of heat evolved on a molar basis. The molar mass of naphthalene is 128.2 g/mol.

$$q = \frac{-41.56 \text{ kJ}}{1.034 \text{ g } C_{10}H_8} \times \frac{128.2 \text{ g } C_{10}H_8}{1 \text{ mol } C_{10}H_8} = -5153 \text{ kJ/mol}$$

q has a negative sign because this is an exothermic reaction.

This reaction is run at constant volume ($\Delta V = 0$); therefore, no work will result from the change.

$$w = -P\Delta V = 0$$

From Equation (6.4) of the text, it follows that the change in energy is equal to the heat change.

$$\Delta E = q + w = q_v = -5153 \text{ kJ/mol}$$

To calculate ΔH, we rearrange Equation (6.10) of the text.

$$\Delta E = \Delta H - RT\Delta n$$

$$\Delta H = \Delta E + RT\Delta n$$

To calculate ΔH, Δn must be determined, which is the difference in moles of *gas* products and moles of *gas* reactants. Looking at the balanced equation for the combustion of naphthalene:

$$C_{10}H_8(s) + 12O_2(g) \rightarrow 10CO_2(g) + 4H_2O(l)$$

$$\Delta n = 10 - 12 = -2$$

$$\Delta H = \Delta E + RT\Delta n$$

$$\Delta H = -5153 \text{ kJ/mol} + (8.314 \text{ J/mol} \cdot \text{K})(298 \text{ K})(-2) \times \frac{1 \text{ kJ}}{1000 \text{ J}}$$

$$\Delta H = -5158 \text{ kJ/mol}$$

Is ΔH equal to q_p in this case?

6.125   Let's write balanced equations for the reactions between Mg and CO_2, and Mg and H_2O. Then, we can calculate ΔH°_{rxn} for each reaction from ΔH°_f values.

   (1) $2Mg(s) + CO_2(g) \rightarrow 2MgO(s) + C(s)$

   (2) $Mg(s) + 2H_2O(l) \rightarrow Mg(OH)_2(s) + H_2(g)$

For reaction (1), ΔH°_{rxn} is:

$$\Delta H^{\circ}_{rxn} = 2\Delta H^{\circ}_f[MgO(s)] + \Delta H^{\circ}_f[C(s)] - \{2\Delta H^{\circ}_f[Mg(s)] + \Delta H^{\circ}_f[CO_2(g)]\}$$

$$\Delta H^{\circ}_{rxn} = (2)(-601.8 \text{ kJ/mol}) + (1)(0) - [(2)(0) + (1)(-393.5 \text{ kJ/mol})] = -810.1 \text{ kJ/mol}$$

For reaction (2), ΔH°_{rxn} is:

$$\Delta H^{\circ}_{rxn} = \Delta H^{\circ}_f[Mg(OH)_2(s)] + \Delta H^{\circ}_f[H_2(g)] - \{\Delta H^{\circ}_f[Mg(s)] + 2\Delta H^{\circ}_f[H_2O(l)]\}$$

$$\Delta H^{\circ}_{rxn} = (1)(-924.66 \text{ kJ/mol}) + (1)(0) - [(1)(0) + (2)(-285.8 \text{ kJ/mol})] = -353.1 \text{ kJ/mol}$$

Both of these reactions are highly exothermic, which will promote the fire rather than extinguishing it.

CHAPTER 7
QUANTUM THEORY AND THE ELECTRONIC STRUCTURE OF ATOMS

7.7 **(a)** $\lambda = \dfrac{c}{\nu} = \dfrac{3.00 \times 10^8 \text{ m/s}}{8.6 \times 10^{13} \text{ /s}} = 3.5 \times 10^{-6} \text{ m} = \mathbf{3.5 \times 10^3 \text{ nm}}$

(b) $\nu = \dfrac{c}{\lambda} = \dfrac{3.00 \times 10^8 \text{ m/s}}{566 \times 10^{-9} \text{ m}} = 5.30 \times 10^{14} \text{ /s} = \mathbf{5.30 \times 10^{14} \text{ Hz}}$

7.8 **(a)**
Strategy: We are given the wavelength of an electromagnetic wave and asked to calculate the frequency. Rearranging Equation (7.1) of the text and replacing u with c (the speed of light) gives:

$$\nu = \frac{c}{\lambda}$$

Solution: Because the speed of light is given in meters per second, it is convenient to first convert wavelength to units of meters. Recall that 1 nm = 1×10^{-9} m (see Table 1.3 of the text). We write:

$$456 \text{ nm} \times \frac{1 \times 10^{-9} \text{ m}}{1 \text{ nm}} = 456 \times 10^{-9} \text{ m} = 4.56 \times 10^{-7} \text{ m}$$

Substituting in the wavelength and the speed of light (3.00×10^8 m/s), the frequency is:

$$\nu = \frac{c}{\lambda} = \frac{3.00 \times 10^8 \dfrac{\text{m}}{\text{s}}}{4.56 \times 10^{-7} \text{ m}} = 6.58 \times 10^{14} \text{ s}^{-1} \text{ or } \mathbf{6.58 \times 10^{14} \text{ Hz}}$$

Check: The answer shows that 6.58×10^{14} waves pass a fixed point every second. This very high frequency is in accordance with the very high speed of light.

(b)
Strategy: We are given the frequency of an electromagnetic wave and asked to calculate the wavelength. Rearranging Equation (7.1) of the text and replacing u with c (the speed of light) gives:

$$\lambda = \frac{c}{\nu}$$

Solution: Substituting in the frequency and the speed of light (3.00×10^8 m/s) into the above equation, the wavelength is:

$$\lambda = \frac{c}{\nu} = \frac{3.00 \times 10^8 \dfrac{\text{m}}{\text{s}}}{2.45 \times 10^9 \dfrac{1}{\text{s}}} = 0.122 \text{ m}$$

The problem asks for the wavelength in units of nanometers. Recall that 1 nm = 1×10^{-9} m.

$$\lambda = 0.122 \text{ m} \times \frac{1 \text{ nm}}{1 \times 10^{-9} \text{ m}} = \mathbf{1.22 \times 10^8 \text{ nm}}$$

7.9   Since the speed of light is 3.00×10^8 m/s, we can write

$$(1.3 \times 10^8 \text{ mi}) \times \frac{1.61 \text{ km}}{1 \text{ mi}} \times \frac{1000 \text{ m}}{1 \text{ km}} \times \frac{1 \text{ s}}{3.00 \times 10^8 \text{ m}} = \mathbf{7.0 \times 10^2 \text{ s}}$$

Would the time be different for other types of electromagnetic radiation?

7.10   A radio wave is an electromagnetic wave, which travels at the speed of light. The speed of light is in units of m/s, so let's convert distance from units of miles to meters. (28 million mi = 2.8×10^7 mi)

$$? \text{ distance (m)} = (2.8 \times 10^7 \text{ mi}) \times \frac{1.61 \text{ km}}{1 \text{ mi}} \times \frac{1000 \text{ m}}{1 \text{ km}} = 4.5 \times 10^{10} \text{ m}$$

Now, we can use the speed of light as a conversion factor to convert from meters to seconds ($c = 3.00 \times 10^8$ m/s).

$$? \text{ min} = (4.5 \times 10^{10} \text{ m}) \times \frac{1 \text{ s}}{3.00 \times 10^8 \text{ m}} = 1.5 \times 10^2 \text{ s} = \mathbf{2.5 \text{ min}}$$

7.11 $\lambda = \dfrac{c}{\nu} = \dfrac{3.00 \times 10^8 \text{ m/s}}{9192631770 \text{ s}^{-1}} = 3.26 \times 10^{-2} \text{ m} = \mathbf{3.26 \times 10^7 \text{ nm}}$

This radiation falls in the microwave region of the spectrum. (See Figure 7.4 of the text.)

7.12   The wavelength is:

$$\lambda = \frac{1 \text{ m}}{1,650,763.73 \text{ wavelengths}} = 6.05780211 \times 10^{-7} \text{ m}$$

$$\nu = \frac{c}{\lambda} = \frac{3.00 \times 10^8 \text{ m/s}}{6.05780211 \times 10^{-7} \text{ m}} = \mathbf{4.95 \times 10^{14} \text{ s}^{-1}}$$

7.15 $E = h\nu = \dfrac{hc}{\lambda} = \dfrac{(6.63 \times 10^{-34} \text{ J·s})(3.00 \times 10^8 \text{ m/s})}{624 \times 10^{-9} \text{ m}} = \mathbf{3.19 \times 10^{-19} \text{ J}}$

7.16   **(a)**
Strategy: We are given the frequency of an electromagnetic wave and asked to calculate the wavelength. Rearranging Equation (7.1) of the text and replacing u with c (the speed of light) gives:

$$\lambda = \frac{c}{\nu}$$

Solution: Substituting in the frequency and the speed of light (3.00×10^8 m/s) into the above equation, the wavelength is:

$$\lambda = \frac{3.00 \times 10^8 \frac{m}{s}}{7.5 \times 10^{14} \frac{1}{s}} = 4.0 \times 10^{-7} \text{ m} = \mathbf{4.0 \times 10^2 \text{ nm}}$$

Check: The wavelength of 400 nm calculated is in the blue region of the visible spectrum as expected.

(b)
Strategy: We are given the frequency of an electromagnetic wave and asked to calculate its energy. Equation (7.2) of the text relates the energy and frequency of an electromagnetic wave.

$$E = h\nu$$

Solution: Substituting in the frequency and Planck's constant (6.63×10^{-34} J·s) into the above equation, the energy of a single photon associated with this frequency is:

$$E = h\nu = (6.63 \times 10^{-34} \text{ J·s})\left(7.5 \times 10^{14} \frac{1}{s}\right) = \mathbf{5.0 \times 10^{-19} \text{ J}}$$

Check: We expect the energy of a single photon to be a very small energy as calculated above, 5.0×10^{-19} J.

7.17 **(a)** $\lambda = \dfrac{c}{\nu} = \dfrac{3.00 \times 10^8 \text{ m/s}}{6.0 \times 10^4 \text{ /s}} = 5.0 \times 10^3 \text{ m} = \mathbf{5.0 \times 10^{12} \text{ nm}}$

The radiation does not fall in the visible region; it is **radio** radiation. (See Figure 7.4 of the text.)

(b) $E = h\nu = (6.63 \times 10^{-34} \text{ J·s})(6.0 \times 10^4 \text{ /s}) = \mathbf{4.0 \times 10^{-29} \text{ J}}$

(c) Converting to J/mol: $E = \dfrac{4.0 \times 10^{-29} \text{ J}}{1 \text{ photon}} \times \dfrac{6.022 \times 10^{23} \text{ photons}}{1 \text{ mol}} = \mathbf{2.4 \times 10^{-5} \text{ J/mol}}$

7.18 The energy given in this problem is for *1 mole* of photons. To apply $E = h\nu$, we must divide the energy by Avogadro's number. The energy of one photon is:

$$E = \frac{1.0 \times 10^3 \text{ kJ}}{1 \text{ mol}} \times \frac{1 \text{ mol}}{6.022 \times 10^{23} \text{ photons}} \times \frac{1000 \text{ J}}{1 \text{ kJ}} = 1.7 \times 10^{-18} \text{ J/photon}$$

The wavelength of this photon can be found using the relationship, $E = \dfrac{hc}{\lambda}$.

$$\lambda = \frac{hc}{E} = \frac{(6.63 \times 10^{-34} \text{ J·s})\left(3.00 \times 10^8 \frac{m}{s}\right)}{1.7 \times 10^{-18} \text{ J}} = 1.2 \times 10^{-7} \text{ m} \times \frac{1 \text{ nm}}{1 \times 10^{-9} \text{ m}} = \mathbf{1.2 \times 10^2 \text{ nm}}$$

The radiation is in the **ultraviolet** region (see Figure 7.4 of the text).

7.19 $E = h\nu = \dfrac{hc}{\lambda} = \dfrac{(6.63 \times 10^{-34} \text{ J·s})(3.00 \times 10^8 \text{ m/s})}{(0.154 \times 10^{-9} \text{ m})} = \mathbf{1.29 \times 10^{-15} \text{ J}}$

7.20   **(a)** $\lambda = \dfrac{c}{\nu}$

$$\lambda = \frac{3.00 \times 10^8 \, \dfrac{m}{s}}{8.11 \times 10^{14} \, \dfrac{1}{s}} = 3.70 \times 10^{-7} \, m = \mathbf{3.70 \times 10^2 \, nm}$$

(b)   Checking Figure 7.4 of the text, you should find that the visible region of the spectrum runs from 400 to 700 nm. 370 nm is in the **ultraviolet** region of the spectrum.

(c) $E = h\nu$. Substitute the frequency (ν) into this equation to solve for the energy of one quantum associated with this frequency.

$$E = h\nu = (6.63 \times 10^{-34} \, J \cdot s)\left(8.11 \times 10^{14} \, \frac{1}{s}\right) = \mathbf{5.38 \times 10^{-19} \, J}$$

7.25   The arrangement of energy levels for each element is unique. The frequencies of light emitted by an element are characteristic of that element. Even the frequencies emitted by isotopes of the same element are very slightly different.

7.26   The emitted light could be analyzed by passing it through a prism.

7.27   Light emitted by fluorescent materials always has lower energy than the light striking the fluorescent substance. Absorption of visible light could not give rise to emitted ultraviolet light because the latter has higher energy.

The reverse process, ultraviolet light producing visible light by fluorescence, is very common. Certain brands of laundry detergents contain materials called "optical brighteners" which, for example, can make a white shirt look much whiter and brighter than a similar shirt washed in ordinary detergent.

7.28   Excited atoms of the chemical elements emit the same characteristic frequencies or lines in a terrestrial laboratory, in the sun, or in a star many light-years distant from earth.

7.29   **(a)**   The energy difference between states E_1 and E_4 is:

$$E_4 - E_1 = (-1.0 \times 10^{-19}) J - (-15 \times 10^{-19}) J = 14 \times 10^{-19} \, J$$

$$\lambda = \frac{hc}{\Delta E} = \frac{(6.63 \times 10^{-34} \, J \cdot s)(3.00 \times 10^8 \, m/s)}{14 \times 10^{-19} \, J} = 1.4 \times 10^{-7} \, m = \mathbf{1.4 \times 10^2 \, nm}$$

(b)   The energy difference between the states E_2 and E_3 is:

$$E_3 - E_2 = (-5.0 \times 10^{-19} \, J) - (-10.0 \times 10^{-19} \, J) = \mathbf{5 \times 10^{-19} \, J}$$

(c)   The energy difference between the states E_1 and E_3 is:

$$E_1 - E_3 = (-15 \times 10^{-19} \, J) - (-5.0 \times 10^{-19} \, J) = -10 \times 10^{-19} \, J$$

Ignoring the negative sign of ΔE, the wavelength is found as in part (a).

$$\lambda = \frac{hc}{\Delta E} = \frac{(6.63 \times 10^{-34} \text{ J} \cdot \text{s})(3.00 \times 10^8 \text{ m/s})}{10 \times 10^{-19} \text{ J}} = 2.0 \times 10^{-7} \text{ m} = 2.0 \times 10^2 \text{ nm}$$

7.30   We use more accurate values of h and c for this problem.

$$E = \frac{hc}{\lambda} = \frac{(6.6256 \times 10^{-34} \text{ J} \cdot \text{s})(2.998 \times 10^8 \text{ m/s})}{656.3 \times 10^{-9} \text{ m}} = 3.027 \times 10^{-19} \text{ J}$$

7.31   In this problem $n_i = 5$ and $n_f = 3$.

$$\Delta E = R_H \left(\frac{1}{n_i^2} - \frac{1}{n_f^2} \right) = (2.18 \times 10^{-18} \text{ J}) \left(\frac{1}{5^2} - \frac{1}{3^2} \right) = -1.55 \times 10^{-19} \text{ J}$$

The sign of ΔE means that this is energy associated with an emission process.

$$\lambda = \frac{hc}{\Delta E} = \frac{(6.63 \times 10^{-34} \text{ J} \cdot \text{s})(3.00 \times 10^8 \text{ m/s})}{1.55 \times 10^{-19} \text{ J}} = 1.28 \times 10^{-6} \text{ m} = 1.28 \times 10^3 \text{ nm}$$

Is the sign of the energy change consistent with the sign conventions for exo- and endothermic processes?

7.32   **Strategy:** We are given the initial and final states in the emission process. We can calculate the energy of the emitted photon using Equation (7.5) of the text. Then, from this energy, we can solve for the frequency of the photon, and from the frequency we can solve for the wavelength. The value of Rydberg's constant is 2.18×10^{-18} J.

Solution: From Equation (7.5) we write:

$$\Delta E = R_H \left(\frac{1}{n_i^2} - \frac{1}{n_f^2} \right)$$

$$\Delta E = (2.18 \times 10^{-18} \text{ J}) \left(\frac{1}{4^2} - \frac{1}{2^2} \right)$$

$$\Delta E = -4.09 \times 10^{-19} \text{ J}$$

The negative sign for ΔE indicates that this is energy associated with an emission process. To calculate the frequency, we will omit the minus sign for ΔE because the frequency of the photon must be positive. We know that

$$\Delta E = h\nu$$

Rearranging the equation and substituting in the known values,

$$\nu = \frac{\Delta E}{h} = \frac{(4.09 \times 10^{-19} \text{ J})}{(6.63 \times 10^{-34} \text{ J} \cdot \text{s})} = 6.17 \times 10^{14} \text{ s}^{-1} \text{ or } 6.17 \times 10^{14} \text{ Hz}$$

We also know that $\lambda = \dfrac{c}{v}$. Substituting the frequency calculated above into this equation gives:

$$\lambda = \frac{3.00 \times 10^8 \, \frac{m}{s}}{\left(6.17 \times 10^{14} \, \frac{1}{s}\right)} = 4.86 \times 10^{-7} \, m = \mathbf{486 \ nm}$$

Check: This wavelength is in the visible region of the electromagnetic region (see Figure 7.4 of the text). This is consistent with the fact that because $n_i = 4$ and $n_f = 2$, this transition gives rise to a spectral line in the Balmer series (see Figure 7.6 of the text).

7.33   This problem must be worked to four significant figure accuracy. We use 6.6256×10^{-34} J·s for Planck's constant and 2.998×10^8 m/s for the speed of light. First calculate the energy of each of the photons.

$$E = \frac{hc}{\lambda} = \frac{(6.6256 \times 10^{-34} \, J \cdot s)(2.998 \times 10^8 \, m/s)}{589.0 \times 10^{-9} \, m} = 3.372 \times 10^{-19} \, J$$

$$E = \frac{hc}{\lambda} = \frac{(6.6256 \times 10^{-34} \, J \cdot s)(2.998 \times 10^8 \, m/s)}{589.6 \times 10^{-9} \, m} = 3.369 \times 10^{-19} \, J$$

For *one* photon the energy difference is:

$$\Delta E = (3.372 \times 10^{-19} \, J) - (3.369 \times 10^{-19} \, J) = \mathbf{3 \times 10^{-22} \ J}$$

For *one mole* of photons the energy difference is:

$$\frac{3 \times 10^{-22} \, J}{1 \, photon} \times \frac{6.022 \times 10^{23} \, photons}{1 \, mol} = \mathbf{2 \times 10^2 \ J/mol}$$

7.34 $\Delta E = R_H \left(\dfrac{1}{n_i^2} - \dfrac{1}{n_f^2} \right)$

n_f is given in the problem and R_H is a constant, but we need to calculate ΔE. The photon energy is:

$$E = \frac{hc}{\lambda} = \frac{(6.63 \times 10^{-34} \, J \cdot s)(3.00 \times 10^8 \, m/s)}{434 \times 10^{-9} \, m} = 4.58 \times 10^{-19} \, J$$

Since this is an emission process, the energy change ΔE must be negative, or -4.58×10^{-19} J.

Substitute ΔE into the following equation, and solve for n_i.

$$\Delta E = R_H \left(\frac{1}{n_i^2} - \frac{1}{n_f^2} \right)$$

$$-4.58 \times 10^{-19} \, J = (2.18 \times 10^{-18} \, J) \left(\frac{1}{n_i^2} - \frac{1}{2^2} \right)$$

$$\frac{1}{n_i^2} = \left(\frac{-4.58 \times 10^{-19} \text{ J}}{2.18 \times 10^{-18} \text{ J}}\right) + \frac{1}{2^2} = -0.210 + 0.250 = 0.040$$

$$n_i = \frac{1}{\sqrt{0.040}} = \mathbf{5}$$

7.39 $\lambda = \dfrac{h}{mu} = \dfrac{6.63 \times 10^{-34} \text{ J} \cdot \text{s}}{(1.675 \times 10^{-27} \text{ kg})(7.00 \times 10^2 \text{ m/s})} = 5.65 \times 10^{-10} \text{ m} = \mathbf{0.565 \text{ nm}}$

7.40   **Strategy:** We are given the mass and the speed of the proton and asked to calculate the wavelength. We need the de Broglie equation, which is Equation (7.7) of the text. Note that because the units of Planck's constant are J·s, m must be in kg and u must be in m/s (1 J = 1 kg·m^2/s^2).

Solution: Using Equation (7.7) we write:

$$\lambda = \frac{h}{mu}$$

$$\lambda = \frac{h}{mu} = \frac{\left(6.63 \times 10^{-34} \dfrac{\text{kg} \cdot \text{m}^2}{\text{s}^2} \cdot \text{s}\right)}{(1.673 \times 10^{-27} \text{ kg})(2.90 \times 10^8 \text{ m/s})} = 1.37 \times 10^{-15} \text{ m}$$

The problem asks to express the wavelength in nanometers.

$$\lambda = (1.37 \times 10^{-15} \text{ m}) \times \frac{1 \text{ nm}}{1 \times 10^{-9} \text{ m}} = \mathbf{1.37 \times 10^{-6} \text{ nm}}$$

7.41   Converting the velocity to units of m/s::

$$\frac{1.20 \times 10^2 \text{ mi}}{1 \text{ hr}} \times \frac{1.61 \text{ km}}{1 \text{ mi}} \times \frac{1000 \text{ m}}{1 \text{ km}} \times \frac{1 \text{ hr}}{3600 \text{ s}} = 53.7 \text{ m/s}$$

$$\lambda = \frac{h}{mu} = \frac{6.63 \times 10^{-34} \text{ J} \cdot \text{s}}{(0.0124 \text{ kg})(53.7 \text{ m/s})} = 9.96 \times 10^{-34} \text{ m} = \mathbf{9.96 \times 10^{-32} \text{ cm}}$$

7.42   First, we convert mph to m/s.

$$\frac{35 \text{ mi}}{1 \text{ h}} \times \frac{1.61 \text{ km}}{1 \text{ mi}} \times \frac{1000 \text{ m}}{1 \text{ km}} \times \frac{1 \text{ h}}{3600 \text{ s}} = 16 \text{ m/s}$$

$$\lambda = \frac{h}{mu} = \frac{\left(6.63 \times 10^{-34} \dfrac{\text{kg} \cdot \text{m}^2}{\text{s}^2} \cdot \text{s}\right)}{(2.5 \times 10^{-3} \text{ kg})(16 \text{ m/s})} = 1.7 \times 10^{-32} \text{ m} = \mathbf{1.7 \times 10^{-23} \text{ nm}}$$

7.53   The angular momentum quantum number l can have integral (i.e. whole number) values from 0 to $n - 1$. In this case $n = 2$, so the allowed values of the angular momentum quantum number, l, are **0** and **1**.

Each allowed value of the angular momentum quantum number labels a subshell. Within a given subshell (label l) there are $2l + 1$ allowed energy states (orbitals) each labeled by a different value of the magnetic quantum number. The allowed values run from $-l$ through 0 to $+l$ (whole numbers only). For the subshell labeled by the angular momentum quantum number $l = 1$, the allowed values of the magnetic quantum number, m_l, are **−1, 0, and 1**. For the other subshell in this problem labeled by the angular momentum quantum number $l = 0$, the allowed value of the magnetic quantum number is **0**.

If the allowed whole number values run from −1 to +1, are there always $2l + 1$ values? Why?

7.54   Strategy: What are the relationships among n, l, and m_l?

Solution: We are given the principal quantum number, $n = 3$. The possible l values range from 0 to $(n - 1)$. Thus, there are three possible values of l: 0, 1, and 2, corresponding to the s, p, and d orbitals, respectively. The values of m_l can vary from $-l$ to l. The values of m_l for each l value are:

$l = 0$: $m_l = 0$ $l = 1$: $m_l = -1, 0, 1$ $l = 2$: $m_l = -2, -1, 0, 1, 2$

7.55   (a) $2p$: $n = 2$, $l = 1$, $m_l = 1, 0,$ or -1
      (b) $3s$: $n = 3$, $l = 0$, $m_l = 0$ (only allowed value)
      (c) $5d$: $n = 5$, $l = 2$, $m_l = 2, 1, 0, -1,$ or -2

An orbital in a subshell can have any of the allowed values of the magnetic quantum number for that subshell. All the orbitals in a subshell have exactly the same energy.

7.56   (a)   The number given in the designation of the subshell is the principal quantum number, so in this case $n = 3$. For s orbitals, $l = 0$. m_l can have integer values from $-l$ to $+l$, therefore, $m_l = 0$. The electron spin quantum number, m_s, can be either **+1/2** or **−1/2**.

Following the same reasoning as part **(a)**

(b) $4p$: $n = 4$; $l = 1$; $m_l = -1, 0, 1$; $m_s = +1/2, -1/2$

(c) $3d$: $n = 3$; $l = 2$; $m_l = -2, -1, 0, 1, 2$; $m_s = +1/2, -1/2$

7.57   A $2s$ orbital is larger than a $1s$ orbital. Both have the same spherical shape. The $1s$ orbital is lower in energy than the $2s$.

7.58   The two orbitals are identical in size, shape, and energy. They differ only in their orientation with respect to each other.

Can you assign a specific value of the magnetic quantum number to these orbitals? What are the allowed values of the magnetic quantum number for the $2p$ subshell?

7.59   The allowed values of l are 0, 1, 2, 3, and 4. These correspond to the $5s$, $5p$, $5d$, $5f$, and $5g$ subshells. These subshells each have one, three, five, seven, and nine orbitals, respectively.

7.60   For $n = 6$, the allowed values of l are 0, 1, 2, 3, 4, and 5 [$l = 0$ to $(n - 1)$, integer values]. These l values correspond to the $6s$, $6p$, $6d$, $6f$, $6g$, and $6h$ subshells. These subshells each have 1, 3, 5, 7, 9, and 11 orbitals, respectively (number of orbitals = $2l + 1$).

7.61   There can be a maximum of two electrons occupying one orbital.

    (a)   two   **(b)**   six   **(c)**   ten   **(d)**   fourteen

    What rule of nature demands a maximum of two electrons per orbital? Do they have the same energy? How are they different? Would five 4*d* orbitals hold as many electrons as five 3*d* orbitals? In other words, does the principal quantum number *n* affect the number of electrons in a given subshell?

7.62

n value	orbital sum	total number of electrons
1	1	2
2	$1 + 3 = 4$	8
3	$1 + 3 + 5 = 9$	18
4	$1 + 3 + 5 + 7 = 16$	32
5	$1 + 3 + 5 + 7 + 9 = 25$	50
6	$1 + 3 + 5 + 7 + 9 + 11 = 36$	72

    In each case the total number of orbitals is just the square of the *n* value (n^2). The total number of electrons is $2n^2$.

7.63   3*s*: two   3*d*: ten   4*p*: six   4*f*: fourteen   5*f*: fourteen

7.64   The electron configurations for the elements are

    (a)   N: $1s^2 2s^2 2p^3$ There are three *p*-type electrons.

    (b)   Si: $1s^2 2s^2 2p^6 3s^2 3p^2$ There are six *s*-type electrons.

    (c)   S: $1s^2 2s^2 2p^6 3s^2 3p^4$ There are no *d*-type electrons.

7.65   See Figure 7.22 in your textbook.

7.66   In the many-electron atom, the 3*p* orbital electrons are more effectively shielded by the inner electrons of the atom (that is, the 1*s*, 2*s*, and 2*p* electrons) than the 3*s* electrons. The 3*s* orbital is said to be more "penetrating" than the 3*p* and 3*d* orbitals. In the hydrogen atom there is only one electron, so the 3*s*, 3*p*, and 3*d* orbitals have the same energy.

7.67   Equation (7.4) of the text gives the orbital energy in terms of the principal quantum number, *n*, alone (for the hydrogen atom). The energy does not depend on any of the other quantum numbers. If two orbitals in the hydrogen atom have the same value of *n*, they have equal energy.

    (a)   2*s* > 1*s*   **(b)**   3*p* > 2*p*   **(c)**   equal   **(d)**   equal   **(e)**   5*s* > 4*f*

7.68   **(a)** 2*s* < 2*p*   **(b)** 3*p* < 3*d*   **(c)** 3*s* < 4*s*   **(d)** 4*d* < 5*f*

7.79   **(a)**   is wrong because the magnetic quantum number m_l can have only whole number values.

    (c)   is wrong because the maximum value of the angular momentum quantum number *l* is $n - 1$.

    (e)   is wrong because the electron spin quantum number m_s can have only half-integral values.

7.80   For aluminum, there are not enough electrons in the $2p$ subshell. (The $2p$ subshell holds six electrons.) The number of electrons (13) is correct. The electron configuration should be $1s^2 2s^2 2p^6 3s^2 3p^1$. The configuration shown might be an excited state of an aluminum atom.

For boron, there are too many electrons. (Boron only has five electrons.) The electron configuration should be $1s^2 2s^2 2p^1$. What would be the electric charge of a boron ion with the electron arrangement given in the problem?

For fluorine, there are also too many electrons. (Fluorine only has nine electrons.) The configuration shown is that of the F^- ion. The correct electron configuration is $1s^2 2s^2 2p^5$.

7.81   Since the atomic number is odd, it is mathematically impossible for all the electrons to be paired. There must be at least one that is unpaired. The element would be paramagnetic.

7.82   You should write the electron configurations for each of these elements to answer this question. In some cases, an orbital diagram may be helpful.

B: $[He]2s^2 2p^1$ (1 unpaired electron)   Ne: (0 unpaired electrons, Why?)

P: $[Ne]3s^2 3p^3$ (3 unpaired electrons)   Sc: $[Ar]4s^2 3d^1$ (1 unpaired electron)

Mn: $[Ar]4s^2 3d^5$ (5 unpaired electrons)   Se: $[Ar]4s^2 3d^{10} 4p^4$ (2 unpaired electrons)

Kr: (0 unpaired electrons)   Fe: $[Ar]4s^2 3d^6$ (4 unpaired electrons)

Cd: $[Kr]5s^2 4d^{10}$ (0 unpaired electrons)   I: $[Kr]5s^2 4d^{10} 5p^5$ (1 unpaired electron)

Pb: $[Xe]6s^2 4f^{14} 5d^{10} 6p^2$ (2 unpaired electrons)

7.83   B: $1s^2 2s^2 2p^1$ As: $[Ar]4s^2 3d^{10} 4p^3$

V: $[Ar]4s^2 3d^3$ I: $[Kr]5s^2 4d^{10} 5p^5$

Ni: $[Ar]4s^2 3d^8$ Au: $[Xe]6s^1 4f^{14} 5d^{10}$

What is the meaning of "[Ar]"? of "[Kr]"? of "[Xe]"?

7.84   **Strategy:** How many electrons are in the Ge atom ($Z = 32$)? We start with $n = 1$ and proceed to fill orbitals in the order shown in Figure 7.23 of the text. Remember that any given orbital can hold at most 2 electrons. However, don't forget about degenerate orbitals. Starting with $n = 2$, there are three p orbitals of equal energy, corresponding to $m_l = -1, 0, 1$. Starting with $n = 3$, there are five d orbitals of equal energy, corresponding to $m_l = -2, -1, 0, 1, 2$. We can place electrons in the orbitals according to the Pauli exclusion principle and Hund's rule. The task is simplified if we use the noble gas core preceding Ge for the inner electrons.

Solution: Germanium has 32 electrons. The noble gas core in this case is [Ar]. (Ar is the noble gas in the period preceding germanium.) [Ar] represents $1s^2 2s^2 2p^6 3s^2 3p^6$. This core accounts for 18 electrons, which leaves 14 electrons to place.

See Figure 7.23 of your text to check the order of filling subshells past the Ar noble gas core. You should find that the order of filling is $4s$, $3d$, $4p$. There are 14 remaining electrons to distribute among these orbitals. The $4s$ orbital can hold two electrons. Each of the five $3d$ orbitals can hold two electrons for a total of 10 electrons. This leaves two electrons to place in the $4p$ orbitals.

The electrons configuration for Ge is:

$$[Ar]4s^2 3d^{10} 4p^2$$

You should follow the same reasoning for the remaining atoms.

Fe: $[Ar]4s^2 3d^6$ Zn: $[Ar]4s^2 3d^{10}$ Ni: $[Ar]4s^2 3d^8$

W: $[Xe]6s^2 4f^{14} 5d^4$ Tl: $[Xe]6s^2 4f^{14} 5d^{10} 6p^1$

7.85 There are a total of twelve electrons:

Orbital	n	l	m_l	m_s
$1s$	1	0	0	$+\frac{1}{2}$
$1s$	1	0	0	$-\frac{1}{2}$
$2s$	2	0	0	$+\frac{1}{2}$
$2s$	2	0	0	$-\frac{1}{2}$
$2p$	2	1	1	$+\frac{1}{2}$
$2p$	2	1	1	$-\frac{1}{2}$
$2p$	2	1	0	$+\frac{1}{2}$
$2p$	2	1	0	$-\frac{1}{2}$
$2p$	2	1	-1	$+\frac{1}{2}$
$2p$	2	1	-1	$-\frac{1}{2}$
$3s$	3	0	0	$+\frac{1}{2}$
$3s$	3	0	0	$-\frac{1}{2}$

The element is magnesium.

7.86

$\uparrow\downarrow$	$\uparrow \quad \uparrow \quad \uparrow$
$3s^2$	$3p^3$

S$^+$ (5 valence electrons)
3 unpaired electrons

$\uparrow\downarrow$	$\uparrow\downarrow \quad \uparrow \quad \uparrow$
$3s^2$	$3p^4$

S (6 valence electrons)
2 unpaired electrons

$\uparrow\downarrow$	$\uparrow\downarrow \quad \uparrow\downarrow \quad \uparrow$
$3s^2$	$3p^5$

S$^-$ (7 valence electrons)
1 unpaired electron

S$^+$ has the most unpaired electrons

7.91 $[Ar]4s^2 3d^{10} 4p^4$

7.92 The ground state electron configuration of Tc is: $[Kr]5s^2 4d^5$.

7.93 We first calculate the wavelength, then we find the color using Figure 7.4 of the text.

$$\lambda = \frac{hc}{E} = \frac{(6.63 \times 10^{-34}\,\text{J}\cdot\text{s})(3.00 \times 10^8\,\text{m/s})}{4.30 \times 10^{-19}\,\text{J}} = 4.63 \times 10^{-7}\,\text{m} = 463\,\text{nm, which is \textbf{blue}.}$$

7.94 Part **(b)** is correct in the view of contemporary quantum theory. Bohr's explanation of emission and absorption line spectra appears to have universal validity. Parts **(a)** and **(c)** are artifacts of Bohr's early planetary model of the hydrogen atom and are *not* considered to be valid today.

7.95 **(a)** Wavelength and frequency are reciprocally related properties of any wave. The two are connected through Equation (7.1) of the text. See Example 7.2 of the text for a simple application of the relationship to a light wave.

(b) Typical wave properties: wavelength, frequency, characteristic wave speed (sound, light, etc.). Typical particle properties: mass, speed or velocity, momentum (mass × velocity), kinetic energy. For phenomena that we normally perceive in everyday life (macroscopic world) these properties are mutually exclusive. At the atomic level (microscopic world) objects can exhibit characteristic properties of both particles and waves. This is completely outside the realm of our everyday common sense experience and is extremely difficult to visualize.

(c) Quantization of energy means that emission or absorption of only descrete energies is allowed (e.g., atomic line spectra). Continuous variation in energy means that all energy changes are allowed (e.g., continuous spectra).

7.96 **(a)** With $n = 2$, there are n^2 orbitals $= 2^2 = 4$. $m_s = +1/2$, specifies 1 electron per orbital, for a total of **4 electrons**.

(b) $n = 4$ and $m_l = +1$, specifies one orbital in each subshell with $l = 1, 2,$ or 3 (i.e., a $4p$, $4d$, and $4f$ orbital). Each of the three orbitals holds 2 electrons for a total of **6 electrons**.

(c) If $n = 3$ and $l = 2$, m_l has the values 2, 1, 0, –1, or –2. Each of the five orbitals can hold 2 electrons for a total of **10 electrons** (2 e⁻ in each of the five $3d$ orbitals).

(d) If $n = 2$ and $l = 0$, then m_l can only be zero. $m_s = -1/2$ specifies 1 electron in this orbital for a total of **1 electron** (one e⁻ in the $2s$ orbital).

(e) $n = 4$, $l = 3$ and $m_l = -2$, specifies one $4f$ orbital. This orbital can hold **2 electrons**.

7.97 See the appropriate sections of the textbook in Chapter 7.

7.98 The wave properties of electrons are used in the operation of an electron microscope.

7.99 In the photoelectric effect, light of sufficient energy shining on a metal surface causes electrons to be ejected (photoelectrons). Since the electrons are charged particles, the metal surface becomes positively charged as more electrons are lost. After a long enough period of time, the positive surface charge becomes large enough to start attracting the ejected electrons back toward the metal with the result that the kinetic energy of the departing electrons becomes smaller.

7.100 **(a)** First convert 100 mph to units of m/s.

$$\frac{100 \text{ mi}}{1 \text{ h}} \times \frac{1 \text{ h}}{3600 \text{ s}} \times \frac{1.609 \text{ km}}{1 \text{ mi}} \times \frac{1000 \text{ m}}{1 \text{ km}} = 44.7 \text{ m/s}$$

Using the de Broglie equation:

$$\lambda = \frac{h}{mu} = \frac{\left(6.63 \times 10^{-34} \frac{\text{kg} \cdot \text{m}^2}{\text{s}^2} \cdot \text{s} \right)}{(0.141 \text{ kg})(44.7 \text{ m/s})} = 1.05 \times 10^{-34} \text{ m} = \mathbf{1.05 \times 10^{-25} \text{ nm}}$$

(b)   The average mass of a hydrogen atom is:

$$\frac{1.008 \text{ g}}{1 \text{ mol}} \times \frac{1 \text{ mol}}{6.022 \times 10^{23} \text{ atoms}} = 1.674 \times 10^{-24} \text{ g/H atom} = 1.674 \times 10^{-27} \text{ kg}$$

$$\lambda = \frac{h}{mu} = \frac{\left(6.63 \times 10^{-34} \frac{\text{kg} \cdot \text{m}^2}{\text{s}^2} \cdot \text{s} \right)}{(1.674 \times 10^{-27} \text{ kg})(44.7 \text{ m/s})} = 8.86 \times 10^{-9} \text{ m} = \textbf{8.86 nm}$$

7.101   There are many more paramagnetic elements than diamagnetic elements because of Hund's rule.

7.102   **(a)**   First, we can calculate the energy of a single photon with a wavelength of 633 nm.

$$E = \frac{hc}{\lambda} = \frac{(6.63 \times 10^{-34} \text{ J} \cdot \text{s})(3.00 \times 10^8 \text{ m/s})}{633 \times 10^{-9} \text{ m}} = 3.14 \times 10^{-19} \text{ J}$$

The number of photons produced in a 0.376 J pulse is:

$$0.376 \text{ J} \times \frac{1 \text{ photon}}{3.14 \times 10^{-19} \text{ J}} = \textbf{1.20} \times \textbf{10}^{\textbf{18}} \textbf{ photons}$$

(b)   Since a 1 W = 1 J/s, the power delivered per a 1.00×10^{-9} s pulse is:

$$\frac{0.376 \text{ J}}{1.00 \times 10^{-9} \text{ s}} = 3.76 \times 10^8 \text{ J/s} = \textbf{3.76} \times \textbf{10}^{\textbf{8}} \textbf{ W}$$

Compare this with the power delivered by a 100-W light bulb!

7.103   The energy required to heat the water is: $ms\Delta t = (368 \text{ g})(4.184 \text{ J/g} \cdot ^\circ\text{C})(5.00^\circ\text{C}) = 7.70 \times 10^3 \text{ J}$

Energy of a photon with a wavelength = 1.06×10^4 nm:

$$E = h\nu = \frac{hc}{\lambda} = \frac{(6.63 \times 10^{-34} \text{ J} \cdot \text{s})(3.00 \times 10^8 \text{ m/s})}{1.06 \times 10^{-5} \text{ m}} = 1.88 \times 10^{-20} \text{ J/photon}$$

The number of photons required is:

$$(7.70 \times 10^3 \text{ J}) \times \frac{1 \text{ photon}}{1.88 \times 10^{-20} \text{ J}} = \textbf{4.10} \times \textbf{10}^{\textbf{23}} \textbf{ photons}$$

7.104   First, let's find the energy needed to photodissociate one water molecule.

$$\frac{285.8 \text{ kJ}}{1 \text{ mol}} \times \frac{1 \text{ mol}}{6.022 \times 10^{23} \text{ molecules}} = 4.746 \times 10^{-22} \text{ kJ/molecule} = 4.746 \times 10^{-19} \text{ J/molecule}$$

The maximum wavelength of a photon that would provide the above energy is:

$$\lambda = \frac{hc}{E} = \frac{(6.63 \times 10^{-34} \text{ J} \cdot \text{s})(3.00 \times 10^8 \text{ m/s})}{4.746 \times 10^{-19} \text{ J}} = \textbf{4.19} \times \textbf{10}^{\textbf{-7}} \textbf{ m} = \textbf{419 nm}$$

This wavelength is in the visible region of the electromagnetic spectrum. Since water is continuously being struck by visible radiation *without* decomposition, it seems unlikely that photodissociation of water by this method is feasible.

7.105   For the Lyman series, we want the longest wavelength (smallest energy), with $n_i = 2$ and $n_f = 1$. Using Equation (7.5) of the text:

$$\Delta E = R_H \left(\frac{1}{n_i^2} - \frac{1}{n_f^2} \right) = (2.18 \times 10^{-18} \text{ J}) \left(\frac{1}{2^2} - \frac{1}{1^2} \right) = -1.64 \times 10^{-18} \text{ J}$$

$$\lambda = \frac{hc}{\Delta E} = \frac{(6.63 \times 10^{-34} \text{ J} \cdot \text{s})(3.00 \times 10^8 \text{ m/s})}{1.64 \times 10^{-18} \text{ J}} = 1.21 \times 10^{-7} \text{ m} = \textbf{121 nm}$$

For the Balmer series, we want the shortest wavelength (highest energy), with $n_i = \infty$ and $n_f = 2$.

$$\Delta E = R_H \left(\frac{1}{n_i^2} - \frac{1}{n_f^2} \right) = (2.18 \times 10^{-18} \text{ J}) \left(\frac{1}{\infty^2} - \frac{1}{2^2} \right) = -5.45 \times 10^{-19} \text{ J}$$

$$\lambda = \frac{hc}{\Delta E} = \frac{(6.63 \times 10^{-34} \text{ J} \cdot \text{s})(3.00 \times 10^8 \text{ m/s})}{5.45 \times 10^{-19} \text{ J}} = 3.65 \times 10^{-7} \text{ m} = \textbf{365 nm}$$

Therefore the two series do not overlap.

7.106   Since 1 W = 1 J/s, the energy output of the light bulb in 1 second is 75 J. The actual energy converted to visible light is 15 percent of this value or 11 J.

First, we need to calculate the energy of one 550 nm photon. Then, we can determine how many photons are needed to provide 11 J of energy.

The energy of one 550 nm photon is:

$$E = \frac{hc}{\lambda} = \frac{(6.63 \times 10^{-34} \text{ J} \cdot \text{s})(3.00 \times 10^8 \text{ m/s})}{550 \times 10^{-9} \text{ m}} = 3.62 \times 10^{-19} \text{ J/photon}$$

The number of photons needed to produce 11 J of energy is:

$$11 \text{ J} \times \frac{1 \text{ photon}}{3.62 \times 10^{-19} \text{ J}} = \textbf{3.0} \times \textbf{10}^{\textbf{19}} \textbf{ photons}$$

7.107   The energy needed per photon for the process is:

$$\frac{248 \times 10^3 \text{ J}}{1 \text{ mol}} \times \frac{1 \text{ mol}}{6.022 \times 10^{23} \text{ photons}} = 4.12 \times 10^{-19} \text{ J/photon}$$

$$\lambda = \frac{hc}{E} = \frac{(6.63 \times 10^{-34} \text{ J} \cdot \text{s})(3.00 \times 10^8 \text{ m/s})}{(4.12 \times 10^{-19} \text{ J})} = 4.83 \times 10^{-7} \text{ m} = \textbf{483 nm}$$

Any wavelength shorter than 483 nm will also promote this reaction. Once a person goes indoors, the reverse reaction $Ag + Cl \rightarrow AgCl$ takes place.

7.108   The Balmer series corresponds to transitions to the $n = 2$ level.

For He$^+$:

$$\Delta E = R_{He^+}\left(\frac{1}{n_i^2} - \frac{1}{n_f^2}\right) \qquad \lambda = \frac{hc}{\Delta E} = \frac{(6.63 \times 10^{-34}\ \text{J}\cdot\text{s})(3.00 \times 10^8\ \text{m/s})}{\Delta E}$$

For the transition, $n = 3 \rightarrow 2$

$$\Delta E = (8.72 \times 10^{-18}\ \text{J})\left(\frac{1}{3^2} - \frac{1}{2^2}\right) = -1.21 \times 10^{-18}\ \text{J} \qquad \lambda = \frac{1.99 \times 10^{-25}\ \text{J}\cdot\text{m}}{1.21 \times 10^{-18}\ \text{J}} = 1.64 \times 10^{-7}\ \text{m} = \textbf{164 nm}$$

For the transition, $n = 4 \rightarrow 2$, $\Delta E = -1.64 \times 10^{-18}$ J $\lambda = \textbf{121 nm}$

For the transition, $n = 5 \rightarrow 2$, $\Delta E = -1.83 \times 10^{-18}$ J $\lambda = \textbf{109 nm}$

For the transition, $n = 6 \rightarrow 2$, $\Delta E = -1.94 \times 10^{-18}$ J $\lambda = \textbf{103 nm}$

For H, the calculations are identical to those above, except the Rydberg constant for H is 2.18×10^{-18} J.

For the transition, $n = 3 \rightarrow 2$, $\Delta E = -3.03 \times 10^{-19}$ J $\lambda = \textbf{657 nm}$

For the transition, $n = 4 \rightarrow 2$, $\Delta E = -4.09 \times 10^{-19}$ J $\lambda = \textbf{487 nm}$

For the transition, $n = 5 \rightarrow 2$, $\Delta E = -4.58 \times 10^{-19}$ J $\lambda = \textbf{434 nm}$

For the transition, $n = 6 \rightarrow 2$, $\Delta E = -4.84 \times 10^{-19}$ J $\lambda = \textbf{411 nm}$

All the Balmer transitions for He$^+$ are in the ultraviolet region; whereas, the transitions for H are all in the visible region. Note the negative sign for energy indicating that a photon has been emitted.

7.109   **(a)** $\Delta H° = \Delta H_f°(O) + \Delta H_f°(O_2) - \Delta H_f°(O_3) = 249.4\ \text{kJ/mol} + (0) - 142.2\ \text{kJ/mol} = \textbf{107.2 kJ/mol}$

(b)   The energy in part (a) is for *one mole* of photons. To apply $E = h\nu$ we must divide by Avogadro's number. The energy of one photon is:

$$E = \frac{107.2\ \text{kJ}}{1\ \text{mol}} \times \frac{1\ \text{mol}}{6.022 \times 10^{23}\ \text{photons}} \times \frac{1000\ \text{J}}{1\ \text{kJ}} = 1.780 \times 10^{-19}\ \text{J/photon}$$

The wavelength of this photon can be found using the relationship $E = hc/\lambda$.

$$\lambda = \frac{hc}{E} = \frac{(6.63 \times 10^{-34}\ \text{J}\cdot\text{s})(3.00 \times 10^8\ \text{m/s})}{1.780 \times 10^{-19}\ \text{J}} \times \frac{1\ \text{nm}}{1 \times 10^{-9}\ \text{m}} = \textbf{1.12} \times \textbf{10}^3\ \textbf{nm}$$

7.110   First, we need to calculate the energy of one 600 nm photon. Then, we can determine how many photons are needed to provide 4.0×10^{-17} J of energy.

The energy of one 600 nm photon is:

$$E = \frac{hc}{\lambda} = \frac{(6.63 \times 10^{-34}\ \text{J}\cdot\text{s})(3.00 \times 10^8\ \text{m/s})}{600 \times 10^{-9}\ \text{m}} = 3.32 \times 10^{-19}\ \text{J/photon}$$

The number of photons needed to produce 4.0×10^{-17} J of energy is:

$$(4.0 \times 10^{-17} \text{ J}) \times \frac{1 \text{ photon}}{3.32 \times 10^{-19} \text{ J}} = \textbf{1.2} \times \textbf{10}^{\textbf{2}} \textbf{ photons}$$

7.111   Since the energy corresponding to a photon of wavelength λ_1 equals the energy of photon of wavelength λ_2 plus the energy of photon of wavelength λ_3, then the equation must relate the wavelength to energy.

energy of photon 1 = (energy of photon 2 + energy of photon 3)

Since $E = \dfrac{hc}{\lambda}$, then:

$$\frac{hc}{\lambda_1} = \frac{hc}{\lambda_2} + \frac{hc}{\lambda_3}$$

Dividing by hc:

$$\frac{1}{\lambda_1} = \frac{1}{\lambda_2} + \frac{1}{\lambda_3}$$

7.112   A "blue" photon (shorter wavelength) is higher energy than a "yellow" photon. For the same amount of energy delivered to the metal surface, there must be fewer "blue" photons than "yellow" photons. Thus, the yellow light would eject more electrons since there are more "yellow" photons. Since the "blue" photons are of higher energy, blue light will eject electrons with greater kinetic energy.

7.113   Refer to Figures 7.20 and 7.21 in the textbook.

7.114   The excited atoms are still neutral, so the total number of electrons is the same as the atomic number of the element.

(a)   He (2 electrons), $1s^2$

(b)   N (7 electrons), $1s^2 2s^2 2p^3$

(c)   Na (11 electrons), $1s^2 2s^2 2p^6 3s^1$

(d)   As (33 electrons), $[\text{Ar}]4s^2 3d^{10} 4p^3$

(e)   Cl (17 electrons), $[\text{Ne}]3s^2 3p^5$

7.115   Applying the Pauli exclusion principle and Hund's rule:

(a) $\underset{1s^2}{\uparrow\downarrow}$ $\underset{2s^2}{\uparrow\downarrow}$ $\underset{2p^5}{\uparrow\downarrow \ \uparrow\downarrow \ \uparrow}$

(b)   [Ne] $\underset{3s^2}{\uparrow\downarrow}$ $\underset{3p^3}{\uparrow \ \ \uparrow \ \ \uparrow}$

(c)   [Ar] $\underset{4s^2}{\uparrow\downarrow}$ $\underset{3d^7}{\uparrow\downarrow \ \uparrow\downarrow \ \uparrow \ \ \uparrow \ \ \uparrow}$

7.116   Rutherford and his coworkers might have discovered the wave properties of electrons.

7.117 $n_i = 236$, $n_f = 235$

$$\Delta E = (2.18 \times 10^{-18} \text{ J})\left(\frac{1}{236^2} - \frac{1}{235^2}\right) = -3.34 \times 10^{-25} \text{ J}$$

$$\lambda = \frac{hc}{\Delta E} = \frac{(6.63 \times 10^{-34} \text{ J} \cdot \text{s})(3.00 \times 10^8 \text{ m/s})}{3.34 \times 10^{-25} \text{ J}} = \mathbf{0.596 \text{ m}}$$

This wavelength is in the *microwave* region. (See Figure 7.4 of the text.)

7.118 The wavelength of a He atom can be calculated using the de Broglie equation. First, we need to calculate the root-mean-square speed using Equation (5.16) from the text.

$$u_{rms} = \sqrt{\frac{3\left(8.314\dfrac{\text{J}}{\text{K} \cdot \text{mol}}\right)(273 + 20)\text{K}}{4.003 \times 10^{-3} \text{ kg/mol}}} = 1.35 \times 10^3 \text{ m/s}$$

To calculate the wavelength, we also need the mass of a He atom in kg.

$$\frac{4.003 \times 10^{-3} \text{ kg He}}{1 \text{ mol He}} \times \frac{1 \text{ mol He}}{6.022 \times 10^{23} \text{ He atoms}} = 6.647 \times 10^{-27} \text{ kg/atom}$$

Finally, the wavelength of a He atom is:

$$\lambda = \frac{h}{mu} = \frac{(6.63 \times 10^{-34} \text{ J} \cdot \text{s})}{(6.647 \times 10^{-27} \text{ kg})(1.35 \times 10^3 \text{ m/s})} = \mathbf{7.39 \times 10^{-11} \text{ m}} = \mathbf{7.39 \times 10^{-2} \text{ nm}}$$

7.119 **(a)** Treating this as an absorption process:

$n_i = 1$, $n_f = \infty$

$$\Delta E = (2.18 \times 10^{-18} \text{ J})\left(\frac{1}{1^2} - \frac{1}{\infty^2}\right) = 2.18 \times 10^{-18} \text{ J}$$

For a mole of hydrogen atoms:

$$\textbf{Ionization energy} = \frac{2.18 \times 10^{-18} \text{ J}}{1 \text{ atom}} \times \frac{6.022 \times 10^{23} \text{ atoms}}{1 \text{ mol}} = 1.31 \times 10^6 \text{ J/mol} = \mathbf{1.31 \times 10^3 \text{ kJ/mol}}$$

(b) $$\Delta E = (2.18 \times 10^{-18} \text{ J})\left(\frac{1}{2^2} - \frac{1}{\infty^2}\right) = 5.45 \times 10^{-19} \text{ J}$$

$$\textbf{Ionization energy} = \frac{5.45 \times 10^{-19} \text{ J}}{1 \text{ atom}} \times \frac{6.022 \times 10^{23} \text{ atoms}}{1 \text{ mol}} = 3.28 \times 10^5 \text{ J/mol} = \mathbf{328 \text{ kJ/mol}}$$

It takes considerably less energy to remove the electron from an excited state.

7.120   **(a)**   **False**. $n = 2$ is the first excited state.

        (b)   **False**. In the $n = 4$ state, the electron is (on average) further from the nucleus and hence easier to remove.

        (c)   **True**.

        (d)   **False**. The $n = 4$ to $n = 1$ transition is a higher energy transition, which corresponds to a *shorter* wavelength.

        (e)   **True**.

7.121   The difference in ionization energy is:

$$(412 - 126)\text{kJ/mol} = 286 \text{ kJ/mol}.$$

In terms of one atom:

$$\frac{286 \times 10^3 \text{ J}}{1 \text{ mol}} \times \frac{1 \text{ mol}}{6.022 \times 10^{23} \text{ atoms}} = 4.75 \times 10^{-19} \text{ J/atom}$$

$$\lambda = \frac{hc}{\Delta E} = \frac{(6.63 \times 10^{-34} \text{ J} \cdot \text{s})(3.00 \times 10^8 \text{ m/s})}{4.75 \times 10^{-19} \text{ J}} = \textbf{4.19} \times \textbf{10}^{-7} \textbf{ m} = \textbf{419 nm}$$

7.122   We use Heisenberg's uncertainty principle with the equality sign to calculate the minimum uncertainty.

$$\Delta x \Delta p = \frac{h}{4\pi}$$

The momentum (p) is equal to the mass times the velocity.

$$p = mu \qquad \text{or} \qquad \Delta p = m\Delta u$$

We can write:

$$\Delta p = m\Delta u = \frac{h}{4\pi \Delta x}$$

Finally, the uncertainty in the velocity of the oxygen molecule is:

$$\Delta u = \frac{h}{4\pi m \Delta x} = \frac{(6.63 \times 10^{-34} \text{ J} \cdot \text{s})}{4\pi(5.3 \times 10^{-26} \text{ kg})(5.0 \times 10^{-5} \text{ m})} = \textbf{2.0} \times \textbf{10}^{-5} \textbf{ m/s}$$

7.123   It takes:

$$(5.0 \times 10^2 \text{ g ice}) \times \frac{334 \text{ J}}{1 \text{ g ice}} = 1.67 \times 10^5 \text{ J to melt } 5.0 \times 10^2 \text{ g of ice.}$$

Energy of a photon with a wavelength of 660 nm:

$$E = \frac{hc}{\lambda} = \frac{(6.63 \times 10^{-34} \text{ J} \cdot \text{s})(3.00 \times 10^8 \text{ m/s})}{660 \times 10^{-9} \text{ m}} = 3.01 \times 10^{-19} \text{ J}$$

Number of photons needed to melt 5.0×10^2 g of ice:

$$(1.67 \times 10^5 \text{ J}) \times \frac{1 \text{ photon}}{3.01 \times 10^{-19} \text{ J}} = \textbf{5.5} \times \textbf{10}^{\textbf{23}} \textbf{ photons}$$

The number of water molecules is:

$$(5.0 \times 10^2 \text{ g H}_2\text{O}) \times \frac{1 \text{ mol H}_2\text{O}}{18.02 \text{ g H}_2\text{O}} \times \frac{6.022 \times 10^{23} \text{ H}_2\text{O molecules}}{1 \text{ mol H}_2\text{O}} = 1.7 \times 10^{25} \text{ H}_2\text{O molecules}$$

The number of water molecules converted from ice to water by one photon is:

$$\frac{1.7 \times 10^{25} \text{ H}_2\text{O molecules}}{5.5 \times 10^{23} \text{ photons}} = \textbf{31 H}_2\textbf{O molecules/photon}$$

7.124   The Pauli exclusion principle states that no two electrons in an atom can have the same four quantum numbers. In other words, only two electrons may exist in the same atomic orbital, and these electrons must have opposite spins. **(a)** and **(f)** violate the Pauli exclusion principle.

Hund's rule states that the most stable arrangement of electrons in subshells is the one with the greatest number of parallel spins. **(b)**, **(d)**, and **(e)** violate Hund's rule.

7.125   Energy of a photon at 360 nm:

$$E = h\nu = \frac{hc}{\lambda} = \frac{(6.63 \times 10^{-34} \text{ J} \cdot \text{s})(3.00 \times 10^8 \text{ m/s})}{360 \times 10^{-9} \text{ m}} = 5.53 \times 10^{-19} \text{ J}$$

Area of exposed body in cm^2:

$$0.45 \text{ m}^2 \times \left(\frac{1 \text{ cm}}{1 \times 10^{-2} \text{ m}} \right)^2 = 4.5 \times 10^3 \text{ cm}^2$$

The number of photons absorbed by the body in 2 hours is:

$$0.5 \times \frac{2.0 \times 10^{16} \text{ photons}}{\text{cm}^2 \cdot \text{s}} \times (4.5 \times 10^3 \text{ cm}^2) \times \frac{7200 \text{ s}}{2 \text{ hr}} = 3.2 \times 10^{23} \text{ photons/2 hr}$$

The factor of 0.5 is used above because only 50% of the radiation is absorbed.

3.2×10^{23} photons with a wavelength of 360 nm correspond to an energy of:

$$(3.2 \times 10^{23} \text{ photons}) \times \frac{5.53 \times 10^{-19} \text{ J}}{1 \text{ photon}} = \textbf{1.8} \times \textbf{10}^{\textbf{5}} \textbf{ J}$$

7.126   As an estimate, we can equate the energy for ionization ($\text{Fe}^{13+} \rightarrow \text{Fe}^{14+}$) to the average kinetic energy $\left(\frac{3}{2} RT \right)$ of the ions.

$$\frac{3.5 \times 10^4 \text{ kJ}}{1 \text{ mol}} \times \frac{1000 \text{ J}}{1 \text{ kJ}} = 3.5 \times 10^7 \text{ J}$$

$$IE = \frac{3}{2}RT$$

$$3.5 \times 10^7 \text{ J/mol} = \frac{3}{2}(8.314 \text{ J/mol·K})T$$

$$T = 2.8 \times 10^6 \text{ K}$$

The actual temperature can be, and most probably is, higher than this.

7.127   The anti-atom of hydrogen should show the same characteristics as a hydrogen atom. Should an anti-atom of hydrogen collide with a hydrogen atom, they would be annihilated and energy would be given off.

7.128   Looking at the de Broglie equation $\lambda = \frac{h}{mu}$, the mass of an N_2 molecule (in kg) and the velocity of an N_2 molecule (in m/s) is needed to calculate the de Broglie wavelength of N_2.

First, calculate the root-mean-square velocity of N_2.

$\mathcal{M}(N_2) = 28.02$ g/mol $= 0.02802$ kg/mol

$$u_{rms}(N_2) = \sqrt{\frac{(3)\left(8.314\dfrac{\text{J}}{\text{mol·K}}\right)(300 \text{ K})}{\left(0.02802\dfrac{\text{kg}}{\text{mol}}\right)}} = 516.8 \text{ m/s}$$

Second, calculate the mass of one N_2 molecule in kilograms.

$$\frac{28.02 \text{ g } N_2}{1 \text{ mol } N_2} \times \frac{1 \text{ mol } N_2}{6.022 \times 10^{23} \text{ } N_2 \text{ molecules}} \times \frac{1 \text{ kg}}{1000 \text{ g}} = 4.653 \times 10^{-26} \text{ kg/molecule}$$

Now, substitute the mass of an N_2 molecule and the root-mean-square velocity into the de Broglie equation to solve for the de Broglie wavelength of an N_2 molecule.

$$\lambda = \frac{h}{mu} = \frac{(6.63 \times 10^{-34} \text{ J·s})}{(4.653 \times 10^{-26} \text{ kg})(516.8 \text{ m/s})} = 2.76 \times 10^{-11} \text{ m}$$

7.129   Based on the *selection rule*, which states that $\Delta l = \pm 1$, only **(b)** and **(d)** are allowed transitions.

7.130   The kinetic energy acquired by the electrons is equal to the voltage times the charge on the electron. After calculating the kinetic energy, we can calculate the velocity of the electrons (KE = $1/2mu^2$). Finally, we can calculate the wavelength associated with the electrons using the de Broglie equation.

$$KE = (5.00 \times 10^3 \text{ V}) \times \frac{1.602 \times 10^{-19} \text{ J}}{1 \text{ V}} = 8.01 \times 10^{-16} \text{ J}$$

We can now calculate the velocity of the electrons.

$$KE = \frac{1}{2}mu^2$$

$$8.01 \times 10^{-16} \text{ J} = \frac{1}{2}(9.1094 \times 10^{-31} \text{ kg})u^2$$

$$u = 4.19 \times 10^7 \text{ m/s}$$

Finally, we can calculate the wavelength associated with the electrons using the de Broglie equation.

$$\lambda = \frac{h}{mu}$$

$$\lambda = \frac{(6.63 \times 10^{-34} \text{ J} \cdot \text{s})}{(9.1094 \times 10^{-31} \text{ kg})(4.19 \times 10^7 \text{ m/s})} = 1.74 \times 10^{-11} \text{ m} = 17.4 \text{ pm}$$

7.131   The heat needed to raise the temperature of 150 mL of water from 20°C to 100°C is:

$$q = ms\Delta t = (150 \text{ g})(4.184 \text{ J/g} \cdot \text{°C})(100 - 20)\text{°C} = 5.0 \times 10^4 \text{ J}$$

The microwave will need to supply more energy than this because only 92.0% of microwave energy is converted to thermal energy of water. The energy that needs to be supplied by the microwave is:

$$\frac{5.0 \times 10^4 \text{ J}}{0.920} = 5.4 \times 10^4 \text{ J}$$

The energy supplied by one photon with a wavelength of 1.22×10^8 nm (0.122 m) is:

$$E = \frac{hc}{\lambda} = \frac{(6.63 \times 10^{-34} \text{ J} \cdot \text{s})(3.00 \times 10^8 \text{ m/s})}{(0.122 \text{ m})} = 1.63 \times 10^{-24} \text{ J}$$

The number of photons needed to supply 5.4×10^4 J of energy is:

$$(5.4 \times 10^4 \text{ J}) \times \frac{1 \text{ photon}}{1.63 \times 10^{-24} \text{ J}} = 3.3 \times 10^{28} \text{ photons}$$

7.132   The energy given in the problem is the energy of 1 mole of gamma rays. We need to convert this to the energy of one gamma ray, then we can calculate the wavelength and frequency of this gamma ray.

$$\frac{1.29 \times 10^{11} \text{ J}}{1 \text{ mol}} \times \frac{1 \text{ mol}}{6.022 \times 10^{23} \text{ gamma rays}} = 2.14 \times 10^{-13} \text{ J/gamma ray}$$

Now, we can calculate the wavelength and frequency from this energy.

$$E = \frac{hc}{\lambda}$$

$$\lambda = \frac{hc}{E} = \frac{(6.63 \times 10^{-34} \text{ J} \cdot \text{s})(3.00 \times 10^8 \text{ m/s})}{2.14 \times 10^{-13} \text{ J}} = 9.29 \times 10^{-13} \text{ m} = 0.929 \text{ pm}$$

and

$$E = h\nu$$

$$\nu = \frac{E}{h} = \frac{2.14 \times 10^{-13} \text{ J}}{6.63 \times 10^{-34} \text{ J} \cdot \text{s}} = 3.23 \times 10^{20} \text{ s}^{-1}$$

7.133 **(a)** Line A corresponds to the longest wavelength or lowest energy transition, which is the $3 \rightarrow 2$ transition. Therefore, line B corresponds to the $4 \rightarrow 2$ transition, and line C corresponds to the $5 \rightarrow 2$ transition.

(b) We can derive an equation for the energy change (ΔE) for an electronic transition.

$$E_f = -R_H Z^2 \left(\frac{1}{n_f^2} \right) \quad \text{and} \quad E_i = -R_H Z^2 \left(\frac{1}{n_i^2} \right)$$

$$\Delta E = E_f - E_i = -R_H Z^2 \left(\frac{1}{n_f^2} \right) - \left(-R_H Z^2 \left(\frac{1}{n_i^2} \right) \right)$$

$$\Delta E = R_H Z^2 \left(\frac{1}{n_i^2} - \frac{1}{n_f^2} \right)$$

Line C corresponds to the $5 \rightarrow 2$ transition. The energy change associated with this transition can be calculated from the wavelength (27.1 nm).

$$E = \frac{hc}{\lambda} = \frac{(6.63 \times 10^{-34} \text{ J·s})(3.00 \times 10^8 \text{ m/s})}{(27.1 \times 10^{-9} \text{ m})} = 7.34 \times 10^{-18} \text{ J}$$

For the $5 \rightarrow 2$ transition, we now know ΔE, n_i, n_f, and R_H ($R_H = 2.18 \times 10^{-18}$ J). Since this transition corresponds to an emission process, energy is released and ΔE is negative. ($\Delta E = -7.34 \times 10^{-18}$ J). We can now substitute these values into the equation above to solve for Z.

$$\Delta E = R_H Z^2 \left(\frac{1}{n_i^2} - \frac{1}{n_f^2} \right)$$

$$-7.34 \times 10^{-18} \text{ J} = (2.18 \times 10^{-18} \text{ J})Z^2 \left(\frac{1}{5^2} - \frac{1}{2^2} \right)$$

$$-7.34 \times 10^{-18} \text{ J} = (-4.58 \times 10^{-19})Z^2$$

$$Z^2 = 16.0$$

$$Z = 4$$

Z must be an integer because it represents the atomic number of the parent atom.

Now, knowing the value of Z, we can substitute in n_i and n_f for the $3 \rightarrow 2$ (Line A) and the $4 \rightarrow 2$ (Line B) transitions to solve for ΔE. We can then calculate the wavelength from the energy.

For Line A ($3 \rightarrow 2$)

$$\Delta E = R_H Z^2 \left(\frac{1}{n_i^2} - \frac{1}{n_f^2} \right) = (2.18 \times 10^{-18} \text{ J})(4)^2 \left(\frac{1}{3^2} - \frac{1}{2^2} \right)$$

$$\Delta E = -4.84 \times 10^{-18} \text{ J}$$

$$\lambda = \frac{hc}{E} = \frac{(6.63 \times 10^{-34} \text{ J·s})(3.00 \times 10^8 \text{ m/s})}{(4.84 \times 10^{-18} \text{ J})} = \textbf{4.11} \times \textbf{10}^{-8} \textbf{ m} = \textbf{41.1 nm}$$

For Line B $(4 \rightarrow 2)$

$$\Delta E = R_H Z^2 \left(\frac{1}{n_i^2} - \frac{1}{n_f^2} \right) = (2.18 \times 10^{-18} \text{ J})(4)^2 \left(\frac{1}{4^2} - \frac{1}{2^2} \right)$$

$$\Delta E = -6.54 \times 10^{-18} \text{ J}$$

$$\lambda = \frac{hc}{E} = \frac{(6.63 \times 10^{-34} \text{ J} \cdot \text{s})(3.00 \times 10^8 \text{ m/s})}{(6.54 \times 10^{-18} \text{ J})} = 3.04 \times 10^{-8} \text{ m} = 30.4 \text{ nm}$$

(c) The value of the final energy state is $n_f = \infty$. Use the equation derived in part (b) to solve for ΔE.

$$\Delta E = R_H Z^2 \left(\frac{1}{n_i^2} - \frac{1}{n_f^2} \right) = (2.18 \times 10^{-18} \text{ J})(4)^2 \left(\frac{1}{4^2} - \frac{1}{\infty^2} \right)$$

$$\Delta E = 2.18 \times 10^{-18} \text{ J}$$

(d) As we move to higher energy levels in an atom or ion, the energy levels get closer together. See Figure 7.11 of the text, which represents the energy levels for the hydrogen atom. Transitions from higher energy levels to the $n = 2$ level will be very close in energy and hence will have similar wavelengths. The lines are so close together that they overlap, forming a continuum. The continuum shows that the electron has been removed from the ion, and we no longer have quantized energy levels associated with the electron. In other words, the energy of the electron can now vary continuously.

CHAPTER 8
PERIODIC RELATIONSHIPS
AMONG THE ELEMENTS

8.19 Hydrogen forms the H^+ ion (resembles the alkali metals) and the H^- ion (resembles the halogens).

8.20 **Strategy:** **(a)** We refer to the building-up principle discussed in Section 7.9 of the text. We start writing the electron configuration with principal quantum number $n = 1$ and continue upward in energy until all electrons are accounted for. **(b)** What are the electron configuration characteristics of representative elements, transition elements, and noble gases? **(c)** Examine the pairing scheme of the electrons in the outermost shell. What determines whether an element is diamagnetic or paramagnetic?

Solution:

(a) We know that for $n = 1$, we have a $1s$ orbital (2 electrons). For $n = 2$, we have a $2s$ orbital (2 electrons) and three $2p$ orbitals (6 electrons). For $n = 3$, we have a $3s$ orbital (2 electrons). The number of electrons left to place is $17 - 12 = 5$. These five electrons are placed in the $3p$ orbitals. The electron configuration is $1s^2 2s^2 2p^6 3s^2 3p^5$ or $[Ne]3s^2 3p^5$.

(b) Because the $3p$ subshell is not completely filled, this is a *representative element*. Without consulting a periodic table, you might know that the halogen family has seven valence electrons. You could then further classify this element as a *halogen*. In addition, all halogens are *nonmetals*.

(c) If you were to write an orbital diagram for this electron configuration, you would see that there is *one* unpaired electron in the p subshell. Remember, the three $3p$ orbitals can hold a total of six electrons. Therefore, the atoms of this element are paramagnetic.

Check: For (b), note that a transition metal possesses an incompletely filled d subshell, and a noble gas has a completely filled outer-shell. For (c), recall that if the atoms of an element contain an odd number of electrons, the element must be paramagnetic.

8.21 **(a)** and **(d)**;   **(b)** and **(f)**;   **(c)** and **(e)**.

8.22 Elements that have the same number of valence electrons will have similarities in chemical behavior. Looking at the periodic table, elements with the same number of valence electrons are in the same group. Therefore, the pairs that would represent similar chemical properties of their atoms are:

    (a) and **(d)**   **(b)** and **(e)**   **(c)** and **(f)**.

8.23 **(a)** $1s^2 2s^2 2p^5$ (halogen)   **(c)** $[Ar]4s^2 3d^6$ (transition metal)

          (b) $[Ar]4s^2$ (alkaline earth metal)   **(d)** $[Ar]4s^2 3d^{10} 4p^3$ (Group 5A)

8.24 **(a)** Group 1A   **(b)** Group 5A   **(c)** Group 8A   **(d)** Group 8B

Identify the elements.

8.25   There are no electrons in the 4s subshell because transition metals lose electrons from the ns valence subshell before they are lost from the $(n-1)d$ subshell. For the neutral atom there are only six valence electrons. The element can be identified as Cr (chromium) simply by counting six across starting with potassium (K, atomic number 19).

What is the electron configuration of neutral chromium?

8.26   You should realize that the metal ion in question is a transition metal ion because it has five electrons in the 3d subshell. Remember that in a transition metal ion, the $(n-1)d$ orbitals are more stable than the ns orbital. Hence, when a cation is formed from an atom of a transition metal, electrons are *always* removed first from the ns orbital and then from the $(n-1)d$ orbitals if necessary. Since the metal ion has a +3 charge, three electrons have been removed. Since the 4s subshell is less stable than the 3d, two electrons would have been lost from the 4s and one electron from the 3d. Therefore, the electron configuration of the neutral atom is $[Ar]4s^2 3d^6$. This is the electron configuration of iron. Thus, the metal is **iron**.

8.27   Determine the number of electrons, and then "fill in" the electrons as you learned (Figure 7.23 and Table 7.3 of the text).

(a) $1s^2$	**(g)** $[Ar]4s^2 3d^{10} 4p^6$	**(m)** $[Xe]$
(b) $1s^2$	**(h)** $[Ar]4s^2 3d^{10} 4p^6$	**(n)** $[Xe]6s^2 4f^{14} 5d^{10}$
(c) $1s^2 2s^2 2p^6$	**(i)** $[Kr]$	**(o)** $[Kr]5d^{10}$
(d) $1s^2 2s^2 2p^6$	**(j)** $[Kr]$	**(p)** $[Xe]6s^2 4f^{14} 5d^{10}$
(e) $[Ne]3s^2 3p^6$	**(k)** $[Kr]5s^2 4d^{10}$	**(q)** $[Xe]4f^{14} 5d^{10}$
(f) $[Ne]$	**(l)** $[Kr]5s^2 4d^{10} 5p^6$	

8.28   **Strategy:** In the formation of a **cation** from the neutral atom of a representative element, one or more electrons are *removed* from the highest occupied n shell. In the formation of an **anion** from the neutral atom of a representative element, one or more electrons are *added* to the highest partially filled n shell. Representative elements typically gain or lose electrons to achieve a stable noble gas electron configuration. When a cation is formed from an atom of a transition metal, electrons are *always* removed first from the ns orbital and then from the $(n-1)d$ orbitals if necessary.

Solution:

(a) $[Ne]$	**(e)** Same as (c)	
(b) same as (a). Do you see why?	**(f)** $[Ar]3d^6$. Why isn't it $[Ar]4s^2 3d^4$?	
(c) $[Ar]$	**(g)** $[Ar]3d^9$. Why not $[Ar]4s^2 3d^7$?	
(d) Same as (c). Do you see why?	**(h)** $[Ar]3d^{10}$. Why not $[Ar]4s^2 3d^8$?	

8.29   This exercise simply depends on determining the total number of electrons and using Figure 7.23 and Table 7.3 of the text.

(a) $[Ar]$	**(f)** $[Ar]3d^6$	**(k)** $[Ar]3d^9$
(b) $[Ar]$	**(g)** $[Ar]3d^5$	**(l)** $[Kr]4d^{10}$
(c) $[Ar]$	**(h)** $[Ar]3d^7$	**(m)** $[Xe]4f^{14} 5d^{10}$
(d) $[Ar]3d^3$	**(i)** $[Ar]3d^8$	**(n)** $[Xe]4f^{14} 5d^8$
(e) $[Ar]3d^5$	**(j)** $[Ar]3d^{10}$	**(o)** $[Xe]4f^{14} 5d^8$

8.30   **(a)** Cr^{3+} **(b)** Sc^{3+} **(c)** Rh^{3+} **(d)** Ir^{3+}

8.31   Two species are isoelectronic if they have the same number of electrons. Can two neutral atoms of different elements be isoelectronic?

    (a)   C and B^- are isoelectronic.   **(b)** Mn^{2+} and Fe^{3+} are isoelectronic.

    (c)   Ar and Cl^- are isoelectronic.   **(d)**   Zn and Ge^{2+} are isoelectronic.

    With which neutral atom are the positive ions in (b) isoelectronic?

8.32   Isoelectronic means that the species have the same number of electrons and the same electron configuration.

    Be^{2+} and He $(2\ e^-)$ F^- and N^{3-} $(10\ e^-)$ Fe^{2+} and Co^{3+} $(24\ e^-)$ S^{2-} and Ar $(18\ e^-)$

8.37   **(a)**   Cs is larger. It is below Na in Group 1A.   **(d)**   Br is larger. It is below F in Group 7A.

    (b)   Ba is larger. It is below Be in Group 2A.   **(e)**   Xe is larger. It is below Ne in Group 8A.

    (c)   Sb is larger. It is below N in Group 5A.

8.38   **Strategy:** What are the trends in atomic radii in a periodic group and in a particular period. Which of the above elements are in the same group and which are in the same period?

    Solution: Recall that the general periodic trends in atomic size are:

    (1)   Moving from left to right across a row (period) of the periodic table, the atomic radius *decreases* due to an increase in effective nuclear charge.

    (2)   Moving down a column (group) of the periodic table, the atomic radius *increases* since the orbital size increases with increasing principal quantum number.

    The atoms that we are considering are all in the same period of the periodic table. Hence, the atom furthest to the left in the row will have the largest atomic radius, and the atom furthest to the right in the row will have the smallest atomic radius. Arranged in order of decreasing atomic radius, we have:

$$Na > Mg > Al > P > Cl$$

    Check: See Figure 8.5 of the text to confirm that the above is the correct order of decreasing atomic radius.

8.39   Pb, as can be seen in Figure 8.5 of the text.

8.40   **Fluorine** is the smallest atom in Group 7A. Atomic radius increases moving down a group since the orbital size increases with increasing principal quantum number, n.

8.41   The electron configuration of lithium is $1s^2 2s^1$. The two $1s$ electrons shield the $2s$ electron effectively from the nucleus. Consequently, the lithium atom is considerably larger than the hydrogen atom.

8.42   The atomic radius is largely determined by how strongly the outer-shell electrons are held by the nucleus. The larger the effective nuclear charge, the more strongly the electrons are held and the smaller the atomic radius. For the second period, the atomic radius of Li is largest because the $2s$ electron is well shielded by the filled $1s$ shell. The effective nuclear charge that the outermost electrons feel increases across the period as a result of incomplete shielding by electrons in the same shell. Consequently, the orbital containing the electrons is compressed and the atomic radius decreases.

8.43   **(a)**   Cl is smaller than Cl$^-$. An atom gets bigger when more electrons are added.

   (b)   Na$^+$ is smaller than Na. An atom gets smaller when electrons are removed.

   (c)   O^{2-} is smaller than S^{2-}. Both elements belong to the same group, and ionic radius increases going down a group.

   (d)   Al^{3+} is smaller than Mg^{2+}. The two ions are isoelectronic (What does that mean? See Section 8.2 of the text) and in such cases the radius gets smaller as the charge becomes more positive.

   (e)   Au^{3+} is smaller than Au$^+$ for the same reason as part (b).

In each of the above cases from which atom would it be harder to remove an electron?

8.44   **Strategy:** In comparing ionic radii, it is useful to classify the ions into three categories: (1) isoelectronic ions, (2) ions that carry the same charges and are generated from atoms of the same periodic group, and (3) ions that carry different charges but are generated from the same atom. In case (1), ions carrying a greater negative charge are always larger; in case (2), ions from atoms having a greater atomic number are always larger; in case (3), ions have a smaller positive charge are always larger.

Solution: The ions listed are all isoelectronic. They each have ten electrons. The ion with the fewest protons will have the largest ionic radius, and the ion with the most protons will have the smallest ionic radius. The effective nuclear charge increases with increasing number of protons. The electrons are attracted more strongly by the nucleus, decreasing the ionic radius. N^{3-} has only 7 protons resulting in the smallest attraction exerted by the nucleus on the 10 electrons. N^{3-} is the largest ion of the group. Mg^{2+} has 12 protons resulting in the largest attraction exerted by the nucleus on the 10 electrons. Mg^{2+} is the smallest ion of the group. The order of increasing atomic radius is:

$$\textbf{Mg}^{2+} \; < \; \textbf{Na}^+ \; < \textbf{F}^- \; < \; \textbf{O}^{2-} \; < \; \textbf{N}^{3-}$$

8.45   The Cu$^+$ ion is larger than Cu^{2+} because it has one more electron.

8.46   Both selenium and tellurium are Group 6A elements. Since atomic radius increases going down a column in the periodic table, it follows that **Te^{2-} must be larger than Se^{2-}.**

8.47   Bromine is liquid; all the others are solids.

8.48   We assume the approximate boiling point of argon is the mean of the boiling points of neon and krypton, based on its position in the periodic table being between Ne and Kr in Group 8A.

$$\textbf{b.p.} \; = \; \frac{-245.9°C + (-152.9°C)}{2} \; = \; -\textbf{199.4°C}$$

The actual boiling point of argon is −185.7°C.

8.51   Ionization energy increases across a row of the periodic table and decreases down a column or group. The correct order of increasing ionization energy is:

$$\textbf{Cs} \; < \; \textbf{Na} \; < \; \textbf{Al} \; < \; \textbf{S} \; < \; \textbf{Cl}$$

8.52   The general periodic trend for first ionization energy is that it increases across a period (row) of the periodic table and it decreases down a group (column). Of the choices, K will have the smallest ionization energy. Ca, just to the right of K, will have a higher first ionization energy. Moving to the right across the periodic table, the ionization energies will continue to increase as we move to P. Continuing across to Cl and moving up the halogen group, F will have a higher ionization energy than P. Finally, Ne is to the right of F in period two, thus it will have a higher ionization energy. The correct order of increasing first ionization energy is:

$$K < Ca < P < F < Ne$$

You can check the above answer by looking up the first ionization energies for these elements in Table 8.2 of the text.

8.53   Apart from the small irregularities, the ionization energies of elements in a period increase with increasing atomic number. We can explain this trend by referring to the increase in effective nuclear charge from left to right. A larger effective nuclear charge means a more tightly held outer electron, and hence a higher first ionization energy. Thus, in the third period, sodium has the lowest and neon has the highest first ionization energy.

8.54   The Group 3A elements (such as Al) all have a single electron in the outermost p subshell, which is well shielded from the nuclear charge by the inner electrons and the ns^2 electrons. Therefore, less energy is needed to remove a single p electron than to remove a paired s electron from the same principal energy level (such as for Mg).

8.55   To form the +2 ion of calcium, it is only necessary to remove two valence electrons. For potassium, however, the second electron must come from the atom's noble gas core which accounts for the much higher second ionization energy. Would you expect a similar effect if you tried to form the +3 ion of calcium?

8.56   **Strategy:** Removal of the outermost electron requires less energy if it is shielded by a filled inner shell.

Solution: The lone electron in the $3s$ orbital will be much easier to remove. This lone electron is shielded from the nuclear charge by the filled inner shell. Therefore, the ionization energy of 496 kJ/mol is paired with the electron configuration $1s^2 2s^2 2p^6 3s^1$.

A noble gas electron configuration, such as $1s^2 2s^2 2p^6$, is a very stable configuration, making it extremely difficult to remove an electron. The $2p$ electron is not as effectively shielded by electrons in the same energy level. The high ionization energy of 2080 kJ/mol would be associated with the element having this noble gas electron configuration.

Check: Compare this answer to the data in Table 8.2. The electron configuration of $1s^2 2s^2 2p^6 3s^1$ corresponds to a Na atom, and the electron configuration of $1s^2 2s^2 2p^6$ corresponds to a Ne atom.

8.57   The ionization energy is the difference between the $n = \infty$ state (final) and the $n = 1$ state (initial).

$$\Delta E = E_\infty - E_1 = (-2.18 \times 10^{-18} \text{ J})(2)^2 \left(\frac{1}{\infty}\right)^2 - (-2.18 \times 10^{-18} \text{ J})(2)^2 \left(\frac{1}{1}\right)^2$$

$$\Delta E = 0 + (2.18 \times 10^{-18} \text{ J})(2)^2 \left(\frac{1}{1}\right)^2 = 8.72 \times 10^{-18} \text{ J}$$

In units of kJ/mol: $(8.72 \times 10^{-18} \text{ J}) \times \dfrac{1 \text{ kJ}}{1000 \text{ J}} \times \dfrac{6.022 \times 10^{23}}{1 \text{ mol}} = \mathbf{5.25 \times 10^3 \text{ kJ/mol}}$

Should this be larger than the first ionization energy of helium (see Table 8.2 of the text)?

8.58   The atomic number of mercury is 80. We write:

$$\Delta E = (2.18 \times 10^{-18} \text{ J})(80^2)\left(\frac{1}{1^2} - \frac{1}{\infty^2}\right) = 1.40 \times 10^{-14} \text{ J/ion}$$

$$\Delta E = \frac{1.40 \times 10^{-14} \text{ J}}{1 \text{ ion}} \times \frac{6.022 \times 10^{23} \text{ ions}}{1 \text{ mol}} \times \frac{1 \text{ kJ}}{1000 \text{ J}} = \mathbf{8.43 \times 10^6 \text{ kJ/mol}}$$

8.61   **(a)**   K < Na < Li   **(b)**   I < Br < F < Cl

8.62   **Strategy:** What are the trends in electron affinity in a periodic group and in a particular period. Which of the above elements are in the same group and which are in the same period?

Solution: One of the general periodic trends for electron affinity is that the tendency to accept electrons increases (that is, electron affinity values become more positive) as we move from left to right across a period. However, this trend does not include the noble gases. We know that noble gases are extremely stable, and they do not want to gain or lose electrons.

Based on the above periodic trend, **Cl** would be expected to have the highest electron affinity. Addition of an electron to Cl forms Cl^-, which has a stable noble gas electron configuration.

8.63   Based on electron affinity values, we would not expect the alkali metals to form anions. A few years ago most chemists would have answered this question with a loud "No"! In the early seventies a chemist named J.L. Dye at Michigan State University discovered that under very special circumstances alkali metals could be coaxed into accepting an electron to form negative ions! These ions are called alkalide ions.

8.64   Alkali metals have a valence electron configuration of ns^1 so they can accept another electron in the ns orbital. On the other hand, alkaline earth metals have a valence electron configuration of ns^2. Alkaline earth metals have little tendency to accept another electron, as it would have to go into a higher energy p orbital.

8.67   Basically, we look for the process that will result in forming a cation of the metal that will be isoelectronic with the noble gas preceding the metal in the periodic table. Since all alkali metals have the ns^1 outer electron configuration, we predict that they will form unipositive ions: M^+. Similarly, the alkaline earth metals, which have the ns^2 outer electron configuration, will form M^{2+} ions.

8.68   Since ionization energies decrease going down a column in the periodic table, francium should have the lowest first ionization energy of all the alkali metals. As a result, Fr should be the most reactive of all the Group 1A elements toward water and oxygen. The reaction with oxygen would probably be similar to that of K, Rb, or Cs.

What would you expect the formula of the oxide to be? The chloride?

8.69   The electron configuration of helium is $1s^2$ and that of the other noble gases is ns^2np^6. The completely filled subshell represents great stability. Consequently, these elements are chemically unreactive.

8.70   The Group 1B elements are much less reactive than the Group 1A elements. The 1B elements are more stable because they have much higher ionization energies resulting from incomplete shielding of the nuclear charge by the inner d electrons. The ns^1 electron of a Group 1A element is shielded from the nucleus more effectively by the completely filled noble gas core. Consequently, the outer s electrons of 1B elements are more strongly attracted by the nucleus.

8.71   Across a period, oxides change from basic to amphoteric to acidic. Going down a group, the oxides become more basic.

8.72   **(a)**   Lithium oxide is a basic oxide. It reacts with water to form the metal hydroxide:

$$Li_2O(s) + H_2O(l) \longrightarrow 2LiOH(aq)$$

(b)   Calcium oxide is a basic oxide. It reacts with water to form the metal hydroxide:

$$CaO(s) + H_2O(l) \longrightarrow Ca(OH)_2(aq)$$

(c)   Sulfur trioxide is an acidic oxide. It reacts with water to form sulfuric acid:

$$SO_3(g) + H_2O(l) \longrightarrow H_2SO_4(aq)$$

8.73   LiH (lithium hydride): ionic compound; BeH_2 (beryllium hydride): covalent compound; B_2H_6 (diborane, you aren't expected to know that name): molecular compound; CH_4 (methane, do you know that one?): molecular compound; NH_3 (ammonia, you should know that one): molecular compound; H_2O (water, if you didn't know that one, you should be ashamed): molecular compound; HF (hydrogen fluoride): molecular compound. LiH and BeH_2 are solids, B_2H_6, CH_4, NH_3, and HF are gases, and H_2O is a liquid.

8.74   As we move down a column, the metallic character of the elements increases. Since magnesium and barium are both Group 2A elements, we expect barium to be more metallic than magnesium and **BaO** to be more basic than MgO.

8.75   **(a)**   Metallic character decreases moving left to right across a period and increases moving down a column (Group).

(b)   Atomic size decreases moving left to right across a period and increases moving down a column (Group).

(c)   Ionization energy increases (with some exceptions) moving left to right across a period and decreases moving down a column.

(d)   Acidity of oxides increases moving left to right across a period and decreases moving down a column.

8.76   **(a)**   bromine   **(b)**   nitrogen   **(c)**   rubidium   **(d)**   magnesium

8.77   **(a)** $S^- + e^- \rightarrow S^{2-}$
          (b) $Ti^{2+} \rightarrow Ti^{3+} + e^-$
          (c) $Mg^{2+} + e^- \rightarrow Mg^+$
          (d) $O^{2-} \rightarrow O^- + e^-$

8.78   This is an isoelectronic series with ten electrons in each species. The nuclear charge interacting with these ten electrons ranges from +8 for oxygen to +12 for magnesium. Therefore the +12 charge in Mg^{2+} will draw in the ten electrons more tightly than the +11 charge in Na^+, than the +9 charge in F^-, than the +8 charge in O^{2-}. Recall that the largest species will be the *easiest* to ionize.

   (a) increasing ionic radius: $Mg^{2+} < Na^+ < F^- < O^{2-}$

   (b) increasing ionization energy: $O^{2-} < F^- < Na^+ < Mg^{2+}$

8.79   Ionic compounds are usually combinations of a metal and a nonmetal. Molecular compounds are usually nonmetal–nonmetal combinations.

   (a) Na_2O (ionic); MgO (ionic); Al_2O_3 (ionic); SiO_2 (molecular);
      P_4O_6 and P_4O_{10} (both molecular); SO_2 or SO_3 (molecular);

      Cl_2O and several others (all molecular).

   (b) $NaCl$ (ionic); $MgCl_2$ (ionic); $AlCl_3$ (ionic); $SiCl_4$ (molecular);
      PCl_3 and PCl_5 (both molecular); SCl_2 (molecular).

8.80   According to the *Handbook of Chemistry and Physics* (1966-67 edition), potassium metal has a melting point of 63.6°C, bromine is a reddish brown liquid with a melting point of −7.2°C, and potassium bromide (KBr) is a colorless solid with a melting point of 730°C. **M** is **potassium** (K) and **X** is **bromine** (Br).

8.81   **(a)** matches bromine (Br_2),   **(b)** matches hydrogen (H_2),   **(c)** matches calcium (Ca),
      (d) matches gold (Au),   **(e)** matches argon (Ar)

8.82 O^+ and N   Ar and S^{2-} Ne and N^{3-} Zn and As^{3+} Cs^+ and Xe

8.83   Only **(b)** is listed in order of decreasing radius. Answer (a) is listed in increasing size because the radius increases down a group. Answer (c) is listed in increasing size because the number of electrons is increasing.

8.84   **(a)** and **(d)**

8.85   The equation is: $CO_2(g) + Ca(OH)_2(aq) \rightarrow CaCO_3(s) + H_2O(l)$

   The milky white color is due to calcium carbonate. Calcium hydroxide is a base and carbon dioxide is an acidic oxide. The products are a salt and water.

8.86   Fluorine is a yellow-green gas that attacks glass; chlorine is a pale yellow gas; bromine is a fuming red liquid; and iodine is a dark, metallic-looking solid.

8.87   **(a)**   (i)   Both react with water to produce hydrogen;
            (ii)   Their oxides are basic;
            (iii) Their halides are ionic.

   (b)   (i)   Both are strong oxidizing agents;
            (ii)   Both react with hydrogen to form HX (where X is Cl or Br);
            (iii) Both form halide ions (Cl^- or Br^-) when combined with electropositive metals (Na, K, Ca, Ba).

8.88   Fluorine

8.89   Sulfur has a ground state electron configuration of $[Ne]3s^2 3p^4$. Therefore, it has a tendency to accept one electron to become S^-. Although adding another electron makes S^{2-}, which is isoelectronic with Ar, the increase in electron repulsion makes the process unfavorable.

8.90 H^- and He are isoelectronic species with two electrons. Since H^- has only one proton compared to two protons for He, the nucleus of H^- will attract the two electrons less strongly compared to He. Therefore, **H^- is larger**.

8.91 Na_2O (basic oxide) $Na_2O + H_2O \rightarrow 2NaOH$

       BaO (basic oxide) $BaO + H_2O \rightarrow Ba(OH)_2$

       CO_2 (acidic oxide) $CO_2 + H_2O \rightarrow H_2CO_3$

       N_2O_5 (acidic oxide) $N_2O_5 + H_2O \rightarrow 2HNO_3$

       P_4O_{10} (acidic oxide) $P_4O_{10} + 6H_2O \rightarrow 4H_3PO_4$

       SO_3 (acidic oxide) $SO_3 + H_2O \rightarrow H_2SO_4$

8.92

Oxide	Name	Property
Li_2O	lithium oxide	basic
BeO	beryllium oxide	amphoteric
B_2O_3	boron oxide	acidic
CO_2	carbon dioxide	acidic
N_2O_5	dinitrogen pentoxide	acidic

Note that only the highest oxidation states are considered.

8.93

Element	State	Form
Mg	solid	three dimensional
Cl	gas	diatomic molecules
Si	solid	three dimensional
Kr	gas	monatomic
O	gas	diatomic molecules
I	solid	diatomic molecules
Hg	liquid	liquid (metallic)
Br	liquid	diatomic molecules

8.94   In its chemistry, hydrogen can behave like an alkali metal (H^+) and like a halogen (H^-). H^+ is a single proton.

8.95   The reactions are:

(a) $Li_2O + CO_2 \rightarrow Li_2CO_3$

(b) $2Na_2O_2 + 2CO_2 \rightarrow 2Na_2CO_3 + O_2$

(c) $4KO_2 + 2CO_2 \rightarrow 2K_2CO_3 + 3O_2$

8.96   Replacing Z in the equation given in Problem 8.57 with $(Z - \sigma)$ gives:

$$E_n = (2.18 \times 10^{-18} \text{ J})(Z - \sigma)^2 \left(\frac{1}{n^2} \right)$$

For helium, the atomic number (Z) is 2, and in the ground state, its two electrons are in the first energy level, so $n = 1$. Substitute Z, n, and the first ionization energy into the above equation to solve for σ.

$$E_1 = 3.94 \times 10^{-18} \text{ J} = (2.18 \times 10^{-18} \text{ J})(2 - \sigma)^2 \left(\frac{1}{1^2} \right)$$

$$(2 - \sigma)^2 = \frac{3.94 \times 10^{-18} \text{ J}}{2.18 \times 10^{-18} \text{ J}}$$

$$2 - \sigma = \sqrt{1.81}$$

$$\sigma = 2 - 1.35 = \mathbf{0.65}$$

8.97   Noble gases have filled shells or subshells. Therefore, they have little tendency to accept electrons (endothermic).

8.98   The percentage of volume occupied by K^+ compared to K is:

$$\frac{\text{volume of } K^+ \text{ ion}}{\text{volume of K atom}} \times 100\% = \frac{\frac{4}{3}\pi(133 \text{ pm})^3}{\frac{4}{3}\pi(216 \text{ pm})^3} \times 100\% = 23.3\%$$

Therefore, there is a decrease in volume of $(100 - 23.3)\% = \mathbf{76.7\%}$ when K^+ is formed from K.

8.99   The volume of a sphere is $4/3\pi r^3$.

The percent change in volume from F to F^- is:

$$\frac{\text{volume of } F^- \text{ ion}}{\text{volume of F atom}} \times 100\% = \frac{\frac{4}{3}\pi(136 \text{ pm})^3}{\frac{4}{3}\pi(72 \text{ pm})^3} \times 100\% = 674\%$$

Therefore, there is an increase in volume of $(674 - 100)\%$ or **574%** as a result of the formation of the F^- ion.

8.100   Rearrange the given equation to solve for ionization energy.

$$IE = h\nu - \frac{1}{2}mu^2$$

or,

$$IE = \frac{hc}{\lambda} - KE$$

The kinetic energy of the ejected electron is given in the problem. Substitute h, c, and λ into the above equation to solve for the ionization energy.

$$IE = \frac{(6.63 \times 10^{-34} \text{ J} \cdot \text{s})(3.00 \times 10^8 \text{ m/s})}{162 \times 10^{-9} \text{ m}} - (5.34 \times 10^{-19} \text{ J})$$

$$IE = 6.94 \times 10^{-19} \text{ J}$$

We might also want to express the ionization energy in kJ/mol.

$$\frac{6.94 \times 10^{-19} \text{ J}}{1 \text{ photon}} \times \frac{6.022 \times 10^{23} \text{ photons}}{1 \text{ mol}} \times \frac{1 \text{ kJ}}{1000 \text{ J}} = \textbf{418 kJ/mol}$$

To ensure that the ejected electron is the valence electron, UV light of the *longest* wavelength (lowest energy) should be used that can still eject electrons.

8.101   **(a)**   Because of argon's lack of reactivity.
  (b)   Once Ar was discovered, scientists began to look for other unreactive elements.
  (c)   Atmosphere's content of helium is too low to be detected.

8.102   We want to determine the second ionization energy of lithium.

$$Li^+ \longrightarrow Li^{2+} + e^- \qquad I_2 = ?$$

The equation given in Problem 8.57 allows us to determine the third ionization energy for Li. Knowing the total energy needed to remove all three electrons from Li, we can calculate the second ionization energy by difference.

Energy needed to remove three electrons $= I_1 + I_2 + I_3$

First, let's calculate I_3. For Li, Z = 3, and n = 1 because the third electron will come from the 1s orbital.

$$I_3 = \Delta E = E_\infty - E_3$$

$$I_3 = -(2.18 \times 10^{-18} \text{ J})(3)^2 \left(\frac{1}{\infty^2}\right) + (2.18 \times 10^{-18} \text{ J})(3)^2 \left(\frac{1}{1^2}\right)$$

$$I_3 = +1.96 \times 10^{-17} \text{ J}$$

Converting to units of kJ/mol:

$$I_3 = (1.96 \times 10^{-17} \text{ J}) \times \frac{6.022 \times 10^{23} \text{ ions}}{1 \text{ mol}} = 1.18 \times 10^7 \text{ J/mol} = 1.18 \times 10^4 \text{ kJ/mol}$$

Energy needed to remove three electrons $= I_1 + I_2 + I_3$

$$1.96 \times 10^4 \text{ kJ/mol} = 520 \text{ kJ/mol} + I_2 + (1.18 \times 10^4 \text{ kJ/mol})$$

$$I_2 = \textbf{7.28} \times \textbf{10}^3 \textbf{ kJ/mol}$$

8.103   The first equation is: $X + H_2 \rightarrow Y$. We are given sufficient information from the decomposition reaction (the reverse reaction) to calculate the relative number of moles of X and H. At STP, 1 mole of a gas occupies a volume of 22.4 L.

$$0.559 \text{ L} \times \frac{1 \text{ mol}}{22.4 \text{ L}} = 0.0250 \text{ mol } H_2$$

$$0.0250 \text{ mol } H_2 \times \frac{2 \text{ mol H}}{1 \text{ mol } H_2} = 0.0500 \text{ mol H}$$

Let $\mathcal{M}$ be the molar mass of X. If we assume that the formula for Y is either XH, XH_2, or XH_3, then if $Y = XH$, then

$$\frac{\text{mol H}}{\text{mol X}} = 1 = \frac{0.0500 \text{ mol}}{1.00 \text{ g} \times \dfrac{1}{\mathcal{M}(\text{g/mol})}}$$

$$\mathcal{M} = 20.0 \text{ g/mol} = \text{the element Ne (closest mass)}$$

if $Y = XH_2$, then

$$\frac{\text{mol H}}{\text{mol X}} = 2 = \frac{0.0500 \text{ mol}}{1.00 \text{ g} \times \dfrac{1}{\mathcal{M}(\text{g/mol})}}$$

$$\mathcal{M} = 40.0 \text{ g/mol} = \text{the element Ca (closest mass)}$$

if $Y = XH_3$, then

$$\frac{\text{mol H}}{\text{mol X}} = 3 = \frac{0.0500 \text{ mol}}{1.00 \text{ g} \times \dfrac{1}{\mathcal{M}(\text{g/mol})}}$$

$$\mathcal{M} = 60.0 \text{ g/mol} = ? \text{ (no element of close mass)}$$

If we deduce that the element $X = Ca$, then the formula for the chloride Z is $CaCl_2$ (why?). (Why couldn't X be Ne?) Calculating the mass percent of chlorine in $CaCl_2$ to compare with the known results.

$$\%Cl = \frac{(2)(35.45)}{[40.08 + (2)(35.45)]} \times 100\% = 63.89\%$$

Therefore X is **calcium**.

8.104   X must belong to Group 4A; it is probably **Sn** or **Pb** because it is not a very reactive metal (it is certainly not reactive like an alkali metal).

Y is a nonmetal since it does *not* conduct electricity. Since it is a light yellow solid, it is probably **phosphorus** (Group 5A).

Z is an **alkali metal** since it reacts with air to form a basic oxide or peroxide.

8.105 Plotting the boiling point versus the atomic number and extrapolating the curve to francium, the estimated boiling point is 670°C.

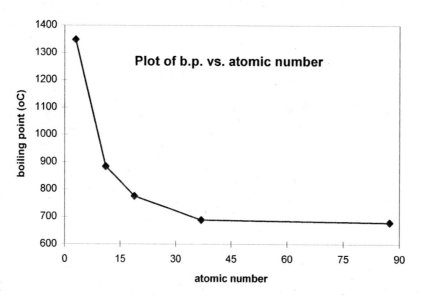

8.106 $Na \longrightarrow Na^+ + e^-$ $I_1 = 495.9$ kJ/mol

This equation is the reverse of the electron affinity for Na^+. Therefore, the electron affinity of Na^+ is **+495.9 kJ/mol**. Note that the electron affinity is positive, indicating that energy is liberated when an electron is added to an atom or ion. You should expect this since we are adding an electron to a positive ion.

8.107 The plot is:

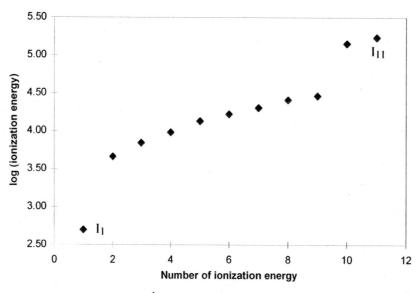

(a) I_1 corresponds to the electron in $3s^1$ I_7 corresponds to the electron in $2p^1$
I_2 corresponds to the first electron in $2p^6$ I_8 corresponds to the first electron in $2s^2$
I_3 corresponds to the first electron in $2p^5$ I_9 corresponds to the electron in $2s^1$
I_4 corresponds to the first electron in $2p^4$ I_{10} corresponds to the first electron in $1s^2$
I_5 corresponds to the first electron in $2p^3$ I_{11} corresponds to the electron in $1s^1$
I_6 corresponds to the first electron in $2p^2$

(b)   It requires more energy to remove an electron from a closed shell. The breaks indicate electrons in different shells and subshells.

8.108   The reaction representing the electron affinity of chlorine is:

$$Cl(g) + e^- \longrightarrow Cl^-(g) \qquad \Delta H° = +349 \text{ kJ/mol}$$

It follows that the energy needed for the reverse process is also +349 kJ/mol.

$$Cl^-(g) + h\nu \longrightarrow Cl(g) + e^- \qquad \Delta H° = +349 \text{ kJ/mol}$$

The energy above is the energy of one mole of photons. We need to convert to the energy of one photon in order to calculate the wavelength of the photon.

$$\frac{349 \text{ kJ}}{1 \text{ mol photons}} \times \frac{1 \text{ mol photons}}{6.022 \times 10^{23} \text{ photons}} \times \frac{1000 \text{ J}}{1 \text{ kJ}} = 5.80 \times 10^{-19} \text{ J/photon}$$

Now, we can calculate the wavelength of a photon with this energy.

$$\lambda = \frac{hc}{E} = \frac{(6.63 \times 10^{-34} \text{ J·s})(3.00 \times 10^8 \text{ m/s})}{5.80 \times 10^{-19} \text{ J}} = 3.43 \times 10^{-7} \text{ m} = \textbf{343 nm}$$

The radiation is in the **ultraviolet** region of the electromagnetic spectrum.

8.109   Considering electron configurations, $Fe^{2+}[Ar]3d^6 \rightarrow Fe^{3+}[Ar]3d^5$

$$Mn^{2+}[Ar]3d^5 \rightarrow Mn^{3+}[Ar]3d^4$$

A half-filled shell has extra stability. In oxidizing Fe^{2+} the product is a d^5-half-filled shell. In oxidizing Mn^{2+}, a d^5-half-filled shell electron is being lost, which requires more energy.

8.110   The equation that we want to calculate the energy change for is:

$$Na(s) \longrightarrow Na^+(g) + e^- \qquad \Delta H° = ?$$

Can we take information given in the problem and other knowledge to end up with the above equation? This is a Hess's law problem (see Chapter 6).

In the problem we are given:	$Na(s) \longrightarrow Na(g)$	$\Delta H° = 108.4$ kJ/mol
We also know the ionization energy of Na (g).	$Na(g) \longrightarrow Na^+(g) + e^-$	$\Delta H° = 495.9$ kJ/mol
Adding the two equations:	$Na(s) \longrightarrow Na^+(g) + e^-$	$\pmb{\Delta H° = 604.3 \text{ kJ/mol}}$

8.111   The hydrides are: LiH (lithium hydride), CH_4 (methane), NH_3 (ammonia), H_2O (water), and HF (hydrogen fluoride).

The reactions with water: $LiH + H_2O \rightarrow LiOH + H_2$

$$CH_4 + H_2O \rightarrow \text{ no reaction at room temperature.}$$

$$NH_3 + H_2O \rightarrow NH_4^+ + OH^-$$

$$H_2O + H_2O \rightarrow H_3O^+ + OH^-$$

$$HF + H_2O \rightarrow H_3O^+ + F^-$$

The last three reactions involve *equilibria* that will be discussed in later chapters.

8.112   The electron configuration of titanium is: $[Ar]4s^2 3d^2$. Titanium has four valence electrons, so the maximum oxidation number it is likely to have in a compound is +4. The compounds followed by the oxidation state of titanium are: K_3TiF_6, +3; $K_2Ti_2O_5$, +4; $TiCl_3$, +3; K_2TiO_4, +6; and K_2TiF_6, +4. **K_2TiO_4** is unlikely to exist because of the oxidation state of Ti of +6. Titanium in an oxidation state greater than +4 is unlikely because of the very high ionization energies needed to remove the fifth and sixth electrons.

8.113   **(a)** Mg in $Mg(OH)_2$ **(d)** Na in $NaHCO_3$ **(g)** Ca in CaO

(b) Na, liquid   **(e)** K in KNO_3 **(h)** Ca

(c) Mg in $MgSO_4 \cdot 7H_2O$ **(f)** Mg   **(i)** Na in NaCl; Ca in $CaCl_2$

8.114   The unbalanced ionic equation is: $MnF_6^{2-} + SbF_5 \longrightarrow SbF_6^- + MnF_3 + F_2$

In this redox reaction, Mn^{4+} is reduced to Mn^{3+}, and F^- from both MnF_6^{2-} and SbF_5 is oxidized to F_2.

We can simplify the half-reactions. $Mn^{4+} \xrightarrow{\text{reduction}} Mn^{3+}$

$F^- \xrightarrow{\text{oxidation}} F_2$

Balancing the two half-reactions: $Mn^{4+} + e^- \longrightarrow Mn^{3+}$

$2F^- \longrightarrow F_2 + 2e^-$

Adding the two half-reactions: $2Mn^{4+} + 2F^- \longrightarrow 2Mn^{3+} + F_2$

We can now reconstruct the complete balanced equation. In the balanced equation, we have 2 moles of Mn ions and 1 mole of F_2 on the products side.

$$2K_2MnF_6 + SbF_5 \longrightarrow KSbF_6 + 2MnF_3 + 1F_2$$

We can now balance the remainder of the equation by inspection. Notice that there are 4 moles of K^+ on the left, but only 1 mole of K^+ on the right. The balanced equation is:

$$2K_2MnF_6 + 4SbF_5 \longrightarrow 4KSbF_6 + 2MnF_3 + F_2$$

8.115   **(a)** $2KClO_3(s) \rightarrow 2KCl(s) + 3O_2(g)$

(b) $N_2(g) + 3H_2(g) \rightarrow 2NH_3(g)$ (industrial)

$NH_4Cl(s) + NaOH(aq) \rightarrow NH_3(g) + NaCl(aq) + H_2O(l)$

(c) $CaCO_3(s) \rightarrow CaO(s) + CO_2(g)$ (industrial)

$CaCO_3(s) + 2HCl(aq) \rightarrow CaCl_2(aq) + H_2O(l) + CO_2(g)$

(d) $Zn(s) + H_2SO_4(aq) \rightarrow ZnSO_4(aq) + H_2(g)$

(e)   Same as **(c)**, (first equation)

8.116   To work this problem, assume that the oxidation number of oxygen is -2.

Oxidation number	Chemical formula
+1	N_2O
+2	NO
+3	N_2O_3
+4	NO_2, N_2O_4
+5	N_2O_5

8.117   Examine a solution of Na_2SO_4 which is colorless. This shows that the SO_4^{2-} ion is colorless. Thus the blue color is due to $Cu^{2+}(aq)$.

8.118   The larger the effective nuclear charge, the more tightly held are the electrons. Thus, the atomic radius will be small, and the ionization energy will be large. The quantities show an opposite periodic trend.

8.119 Z_{eff} increases from left to right across the table, so electrons are held more tightly. (This explains the electron affinity values of C and O.) Nitrogen has a zero value of electron affinity because of the stability of the half-filled $2p$ subshell (that is, N has little tendency to accept another electron).

8.120   We assume that the m.p. and b.p. of bromine will be between those of chlorine and iodine.

Taking the average of the melting points and boiling points:

$$\textbf{m.p.} = \frac{-101.0°C + 113.5°C}{2} = \textbf{6.3°C} \qquad \text{(Handbook: } -7.2°C\text{)}$$

$$\textbf{b.p.} = \frac{-34.6°C + 184.4°C}{2} = \textbf{74.9°C} \qquad \text{(Handbook: } 58.8 \ °C\text{)}$$

The estimated values do not agree very closely with the actual values because $Cl_2(g)$, $Br_2(l)$, and $I_2(s)$ are in different physical states. If you were to perform the same calculations for the noble gases, your calculations would be much closer to the actual values.

8.121   Once an atom gains an electron forming a negative ion, adding additional electrons is typically an unfavorable process due to electron-electron repulsions. 2nd and 3rd electron affinities do not occur spontaneously and are therefore difficult to measure.

8.122   The heat generated from the radioactive decay can break bonds; therefore, few radon compounds exist.

8.123   Physical characteristics: Solid; metallic appearance like iodine; melting point greater than 114°C.

Reaction with sulfuric acid:

$2NaAt + 2H_2SO_4 \rightarrow At_2 + SO_2 + Na_2SO_4 + 2H_2O$

8.124   **(a)**   It was determined that the periodic table was based on atomic number, not atomic mass.

(b)   Argon:

$(0.00337 \times 35.9675 \text{ amu}) + (0.00063 \times 37.9627 \text{ amu}) + (0.9960 \times 39.9624 \text{ amu}) = \textbf{39.95 amu}$

Potassium:

$(0.93258 \times 38.9637 \text{ amu}) + (0.000117 \times 39.9640 \text{ amu}) + (0.0673 \times 40.9618 \text{ amu}) = \textbf{39.10 amu}$

8.125 $Na(g) \rightarrow Na^+(g) + e^-$ $I_1 = 495.9$ kJ/mol

Energy needed to ionize one Na atom:

$$\frac{495.9 \times 10^3 \text{ J}}{1 \text{ mol}} \times \frac{1 \text{ mol}}{6.022 \times 10^{23} \text{ atoms}} = 8.235 \times 10^{-19} \text{ J/atom}$$

The corresponding wavelength is:

$$\lambda = \frac{hc}{I_1} = \frac{(6.63 \times 10^{-34} \text{ J} \cdot \text{s})(3.00 \times 10^8 \text{ m/s})}{8.235 \times 10^{-19} \text{ J}} = 2.42 \times 10^{-7} \text{ m} = \mathbf{242 \text{ nm}}$$

8.126 $Z = 119$

Electron configuration: $[Rn]7s^2 5f^{14} 6d^{10} 7p^6 8s^1$

8.127   Both ionization energy and electron affinity are affected by atomic size – the smaller the atom, the greater the attraction between the electrons and the nucleus. If it is difficult to remove an electron from an atom (that is, high ionization energy), then it follows that it would also be favorable to add an electron to the atom (large electron affinity).

Noble gases would be an exception to this generalization.

8.128   There is a large jump from the second to the third ionization energy, indicating a change in the principal quantum number n. In other words, the third electron removed is an inner, noble gas core electron, which is difficult to remove. Therefore, the element is in **Group 2A**.

8.129   Helium should be named helon to match the other noble gases: neon, argon, xenon, krypton, and radon. In addition, the ending, -ium, suggests that helium has properties similar to some metals (i.e., sodium, magnesium, barium, etc.). Since helium in an unreactive gas, this ending in not appropriate.

8.130   **(a)** SiH_4, GeH_4, SnH_4, PbH_4
   (b)   Metallic character increases going down a family of the periodic table. Therefore, RbH would be more ionic than NaH.
   (c)   Since Ra is in Group 2A, we would expect the reaction to be the same as other alkaline earth metals with water.

$$Ra(s) + 2H_2O(l) \rightarrow Ra(OH)_2(aq) + H_2(g)$$

   (d)   Beryllium (diagonal relationship)

8.131   **(a)** F_2 **(b)** Na   **(c)** B   **(d)** N_2 **(e)** Al

8.132   The importance and usefulness of the periodic table lie in the fact that we can use our understanding of the general properties and trends within a group or a period to predict with considerable accuracy the properties of any element, even though the element may be unfamiliar to us. For example, elements in the same group or family have the same valence electron configurations. Due to the same number of valence electrons occupying similar orbitals, elements in the same family have similar chemical properties. In addition, trends in properties such as ionization energy, atomic radius, electron affinity, and metallic character can be predicted based on an element's position in the periodic table. Ionization energy typically increases across a period of the period table and decreases down a group. Atomic radius typically decreases across a period and increases down a group. Electron affinity typically increases across a period and decreases down a group.

Metallic character typically decreases across a period and increases down a group. The periodic table is an extremely useful tool for a scientist. Without having to look in a reference book for a particular element's properties, one can look at its position in the periodic table and make educated predictions as to its many properties such as those mentioned above.

8.133

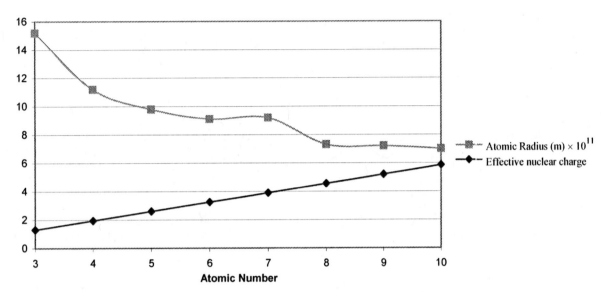

Note that the atomic radius values (in meters) have been multiplied by 1×10^{11}, so that the effective nuclear charge and radius data would fit better on the same graph. In general, as the effective nuclear charge increases, the outer-shell electrons are held more strongly, and hence the atomic radius decreases.

8.134   The first statement that an allotropic form of the element is a colorless crystalline solid, might lead you to think about diamond, a form of carbon. When carbon is reacted with excess oxygen, the colorless gas, carbon dioxide is produced.

$$C(s) + O_2(g) \rightarrow CO_2(g)$$

When $CO_2(g)$ is dissolved in water, carbonic acid is produced.

$$CO_2(g) + H_2O(l) \rightarrow H_2CO_3(aq)$$

Element X is most likely carbon, choice **(c)**.

8.135   Referring to the Chemistry in Action on p. 336 of the text, Mg will react with air (O_2 and N_2) to produce $MgO(s)$ and $Mg_3N_2(s)$. The reaction is:

$$5Mg(s) + O_2(g) + N_2(g) \rightarrow 2MgO(s) + Mg_3N_2(s)$$

$MgO(s)$ will react with water to produce the basic solution, $Mg(OH)_2(aq)$. The reaction is:

$$MgO(s) + H_2O(l) \rightarrow Mg(OH)_2(aq)$$

The problem states that B forms a similar solution to A, plus a gas with a pungent odor. This gas is ammonia, NH_3. The reaction is:

$$Mg_3N_2(s) + 6H_2O(l) \rightarrow 3Mg(OH)_2(aq) + 2NH_3(g)$$

A is **MgO**, and **B** is **Mg₃N₂**.

CHAPTER 9
CHEMICAL BONDING I:
BASIC CONCEPTS

9.15 We use Coulomb's law to answer this question: $E = k\dfrac{Q_{cation} Q_{anion}}{r}$

    (a) Doubling the radius of the cation would increase the distance, r, between the centers of the ions. A larger value of r results in a smaller energy, E, of the ionic bond. Is it possible to say how much smaller E will be?

    (b) Tripling the charge on the cation will result in tripling of the energy, E, of the ionic bond, since the energy of the bond is directly proportional to the charge on the cation, Q_{cation}.

    (c) Doubling the charge on both the cation and anion will result in quadrupling the energy, E, of the ionic bond.

    (d) Decreasing the radius of both the cation and the anion to half of their original values is the same as halving the distance, r, between the centers of the ions. Halving the distance results in doubling the energy.

9.16 **(a)** RbI, rubidium iodide   **(b)** Cs_2SO_4, cesium sulfate

    (c) Sr_3N_2, strontium nitride   **(d)** Al_2S_3, aluminum sulfide

9.17 Lewis representations for the ionic reactions are as follows.

    (a) $Na\cdot \;+\; :\!\overset{\cdot\cdot}{\underset{\cdot\cdot}{F}}\!\cdot \;\longrightarrow\; Na^{+}\;:\!\overset{\cdot\cdot}{\underset{\cdot\cdot}{F}}\!:^{-}$

    (b) $2K\cdot \;+\; \cdot\!\overset{\cdot\cdot}{\underset{\cdot\cdot}{S}}\!\cdot \;\longrightarrow\; 2K^{+}\;:\!\overset{\cdot\cdot}{\underset{\cdot\cdot}{S}}\!:^{2-}$

    (c) $\overset{\cdot}{\underset{\cdot}{Ba}} \;+\; \cdot\!\overset{\cdot\cdot}{\underset{\cdot\cdot}{O}}\!\cdot \;\longrightarrow\; Ba^{2+}\;:\!\overset{\cdot\cdot}{\underset{\cdot\cdot}{O}}\!:^{2-}$

    (d) $\overset{\cdot}{\underset{\cdot}{Al}}\!\cdot \;+\; \cdot\!\overset{\cdot\cdot}{N}\!\cdot \;\longrightarrow\; Al^{3+}\;:\!\overset{\cdot\cdot}{\underset{\cdot\cdot}{N}}\!:^{3-}$

9.18 The Lewis representations for the reactions are:

    (a) $\overset{\cdot}{\underset{\cdot}{Sr}} \;+\; \cdot\!\overset{\cdot\cdot}{\underset{\cdot\cdot}{Se}}\!\cdot \;\longrightarrow\; Sr^{2+}\;:\!\overset{\cdot\cdot}{\underset{\cdot\cdot}{Se}}\!:^{2-}$

    (b) $\overset{\cdot}{\underset{\cdot}{Ca}} \;+\; 2\,H\cdot \;\longrightarrow\; Ca^{2+}\;2H\!:^{-}$

    (c) $3Li\cdot \;+\; \cdot\!\overset{\cdot\cdot}{N}\!\cdot \;\longrightarrow\; 3Li^{+}\;:\!\overset{\cdot\cdot}{\underset{\cdot\cdot}{N}}\!:^{3-}$

    (d) $2\overset{\cdot}{\underset{\cdot}{Al}}\!\cdot \;+\; 3\cdot\!\overset{\cdot\cdot}{\underset{\cdot\cdot}{S}}\!: \;\longrightarrow\; 2Al^{3+}\;3:\!\overset{\cdot\cdot}{\underset{\cdot\cdot}{S}}\!:^{2-}$

9.19   **(a)**   I and Cl should form a molecular compound; both elements are nonmetals. One possibility would be ICl, iodine chloride.

(b)   Mg and F will form an ionic compound; Mg is a metal while F is a nonmetal. The substance will be MgF_2, magnesium fluoride.

9.20   **(a)** Covalent (BF_3, boron trifluoride)   **(b)** ionic (KBr, potassium bromide)

9.25   (1) $Na(s) \rightarrow Na(g)$ $\Delta H_1^{\circ} = 108$ kJ/mol

(2) $\frac{1}{2} Cl_2(g) \rightarrow Cl(g)$ $\Delta H_2^{\circ} = 121.4$ kJ/mol

(3) $Na(g) \rightarrow Na^+(g) + e^-$ $\Delta H_3^{\circ} = 495.9$ kJ/mol

(4) $Cl(g) + e^- \rightarrow Cl^-(g)$ $\Delta H_4^{\circ} = -349$ kJ/mol

(5) $Na^+(g) + Cl^-(g) \rightarrow NaCl(s)$ $\Delta H_5^{\circ} = ?$

$Na(s) + \frac{1}{2} Cl_2(g) \rightarrow NaCl(s)$ $\Delta H_{overall}^{\circ} = -411$ kJ/mol

$$\Delta H_5^{\circ} = \Delta H_{overall}^{\circ} - \Delta H_1^{\circ} - \Delta H_2^{\circ} - \Delta H_3^{\circ} - \Delta H_4^{\circ} = (-411) - (108) - (121.4) - (495.9) - (-349) = -787 \text{ kJ/mol}$$

The lattice energy of NaCl is **787 kJ/mol**.

9.26   (1) $Ca(s) \rightarrow Ca(g)$ $\Delta H_1^{\circ} = 121$ kJ/mol

(2) $Cl_2(g) \rightarrow 2Cl(g)$ $\Delta H_2^{\circ} = 242.8$ kJ/mol

(3) $Ca(g) \rightarrow Ca^+(g) + e^-$ $\Delta H_3^{\circ \prime} = 589.5$ kJ/mol

$Ca^+(g) \rightarrow Ca^{2+}(g) + e^-$ $\Delta H_3^{\circ \prime\prime} = 1145$ kJ/mol

(4) $2[Cl(g) + e^- \rightarrow Cl^-(g)]$ $\Delta H_4^{\circ} = 2(-349 \text{ kJ/mol}) = -698$ kJ/mol

(5) $Ca^{2+}(g) + 2Cl^-(g) \rightarrow CaCl_2(s)$ $\Delta H_5^{\circ} = ?$

$Ca(s) + Cl_2(g) \rightarrow CaCl_2(s)$ $\Delta H_{overall}^{\circ} = -795$ kJ/mol

Thus we write:

$$\Delta H_{overall}^{\circ} = \Delta H_1^{\circ} + \Delta H_2^{\circ} + \Delta H_3^{\circ \prime} + \Delta H_3^{\circ \prime\prime} + \Delta H_4^{\circ} + \Delta H_5^{\circ}$$

$$\Delta H_5^{\circ} = (-795 - 121 - 242.8 - 589.5 - 1145 + 698) \text{kJ/mol} = -2195 \text{ kJ/mol}$$

The lattice energy is represented by the reverse of equation (5); therefore, the lattice energy is **+2195 kJ/mol**.

9.35   The degree of ionic character in a bond is a function of the difference in electronegativity between the two bonded atoms. Figure 9.5 lists electronegativity values of the elements. The bonds in order of increasing ionic character are: N–N (zero difference in electronegativity) < S–O (difference 1.0) = Cl–F (difference 1.0) < K–O (difference 2.7) < Li–F (difference 3.0).

9.36   **Strategy:** We can look up electronegativity values in Figure 9.5 of the text. The amount of ionic character is based on the electronegativity difference between the two atoms. The larger the electronegativity difference, the greater the ionic character.

Solution: Let ΔEN = electronegativity difference. The bonds arranged in order of increasing ionic character are:

C–H ($\Delta EN = 0.4$) < Br–H ($\Delta EN = 0.7$) < F–H ($\Delta EN = 1.9$) < Li–Cl ($\Delta EN = 2.0$)

< Na–Cl ($\Delta EN = 2.1$) < K–F ($\Delta EN = 3.2$)

9.37   We calculate the electronegativity differences for each pair of atoms:

DE: $3.8 - 3.3 = 0.5$ DG: $3.8 - 1.3 = 2.5$ EG: $3.3 - 1.3 = 2.0$ DF: $3.8 - 2.8 = 1.0$

The order of increasing covalent bond character is: **DG < EG < DF < DE**

9.38   The order of increasing ionic character is:

Cl–Cl (zero difference in electronegativity) < Br–Cl (difference 0.2) < Si–C (difference 0.7)

< Cs–F (difference 3.3).

9.39   **(a)**   The two carbon atoms are the same. The bond is covalent.

(b)   The elelctronegativity difference between K and I is $2.5 - 0.8 = 1.7$. The bond is polar covalent.

(c)   The electronegativity difference between N and B is $3.0 - 2.0 = 1.0$. The bond is polar covalent.

(d)   The electronegativity difference between C and F is $4.0 - 2.5 = 1.5$. The bond is polar covalent.

9.40   **(a)**   The two silicon atoms are the same. The bond is covalent.

(b)   The electronegativity difference between Cl and Si is $3.0 - 1.8 = 1.2$. The bond is polar covalent.

(c)   The electronegativity difference between F and Ca is $4.0 - 1.0 = 3.0$. The bond is ionic.

(d)   The electronegativity difference between N and H is $3.0 - 2.1 = 0.9$. The bond is polar covalent.

9.43   **(a)** 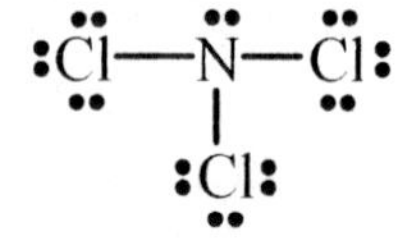 :Cl—N—Cl: **(b)** 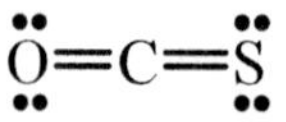 O=C=S   **(c)** H—O—O—H
                  |
                 :Cl:

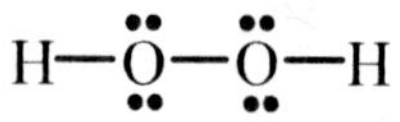

(d) 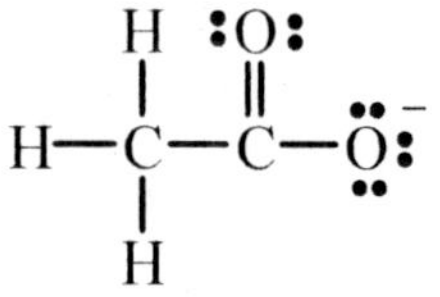   **(e)** 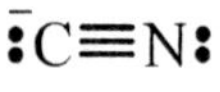 :C≡N: **(f)** 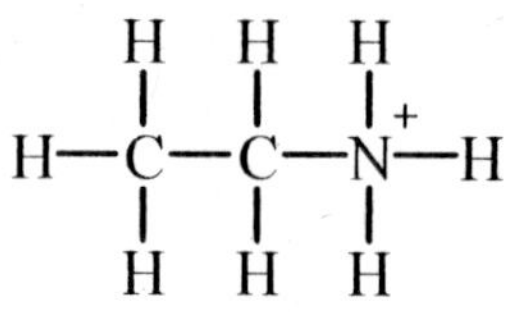
      H :O: H H H
      | || | | |+
H—C—C—O:⁻   H—C—C—N—H
      | | | |
      H   H H H

9.44   **(a)** 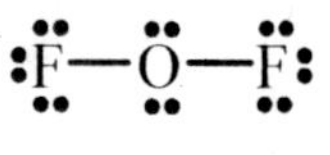 :F—O—F: **(b)** 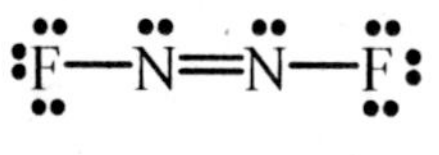 :F—N=N—F: **(c)** 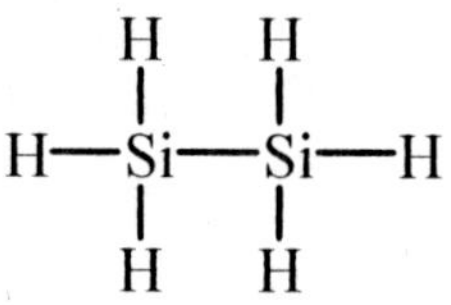
                                                              H   H
                                                              | |
                                                        H—Si—Si—H
                                                              | |
                                                              H   H

(d) 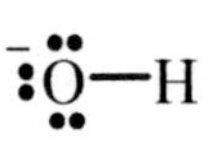 $:\ddot{\underset{..}{O}}{-}H$ **(e)** 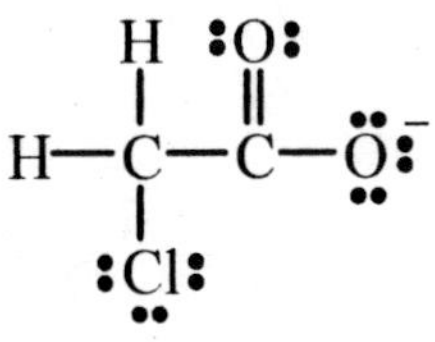 (structure)   **(f)** 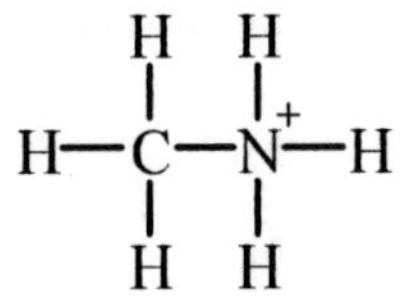 (structure)

9.45 The lewis dot structures are:

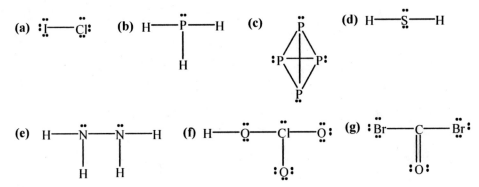

9.46 **Strategy:** We follow the procedure for drawing Lewis structures outlined in Section 9.6 of the text.

Solution:

(a)

Step 1: It is obvious that the skeletal structure is: O   O

Step 2: The outer-shell electron configuration of O is $2s^2 2p^4$. Also, we must add the negative charges to the number of valence electrons, Thus, there are

$$(2 \times 6) + 2 = 14 \text{ valence electrons}$$

Step 3: We draw a single covalent bond between each O, and then attempt to complete the octets for the O atoms.

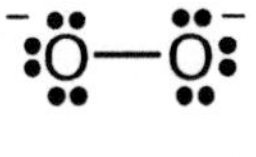

$$:\ddot{\underset{..}{O}}{-}\ddot{\underset{..}{O}}:$$

Because this structure satisfies the octet rule for both oxygen atoms, step 4 outlined in the text is not required.

Check: As a final check, we verify that there are 14 valence electrons in the Lewis structure of O_2^-.

Follow the same procedure as part (a) for parts (b), (c), and (d). The appropriate Lewis structures are:

(b) 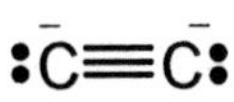 $:C{\equiv}C:$ **(c)** $:N{\equiv}\overset{+}{O}:$ **(d)** 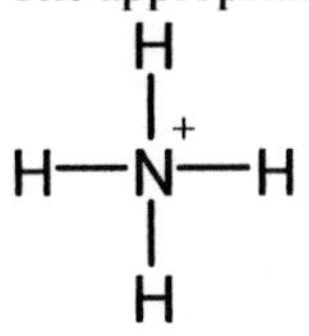 $H{-}\overset{+}{N}{-}H$ (with H above and H below)

9.47 **(a)** Too many electrons. The correct structure is

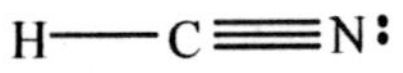

$$H{-}C{\equiv}N:$$

(b)   Hydrogen atoms do not form double bonds. The correct structure is

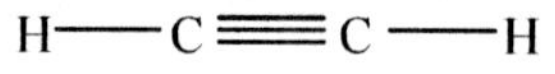

$$H\!\!-\!\!-\!\!-C\equiv C\!\!-\!\!-\!\!H$$

(c)   Too few electrons.

$$\ddot{:}\!\ddot{O}\!=\!\!=\!Sn\!=\!\!=\!\ddot{O}\ddot{:}$$

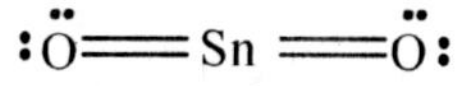

(d)   Too many electrons. The correct structure is

$$\begin{array}{c} :\ddot{F}: \quad :\ddot{F}: \\ \diagdown \quad \diagup \\ B \\ | \\ :\ddot{F}: \end{array}$$

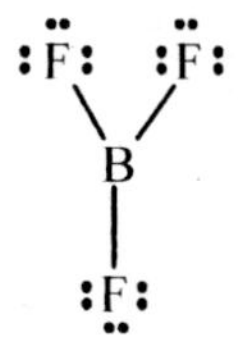

(e)   Fluorine has more than an octet. The correct structure is

$$H\!\!-\!\!-\!\ddot{O}\!\!-\!\!-\!\ddot{F}:$$

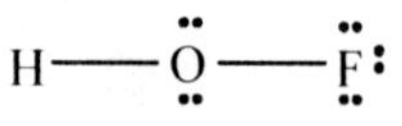

(f)   Oxygen does not have an octet. The correct structure is

$$\begin{array}{c} :\ddot{O}: \\ \| \\ H\!\!-\!\!-\!C\!\!-\!\!-\!\ddot{F}: \end{array}$$

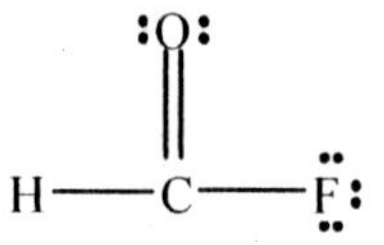

(g)   Too few electrons. The correct structure is

$$\begin{array}{c} :\ddot{F}\!\!-\!\!N\!\!-\!\!\ddot{F}: \\ | \\ :\ddot{F}: \end{array}$$

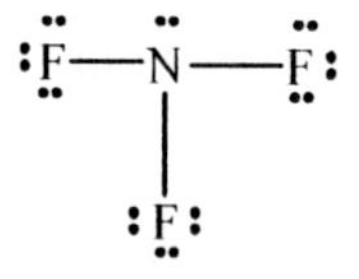

9.48   **(a)**   Neither oxygen atom has a complete octet. The left-most hydrogen atom is forming two bonds (4 e⁻). Hydrogen can only be surrounded by at most two electrons.

(b)   The correct structure is:

$$\begin{array}{c} H \quad :\ddot{O}: \\ | \quad\quad \| \\ H\!-\!C\!-\!C\!-\!\ddot{O}\!-\!H \\ | \\ H \end{array}$$

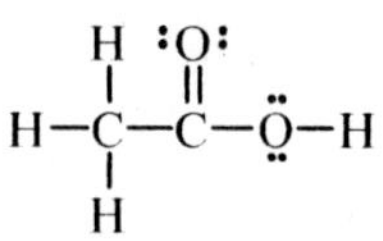

Do the two structures have the same number of electrons? Is the octet rule satisfied for all atoms other than hydrogen, which should have a duet of electrons?

9.51   The resonance structures are:

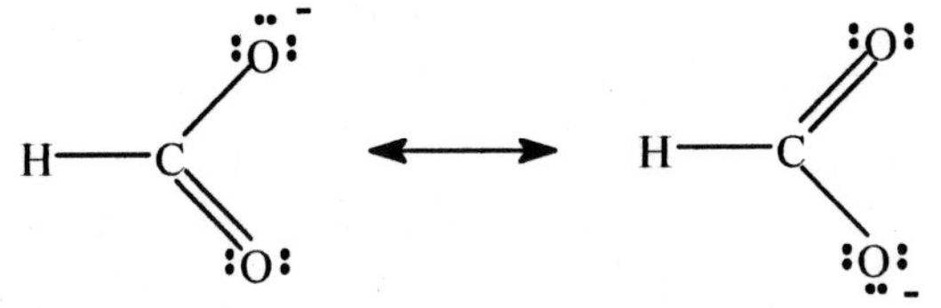

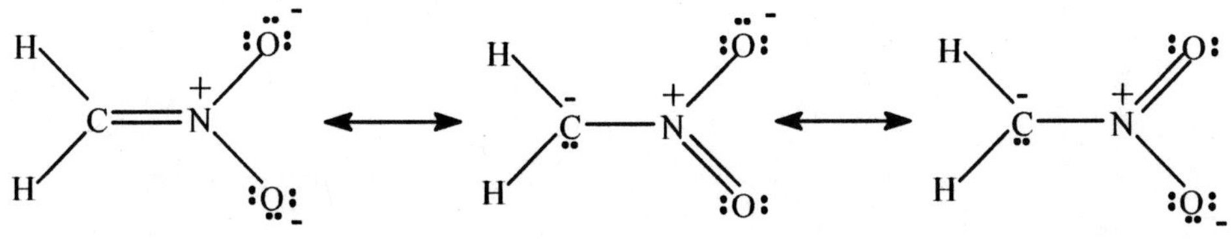

9.52   **Strategy:** We follow the procedure for drawing Lewis structures outlined in Section 9.6 of the text. After we complete the Lewis structure, we draw the resonance structures.

Solution: Following the procedure in Section 9.6 of the text, we come up with the following Lewis structure for ClO_3^-.

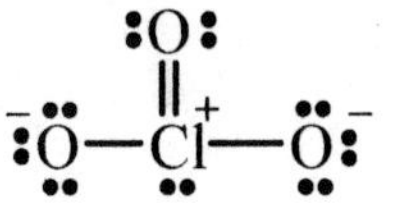

We can draw two more equivalent Lewis structures with the double bond between Cl and a different oxygen atom.

The resonance structures with formal charges are as follows:

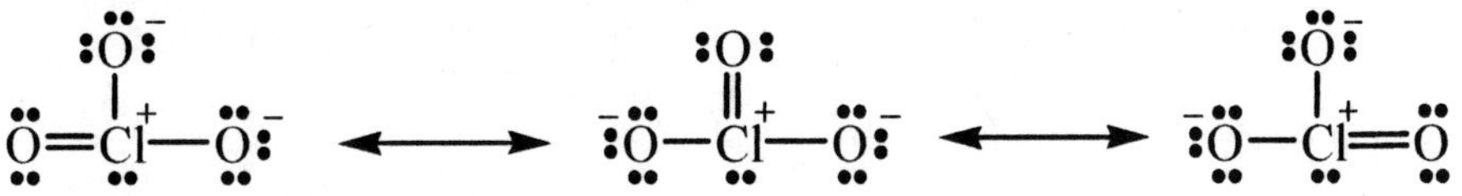

9.53   The structures of the most important resonance forms are:

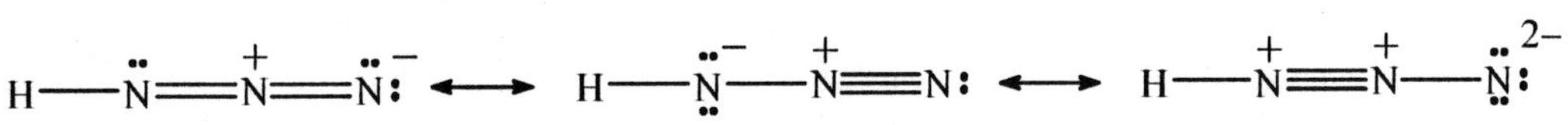

9.54   The structures of the most important resonance forms are:

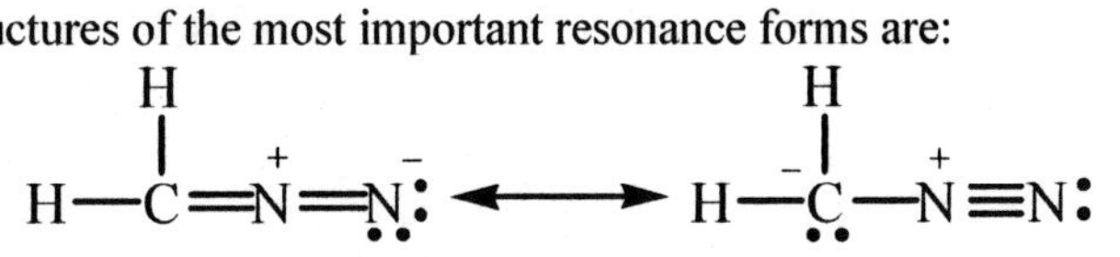

9.55   Three reasonable resonance structures for OCN^- are:

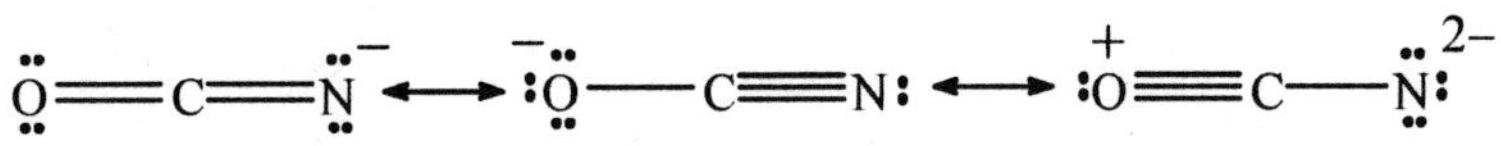

9.56   Three reasonable resonance structures with the formal charges indicated are

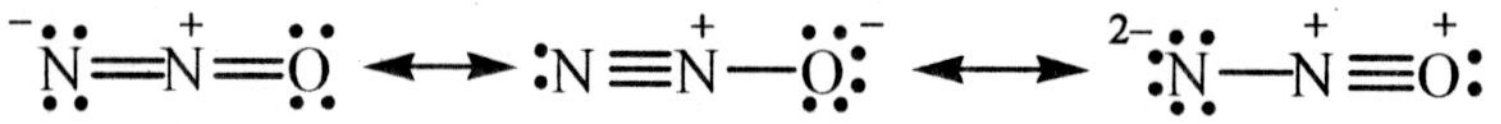

$$\overset{-}{:}\overset{..}{N}=\overset{+}{N}=\overset{..}{O} \longleftrightarrow :N\equiv\overset{+}{N}-\overset{..}{\underset{..}{O}}\overset{-}{:} \longleftrightarrow \overset{2-}{:}\overset{..}{\underset{..}{N}}-\overset{+}{N}\equiv\overset{+}{O}:$$

9.61   The resonance structures are

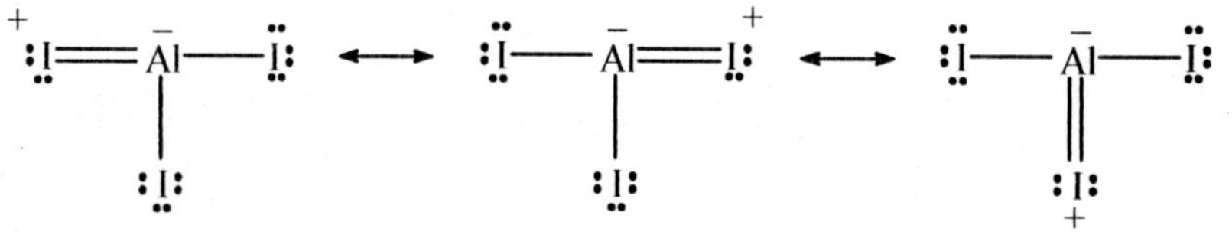

9.62   **Strategy:** We follow the procedure outlined in Section 9.6 of the text for drawing Lewis structures. We assign formal charges as discussed in Section 9.7 of the text.

Solution: Drawing the structure with single bonds between Be and each of the Cl atoms, the octet rule for Be is *not* satisfied. The Lewis structure is:

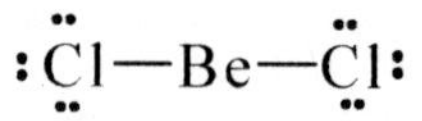

$$:\overset{..}{\underset{..}{Cl}}-Be-\overset{..}{\underset{..}{Cl}}:$$

An octet of electrons on Be can only be formed by making two double bonds as shown below:

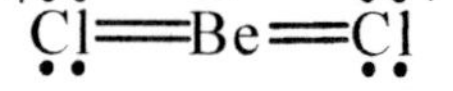

$$\overset{+}{\underset{..}{\overset{..}{Cl}}}=\overset{2-}{Be}=\overset{..}{\underset{..}{Cl}}\overset{+}{}$$

This places a high negative formal charge on Be and positive formal charges on the Cl atoms. This structure distributes the formal charges counter to the electronegativities of the elements. It is not a plausible Lewis structure.

9.63   For simplicity, the three, nonbonding pairs of electrons around the fluorine atoms are omitted.

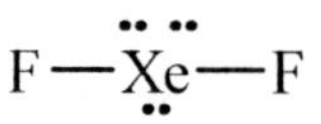

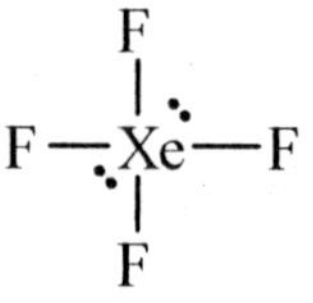

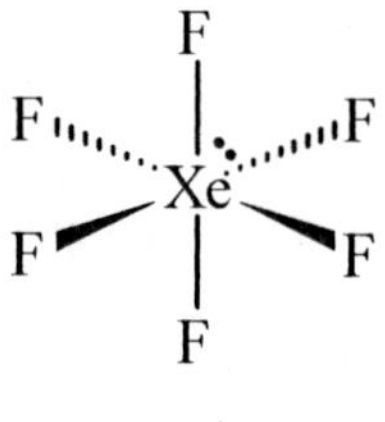

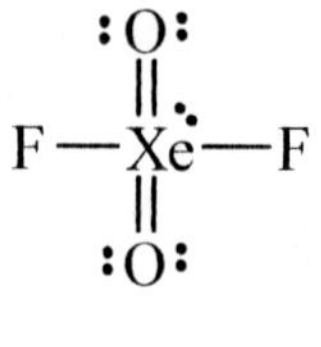

The octet rule is exceeded in each case.

9.64   The outer electron configuration of antimony is $5s^2 5p^3$. The Lewis structure is shown below. All five valence electrons are shared in the five covalent bonds. The octet rule is not obeyed. (The electrons on the chlorine atoms have been omitted for clarity.)

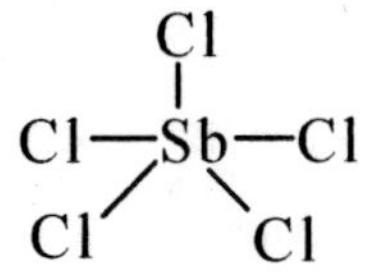

Can Sb have an expanded octet?

9.65   For simplicity, the three, nonbonding pairs of electrons around the fluorine are omitted.

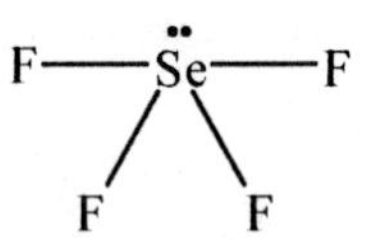

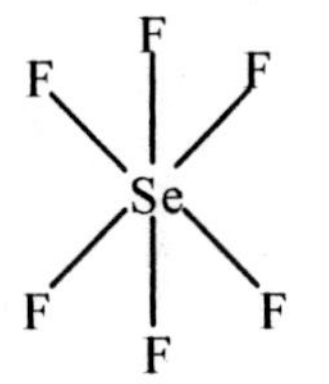

The octet rule is not satisfied for Se in both compounds (why not?).

9.66   The reaction can be represented as:

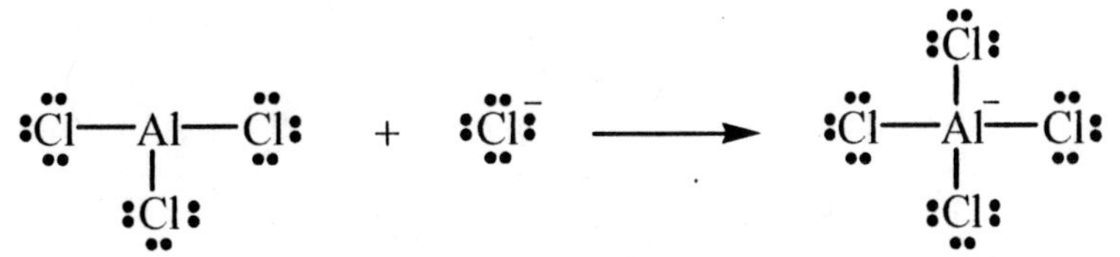

The new bond formed is called a **coordinate covalent bond**.

9.69   The enthalpy change for the equation showing ammonia dissociating into a nitrogen atom and three hydrogen atoms is equal to three times the average bond energy of the N–H bond (Why three?).

$$NH_3(g) \rightarrow N(g) + 3H(g) \qquad \Delta H° = 3\Delta H°(N-H)$$

The equation is the sum of the three equations given in the problem, and by Hess's law (Section 6.6 of the text) the enthalpy change is just the sum of the enthalpies of the individual steps.

$NH_3(g) \rightarrow NH_2(g) + H(g)$	$\Delta H° = 435 \text{ kJ/mol}$
$NH_2(g) \rightarrow NH(g) + H(g)$	$\Delta H° = 381 \text{ kJ/mol}$
$NH(g) \rightarrow N(g) + H(g)$	$\Delta H° = 360 \text{ kJ/mol}$
$NH_3(g) \rightarrow N(g) + 3H(g)$	$\Delta H° = 1176 \text{ kJ/mol}$

$$\Delta H°(N-H) = \frac{1176 \text{ kJ/mol}}{3} = \textbf{392 kJ/mol}$$

9.70   **Strategy:** Keep in mind that bond breaking is an energy absorbing (endothermic) process and bond making is an energy releasing (exothermic) process. Therefore, the overall energy change is the difference between these two opposing processes, as described in Equation (9.3) of the text.

Solution: There are two oxygen-to-oxygen bonds in ozone. We will represent these bonds as O–O. However, these bonds might not be true oxygen-to-oxygen single bonds. Using Equation (9.3) of the text, we write:

$$\Delta H° = \Sigma BE(\text{reactants}) - \Sigma BE(\text{products})$$

$$\Delta H° = BE(\text{O=O}) - 2BE(\text{O–O})$$

In the problem, we are given $\Delta H°$ for the reaction, and we can look up the O=O bond energy in Table 9.4 of the text. Solving for the average bond energy in ozone,

$$-2BE(\text{O–O}) = \Delta H° - BE(\text{O=O})$$

$$BE(\text{O–O}) = \frac{BE(\text{O=O}) - \Delta H°}{2} = \frac{498.7 \text{ kJ/mol} + 107.2 \text{ kJ/mol}}{2} = \textbf{303.0 kJ/mol}$$

Considering the resonance structures for ozone, is it expected that the O–O bond energy in ozone is between the single O–O bond energy (142 kJ) and the double O=O bond energy (498.7 kJ)?

9.71   When molecular fluorine dissociates, two fluorine atoms are produced. Since the enthalpy of formation of atomic fluorine is in units of kJ/mol, this number is half the bond dissociation energy of the fluorine molecule.

$$F_2(g) \rightarrow 2F(g) \qquad \Delta H° = 156.9 \text{ kJ/mol}$$

$$\Delta H° = 2\Delta H_f°(\text{F}) - \Delta H_f°(\text{F}_2)$$

$$156.9 \text{ kJ/mol} = 2\Delta H_f°(\text{F}) - (1)(0)$$

$$\Delta H_f°(\text{F}) = \frac{156.9 \text{ kJ/mol}}{2} = \textbf{78.5 kJ/mol}$$

9.72   **(a)**

Bonds Broken	Number Broken	Bond Energy (kJ/mol)	Energy Change (kJ)
C – H	12	414	4968
C – C	2	347	694
O = O	7	498.7	3491

Bonds Formed	Number Formed	Bond Energy (kJ/mol)	Energy Change (kJ)
C = O	8	799	6392
O – H	12	460	5520

$$\Delta H° = \text{total energy input} - \text{total energy released}$$

$$= (4968 + 694 + 3491) - (6392 + 5520) = \textbf{-2759 kJ/mol}$$

(b) $\Delta H° = 4\Delta H_f°(\text{CO}_2) + 6\Delta H_f°(\text{H}_2\text{O}) - [2\Delta H_f°(\text{C}_2\text{H}_6) + 7\Delta H_f°(\text{O}_2)]$

$\Delta H° = (4)(-393.5 \text{ kJ/mol}) + (6)(-241.8 \text{ kJ/mol}) - [(2)(-84.7 \text{ kJ/mol}) + (7)(0)] = \textbf{-2855 kJ/mol}$

The answers for part (a) and (b) are different, because *average* bond energies are used for part (a).

9.73 CH_4, CO, and $SiCl_4$ are covalent compounds. KF and $BaCl_2$ are ionic compounds.

9.74   Typically, ionic compounds are composed of a metal cation and a nonmetal anion. RbCl and KO_2 are ionic compounds.

Typically, covalent compounds are composed of two nonmetals. PF_5, BrF_3, and CI_4 are covalent compounds.

9.75   **(a)**   electron affinity of fluorine   **(b)**   bond energy of molecular fluorine

   (c)   ionization energy of sodium   **(d)**   standard enthalpy of formation of sodium fluoride

9.76   Recall that you can classify bonds as ionic or covalent based on electronegativity difference.

The melting points (°C) are shown in parentheses following the formulas.

Ionic: NaF (993) MgF_2 (1261) AlF_3 (1291)

Covalent: SiF_4 (−90.2) PF_5 (−83) SF_6 (−121) ClF_3 (−83)

Is there any correlation between ionic character and melting point?

9.77   By Hess's law, the overall enthalpy (energy) change in a reaction is equal to the sum of the enthalpy (energy) changes for the individual steps. The reactions shown in the problem are just the sums of the ionization energy of the alkali metal and the electron affinity of the halogen.

(a)   Taking data from the referenced figures we have:

$$Li(g) \rightarrow Li^+(g) + e^- \qquad \Delta H^\circ = 520 \text{ kJ/mol}$$
$$\underline{I(g) + e^- \rightarrow I^-(g) \qquad \Delta H^\circ = -295 \text{ kJ/mol}}$$
$$Li(g) + I(g) \rightarrow Li^+(g) + I^-(g) \qquad \Delta H^\circ = \textbf{225 kJ/mol}$$

Parts (b) and (c) are solved in an analogous manner.

(b) $Na(g) + F(g) \rightarrow Na^+(g) + F^-(g)$ $\Delta H^\circ = \textbf{163 kJ/mol}$

(c) $K(g) + Cl(g) \rightarrow K^+(g) + Cl^-(g)$ $\Delta H^\circ = \textbf{71 kJ/mol}$

9.78   KF is an ionic compound. It is a solid at room temperature made up of K^+ and F^- ions. It has a high melting point, and it is a strong electrolyte. Benzene, C_6H_6, is a covalent compound that exists as discrete molecules. It is a liquid at room temperature. It has a low melting point, is insoluble in water, and is a nonelectrolyte.

9.79   The three pairs of nonbonding electrons around each fluorine have been omitted for simplicity.

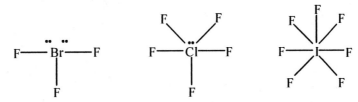

The octet rule is not obeyed in any of the compounds. In order for the octet rule to be obeyed, what would the value of n in the compound ICl_n have to be? [Hint: see Problem 9.45 (a)]

9.80   The resonance structures are:

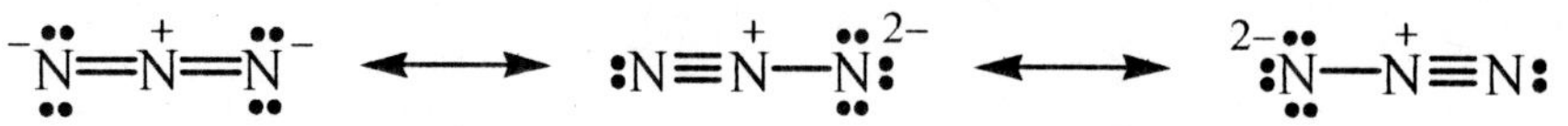

Which is the most plausible structure based on a formal charge argument?

9.81   A resonance structure is:

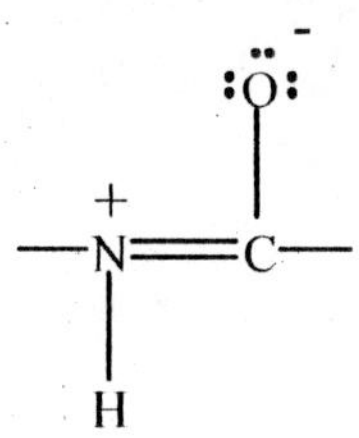

9.82   **(a)**   An example of an aluminum species that satisfies the octet rule is the anion $AlCl_4^-$. The Lewis dot structure is drawn in Problem 9.66.

    (b)   An example of an aluminum species containing an expanded octet is anion AlF_6^{3-}. (How many pairs of electrons surround the central atom?)

    (c)   An aluminum species that has an incomplete octet is the compound $AlCl_3$. The dot structure is given in Problem 9.66.

9.83   Four resonance structures together with the formal charges are

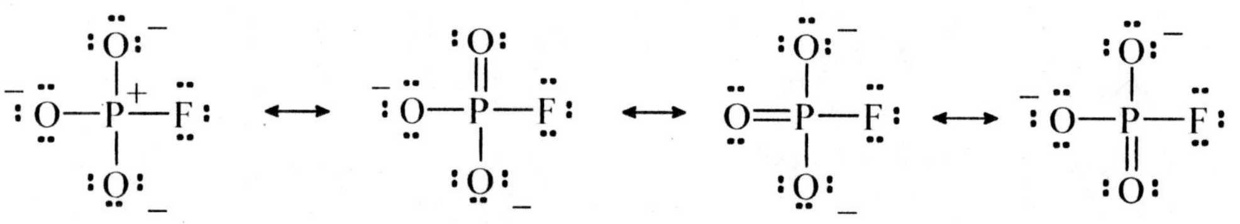

9.84 CF_2 would be very unstable because carbon does not have an octet. (How many electrons does it have?)

    LiO_2 would not be stable because the lattice energy between Li^+ and superoxide O_2^- would be too low to stabilize the solid.

    $CsCl_2$ requires a Cs^{2+} cation. The second ionization energy is too large to be compensated by the increase in lattice energy.

    PI_5 appears to be a reasonable species (compared to PF_5 in Example 9.10 of the text). However, the iodine atoms are too large to have five of them "fit" around a single P atom.

9.85   Reasonable resonance structures are:

(a)

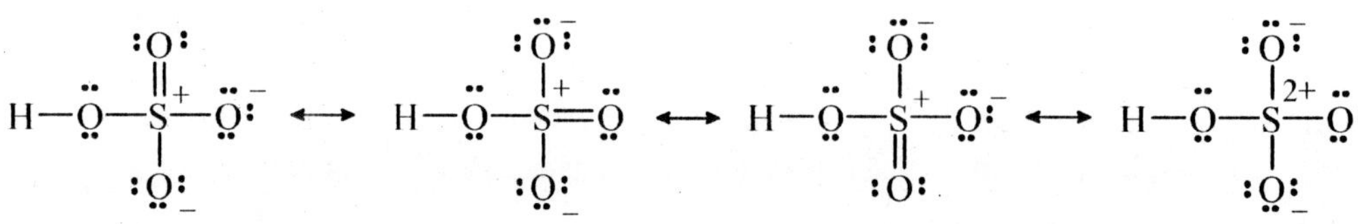

(b)

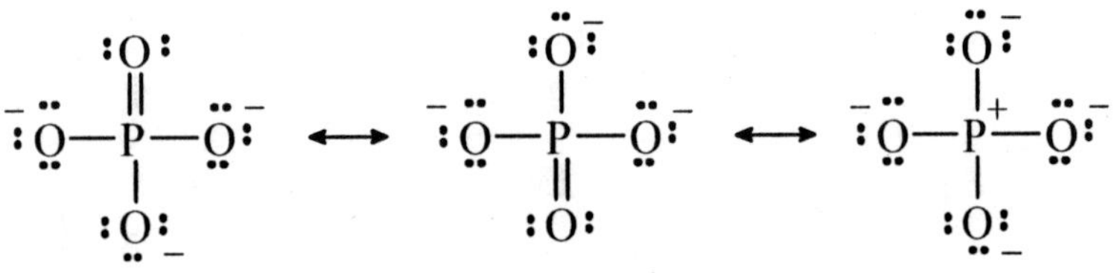

$$\text{:O:} \quad\quad\quad \text{:O:}^- \quad\quad\quad \text{:O:}^-$$

There are two more equivalent resonance structures to the first structure above.

(c)

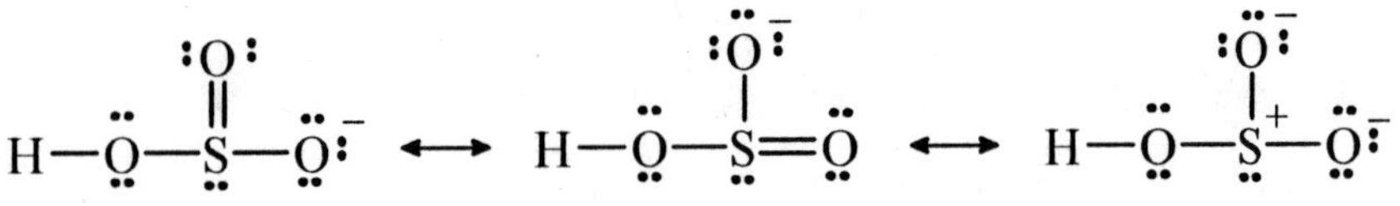

There are two more equivalent resonance structures to the first structure.

(d)

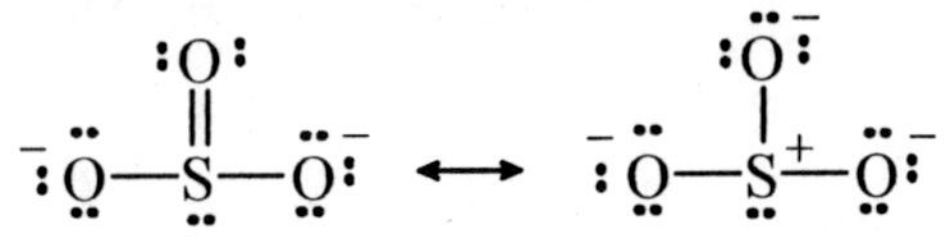

There are two more equivalent resonance structures to the first structure.

9.86   **(a)** false   **(b)** true   **(c)** false   **(d)** false

For question (c), what is an example of a second-period species that violates the octet rule?

9.87   If the central atom were more electronegative, there would be a concentration of negative charges at the central atom. This would lead to instability.

9.88   The formation of CH_4 from its elements is:

$$C(s) + 2H_2(g) \longrightarrow CH_4(g)$$

The reaction could take place in two steps:

Step 1: $C(s) + 2H_2(g) \longrightarrow C(g) + 4H(g)$ $\Delta H^\circ_{rxn} = (716 + 872.8)\text{kJ/mol} = 1589$ kJ/mol

Step 2: $C(g) + 4H(g) \longrightarrow CH_4(g)$ $\Delta H^\circ_{rxn} \approx -4 \times$ (bond energy of C–H bond)

$$= -4 \times 414 \text{ kJ/mol} = -1656 \text{ kJ/mol}$$

Therefore, $\Delta H^\circ_f(CH_4)$ would be approximately the sum of the enthalpy changes for the two steps. See Section 6.6 of the text (Hess's law).

$$\Delta H^\circ_f(CH_4) = \Delta H^\circ_{rxn}(1) + \Delta H^\circ_{rxn}(2)$$

$$\Delta H^\circ_f(CH_4) = (1589 - 1656)\text{kJ/mol} = -67 \text{ kJ/mol}$$

The actual value of $\Delta H^\circ_f(CH_4) = -74.85$ kJ/mol.

9.89   **(a)**   Bond broken: C–H $\Delta H^\circ = 414$ kJ/mol
                        Bond made: C–Cl $\Delta H^\circ = -338$ kJ/mol

$$\Delta H^\circ_{rxn} = 414 - 338 = 76 \text{ kJ/mol}$$

(b) Bond broken: C–H $\Delta H° = 414$ kJ/mol
Bond made: H–Cl $\Delta H° = -431.9$ kJ/mol

$$\Delta H^{\circ}_{\text{rxn}} = 414 - 431.9 = -18 \text{ kJ/mol}$$

Based on energy considerations, reaction **(b)** will occur readily since it is exothermic. Reaction (a) is endothermic.

9.90 Only $\mathbf{N_2}$ has a triple bond. Therefore, it has the shortest bond length.

9.91 The rest of the molecule (in this problem, unidentified) would be attached at the end of the free bond.

(a)

(b)

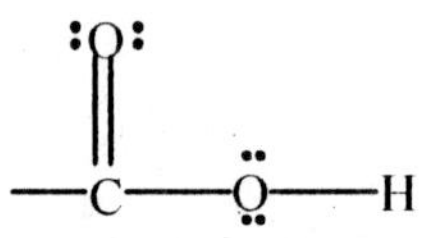

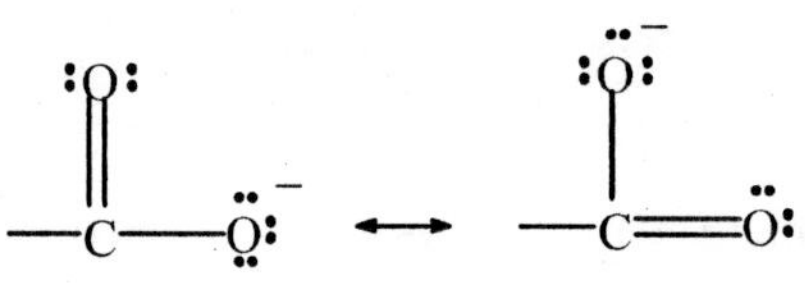

9.92 To be isoelectronic, molecules must have the same number and arrangement of valence electrons. NH_4^+ and CH_4 are isoelectronic (8 valence electrons), as are CO and N_2 (10 valence electrons), as are $B_3N_3H_6$ and C_6H_6 (30 valence electrons). Draw Lewis structures to convince yourself that the electron arrangements are the same in each isoelectronic pair.

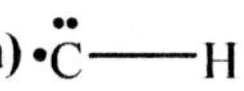

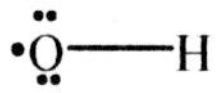

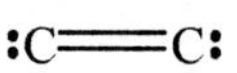

9.93 The Lewis structures are:

(a) paramagnetic **(b)** paramagnetic **(c)** diamagnetic

(d) diamagnetic **(e)** paramagnetic

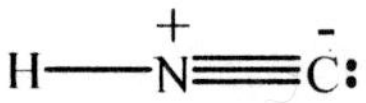

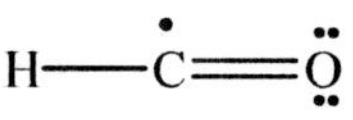

9.94 The reaction can be represented as:

9.95 The Lewis structures are:

(a)

(b)

(c)

(d)

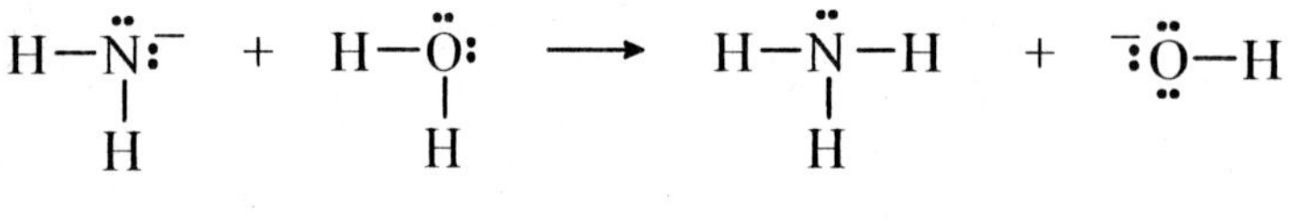

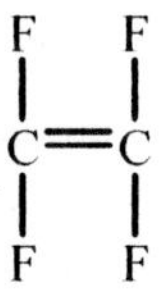

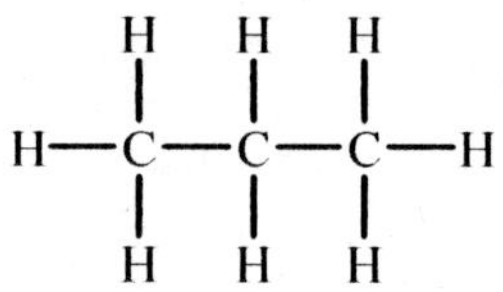

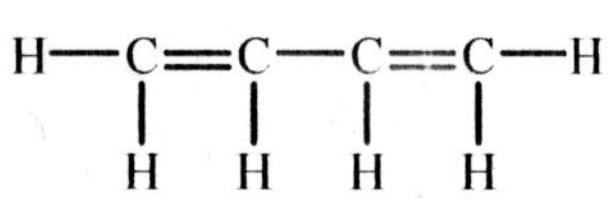

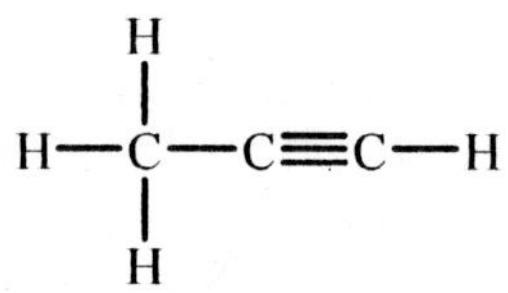

(e)

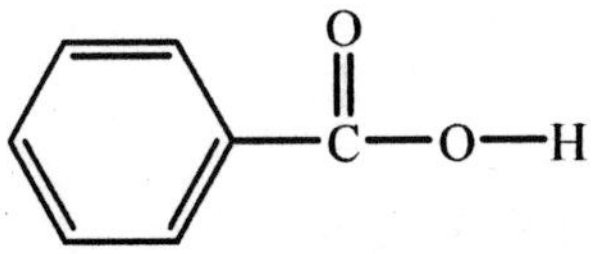

9.96   The central iodine atom in I_3^- has *ten* electrons surrounding it: two bonding pairs and three lone pairs. The central iodine has an expanded octet. Elements in the second period such as fluorine cannot have an expanded octet as would be required for F_3^-.

9.97   See Problem 9.71. The bond energy for F_2 is:

$$F_2(g) \rightarrow F(g) + F(g) \qquad \Delta H^\circ = 156.9 \text{ kJ/mol}$$

The energy for the process $F_2(g) \rightarrow F^+(g) + F^-(g)$ can be found by Hess's law. Thus,

$F_2(g) \rightarrow F(g) + F(g)$	$\Delta H^\circ = 156.9$ kJ/mol
$F(g) \rightarrow F^+(g) + e^-$	$\Delta H^\circ = 1680$ kJ/mol [I_1(F), see Table 8.2 of the text]
$F(g) + e^- \rightarrow F^-(g)$	$\Delta H^\circ = -328$ kJ/mol [EA_1(F), see Table 8.3 of the text]
$F_2(g) \rightarrow F^+(g) + F^-(g)$	$\Delta H^\circ = 1509$ kJ/mol

It is much easier to dissociate F_2 into two neutral F atoms than it is to dissociate it into a fluorine cation and anion.

9.98   The skeletal structure is:

```
          H
          |
  H   C   N   C   O
          |
          H
```

The number of valence electron is: $(1 \times 3) + (2 \times 4) + 5 + 6 = 22$ valence electrons

We can draw two resonance structures for methyl isocyanate.

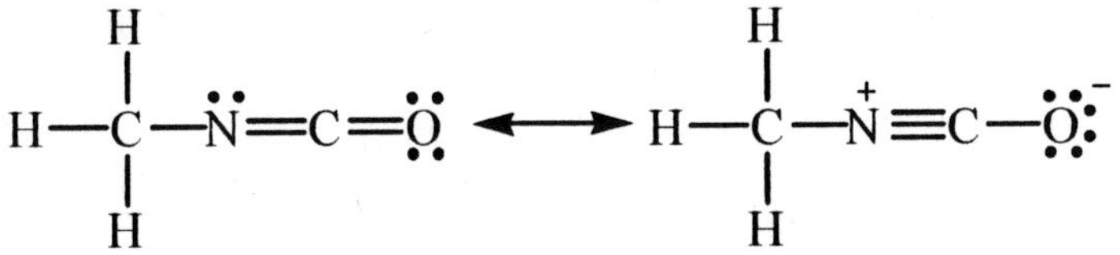

9.99   A reasonable Lewis structure is:

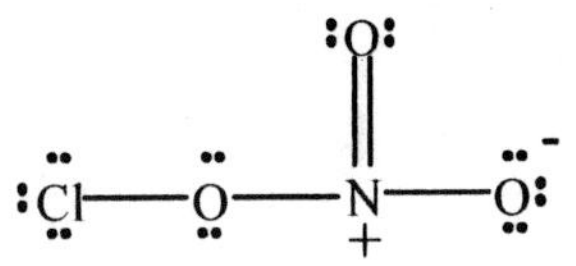

9.100   **(a)**   This is a very good resonance form; there are no formal charges and each atom satisfies the octet rule.

(b)   This is a second choice after (a) because of the positive formal charge on the oxygen (high electronegativity).

(c)   This is a poor choice for several reasons. The formal charges are placed counter to the electronegativities of C and O, the oxygen atom does not have an octet, and there is no bond between that oxygen and carbon!

(d)   This is a mediocre choice because of the large formal charge and lack of an octet on carbon.

9.101   For C_4H_{10} and C_5H_{12} there are a number of structural isomers.

C_2H_6 C_4H_{10}

C_5H_{12}

9.102   The nonbonding electron pairs around Cl and F are omitted for simplicity.

9.103   The structures are (the nonbonding electron pairs on fluorine have been omitted for simplicity):

9.104   (a)   Using Equation (9.3) of the text,

$\Delta H = \Sigma BE(\text{reactants}) - \Sigma BE(\text{products})$

$\Delta H = [(436.4 + 151.0) - 2(298.3)] = -9.2 \text{ kJ/mol}$

(b) Using Equation (6.17) of the text,

$$\Delta H° = 2\Delta H_f°[HI(g)] - \{\Delta H_f°[H_2(g)] + \Delta H_f°[I_2(g)]\}$$

$$\Delta H° = (2)(25.9 \text{ kJ/mol}) - [(0) + (1)(61.0 \text{ kJ/mol})] = -9.2 \text{ kJ/mol}$$

9.105 Note that the nonbonding electron pairs have been deleted from oxygen, nitrogen, sulfur, and chlorine for simplicity.

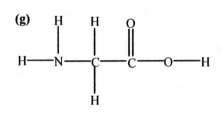

Note: in part (c) above, ethyl = C_2H_5 =

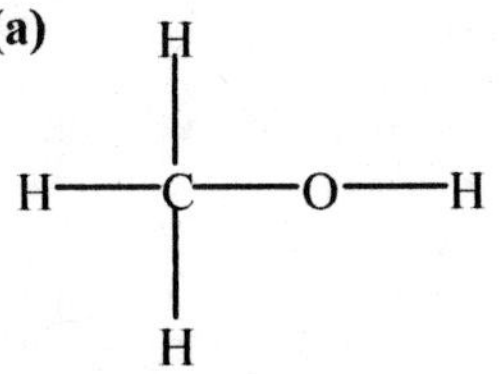

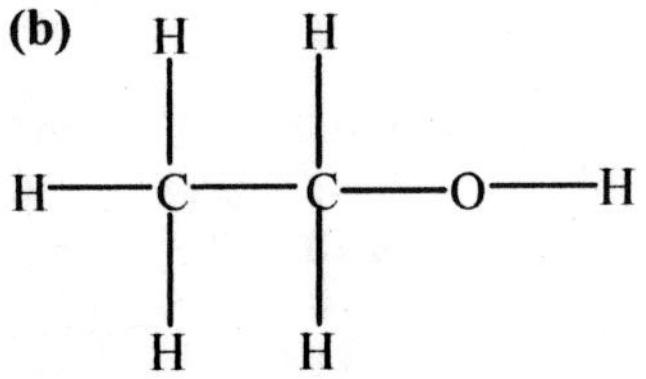

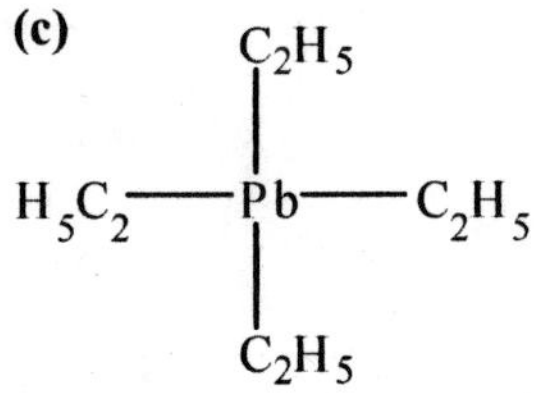

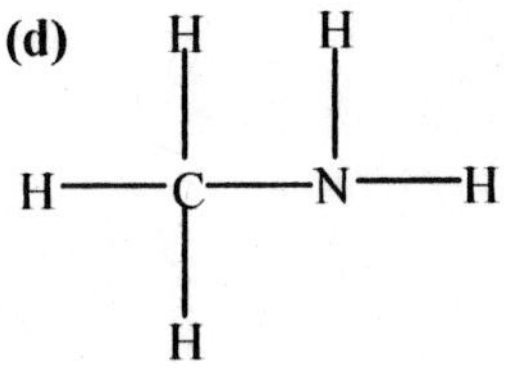

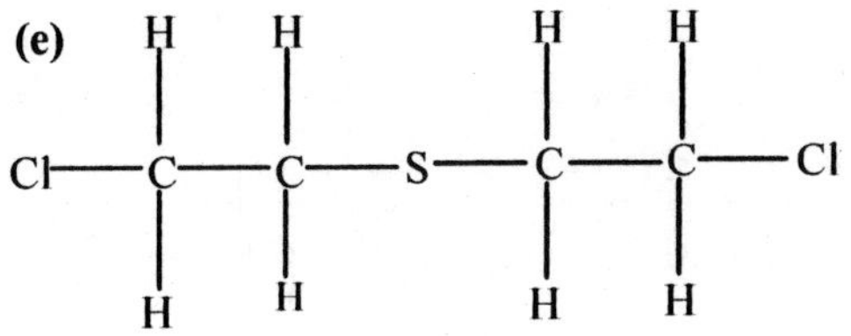

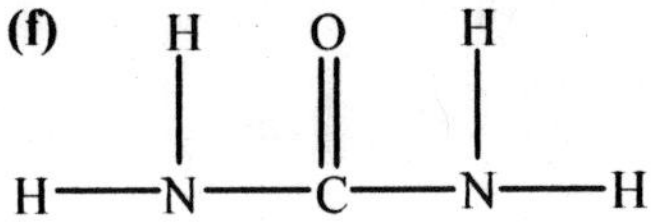

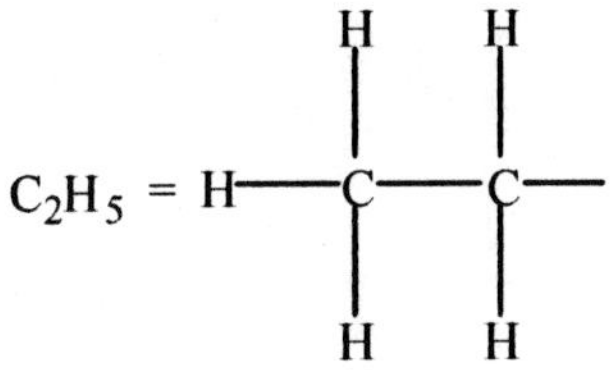

9.106 The Lewis structures are:

(a) :C≡O: ⁻ ⁺   (b) :N≡O: ⁺   (c) :C≡N: ⁻   (d) :N≡N:

9.107

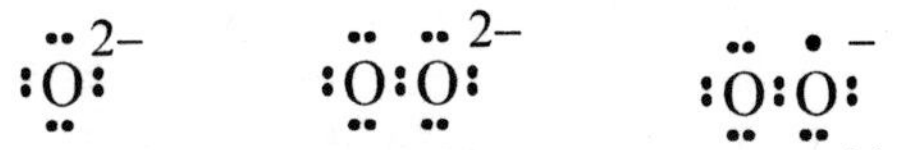

:O: ²⁻   :O:O: ²⁻   :O:O: ·⁻

oxide   peroxide   superoxide

9.108 True. Each noble gas atom already has completely filled ns and np subshells.

9.109   The resonance structures are:

(a) 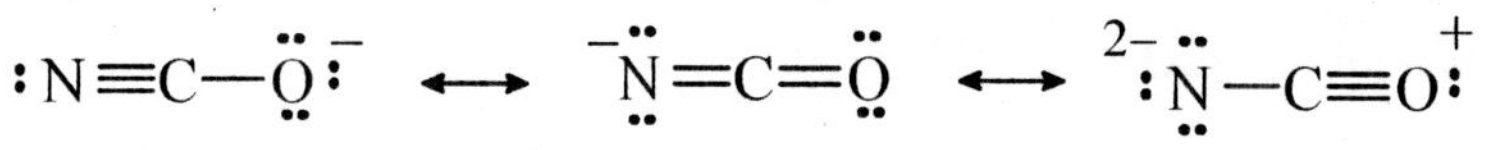 $:N\equiv C—\overset{..}{\underset{..}{O}}:^{-}$ ⟷ $^{-}\overset{..}{N}=C=\overset{..}{\underset{..}{O}}$ ⟷ $^{2-}:\overset{..}{\underset{..}{N}}—C\equiv O:^{+}$

(b) 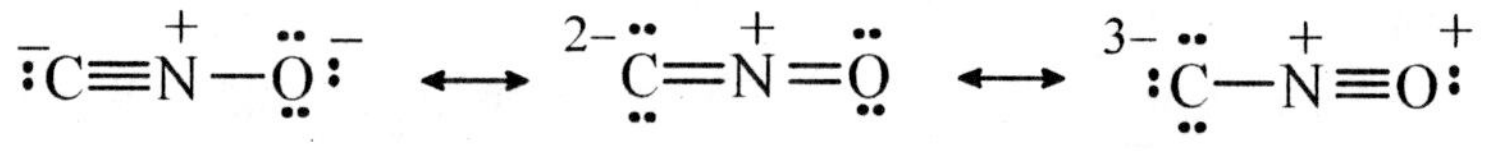 $^{-}:C\equiv \overset{+}{N}—\overset{..}{\underset{..}{O}}:^{-}$ ⟷ $^{2-}\overset{..}{C}=\overset{+}{N}=\overset{..}{\underset{..}{O}}$ ⟷ $^{3-}:\overset{..}{\underset{..}{C}}—\overset{+}{N}\equiv O:^{+}$

In both cases, the most likely structure is on the left and the least likely structure is on the right.

9.110   (a)   The bond energy of F_2^- is the energy required to break up F_2^- into an F atom and an F^- ion.

$$F_2^-(g) \longrightarrow F(g) + F^-(g)$$

We can arrange the equations given in the problem so that they add up to the above equation. See Section 6.6 of the text (Hess's law).

$F_2^-(g) \longrightarrow F_2(g) + e^-$	$\Delta H° = 290$ kJ/mol
$F_2(g) \longrightarrow 2F(g)$	$\Delta H° = 156.9$ kJ/mol
$F(g) + e^- \longrightarrow F^-(g)$	$\Delta H° = -333$ kJ/mol
$F_2^-(g) \longrightarrow F(g) + F^-(g)$	

The bond energy of F_2^- is the sum of the enthalpies of reaction.

$$BE(F_2^-) = [290 + 156.9 + (-333 \text{ kJ})]\text{kJ/mol} = \textbf{114 kJ/mol}$$

(b)   The bond in F_2^- is weaker (114 kJ/mol) than the bond in F_2 (156.9 kJ/mol), because the extra electron increases repulsion between the F atoms.

9.111   The description involving a griffin and a unicorn is more appropriate. Both mule and donkey are real animals whereas resonance structures are nonexistent.

9.112   In (a) there is a lone pair on the C atom and the negative formal charge is on the less electronegative C atom.

9.113   Except for nitric oxide, lone pairs of electrons have been left off for simplicity. H–H (hydrogen),

O=O (oxygen), N≡N (nitrogen), F–F (fluorine), Cl–Cl (chlorine), Br–Br (bromine), I–I (iodine),

C≡O (carbon monoxide), H–F (hydrogen fluoride), H–Cl (hydrogen chloride), H–Br (hydrogen bromide),

H–I (hydrogen iodide), and $\overset{\bullet}{N}=\overset{..}{\underset{..}{O}}$ (nitric oxide).

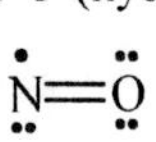

9.114   (a)  $:\overset{\bullet}{N}=\overset{..}{\underset{..}{O}}$ ⟷ $^{-}:\overset{..}{N}=\overset{\bullet}{\underset{..}{O}}{}^{+}$

The first structure is the most important. Both N and O have formal charges of zero. In the second structure, the more electronegative oxygen atom has a formal charge of +1. Having a positive formal charge on an highly electronegative atom is not favorable. In addition, both structures leave one atom with an incomplete octet. This cannot be avoided due to the odd number of electrons.

(b)   It is not possible to draw a structure with a triple bond between N and O.

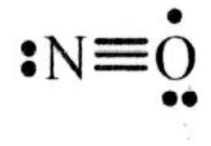

$:N\equiv \overset{\bullet}{\underset{..}{O}}$

Any structure drawn with a triple bond will lead to an expanded octet. Elements in the second row of the period table cannot exceed the octet rule.

9.115   The unpaired electron on each N atom will be shared to form a covalent bond:

$$O_2N^\bullet + {}^\bullet NO_2 \longrightarrow N_2O_4$$

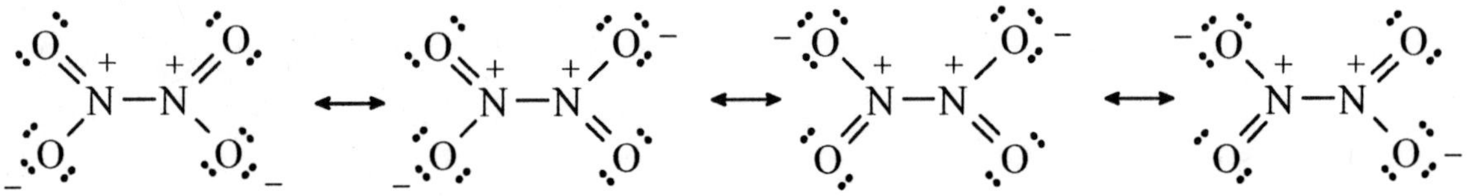

9.116   The OCOO structure violates the octet rule (expanded octet). The structure shown below satisfies the octet rule with 22 valence electrons. However, CO_3^{2-} has 24 valence electrons. Adding two more electrons to the structure would cause at least one atom to exceed the octet rule.

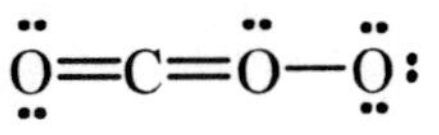

9.117   One resonance form showing formal charges is:

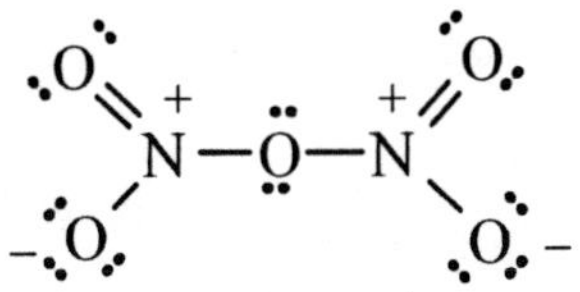

9.118

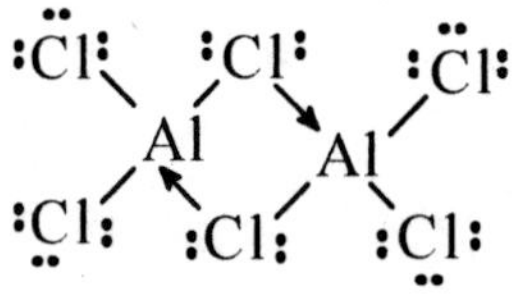

The arrows indicate coordinate covalent bonds.

9.119   **(a)**

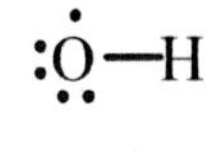

(b)   The O–H bond is quite strong (460 kJ/mol). To complete its octet, the OH radical has a strong tendency to form a bond with a H atom.

(c)   One C–H bond is being broken, and an O–H bond is being formed.

$$\Delta H = \Sigma BE(\text{reactants}) - \Sigma BE(\text{products})$$

$$\boldsymbol{\Delta H = 414 - 460 = -46 \text{ kJ/mol}}$$

(d)   Energy of one O–H bond $= \dfrac{460 \times 10^3 \text{ J}}{1 \text{ mol}} \times \dfrac{1 \text{ mol}}{6.022 \times 10^{23} \text{ bonds}} = 7.64 \times 10^{-19} \text{ J/bond}$

$$\lambda = \frac{hc}{\Delta H} = \frac{(6.63 \times 10^{-34}\,\text{J}\cdot\text{s})(3.00 \times 10^8\,\text{m/s})}{7.64 \times 10^{-19}\,\text{J}}$$

$$\lambda = 2.60 \times 10^{-7}\,\text{m} = \textbf{260 nm}$$

9.120 There are four C–H bonds in CH_4, so the average bond energy of a C–H bond is:

$$\frac{1656\,\text{kJ/mol}}{4} = 414\,\text{kJ/mol}$$

The Lewis structure of propane is:

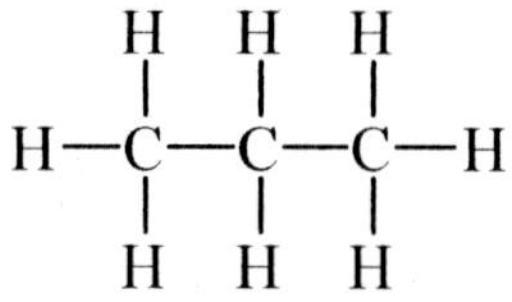

There are eight C–H bonds and two C–C bonds. We write:

$$8(\text{C–H}) + 2(\text{C–C}) = 4006\,\text{kJ/mol}$$

$$8(414\,\text{kJ/mol}) + 2(\text{C–C}) = 4006\,\text{kJ/mol}$$

$$2(\text{C–C}) = 694\,\text{kJ/mol}$$

So, the average bond energy of a C–C bond is: $\dfrac{694}{2}\,\text{kJ/mol} = \textbf{347 kJ/mol}$

9.121 Three resonance structures with formal charges are:

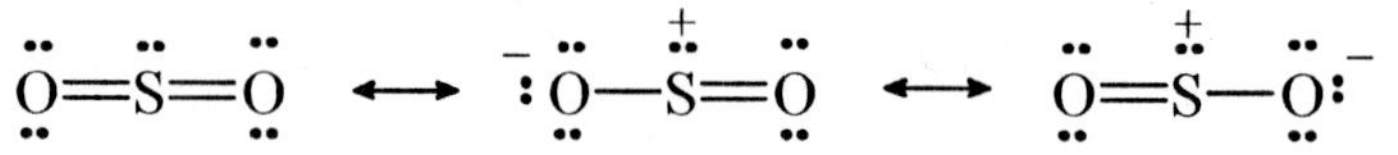

According to the comments in Example 9.11 of the text, the *second* and *third* structures above are more important.

9.122

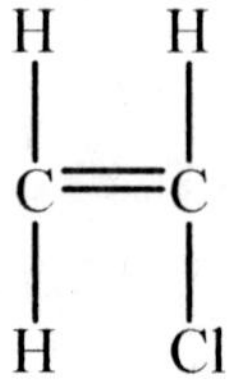

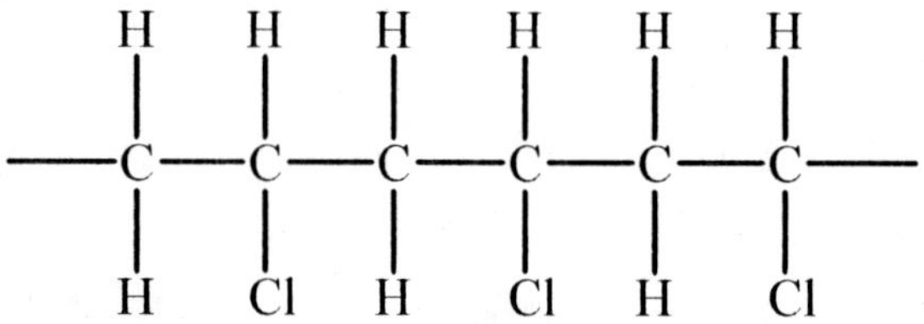

(c) In the formation of poly(vinyl chloride) form vinyl chloride, for every C=C double bond broken, 2 C–C single bonds are formed. No other bonds are broken or formed. The energy changes for 1 mole of vinyl chloride reacted are:

total energy input (breaking C=C bonds) = 620 kJ

total energy released (forming C–C bonds) = 2×347 kJ = 694 kJ

$$\Delta H° = 620\,\text{kJ} - 694\,\text{kJ} = -74\,\text{kJ}$$

The negative sign shows that this is an exothermic reaction. To find the total heat released when 1.0×10^3 kg of vinyl chloride react, we proceed as follows:

$$\text{heat released} = (1.0 \times 10^6 \text{ g C}_2\text{H}_3\text{Cl}) \times \frac{1 \text{ mol C}_2\text{H}_3\text{Cl}}{62.49 \text{ g C}_2\text{H}_3\text{Cl}} \times \frac{-74 \text{ kJ}}{1 \text{ mol C}_2\text{H}_3\text{Cl}} = -1.2 \times 10^6 \text{ kJ}$$

9.123   Work done = force × distance

$$= (2.0 \times 10^{-9} \text{ N}) \times (2 \times 10^{-10} \text{ m})$$

$$= 4 \times 10^{-19} \text{ N·m}$$

$$= 4 \times 10^{-19} \text{ J to break one bond}$$

Expressing the bond energy in kJ/mol:

$$\frac{4 \times 10^{-19} \text{ J}}{1 \text{ bond}} \times \frac{1 \text{ kJ}}{1000 \text{ J}} \times \frac{6.022 \times 10^{23} \text{ bonds}}{1 \text{ mol}} = \mathbf{2 \times 10^2 \text{ kJ/mol}}$$

9.124 $\text{EN(O)} = \dfrac{1314 + 141}{2} = 727.5 \qquad \text{EN(F)} = \dfrac{1680 + 328}{2} = 1004 \qquad \text{EN(Cl)} = \dfrac{1251 + 349}{2} = 800$

Using Mulliken's definition, the electronegativity of chlorine is greater than that of oxygen, and fluorine is still the most electronegative element. We can convert to the Pauling scale by dividing each of the above by 230 kJ/mol.

$$\text{EN(O)} = \frac{727.5}{230} = \mathbf{3.16} \qquad\qquad \text{EN(F)} = \frac{1004}{230} = \mathbf{4.37} \qquad\qquad \text{EN(Cl)} = \frac{800}{230} = \mathbf{3.48}$$

These values compare to the Pauling values for oxygen of 3.5, fluorine of 4.0, and chlorine of 3.0.

9.125

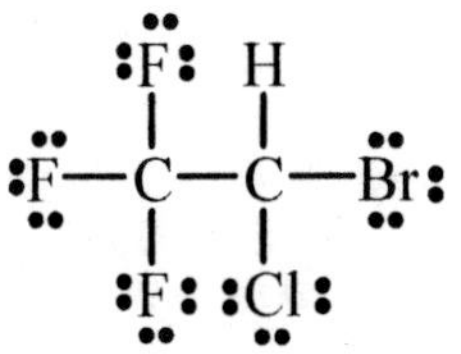

halothane

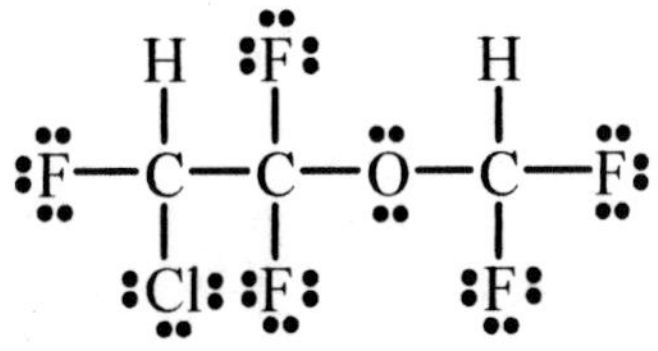

enflurane

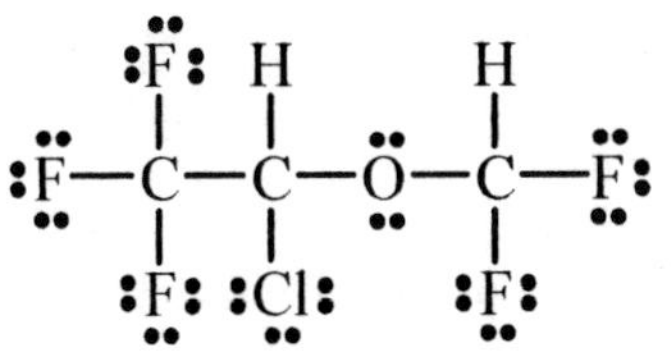

isoflurane

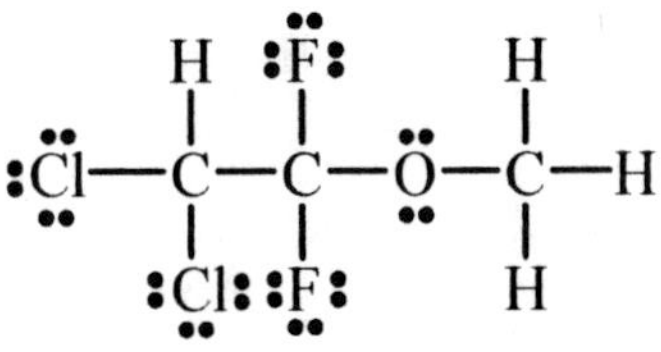

methoxyflurane

9.126   (1)   You could determine the magnetic properties of the solid. An Mg^+O^- solid would be paramagnetic while $Mg^{2+}O^{2-}$ solid is diamagnetic.

(2)   You could determine the lattice energy of the solid. Mg^+O^- would have a lattice energy similar to Na^+Cl^-. This lattice energy is much lower than the lattice energy of $Mg^{2+}O^{2-}$.

CHAPTER 10
CHEMICAL BONDING II: MOLECULAR GEOMETRY AND HYBRIDIZATION OF ATOMIC ORBITALS

10.7 **(a)** The Lewis structure of PCl_3 is shown below. Since in the VSEPR method the number of bonding pairs and lone pairs of electrons around the *central atom* (phosphorus, in this case) is important in determining the structure, the lone pairs of electrons around the chlorine atoms have been omitted for simplicity. There are three bonds and one lone electron pair around the central atom, phosphorus, which makes this an AB_3E case. The information in Table 10.2 shows that the structure is a trigonal pyramid like ammonia.

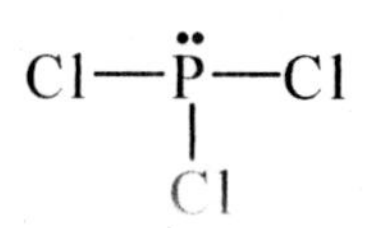

$$Cl—\overset{\displaystyle ..}{P}—Cl$$
$$|$$
$$Cl$$

What would be the structure of the molecule if there were no lone pairs and only three bonds?

(b) The Lewis structure of $CHCl_3$ is shown below. There are four bonds and no lone pairs around carbon which makes this an AB_4 case. The molecule should be tetrahedral like methane (Table 10.1).

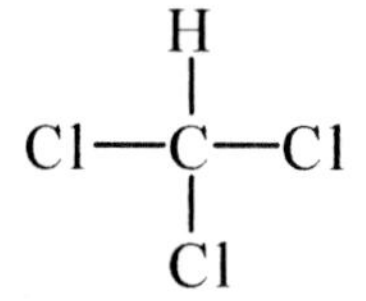

$$H$$
$$|$$
$$Cl—C—Cl$$
$$|$$
$$Cl$$

(c) The Lewis structure of SiH_4 is shown below. Like part (b), it is a tetrahedral AB_4 molecule.

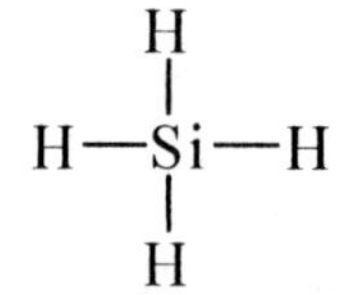

$$H$$
$$|$$
$$H—Si—H$$
$$|$$
$$H$$

(d) The Lewis structure of $TeCl_4$ is shown below. There are four bonds and one lone pair which make this an AB_4E case. Consulting Table 10.2 shows that the structure should be that of a distorted tetrahedron like SF_4.

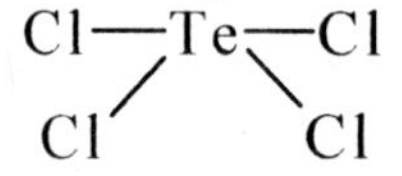

$$Cl—\overset{\displaystyle ..}{Te}—Cl$$
$$Cl \diagup \qquad \diagdown Cl$$

Are $TeCl_4$ and SF_4 isoelectronic? Should isoelectronic molecules have similar VSEPR structures?

10.8   **Strategy:** The sequence of steps in determining molecular geometry is as follows:

draw Lewis ⟶ find arrangement of ⟶ find arrangement ⟶ determine geometry
structure   electrons pairs   of bonding pairs   based on bonding pairs

Solution:

Lewis structure	Electron pairs on central atom	Electron arrangement	Lone pairs	Geometry
(a) Cl—Al—Cl with Cl above Al	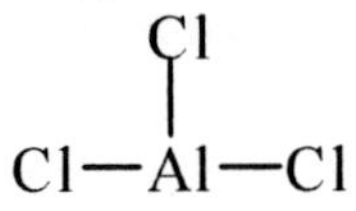 3	trigonal planar	0	trigonal planar, AB$_3$
(b) Cl—Zn—Cl	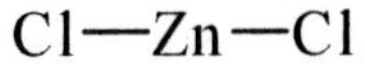 2	linear	0	linear, AB$_2$
(c) [Cl—Zn—Cl with Cl above and Cl below]$^{2-}$	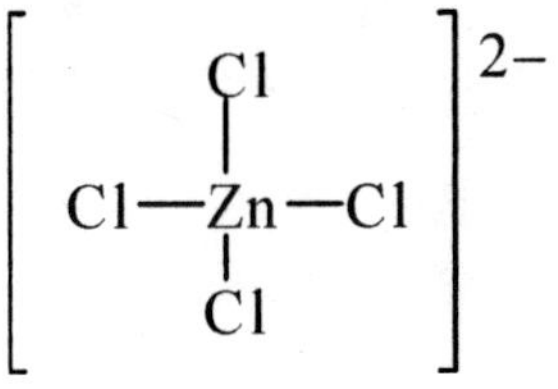 4	tetrahedral	0	tetrahedral, AB$_4$

10.9

	Lewis Structure	e$^-$ pair arrangement	geometry
(a)	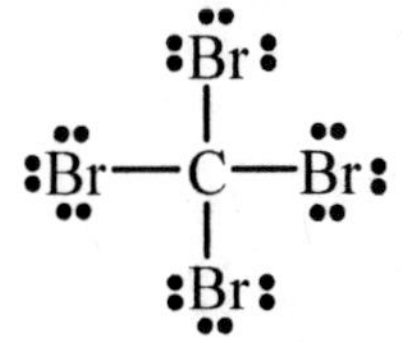 :Br: above, :Br—C—Br: , :Br: below	tetrahedral	tetrahedral
(b)	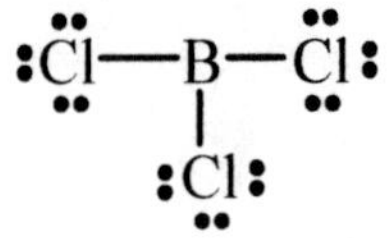 :Cl—B—Cl: , :Cl: below	trigonal planar	trigonal planar
(c)	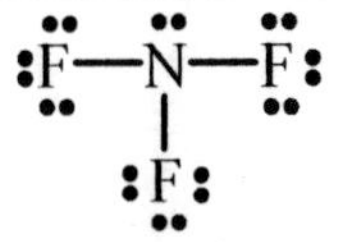 :F—N—F: , :F: below	tetrahedral	trigonal pyramidal
(d)	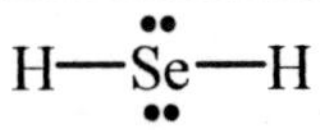 H—Se—H	tetrahedral	bent
(e)	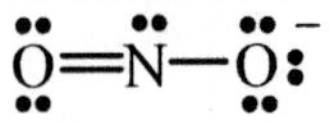 O=N—O:$^-$	trigonal planar	bent

10.10   We use the following sequence of steps to determine the geometry of the molecules.

draw Lewis ⟶ find arrangement of ⟶ find arrangement ⟶ determine geometry
structure   electrons pairs   of bonding pairs   based on bonding pairs

(a) Looking at the Lewis structure we find 4 pairs of electrons around the central atom. The electron pair arrangement is tetrahedral. Since there are no lone pairs on the central atom, the geometry is also **tetrahedral**.

$$
\begin{array}{c}
H \\
| \\
H-C-\ddot{\underset{\displaystyle\cdot\cdot}{I}}\colon \\
| \\
H
\end{array}
$$

(b) Looking at the Lewis structure we find 5 pairs of electrons around the central atom. The electron pair arrangement is trigonal bipyramidal. There are two lone pairs on the central atom, which occupy positions in the trigonal plane. The geometry is **t-shaped**.

$$
\begin{array}{c}
\colon\!\ddot{F}-\overset{\cdot\cdot}{Cl}-\ddot{F}\colon \\
| \\
\colon\!\ddot{F}\colon
\end{array}
$$

(c) Looking at the Lewis structure we find 4 pairs of electrons around the central atom. The electron pair arrangement is tetrahedral. There are two lone pairs on the central atom. The geometry is **bent**.

$$
H-\overset{\cdot\cdot}{\underset{\cdot\cdot}{S}}-H
$$

(d) Looking at the Lewis structure, there are 3 VSEPR pairs of electrons around the central atom. Recall that a double bond counts as one VSEPR pair. The electron pair arrangement is trigonal planar. Since there are no lone pairs on the central atom, the geometry is also **trigonal planar**.

$$
\begin{array}{c}
\colon\!O\colon \\
\parallel \\
\colon\!\ddot{O}-S-\ddot{O}\colon
\end{array}
$$

(e) Looking at the Lewis structure, there are 4 pairs of electrons around the central atom. The electron pair arrangement is tetrahedral. Since there are no lone pairs on the central atom, the geometry is also **tetrahedral**.

$$
\left[\begin{array}{c}
\colon\!\ddot{O}\colon \\
| \\
\colon\!\ddot{O}-S-\ddot{O}\colon \\
| \\
\colon\!\ddot{O}\colon
\end{array}\right]^{2-}
$$

10.11 The lone pairs of electrons on the bromine atoms have been omitted for simplicity.

$$
\begin{array}{ccc}
Br-Hg-Br & \colon\!N\!\equiv\!\overset{+}{N}\!-\!\ddot{O}\colon^{-} & \colon\!\ddot{S}\!-\!C\!\equiv\!N\colon^{-} \\
\text{linear} & \text{linear} & \text{linear}
\end{array}
$$

10.12 **(a)** AB_4 tetrahedral   **(f)** AB_4 tetrahedral

  (b) AB_2E_2 bent   **(g)** AB_5 trigonal bipyramidal

  (c) AB_3 trigonal planar   **(h)** AB_3E trigonal pyramidal

  (d) AB_2E_3 linear   **(i)** AB_4 tetrahedral

  (e) AB_4E_2 square planar

10.13   The Lewis structure is:

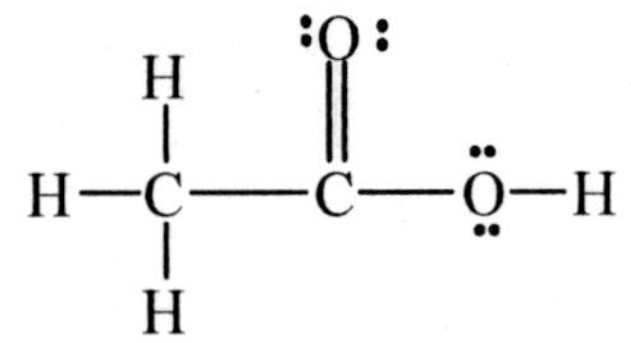

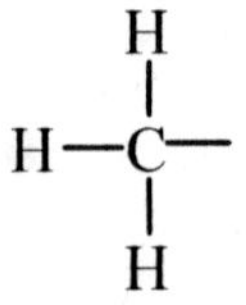

AB_4 tetrahedral AB_3 trigonal planar

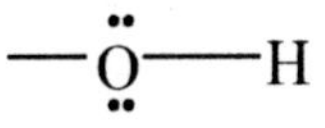

AB_2E_2 bent

10.14   Only molecules with four bonds to the central atom and no lone pairs are tetrahedral (AB_4).

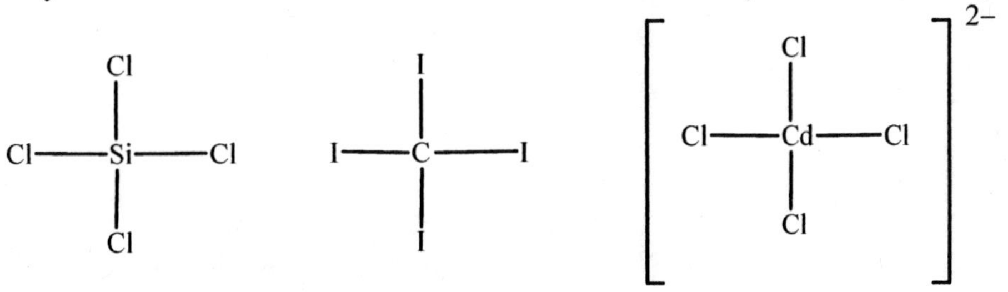

What are the Lewis structures and shapes for XeF_4 and SeF_4?

10.19   All four molecules have two bonds and two lone pairs (AB_2E_2) and therefore the bond angles are not linear. Since electronegativity decreases going down a column (group) in the periodic table, the electronegativity differences between hydrogen and the other Group 6 element will increase in the order Te < Se < S < O. The dipole moments will increase in the same order. Would this conclusion be as easy if the elements were in different groups?

10.20   The electronegativity of the halogens decreases from F to I. Thus, the polarity of the H–X bond (where X denotes a halogen atom) also decreases from HF to HI. This difference in electronegativity accounts for the decrease in dipole moment.

10.21 CO_2 = CBr_4 (μ = 0 for both) < H_2S < NH_3 < H_2O < HF

10.22   Draw the Lewis structures. Both molecules are linear (AB_2). In CS_2, the two C–S bond moments are equal in magnitude and opposite in direction. The sum or resultant dipole moment will be *zero*. Hence, CS_2 is a nonpolar molecule. Even though OCS is linear, the C–O and C–S bond moments are not exactly equal, and there will be a small net dipole moment. Hence, OCS has a **larger** dipole moment than CS_2 (zero).

10.23   Molecule **(b)** will have a higher dipole moment. In molecule (a), the *trans* arrangement cancels the bond dipoles and the molecule is nonpolar.

10.24   **Strategy:** Keep in mind that the dipole moment of a molecule depends on both the difference in electronegativities of the elements present and its geometry. A molecule can have polar bonds (if the bonded atoms have different electronegativities), but it may not possess a dipole moment if it has a highly symmetrical geometry.

Solution: Each vertex of the hexagonal structure of benzene represents the location of a C atom. Around the ring, there is no difference in electronegativity between C atoms, so the only bonds we need to consider are the polar C–Cl bonds.

The molecules shown in **(b)** and **(d)** are nonpolar. Due to the high symmetry of the molecules and the equal magnitude of the bond moments, the bond moments in each molecule cancel one another. The resultant dipole moment will be *zero*. For the molecules shown in **(a)** and **(c)**, the bond moments do not cancel and there will be net dipole moments. The dipole moment of the molecule in **(a)** is larger than that in **(c)**, because in **(a)** all the bond moments point in the same relative direction, reinforcing each other (see Lewis structure below). Therefore, the order of increasing dipole moments is:

$$(b) = (d) < (c) < (a).$$

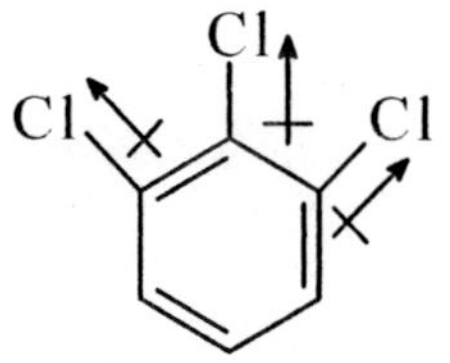

(a)

10.33 AsH_3 has the Lewis structure shown below. There are three bond pairs and one lone pair. The four electron pairs have a tetrahedral arrangement, and the molecular geometry is trigonal pyramidal (AB_3E) like ammonia (See Table 10.2). The As (arsenic) atom is in an sp^3 hybridization state.

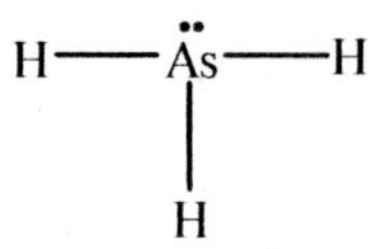

Three of the sp^3 hybrid orbitals form bonds to the hydrogen atoms by overlapping with the hydrogen $1s$ orbitals. The fourth sp^3 hybrid orbital holds the lone pair.

10.34   **Strategy:** The steps for determining the hybridization of the central atom in a molecule are:

draw Lewis Structure of the molecule ⟶ use VSEPR to determine the electron pair arrangement surrounding the central atom (Table 10.1 of the text) ⟶ use Table 10.4 of the text to determine the hybridization state of the central atom

Solution:

(a)   Write the Lewis structure of the molecule.

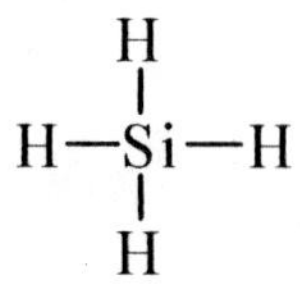

Count the number of electron pairs around the central atom. Since there are four electron pairs around Si, the electron arrangement that minimizes electron-pair repulsion is **tetrahedral**.

We conclude that Si is sp^3 **hybridized** because it has the electron arrangement of four sp^3 hybrid orbitals.

(b)   Write the Lewis structure of the molecule.

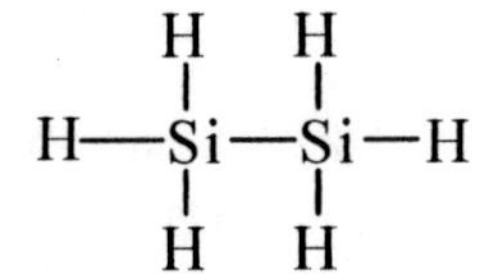

$$\begin{array}{c}
\quad\;\; \text{H} \quad\;\, \text{H} \\
\quad\;\; | \quad\quad\; | \\
\text{H}-\text{Si}-\text{Si}-\text{H} \\
\quad\;\; | \quad\quad\; | \\
\quad\;\; \text{H} \quad\;\, \text{H}
\end{array}$$

Count the number of electron pairs around the "central atoms". Since there are four electron pairs around each Si, the electron arrangement that minimizes electron-pair repulsion for each Si is **tetrahedral**.

We conclude that each Si is sp^3 **hybridized** because it has the electron arrangement of four sp^3 hybrid orbitals.

10.35   The Lewis structures of $AlCl_3$ and $AlCl_4^-$ are shown below. By the reasoning of the two problems above, the hybridization changes from sp^2 to sp^3.

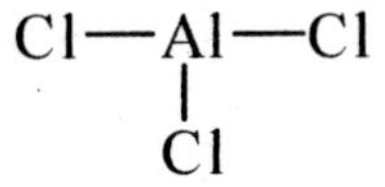

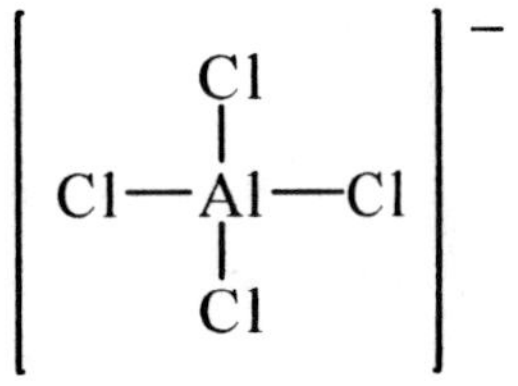

$$\begin{array}{c}
\text{Cl}-\text{Al}-\text{Cl} \\
\quad\;\; | \\
\quad\;\; \text{Cl}
\end{array}
\qquad\qquad
\left[\begin{array}{c}
\quad\;\; \text{Cl} \\
\quad\;\; | \\
\text{Cl}-\text{Al}-\text{Cl} \\
\quad\;\; | \\
\quad\;\; \text{Cl}
\end{array}\right]^-$$

What are the geometries of these molecules?

10.36   Draw the Lewis structures. Before the reaction, boron is sp^2 hybridized (trigonal planar electron arrangement) in BF_3 and nitrogen is sp^3 hybridized (tetrahedral electron arrangement) in NH_3. After the reaction, boron and nitrogen are both sp^3 hybridized (tetrahedral electron arrangement).

10.37   **(a)** NH_3 is an AB_3E type molecule just as AsH_3 in Problem 10.33. Referring to Table 10.4 of the text, the nitrogen is sp^3 hybridized.

(b) N_2H_4 has two equivalent nitrogen atoms. Centering attention on just one nitrogen atom shows that it is an AB_3E molecule, so the nitrogen atoms are sp^3 hybridized. From structural considerations, how can N_2H_4 be considered to be a derivative of NH_3?

(c)   The nitrate anion NO_3^- is isoelectronic and isostructural with the carbonate anion CO_3^{2-} that is discussed in Example 9.5 of the text. There are three resonance structures, and the ion is of type AB_3; thus, the nitrogen is sp^2 hybridized.

10.38   **(a)**   Each carbon has four bond pairs and no lone pairs and therefore has a tetrahedral electron pair arrangement. This implies sp^3 hybrid orbitals.

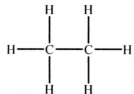

(b)   The left-most carbon is tetrahedral and therefore has sp^3 hybrid orbitals. The two carbon atoms connected by the double bond are trigonal planar with sp^2 hybrid orbitals.

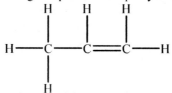

(c)   Carbons 1 and 4 have sp^3 hybrid orbitals. Carbons 2 and 3 have sp hybrid orbitals.

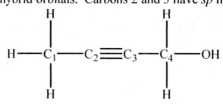

(d)   The left-most carbon is tetrahedral (sp^3 hybrid orbitals). The carbon connected to oxygen is trigonal planar (why?) and has sp^2 hybrid orbitals.

$$H-\overset{\overset{\displaystyle H}{|}}{\underset{\underset{\displaystyle H}{|}}{C}}-\overset{\overset{\displaystyle H}{|}}{C}=O$$

(e)   The left-most carbon is tetrahedral (sp^3 hybrid orbitals). The other carbon is trigonal planar with sp^2 hybridized orbitals.

$$H-\overset{\overset{\displaystyle H}{|}}{\underset{\underset{\displaystyle H}{|}}{C}}-\overset{\overset{\displaystyle O}{||}}{C}-O-H$$

10.39   **(a)** sp **(b)** sp **(c)** sp

10.40   **Strategy:** The steps for determining the hybridization of the central atom in a molecule are:

| draw Lewis Structure of the molecule | $\longrightarrow$ | use VSEPR to determine the electron pair arrangement surrounding the central atom (Table 10.1 of the text) | $\longrightarrow$ | use Table 10.4 of the text to determine the hybridization state of the central atom |

Solution:

Write the Lewis structure of the molecule. Several resonance forms with formal charges are shown.

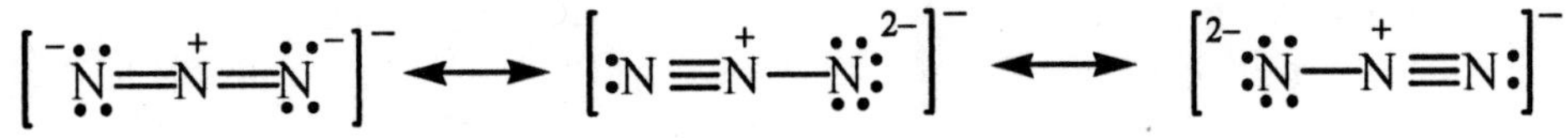

$$\left[\bar{\ddot{N}}=\overset{+}{N}=\bar{\ddot{N}} \right]^{-} \longleftrightarrow \left[:N\equiv\overset{+}{N}-\ddot{\ddot{N}}\text{:}^{2-} \right]^{-} \longleftrightarrow \left[^{2-}\ddot{\ddot{N}}-\overset{+}{N}\equiv N\text{:} \right]^{-}$$

Count the number of electron pairs around the central atom. Since there are two electron pairs around N, the electron arrangement that minimizes electron-pair repulsion is **linear** (AB_2). Remember, for VSEPR purposes a multiple bond counts the same as a single bond.

We conclude that N is **sp hybridized** because it has the electron arrangement of two *sp* hybrid orbitals.

10.41 The Lewis structure is shown below. The two end carbons are trigonal planar and therefore use sp^2 hybrid orbitals. The central carbon is linear and must use *sp* hybrid orbitals.

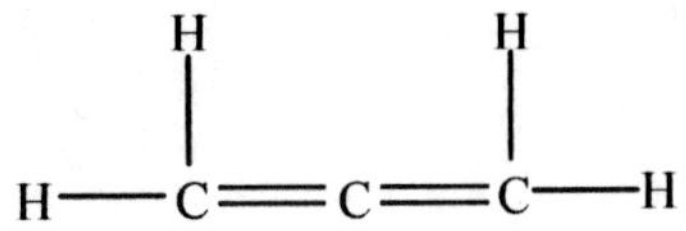

$$\begin{array}{ccccc} & H & & H & \\ & | & & | & \\ H- & C & = C = & C & -H \end{array}$$

A Lewis drawing does not necessarily show actual molecular geometry. Notice that the two CH_2 groups at the ends of the molecule must be perpendicular. This is because the two double bonds must use different *2p* orbitals on the middle carbon, and these two *2p* orbitals are perpendicular. The overlap of the *2p* orbitals on each carbon is shown below.

Is the allene molecule polar?

10.42 **Strategy:** The steps for determining the hybridization of the central atom in a molecule are:

draw Lewis Structure of the molecule	$\longrightarrow$	use VSEPR to determine the electron pair arrangement surrounding the central atom (Table 10.1 of the text)	$\longrightarrow$	use Table 10.4 of the text to determine the hybridization state of the central atom

Solution:

Write the Lewis structure of the molecule.

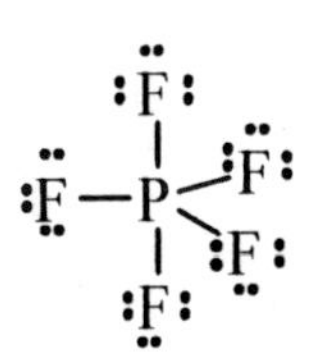

$$\begin{array}{c} \ddot{:}\ddot{F}\text{:} \\ | \\ \ddot{:}\ddot{F}-P\overset{\ddot{:}\ddot{F}\text{:}}{\underset{:\ddot{F}\text{:}}{<}} \\ | \\ :\ddot{F}\text{:} \end{array}$$

Count the number of electron pairs around the central atom. Since there are five electron pairs around P, the electron arrangement that minimizes electron-pair repulsion is **trigonal bipyramidal** (AB_5).

We conclude that P is **sp^3d hybridized** because it has the electron arrangement of five sp^3d hybrid orbitals.

10.43   It is almost always true that a single bond is a sigma bond, that a double bond is a sigma bond and a pi bond, and that a triple bond is *always* a sigma bond and two pi bonds.

    (a)   sigma bonds: 4; pi bonds: 0   **(b)**   sigma bonds: 5; pi bonds: 1

    (c)   sigma bonds: 10; pi bonds: 3

10.44   A single bond is usually a sigma bond, a double bond is usually a sigma bond and a pi bond, and a triple bond is always a sigma bond and two pi bonds. Therefore, there are **nine pi bonds** and **nine sigma bonds** in the molecule.

10.49   The molecular orbital electron configuration and bond order of each species is shown below.

$$H_2 \qquad\qquad H_2^+ \qquad\qquad H_2^{2+}$$

$$\sigma_{1s}^\star \;\underline{\quad\quad} \qquad \sigma_{1s}^\star \;\underline{\quad\quad} \qquad \sigma_{1s}^\star \;\underline{\quad\quad}$$

$$\sigma_{1s} \;\underline{\uparrow\downarrow} \qquad \sigma_{1s} \;\underline{\uparrow\quad} \qquad \sigma_{1s} \;\underline{\quad\quad}$$

$$\text{bond order} = 1 \qquad\qquad \text{bond order} = \frac{1}{2} \qquad\qquad \text{bond order} = 0$$

The internuclear distance in the +1 ion should be greater than that in the neutral hydrogen molecule. The distance in the +2 ion will be arbitrarly large because there is no bond (bond order zero).

10.50   In order for the two hydrogen atoms to combine to form a H_2 molecule, the electrons must have opposite spins. Furthermore, the combined energy of the two atoms must not be too great. Otherwise, the H_2 molecule will possess too much energy and will break apart into two hydrogen atoms.

10.51   The energy level diagrams are shown below.

$$He_2 \qquad\qquad HHe \qquad\qquad He_2^+$$

$$\sigma_{1s}^\star \;\underline{\uparrow\downarrow} \qquad \sigma_{1s}^\star \;\underline{\uparrow\quad} \qquad \sigma_{1s}^\star \;\underline{\uparrow\quad}$$

$$\sigma_{1s} \;\underline{\uparrow\downarrow} \qquad \sigma_{1s} \;\underline{\uparrow\downarrow} \qquad \sigma_{1s} \;\underline{\uparrow\downarrow}$$

$$\text{bond order} = 0 \qquad\qquad \text{bond order} = \frac{1}{2} \qquad\qquad \text{bond order} = \frac{1}{2}$$

He_2 has a bond order of zero; the other two have bond orders of 1/2. Based on bond orders alone, He_2 has no stability, while the other two have roughly equal stabilities.

10.52   The electron configurations are listed. Refer to Table 10.5 of the text for the molecular orbital diagram.

    Li_2: $(\sigma_{1s})^2(\sigma_{1s}^\star)^2(\sigma_{2s})^2$ bond order = 1

    Li_2^+: $(\sigma_{1s})^2(\sigma_{1s}^\star)^2(\sigma_{2s})^1$ bond order = $\dfrac{1}{2}$

    Li_2^-: $(\sigma_{1s})^2(\sigma_{1s}^\star)^2(\sigma_{2s})^2(\sigma_{2s}^\star)^1$ bond order = $\dfrac{1}{2}$

    Order of increasing stability: **Li_2^- = Li_2^+ < Li_2**

In reality, Li_2^+ is more stable than Li_2^- because there is less electrostatic repulsion in Li_2^+.

10.53   The Be_2 molecule does not exist because there are equal numbers of electrons in bonding and antibonding molecular orbitals, making the bond order zero.

$$
\begin{array}{ll}
\sigma_{2s}^{\star} & \underline{\uparrow\downarrow} \\
\sigma_{2s} & \underline{\uparrow\downarrow} \\[1em]
\sigma_{1s}^{\star} & \underline{\uparrow\downarrow} \\
\sigma_{1s} & \underline{\uparrow\downarrow}
\end{array}
$$

$$\text{bond order } = 0$$

10.54   See Table 10.5 of the text. Removing an electron from B_2 (bond order = 1) gives B_2^{+}, which has a bond order of (1/2). Therefore, **B_2^{+}** has a weaker and longer bond than B_2.

10.55   The energy level diagrams are shown below.

$$C_2^{2-} \qquad\qquad\qquad\qquad C_2$$

C_2^{2-}		C_2	
$\sigma_{2p_x}^{\star}$	———	$\sigma_{2p_x}^{\star}$	
$\pi_{2p_y}^{\star}, \pi_{2p_z}^{\star}$	——— ———	$\pi_{2p_y}^{\star}, \pi_{2p_z}^{\star}$	
σ_{2p_x}	$\uparrow\downarrow$	σ_{2p_x}	
π_{2p_y}, π_{2p_z}	$\uparrow\downarrow \quad \uparrow\downarrow$	π_{2p_y}, π_{2p_z}	$\uparrow\downarrow \quad \uparrow\downarrow$
$\sigma_{2s}^{\star}$	$\uparrow\downarrow$	$\sigma_{2s}^{\star}$	$\uparrow\downarrow$
σ_{2s}	$\uparrow\downarrow$	σ_{2s}	$\uparrow\downarrow$
$\sigma_{1s}^{\star}$	$\uparrow\downarrow$	$\sigma_{1s}^{\star}$	$\uparrow\downarrow$
σ_{1s}	$\uparrow\downarrow$	σ_{1s}	$\uparrow\downarrow$

The bond order of the carbide ion is 3 and that of C_2 is only 2. With what homonuclear diatomic molecule is the carbide ion isoelectronic?

10.56   In both the Lewis structure and the molecular orbital energy level diagram (Table 10.5 of the text), the oxygen molecule has a double bond (bond order = 2). The principal difference is that the molecular orbital treatment predicts that the molecule will have two unpaired electrons (paramagnetic). Experimentally this is found to be true.

10.57   In forming the N_2^{+} from N_2, an electron is removed from the sigma *bonding* molecular orbital. Consequently, the bond order decreases to 2.5 from 3.0. In forming the O_2^{+} ion from O_2, an electron is removed from the pi *antibonding* molecular orbital. Consequently, the bond order increases to 2.5 from 2.0.

10.58   We refer to Table 10.5 of the text.

   O_2 has a bond order of 2 and is paramagnetic (two unpaired electrons).

   O_2^{+} has a bond order of 2.5 and is paramagnetic (one unpaired electron).

   O_2^{-} has a bond order of 1.5 and is paramagnetic (one unpaired electron).

O_2^{2-} has a bond order of 1 and is diamagnetic.

Based on molecular orbital theory, the stability of these molecules increases as follows:

$$O_2^{2-} \; < \; O_2^{-} \; < \; O_2 \; < \; O_2^{+}$$

10.59   From Table 10.5 of the text, we see that the bond order of F_2^{+} is 1.5 compared to 1 for F_2. Therefore, F_2^{+} should be more stable than F_2 (stronger bond) and should also have a shorter bond length.

10.60   As discussed in the text (see Table 10.5), the single bond in B_2 is a pi bond (the electrons are in a pi *bonding* molecular orbital) and the double bond in C_2 is made up of two pi bonds (the electrons are in the pi *bonding* molecular orbitals).

10.63   Benzene is stabilized by delocalized molecular orbitals. The C–C bonds are equivalent, rather than alternating single and double bonds. The additional stabilization makes the bonds in benzene much less reactive chemically than isolated double bonds such as those in ethylene.

10.64   The symbol on the left shows the pi bond delocalized over the entire molecule. The symbol on the right shows only one of the two resonance structures of benzene; it is an incomplete representation.

10.65   If the two rings happen to be perpendicular in biphenyl, the pi molecular orbitals are less delocalized. In naphthalene the pi molecular orbital is always delocalized over the entire molecule. What do you think is the most stable structure for biphenyl: both rings in the same plane or both rings perpendicular?

10.66   **(a)**   Two Lewis resonance forms are shown below. Formal charges different than zero are indicated.

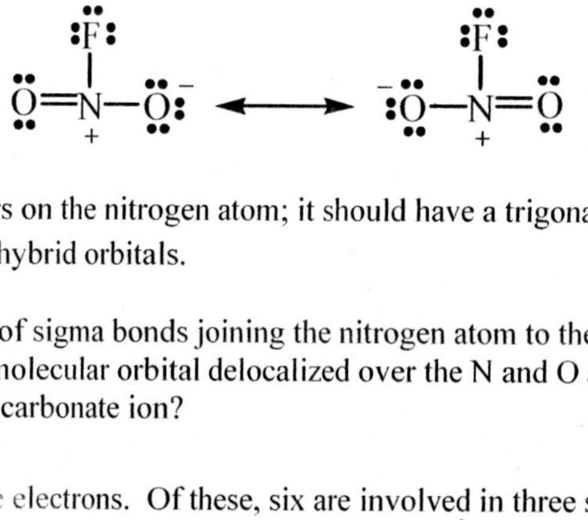

(b)   There are no lone pairs on the nitrogen atom; it should have a trigonal planar electron pair arrangement and therefore use sp^2 hybrid orbitals.

(c)   The bonding consists of sigma bonds joining the nitrogen atom to the fluorine and oxygen atoms. In addition there is a pi molecular orbital delocalized over the N and O atoms. Is nitryl fluoride isoelectronic with the carbonate ion?

10.67   The ion contains 24 valence electrons. Of these, six are involved in three sigma bonds between the nitrogen and oxygen atoms. The hybridization of the nitrogen atom is sp^2. There are 16 non-bonding electrons on the oxygen atoms. The remaining two electrons are in a delocalized pi molecular orbital which results from the overlap of the p_z orbital of nitrogen and the p_z orbitals of the three oxygen atoms. The molecular orbitals are similar to those of the carbonate ion (See Section 10.8 of the text).

10.68   The Lewis structures of ozone are:

The central oxygen atom is sp^2 hybridized (AB_2E). The unhybridized $2p_z$ orbital on the central oxygen overlaps with the $2p_z$ orbitals on the two end atoms.

10.69   Only **(c)** will not be tetrahedral. All the others have AB_4–type Lewis structures and will therefore be tetrahedral. For SF_4 the Lewis structure is of the AB_4E type which gives rise to a distorted tetrahedral geometry (Table 10.2 of the text).

10.70   **Strategy:** The sequence of steps in determining molecular geometry is as follows:

draw Lewis $\longrightarrow$ find arrangement of $\longrightarrow$ find arrangement $\longrightarrow$ determine geometry
structure   electrons pairs   of bonding pairs   based on bonding pairs

Solution:

Write the Lewis structure of the molecule.

$$:\!\overset{..}{\underset{..}{Br}}\!-\!Hg\!-\!\overset{..}{\underset{..}{Br}}\!:$$

Count the number of electron pairs around the central atom. There are two electron pairs around Hg.

Since there are two electron pairs around Hg, the electron-pair arrangement that minimizes electron-pair repulsion is **linear**.

In addition, since there are no lone pairs around the central atom, the geometry is also **linear** (AB_2).

You could establish the geometry of $HgBr_2$ by measuring its dipole moment. If mercury(II) bromide were bent, it would have a measurable dipole moment. Experimentally, it has no dipole moment and therefore must be linear.

10.71

Dot structure	*Label*	*Shape*	*Bond dipole*	*Resultant*	*dipole moment*
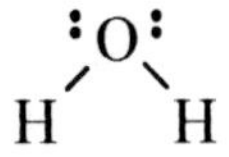 (H₂O structure)	AB_2E_2	bent			$\mu > 0$
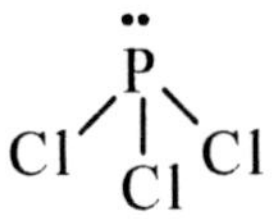 (PCl₃ structure)	AB_3E	trigonal pyramidal			$\mu > 0$
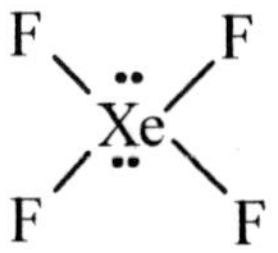 (XeF₄ structure)	AB_4E_2	square planar			$\mu = 0$
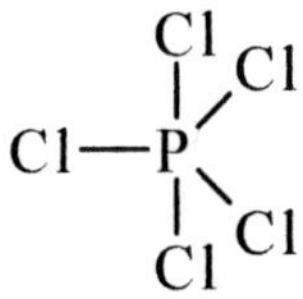 (PCl₅ structure)	AB_5	trigonal bipyramid	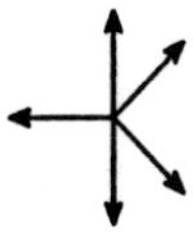		$\mu = 0$

AB₆   octahedral   μ = 0

Why do the bond dipoles add to zero in PCl_5?

10.72   According to valence bond theory, a pi bond is formed through the side-to-side overlap of a pair of *p* orbitals. As atomic size increases, the distance between atoms is too large for *p* orbitals to overlap effectively in a side-to-side fashion. If two orbitals overlap poorly, that is, they share very little space in common, then the resulting bond will be very weak. This situation applies in the case of pi bonds between silicon atoms as well as between any other elements not found in the second period. It is usually far more energetically favorable for silicon, or any other heavy element, to form two single (sigma) bonds to two other atoms than to form a double bond (sigma + pi) to only one other atom.

10.73   Geometry: bent; hybridization: sp^3.

10.74   The Lewis structures and VSEPR geometries of these species are shown below. The three nonbonding pairs of electrons on each fluorine atom have been omitted for simplicity.

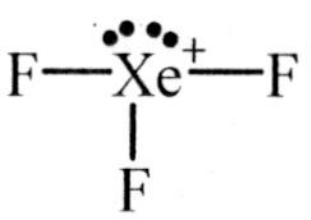

AB₃E₂   AB₅E   AB₆
T-shaped   Square Pyramidal   Octahedral

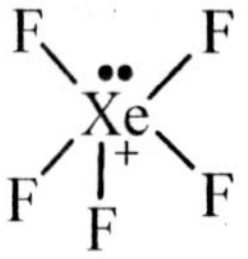

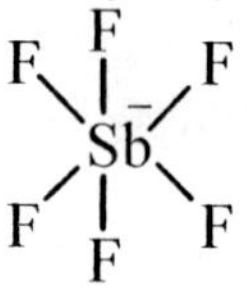

10.75   **(a)**   The Lewis structure is:

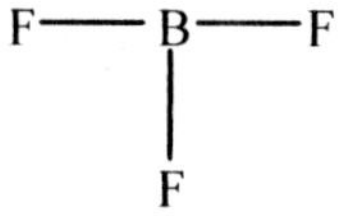

The shape will be trigonal *planar* (AB₃)

(b)   The Lewis structure is:

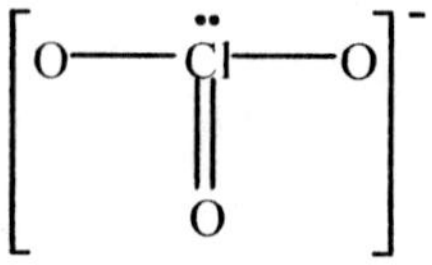

The molecule will be a trigonal pyramid (*nonplanar*).

(c)   The Lewis structure and the dipole moment for H_2O is presented in Problem 10.71. The dipole moment is directed from the positive hydrogen end to the more negative oxygen.

(d)   The Lewis structure is:

The molecule is bent and therefore *polar*.

(e)   The Lewis structure is:

The nitrogen atom is of the AB₂E type, but there is only one unshared electron rather than the usual pair. As a result, the repulsion will not be as great and the O–N–O angle will be *greater than 120°* expected for AB₂E geometry. Experiment shows the angle to be around 135°.

Which of the species in this problem has resonance structures?

10.76   To predict the bond angles for the molecules, you would have to draw the Lewis structure and determine the geometry using the VSEPR model. From the geometry, you can predict the bond angles.

(a) $BeCl_2$:   AB₂ type, 180° (linear).

(b) BCl_3:   AB₃ type, 120° (trigonal planar).

(c) CCl_4:   AB₄ type, 109.5° (tetrahedral).

(d) CH_3Cl: AB₄ type, 109.5° (tetrahedral with a possible slight distortion resulting from the different sizes of the chlorine and hydrogen atoms).

(e) Hg_2Cl_2:   Each mercury atom is of the AB₂ type. The entire molecule is linear, 180° bond angles.

(f) $SnCl_2$:   AB₂E type, roughly 120° (bent).

(g) H_2O_2:   The atom arrangement is HOOH. Each oxygen atom is of the AB₂E₂ type and the H–O–O angles will be roughly 109.5°.

(h) SnH_4:   AB₄ type, 109.5° (tetrahedral).

10.77   The two approaches are discussed in Sections 10.1, 10.3, and 10.4 of the textbook.

10.78   Since arsenic and phosphorus are both in the same group of the periodic table, this problem is exactly like Problem 10.42. AsF_5 is an AB₅ type molecule, so the geometry is trigonal bipyramidal. We conclude that As is *sp³d hybridized* because it has the electron arrangement of five sp^3d hybrid orbitals.

10.79   **(a)**   The Lewis structure is:

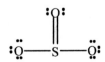

The geometry is trigonal planar; the molecule is *nonpolar*.

(b)   The Lewis structure is:

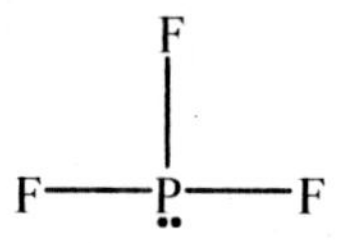

$$
\begin{array}{c}
\text{F} \\
| \\
\text{F}\!-\!\overset{}{\underset{\textbf{..}}{\text{P}}}\!-\!\text{F}
\end{array}
$$

The molecule has trigonal pyramidal geometry. It is ***polar***.

(c)   The Lewis structure is:

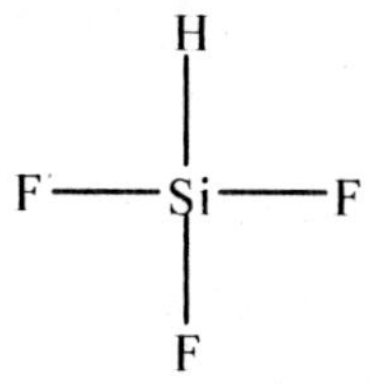

$$
\begin{array}{c}
\text{H} \\
| \\
\text{F}\!-\!\text{Si}\!-\!\text{F} \\
| \\
\text{F}
\end{array}
$$

The molecule will be tetrahedral (AB_4). Both fluorine and hydrogen are more electronegative than silicon, but fluorine is the most electronegative element, so the molecule will be polar (fluorine side negative).

(d)   The Lewis structure is:

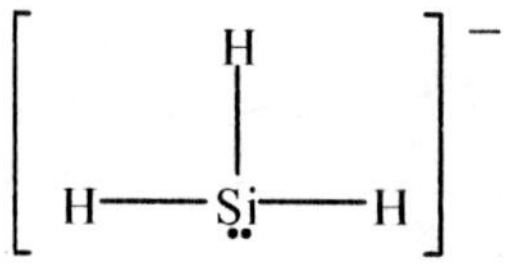

$$
\left[
\begin{array}{c}
\text{H} \\
| \\
\text{H}\!-\!\overset{}{\underset{\textbf{..}}{\text{Si}}}\!-\!\text{H}
\end{array}
\right]^{-}
$$

The ion has a ***trigonal pyramidal*** geometry (AB_3E).

(e)   The Lewis structure is:

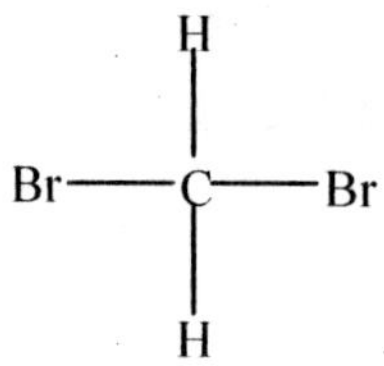

$$
\begin{array}{c}
\text{H} \\
| \\
\text{Br}\!-\!\text{C}\!-\!\text{Br} \\
| \\
\text{H}
\end{array}
$$

The molecule will be tetrahedral (AB_4) but still ***polar***. The negative end of the dipole moment will be on the side with the two bromine atoms; the positive end will be on the hydrogen side.

10.80   Only ICl_2^- and $CdBr_2$ will be linear. The rest are bent.

10.81   The Lewis structure is shown below.

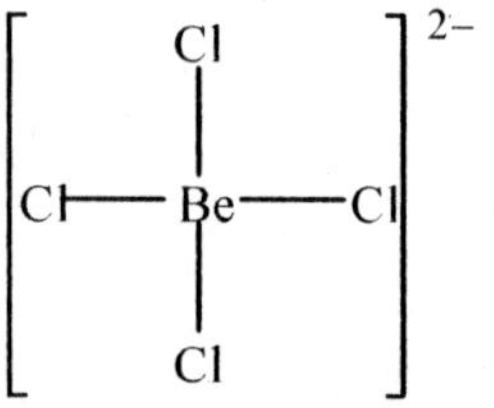

$$
\left[
\begin{array}{c}
\text{Cl} \\
| \\
\text{Cl}\!-\!\text{Be}\!-\!\text{Cl} \\
| \\
\text{Cl}
\end{array}
\right]^{2-}
$$

The molecule is of the AB_4 type and should therefore be ***tetrahedral***. The hybridization of the Be atom should be ***sp^3***.

10.82   (a)

Strategy: The steps for determining the hybridization of the central atom in a molecule are:

draw Lewis Structure of the molecule ⟶ use VSEPR to determine the electron pair arrangement surrounding the central atom (Table 10.1 of the text) ⟶ use Table 10.4 of the text to determine the hybridization state of the central atom

Solution:

The geometry around each nitrogen is identical. To complete an octet of electrons around N, you must add a lone pair of electrons. Count the number of electron pairs around N. There are three electron pairs around each N.

Since there are three electron pairs around N, the electron-pair arrangement that minimizes electron-pair repulsion is **trigonal planar**.

We conclude that each N is sp^2 **hybridized** because it has the electron arrangement of three sp^2 hybrid orbitals.

(b)

Strategy: Keep in mind that the dipole moment of a molecule depends on both the difference in electronegativities of the elements present and its geometry. A molecule can have polar bonds (if the bonded atoms have different electronegativities), but it may not possess a dipole moment if it has a highly symmetrical geometry.

Solution: An N–F bond is polar because F is more electronegative than N. The structure on the right has a dipole moment because the two N–F bond moments do not cancel each other out and so the molecule has a net dipole moment. On the other hand, the two N–F bond moments in the left-hand structure cancel. The sum or resultant dipole moment will be *zero*.

10.83   (a)   The structures for cyclopropane and cubane are

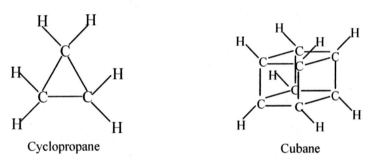

Cyclopropane   Cubane

(b)   The C–C–C bond in cyclopropane is 60° and in cubane is 90°. Both are smaller than the 109.5° expected for sp^3 hybridized carbon. Consequently, there is considerable strain on the molecules.

(c)   They would be more difficult to make than an sp^3 hybridized carbon in an unconstrained (that is, not in a small ring) system in which the carbons can adopt bond angles closer to 109.5°.

10.84   In 1,2-dichloroethane, the two C atoms are joined by a sigma bond. Rotation about a sigma bond does not destroy the bond, and the bond is therefore free (or relatively free) to rotate. Thus, all angles are permitted and the molecule is nonpolar because the C–Cl bond moments cancel each other because of the averaging effect brought about by rotation. In *cis*-dichloroethylene the two C–Cl bonds are locked in position. The π bond between the C atoms prevents rotation (in order to rotate, the π bond must be broken, using an energy source such as light or heat). Therefore, there is no rotation about the C=C in *cis*-dichloroethylene, and the molecule is polar.

10.85   Consider the overlap of the 2*p* orbitals on each carbon atom.

The geometric planes containing the CHCl groups at each end of the molecule are mutually perpendicular. This is because the two carbon-carbon double bonds must use different 2*p* orbitals on the middle carbon, and these two 2*p* orbitals are perpendicular. This means that the two chlorine atoms can be considered to be on one side of the molecule and the two hydrogen atoms on the other. The molecule has a dipole moment. Draw the end-on view if you aren't convinced.

10.86 O_3, CO, CO_2, NO_2, N_2O, CH_4, and $CFCl_3$ are greenhouse gases.

10.87   Both sulfur–to–oxygen bonds are double bonds which occupy more space than single bonds and the repulsion between the double bonds spread out the angle close to 120°.

10.88   The Lewis structure is:

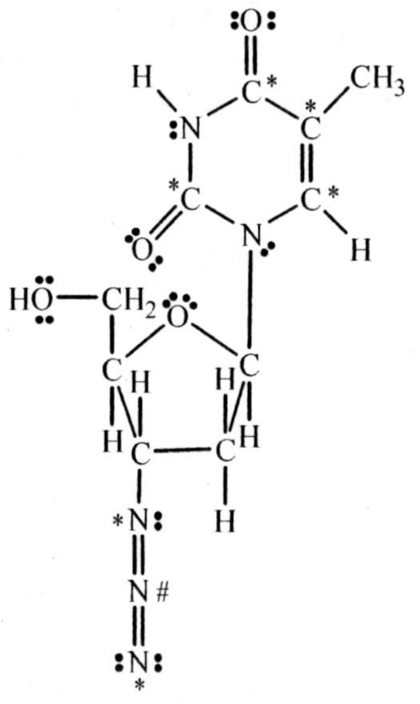

The carbon atoms and nitrogen atoms marked with an asterisk (C* and N*) are sp^2 hybridized; unmarked carbon atoms and nitrogen atoms are sp^3 hybridized; and the nitrogen atom marked with (#) is sp hybridized.

10.89   For an octahedral AX_4Y_2 molecule only two different structures are possible: one with the two Y's next to each other like **(b)** and **(d)**, and one with the two Y's on opposite sides of the molecule like **(a)** and **(c)**. The different looking drawings simply depict the same molecule seen from a different angle or side.

It would help to develop your power of spatial visualization to make some simple models and convince yourself of the validity of these answers. How many different structures are possible for octahedral AX_5Y or AX_3Y_3 molecules? Would an octahedral AX_2Y_4 molecule have a different number of structures from AX_4Y_2? Ask your instructor if you aren't sure.

10.90   C has no *d* orbitals but Si does (3*d*). Thus, H_2O molecules can add to Si in hydrolysis (valence-shell expansion).

10.91 B_2 is $(\sigma_{1s})^2(\sigma_{1s}^{\star})^2(\sigma_{2s})^2(\sigma_{2s}^{\star})^2(\pi_{2p_y})^1(\pi_{2p_z})^1$. It is ***paramagnetic***.

10.92   The carbons are in sp^2 hybridization states. The nitrogens are in the sp^3 hybridization state, except for the ring nitrogen double-bonded to a carbon that is sp^2 hybridized. The oxygen atom is sp^2 hybridized.

10.93   Referring to Table 10.5, we see that F_2^- has an extra electron in $\sigma_{2p_x}^{\star}$. Therefore, it only has a bond order of $\frac{1}{2}$ (compared with a bond order of one for F_2).

10.94   **(a)**   Use a conventional oven. A microwave oven would not cook the meat from the outside toward the center (it penetrates).

      (b)   Polar molecules absorb microwaves and would interfere with the operation of radar.

      (c)   Too much water vapor (polar molecules) absorbed the microwaves, interfering with the operation of radar.

10.95 CCl_4 can be represented by:

Let ρ be a C–Cl bond moment. Thus,

$$\rho = 3\rho \cos \theta$$

$$\cos \theta = \frac{1}{3} \qquad \theta = 70.5°$$

Tetrahedral angle $= 180° - 70.5° = $ **109.5°**

10.96   The smaller size of F compared to Cl results in a shorter F–F bond than a Cl–Cl bond. The closer proximity of the lone pairs of electrons on the F atoms results in greater electron-electron repulsions that weaken the bond.

10.97   Since nitrogen is a second row element, it cannot exceed an octet of electrons. Since there are no lone pairs on the central nitrogen, the molecule must be linear and sp hybrid orbitals must be used.

$$:N\equiv\overset{+}{N}-\overset{\cdot\cdot}{\underset{\cdot\cdot}{N}}:^{2-} \longleftrightarrow \overset{\cdot\cdot}{\underset{\cdot\cdot}{N}}{}^{-}=\overset{+}{N}=\overset{\cdot\cdot}{\underset{\cdot\cdot}{N}}{}^{-} \longleftrightarrow {}^{2-}\overset{\cdot\cdot}{\underset{\cdot\cdot}{N}}:-\overset{+}{N}\equiv N:$$

The $2p_y$ orbital on the central nitrogen atom overlaps with the $2p_y$ on the terminal nitrogen atoms, and the $2p_z$ orbital on the central nitrogen overlaps with the $2p_z$ orbitals on the terminal nitrogen atoms to form delocalized molecular orbitals.

10.98 $1 \text{ D} = 3.336 \times 10^{-30} \text{ C·m}$

electronic charge $(e) = 1.6022 \times 10^{-19} \text{ C}$

$$\frac{\mu}{ed} \times 100\% = \frac{1.92 \text{ D} \times \dfrac{3.336 \times 10^{-30} \text{ C·m}}{1 \text{ D}}}{(1.6022 \times 10^{-19} \text{ C}) \times (91.7 \times 10^{-12} \text{ m})} \times 100\% = \textbf{43.6\%} \text{ ionic character}$$

10.99   In the structure on the left, the bonds are too strained (bond angles too small). This molecule does ***not*** exist.

The structure on the right ***does*** exist, although it is highly reactive. This molecule is called benzyne.

10.100   The second and third vibrational motions are responsible for CO_2 to behave as a greenhouse gas. CO_2 is a nonpolar molecule. The second and third vibrational motions, create a changing dipole moment. The first vibration, a symmetric stretch, does *not* create a dipole moment. Since CO, NO_2, and N_2O are all polar molecules, they will also act as greenhouse gases.

10.101 (a)

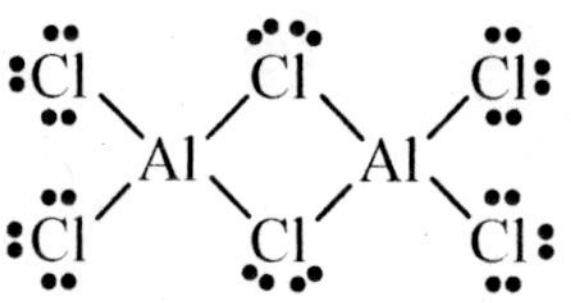

(b)   The hybridization of Al in $AlCl_3$ is sp^2. The molecule is trigonal planar. The hybridization of Al in Al_2Cl_6 is sp^3.

(c)   The geometry about each Al atom is tetrahedral.

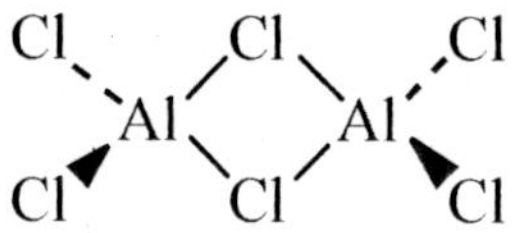

(d)   The molecules are nonpolar; they do not possess a dipole moment.

10.102 (a)   Looking at the electronic configuration for N_2 shown in Table 10.5 of the text, we write the electronic configuration for P_2.

$$[\text{Ne}_2](\sigma_{3s})^2(\sigma_{3s}^\star)^2(\pi_{3p_y})^2(\pi_{3p_z})^2(\sigma_{3p_x})^2$$

(b)   Past the Ne_2 core configuration, there are 8 bonding electrons and 2 antibonding electrons. The bond order is:

bond order $= \frac{1}{2}(8-2) = \textbf{3}$

(c)   All the electrons in the electronic configuration are paired. P_2 is **diamagnetic**.

10.103   The complete structure of progesterone is shown below.

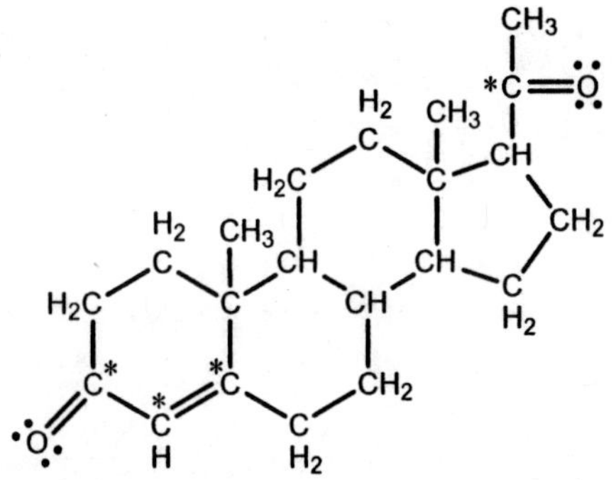

The four carbons marked with an asterisk are sp^2 hybridized. The remaining carbons are sp^3 hybridized.

10.104   **(a)**   A σ bond is formed by orbitals overlapping end-to-end. Rotation will not break this end-to-end overlap. A π bond is formed by the sideways overlapping of orbitals. The two 90° rotations (180° total) will break and then reform the pi bond, thereby converting *cis*-dichloroethylene to *trans*-dichloroethylene.

(b)   The pi bond is weaker because of the lesser extent of sideways orbital overlap, compared to the end-to-end overlap in a sigma bond.

(c)   The bond energy is given in the unit, kJ/mol. To find the longest wavelength of light needed to bring about the conversion from *cis* to *trans*, we need the energy to break a pi bond in a single molecule. We convert from kJ/mol to J/molecule.

$$\frac{270 \text{ kJ}}{1 \text{ mol}} \times \frac{1 \text{ mol}}{6.022 \times 10^{23} \text{ molecules}} = 4.48 \times 10^{-22} \text{ kJ/molecule} = 4.48 \times 10^{-19} \text{ J/molecule}$$

Now that we have the energy needed to cause the conversion from *cis* to *trans* in one molecule, we can calculate the wavelength from this energy.

$$E = \frac{hc}{\lambda}$$

$$\lambda = \frac{hc}{E} = \frac{(6.63 \times 10^{-34} \text{ J} \cdot \text{s})(3.00 \times 10^8 \text{ m/s})}{4.48 \times 10^{-19} \text{ J}}$$

$$\lambda = 4.44 \times 10^{-7} \text{ m} = 444 \text{ nm}$$

CHAPTER 11
INTERMOLECULAR FORCES AND LIQUIDS AND SOLIDS

11.7 ICl has a dipole moment and Br_2 does not. The dipole moment increases the intermolecular attractions between ICl molecules and causes that substance to have a higher melting point than bromine.

11.8 **Strategy:** Classify the species into three categories: ionic, polar (possessing a dipole moment), and nonpolar. Keep in mind that dispersion forces exist between *all* species.

Solution: The three molecules are essentially nonpolar. There is little difference in electronegativity between carbon and hydrogen. Thus, the only type of intermolecular attraction in these molecules is dispersion forces. Other factors being equal, the molecule with the greater number of electrons will exert greater intermolecular attractions. By looking at the molecular formulas you can predict that the order of increasing boiling points will be $CH_4 < C_3H_8 < C_4H_{10}$.

On a very cold day, propane and butane would be liquids (boiling points $-44.5°C$ and $-0.5°C$, respectively); only **methane** would still be a gas (boiling point $-161.6°C$).

11.9 All are tetrahedral (AB_4 type) and are nonpolar. Therefore, the only intermolecular forces possible are dispersion forces. Without worrying about what causes dispersion forces, you only need to know that the strength of the dispersion force increases with the number of electrons in the molecule (all other things being equal). As a consequence, the magnitude of the intermolecular attractions and of the boiling points should increase with increasing molar mass.

11.10 **(a)** Benzene (C_6H_6) molecules are nonpolar. Only dispersion forces will be present.

(b) Chloroform (CH_3Cl) molecules are polar (why?). Dispersion and dipole-dipole forces will be present.

(c) Phosphorus trifluoride (PF_3) molecules are polar. Dispersion and dipole-dipole forces will be present.

(d) Sodium chloride (NaCl) is an ionic compound. Ion-ion (and dispersion) forces will be present.

(e) Carbon disulfide (CS_2) molecules are nonpolar. Only dispersion forces will be present.

11.11 The center ammonia molecule is hydrogen–bonded to two other ammonia molecules.

11.12 In this problem you must identify the species capable of hydrogen bonding among themselves, not with water. In order for a molecule to be capable of hydrogen bonding with another molecule like itself, it must have at least one hydrogen atom bonded to N, O, or F. Of the choices, only **(e)** CH_3COOH (acetic acid) shows this structural feature. The others cannot form hydrogen bonds among themselves.

11.13 CO_2 is a nonpolar molecular compound. The only intermolecular force present is a relatively weak dispersion force (small molar mass). CO_2 will have the lowest boiling point.

CH$_3$Br is a polar molecule. Dispersion forces (present in all matter) and dipole–dipole forces will be present. This compound has the next highest boiling point.

CH$_3$OH is polar and can form hydrogen bonds, which are especially strong dipole-dipole attractions. Dispersion forces and hydrogen bonding are present to give this substance the next highest boiling point.

RbF is an ionic compound (Why?). Ion–ion attractions are much stronger than any intermolecular force. RbF has the highest boiling point.

11.14 **Strategy:** The molecule with the stronger intermolecular forces will have the higher boiling point. If a molecule contains an N–H, O–H, or F–H bond it can form intermolecular hydrogen bonds. A hydrogen bond is a particularly strong dipole-dipole intermolecular attraction.

Solution: 1-butanol has the higher boiling point because the molecules can form hydrogen bonds with each other (It contains an O–H bond). Diethyl ether molecules do contain both oxygen atoms and hydrogen atoms. However, all the hydrogen atoms are bonded to carbon, not oxygen. There is no hydrogen bonding in diethyl ether, because carbon is not electronegative enough.

11.15 **(a)** Cl_2: it has more electrons the O_2 (both are nonpolar) and therefore has stronger dispersion forces.

(b) SO_2: it is polar (most important) and also has more electrons than CO_2 (nonpolar). More electrons imply stronger dispersion forces.

(c) HF: although HI has more electrons and should therefore exert stronger dispersion forces, HF is capable of hydrogen bonding and HI is not. Hydrogen bonding is the stronger attractive force.

11.16 **(a)** Xe: it has more electrons and therefore stronger dispersion forces.

(b) CS_2:   it has more electrons (both molecules nonpolar) and therefore stronger dispersion forces.

(c) Cl_2:   it has more electrons (both molecules nonpolar) and therefore stronger dispersion forces.

(d) LiF: it is an ionic compound, and the ion-ion attractions are much stronger than the dispersion forces between F_2 molecules.

(e) NH_3:   it can form hydrogen bonds and PH_3 cannot.

11.17 **(a)** CH_4 has a lower boiling point because NH_3 is polar and can form hydrogen bonds; CH_4 is nonpolar and can only form weak attractions through dispersion forces.

(b) KCl is an ionic compound. Ion–Ion forces are much stronger than any intermolecular forces. I_2 is a nonpolar molecular substance; only weak dispersion forces are possible.

11.18 **Strategy:** Classify the species into three categories: ionic, polar (possessing a dipole moment), and nonpolar. Also look for molecules that contain an N–H, O–H, or F–H bond, which are capable of forming intermolecular hydrogen bonds. Keep in mind that dispersion forces exist between *all* species.

Solution:
(a) Water has O–H bonds. Therefore, water molecules can form hydrogen bonds. The attractive forces that must be overcome are hydrogen bonding and dispersion forces.

(b) Bromine (Br_2) molecules are nonpolar. Only dispersion forces must be overcome.

(c)   Iodine (I_2) molecules are nonpolar. Only dispersion forces must be overcome.

(d)   In this case, the F–F bond must be broken. This is an *intra*molecular force between two F atoms, not an *inter*molecular force between F_2 molecules. The attractive forces of the covalent bond must be overcome.

11.19   Both molecules are nonpolar, so the only intermolecular forces are dispersion forces. The linear structure (*n*–butane) has a higher boiling point (–0.5°C) than the branched structure (2–methylpropane, boiling point –11.7°C) because the linear form can be stacked together more easily.

11.20   The lower melting compound (shown below) can form hydrogen bonds only with itself (*intra*molecular hydrogen bonds), as shown in the figure. Such bonds do not contribute to *inter*molecular attraction and do not help raise the melting point of the compound. The other compound can form *inter*molecular hydrogen bonds; therefore, it will take a higher temperature to provide molecules of the liquid with enough kinetic energy to overcome these attractive forces to escape into the gas phase.

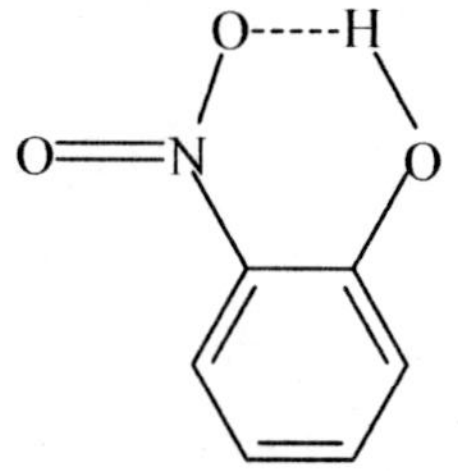

11.31   Ethanol molecules can attract each other with strong hydrogen bonds; dimethyl ether molecules cannot (why?). The surface tension of ethanol is greater than that of dimethyl ether because of stronger intermolecular forces (the hydrogen bonds). Note that ethanol and dimethyl ether have identical molar masses and molecular formulas so attractions resulting from dispersion forces will be equal.

11.32   Ethylene glycol has two –OH groups, allowing it to exert strong intermolecular forces through hydrogen bonding. Its viscosity should fall between ethanol (1 OH group) and glycerol (3 OH groups).

11.37   **(a)**   In a simple cubic structure each sphere touches **six** others on the $\pm x$, $\pm y$ and $\pm z$ axes.

(b)   In a body-centered cubic lattice each sphere touches **eight** others. Visualize the body-center sphere touching the eight corner spheres.

(c)   In a face–centered cubic lattice each sphere touches **twelve** others.

11.38   A corner sphere is shared equally among eight unit cells, so only one-eighth of each corner sphere "belongs" to any one unit cell. A face-centered sphere is divided equally between the two unit cells sharing the face. A body-centered sphere belongs entirely to its own unit cell.

In a *simple cubic cell* there are eight corner spheres. One-eighth of each belongs to the individual cell giving a total of **one** whole sphere per cell. In a *body-centered cubic cell*, there are eight corner spheres and one body-center sphere giving a total of **two** spheres per unit cell (one from the corners and one from the body-center). In a *face-center* sphere, there are eight corner spheres and six face-centered spheres (six faces). The total number of spheres would be **four**: one from the corners and three from the faces.

11.39   The mass of one cube of edge 287 pm can be found easily from the mass of one cube of edge 1.00 cm (7.87 g):

$$\frac{7.87 \text{ g Fe}}{1 \text{ cm}^3} \times \left(\frac{1 \text{ cm}}{0.01 \text{ m}}\right)^3 \times \left(\frac{1 \times 10^{-12} \text{ m}}{1 \text{ pm}}\right)^3 \times (287 \text{ pm})^3 = 1.86 \times 10^{-22} \text{ g Fe/unit cell}$$

The mass of one iron atom can be found by dividing the molar mass of iron (55.85 g) by Avogadro's number:

$$\frac{55.85 \text{ g Fe}}{1 \text{ mol Fe}} \times \frac{1 \text{ mol Fe}}{6.022 \times 10^{23} \text{ Fe atoms}} = 9.27 \times 10^{-23} \text{ g Fe/atom}$$

Converting to atoms/unit cell:

$$\frac{1 \text{ atom Fe}}{9.27 \times 10^{-23} \text{ g Fe}} \times \frac{1.86 \times 10^{-22} \text{ g Fe}}{1 \text{ unit cell}} = 2.01 \approx \textbf{2 atoms/unit cell}$$

What type of cubic cell is this?

11.40   **Strategy:** The problem gives a generous hint. First, we need to calculate the volume (in cm^3) occupied by 1 mole of Ba atoms. Next, we calculate the volume that a Ba atom occupies. Once we have these two pieces of information, we can multiply them together to end up with the number of Ba atoms per mole of Ba.

$$\frac{\text{number of Ba atoms}}{\text{cm}^3} \times \frac{\text{cm}^3}{1 \text{ mol Ba}} = \frac{\text{number of Ba atoms}}{1 \text{ mol Ba}}$$

Solution: The volume that contains one mole of barium atoms can be calculated from the density using the following strategy:

$$\frac{\text{volume}}{\text{mass of Ba}} \rightarrow \frac{\text{volume}}{\text{mol Ba}}$$

$$\frac{1 \text{ cm}^3}{3.50 \text{ g Ba}} \times \frac{137.3 \text{ g Ba}}{1 \text{ mol Ba}} = \frac{39.2 \text{ cm}^3}{1 \text{ mol Ba}}$$

Next, the volume that contains two barium atoms is the volume of the body-centered cubic unit cell. Some of this volume is empty space because packing is only 68.0 percent efficient. But, this will not affect our calculation.

$$V = a^3$$

Let's also convert to cm^3.

$$V = (502 \text{ pm})^3 \times \left(\frac{1 \times 10^{-12} \text{ m}}{1 \text{ pm}}\right)^3 \times \left(\frac{1 \text{ cm}}{0.01 \text{ m}}\right)^3 = \frac{1.27 \times 10^{-22} \text{ cm}^3}{2 \text{ Ba atoms}}$$

We can now calculate the number of barium atoms in one mole using the strategy presented above.

$$\frac{\text{number of Ba atoms}}{\text{cm}^3} \times \frac{\text{cm}^3}{1 \text{ mol Ba}} = \frac{\text{number of Ba atoms}}{1 \text{ mol Ba}}$$

$$\frac{2 \text{ Ba atoms}}{1.27 \times 10^{-22} \text{ cm}^3} \times \frac{39.2 \text{ cm}^3}{1 \text{ mol Ba}} = \textbf{6.17} \times \textbf{10}^{\textbf{23}} \textbf{ atoms/mol}$$

This is close to Avogadro's number, 6.022×10^{23} particles/mol.

11.41   In a body–centered cubic cell, there is one sphere at the cubic center and one at each of the eight corners. Each corner sphere is shared among eight adjacent unit cells. We have:

$$1 \text{ center sphere} + \left(\frac{1}{8} \times 8 \text{ corner spheres} \right) = 2 \text{ spheres per cell}$$

There are **two** vanadium atoms per unit cell.

11.42   The mass of the unit cell is the mass in grams of two europium atoms.

$$m = \frac{2 \text{ Eu atoms}}{1 \text{ unit cell}} \times \frac{1 \text{ mol Eu}}{6.022 \times 10^{23} \text{ Eu atoms}} \times \frac{152.0 \text{ g Eu}}{1 \text{ mol Eu}} = 5.048 \times 10^{-22} \text{ g Eu/unit cell}$$

$$V = \frac{5.048 \times 10^{-22} \text{ g}}{1 \text{ unit cell}} \times \frac{1 \text{ cm}^3}{5.26 \text{ g}} = 9.60 \times 10^{-23} \text{ cm}^3/\text{unit cell}$$

The edge length (a) is:

$$a = V^{1/3} = (9.60 \times 10^{-23} \text{ cm}^3)^{1/3} = 4.58 \times 10^{-8} \text{ cm} = \textbf{458 pm}$$

11.43   The volume of the unit cell is:

$$V = a^3 = (543 \text{ pm})^3 \times \left(\frac{1 \times 10^{-12} \text{ m}}{1 \text{ pm}} \right)^3 \times \left(\frac{1 \text{ cm}}{0.01 \text{ m}} \right)^3 = 1.60 \times 10^{-22} \text{ cm}^3$$

$$m = dV = \frac{2.33 \text{ g}}{1 \text{ cm}^3} \times (1.60 \times 10^{-22} \text{ cm}^3) = 3.73 \times 10^{-22} \text{ g}$$

The mass of one silicon atom is: $$\frac{28.09 \text{ g Si}}{1 \text{ mol Si}} \times \frac{1 \text{ mol Si}}{6.022 \times 10^{23} \text{ atoms Si}} = 4.665 \times 10^{-23} \text{ g/atom}$$

The number of silicon atoms in one unit cell is:

$$\frac{1 \text{ atom Si}}{4.665 \times 10^{-23} \text{ g Si}} \times \frac{3.73 \times 10^{-22} \text{ g Si}}{1 \text{ unit cell}} = \textbf{8 atoms/unit cell}$$

11.44   **Strategy:** Recall that a corner atom is shared with 8 unit cells and therefore only 1/8 of corner atom is within a given unit cell. Also recall that a face atom is shared with 2 unit cells and therefore 1/2 of a face atom is within a given unit cell. See Figure 11.19 of the text.

Solution: In a face-centered cubic unit cell, there are atoms at each of the eight corners, and there is one atom in each of the six faces. Only one-half of each face-centered atom and one-eighth of each corner atom belongs to the unit cell.

X atoms/unit cell = (8 corner atoms)(1/8 atom per corner) = 1 X atom/unit cell

Y atoms/unit cell = (6 face-centered atoms)(1/2 atom per face) = 3 Y atoms/unit cell

The unit cell is the smallest repeating unit in the crystal; therefore, the empirical formula is **XY₃**.

11.47   From Equation (11.1) of the text we can write

$$d = \frac{n\lambda}{2\sin\theta} = \frac{\lambda}{2\sin\theta} = \frac{0.090 \text{ nm} \times \dfrac{1000 \text{ pm}}{1 \text{ nm}}}{2\sin(15.2°)} = \mathbf{172 \text{ pm}}$$

11.48   Rearranging the Bragg equation, we have:

$$\lambda = \frac{2d\sin\theta}{n} = \frac{2(282 \text{ pm})(\sin 23.0°)}{1} = 220 \text{ pm} = \mathbf{0.220 \text{ nm}}$$

11.51   See Table 11.4 of the text. The properties listed are those of an **ionic solid**.

11.52   See Table 11.4 of the text. The properties listed are those of a **molecular solid**.

11.53   See Table 11.4 of the text. The properties listed are those of a **covalent solid**.

11.54   In a molecular crystal the lattice points are occupied by molecules. Of the solids listed, the ones that are composed of molecules are Se_8, HBr, CO_2, P_4O_6, and SiH_4. In covalent crystals, atoms are held together in an extensive three-dimensional network entirely by covalent bonds. Of the solids listed, the ones that are composed of atoms held together by covalent bonds are Si and C.

11.55   **(a)**   Carbon dioxide forms molecular crystals; it is a molecular compound and can only exert weak dispersion type intermolecular attractions because of its lack of polarity.

   (b)   Boron is a nonmetal with an extremely high melting point. It forms covalent crystals like carbon (diamond).

   (c)   Sulfur forms molecular crystals; it is a molecular substance (S_8) and can only exert weak dispersion type intermolecular attractions because of its lack of polarity.

   (d)   KBr forms ionic crystals because it is an ionic compound.

   (e)   Mg is a metal; it forms metallic crystals.

   (f) SiO_2 (quartz) is a hard, high melting nonmetallic compound; it forms covalent crystals like boron and C (diamond).

   (g)   LiCl is an ionic compound; it forms ionic crystals.

   (h)   Cr (chromium) is a metal and forms metallic crystals.

11.56   In diamond, each carbon atom is covalently bonded to four other carbon atoms. Because these bonds are strong and uniform, diamond is a very hard substance. In graphite, the carbon atoms in each layer are linked by strong bonds, but the layers are bound by weak dispersion forces. As a result, graphite may be cleaved easily between layers and is not hard.

11.77   The molar heat of vaporization of water is 40.79 kJ/mol. One must find the number of moles of water in the sample:

$$\text{Moles } H_2O = 74.6 \text{ g } H_2O \times \frac{1 \text{ mol } H_2O}{18.02 \text{ g } H_2O} = 4.14 \text{ mol } H_2O$$

We can then calculate the amount of heat.

$$q = 4.14 \text{ mol H}_2\text{O} \times \frac{40.79 \text{ kJ}}{1 \text{ mol H}_2\text{O}} = \textbf{169 kJ}$$

11.78   *Step 1:* Warming ice to the melting point.

$$q_1 = ms\Delta t = (866 \text{ g H}_2\text{O})(2.03 \text{ J/g°C})[0 - (-10)°\text{C}] = 17.6 \text{ kJ}$$

Step 2: Converting ice at the melting point to liquid water at 0°C. (See Table 11.8 of the text for the heat of fusion of water.)

$$q_2 = 866 \text{ g H}_2\text{O} \times \frac{1 \text{ mol}}{18.02 \text{ g H}_2\text{O}} \times \frac{6.01 \text{ kJ}}{1 \text{ mol}} = 289 \text{ kJ}$$

Step 3: Heating water from 0°C to 100°C.

$$q_3 = ms\Delta t = (866 \text{ g H}_2\text{O})(4.184 \text{ J/g°C})[(100 - 0)°\text{C}] = 362 \text{ kJ}$$

Step 4: Converting water at 100°C to steam at 100°C. (See Table 11.6 of the text for the heat of vaporization of water.)

$$q_4 = 866 \text{ g H}_2\text{O} \times \frac{1 \text{ mol}}{18.02 \text{ g H}_2\text{O}} \times \frac{40.79 \text{ kJ}}{1 \text{ mol}} = 1.96 \times 10^3 \text{ kJ}$$

Step 5: Heating steam from 100°C to 126°C.

$$q_5 = ms\Delta t = (866 \text{ g H}_2\text{O})(1.99 \text{ J/g°C})[(126 - 100)°\text{C}] = 44.8 \text{ kJ}$$

$$\textbf{q}_{\textbf{total}} = q_1 + q_2 + q_3 + q_4 + q_5 = \textbf{2.67} \times \textbf{10}^{\textbf{3}} \textbf{ kJ}$$

How would you set up and work this problem if you were computing the heat lost in cooling steam from 126°C to ice at −10°C?

11.79   **(a)**   Other factors being equal, liquids evaporate faster at higher temperatures.

   (b)   The greater the surface area, the greater the rate of evaporation.

   (c)   Weak intermolecular forces imply a high vapor pressure and rapid evaporation.

11.80 $\Delta H_{\text{vap}} = \Delta H_{\text{sub}} - \Delta H_{\text{fus}} = 62.30 \text{ kJ/mol} - 15.27 \text{ kJ/mol} = \textbf{47.03 kJ/mol}$

11.81   The substance with the lowest boiling point will have the highest vapor pressure at some particular temperature. Thus, butane will have the highest vapor pressure at −10°C and toluene the lowest.

11.82   Two phase changes occur in this process. First, the liquid is turned to solid (freezing), then the solid ice is turned to gas (sublimation).

11.83   The solid ice turns to vapor (sublimation). The temperature is too low for melting to occur.

11.84   When steam condenses to liquid water at 100°C, it releases a large amount of heat equal to the enthalpy of vaporization. Thus steam at 100°C exposes one to more heat than an equal amount of water at 100°C.

11.85   The graph is shown below:

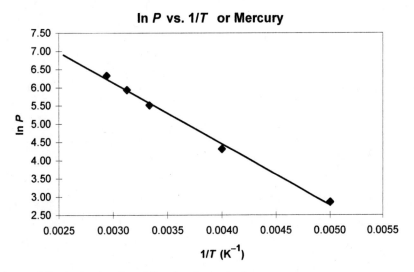

ln *P* vs. 1/*T* or Mercury

Using the first and last points to determine the slope, we have:

$$\text{Slope} = \frac{\ln P_i - \ln P_f}{\dfrac{1}{T_i} - \dfrac{1}{T_f}} = \frac{6.32 - 2.85}{(1.63 \times 10^{-3} - 2.11 \times 10^{-3})K^{-1}} = -7230 \text{ K}$$

$$-7230 \text{ K} = \frac{-\Delta H_{vap}}{R} = \frac{-\Delta H_{vap}}{8.314 \text{ J/K} \cdot \text{mol}}$$

$$\Delta H_{vap} = \textbf{60.1 \text{ kJ/mol}}$$

11.86   We can use a modified form of the Clausius-Clapeyron equation to solve this problem. See Equation (11.5) in the text.

$P_1 = 40.1$ mmHg $P_2 = ?$
$T_1 = 7.6°C = 280.6 \text{ K}$ $T_2 = 60.6°C = 333.6 \text{ K}$

$$\ln\frac{P_1}{P_2} = \frac{\Delta H_{vap}}{R}\left(\frac{1}{T_2} - \frac{1}{T_1}\right)$$

$$\ln\frac{40.1}{P_2} = \frac{31000 \text{ J/mol}}{8.314 \text{ J/K} \cdot \text{mol}}\left(\frac{1}{333.6 \text{ K}} - \frac{1}{280.6 \text{ K}}\right)$$

$$\ln\frac{40.1}{P_2} = -2.11$$

Taking the antilog of both sides, we have:

$$\frac{40.1}{P_2} = 0.121$$

$$P_2 = \textbf{331 \text{ mmHg}}$$

11.87   Application of the Clausius-Clapeyron, Equation (11.5) of the text, predicts that the more the vapor pressure rises over a temperature range, the smaller the heat of vaporization will be. Considering the equation below, if the vapor pressure change is greater, then $\dfrac{P_1}{P_2}$ is a smaller number and therefore ΔH is smaller. Thus, the molar heat of vaporization of **X** < **Y**.

$$\ln\frac{P_1}{P_2} = \frac{\Delta H_{vap}}{R}\left(\frac{1}{T_2} - \frac{1}{T_1}\right)$$

11.88   Using Equation (11.5) of the text:

$$\ln\frac{P_1}{P_2} = \frac{\Delta H_{vap}}{R}\left(\frac{1}{T_2} - \frac{1}{T_1}\right)$$

$$\ln\left(\frac{1}{2}\right) = \left(\frac{\Delta H_{vap}}{8.314 \text{ J/K}\cdot\text{mol}}\right)\left(\frac{1}{368 \text{ K}} - \frac{1}{358 \text{ K}}\right) = \Delta H_{vap}\left(\frac{-7.59 \times 10^{-5}}{8.314 \text{ J/mol}}\right)$$

$$\Delta H_{vap} = 7.59 \times 10^{4} \text{ J/mol} = \textbf{75.9 kJ/mol}$$

11.91   The pressure exerted by the blades on the ice lowers the melting point of the ice. A film of liquid water between the blades and the solid ice provides lubrication for the motion of the skater. The main mechanism for ice skating, however, is due to friction. See Chemistry in Action on p. 473 of the text.

11.92   Initially, the ice melts because of the increase in pressure. As the wire sinks into the ice, the water above the wire refreezes. Eventually the wire actually moves completely through the ice block without cutting it in half.

11.93

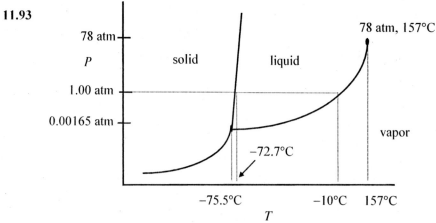

11.94   Region labels: The region containing point A is the solid region. The region containing point B is the liquid region. The region containing point C is the gas region.

(a)   Raising the temperature at constant pressure beginning at A implies starting with solid ice and warming until melting occurs. If the warming continued, the liquid water would eventually boil and change to steam. Further warming would increase the temperature of the steam.

(b)   At point C water is in the gas phase. Cooling without changing the pressure would eventually result in the formation of solid ice. Liquid water would never form.

(c)   At B the water is in the liquid phase. Lowering the pressure without changing the temperature would eventually result in boiling and conversion to water in the gas phase.

11.95 **(a)** Boiling liquid ammonia requires breaking hydrogen bonds between molecules. Dipole–dipole and dispersion forces must also be overcome.

(b) P_4 is a nonpolar molecule, so the only intermolecular forces are of the dispersion type.

(c) CsI is an ionic solid. To dissolve in any solvent ion–ion interparticle forces must be overcome.

(d) Metallic bonds must be broken.

11.96 **(a)** A low surface tension means the attraction between molecules making up the surface is weak. Water has a high surface tension; water bugs could not "walk" on the surface of a liquid with a low surface tension.

(b) A low critical temperature means a gas is very difficult to liquefy by cooling. This is the result of weak intermolecular attractions. Helium has the lowest known critical temperature (5.3 K).

(c) A low boiling point means weak intermolecular attractions. It takes little energy to separate the particles. All ionic compounds have extremely high boiling points.

(d) A low vapor pressure means it is difficult to remove molecules from the liquid phase because of high intermolecular attractions. Substances with low vapor pressures have high boiling points (why?).

Thus, only choice **(d)** indicates strong intermolecular forces in a liquid. The other choices indicate weak intermolecular forces in a liquid.

11.97 The HF molecules are held together by strong intermolecular hydrogen bonds. Therefore, liquid HF has a lower vapor pressure than liquid HI. (The HI molecules do not form hydrogen bonds with each other.)

11.98 The properties of hardness, high melting point, poor conductivity, and so on, could place boron in either the ionic or covalent categories. However, boron atoms will not alternately form positive and negative ions to achieve an ionic crystal. The structure is **covalent** because the units are single boron atoms.

11.99 Reading directly from the graph: **(a)** solid; **(b)** vapor.

11.100 **CCl_4.** Generally, the larger the number of electrons and the more diffuse the electron cloud in an atom or a molecule, the greater its polarizability. Recall that polarizability is the ease with which the electron distribution in an atom or molecule can be distorted.

11.101 Because the critical temperature of CO_2 is only 31°C, the liquid CO_2 in the fire extinguisher vaporizes above this temperature, no matter the applied pressure inside the extinguisher. 31°C is approximately 88°F, so on a hot summer day, no liquid CO_2 will exist inside the extinguisher, and hence no sloshing sound would be heard.

11.102 The vapor pressure of mercury (as well as all other substances) is 760 mmHg at its normal boiling point.

11.103 As the vacuum pump is turned on and the pressure is reduced, the liquid will begin to boil because the vapor pressure of the liquid is greater than the external pressure (approximately zero). The heat of vaporization is supplied by the water, and thus the water cools. Soon the water loses sufficient heat to drop the temperature below the freezing point. Finally the ice sublimes under reduced pressure.

11.104 It has reached the critical point; the point of critical temperature (T_c) and critical pressure (P_c).

11.105   The graph is shown below. See Table 9.1 of the text for the lattice energies.

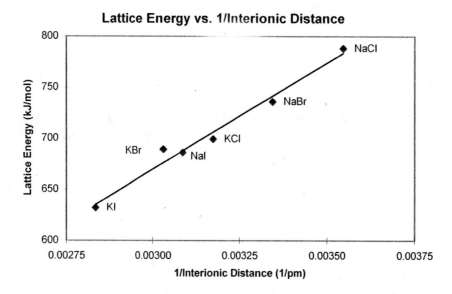

This plot is fairly linear. The energy required to separate two opposite charges is given by:

$$E = k\frac{Q_+Q_-}{r}$$

As the separation increases, less work is needed to pull the ions apart; therefore, the lattice energies become smaller as the interionic distances become larger. This is in accordance with Coulomb's law.

From these data what can you conclude about the relationship between lattice energy and the size of the negative ion? What about lattice energy versus positive ion size (compare KCl with NaCl, KBr with NaBr, etc.)?

11.106   Crystalline SiO_2. Its regular structure results in a more efficient packing.

11.107   W must be a reasonably non-reactive metal. It conducts electricity and is malleable, but doesn't react with nitric acid. Of the choices, it must be gold.

X is nonconducting (and therefore isn't a metal), is brittle, is high melting, and reacts with nitric acid. Of the choices, it must be lead sulfide.

Y doesn't conduct and is soft (and therefore is a nonmetal). It melts at a low temperature with sublimation. Of the choices, it must be iodine.

Z doesn't conduct, is chemically inert, and is high melting (network solid). Of the choices, it must be mica (SiO_2).

Would the colors of the species have been any help in determining their identity?

11.108   **(a)**   **False.** Permanent dipoles are usually much stronger than temporary dipoles.

(b)   **False.** The hydrogen atom must be bonded to N, O, or F.

(c)   **True.**

(d)   **False.** The magnitude of the attraction depends on both the ion charge and the polarizability of the neutral atom or molecule.

11.109 A: Steam   B: Water vapor.

(Most people would call the mist "steam". Steam is invisible.)

11.110 Sublimation temperature is −78°C or 195 K at a pressure of 1 atm.

$$\ln \frac{P_1}{P_2} = \frac{\Delta H_{sub}}{R} \left(\frac{1}{T_2} - \frac{1}{T_1} \right)$$

$$\ln \frac{1}{P_2} = \frac{25.9 \times 10^3 \text{ J/mol}}{8.314 \text{ J/mol} \cdot \text{K}} \left(\frac{1}{150 \text{ K}} - \frac{1}{195 \text{ K}} \right)$$

$$\ln \frac{1}{P_2} = 4.79$$

Taking the antilog of both sides gives:

$$P_2 = 8.3 \times 10^{-3} \text{ atm}$$

11.111 **(a)** The average separation between particles decreases from gases to liquids to solids, so the ease of compressibility decreases in the same order.

(b) In solids, the molecules or atoms are usually locked in a rigid 3-dimensional structure which determines the shape of the crystal. In liquids and gases the particles are free to move relative to each other.

(c) The trend in volume is due to the same effect as part (a).

11.112 **(a)** K_2S: Ionic forces are much stronger than the dipole-dipole forces in $(CH_3)_3N$.

(b) Br_2: Both molecules are nonpolar; but Br_2 has more electrons. (The boiling point of Br_2 is 50°C and that of C_4H_{10} is −0.5°C.)

11.113 Oil is made up of nonpolar molecules and therefore does not mix with water. To minimize contact, the oil drop assumes a spherical shape. (For a given volume the sphere has the smallest surface area.)

11.114 CH_4 is a tetrahedral, nonpolar molecule that can only exert weak dispersion type attractive forces. SO_2 is bent (why?) and possesses a dipole moment, which gives rise to stronger dipole-dipole attractions. Sulfur dioxide will have a larger value of "a" in the van der Waals equation (a is a measure of the strength of the interparticle attraction) and will behave less like an ideal gas than methane.

11.115 LiF, ionic bonding and dispersion forces; BeF_2, ionic bonding and dispersion forces; BF_3, dispersion forces; CF_4, dispersion forces; NF_3, dipole-dipole interaction and dispersion forces; OF_2, dipole-dipole interaction and dispersion forces; F_2, dispersion forces.

11.116 The standard enthalpy change for the formation of gaseous iodine from solid iodine is simply the difference between the standard enthalpies of formation of the products and the reactants in the equation:

$$I_2(s) \rightarrow I_2(g)$$

$$\Delta H_{vap} = \Delta H_f^\circ[I_2(g)] - \Delta H_f^\circ[I_2(s)] = 62.4 \text{ kJ/mol} - 0 \text{ kJ/mol} = \textbf{62.4 kJ/mol}$$

11.117 The Li-Cl bond length is longer in the solid phase because each Li^+ is shared among several Cl^- ions. In the gas phase the ion pairs (Li^+ and Cl^-) tend to get as close as possible for maximum net attraction.

11.118 Smaller ions have more concentrated charges (charge densities) and are more effective in ion-dipole interaction. The greater the ion-dipole interaction, the larger is the heat of hydration.

11.119 (a) If water were linear, the two O–H bond dipoles would cancel each other as in CO_2. Thus a linear water molecule would not be polar.

(b) Hydrogen bonding would still occur between water molecules even if they were linear.

11.120 (a) For the process: $Br_2(l) \rightarrow Br_2(g)$

$\Delta H° = \Delta H_f°[Br_2(g)] - \Delta H_f°[Br_2(l)] = (1)(30.7 \text{ kJ/mol}) - 0 = \mathbf{30.7 \text{ kJ/mol}}$

(b) For the process: $Br_2(g) \rightarrow 2Br(g)$

$\Delta H° = \mathbf{192.5 \text{ kJ/mol}}$ (from Table 9.4 of the text)

As expected, the bond energy represented in part (b) is much greater than the energy of vaporization represented in part (a). It requires more energy to break the bond than to vaporize the molecule.

11.121 Water molecules can attract each other with strong hydrogen bonds; diethyl ether molecules cannot (why?). The surface tension of water is greater than that of diethyl ether because of stronger intermolecular forces (the hydrogen bonds).

11.122 (a) Decreases   **(b)** No change   **(c)** No change

11.123 $3Hg(l) + O_3(g) \rightarrow 3HgO(s)$

Conversion to solid HgO changes its surface tension.

11.124 $CaCO_3(s) \rightarrow CaO(s) + CO_2(g)$

Three phases (two solid and one gas). $CaCO_3$ and CaO constitute two separate solid phases because they are separated by well-defined boundaries.

11.125 (a) To calculate the boiling point of trichlorosilane, we rearrange Equation (11.5) of the text to get:

$$\frac{1}{T_2} = \frac{R}{\Delta H_{vap}} \ln \frac{P_1}{P_2} + \frac{1}{T_1}$$

where T_2 is the normal boiling point of trichlorosilane. Setting $P_1 = 0.258$ atm, $T_1 = (-2 + 273)K = 271$ K, $P_2 = 1.00$ atm, we write:

$$\frac{1}{T_2} = \frac{8.314 \text{ J/K} \cdot \text{mol}}{28.8 \times 10^3 \text{ J/mol}} \ln \frac{0.258}{1.00} + \frac{1}{271 \text{ K}}$$

$T_2 = \mathbf{303 \text{ K}} = \mathbf{30°C}$

The Si atom is sp^3-hybridized and the $SiCl_3H$ molecule has a tetrahedral geometry and a dipole moment. Thus, trichlorosilane is polar and the predominant forces among its molecules are dipole-dipole forces. Since dipole-dipole forces are normally fairly weak, we expect trichlorosilane to have a low boiling point, which is consistent with the calculated value of 30°C.

(b) From Section 11.6 of the text, we see that SiO_2 forms a covalent crystal. Silicon, like carbon in Group 4A, also forms a covalent crystal. The strong covalent bonds between Si atoms (in silicon) and between Si and O atoms (in quartz) account for their high melting points and boiling points.

(c) To test the 10^{-9} purity requirement, we need to calculate the number of Si atoms in 1 cm^3. We can arrive at the answer by carrying out the following three steps: (1) Determine the volume of an Si unit cell in cubic centimeters, (2) determine the number of Si unit cells in 1 cm^3, and (3) multiply the number of unit cells in 1 cm^3 by 8, the number of Si atoms in a unit cell.

Step 1: The volume of the unit cell, V, is

$$V = a^3$$

$$V = (543 \text{ pm})^3 \times \left(\frac{1 \times 10^{-12} \text{ m}}{1 \text{ pm}} \times \frac{1 \text{ cm}}{1 \times 10^{-2} \text{ m}} \right)^3$$

$$V = 1.60 \times 10^{-22} \text{ cm}^3$$

Step 2: The number of cells per cubic centimeter is given by:

$$\text{number of unit cells} = 1 \text{ cm}^3 \times \frac{1 \text{ unit cell}}{1.60 \times 10^{-22} \text{ cm}^3} = 6.25 \times 10^{21} \text{ unit cells}$$

Step 3: Since there are 8 Si atoms per unit cell, the total number of Si atoms is:

$$\text{number of Si atoms} = \frac{8 \text{ Si atoms}}{1 \text{ unit cell}} \times (6.25 \times 10^{21} \text{ unit cells}) = 5.00 \times 10^{22} \text{ Si atoms}$$

Finally, to calculate the purity of the Si crystal, we write:

$$\frac{\text{B atoms}}{\text{Si atoms}} = \frac{1.0 \times 10^{13} \text{ B atoms}}{5.00 \times 10^{22} \text{ Si atoms}} = \mathbf{2.0 \times 10^{-10}}$$

Since this number is smaller than 10^{-9}, the purity requirement is satisfied.

11.126 SiO_2 has an extensive three-dimensional structure. CO_2 exists as discrete molecules. It will take much more energy to break the strong network covalent bonds of SiO_2; therefore, SiO_2 has a much higher boiling point than CO_2.

11.127 The pressure inside the cooker increases and so does the boiling point of water.

11.128 The moles of water vapor can be calculated using the ideal gas equation.

$$n = \frac{PV}{RT} = \frac{\left(187.5 \text{ mmHg} \times \dfrac{1 \text{ atm}}{760 \text{ mmHg}}\right)(5.00 \text{ L})}{\left(0.0821 \dfrac{\text{L} \cdot \text{atm}}{\text{mol} \cdot \text{K}}\right)(338 \text{ K})} = 0.0445 \text{ mol}$$

mass of water vapor $= 0.0445 \text{ mol} \times 18.02 \text{ g/mol} = 0.802 \text{ g}$

Now, we can calculate the percentage of the 1.20 g sample of water that is vapor.

$$\textbf{\% of H}_2\textbf{O vaporized} = \frac{0.802 \text{ g}}{1.20 \text{ g}} \times 100\% = \textbf{66.8\%}$$

11.129 **(a)** Extra heat produced when steam condenses at 100°C.

(b) Avoids extraction of ingredients by boiling in water.

11.130 The packing efficiency is: $\dfrac{\text{volume of atoms in unit cell}}{\text{volume of unit cell}} \times 100\%$

An atom is assumed to be spherical, so the volume of an atom is $(4/3)\pi r^3$. The volume of a cubic unit cell is a^3 (a is the length of the cube edge). The packing efficiencies are calculated below:

(a) Simple cubic cell: cell edge $(a) = 2r$

$$\text{Packing efficiency} = \frac{\left(\dfrac{4\pi r^3}{3}\right) \times 100\%}{(2r)^3} = \frac{4\pi r^3 \times 100\%}{24r^3} = \frac{\pi}{6} \times 100\% = \textbf{52.4\%}$$

(b) Body-centered cubic cell: cell edge $= \dfrac{4r}{\sqrt{3}}$

$$\text{Packing efficiency} = \frac{2 \times \left(\dfrac{4\pi r^3}{3}\right) \times 100\%}{\left(\dfrac{4r}{\sqrt{3}}\right)^3} = \frac{2 \times \left(\dfrac{4\pi r^3}{3}\right) \times 100\%}{\left(\dfrac{64r^3}{3\sqrt{3}}\right)} = \frac{2\pi\sqrt{3}}{16} \times 100\% = \textbf{68.0\%}$$

Remember, there are two atoms per body-centered cubic unit cell.

(c) Face-centered cubic cell: cell edge $= \sqrt{8}r$

$$\text{Packing efficiency} = \frac{4 \times \left(\dfrac{4\pi r^3}{3}\right) \times 100\%}{\left(\sqrt{8}r\right)^3} = \frac{\left(\dfrac{16\pi r^3}{3}\right) \times 100\%}{8r^3\sqrt{8}} = \frac{2\pi}{3\sqrt{8}} \times 100\% = \textbf{74.0\%}$$

Remember, there are four atoms per face-centered cubic unit cell.

11.131 **(a)** Pumping allows Ar atoms to escape, thus removing heat from the liquid phase. Eventually the liquid freezes.

(b) The solid-liquid line of cyclohexane is positive. Therefore, its melting point increases with pressure.

(c) These droplets are super-cooled liquids.

(d) When the dry ice is added to water, it sublimes. The cold CO_2 gas generated causes nearby water vapor to condense, hence the appearance of fog.

11.132 For a face-centered cubic unit cell, the length of an edge (a) is given by:

$$a = \sqrt{8}r$$

$$a = \sqrt{8}\,(191\,\text{pm}) = 5.40 \times 10^2\,\text{pm}$$

The volume of a cube equals the edge length cubed (a^3).

$$V = a^3 = (5.40 \times 10^2\,\text{pm})^3 \times \left(\frac{1 \times 10^{-12}\,\text{m}}{1\,\text{pm}}\right)^3 \times \left(\frac{1\,\text{cm}}{1 \times 10^{-2}\,\text{m}}\right)^3 = 1.57 \times 10^{-22}\,\text{cm}^3$$

Now that we have the volume of the unit cell, we need to calculate the mass of the unit cell in order to calculate the density of Ar. The number of atoms in one face centered cubic unit cell is four.

$$m = \frac{4\,\text{atoms}}{1\,\text{unit cell}} \times \frac{1\,\text{mol}}{6.022 \times 10^{23}\,\text{atoms}} \times \frac{39.95\,\text{g}}{1\,\text{mol}} = \frac{2.65 \times 10^{-22}\,\text{g}}{1\,\text{unit cell}}$$

$$d = \frac{m}{V} = \frac{2.65 \times 10^{-22}\,\text{g}}{1.57 \times 10^{-22}\,\text{cm}^3} = \textbf{1.69 g/cm}^\textbf{3}$$

11.133 The ice condenses the water vapor inside. Since the water is still hot, it will begin to boil at reduced pressure. (Be sure to drive out as much air in the beginning as possible.)

11.134 **(a)** Two triple points: Diamond/graphite/liquid and graphite/liquid/vapor.

(b) Diamond.

(c) Apply high pressure at high temperature.

11.135 Ethanol mixes well with water. The mixture has a lower surface tension and readily flows out of the ear channel.

11.136 The cane is made of many molecules held together by intermolecular forces. The forces are strong and the molecules are packed tightly. Thus, when the handle is raised, all the molecules are raised because they are held together.

11.137 The two main reasons for spraying the trees with water are:

1) As water freezes, heat is released.

$$H_2O(l) \rightarrow H_2O(s) \qquad \Delta H_{\text{fus}} = -6.01\,\text{kJ/mol}$$

The heat released protects the fruit. Of course, spraying the trees with warm water is even more helpful.

2)   The ice forms an insulating layer to protect the fruit.

11.138   When the tungsten filament inside the bulb is heated to a high temperature (about 3000°C), the tungsten sublimes (solid → gas phase transition) and then it condenses on the inside walls of the bulb. The inert, pressurized Ar gas retards sublimation and oxidation of the tungsten filament.

11.139

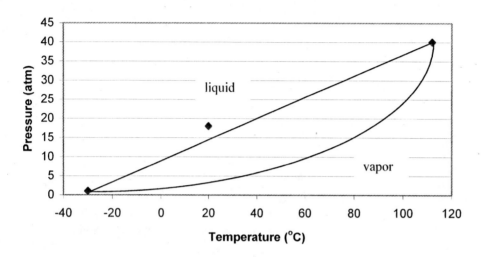

Plotting the three points, and connecting the boiling point to the critical point with both a straight line and a curved line, we see that the point (20°C, 18 atm) lies on the liquid side of the phase boundary. The gas will condense under these conditions. The curved line better represents the liquid/vapor boundary for a typical phase diagram. See Figures 11.40 and 11.41 of the text.

11.140   The fuel source for the Bunsen burner is most likely methane gas. When methane burns in air, carbon dioxide and water are produced.

$$CH_4(g) + 2O_2(g) \rightarrow CO_2(g) + 2H_2O(g)$$

The water vapor produced during the combustion condenses to liquid water when it comes in contact with the outside of the cold beaker.

11.141   The original diagram shows that as heat is supplied to the water, its temperature rises. At the boiling point (represented by the horizontal line), water is converted to steam. Beyond this point the temperature of the steam rises above 100°C.

Choice (a) is eliminated because it shows no change from the original diagram even though the mass of water is doubled.

Choice (b) is eliminated because the rate of heating is greater than that for the original system. Also, it shows water boiling at a higher temperature, which is not possible.

Choice (c) is eliminated because it shows that water now boils at a temperature below 100°C, which is not possible.

Choice **(d)** therefore represents what actually happens. The heat supplied is enough to bring the water to its boiling point, but not raise the temperature of the steam.

CHAPTER 12
PHYSICAL PROPERTIES OF SOLUTIONS

12.9 CsF is an ionic solid; the ion–ion attractions are too strong to be overcome in the dissolving process in benzene. The ion–induced dipole interaction is too weak to stabilize the ion. Nonpolar naphthalene molecules form a molecular solid in which the only interparticle forces are of the weak dispersion type. The same forces operate in liquid benzene causing naphthalene to dissolve with relative ease. Like dissolves like.

12.10 **Strategy:** In predicting solubility, remember the saying: Like dissolves like. A nonpolar solute will dissolve in a nonpolar solvent; ionic compounds will generally dissolve in polar solvents due to favorable ion-dipole interactions; solutes that can form hydrogen bonds with a solvent will have high solubility in the solvent.

Solution: Strong hydrogen bonding (dipole-dipole attraction) is the principal intermolecular attraction in liquid ethanol, but in liquid cyclohexane the intermolecular forces are dispersion forces because cyclohexane is nonpolar. Cyclohexane cannot form hydrogen bonds with ethanol, and therefore cannot attract ethanol molecules strongly enough to form a solution.

12.11 The order of increasing solubility is: $O_2 < Br_2 < LiCl < CH_3OH$. Methanol is miscible with water because of strong hydrogen bonding. LiCl is an ionic solid and is very soluble because of the high polarity of the water molecules. Both oxygen and bromine are nonpolar and exert only weak dispersion forces. Bromine is a larger molecule and is therefore more polarizable and susceptible to dipole–induced dipole attractions.

12.12 The longer the C–C chain, the more the molecule "looks like" a hydrocarbon and the less important the –OH group becomes. Hence, as the C–C chain length increases, the molecule becomes less polar. Since "like dissolves like", as the molecules become more nonpolar, the solubility in polar water decreases. The –OH group of the alcohols can form strong hydrogen bonds with water molecules, but this property decreases as the chain length increases.

12.15 Percent mass equals the mass of solute divided by the mass of the solution (that is, solute plus solvent) times 100 (to convert to percentage).

(a) $\dfrac{5.50 \text{ g NaBr}}{78.2 \text{ g soln}} \times 100\% = \textbf{7.03\%}$

(b) $\dfrac{31.0 \text{ g KCl}}{(31.0 + 152) \text{ g soln}} \times 100\% = \textbf{16.9\%}$

(c) $\dfrac{4.5 \text{ g toluene}}{(4.5 + 29) \text{ g soln}} \times 100\% = \textbf{13\%}$

12.16 **Strategy:** We are given the percent by mass of the solute and the mass of the solute. We can use Equation (12.1) of the text to solve for the mass of the solvent (water).

Solution:

(a)   The percent by mass is defined as

$$\text{percent by mass of solute} = \frac{\text{mass of solute}}{\text{mass of solute} + \text{mass of solvent}} \times 100\%$$

Substituting in the percent by mass of solute and the mass of solute, we can solve for the mass of solvent (water).

$$16.2\% = \frac{5.00 \text{ g urea}}{5.00 \text{ g urea} + \text{mass of water}} \times 100\%$$

$$(0.162)(\text{mass of water}) = 5.00 \text{ g} - (0.162)(5.00\text{g})$$

mass of water = 25.9 g

(b)   Similar to part (a),

$$1.5\% = \frac{26.2 \text{ g MgCl}_2}{26.2 \text{ g MgCl}_2 + \text{mass of water}} \times 100\%$$

mass of water = 1.72×10^3 g

12.17   (a)   The molality is the number of moles of sucrose (molar mass 342.3 g/mol) divided by the mass of the solvent (water) in kg.

$$\text{mol sucrose} = 14.3 \text{ g sucrose} \times \frac{1 \text{ mol}}{342.3 \text{ g sucrose}} = 0.0418 \text{ mol}$$

$$\textbf{Molality} = \frac{0.0418 \text{ mol sucrose}}{0.676 \text{ kg H}_2\text{O}} = \textbf{0.0618 } \textit{m}$$

(b)   **Molality** $= \dfrac{7.20 \text{ mol ethylene glycol}}{3.546 \text{ kg H}_2\text{O}} = \textbf{2.03 } \textit{m}$

12.18 $\text{molality} = \dfrac{\text{moles of solute}}{\text{mass of solvent (kg)}}$

(a) $\text{mass of 1 L soln} = 1000 \text{ mL} \times \dfrac{1.08 \text{ g}}{1 \text{ mL}} = 1080 \text{ g}$

$$\text{mass of water} = 1080 \text{ g} - \left(2.50 \text{ mol NaCl} \times \frac{58.44 \text{ g NaCl}}{1 \text{ mol NaCl}} \right) = 934 \text{ g} = 0.934 \text{ kg}$$

$$\textit{m} = \frac{2.50 \text{ mol NaCl}}{0.934 \text{ kg H}_2\text{O}} = \textbf{2.68 } \textit{m}$$

(b)   100 g of the solution contains 48.2 g KBr and 51.8 g H$_2$O.

$$\text{mol of KBr} = 48.2 \text{ g KBr} \times \frac{1 \text{ mol KBr}}{119.0 \text{ g KBr}} = 0.405 \text{ mol KBr}$$

$$\text{mass of } H_2O \text{ (in kg)} = 51.8 \text{ g } H_2O \times \frac{1 \text{ kg}}{1000 \text{ g}} = 0.0518 \text{ kg } H_2O$$

$$m = \frac{0.405 \text{ mol KBr}}{0.0518 \text{ kg } H_2O} = \mathbf{7.82 \text{ } m}$$

12.19 In each case we consider one liter of solution. mass of solution = volume × density

(a) $$\text{mass of sugar} = 1.22 \text{ mol sugar} \times \frac{342.3 \text{ g sugar}}{1 \text{ mol sugar}} = 418 \text{ g sugar} \times \frac{1 \text{ kg}}{1000 \text{ g}} = 0.418 \text{ kg sugar}$$

$$\text{mass of soln} = 1000 \text{ mL} \times \frac{1.12 \text{ g}}{1 \text{ mL}} = 1120 \text{ g} \times \frac{1 \text{ kg}}{1000 \text{ g}} = 1.120 \text{ kg}$$

$$\text{molality} = \frac{1.22 \text{ mol sugar}}{(1.120 - 0.418) \text{kg } H_2O} = \mathbf{1.74 \text{ } m}$$

(b) $$\text{mass of NaOH} = 0.87 \text{ mol NaOH} \times \frac{40.00 \text{ g NaOH}}{1 \text{ mol NaOH}} = 35 \text{ g NaOH}$$

$$\text{mass solvent } (H_2O) = 1040 \text{ g} - 35 \text{ g} = 1005 \text{ g} = 1.005 \text{ kg}$$

$$\text{molality} = \frac{0.87 \text{ mol NaOH}}{1.005 \text{ kg } H_2O} = \mathbf{0.87 \text{ } m}$$

(c) $$\text{mass of NaHCO}_3 = 5.24 \text{ mol NaHCO}_3 \times \frac{84.01 \text{ g NaHCO}_3}{1 \text{ mol NaHCO}_3} = 440 \text{ g NaHCO}_3$$

$$\text{mass solvent } (H_2O) = 1190 \text{ g} - 440 \text{ g} = 750 \text{ g} = 0.750 \text{ kg}$$

$$\text{molality} = \frac{5.24 \text{ mol NaHCO}_3}{0.750 \text{ kg } H_2O} = \mathbf{6.99 \text{ } m}$$

12.20 Let's assume that we have 1.0 L of a 0.010 M solution.

Assuming a solution density of 1.0 g/mL, the mass of 1.0 L (1000 mL) of the solution is 1000 g or 1.0×10^3 g.

The mass of 0.010 mole of urea is:

$$0.010 \text{ mol urea} \times \frac{60.06 \text{ g urea}}{1 \text{ mol urea}} = 0.60 \text{ g urea}$$

The mass of the solvent is:

$$\text{(solution mass)} - \text{(solute mass)} = (1.0 \times 10^3 \text{ g}) - (0.60 \text{ g}) = 1.0 \times 10^3 \text{ g} = 1.0 \text{ kg}$$

$$m = \frac{\text{moles solute}}{\text{mass solvent}} = \frac{0.010 \text{ mol}}{1.0 \text{ kg}} = \mathbf{0.010 \text{ } m}$$

12.21   We find the volume of ethanol in 1.00 L of 75 proof gin. Note that 75 proof means $\left(\dfrac{75}{2}\right)\%$.

$$\text{Volume} = 1.00 \text{ L} \times \left(\frac{75}{2}\right)\% = 0.38 \text{ L} = 3.8 \times 10^2 \text{ mL}$$

$$\textbf{Ethanol mass} = (3.8 \times 10^2 \text{ mL}) \times \frac{0.798 \text{ g}}{1 \text{ mL}} = \textbf{3.0} \times \textbf{10}^2 \textbf{ g}$$

12.22   **(a)**   Converting mass percent to molality.

Strategy: In solving this type of problem, it is convenient to assume that we start with 100.0 grams of the solution. If the mass of sulfuric acid is 98.0% of 100.0 g, or 98.0 g, the percent by mass of water must be 100.0% – 98.0% = 2.0%. The mass of water in 100.0 g of solution would be 2.0 g. From the definition of molality, we need to find moles of solute (sulfuric acid) and kilograms of solvent (water).

Solution: Since the definition of molality is

$$\text{molality} = \frac{\text{moles of solute}}{\text{mass of solvent (kg)}}$$

we first convert 98.0 g H_2SO_4 to moles of H_2SO_4 using its molar mass, then we convert 2.0 g of H_2O to units of kilograms.

$$98.0 \text{ g } H_2SO_4 \times \frac{1 \text{ mol } H_2SO_4}{98.09 \text{ g } H_2SO_4} = 0.999 \text{ mol } H_2SO_4$$

$$2.0 \text{ g } H_2O \times \frac{1 \text{ kg}}{1000 \text{ g}} = 2.0 \times 10^{-3} \text{ kg } H_2O$$

Lastly, we divide moles of solute by mass of solvent in kg to calculate the molality of the solution.

$$m = \frac{\text{mol of solute}}{\text{kg of solvent}} = \frac{0.999 \text{ mol}}{2.0 \times 10^{-3} \text{ kg}} = \textbf{5.0} \times \textbf{10}^2 \textbf{ m}$$

(b)   Converting molality to molarity.

Strategy: From part (a), we know the moles of solute (0.999 mole H_2SO_4) and the mass of the solution (100.0 g). To solve for molarity, we need the volume of the solution, which we can calculate from its mass and density.

Solution: First, we use the solution density as a conversion factor to convert to volume of solution.

$$? \text{ volume of solution} = 100.0 \text{ g} \times \frac{1 \text{ mL}}{1.83 \text{ g}} = 54.6 \text{ mL} = 0.0546 \text{ L}$$

Since we already know moles of solute from part (a), 0.999 mole H_2SO_4, we divide moles of solute by liters of solution to calculate the molarity of the solution.

$$M = \frac{\text{mol of solute}}{\text{L of soln}} = \frac{0.999 \text{ mol}}{0.0546 \text{ L}} = \textbf{18.3 } M$$

12.23 $mol\ NH_3\ =\ 30.0\ g\ NH_3 \times \dfrac{1\ mol\ NH_3}{17.03\ g\ NH_3}\ =\ 1.76\ mol\ NH_3$

Volume of the solution $=\ 100.0\ g\ soln \times \dfrac{1\ mL}{0.982\ g} \times \dfrac{1\ L}{1000\ mL}\ =\ 0.102\ L$

molarity $=\ \dfrac{1.76\ mol\ NH_3}{0.102\ L\ soln}\ =\ \textbf{17.3\ }\textbf{\textit{M}}$

kg of solvent (H_2O) $=\ 70.0\ g\ H_2O \times \dfrac{1\ kg}{1000\ g}\ =\ 0.0700\ kg\ H_2O$

molality $=\ \dfrac{1.76\ mol\ NH_3}{0.0700\ kg\ H_2O}\ =\ \textbf{25.1\ }\textbf{\textit{m}}$

12.24   Assume 100.0 g of solution.

(a)   The mass of ethanol in the solution is $0.100 \times 100.0\ g = 10.0\ g$. The mass of the water is $100.0\ g - 10.0\ g = 90.0\ g = 0.0900\ kg$. The amount of ethanol in moles is:

$$10.0\ g\ ethanol \times \dfrac{1\ mol}{46.07\ g}\ =\ 0.217\ mol\ ethanol$$

$$m\ =\ \dfrac{mol\ solute}{kg\ solvent}\ =\ \dfrac{0.217\ mol}{0.0900\ kg}\ =\ \textbf{2.41\ }\textbf{\textit{m}}$$

(b)   The volume of the solution is:

$$100.0\ g \times \dfrac{1\ mL}{0.984\ g}\ =\ 102\ mL\ =\ 0.102\ L$$

The amount of ethanol in moles is 0.217 mole [part (a)].

$$M\ =\ \dfrac{mol\ solute}{liters\ of\ soln}\ =\ \dfrac{0.217\ mol}{0.102\ L}\ =\ \textbf{2.13\ }\textbf{\textit{M}}$$

(c)   **Solution volume** $=\ 0.125\ mol \times \dfrac{1\ L}{2.13\ mol}\ =\ \textbf{0.0587\ L}\ =\ \textbf{58.7\ mL}$

12.27   The amount of salt dissolved in 100 g of water is:

$$\dfrac{3.20\ g\ salt}{9.10\ g\ H_2O} \times 100\ g\ H_2O\ =\ 35.2\ g\ salt$$

Therefore, the solubility of the salt is **35.2 g salt/100 g H_2O**.

12.28   At 75°C, 155 g of KNO_3 dissolves in 100 g of water to form 255 g of solution. When cooled to 25°C, only 38.0 g of KNO_3 remain dissolved. This means that $(155 - 38.0)\ g = 117\ g$ of KNO_3 will crystallize.

The amount of KNO_3 formed when 100 g of saturated solution at 75°C is cooled to 25°C can be found by a simple unit conversion.

$$100 \text{ g saturated soln} \times \frac{117 \text{ g } KNO_3 \text{ crystallized}}{255 \text{ g saturated soln}} = \textbf{45.9 g } KNO_3$$

12.29 The mass of KCl is 10% of the mass of the whole sample or 5.0 g. The $KClO_3$ mass is 45 g. If 100 g of water will dissolve 25.5 g of KCl, then the amount of water to dissolve 5.0 g KCl is:

$$5.0 \text{ g KCl} \times \frac{100 \text{ g } H_2O}{25.5 \text{ g KCl}} = 20 \text{ g } H_2O$$

The 20 g of water will dissolve:

$$20 \text{ g } H_2O \times \frac{7.1 \text{ g } KClO_3}{100 \text{ g } H_2O} = 1.4 \text{ g } KClO_3$$

The $KClO_3$ remaining undissolved will be:

$$(45 - 1.4) \text{ g } KClO_3 = \textbf{44 g } KClO_3$$

12.35 When a dissolved gas is in dynamic equilibrium with its surroundings, the number of gas molecules entering the solution (dissolving) is equal to the number of dissolved gas molecules leaving and entering the gas phase. When the surrounding air is replaced by helium, the number of air molecules leaving the solution is greater than the number dissolving. As time passes the concentration of dissolved air becomes very small or zero, and the concentration of dissolved helium increases to a maximum.

12.36 According to Henry's law, the solubility of a gas in a liquid increases as the pressure increases ($c = kP$). The soft drink tastes flat at the bottom of the mine because the carbon dioxide pressure is greater and the dissolved gas is not released from the solution. As the miner goes up in the elevator, the atmospheric carbon dioxide pressure decreases and dissolved gas is released from his stomach.

12.37 We first find the value of k for Henry's law

$$k = \frac{c}{P} = \frac{0.034 \text{ mol/L}}{1 \text{ atm}} = 0.034 \text{ mol/L} \cdot \text{atm}$$

For atmospheric conditions we write:

$$c = kP = (0.034 \text{ mol/L} \cdot \text{atm})(0.00030 \text{ atm}) = \textbf{1.0} \times \textbf{10}^{-5} \textbf{ mol/L}$$

12.38 **Strategy:** The given solubility allows us to calculate Henry's law constant (k), which can then be used to determine the concentration of N_2 at 4.0 atm. We can then compare the solubilities of N_2 in blood under normal pressure (0.80 atm) and under a greater pressure that a deep-sea diver might experience (4.0 atm) to determine the moles of N_2 released when the diver returns to the surface. From the moles of N_2 released, we can calculate the volume of N_2 released.

Solution: First, calculate the Henry's law constant, k, using the concentration of N_2 in blood at 0.80 atm.

$$k = \frac{c}{P}$$

$$k = \frac{5.6 \times 10^{-4} \text{ mol/L}}{0.80 \text{ atm}} = 7.0 \times 10^{-4} \text{ mol/L} \cdot \text{atm}$$

Next, we can calculate the concentration of N_2 in blood at 4.0 atm using k calculated above.

$$c = kP$$

$$c = (7.0 \times 10^{-4} \text{ mol/L} \cdot \text{atm})(4.0 \text{ atm}) = 2.8 \times 10^{-3} \text{ mol/L}$$

From each of the concentrations of N_2 in blood, we can calculate the number of moles of N_2 dissolved by multiplying by the total blood volume of 5.0 L. Then, we can calculate the number of moles of N_2 released when the diver returns to the surface.

The number of moles of N_2 in 5.0 L of blood at 0.80 atm is:

$$(5.6 \times 10^{-4} \text{ mol/L})(5.0 \text{ L}) = 2.8 \times 10^{-3} \text{ mol}$$

The number of moles of N_2 in 5.0 L of blood at 4.0 atm is:

$$(2.8 \times 10^{-3} \text{ mol/L})(5.0 \text{ L}) = 1.4 \times 10^{-2} \text{ mol}$$

The amount of N_2 released in moles when the diver returns to the surface is:

$$(1.4 \times 10^{-2} \text{ mol}) - (2.8 \times 10^{-3} \text{ mol}) = 1.1 \times 10^{-2} \text{ mol}$$

Finally, we can now calculate the volume of N_2 released using the ideal gas equation. The total pressure pushing on the N_2 that is released is atmospheric pressure (1 atm).

The volume of N_2 released is:

$$V_{N_2} = \frac{nRT}{P}$$

$$V_{N_2} = \frac{(1.1 \times 10^{-2} \text{ mol})(273 + 37)K}{(1.0 \text{ atm})} \times \frac{0.0821 \text{ L} \cdot \text{atm}}{\text{mol} \cdot \text{K}} = \mathbf{0.28 \text{ L}}$$

12.51   The first step is to find the number of moles of sucrose and of water.

$$\text{Moles sucrose} = 396 \text{ g} \times \frac{1 \text{ mol}}{342.3 \text{ g}} = 1.16 \text{ mol sucrose}$$

$$\text{Moles water} = 624 \text{ g} \times \frac{1 \text{ mol}}{18.02 \text{ g}} = 34.6 \text{ mol water}$$

The mole fraction of water is:

$$X_{H_2O} = \frac{34.6 \text{ mol}}{34.6 \text{ mol} + 1.16 \text{ mol}} = 0.968$$

The vapor pressure of the solution is found as follows:

$$\boldsymbol{P_{\text{solution}}} = X_{H_2O} \times P^{\circ}_{H_2O} = (0.968)(31.8 \text{ mmHg}) = \mathbf{30.8 \text{ mmHg}}$$

12.52   **Strategy:** From the vapor pressure of water at 20°C and the change in vapor pressure for the solution (2.0 mmHg), we can solve for the mole fraction of sucrose using Equation (12.5) of the text. From the mole fraction of sucrose, we can solve for moles of sucrose. Lastly, we convert form moles to grams of sucrose.

Solution: Using Equation (12.5) of the text, we can calculate the mole fraction of sucrose that causes a 2.0 mmHg drop in vapor pressure.

$$\Delta P = X_2 P_1^{\circ}$$

$$\Delta P = X_{sucrose} P_{water}^{\circ}$$

$$X_{sucrose} = \frac{\Delta P}{P_{water}^{\circ}} = \frac{2.0 \text{ mmHg}}{17.5 \text{ mmHg}} = 0.11$$

From the definition of mole fraction, we can calculate moles of sucrose.

$$X_{sucrose} = \frac{n_{sucrose}}{n_{water} + n_{sucrose}}$$

$$\text{moles of water} = 552 \text{ g} \times \frac{1 \text{ mol}}{18.02 \text{ g}} = 30.6 \text{ mol H}_2\text{O}$$

$$X_{sucrose} = 0.11 = \frac{n_{sucrose}}{30.6 + n_{sucrose}}$$

$$n_{sucrose} = 3.8 \text{ mol sucrose}$$

Using the molar mass of sucrose as a conversion factor, we can calculate the mass of sucrose.

$$\textbf{mass of sucrose} = 3.8 \text{ mol sucrose} \times \frac{342.3 \text{ g sucrose}}{1 \text{ mol sucrose}} = \textbf{1.3} \times \textbf{10}^3 \textbf{ g sucrose}$$

12.53   Let us call benzene component 1 and camphor component 2.

$$P_1 = X_1 P_1^{\circ} = \left(\frac{n_1}{n_1 + n_2} \right) P_1^{\circ}$$

$$n_1 = 98.5 \text{ g benzene} \times \frac{1 \text{ mol}}{78.11 \text{ g}} = 1.26 \text{ mol benzene}$$

$$n_2 = 24.6 \text{ g camphor} \times \frac{1 \text{ mol}}{152.2 \text{ g}} = 0.162 \text{ mol camphor}$$

$$P_1 = \frac{1.26 \text{ mol}}{(1.26 + 0.162) \text{ mol}} \times 100.0 \text{ mmHg} = \textbf{88.6 mmHg}$$

12.54   For any solution the sum of the mole fractions of the components is always 1.00, so the mole fraction of 1–propanol is 0.700. The partial pressures are:

$$P_{ethanol} = X_{ethanol} \times P_{ethanol}^{\circ} = (0.300)(100 \text{ mmHg}) = \textbf{30.0 mmHg}$$

$$P_{1-propanol} = X_{1-propanol} \times P_{1-propanol}^{\circ} = (0.700)(37.6 \text{ mmHg}) = \textbf{26.3 mmHg}$$

Is the vapor phase richer in one of the components than the solution? Which component? Should this always be true for ideal solutions?

12.55 **(a)** First find the mole fractions of the solution components.

$$\text{Moles methanol} = 30.0 \text{ g} \times \frac{1 \text{ mol}}{32.04 \text{ g}} = 0.936 \text{ mol CH}_3\text{OH}$$

$$\text{Moles ethanol} = 45.0 \text{ g} \times \frac{1 \text{ mol}}{46.07 \text{ g}} = 0.977 \text{ mol C}_2\text{H}_5\text{OH}$$

$$X_{\text{methanol}} = \frac{0.936 \text{ mol}}{0.936 \text{ mol} + 0.977 \text{ mol}} = 0.489$$

$$X_{\text{ethanol}} = 1 - X_{\text{methanol}} = 0.511$$

The vapor pressures of the methanol and ethanol are:

$$P_{\text{methanol}} = (0.489)(94 \text{ mmHg}) = \textbf{46 mmHg}$$

$$P_{\text{ethanol}} = (0.511)(44 \text{ mmHg}) = \textbf{22 mmHg}$$

(b) Since $n = PV/RT$ and V and T are the same for both vapors, the number of moles of each substance is proportional to the partial pressure. We can then write for the mole fractions:

$$X_{\textbf{methanol}} = \frac{P_{\text{methanol}}}{P_{\text{methanol}} + P_{\text{ethanol}}} = \frac{46 \text{ mmHg}}{46 \text{ mmHg} + 22 \text{ mmHg}} = \textbf{0.68}$$

$$X_{\textbf{ethanol}} = 1 - X_{\text{methanol}} = \textbf{0.32}$$

(c) The two components could be separated by fractional distillation. See Section 12.6 of the text.

12.56 This problem is very similar to Problem 12.52.

$$\Delta P = X_{\text{urea}} P°_{\text{water}}$$

$$2.50 \text{ mmHg} = X_{\text{urea}}(31.8 \text{ mmHg})$$

$$X_{\text{urea}} = 0.0786$$

The number of moles of water is:

$$n_{\text{water}} = 450 \text{ g H}_2\text{O} \times \frac{1 \text{ mol H}_2\text{O}}{18.02 \text{ g H}_2\text{O}} = 25.0 \text{ mol H}_2\text{O}$$

$$X_{\text{urea}} = \frac{n_{\text{urea}}}{n_{\text{water}} + n_{\text{urea}}}$$

$$0.0786 = \frac{n_{\text{urea}}}{25.0 + n_{\text{urea}}}$$

$$n_{\text{urea}} = 2.13 \text{ mol}$$

$$\textbf{mass of urea} = 2.13 \text{ mol urea} \times \frac{60.06 \text{ g urea}}{1 \text{ mol urea}} = \textbf{128 g of urea}$$

12.57 $\Delta T_b = K_b m = (2.53°C/m)(2.47\ m) = 6.25°C$

The new **boiling point** is $80.1°C + 6.25°C = \textbf{86.4°C}$

$\Delta T_f = K_f m = (5.12°C/m)(2.47\ m) = 12.6°C$

The new **freezing point** is $5.5°C - 12.6°C = \textbf{-7.1°C}$

12.58 $m = \dfrac{\Delta T_f}{K_f} = \dfrac{1.1°C}{1.86°C/m} = \textbf{0.59}\ \textbf{\textit{m}}$

12.59   **METHOD 1:** The empirical formula can be found from the percent by mass data assuming a 100.0 g sample.

$$\text{Moles C} = 80.78 \text{ g} \times \frac{1 \text{ mol}}{12.01 \text{ g}} = 6.726 \text{ mol C}$$

$$\text{Moles H} = 13.56 \text{ g} \times \frac{1 \text{ mol}}{1.008 \text{ g}} = 13.45 \text{ mol H}$$

$$\text{Moles O} = 5.66 \text{ g} \times \frac{1 \text{ mol}}{16.00 \text{ g}} = 0.354 \text{ mol O}$$

This gives the formula: $C_{6.726}H_{13.45}O_{0.354}$. Dividing through by the smallest subscript (0.354) gives the empirical formula, $C_{19}H_{38}O$.

The freezing point depression is $\Delta T_f = 5.5°C - 3.37°C = 2.1°C$. This implies a solution molality of:

$$m = \frac{\Delta T_f}{K_f} = \frac{2.1°C}{5.12°C/m} = 0.41\ m$$

Since the solvent mass is 8.50 g or 0.00850 kg, the amount of solute is:

$$\frac{0.41 \text{ mol}}{1 \text{ kg benzene}} \times 0.00850 \text{ kg benzene} = 3.5 \times 10^{-3} \text{ mol}$$

Since 1.00 g of the sample represents 3.5×10^{-3} mol, the molar mass is:

$$\textbf{molar mass} = \frac{1.00 \text{ g}}{3.5 \times 10^{-3} \text{ mol}} = \textbf{286 g/mol}$$

The mass of the empirical formula is 282 g/mol, so the molecular formula is the same as the empirical formula, $\textbf{C}_{\textbf{19}}\textbf{H}_{\textbf{38}}\textbf{O}$.

METHOD 2: Use the freezing point data as above to determine the molar mass.

$$\textbf{molar mass} = \textbf{286 g/mol}$$

Multiply the mass % (converted to a decimal) of each element by the molar mass to convert to grams of each element. Then, use the molar mass to convert to moles of each element.

$$n_C = (0.8078) \times (286 \text{ g}) \times \frac{1 \text{ mol C}}{12.01 \text{ g C}} = \textbf{19.2 mol C}$$

$$n_H = (0.1356) \times (286 \text{ g}) \times \frac{1 \text{ mol H}}{1.008 \text{ g H}} = \textbf{38.5 mol H}$$

$$n_O = (0.0566) \times (286 \text{ g}) \times \frac{1 \text{ mol O}}{16.00 \text{ g O}} = \textbf{1.01 mol O}$$

Since we used the molar mass to calculate the moles of each element present in the compound, this method directly gives the molecular formula. The formula is $\textbf{C}_{19}\textbf{H}_{38}\textbf{O}$.

12.60 **METHOD 1:**

Strategy: First, we can determine the empirical formula from mass percent data. Then, we can determine the molar mass from the freezing-point depression. Finally, from the empirical formula and the molar mass, we can find the molecular formula.

Solution: If we assume that we have 100 g of the compound, then each percentage can be converted directly to grams. In this sample, there will be 40.0 g of C, 6.7 g of H, and 53.3 g of O. Because the subscripts in the formula represent a mole ratio, we need to convert the grams of each element to moles. The conversion factor needed is the molar mass of each element. Let n represent the number of moles of each element so that

$$n_C = 40.0 \text{ g C} \times \frac{1 \text{ mol C}}{12.01 \text{ g C}} = 3.33 \text{ mol C}$$

$$n_H = 6.7 \text{ g H} \times \frac{1 \text{ mol H}}{1.008 \text{ g H}} = 6.6 \text{ mol H}$$

$$n_O = 53.3 \text{ g O} \times \frac{1 \text{ mol O}}{16.00 \text{ g O}} = 3.33 \text{ mol O}$$

Thus, we arrive at the formula $C_{3.33}H_{6.6}O_{3.3}$, which gives the identity and the ratios of atoms present. However, chemical formulas are written with whole numbers. Try to convert to whole numbers by dividing all the subscripts by the smallest subscript.

$$\text{C}: \frac{3.33}{3.33} = 1.00 \qquad \text{H}: \frac{6.6}{3.33} = 2.0 \qquad \text{O}: \frac{3.33}{3.33} = 1.00$$

This gives us the empirical, CH_2O.

Now, we can use the freezing point data to determine the molar mass. First, calculate the molality of the solution.

$$m = \frac{\Delta T_f}{K_f} = \frac{1.56°C}{8.00°C/m} = 0.195 \, m$$

Multiplying the molality by the mass of solvent (in kg) gives moles of unknown solute. Then, dividing the mass of solute (in g) by the moles of solute, gives the molar mass of the unknown solute.

$$? \text{ mol of unknown solute} = \frac{0.195 \text{ mol solute}}{1 \text{ kg diphenyl}} \times 0.0278 \text{ kg diphenyl}$$

$$= 0.00542 \text{ mol solute}$$

$$\text{molar mass of unknown} = \frac{0.650 \text{ g}}{0.00542 \text{ mol}} = 1.20 \times 10^2 \text{ g/mol}$$

Finally, we compare the empirical molar mass to the molar mass above.

$$\text{empirical molar mass} = 12.01 \text{ g} + 2(1.008 \text{ g}) + 16.00 \text{ g} = 30.03 \text{ g/mol}$$

The number of (CH_2O) units present in the molecular formula is:

$$\frac{\text{molar mass}}{\text{empirical molar mass}} = \frac{1.20 \times 10^2 \text{ g}}{30.03 \text{ g}} = 4.00$$

Thus, there are four CH_2O units in each molecule of the compound, so the molecular formula is ($CH_2O)_4$, or **$C_4H_8O_4$**.

METHOD 2:

Strategy: As in Method 1, we determine the molar mass of the unknown from the freezing point data. Once the molar mass is known, we can multiply the mass % of each element (converted to a decimal) by the molar mass to convert to grams of each element. From the grams of each element, the moles of each element can be determined and hence the mole ratio in which the elements combine.

Solution: We use the freezing point data to determine the molar mass. First, calculate the molality of the solution.

$$m = \frac{\Delta T_f}{K_f} = \frac{1.56°C}{8.00°C/m} = 0.195 \ m$$

Multiplying the molality by the mass of solvent (in kg) gives moles of unknown solute. Then, dividing the mass of solute (in g) by the moles of solute, gives the molar mass of the unknown solute.

$$? \text{ mol of unknown solute} = \frac{0.195 \text{ mol solute}}{1 \text{ kg diphenyl}} \times 0.0278 \text{ kg diphenyl}$$

$$= 0.00542 \text{ mol solute}$$

$$\text{molar mass of unknown} = \frac{0.650 \text{ g}}{0.00542 \text{ mol}} = 1.20 \times 10^2 \text{ g/mol}$$

Next, we multiply the mass % (converted to a decimal) of each element by the molar mass to convert to grams of each element. Then, we use the molar mass to convert to moles of each element.

$$n_C = (0.400) \times (1.20 \times 10^2 \text{ g}) \times \frac{1 \text{ mol C}}{12.01 \text{ g C}} = 4.00 \text{ mol C}$$

$$n_H = (0.067) \times (1.20 \times 10^2 \text{ g}) \times \frac{1 \text{ mol H}}{1.008 \text{ g H}} = 7.98 \text{ mol H}$$

$$n_O = (0.533) \times (1.20 \times 10^2 \text{ g}) \times \frac{1 \text{ mol O}}{16.00 \text{ g O}} = 4.00 \text{ mol O}$$

Since we used the molar mass to calculate the moles of each element present in the compound, this method directly gives the molecular formula. The formula is **$C_4H_8O_4$**.

12.61 We want a freezing point depression of 20°C.

$$m = \frac{\Delta T_f}{K_f} = \frac{20°C}{1.86°C/m} = 10.8\ m$$

The mass of ethylene glycol (EG) in 6.5 L or 6.5 kg of water is:

$$\text{mass EG} = 6.50\ \text{kg H}_2\text{O} \times \frac{10.8\ \text{mol EG}}{1\ \text{kg H}_2\text{O}} \times \frac{62.07\ \text{g EG}}{1\ \text{mol EG}} = 4.36 \times 10^3\ \text{g EG}$$

The volume of EG needed is:

$$V = (4.36 \times 10^3\ \text{g EG}) \times \frac{1\ \text{mL EG}}{1.11\ \text{g EG}} \times \frac{1\ \text{L}}{1000\ \text{mL}} = \textbf{3.93 L}$$

Finally, we calculate the boiling point:

$$\Delta T_b = mK_b = (10.8\ m)(0.52°C/m) = 5.6°C$$

The **boiling point** of the solution will be 100.0°C + 5.6°C = **105.6°C**.

12.62 We first find the number of moles of gas using the ideal gas equation.

$$n = \frac{PV}{RT} = \frac{\left(748\ \text{mmHg} \times \dfrac{1\ \text{atm}}{760\ \text{mmHg}}\right)(4.00\ \text{L})}{(27 + 273)\,\text{K}} \times \frac{\text{mol}\cdot\text{K}}{0.0821\ \text{L}\cdot\text{atm}} = 0.160\ \text{mol}$$

$$\text{molality} = \frac{0.160\ \text{mol}}{0.0580\ \text{kg benzene}} = 2.76\ m$$

$$\Delta T_f = K_f m = (5.12°C/m)(2.76\ m) = 14.1°C$$

freezing point = 5.5°C − 14.1°C = **−8.6°C**

12.63 The experimental data indicate that the benzoic acid molecules are associated together in pairs in solution due to hydrogen bonding.

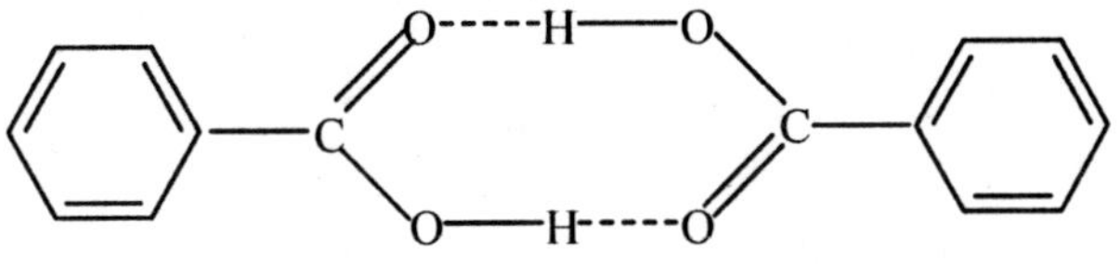

12.64 First, from the freezing point depression we can calculate the molality of the solution. See Table 12.2 of the text for the normal freezing point and K_f value for benzene.

$$\Delta T_f = (5.5 - 4.3)°C = 1.2°C$$

$$m = \frac{\Delta T_f}{K_f} = \frac{1.2°C}{5.12°C/m} = 0.23\ m$$

Multiplying the molality by the mass of solvent (in kg) gives moles of unknown solute. Then, dividing the mass of solute (in g) by the moles of solute, gives the molar mass of the unknown solute.

$$? \text{ mol of unknown solute} = \frac{0.23 \text{ mol solute}}{1 \text{ kg benzene}} \times 0.0250 \text{ kg benzene}$$

$$= 0.0058 \text{ mol solute}$$

$$\textbf{molar mass of unknown} = \frac{2.50 \text{ g}}{0.0058 \text{ mol}} = \textbf{4.3} \times \textbf{10}^2 \textbf{ g/mol}$$

The empirical molar mass of C_6H_5P is 108.1 g/mol. Therefore, the molecular formula is $(C_6H_5P)_4$ or $C_{24}H_{20}P_4$.

12.65 $\pi = MRT = (1.36 \text{ mol/L})(0.0821 \text{ L·atm/K·mol})(22.0 + 273) \text{ K} = \textbf{32.9 atm}$

12.66   **Strategy:** We are asked to calculate the molar mass of the polymer. Grams of the polymer are given in the problem, so we need to solve for moles of polymer.

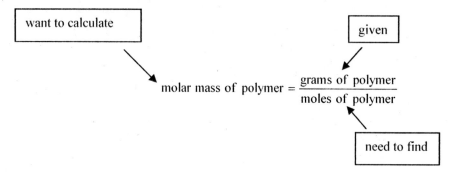

From the osmotic pressure of the solution, we can calculate the molarity of the solution. Then, from the molarity, we can determine the number of moles in 0.8330 g of the polymer. What units should we use for π and temperature?

Solution: First, we calculate the molarity using Equation (12.8) of the text.

$$\pi = MRT$$

$$M = \frac{\pi}{RT} = \frac{\left(5.20 \text{ mmHg} \times \dfrac{1 \text{ atm}}{760 \text{ mmHg}}\right)}{298 \text{ K}} \times \frac{\text{mol} \cdot \text{K}}{0.0821 \text{ L} \cdot \text{atm}} = 2.80 \times 10^{-4} M$$

Multiplying the molarity by the volume of solution (in L) gives moles of solute (polymer).

$$? \text{ mol of polymer} = (2.80 \times 10^{-4} \text{ mol/L})(0.170 \text{ L}) = 4.76 \times 10^{-5} \text{ mol polymer}$$

Lastly, dividing the mass of polymer (in g) by the moles of polymer, gives the molar mass of the polymer.

$$\textbf{molar mass of polymer} = \frac{0.8330 \text{ g polymer}}{4.76 \times 10^{-5} \text{ mol polymer}} = \textbf{1.75} \times \textbf{10}^4 \textbf{ g/mol}$$

12.67 **Method 1:** First, find the concentration of the solution, then work out the molar mass. The concentration is:

$$\text{Molarity} = \frac{\pi}{RT} = \frac{1.43 \text{ atm}}{(0.0821 \text{ L} \cdot \text{atm/K} \cdot \text{mol})(300 \text{ K})} = 0.0581 \text{ mol/L}$$

The solution volume is 0.3000 L so the number of moles of solute is:

$$\frac{0.0581 \text{ mol}}{1 \text{ L}} \times 0.3000 \text{ L} = 0.0174 \text{ mol}$$

The molar mass is then:

$$\frac{7.480 \text{ g}}{0.0174 \text{ mol}} = 430 \text{ g/mol}$$

The empirical formula can be found most easily by assuming a 100.0 g sample of the substance.

$$\text{Moles C} = 41.8 \text{ g} \times \frac{1 \text{ mol}}{12.01 \text{ g}} = 3.48 \text{ mol C}$$

$$\text{Moles H} = 4.7 \text{ g} \times \frac{1 \text{ mol}}{1.008 \text{ g}} = 4.7 \text{ mol H}$$

$$\text{Moles O} = 37.3 \text{ g} \times \frac{1 \text{ mol}}{16.00 \text{ g}} = 2.33 \text{ mol O}$$

$$\text{Moles N} = 16.3 \text{ g} \times \frac{1 \text{ mol}}{14.01 \text{ g}} = 1.16 \text{ mol N}$$

The gives the formula: $C_{3.48}H_{4.7}O_{2.33}N_{1.16}$. Dividing through by the smallest subscript (1.16) gives the empirical formula, $C_3H_4O_2N$, which has a mass of 86.0 g per formula unit. The molar mass is five times this amount (430 ÷ 86.0 = 5.0), so the *molecular formula* is $(C_3H_4O_2N)_5$ or **$C_{15}H_{20}O_{10}N_5$**.

METHOD 2: Use the molarity data as above to determine the molar mass.

molar mass = 430 g/mol

Multiply the mass % (converted to a decimal) of each element by the molar mass to convert to grams of each element. Then, use the molar mass to convert to moles of each element.

$$n_C = (0.418) \times (430 \text{ g}) \times \frac{1 \text{ mol C}}{12.01 \text{ g C}} = \textbf{15.0 mol C}$$

$$n_H = (0.047) \times (430 \text{ g}) \times \frac{1 \text{ mol H}}{1.008 \text{ g H}} = \textbf{20 mol H}$$

$$n_O = (0.373) \times (430 \text{ g}) \times \frac{1 \text{ mol O}}{16.00 \text{ g O}} = \textbf{10.0 mol O}$$

$$n_N = (0.163) \times (430 \text{ g}) \times \frac{1 \text{ mol N}}{14.01 \text{ g N}} = \textbf{5.00 mol N}$$

Since we used the molar mass to calculate the moles of each element present in the compound, this method directly gives the molecular formula. The formula is **$C_{15}H_{20}O_{10}N_5$**.

12.68   We use the osmotic pressure data to determine the molarity.

$$M = \frac{\pi}{RT} = \frac{4.61 \text{ atm}}{(20 + 273)\text{K}} \times \frac{\text{mol} \cdot \text{K}}{0.0821 \text{ L} \cdot \text{atm}} = 0.192 \text{ mol/L}$$

Next we use the density and the solution mass to find the volume of the solution.

$$\text{mass of soln} = 6.85 \text{ g} + 100.0 \text{ g} = 106.9 \text{ g soln}$$

$$\text{volume of soln} = 106.9 \text{ g soln} \times \frac{1 \text{ mL}}{1.024 \text{ g}} = 104.4 \text{ mL} = 0.1044 \text{ L}$$

Multiplying the molarity by the volume (in L) gives moles of solute (carbohydrate).

$$\text{mol of solute} = M \times L = (0.192 \text{ mol/L})(0.1044 \text{ L}) = 0.0200 \text{ mol solute}$$

Finally, dividing mass of carbohydrate by moles of carbohydrate gives the molar mass of the carbohydrate.

$$\textbf{molar mass} = \frac{6.85 \text{ g carbohydrate}}{0.0200 \text{ mol carbohydrate}} = \textbf{343 g/mol}$$

12.73 $CaCl_2$ is an ionic compound (why?) and is therefore an electrolyte in water. Assuming that $CaCl_2$ is a strong electrolyte and completely dissociates (no ion pairs, van't Hoff factor $i = 3$), the total ion concentration will be $3 \times 0.35 = 1.05 \ m$, which is larger than the urea (nonelectrolyte) concentration of $0.90 \ m$.

(a)   The **$CaCl_2$** solution will show a larger boiling point elevation.

(b)   The $CaCl_2$ solution will show a larger freezing point depression. The freezing point of the **urea** solution will be higher.

(c)   The **$CaCl_2$** solution will have a larger vapor pressure lowering.

12.74   Boiling point, vapor pressure, and osmotic pressure all depend on particle concentration. Therefore, these solutions also have the same boiling point, osmotic pressure, and vapor pressure.

12.75   Assume that all the salts are completely dissociated. Calculate the molality of the ions in the solutions.

(a)	$0.10 \ m \ Na_3PO_4$:	$0.10 \ m \times 4$ ions/unit $= 0.40 \ m$
(b)	$0.35 \ m \ NaCl$:	$0.35 \ m \times 2$ ions/unit $= 0.70 \ m$
(c)	$0.20 \ m \ MgCl_2$:	$0.20 \ m \times 3$ ions/unit $= 0.60 \ m$
(d)	$0.15 \ m \ C_6H_{12}O_6$:	nonelectrolyte, $0.15 \ m$
(e)	$0.15 \ m \ CH_3COOH$:	weak electrolyte, slightly greater than $0.15 \ m$

The solution with the lowest molality will have the highest freezing point (smallest freezing point depression): **(d) > (e) > (a) > (c) > (b)**.

12.76   The freezing point will be depressed most by the solution that contains the most solute particles. You should try to classify each solute as a strong electrolyte, a weak electrolyte, or a nonelectrolyte. All three solutions have the same concentration, so comparing the solutions is straightforward. HCl is a strong electrolyte, so under ideal conditions it will completely dissociate into two particles per molecule. The concentration of particles will be $1.00 \ m$. Acetic acid is a weak electrolyte, so it will only dissociate to a small extent. The concentration of particles will be greater than $0.50 \ m$, but less than $1.00 \ m$. Glucose is a nonelectrolyte, so glucose molecules remain as glucose molecules in solution. The concentration of particles will be 0.50 m. For these solutions, the order in which the freezing points become *lower* is:

$$0.50 \ m \ \text{glucose} \ > \ 0.50 \ m \ \text{acetic acid} \ > \ 0.50 \ m \ \text{HCl}$$

In other words, the HCl solution will have the lowest freezing point (greatest freezing point depression).

12.77 **(a)** NaCl is a strong electrolyte. The concentration of particles (ions) is double the concentration of NaCl. Note that 135 mL of water has a mass of 135 g (why?).

The number of moles of NaCl is:

$$21.2 \ \text{g NaCl} \times \frac{1 \ \text{mol}}{58.44 \ \text{g}} = 0.363 \ \text{mol NaCl}$$

Next, we can find the changes in boiling and freezing points ($i = 2$)

$$m = \frac{0.363 \ \text{mol}}{0.135 \ \text{kg}} = 2.70 \ m$$

$$\Delta T_b = iK_b m = 2(0.52°C/m)(2.70 \ m) = 2.8°C$$

$$\Delta T_f = iK_f m = 2(1.86°C/m)(2.70 \ m) = 10.0°C$$

The *boiling point* is **102.8°C**; the *freezing point* is **−10.0°C**.

(b) Urea is a nonelectrolyte. The particle concentration is just equal to the urea concentration.

The molality of the urea solution is:

$$\text{moles urea} = 15.4 \ \text{g urea} \times \frac{1 \ \text{mol urea}}{60.06 \ \text{g urea}} = 0.256 \ \text{mol urea}$$

$$m = \frac{0.256 \ \text{mol urea}}{0.0667 \ \text{kg H}_2\text{O}} = 3.84 \ m$$

$$\Delta T_b = iK_b m = 1(0.52°C/m)(3.84 \ m) = 2.0°C$$

$$\Delta T_f = iK_f m = 1(1.86°C/m)(3.84 \ m) = 7.14°C$$

The *boiling point* is **102.0°C**; the *freezing point* is **−7.14°C**.

12.78 Using Equation (12.5) of the text, we can find the mole fraction of the NaCl. We use subscript 1 for H_2O and subscript 2 for NaCl.

$$\Delta P = X_2 P_1°$$

$$X_2 = \frac{\Delta P}{P_1°}$$

$$X_2 = \frac{23.76 \ \text{mmHg} - 22.98 \ \text{mmHg}}{23.76 \ \text{mmHg}} = 0.03283$$

Let's assume that we have 1000 g (1 kg) of water as the solvent, because the definition of molality is moles of solute per kg of solvent. We can find the number of moles of particles dissolved in the water using the definition of mole fraction.

$$X_2 = \frac{n_2}{n_1 + n_2}$$

$$n_1 = 1000 \text{ g } H_2O \times \frac{1 \text{ mol } H_2O}{18.02 \text{ g } H_2O} = 55.49 \text{ mol } H_2O$$

$$\frac{n_2}{55.49 + n_2} = 0.03283$$

$$n_2 = 1.884 \text{ mol}$$

Since NaCl dissociates to form two particles (ions), the number of moles of NaCl is half of the above result.

$$\text{Moles NaCl} = 1.884 \text{ mol particles} \times \frac{1 \text{ mol NaCl}}{2 \text{ mol particles}} = 0.9420 \text{ mol}$$

The molality of the solution is:

$$\frac{0.9420 \text{ mol}}{1.000 \text{ kg}} = \textbf{0.9420 } \textit{m}$$

12.79   Both NaCl and $CaCl_2$ are strong electrolytes. Urea and sucrose are nonelectrolytes. The NaCl or $CaCl_2$ will yield more particles per mole of the solid dissolved, resulting in greater freezing point depression. Also, sucrose and urea would make a mess when the ice melts.

12.80   **Strategy:** We want to calculate the osmotic pressure of a NaCl solution. Since NaCl is a strong electrolyte, i in the van't Hoff equation is 2.

$$\pi = iMRT$$

Since, R is a constant and T is given, we need to first solve for the molarity of the solution in order to calculate the osmotic pressure (π). If we assume a given volume of solution, we can then use the density of the solution to determine the mass of the solution. The solution is 0.86% by mass NaCl, so we can find grams of NaCl in the solution.

Solution: To calculate molarity, let's assume that we have 1.000 L of solution (1.000×10^3 mL). We can use the solution density as a conversion factor to calculate the mass of 1.000×10^3 mL of solution.

$$(1.000 \times 10^3 \text{ mL soln}) \times \frac{1.005 \text{ g soln}}{1 \text{ mL soln}} = 1005 \text{ g of soln}$$

Since the solution is 0.86% by mass NaCl, the mass of NaCl in the solution is:

$$1005 \text{ g} \times \frac{0.86\%}{100\%} = 8.6 \text{ g NaCl}$$

The molarity of the solution is:

$$\frac{8.6 \text{ g NaCl}}{1.000 \text{ L}} \times \frac{1 \text{ mol NaCl}}{58.44 \text{ g NaCl}} = 0.15 \text{ } M$$

Since NaCl is a strong electrolyte, we assume that the van't Hoff factor is 2. Substituting *i*, *M*, *R*, and *T* into the equation for osmotic pressure gives:

$$\pi = iMRT = (2)\left(\frac{0.15 \text{ mol}}{\text{L}}\right)\left(\frac{0.0821 \text{ L} \cdot \text{atm}}{\text{mol} \cdot \text{K}}\right)(310 \text{ K}) = \textbf{7.6 atm}$$

12.81   The temperature and molarity of the two solutions are the same. If we divide Equation (12.12) of the text for one solution by the same equation for the other, we can find the ratio of the van't Hoff factors in terms of the osmotic pressures (*i* = 1 for urea).

$$\frac{\pi_{CaCl_2}}{\pi_{urea}} = \frac{iMRT}{MRT} = i = \frac{0.605 \text{ atm}}{0.245 \text{ atm}} = \textbf{2.47}$$

12.82   From Table 12.3 of the text, *i* = 1.3

$$\pi = iMRT$$

$$\pi = (1.3)\left(\frac{0.0500 \text{ mol}}{\text{L}}\right)\left(\frac{0.0821 \text{ L} \cdot \text{atm}}{\text{mol} \cdot \text{K}}\right)(298 \text{ K})$$

$$\pi = \textbf{1.6 atm}$$

12.85   For this problem we must find the solution mole fractions, the molality, and the molarity. For molarity, we can assume the solution to be so dilute that its density is 1.00 g/mL. We first find the number of moles of lysozyme and of water.

$$n_{lysozyme} = 0.100 \text{ g} \times \frac{1 \text{ mol}}{13930 \text{ g}} = 7.18 \times 10^{-6} \text{ mol}$$

$$n_{water} = 150 \text{ g} \times \frac{1 \text{ mol}}{18.02 \text{ g}} = 8.32 \text{ mol}$$

Vapor pressure lowering: $\Delta P = X_{lysozyme} P^{\circ}_{water} = \dfrac{n_{lysozyme}}{n_{lysozyme} + n_{water}}(23.76 \text{ mmHg})$

$$\Delta P = \frac{7.18 \times 10^{-6} \text{ mol}}{[(7.18 \times 10^{-6}) + 8.32]\text{mol}}(23.76 \text{ mmHg}) = \textbf{2.05} \times \textbf{10}^{-5} \textbf{ mmHg}$$

Freezing point depression: $\Delta T_f = K_f m = (1.86°C/m)\left(\dfrac{7.18 \times 10^{-6} \text{ mol}}{0.150 \text{ kg}}\right) = \textbf{8.90} \times \textbf{10}^{-5} \textbf{°C}$

Boiling point elevation: $\Delta T_b = K_b m = (0.52°C/m)\left(\dfrac{7.18 \times 10^{-6} \text{ mol}}{0.150 \text{ kg}}\right) = \textbf{2.5} \times \textbf{10}^{-5} \textbf{°C}$

Osmotic pressure: As stated above, we assume the density of the solution is 1.00 g/mL. The volume of the solution will be 150 mL.

$$\pi = MRT = \left(\frac{7.18 \times 10^{-6}\ \text{mol}}{0.150\ \text{L}}\right)(0.0821\ \text{L}\cdot\text{atm/mol}\cdot\text{K})(298\ \text{K}) = \mathbf{1.17 \times 10^{-3}\ atm} = \mathbf{0.889\ mmHg}$$

Note that only the osmotic pressure is large enough to measure.

12.86   At constant temperature, the osmotic pressure of a solution is proportional to the molarity. When equal volumes of the two solutions are mixed, the molarity will just be the mean of the molarities of the two solutions (assuming additive volumes). Since the osmotic pressure is proportional to the molarity, the osmotic pressure of the solution will be the mean of the osmotic pressure of the two solutions.

$$\pi = \frac{2.4\ \text{atm} + 4.6\ \text{atm}}{2} = \mathbf{3.5\ atm}$$

12.87   Water migrates through the semipermiable cell walls of the cucumber into the concentrated salt solution.

When we go swimming in the ocean, why don't we shrivel up like a cucumber? When we swim in fresh water pool, why don't we swell up and burst?

12.88   (a)   We use Equation (12.4) of the text to calculate the vapor pressure of each component.

$$P_1 = X_1 P_1^{\circ}$$

First, you must calculate the mole fraction of each component.

$$X_A = \frac{n_A}{n_A + n_B} = \frac{1.00\ \text{mol}}{1.00\ \text{mol} + 1.00\ \text{mol}} = 0.500$$

Similarly,

$$X_B = 0.500$$

Substitute the mole fraction calculated above and the vapor pressure of the pure solvent into Equation (12.4) to calculate the vapor pressure of each component of the solution.

$$P_A = X_A P_A^{\circ} = (0.500)(76\ \text{mmHg}) = 38\ \text{mmHg}$$

$$P_B = X_B P_B^{\circ} = (0.500)(132\ \text{mmHg}) = 66\ \text{mmHg}$$

The total vapor pressure is the sum of the vapor pressures of the two components.

$$\mathbf{P_{Total}} = P_A + P_B = 38\ \text{mmHg} + 66\ \text{mmHg} = \mathbf{104\ mmHg}$$

(b)   This problem is solved similarly to part (a).

$$X_A = \frac{n_A}{n_A + n_B} = \frac{2.00\ \text{mol}}{2.00\ \text{mol} + 5.00\ \text{mol}} = 0.286$$

Similarly,

$$X_B = 0.714$$

$$P_A = X_A P_A^\circ = (0.286)(76 \text{ mmHg}) = 22 \text{ mmHg}$$

$$P_B = X_B P_B^\circ = (0.714)(132 \text{ mmHg}) = 94 \text{ mmHg}$$

$$\mathbf{P_{Total}} = P_A + P_B = 22 \text{ mmHg} + 94 \text{ mmHg} = \mathbf{116 \text{ mmHg}}$$

12.89 $\Delta T_f = iK_f m$

$$i = \frac{\Delta T_f}{K_f m} = \frac{2.6}{(1.86)(0.40)} = \mathbf{3.5}$$

12.90   From the osmotic pressure, you can calculate the molarity of the solution.

$$M = \frac{\pi}{RT} = \frac{\left(30.3 \text{ mmHg} \times \dfrac{1 \text{ atm}}{760 \text{ mmHg}}\right)}{308 \text{ K}} \times \frac{\text{mol} \cdot \text{K}}{0.0821 \text{ L} \cdot \text{atm}} = 1.58 \times 10^{-3} \text{ mol/L}$$

Multiplying molarity by the volume of solution in liters gives the moles of solute.

$$(1.58 \times 10^{-3} \text{ mol solute/L soln}) \times (0.262 \text{ L soln}) = 4.14 \times 10^{-4} \text{ mol solute}$$

Divide the grams of solute by the moles of solute to calculate the molar mass.

$$\textbf{molar mass of solute} = \frac{1.22 \text{ g}}{4.14 \times 10^{-4} \text{ mol}} = \mathbf{2.95 \times 10^3 \text{ g/mol}}$$

12.91   One manometer has pure water over the mercury, one manometer has a 1.0 M solution of NaCl and the other manometer has a 1.0 M solution of urea. The pure water will have the highest vapor pressure and will thus force the mercury column down the most; column X. Both the salt and the urea will lower the overall pressure of the water. However, the salt dissociates into sodium and chloride ions (van't Hoff factor $i = 2$), whereas urea is a molecular compound with a van't Hoff factor of 1. Therefore the urea solution will lower the pressure only half as much as the salt solution. Y is the NaCl solution and Z is the urea solution.

Assuming that you knew the temperature, could you actually calculate the distance from the top of the solution to the top of the manometer?

12.92   Solve Equation (12.7) of the text algebraically for molality (m), then substitute ΔT_f and K_f into the equation to calculate the molality. You can find the normal freezing point for benzene and K_f for benzene in Table 12.2 of the text.

$$\Delta T_f = 5.5°C - 3.9°C = 1.6°C$$

$$m = \frac{\Delta T_f}{K_f} = \frac{1.6°C}{5.12°C/m} = 0.31 \ m$$

Multiplying the molality by the mass of solvent (in kg) gives moles of unknown solute. Then, dividing the mass of solute (in g) by the moles of solute, gives the molar mass of the unknown solute.

$$? \text{ mol of unknown solute} = \frac{0.31 \text{ mol solute}}{1 \text{ kg benzene}} \times (8.0 \times 10^{-3} \text{ kg benzene})$$

$$= 2.5 \times 10^{-3} \text{ mol solute}$$

$$\text{molar mass of unknown} = \frac{0.50 \text{ g}}{2.5 \times 10^{-3} \text{ mol}} = 2.0 \times 10^2 \text{ g/mol}$$

The molar mass of cocaine $C_{17}H_{21}NO_4 = 303$ g/mol, so the compound is not cocaine. We assume in our analysis that the compound is a pure, monomeric, nonelectrolyte.

12.93   The pill is in a hypotonic solution. Consequently, by osmosis, water moves across the semipermeable membrane into the pill. The increase in pressure pushes the elastic membrane to the right, causing the drug to exit through the small holes at a constant rate.

12.94   The molality of the solution assuming $AlCl_3$ to be a nonelectrolyte is:

$$\text{mol } AlCl_3 = 1.00 \text{ g } AlCl_3 \times \frac{1 \text{ mol } AlCl_3}{133.3 \text{ g } AlCl_3} = 0.00750 \text{ mol } AlCl_3$$

$$m = \frac{0.00750 \text{ mol}}{0.0500 \text{ kg}} = 0.150 \; m$$

The molality calculated with Equation (12.7) of the text is:

$$m = \frac{\Delta T_f}{K_f} = \frac{1.11°C}{1.86°C/m} = 0.597 \; m$$

The ratio $\dfrac{0.597 \; m}{0.150 \; m}$ is 4. Thus each $AlCl_3$ dissociates as follows:

$$AlCl_3(s) \rightarrow Al^{3+}(aq) + 3Cl^-(aq)$$

12.95   Reverse osmosis uses high pressure to force water from a more concentrated solution to a less concentrated one through a semipermeable membrane. Desalination by reverse osmosis is considerably cheaper than by distillation and avoids the technical difficulties associated with freezing.

To reverse the osmotic migration of water across a semipermeable membrane, an external pressure exceeding the osmotic pressure must be applied. To find the osmotic pressure of 0.70 M NaCl solution, we must use the van't Hoff factor because NaCl is a strong electrolyte and the total ion concentration becomes 2(0.70 M) = 1.4 M.

The osmotic pressure of sea water is:

$$\pi = iMRT = 2(0.70 \text{ mol/L})(0.0821 \text{ L·atm/mol·K})(298 \text{ K}) = \textbf{34 atm}$$

To cause reverse osmosis a pressure in excess of 34 atm must be applied.

12.96   First, we tabulate the concentration of all of the ions. Notice that the chloride concentration comes from more than one source.

$MgCl_2$:	If $[MgCl_2] = 0.054 \; M$,	$[Mg^{2+}] = 0.054 \; M$	$[Cl^-] = 2 \times 0.054 \; M$
Na_2SO_4:	if $[Na_2SO_4] = 0.051 \; M$,	$[Na^+] = 2 \times 0.051 \; M$	$[SO_4^{2-}] = 0.051 \; M$
$CaCl_2$:	if $[CaCl_2] = 0.010 \; M$,	$[Ca^{2+}] = 0.010 \; M$	$[Cl^-] = 2 \times 0.010 \; M$
$NaHCO_3$:	if $[NaHCO_3] = 0.0020 \; M$	$[Na^+] = 0.0020 \; M$	$[HCO_3^-] = 0.0020 \; M$
KCl:	if $[KCl] = 0.0090 \; M$	$[K^+] = 0.0090 \; M$	$[Cl^-] = 0.0090 \; M$

The subtotal of chloride ion concentration is:

$$[Cl^-] = (2 \times 0.0540) + (2 \times 0.010) + (0.0090) = 0.137\ M$$

Since the required $[Cl^-]$ is 2.60 M, the difference (2.6 − 0.137 = 2.46 M) must come from NaCl.

The subtotal of sodium ion concentration is:

$$[Na^+] = (2 \times 0.051) + (0.0020) = 0.104\ M$$

Since the required $[Na^+]$ is 2.56 M, the difference (2.56 − 0.104 = 2.46 M) must come from NaCl.

Now, calculating the mass of the compounds required:

NaCl: $\qquad$ $2.46\ \text{mol} \times \dfrac{58.44\ \text{g NaCl}}{1\ \text{mol NaCl}} = \mathbf{143.8\ g}$

MgCl$_2$: $\qquad$ $0.054\ \text{mol} \times \dfrac{95.21\ \text{g MgCl}_2}{1\ \text{mol MgCl}_2} = \mathbf{5.14\ g}$

Na$_2$SO$_4$: $\qquad$ $0.051\ \text{mol} \times \dfrac{142.1\ \text{g Na}_2\text{SO}_4}{1\ \text{mol Na}_2\text{SO}_4} = \mathbf{7.25\ g}$

CaCl$_2$: $\qquad$ $0.010\ \text{mol} \times \dfrac{111.0\ \text{g CaCl}_2}{1\ \text{mol CaCl}_2} = \mathbf{1.11\ g}$

KCl: $\qquad$ $0.0090\ \text{mol} \times \dfrac{74.55\ \text{g KCl}}{1\ \text{mol KCl}} = \mathbf{0.67\ g}$

NaHCO$_3$: $\qquad$ $0.0020\ \text{mol} \times \dfrac{84.01\ \text{g NaHCO}_3}{1\ \text{mol NaHCO}_3} = \mathbf{0.17\ g}$

12.97 **(a)** Using Equation (12.8) of the text, we find the molarity of the solution.

$$M = \frac{\pi}{RT} = \frac{0.257\ \text{atm}}{(0.0821\ \text{L} \cdot \text{atm/mol} \cdot \text{K})(298\ \text{K})} = 0.0105\ \text{mol/L}$$

This is the combined concentrations of all the ions. The amount dissolved in 10.0 mL (0.01000 L) is

$$?\ \text{moles} = \frac{0.0105\ \text{mol}}{1\ \text{L}} \times 0.0100\ \text{L} = 1.05 \times 10^{-4}\ \text{mol}$$

Since the mass of this amount of protein is 0.225 g, the apparent molar mass is

$$\frac{0.225\ \text{g}}{1.05 \times 10^{-4}\ \text{mol}} = \mathbf{2.14 \times 10^3\ g/mol}$$

(b) We need to use a van't Hoff factor to take into account the fact that the protein is a strong electrolyte. The van't Hoff factor will be $i = 21$ (why?).

$$M = \frac{\pi}{iRT} = \frac{0.257\ \text{atm}}{(21)(0.0821\ \text{L} \cdot \text{atm/mol} \cdot \text{K})(298\ \text{K})} = 5.00 \times 10^{-4}\ \text{mol/L}$$

This is the actual concentration of the protein. The amount in 10.0 mL (0.0100 L) is

$$\frac{5.00 \times 10^{-4} \text{ mol}}{1 \text{ L}} \times 0.0100 \text{ L} = 5.00 \times 10^{-6} \text{ mol}$$

Therefore the actual molar mass is:

$$\frac{0.225 \text{ g}}{5.00 \times 10^{-6} \text{ mol}} = \textbf{4.50} \times \textbf{10}^{\textbf{4}} \textbf{ g/mol}$$

12.98   Solution A: Let molar mass be $\mathcal{M}$.

$$\Delta P = X_A P_A^\circ$$

$$(760 - 754.5) = X_A(760)$$

$$X_A = 7.237 \times 10^{-3}$$

$$n = \frac{\text{mass}}{\text{molar mass}}$$

$$X_A = \frac{n_A}{n_A + n_{\text{water}}} = \frac{5.00/\mathcal{M}}{5.00/\mathcal{M} + 100/18.02} = 7.237 \times 10^{-3}$$

$$\mathcal{M} = \textbf{124 g/mol}$$

Solution B: Let molar mass be $\mathcal{M}$

$$\Delta P = X_B P_B^\circ$$

$$X_B = 7.237 \times 10^{-3}$$

$$n = \frac{\text{mass}}{\text{molar mass}}$$

$$X_B = \frac{n_B}{n_B + n_{\text{benzene}}} = \frac{2.31/\mathcal{M}}{2.31/\mathcal{M} + 100/78.11} = 7.237 \times 10^{-3}$$

$$\mathcal{M} = \textbf{248 g/mol}$$

The molar mass in benzene is about twice that in water. This suggests some sort of dimerization is occurring in a nonpolar solvent such as benzene.

12.99 $2H_2O_2 \rightarrow 2H_2O + O_2$

$$10 \text{ mL} \times \frac{3.0 \text{ g } H_2O_2}{100 \text{ mL}} \times \frac{1 \text{ mol } H_2O_2}{34.02 \text{ g } H_2O_2} \times \frac{1 \text{ mol } O_2}{2 \text{ mol } H_2O_2} = 4.4 \times 10^{-3} \text{ mol } O_2$$

(a)   Using the ideal gas law:

$$V = \frac{nRT}{P} = \frac{(4.4 \times 10^{-3} \text{ mol } O_2)(0.0821 \text{ L} \cdot \text{atm/mol} \cdot \text{K})(273 \text{ K})}{1.0 \text{ atm}} = \textbf{99 mL}$$

(b)   The ratio of the volumes: $\dfrac{99 \text{ mL}}{10 \text{ mL}} = \textbf{9.9}$

Could we have made the calculation in part (a) simpler if we used the fact that 1 mole of all ideal gases at STP occupies a volume of 22.4 L?

12.100   As the chain becomes longer, the alcohols become more like hydrocarbons (nonpolar) in their properties. The alcohol with five carbons (*n*-pentanol) would be the best solvent for iodine (a) and *n*-pentane (c) (why?). Methanol (CH_3OH) is the most water like and is the best solvent for an ionic solid like KBr.

12.101   **(a)**   Boiling under reduced pressure.

(b) CO_2 boils off, expands and cools, condensing water vapor to form fog.

12.102 $I_2 - H_2O$: Dipole - induced dipole.

$I_3^- - H_2O$: Ion - dipole. Stronger interaction causes more I_2 to be converted to I_3^-.

12.103   At equilibrium, the concentration in the 2 beakers is equal. Let *x* mL be the change in volume.

$$(50 - x)(1.0 \text{ } M) = (50 + x)(2.0 \text{ } M)$$

$$3x = 50$$

$$x = 16.7 \text{ mL}$$

The final volumes are:

$$(50 - 16.7) \text{ mL} = \textbf{33.3 mL}$$

$$(50 + 16.7) \text{ mL} = \textbf{66.7 mL}$$

12.104   **(a)**   If the membrane is permeable to all the ions and to the water, the result will be the same as just removing the membrane. You will have two solutions of equal NaCl concentration.

(b)   This part is tricky. The movement of one ion but not the other would result in one side of the apparatus acquiring a positive electric charge and the other side becoming equally negative. This has never been known to happen, so we must conclude that migrating ions always drag other ions of the opposite charge with them. In this hypothetical situation only water would move through the membrane from the dilute to the more concentrated side.

(c)   This is the classic osmosis situation. Water would move through the membrane from the dilute to the concentrated side.

12.105   To protect the red blood cells and other cells from shrinking (in a hypertonic solution) or expanding (in a hypotonic solution).

12.106   First, we calculate the number of moles of HCl in 100 g of solution.

$$n_{HCl} = 100 \text{ g soln} \times \frac{37.7 \text{ g HCl}}{100 \text{ g soln}} \times \frac{1 \text{ mol HCl}}{36.46 \text{ g HCl}} = 1.03 \text{ mol HCl}$$

Next, we calculate the volume of 100 g of solution.

$$V = 100 \text{ g} \times \frac{1 \text{ mL}}{1.19 \text{ g}} \times \frac{1 \text{ L}}{1000 \text{ mL}} = 0.0840 \text{ L}$$

Finally, the molarity of the solution is:

$$\frac{1.03 \text{ mol}}{0.0840 \text{ L}} = \textbf{12.3 } \textbf{\textit{M}}$$

12.107 **(a)** Seawater has a larger number of ionic compounds dissolved in it; thus the boiling point is elevated.

(b) Carbon dioxide escapes from an opened soft drink bottle because gases are less soluble in liquids at lower pressure (Henry's law).

(c) As you proved in Problem 12.20, at dilute concentrations molality and molarity are almost the same because the density of the solution is almost equal to that of the pure solvent.

(d) For colligative properties we are concerned with the number of solute particles in solution relative to the number of solvent particles. Since in colligative particle measurements we frequently are dealing with changes in temperature (and since density varies with temperature), we need a concentration unit that is temperature invariant. We use units of moles per kilogram of mass (molality) rather than moles per liter of solution (molarity).

(e) Methanol is very water soluble (why?) and effectively lowers the freezing point of water. However in the summer, the temperatures are sufficiently high so that most of the methanol would be lost to vaporization.

12.108 Let the mass of NaCl be x g. Then, the mass of sucrose is $(10.2 - x)$g.

We know that the equation representing the osmotic pressure is:

$$\pi = MRT$$

π, R, and T are given. Using this equation and the definition of molarity, we can calculate the percentage of NaCl in the mixture.

$$\text{molarity} = \frac{\text{mol solute}}{\text{L soln}}$$

Remember that NaCl dissociates into two ions in solution; therefore, we multiply the moles of NaCl by two.

$$\text{mol solute} = 2\left(x \text{ g NaCl} \times \frac{1 \text{ mol NaCl}}{58.44 \text{ g NaCl}} \right) + \left((10.2 - x)\text{g sucrose} \times \frac{1 \text{ mol sucrose}}{342.3 \text{ g sucrose}} \right)$$

$$\text{mol solute} = 0.03422x + 0.02980 - 0.002921x$$

$$\text{mol solute} = 0.03130x + 0.02980$$

$$\text{Molarity of solution} = \frac{\text{mol solute}}{\text{L soln}} = \frac{0.03130x + 0.02980}{0.250 \text{ L}}$$

Substitute molarity into the equation for osmotic pressure to solve for x.

$$\pi = MRT$$

$$7.32 \text{ atm} = \left(\frac{(0.03130x + 0.02980)\,\text{mol}}{0.250\,\text{L}}\right)\left(0.0821\frac{\text{L}\cdot\text{atm}}{\text{mol}\cdot\text{K}}\right)(296\text{ K})$$

$$0.0753 = 0.03130x + 0.02980$$

$$x = 1.45\text{ g} = \text{mass of NaCl}$$

$$\textbf{Mass \% NaCl} = \frac{1.45\text{ g}}{10.2\text{ g}} \times 100\% = \textbf{14.2\%}$$

12.109 $\Delta T_f = 5.5 - 2.2 = 3.3°C$ $C_{10}H_8$: 128.2 g/mol

$$m = \frac{\Delta T_f}{K_f} = \frac{3.3}{5.12} = 0.645\ m$$ C_6H_{12}: 84.16 g/mol

Let x = mass of C_6H_{12} (in grams).

Using,

$$m = \frac{\text{mol solute}}{\text{kg solvent}} \quad \text{and} \quad \text{mol} = \frac{\text{mass}}{\text{molar mass}}$$

$$0.645 = \frac{\dfrac{x}{84.16} + \dfrac{1.32 - x}{128.2}}{0.0189\text{ kg}}$$

$$0.0122 = \frac{128.2x + 111.1 - 84.16x}{84.16 \times 128.2}$$

$$x = 0.47\text{ g}$$

$$\%C_6H_{12} = \frac{0.47}{1.32} \times 100\% = \textbf{36\%}$$

$$\%C_{10}H_8 = \frac{0.86}{1.32} \times 100\% = \textbf{65\%}$$

The percentages don't add up to 100% because of rounding procedures.

12.110   **(a)**   Solubility decreases with increasing lattice energy.

   (b)   Ionic compounds are more soluble in a polar solvent.

   (c)   Solubility increases with enthalpy of hydration of the cation and anion.

12.111 The completed table is shown below:

Attractive Forces	Deviation from Raoult's	$\Delta H_{solution}$
A $\leftrightarrow$ A, B $\leftrightarrow$ B > A $\leftrightarrow$ B	Positive	Positive (endothermic)
A $\leftrightarrow$ A, B $\leftrightarrow$ B < A $\leftrightarrow$ B	Negative	Negative (exothermic)
A $\leftrightarrow$ A, B $\leftrightarrow$ B = A $\leftrightarrow$ B	Zero	Zero

The first row represents a Case 1 situation in which A's attract A's and B's attract B's more strongly than A's attract B's. As described in Section 12.6 of the text, this results in positive deviation from Raoult's law (higher vapor pressure than calculated) and positive heat of solution (endothermic).

In the second row a negative deviation from Raoult's law (lower than calculated vapor pressure) means A's attract B's better than A's attract A's and B's attract B's. This causes a negative (exothermic) heat of solution.

In the third row a zero heat of solution means that A–A, B–B, and A–B interparticle attractions are all the same. This corresponds to an ideal solution which obeys Raoult's law exactly.

What sorts of substances form ideal solutions with each other?

12.112 $$\text{molality} = \frac{98.0 \text{ g } H_2SO_4 \times \dfrac{1 \text{ mol } H_2SO_4}{98.09 \text{ g } H_2SO_4}}{2.0 \text{ g } H_2O \times \dfrac{1 \text{ kg } H_2O}{1000 \text{ g } H_2O}} = 5.0 \times 10^2 \, m$$

We can calculate the density of sulfuric acid from the molarity.

$$\text{molarity} = 18 \, M = \frac{18 \text{ mol } H_2SO_4}{1 \text{ L soln}}$$

The 18 mol of H_2SO_4 has a mass of:

$$18 \text{ mol } H_2SO_4 \times \frac{98.0 \text{ g } H_2SO_4}{1 \text{ mol } H_2SO_4} = 1.8 \times 10^3 \text{ g } H_2SO_4$$

$1 \text{ L} = 1000 \text{ mL}$

$$\textbf{density} = \frac{\text{mass } H_2SO_4}{\text{volume}} = \frac{1.8 \times 10^3 \text{ g}}{1000 \text{ mL}} = \textbf{1.80 g/mL}$$

12.113 Let's assume we have 100 g of solution. The 100 g of solution will contain 70.0 g of HNO_3 and 30.0 g of H_2O.

$$\text{mol solute } (HNO_3) = 70.0 \text{ g } HNO_3 \times \frac{1 \text{ mol } HNO_3}{63.02 \text{ g } HNO_3} = 1.11 \text{ mol } HNO_3$$

$$\text{kg solvent } (H_2O) = 30.0 \text{ g } H_2O \times \frac{1 \text{ kg}}{1000 \text{ g}} = 0.0300 \text{ kg } H_2O$$

$$\textbf{molality} = \frac{1.11 \text{ mol } HNO_3}{0.0300 \text{ kg } H_2O} = \textbf{37.0 } \textit{m}$$

To calculate the density, let's again assume we have 100 g of solution. Since,

$$d = \frac{\text{mass}}{\text{volume}}$$

we know the mass (100 g) and therefore need to calculate the volume of the solution. We know from the molarity that 15.9 mol of HNO_3 are dissolved in a solution volume of 1000 mL. In 100 g of solution, there are 1.11 moles HNO_3 (calculated above). What volume will 1.11 moles of HNO_3 occupy?

$$1.11 \text{ mol } HNO_3 \times \frac{1000 \text{ mL soln}}{15.9 \text{ mol } HNO_3} = 69.8 \text{ mL soln}$$

Dividing the mass by the volume gives the density.

$$d = \frac{100 \text{ g}}{69.8 \text{ mL}} = \textbf{1.43 g/mL}$$

12.114 $P_A = X_A P_A^\circ$

$P_{\text{ethanol}} = (0.62)(108 \text{ mmHg}) = 67.0 \text{ mmHg}$

$P_{\text{1-propanol}} = (0.38)(40.0 \text{ mmHg}) = 15.2 \text{ mmHg}$

In the vapor phase:

$$X_{\text{ethanol}} = \frac{67.0}{67.0 + 15.2} = \textbf{0.815}$$

12.115   Since the total volume is less than the sum of the two volumes, the ethanol and water must have an intermolecular attraction that results in an overall smaller volume.

12.116 NH_3 can form hydrogen bonds with water; NCl_3 cannot. (Like dissolves like.)

12.117   In solution, the $Al(H_2O)_6^{3+}$ ions neutralize the charge on the hydrophobic colloidal soil particles, leading to their precipitation from water.

12.118   We can calculate the molality of the solution from the freezing point depression.

$$\Delta T_f = K_f m$$

$$0.203 = 1.86 \, m$$

$$m = \frac{0.203}{1.86} = 0.109 \, m$$

The molality of the original solution was 0.106 m. Some of the solution has ionized to H^+ and CH_3COO^-.

$$CH_3COOH \rightleftharpoons CH_3COO^- + H^+$$

	CH₃COOH	CH₃COO⁻	H⁺
Initial	0.106 m	0	0
Change	$-x$	$+x$	$+x$
Equil.	0.106 $m - x$	x	x

At equilibrium, the total concentration of species in solution is 0.109 m.

$$(0.106 - x) + 2x = 0.109 \ m$$

$$x = 0.003 \ m$$

The percentage of acid that has undergone ionization is:

$$\frac{0.003 \ m}{0.106 \ m} \times 100\% = \textbf{2.8\%}$$

12.119   Egg yolk contains lecithins which solubilize oil in water (See Figure 12.20 of the text). The nonpolar oil becomes soluble in water because the nonpolar tails of lecithin dissolve in the oil, and the polar heads of the lecithin molecules dissolve in polar water (like dissolves like).

12.120   First, we can calculate the molality of the solution from the freezing point depression.

$$\Delta T_f = (5.12)m$$

$$(5.5 - 3.5) = (5.12)m$$

$$m = 0.39$$

Next, from the definition of molality, we can calculate the moles of solute.

$$m = \frac{\text{mol solute}}{\text{kg solvent}}$$

$$0.39 \ m = \frac{\text{mol solute}}{80 \times 10^{-3} \ \text{kg benzene}}$$

$$\text{mol solute} = 0.031 \ \text{mol}$$

The molar mass ($\mathcal{M}$) of the solute is:

$$\frac{3.8 \ \text{g}}{0.031 \ \text{mol}} = \textbf{1.2} \times \textbf{10}^2 \ \textbf{g/mol}$$

The molar mass of CH_3COOH is 60.05 g/mol. Since the molar mass of the solute calculated from the freezing point depression is twice this value, the structure of the solute most likely is a dimer that is held together by hydrogen bonds.

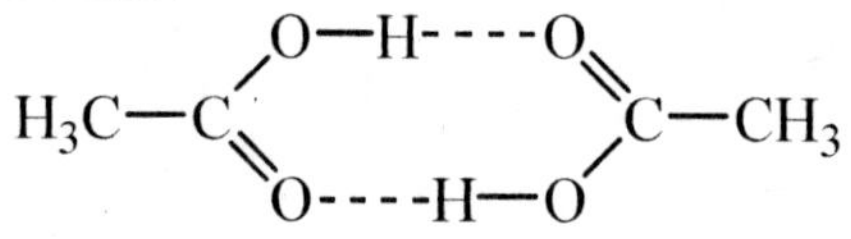

$$H_3C-C\underset{\underset{\textstyle O----H-O}{}}{\overset{\overset{\textstyle O-H----O}{}}{}}C-CH_3 \qquad \text{A dimer}$$

12.121   192 μg = 192×10^{-6} g or 1.92×10^{-4} g

$$\text{mass of lead/L} = \frac{1.92 \times 10^{-4} \ \text{g}}{2.6 \ \text{L}} = 7.4 \times 10^{-5} \ \text{g/L}$$

Safety limit: 0.050 ppm implies a mass of 0.050 g Pb per 1×10^6 g of water. 1 liter of water has a mass of 1000 g.

$$\text{mass of lead} = \frac{0.050 \ \text{g Pb}}{1 \times 10^6 \ \text{g H}_2\text{O}} \times 1000 \ \text{g H}_2\text{O} = 5.0 \times 10^{-5} \ \text{g/L}$$

The concentration of lead calculated above (7.4×10^{-5} g/L) exceeds the safety limit of 5.0×10^{-5} g/L. Don't drink the water!

12.122   (a) $\Delta T_f = K_f m$

$2 = (1.86)(m)$

molality = **1.1 m**

This concentration is too high and is *not* a reasonable physiological concentration.

(b)   Although the protein is present in low concentrations, it can prevent the formation of ice crystals.

12.123   If the can is tapped with a metal object, the vibration releases the bubbles and they move to the top of the can where they join up to form bigger bubbles or mix with the gas at the top of the can. When the can is opened, the gas escapes without dragging the liquid out of the can with it. If the can is not tapped, the bubbles expand when the pressure is released and push the liquid out ahead of them.

12.124   As the water freezes, dissolved minerals in the water precipitate from solution. The minerals refract light and create an opaque appearance.

12.125   Starting with $n = kP$ and substituting into the ideal gas equation ($PV = nRT$), we find:

$$PV = (kP)RT$$
$$V = kRT$$

This equation shows that the volume of a gas that dissolves in a given amount of solvent is dependent on the *temperature*, not the pressure of the gas.

12.126   To solve for the molality of the solution, we need the moles of solute (urea) and the kilograms of solvent (water). If we assume that we have 1 mole of water, we know the mass of water. Using the change in vapor pressure, we can solve for the mole fraction of urea and then the moles of urea.

Using Equation (12.5) of the text, we solve for the mole fraction of urea.

$\Delta P = 23.76$ mmHg $- 22.98$ mmHg $= 0.78$ mmHg

$$\Delta P = X_2 P_1^\circ = X_{urea} P_{water}^\circ$$

$$X_{urea} = \frac{\Delta P}{P_{water}^\circ} = \frac{0.78 \text{ mmHg}}{23.76 \text{ mmHg}} = 0.033$$

Assuming that we have 1 mole of water, we can now solve for moles of urea.

$$X_{urea} = \frac{\text{mol urea}}{\text{mol urea} + \text{mol water}}$$

$$0.033 = \frac{n_{urea}}{n_{urea} + 1}$$

$0.033 n_{urea} + 0.033 = n_{urea}$

$0.033 = 0.967 n_{urea}$

$n_{urea} = 0.034$ mol

1 mole of water has a mass of 18.02 g or 0.01802 kg. We now know the moles of solute (urea) and the kilograms of solvent (water), so we can solve for the molality of the solution.

$$m = \frac{\text{mol solute}}{\text{kg solvent}} = \frac{0.034 \text{ mol}}{0.01802 \text{ kg}} = \textbf{1.9 } \boldsymbol{m}$$

12.127 (a)

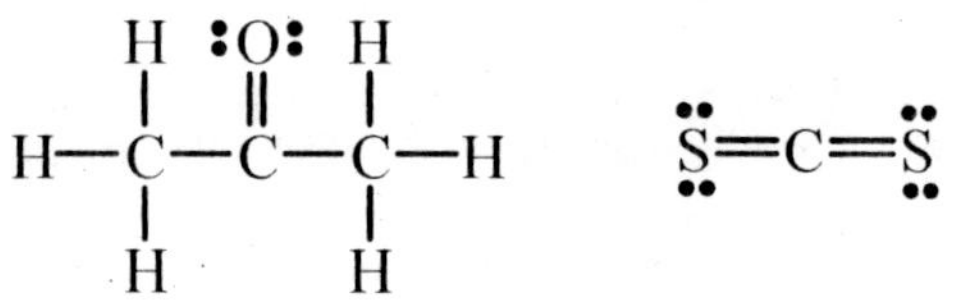

$$\begin{array}{c} \quad\quad\; H \;\; :O: \; H \\ \quad\quad\; | \quad\;\; \| \quad\; | \\ H-C-C-C-H \\ \quad\quad\; | \quad\quad\quad | \\ \quad\quad\; H \quad\quad\; H \end{array} \qquad \ddot{S}{=}C{=}\ddot{S}$$

Acetone is a polar molecule and carbon disulfide is a nonpolar molecule. The intermolecular attractions between acetone and CS_2 will be weaker than those between acetone molecules and those between CS_2 molecules. Because of the weak attractions between acetone and CS_2, there is a greater tendency for these molecules to leave the solution compared to an ideal solution. Consequently, the vapor pressure of the solution is greater than the sum of the vapor pressures as predicted by Raoult's law for the same concentration.

(b)   Let acetone be component A of the solution and carbon disulfide component B. For an ideal solution, $P_A = X_A P_A^\circ$, $P_B = X_B P_B^\circ$, and $P_T = P_A + P_B$.

$$P_{\text{acetone}} = X_A P_A^\circ = (0.60)(349 \text{ mmHg}) = 209.4 \text{ mmHg}$$

$$P_{CS_2} = X_B P_B^\circ = (0.40)(501 \text{ mmHg}) = 200.4 \text{ mmHg}$$

$$P_T = (209.4 + 200.4)\text{mmHg} = \textbf{410 mmHg}$$

Note that the ideal vapor pressure is less than the actual vapor pressure of 615 mmHg.

(c)   The behavior of the solution described in part (a) gives rise to a positive deviation from Raoult's law [See Figure 12.8(a) of the text]. In this case, the heat of solution is **positive** (that is, mixing is an endothermic process).

12.128 (a)   The solution is prepared by mixing equal masses of A and B. Let's assume that we have 100 grams of each component. We can convert to moles of each substance and then solve for the mole fraction of each component.

Since the molar mass of A is 100 g/mol, we have 1.00 mole of A. The moles of B are:

$$100 \text{ g B} \times \frac{1 \text{ mol B}}{110 \text{ g B}} = 0.909 \text{ mol B}$$

The mole fraction of A is:

$$X_A = \frac{n_A}{n_A + n_B} = \frac{1}{1 + 0.909} = \textbf{0.524}$$

Since this is a two component solution, the mole fraction of B is: $X_B = 1 - 0.524 = \textbf{0.476}$

(b)   We can use Equation (12.4) of the text and the mole fractions calculated in part (a) to calculate the partial pressures of A and B over the solution.

$$P_A = X_A P_A^\circ = (0.524)(95 \text{ mmHg}) = \mathbf{50 \text{ mmHg}}$$

$$P_B = X_B P_B^\circ = (0.476)(42 \text{ mmHg}) = \mathbf{20 \text{ mmHg}}$$

(c)   Recall that pressure of a gas is directly proportional to moles of gas ($P \propto n$). The ratio of the partial pressures calculated in part (b) is 50 : 20, and therefore the ratio of moles will also be 50 : 20. Let's assume that we have 50 moles of A and 20 moles of B. We can solve for the mole fraction of each component and then solve for the vapor pressures using Equation (12.4) of the text.

The mole fraction of A is:

$$X_A = \frac{n_A}{n_A + n_B} = \frac{50}{50 + 20} = \mathbf{0.71}$$

Since this is a two component solution, the mole fraction of B is: $X_B = 1 - 0.71 = \mathbf{0.29}$

The vapor pressures of each component above the solution are:

$$P_A = X_A P_A^\circ = (0.71)(95 \text{ mmHg}) = \mathbf{67 \text{ mmHg}}$$

$$P_B = X_B P_B^\circ = (0.29)(42 \text{ mmHg}) = \mathbf{12 \text{ mmHg}}$$

CHAPTER 13
CHEMICAL KINETICS

13.5 In general for a reaction $a\text{A} + b\text{B} \rightarrow c\text{C} + d\text{D}$

$$\text{rate} = -\frac{1}{a}\frac{\Delta[\text{A}]}{\Delta t} = -\frac{1}{b}\frac{\Delta[\text{B}]}{\Delta t} = \frac{1}{c}\frac{\Delta[\text{C}]}{\Delta t} = \frac{1}{d}\frac{\Delta[\text{D}]}{\Delta t}$$

(a) $\quad \text{rate} = -\frac{\Delta[\text{H}_2]}{\Delta t} = -\frac{\Delta[\text{I}_2]}{\Delta t} = \frac{1}{2}\frac{\Delta[\text{HI}]}{\Delta t}$

(b) $\quad \text{rate} = -\frac{1}{5}\frac{\Delta[\text{Br}^-]}{\Delta t} = -\frac{\Delta[\text{BrO}_3^-]}{\Delta t} = -\frac{1}{6}\frac{\Delta[\text{H}^+]}{\Delta t} = \frac{1}{3}\frac{\Delta[\text{Br}_2]}{\Delta t}$

Note that because the reaction is carried out in the aqueous phase, we do not monitor the concentration of water.

13.6 **(a)** $\quad \text{rate} = -\frac{1}{2}\frac{\Delta[\text{H}_2]}{\Delta t} = -\frac{\Delta[\text{O}_2]}{\Delta t} = \frac{1}{2}\frac{\Delta[\text{H}_2\text{O}]}{\Delta t}$

(b) $\quad \text{rate} = -\frac{1}{4}\frac{\Delta[\text{NH}_3]}{\Delta t} = -\frac{1}{5}\frac{\Delta[\text{O}_2]}{\Delta t} = \frac{1}{4}\frac{\Delta[\text{NO}]}{\Delta t} = \frac{1}{6}\frac{\Delta[\text{H}_2\text{O}]}{\Delta t}$

13.7 $\quad \text{Rate} = -\frac{1}{2}\frac{\Delta[\text{NO}]}{\Delta t} \qquad \frac{\Delta[\text{NO}]}{\Delta t} = -0.066 \ M/s$

$$-\frac{1}{2}\frac{\Delta[\text{NO}]}{\Delta t} = \frac{1}{2}\frac{\Delta[\text{NO}_2]}{\Delta t}$$

(a) $\quad \dfrac{\Delta[\text{NO}_2]}{\Delta t} = \textbf{0.066 } \textbf{\textit{M}/s}$

(b) $\quad -\frac{1}{2}\frac{\Delta[\text{NO}]}{\Delta t} = -\frac{\Delta[\text{O}_2]}{\Delta t}$

$$\frac{\Delta[\text{O}_2]}{\Delta t} = \frac{-0.066 \ M/s}{2} = \textbf{-0.033 } \textbf{\textit{M}/s}$$

13.8 **Strategy:** The rate is defined as the change in concentration of a reactant or product with time. Each "change in concentration" term is divided by the corresponding stoichiometric coefficient. Terms involving reactants are preceded by a minus sign.

$$\text{rate} = -\frac{\Delta[\text{N}_2]}{\Delta t} = -\frac{1}{3}\frac{\Delta[\text{H}_2]}{\Delta t} = \frac{1}{2}\frac{\Delta[\text{NH}_3]}{\Delta t}$$

Solution:

(a) If hydrogen is reacting at the rate of -0.074 M/s, the rate at which ammonia is being formed is

$$\frac{1}{2}\frac{\Delta[NH_3]}{\Delta t} = -\frac{1}{3}\frac{\Delta[H_2]}{\Delta t}$$

or

$$\frac{\Delta[NH_3]}{\Delta t} = -\frac{2}{3}\frac{\Delta[H_2]}{\Delta t}$$

$$\frac{\Delta[NH_3]}{\Delta t} = -\frac{2}{3}(-0.074 \ M/s) = \mathbf{0.049 \ \textit{M}/s}$$

(b) The rate at which nitrogen is reacting must be:

$$\frac{\Delta[N_2]}{\Delta t} = \frac{1}{3}\frac{\Delta[H_2]}{\Delta t} = \frac{1}{3}(-0.074 \ M/s) = \mathbf{-0.025 \ \textit{M}/s}$$

Will the rate at which ammonia forms always be twice the rate of reaction of nitrogen, or is this true only at the instant described in this problem?

13.15 $\textbf{rate} = k[NH_4^+][NO_2^-] = (3.0 \times 10^{-4}/M\cdot s)(0.26 \ M)(0.080 \ M) = \mathbf{6.2 \times 10^{-6} \ \textit{M}/s}$

13.16 Assume the rate law has the form:

$$\text{rate} = k[F_2]^x[ClO_2]^y$$

To determine the order of the reaction with respect to F_2, find two experiments in which the $[ClO_2]$ is held constant. Compare the data from experiments 1 and 3. When the concentration of F_2 is doubled, the reaction rate doubles. Thus, the reaction is *first-order* in F_2.

To determine the order with respect to ClO_2, compare experiments 1 and 2. When the ClO_2 concentration is quadrupled, the reaction rate quadruples. Thus, the reaction is *first-order* in ClO_2.

The rate law is:

$$\text{rate} = k[F_2][ClO_2]$$

The value of k can be found using the data from any of the experiments. If we take the numbers from the second experiment we have:

$$k = \frac{\text{rate}}{[F_2][ClO_2]} = \frac{4.8 \times 10^{-3} \ M/s}{(0.10 \ M)(0.040 \ M)} = 1.2 \ M^{-1}s^{-1}$$

Verify that the same value of k can be obtained from the other sets of data.

Since we now know the rate law and the value of the rate constant, we can calculate the rate at any concentration of reactants.

$$\textbf{rate} = k[F_2][ClO_2] = (1.2 \ M^{-1}s^{-1})(0.010 \ M)(0.020 \ M) = \mathbf{2.4 \times 10^{-4} \ \textit{M}/s}$$

13.17 By comparing the first and second sets of data, we see that changing [B] does not affect the rate of the reaction. Therefore, the reaction is zero order in B. By comparing the first and third sets of data, we see that doubling [A] doubles the rate of the reaction. This shows that the reaction is first order in A.

$$\text{rate} = k[A]$$

From the first set of data:

$$3.20 \times 10^{-1} \ M/s = k(1.50 \ M)$$

$$k = 0.213 \ s^{-1}$$

What would be the value of k if you had used the second or third set of data? Should k be constant?

13.18   Strategy: We are given a set of concentrations and rate data and asked to determine the order of the reaction and the initial rate for specific concentrations of X and Y. To determine the order of the reaction, we need to find the rate law for the reaction. We assume that the rate law takes the form

$$\text{rate} = k[X]^x[Y]^y$$

How do we use the data to determine x and y? Once the orders of the reactants are known, we can calculate k for any set of rate and concentrations. Finally, the rate law enables us to calculate the rate at any concentrations of X and Y.

Solution:

(a)   Experiments 2 and 5 show that when we double the concentration of X at constant concentration of Y, the rate quadruples. Taking the ratio of the rates from these two experiments

$$\frac{\text{rate}_5}{\text{rate}_2} = \frac{0.509 \ M/s}{0.127 \ M/s} \approx 4 = \frac{k(0.40)^x(0.30)^y}{k(0.20)^x(0.30)^y}$$

Therefore,

$$\frac{(0.40)^x}{(0.20)^x} = 2^x = 4$$

or, $x = 2$. That is, the reaction is second order in X. Experiments 2 and 4 indicate that doubling [Y] at constant [X] doubles the rate. Here we write the ratio as

$$\frac{\text{rate}_4}{\text{rate}_2} = \frac{0.254 \ M/s}{0.127 \ M/s} = 2 = \frac{k(0.20)^x(0.60)^y}{k(0.20)^x(0.30)^y}$$

Therefore,

$$\frac{(0.60)^y}{(0.30)^y} = 2^y = 2$$

or, $y = 1$. That is, the reaction is first order in Y. Hence, the rate law is given by:

$$\text{rate} = k[X]^2[Y]$$

The order of the reaction is $(2 + 1) = 3$. The reaction is *3rd-order*.

(b)   The rate constant k can be calculated using the data from any one of the experiments. Rearranging the rate law and using the first set of data, we find:

$$k = \frac{\text{rate}}{[X]^2[Y]} = \frac{0.053 \ M/s}{(0.10 \ M)^2(0.50 \ M)} = 10.6 \ M^{-2}s^{-1}$$

Next, using the known rate constant and substituting the concentrations of X and Y into the rate law, we can calculate the initial rate of disappearance of X.

$$\text{rate} = (10.6\ M^{-2}s^{-1})(0.30\ M)^2(0.40\ M) = \textbf{0.38}\ \textbf{\textit{M}/s}$$

13.19   **(a)**   second order,   **(b)**   zero order,   **(c)**   1.5 order,   **(d)**   third order

13.20   **(a)**   For a reaction first-order in A,

$$\text{Rate} = k[A]$$

$$1.6 \times 10^{-2}\ M/s = k(0.35\ M)$$

$$k = \textbf{0.046}\ \textbf{s}^{-1}$$

(b)   For a reaction second-order in A,

$$\text{Rate} = k[A]^2$$

$$1.6 \times 10^{-2}\ M/s = k(0.35\ M)^2$$

$$k = \textbf{0.13}\ \textbf{\textit{M}}^{-1}\textbf{s}^{-1}$$

13.21   The graph below is a plot of ln P vs. time. Since the plot is linear, the reaction is 1st order.

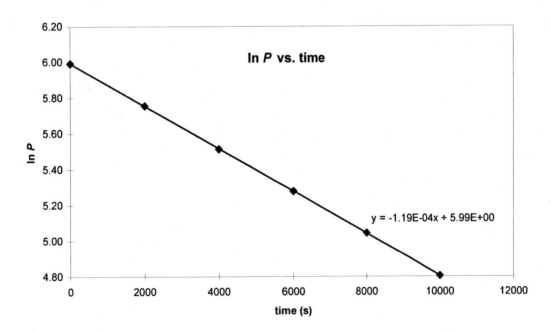

Slope = $-k$

$$k = \textbf{1.19} \times \textbf{10}^{-4}\ \textbf{s}^{-1}$$

13.22   Let P_0 be the pressure of $ClCO_2CCl_3$ at $t = 0$, and let x be the decrease in pressure after time t. Note that from the coefficients in the balanced equation that the loss of 1 atmosphere of $ClCO_2CCl_3$ results in the formation of two atmospheres of $COCl_2$. We write:

$$ClCO_2CCl_3 \rightarrow 2COCl_2$$

Time	$[ClCO_2CCl_3]$	$[COCl_2]$
$t = 0$	P_0	0
$t = t$	$P_0 - x$	$2x$

Thus the change (increase) in pressure (ΔP) is $2x - x = x$. We have:

$t(s)$	P (mmHg)	$\Delta P = x$	$P_{ClCO_2CCl_3}$	$\ln P_{ClCO_2CCl_3}$	$\dfrac{1}{P_{ClCO_2CCl_3}}$
0	15.76	0.00	15.76	2.757	0.0635
181	18.88	3.12	12.64	2.537	0.0791
513	22.79	7.03	8.73	2.167	0.115
1164	27.08	11.32	4.44	1.491	0.225

If the reaction is first order, then a plot of $\ln P_{ClCO_2CCl_3}$ vs. t would be linear. If the reaction is second order, a plot of $1/P_{ClCO_2CCl_3}$ vs. t would be linear. The two plots are shown below.

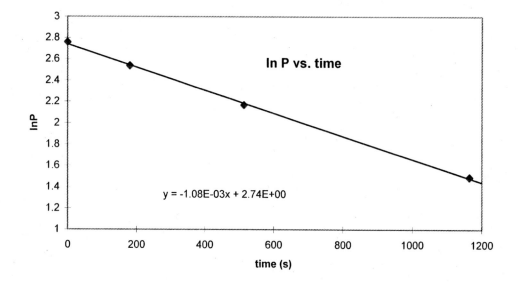

y = -1.08E-03x + 2.74E+00

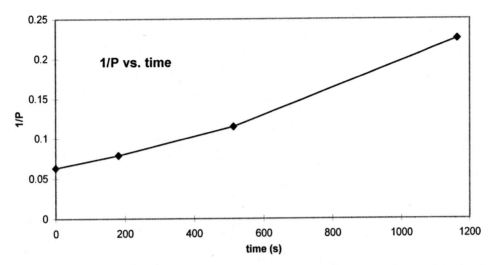

From the graphs we see that the reaction must be **first-order**. For a first-order reaction, the slope is equal to $-k$. The equation of the line is given on the graph. The rate constant is: $k = 1.08 \times 10^{-3} \text{ s}^{-1}$.

13.27   We know that half of the substance decomposes in a time equal to the half-life, $t_{1/2}$. This leaves half of the compound. Half of what is left decomposes in a time equal to another half-life, so that only one quarter of the original compound remains. We see that 75% of the original compound has decomposed after two half-lives. Thus two half-lives equal one hour, or the half-life of the decay is 30 min.

$$100\% \text{ starting compound} \xrightarrow{t_{1/2}} 50\% \text{ starting compound} \xrightarrow{t_{1/2}} 25\% \text{ starting compound}$$

Using first order kinetics, we can solve for k using Equation (13.3) of the text, with $[A]_0 = 100$ and $[A] = 25$,

$$\ln \frac{[A]_t}{[A]_0} = -kt$$

$$\ln \frac{25}{100} = -k(60 \text{ min})$$

$$k = -\frac{\ln(0.25)}{60 \text{ min}} = 0.023 \text{ min}^{-1}$$

Then, substituting k into Equation (13.6) of the text, you arrive at the same answer for $t_{1/2}$.

$$t_{\frac{1}{2}} = \frac{0.693}{k} = \frac{0.693}{0.023 \text{ min}^{-1}} = \textbf{30 min}$$

13.28   **(a)**
Strategy: To calculate the rate constant, k, from the half-life of a first-order reaction, we use Equation (13.6) of the text.

Solution: For a first-order reaction, we only need the half-life to calculate the rate constant. From Equation (13.6)

$$k = \frac{0.693}{t_{\frac{1}{2}}}$$

$$k = \frac{0.693}{35.0 \text{ s}} = 0.0198 \text{ s}^{-1}$$

(b)

Strategy: The relationship between the concentration of a reactant at different times in a first-order reaction is given by Equations (13.3) and (13.4) of the text. We are asked to determine the time required for 95% of the phosphine to decompose. If we initially have 100% of the compound and 95% has reacted, then what is left must be (100% − 95%), or 5%. Thus, the ratio of the percentages will be equal to the ratio of the actual concentrations; that is, $[A]_t/[A]_0 = 5\%/100\%$, or 0.05/1.00.

Solution: The time required for 95% of the phosphine to decompose can be found using Equation (13.3) of the text.

$$\ln \frac{[A]_t}{[A]_0} = -kt$$

$$\ln \frac{(0.05)}{(1.00)} = -(0.0198 \text{ s}^{-1})t$$

$$t = -\frac{\ln(0.0500)}{0.0198 \text{ s}^{-1}} = 151 \text{ s}$$

13.29   **(a)**   Since the reaction is known to be second-order, the relationship between reactant concentration and time is given by Equation (13.7) of the text. The problem supplies the rate constant and the initial (time = 0) concentration of NOBr. The concentration after 22s can be found easily.

$$\frac{1}{[NOBr]_t} = kt + \frac{1}{[NOBr]_0}$$

$$\frac{1}{[NOBr]_t} = (0.80/M \cdot \text{s})(22\text{s}) + \frac{1}{0.086 \ M}$$

$$\frac{1}{[NOBr]_t} = 29 \ M^{-1}$$

[NOBr] = 0.034 M

If the reaction were first order with the same k and initial concentration, could you calculate the concentration after 22 s? If the reaction were first order and you were given the $t_{1/2}$, could you calculate the concentration after 22 s?

(b)   The half-life for a second-order reaction *is* dependent on the initial concentration. The half-lives can be calculated using Equation (13.8) of the text.

$$t_{\frac{1}{2}} = \frac{1}{k[A]_0}$$

$$t_{\frac{1}{2}} = \frac{1}{(0.80/M \cdot \text{s})(0.072 \ M)}$$

$$t_{\frac{1}{2}} = 17 \text{ s}$$

For an initial concentration of 0.054 M, you should find $t_{\frac{1}{2}} = 23\,\text{s}$. Note that the half-life of a second-order reaction is inversely proportional to the initial reactant concentration.

13.30 $\dfrac{1}{[A]} = \dfrac{1}{[A]_0} + kt$

$\dfrac{1}{0.28} = \dfrac{1}{0.62} + 0.54t$

$t = 3.6\,\text{s}$

13.37   Graphing Equation (13.11) of the text requires plotting $\ln k$ versus $1/T$. The graph is shown below.

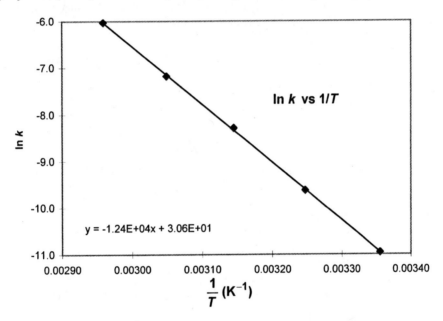

The slope of the line is -1.24×10^4 K, which is $-E_a/R$. The activation energy is:

$$-E_a = \text{slope} \times R = (-1.24 \times 10^4\,\text{K}) \times (8.314\,\text{J/K}\cdot\text{mol})$$

$$E_a = 1.03 \times 10^5\,\text{J/mol} = 103\,\text{kJ/mol}$$

Do you need to know the order of the reaction to find the activation energy? Is it possible to have a negative activation energy? What would a potential energy versus reaction coordinate diagram look like in such a case?

13.38   **Strategy:** A modified form of the Arrhenius equation relates two rate constants at two different temperatures [see Equation (13.12) of the text]. Make sure the units of R and E_a are consistent. Since the rate of the reaction at 250°C is 1.50×10^3 times faster than the rate at 150°C, the ratio of the rate constants, k, is also $1.50 \times 10^3 : 1$, because rate and rate constant are directly proportional.

Solution: The data are: $T_1 = 250°C = 523$ K, $T_2 = 150°C = 423$ K, and $k_1/k_2 = 1.50 \times 10^3$. Substituting into Equation (13.12) of the text,

$$\ln \frac{k_1}{k_2} = \frac{E_a}{R}\left(\frac{T_1 - T_2}{T_1 T_2}\right)$$

$$\ln(1.50 \times 10^3) = \frac{E_a}{8.314 \text{ J/mol·K}}\left(\frac{523 \text{ K} - 423 \text{ K}}{(523 \text{ K})(423 \text{ K})}\right)$$

$$7.31 = \frac{E_a}{8.314 \dfrac{\text{J}}{\text{mol·K}}}\left(4.52 \times 10^{-4} \frac{1}{\text{K}}\right)$$

$$E_a = 1.35 \times 10^5 \text{ J/mol} = 135 \text{ kJ/mol}$$

13.39 The appropriate value of R is 8.314 J/K mol, not 0.0821 L·atm/mol·K. You must also use the activation energy value of 63000 J/mol (why?). Once the temperature has been converted to Kelvin, the rate constant is:

$$k = Ae^{-E_a/RT} = (8.7 \times 10^{12} \text{ s}^{-1})e^{-\left[\frac{63000 \text{ J/mol}}{(8.314 \text{ J/mol·K})(348 \text{ K})}\right]} = (8.7 \times 10^{12} \text{ s}^{-1})(3.5 \times 10^{-10})$$

$$k = 3.0 \times 10^3 \text{ s}^{-1}$$

Can you tell from the units of k what the order of the reaction is?

13.40 Use a modified form of the Arrhenius equation to calculate the temperature at which the rate constant is $8.80 \times 10^{-4} \text{ s}^{-1}$.

$$\ln \frac{k_1}{k_2} = \frac{E_a}{R}\left(\frac{1}{T_2} - \frac{1}{T_1}\right)$$

$$\ln\left(\frac{4.60 \times 10^{-4} \text{ s}^{-1}}{8.80 \times 10^{-4} \text{ s}^{-1}}\right) = \frac{1.04 \times 10^5 \text{ J/mol}}{8.314 \text{ J/mol·K}}\left(\frac{1}{T_2} - \frac{1}{623 \text{ K}}\right)$$

$$\ln(0.523) = (1.25 \times 10^4 \text{ K})\left(\frac{1}{T_2} - \frac{1}{623 \text{ K}}\right)$$

$$-0.648 + 20.1 = \frac{1.25 \times 10^4 \text{ K}}{T_2}$$

$$19.5 T_2 = 1.25 \times 10^4 \text{ K}$$

$$T_2 = 641 \text{ K} = 368°C$$

13.41 Let k_1 be the rate constant at 295 K and $2k_1$ the rate constant at 305 K. We write:

$$\ln \frac{k_1}{2k_1} = \frac{E_a}{R}\left(\frac{T_1 - T_2}{T_1 T_2}\right)$$

$$-0.693 = \frac{E_a}{8.314 \text{ J/K·mol}}\left(\frac{295 \text{ K} - 305 \text{ K}}{(295 \text{ K})(305 \text{ K})}\right)$$

$$E_a = 5.18 \times 10^4 \text{ J/mol} = 51.8 \text{ kJ/mol}$$

13.42   Since the ratio of rates is equal to the ratio of rate constants, we can write:

$$\ln\frac{\text{rate}_1}{\text{rate}_2} = \ln\frac{k_1}{k_2}$$

$$\ln\frac{k_1}{k_2} = \ln\left(\frac{2.0 \times 10^2}{39.6}\right) = \frac{E_a}{8.314 \text{ J/K} \cdot \text{mol}}\left(\frac{(300 \text{ K} - 278 \text{ K})}{(300 \text{ K})(278 \text{ K})}\right)$$

$$E_a = 5.10 \times 10^4 \text{ J/mol} = 51.0 \text{ kJ/mol}$$

13.51   **(a)**   The order of the reaction is simply the sum of the exponents in the rate law (Section 13.2 of the text). The order of this reaction is 2.

(b)   The rate law reveals the identity of the substances participating in the slow or rate-determining step of a reaction mechanism. This rate law implies that the slow step involves the reaction of a molecule of NO with a molecule of Cl_2. If this is the case, then the first reaction shown must be the rate-determining (slow) step, and the second reaction must be much faster.

13.52   **(a)**
Strategy: We are given information as to how the concentrations of X_2, Y, and Z affect the rate of the reaction and are asked to determine the rate law. We assume that the rate law takes the form

$$\text{rate} = k[X_2]^x[Y]^y[Z]^z$$

How do we use the information to determine x, y, and z?

Solution: Since the reaction rate doubles when the X_2 concentration is doubled, the reaction is first-order in X. The reaction rate triples when the concentration of Y is tripled, so the reaction is also first-order in Y. The concentration of Z has no effect on the rate, so the reaction is zero-order in Z.

The rate law is:

$$\text{rate} = k[X_2][Y]$$

(b)   If a change in the concentration of Z has no effect on the rate, the concentration of Z is not a term in the rate law. This implies that Z does not participate in the rate-determining step of the reaction mechanism.

(c)
Strategy: The rate law, determined in part (a), shows that the slow step involves reaction of a molecule of X_2 with a molecule of Y. Since Z is not present in the rate law, it does not take part in the slow step and must appear in a fast step at a later time. (If the fast step involving Z happened before the rate-determining step, the rate law would involve Z in a more complex way.)

Solution: A mechanism that is consistent with the rate law could be:

$$X_2 + Y \longrightarrow XY + X \qquad \text{(slow)}$$

$$X + Z \longrightarrow XZ \qquad \text{(fast)}$$

The rate law only tells us about the slow step. Other mechanisms with different subsequent fast steps are possible. Try to invent one.

Check: The rate law written from the rate-determining step in the proposed mechanism matches the rate law determined in part (a). Also, the two elementary steps add to the overall balanced equation given in the problem.

13.53   The first step involves forward and reverse reactions that are much faster than the second step. The rates of the reaction in the first step are given by:

$$\text{forward rate} = k_1[O_3]$$

$$\text{reverse rate} = k_{-1}[O][O_2]$$

It is assumed that these two processes rapidly reach a state of dynamic equilibrium in which the rates of the forward and reverse reactions are equal:

$$k_1[O_3] = k_{-1}[O][O_2]$$

If we solve this equality for [O] we have:

$$[O] = \frac{k_1[O_3]}{k_{-1}[O_2]}$$

The equation for the rate of the second step is:

$$\text{rate} = k_2[O][O_3]$$

If we substitute the expression for [O] derived from the first step, we have the experimentally verified rate law.

$$\text{overall rate} = \frac{k_1 k_2}{k_{-1}} \frac{[O_3]^2}{[O_2]} = k \frac{[O_3]^2}{[O_2]}$$

The above rate law predicts that higher concentrations of O_2 will decrease the rate. This is because of the reverse reaction in the first step of the mechanism. Notice that if more O_2 molecules are present, they will serve to scavenge free O atoms and thus slow the disappearance of O_3.

13.54   The experimentally determined rate law is first order in H_2 and second order in NO. In Mechanism I the slow step is bimolecular and the rate law would be:

$$\text{rate} = k[H_2][NO]$$

Mechanism I can be discarded.

The rate-determining step in Mechanism II involves the simultaneous collision of two NO molecules with one H_2 molecule. The rate law would be:

$$\text{rate} = k[H_2][NO]^2$$

Mechanism II is a possibility.

In Mechanism III we assume the forward and reverse reactions in the first fast step are in dynamic equilibrium, so their rates are equal:

$$k_f[NO]^2 = k_r[N_2O_2]$$

The slow step is bimolecular and involves collision of a hydrogen molecule with a molecule of N_2O_2. The rate would be:

$$\text{rate} = k_2[H_2][N_2O_2]$$

If we solve the dynamic equilibrium equation of the first step for $[N_2O_2]$ and substitute into the above equation, we have the rate law:

$$\text{rate} = \frac{k_2 k_f}{k_r}[H_2][NO]^2 = k[H_2][NO]^2$$

Mechanism III is also a possibility. Can you suggest an experiment that might help to decide between the two mechanisms?

13.61   Higher temperatures may disrupt the intricate three dimensional structure of the enzyme, thereby reducing or totally destroying its catalytic activity.

13.62   The rate-determining step involves the breakdown of ES to E and P. The rate law for this step is:

$$\text{rate} = k_2[ES]$$

In the first elementary step, the intermediate ES is in equilibrium with E and S. The equilibrium relationship is:

$$\frac{[ES]}{[E][S]} = \frac{k_1}{k_{-1}}$$

or

$$[ES] = \frac{k_1}{k_{-1}}[E][S]$$

Substitute [ES] into the rate law expression.

$$\textbf{rate} = k_2[ES] = \frac{k_1 k_2}{k_{-1}}[E][S]$$

13.63   In each case the gas pressure will either increase or decrease. The pressure can be related to the progress of the reaction through the balanced equation. In (d), an electrical conductance measurement could also be used.

13.64   Temperature, energy of activation, concentration of reactants, and a catalyst.

13.65   Strictly, the temperature must be given whenever the rate or rate constant of a reaction is quoted.

13.66   First, calculate the radius of the 10.0 cm^3 sphere.

$$V = \frac{4}{3}\pi r^3$$

$$10.0 \text{ cm}^3 = \frac{4}{3}\pi r^3$$

$$r = 1.34 \text{ cm}$$

The surface area of the sphere is:

$$\text{area} = 4\pi r^2 = 4\pi(1.34 \text{ cm})^2 = \textbf{22.6 cm}^2$$

Next, calculate the radius of the 1.25 cm^3 sphere.

$$V = \frac{4}{3}\pi r^3$$

$$1.25 \text{ cm}^3 = \frac{4}{3}\pi r^3$$

$$r = 0.668 \text{ cm}$$

The surface area of one sphere is:

$$\text{area} = 4\pi r^2 = 4\pi(0.668 \text{ cm})^2 = 5.61 \text{ cm}^2$$

$$\textbf{The total area of 8 spheres} = 5.61 \text{ cm}^2 \times 8 = \textbf{44.9 cm}^2$$

Obviously, the surface area of the eight spheres (44.9 cm^2) is greater than that of one larger sphere (22.6 cm^2). A greater surface area promotes the catalyzed reaction more effectively.

It can be dangerous to work in grain elevators, because the large surface area of the grain dust can result in a violent explosion.

13.67

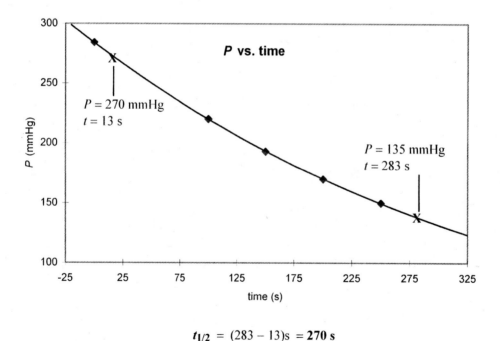

$$t_{1/2} = (283 - 13)\text{s} = \textbf{270 s}$$

13.68   The overall rate law is of the general form: rate $= k[H_2]^x[NO]^y$

(a)   Comparing Experiment #1 and Experiment #2, we see that the concentration of NO is constant and the concentration of H_2 has decreased by one-half. The initial rate has also decreased by one-half. Therefore, the initial rate is directly proportional to the concentration of H_2; $x = 1$.

Comparing Experiment #1 and Experiment #3, we see that the concentration of H_2 is constant and the concentration of NO has decreased by one-half. The initial rate has decreased by one-fourth. Therefore, the initial rate is proportional to the squared concentration of NO; $y = 2$.

The overall rate law is: rate $= k[H_2][NO]^2$, and the order of the reaction is $1 + 2 = 3$.

(b) Using Experiment #1 to calculate the rate constant,

$$\text{rate} = k[H_2][NO]^2$$

$$k = \frac{\text{rate}}{[H_2][NO]^2}$$

$$k = \frac{2.4 \times 10^{-6} \ M/s}{(0.010 \ M)(0.025 \ M)^2} = 0.38 \ M^{-2}s^{-1}$$

(c) Consulting the rate law, we assume that the slow step in the reaction mechanism will probably involve one H_2 molecule and two NO molecules. Additionally the hint tells us that O atoms are an intermediate.

$$
\begin{array}{ll}
H_2 + 2NO \rightarrow N_2 + H_2O + O & \text{slow step} \\
O + H_2 \rightarrow H_2O & \text{fast step} \\
\hline
2H_2 + 2NO \rightarrow N_2 + 2H_2O &
\end{array}
$$

13.69 Since the methanol contains no oxygen–18, the oxygen atom must come from the phosphate group and not the water. The mechanism must involve a bond–breaking process like:

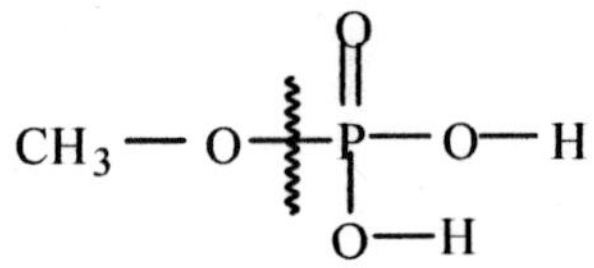

13.70 If water is also the solvent in this reaction, it is present in vast excess over the other reactants and products. Throughout the course of the reaction, the concentration of the water will not change by a measurable amount. As a result, the reaction rate will not appear to depend on the concentration of water.

13.71 Most transition metals have several stable oxidation states. This allows the metal atoms to act as either a source or a receptor of electrons in a broad range of reactions.

13.72 Since the reaction is first order in both A and B, then we can write the rate law expression:

$$\text{rate} = k[A][B]$$

Substituting in the values for the rate, [A], and [B]:

$$4.1 \times 10^{-4} \ M/s = k(1.6 \times 10^{-2})(2.4 \times 10^{-3})$$

$$k = 10.7 \ M^{-1}s^{-1}$$

Knowing that the overall reaction was second order, could you have predicted the units for k?

13.73 **(a)** To determine the rate law, we must determine the exponents in the equation

$$rate = k[CH_3COCH_3]^x[Br_2]^y[H^+]^z$$

To determine the order of the reaction with respect to CH_3COCH_3, find two experiments in which the $[Br_2]$ and $[H^+]$ are held constant. Compare the data from experiments (1) and (5). When the concentration of CH_3COCH_3 is increased by a factor of 1.33, the reaction rate increases by a factor of 1.33. Thus, the reaction is <u>first-order</u> in CH_3COCH_3.

To determine the order with respect to Br_2, compare experiments (1) and (2). When the Br_2 concentration is doubled, the reaction rate does not change. Thus, the reaction is <u>zero-order</u> in Br_2.

To determine the order with respect to H^+, compare experiments (1) and (3). When the H^+ concentration is doubled, the reaction rate doubles. Thus, the reaction is <u>first-order</u> in H^+.

The rate law is:

$$\mathbf{rate = k[CH_3COCH_3][H^+]}$$

(b) Rearrange the rate law from part (a), solving for k.

$$k = \frac{rate}{[CH_3COCH_3][H^+]}$$

Substitute the data from any one of the experiments to calculate k. Using the data from Experiment (1),

$$k = \frac{5.7 \times 10^{-5} \ M/s}{(0.30 \ M)(0.050 \ M)} = 3.8 \times 10^{-3} \ /M \cdot s$$

(c) Let k_2 be the rate constant for the slow step:

$$rate = k_2[CH_3\text{-}\overset{\displaystyle \overset{+}{O}H}{\overset{\|}{C}}\text{---}CH_3][H_2O] \qquad (1)$$

Let k_1 and k_{-1} be the rate constants for the forward and reverse steps in the fast equilibrium.

$$k_1[CH_3COCH_3][H_3O^+] = k_{-1}[CH_3\text{-}\overset{\displaystyle \overset{+}{O}H}{\overset{\|}{C}}\text{---}CH_3][H_2O] \qquad (2)$$

Therefore, Equation (1) becomes

$$rate = \frac{k_1k_2}{k_{-1}}[CH_3COCH_3][H_3O^+]]$$

which is the same as (a), where $k = k_1k_2/k_{-1}$.

13.74 Recall that the pressure of a gas is directly proportional to the number of moles of gas. This comes from the ideal gas equation.

$$P = \frac{nRT}{V}$$

The balanced equation is:

$$2N_2O(g) \longrightarrow 2N_2(g) + O_2(g)$$

From the stoichiometry of the balanced equation, for every one mole of N_2O that decomposes, one mole of N_2 and 0.5 moles of O_2 will be formed. Let's assume that we had 2 moles of N_2O at $t = 0$. After one half-life there will be one mole of N_2O remaining and one mole of N_2 and 0.5 moles of O_2 will be formed. The total number of moles of gas after one half-life will be:

$$n_T = n_{N_2O} + n_{N_2} + n_{O_2} = 1 \text{ mol} + 1 \text{ mol} + 0.5 \text{ mol} = 2.5 \text{ mol}$$

At $t = 0$, there were 2 mol of gas. Now, at $t_{\frac{1}{2}}$, there are 2.5 mol of gas. Since the pressure of a gas is directly proportional to the number of moles of gas, we can write:

$$\frac{2.10 \text{ atm}}{2 \text{ mol gas } (t = 0)} \times 2.5 \text{ mol gas} \left(\text{at } t_{\frac{1}{2}} \right) = \textbf{2.63 atm after one half-life}$$

13.75 Fe^{3+} undergoes a redox cycle: $\quad Fe^{3+} \rightarrow Fe^{2+} \rightarrow Fe^{3+}$

$\quad\quad\quad$ Fe^{3+} oxidizes I^-: $\quad\quad\quad 2Fe^{3+} + 2I^- \rightarrow 2Fe^{2+} + I_2$

$\quad\quad\quad$ Fe^{2+} reduces $S_2O_8^{2-}$: $\quad\quad 2Fe^{2+} + S_2O_8^{2-} \rightarrow 2Fe^{3+} + 2SO_4^{2-}$

$$\overline{\quad\quad\quad\quad\quad\quad\quad\quad 2I^- + S_2O_8^{2-} \rightarrow I_2 + 2SO_4^{2-} \quad\quad\quad\quad}$$

The uncatalyzed reaction is slow because both I^- and $S_2O_8^{2-}$ are negatively charged which makes their mutual approach unfavorable.

13.76 The rate expression for a third order reaction is:

$$\text{rate} = -\frac{\Delta[A]}{\Delta t} = k[A]^3$$

The units for the rate law are:

$$\frac{M}{s} = kM^3$$

$$k = M^{-2}s^{-1}$$

13.77 For a rate law, *zero order* means that the exponent is zero. In other words, the reaction rate is just equal to a constant; it doesn't change as time passes.

(a) The rate law would be:

$$\text{rate} = k[A]^0 = k$$

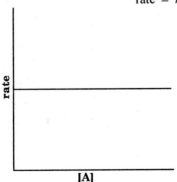

The integrated zero-order rate law is: $[A] = -kt + [A]_0$. Therefore, a plot of $[A]$ versus time should be a straight line with a slope equal to $-k$.

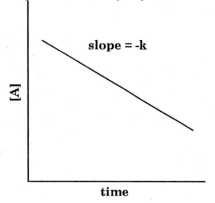

(b) $[A] = [A]_0 - kt$

At $t_{\frac{1}{2}}$, $[A] = \dfrac{[A]_0}{2}$.   Substituting into the above equation:

$$\dfrac{[A]_0}{2} = [A]_0 - kt_{\frac{1}{2}}$$

$$t_{\frac{1}{2}} = \dfrac{[A]_0}{2k}$$

$$k = \dfrac{[A]_0}{2t_{\frac{1}{2}}}$$

(c)   When $[A] = 0$,

$$[A]_0 = kt$$

$$t = \dfrac{[A]_0}{k}$$

Substituting for k,

$$t = \dfrac{[A]_0}{\dfrac{[A]_0}{2t_{\frac{1}{2}}}}$$

$$t = 2t_{\frac{1}{2}}$$

This indicates that the integrated rate law is no longer valid after **two** half-lives.

13.78   Both compounds, A and B, decompose by first-order kinetics. Therefore, we can write a first-order rate equation for A and also one for B.

$$\ln\dfrac{[A]_t}{[A]_0} = -k_A t \qquad\qquad\qquad \ln\dfrac{[B]_t}{[B]_0} = -k_B t$$

$$\frac{[A]_t}{[A]_0} = e^{-k_At} \qquad\qquad \frac{[B]_t}{[B]_0} = e^{-k_Bt}$$

$$[A]_t = [A]_0 e^{-k_At} \qquad\qquad [B]_t = [B]_0 e^{-k_Bt}$$

We can calculate each of the rate constants, k_A and k_B, from their respective half-lives.

$$k_A = \frac{0.693}{50.0 \text{ min}} = 0.0139 \text{ min}^{-1} \qquad k_B = \frac{0.693}{18.0 \text{ min}} = 0.0385 \text{ min}^{-1}$$

The initial concentration of A and B are equal. $[A]_0 = [B]_0$. Therefore, from the first-order rate equations, we can write:

$$\frac{[A]_t}{[B]_t} = 4 = \frac{[A]_0 e^{-k_At}}{[B]_0 e^{-k_Bt}} = \frac{e^{-k_At}}{e^{-k_Bt}} = e^{(k_B - k_A)t} = e^{(0.0385 - 0.0139)t}$$

$$4 = e^{0.0246t}$$

$$\ln 4 = 0.0246t$$

$$t = \mathbf{56.4 \text{ min}}$$

13.79   There are three gases present and we can measure only the total pressure of the gases. To measure the partial pressure of azomethane at a particular time, we must withdraw a sample of the mixture, analyze and determine the mole fractions. Then,

$$P_{\text{azomethane}} = P_T X_{\text{azomethane}}$$

This is a rather tedious process if many measurements are required. A mass spectrometer will help (see Section 3.4 of the text).

13.80   **(a)**   Changing the concentration of a reactant has no effect on k.

　　　　(b)   If a reaction is run in a solvent other than in the gas phase, then the reaction mechanism will probably change and will thus change k.

　　　　(c)   Doubling the pressure simply changes the concentration. No effect on k, as in (a).

　　　　(d)   The rate constant k changes with temperature.

　　　　(e)   A catalyst changes the reaction mechanism and therefore changes k.

13.81

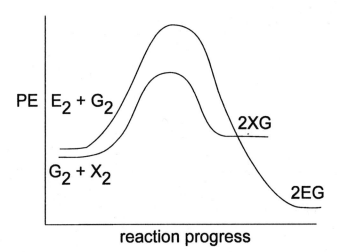

13.82   Mathematically, the amount left after ten half–lives is:

$$\left(\frac{1}{2}\right)^{10} = 9.8 \times 10^{-4}$$

13.83   **(a)**   A catalyst works by changing the reaction mechanism, thus lowering the activation energy.
        (b)   A catalyst changes the reaction mechanism.
        (c)   A catalyst does not change the enthalpy of reaction.
        (d)   A catalyst increases the forward rate of reaction.
        (e)   A catalyst increases the reverse rate of reaction.

13.84   The net ionic equation is:

$$Zn(s) + 2H^+(aq) \longrightarrow Zn^{2+}(aq) + H_2(g)$$

(a)   Changing from the same mass of granulated zinc to powdered zinc **increases** the rate because the surface area of the zinc (and thus its concentration) has increased.

(b)   Decreasing the mass of zinc (in the same granulated form) will **decrease** the rate because the total surface area of zinc has decreased.

(c)   The concentration of protons has decreased in changing from the strong acid (hydrochloric) to the weak acid (acetic); the rate will **decrease**.

(d)   An increase in temperature will **increase** the rate constant k; therefore, the rate of reaction increases.

13.85   At very high $[H_2]$,

$$k_2[H_2] \gg 1$$

$$\text{rate} = \frac{k_1[NO]^2[H_2]}{k_2[H_2]} = \frac{k_1}{k_2}[NO]^2$$

At very low $[H_2]$,

$$k_2[H_2] \ll 1$$

$$\text{rate} = \frac{k_1[NO]^2[H_2]}{1} = k_1[NO]^2[H_2]$$

The result from Problem 13.68 agrees with the rate law determined for low $[H_2]$.

13.86   If the reaction is 35.5% complete, the amount of A remaining is 64.5%. The ratio of $[A]_t/[A]_0$ is 64.5%/100% or 0.645/1.00. Using the first-order integrated rate law, Equation (13.3) of the text, we have

$$\ln\frac{[A]_t}{[A]_0} = -kt$$

$$\ln\frac{0.645}{1.00} = -k(4.90 \text{ min})$$

$$-0.439 = -k(4.90 \text{ min})$$

$$k = \mathbf{0.0896 \text{ min}^{-1}}$$

13.87   First we plot the data for the reaction: $2N_2O_5 \rightarrow 4NO_2 + O_2$

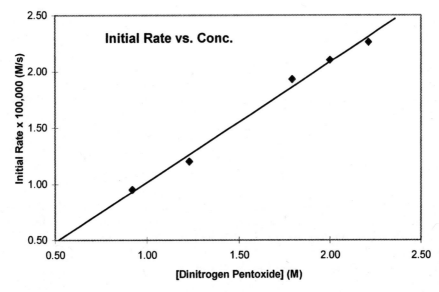

The data in linear, what means that the initial rate is directly proportional to the concentration of N_2O_5.

Thus, the rate law is:

$$\text{Rate} = k[N_2O_5]$$

The rate constant k can be determined from the slope of the graph $\left(\dfrac{\Delta(\text{Initial Rate})}{\Delta[N_2O_5]} \right)$ or by using any set of

data.

$$k = 1.0 \times 10^{-5}\ s^{-1}$$

Note that the rate law is ***not*** Rate $= k[N_2O_5]^2$, as we might expect from the balanced equation. In general, the order of a reaction must be determined by experiment; it cannot be deduced from the coefficients in the balanced equation.

13.88   The first-order rate equation can be arranged to take the form of a straight line.

$$\ln[A] = -kt + \ln[A]_0$$

If a reaction obeys first-order kinetics, a plot of $\ln[A]$ vs. t will be a straight line with a slope of $-k$.

The slope of a plot of $\ln[N_2O_5]$ vs. t is $-6.18 \times 10^{-4}\ min^{-1}$. Thus,

$$k = 6.18 \times 10^{-4}\ min^{-1}$$

The equation for the half-life of a first-order reaction is:

$$t_{\frac{1}{2}} = \frac{0.693}{k}$$

$$t_{\frac{1}{2}} = \frac{0.693}{6.18 \times 10^{-4}\ min^{-1}} = 1.12 \times 10^3\ min$$

13.89   The red bromine vapor absorbs photons of blue light and dissociates to form bromine atoms.

$$Br_2 \rightarrow 2Br\cdot$$

The bromine atoms collide with methane molecules and abstract hydrogen atoms.

$$Br\cdot + CH_4 \rightarrow HBr + \cdot CH_3$$

The methyl radical then reacts with Br_2, giving the observed product and regenerating a bromine atom to start the process over again:

$$\cdot CH_3 + Br_2 \rightarrow CH_3Br + Br\cdot$$

$$Br\cdot + CH_4 \rightarrow HBr + \cdot CH_3 \qquad \text{and so on...}$$

13.90   **(a)**   In the two-step mechanism the rate-determining step is the collision of a hydrogen molecule with two iodine atoms. If visible light increases the concentration of iodine atoms, then the rate must increase. If the true rate-determining step were the collision of a hydrogen molecule with an iodine molecule (the one-step mechanism), then the visible light would have no effect (it might even slow the reaction by depleting the number of available iodine molecules).

(b)   To split hydrogen molecules into atoms, one needs ultraviolet light of much higher energy.

13.91   For a first order reaction: $\ln\left(\dfrac{\text{decay rate at } t = t}{\text{decay rate at } t = 0}\right) = -kt$

$$\ln\left(\frac{0.186}{0.260}\right) = -(1.21 \times 10^{-4} \text{ yr}^{-1})t$$

$$t = 2.77 \times 10^3 \text{ yr}$$

13.92   **(a)**   We can write the rate law for an elementary step directly from the stoichiometry of the balanced reaction. In this rate-determining elementary step three molecules must collide simultaneously (one X and two Y's). This makes the reaction termolecular, and consequently the rate law must be third order: first order in X and second order in Y.

The rate law is:
$$\text{rate} = k[X][Y]^2$$

(b)   The value of the rate constant can be found by solving algebraically for k.

$$k = \frac{\text{rate}}{[X][Y]^2} = \frac{3.8 \times 10^{-3} \text{ M/s}}{(0.26 \text{ M})(0.88 \text{ M})^2} = 1.9 \times 10^{-2} \text{ M}^{-2}\text{s}^{-1}$$

Could you write the rate law if the reaction shown were the overall balanced equation and not an elementary step?

13.93   **(a)** $O + O_3 \rightarrow 2O_2$
(b)   Cl is a catalyst; ClO is an intermediate.
(c)   The C–F bond is stronger than the C–Cl bond.
(d)   Ethane will remove the Cl atoms:
$$Cl + C_2H_6 \rightarrow HCl + C_2H_5$$

(e)

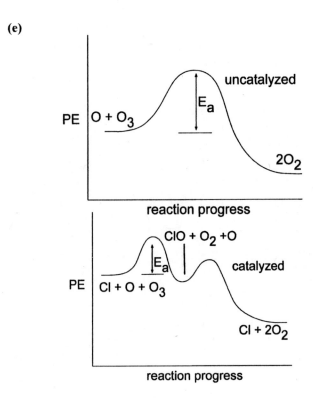

The overall reaction is: $O + O_3 \rightarrow 2O_2$.

$$\Delta H^\circ_{rxn} = 2\Delta H^\circ_f(O_2) - [\Delta H^\circ_f(O) + \Delta H^\circ_f(O_3)]$$

$$\Delta H^\circ_{rxn} = 2(0) - [(1)(249.4 \text{ kJ/mol}) + (1)(142.2 \text{ kJ/mol})]$$

$$\Delta H^\circ_{rxn} = -391.6 \text{ kJ/mol}$$

The reaction is exothermic.

13.94

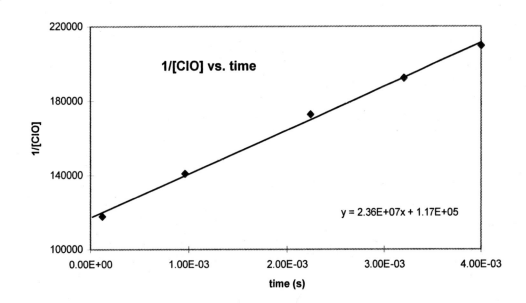

Reaction is **second-order** because a plot of 1/[ClO] vs. time is a straight line. The slope of the line equals the rate constant, k.

$$k = \text{Slope} = 2.4 \times 10^7 \ /M \cdot s$$

13.95   We can calculate the ratio of k_1/k_2 at 40°C using the Arrhenius equation.

$$\frac{k_1}{k_2} = \frac{A e^{-E_{a_1}/RT}}{A e^{-E_{a_2}/RT}} = e^{-(E_{a_1} - E_{a_2})/RT} = e^{-\Delta E_a/RT}$$

$$8.0 = e^{\dfrac{-\Delta E_a}{(8.314 \ \text{J/K·mol})(313 \ \text{K})}}$$

$$\ln(8.0) = \frac{-\Delta E_a}{(8.314 \ \text{J/K·mol})(313 \ \text{K})}$$

$$\Delta E_a = -5.4 \times 10^3 \ \text{J/mol}$$

Having calculated ΔE_a, we can substitute back into the equation to calculate the ratio k_1/k_2 at 300°C (573 K).

$$\frac{k_1}{k_2} = e^{-\dfrac{-5.4 \times 10^3 \ \text{J/mol}}{(8.314 \ \text{J/K·mol})(573 \ \text{K})}} = 3.1$$

13.96   During the first five minutes or so the engine is relatively cold, so the exhaust gases will not fully react with the components of the catalytic converter. Remember, for almost all reactions, the rate of reaction increases with temperature.

13.97

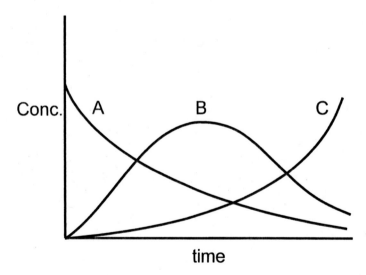

The actual appearance depends on the relative magnitudes of the rate constants for the two steps.

13.98   (a) E_a has a large value.

(b) $E_a \approx 0$. Orientation factor is not important.

13.99 A plausible two-step mechanism is:

$$NO_2 + NO_2 \rightarrow NO_3 + NO \qquad \text{(slow)}$$

$$NO_3 + CO \rightarrow NO_2 + CO_2 \qquad \text{(fast)}$$

13.100 First, solve for the rate constant, k, from the half-life of the decay.

$$t_{\frac{1}{2}} = 2.44 \times 10^5 \text{ yr} = \frac{0.693}{k}$$

$$k = \frac{0.693}{2.44 \times 10^5 \text{ yr}} = 2.84 \times 10^{-6} \text{ yr}^{-1}$$

Now, we can calculate the time for the plutonium to decay from 5.0×10^2 g to 1.0×10^2 g using the equation for a first-order reaction relating concentration and time.

$$\ln \frac{[A]_t}{[A]_0} = -kt$$

$$\ln \frac{1.0 \times 10^2}{5.0 \times 10^2} = -(2.84 \times 10^{-6} \text{ yr}^{-1})t$$

$$-1.61 = -(2.84 \times 10^{-6} \text{ yr}^{-1})t$$

$$t = \mathbf{5.7 \times 10^5 \ yr}$$

13.101 At high pressure of PH_3, all the sites on W are occupied, so the rate is independent of $[PH_3]$.

13.102 **(a)** Catalyst: Mn^{2+}; intermediate: Mn^{3+}

First step is rate-determining.

(b) Without the catalyst, the reaction would be a termolecular one involving 3 cations! (Tl^+ and two Ce^{4+}). The reaction would be slow.

(c) The catalyst is a homogeneous catalyst because it has the same phase (aqueous) as the reactants.

13.103 **(a)**   Since a plot of ln (sucrose) vs. time is linear, the reaction is 1st order.

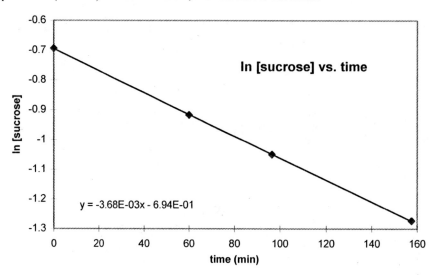

Slope $= -3.68 \times 10^{-3} \text{ min}^{-1} = -k$

$k = 3.68 \times 10^{-3} \text{ min}^{-1}$

(b) $\ln \dfrac{[A]_t}{[A]_0} = -kt$

$\ln \left(\dfrac{0.05}{1} \right) = -(3.68 \times 10^{-3})t$

$t = 814 \text{ min}$

(c) $[H_2O]$ is roughly unchanged. This is a pseudo-first-order reaction.

13.104   Initially, the number of moles of gas in terms of the volume is:

$$n = \frac{PV}{RT} = \frac{(0.350 \text{ atm})V}{\left(0.0821 \dfrac{\text{L} \cdot \text{atm}}{\text{mol} \cdot \text{K}} \right)(450 + 273)\text{K}} = 5.90 \times 10^{-3} \, V$$

We can calculate the concentration of dimethyl ether from the following equation.

$$\ln \frac{[(CH_3)_2O]_t}{[(CH_3)_2O]_0} = -kt$$

$$\frac{[(CH_3)_2O]_t}{[(CH_3)_2O]_0} = e^{-kt}$$

Since, the volume is held constant, it will cancel out of the equation. The concentration of dimethyl ether after 8.0 minutes (480 s) is:

$$[(CH_3)_2O]_t = \left(\frac{5.90 \times 10^{-3} \, V}{V} \right) e^{-\left(3.2 \times 10^{-4} \frac{1}{\text{s}} \right)(480 \text{ s})}$$

$$[(CH_3)_2O]_t = 5.06 \times 10^{-3} \, M$$

After 8.0 min, the concentration of $(CH_3)_2O$ has decreased by $(5.90 \times 10^{-3} - 5.06 \times 10^{-3})M$ or 8.4×10^{-4} M. Since three moles of product form for each mole of dimethyl ether that reacts, the concentrations of the products are $(3)(8.4 \times 10^{-4}$ M$) = 2.5 \times 10^{-3}$ M.

The pressure of the system after 8.0 minutes is:

$$P = \frac{nRT}{V} = \left(\frac{n}{V}\right)RT = MRT$$

$$P = [(5.06 \times 10^{-3}) + (2.5 \times 10^{-3})]M \times (0.0821 \text{ L·atm/mol·K})(723 \text{ K})$$

$$P = 0.45 \text{ atm}$$

13.105 (a) The relationship between half-life and rate constant is given in Equation (13.6) of the text.

$$k = \frac{0.693}{t_{\frac{1}{2}}}$$

$$k = \frac{0.693}{19.8 \text{ min}}$$

$$k = 0.0350 \text{ min}^{-1}$$

(b) Following the same procedure as in part (a), we find the rate constant at 70°C to be 1.58×10^{-3} min^{-1}. We now have two values of rate constants (k_1 and k_2) at two temperatures (T_1 and T_2). This information allows us to calculate the activation energy, E_a, using Equation (13.11) of the text.

$$\ln \frac{k_1}{k_2} = \frac{E_a}{R}\left(\frac{T_1 - T_2}{T_1 T_2}\right)$$

$$\ln\left(\frac{0.0350 \text{ min}^{-1}}{1.58 \times 10^{-3} \text{ min}^{-1}}\right) = \frac{E_a}{(8.314 \text{ J/mol·K})}\left(\frac{373 \text{ K} - 343 \text{ K}}{(373 \text{ K})(343 \text{ K})}\right)$$

$$E_a = 1.10 \times 10^5 \text{ J/mol} = 110 \text{ kJ/mol}$$

(c) Since all the above steps are elementary steps, we can deduce the rate law simply from the equations representing the steps. The rate laws are:

Initiation: rate $= k_i[R_2]$

Propagation: rate $= k_p[M][M_1]$

Termination: rate $= k_t[M'][M'']$

The reactant molecules are the ethylene monomers, and the product is polyethylene. Recalling that intermediates are species that are formed in an early elementary step and consumed in a later step, we see that they are the radicals M'·, M''·, and so on. (The R· species also qualifies as an intermediate.)

(d) The growth of long polymers would be favored by a high rate of propagations and a low rate of termination. Since the rate law of propagation depends on the concentration of monomer, an increase in the concentration of ethylene would increase the propagation (growth) rate. From the rate law for termination we see that a low concentration of the radical fragment M'· or M''· would lead to a slower rate of termination. This can be accomplished by using a low concentration of the initiator, R_2.

13.106 (a) $\dfrac{\Delta[B]}{\Delta t} = k_1[A] - k_2[B]$

(b)   If, $\dfrac{\Delta[B]}{\Delta t} = 0$

Then, from part (a) of this problem:

$$k_1[A] = k_2[B]$$

$$[B] = \frac{k_1}{k_2}[A]$$

13.107 (a)   Drinking too much alcohol too fast means all the alcohol dehydrogenase (ADH) active sites are tied up and the excess alcohol will damage the central nervous system.

(b)   Both ethanol and methanol will compete for the same site at ADH. An excess of ethanol will replace methanol at the active site, leading to methanol's discharge from the body.

13.108 (a)   The first-order rate constant can be determined from the half-life.

$$t_{\frac{1}{2}} = \frac{0.693}{k}$$

$$k = \frac{0.693}{t_{\frac{1}{2}}} = \frac{0.693}{28.1 \text{ yr}} = \mathbf{0.0247 \text{ yr}^{-1}}$$

(b)   See Problem 13.82. Mathematically, the amount left after ten half–lives is:

$$\left(\frac{1}{2}\right)^{10} = \mathbf{9.8 \times 10^{-4}}$$

(c)   If 99.0% has disappeared, then 1.0% remains. The ratio of $[A]_t/[A]_0$ is 1.0%/100% or 0.010/1.00. Substitute into the first-order integrated rate law, Equation (13.3) of the text, to determine the time.

$$\ln\frac{[A]_t}{[A]_0} = -kt$$

$$\ln\frac{0.010}{1.0} = -(0.0247 \text{ yr}^{-1})t$$

$$-4.6 = -(0.0247 \text{ yr}^{-1})t$$

$$t = \mathbf{186 \text{ yr}}$$

13.109 (1)   Assuming the reactions have roughly the same frequency factors, the one with the largest activation energy will be the slowest, and the one with the smallest activation energy will be the fastest. The reactions ranked from slowest to fastest are:

$$\mathbf{(b) < (c) < (a)}$$

(2)   Reaction (a): $\Delta H = -40$ kJ/mol

Reaction (b): $\Delta H = 20$ kJ/mol

Reaction (c): $\Delta H = -20$ kJ/mol

(a) and (c) are exothermic, and (b) is endothermic.

13.110 **(a)**   There are three elementary steps: $A \rightarrow B$, $B \rightarrow C$, and $C \rightarrow D$.

(b)   There are two intermediates: B and C.

(c)   The third step, $C \rightarrow D$, is rate determining because it has the largest activation energy.

(d)   The overall reaction is exothermic.

13.111 The fire should not be doused with water, because titanium acts as a catalyst to decompose steam as follows:

$$2H_2O(g) \rightarrow 2H_2(g) + O_2(g)$$

H_2 gas is flammable and forms an explosive mixture with O_2.

13.112 Let $k_{cat} = k_{uncat}$

Then,

$$Ae^{\frac{-E_a(cat)}{RT_1}} = Ae^{\frac{-E_a(uncat)}{RT_2}}$$

Since the frequency factor is the same, we can write:

$$e^{\frac{-E_a(cat)}{RT_1}} = e^{\frac{-E_a(uncat)}{RT_2}}$$

Taking the natural log (*ln*) of both sides of the equation gives:

$$\frac{-E_a(cat)}{RT_1} = \frac{-E_a(uncat)}{RT_2}$$

or,

$$\frac{E_a(cat)}{T_1} = \frac{E_a(uncat)}{T_2}$$

Substituting in the given values:

$$\frac{7.0 \text{ kJ/mol}}{293 \text{ K}} = \frac{42 \text{ kJ/mol}}{T_2}$$

$$T_2 = 1.8 \times 10^3 \text{ K}$$

This temperature is much too high to be practical.

13.113 First, let's calculate the number of radium nuclei in 1.0 g.

$$1.0 \text{ g} \times \frac{1 \text{ mol Ra}}{226.03 \text{ g Ra}} \times \frac{6.022 \times 10^{23} \text{ Ra nuclei}}{1 \text{ mol Ra}} = 2.7 \times 10^{21} \text{ Ra nuclei}$$

We can now calculate the rate constant, k, from the activity and the number of nuclei, and then we can calculate the half-life from the rate contant.

$$\text{activity} = kN$$

$$k = \frac{\text{activity}}{N} = \frac{3.70 \times 10^{10} \text{ nuclear disintegrations/s}}{2.7 \times 10^{21} \text{ nuclei}} = 1.4 \times 10^{-11} \text{ /s}$$

The half-life is:

$$t_{\frac{1}{2}} = \frac{0.693}{k} = \frac{0.693}{1.4 \times 10^{-11} \text{ 1/s}} = 5.0 \times 10^{10} \text{ s}$$

Next, let's convert 500 years to seconds. Then we can calculate the number of nuclei remaining after 500 years.

$$500 \text{ yr} \times \frac{365 \text{ days}}{1 \text{ yr}} \times \frac{24 \text{ h}}{1 \text{ day}} \times \frac{3600 \text{ s}}{1 \text{ h}} = 1.58 \times 10^{10} \text{ s}$$

Use the first-order integrated rate law to determine the number of nuclei remaining after 500 years.

$$\ln \frac{N_t}{N_0} = -kt$$

$$\ln \left(\frac{N_t}{2.7 \times 10^{21}} \right) = -(1.4 \times 10^{-11} \text{ 1/s})(1.58 \times 10^{10} \text{ s})$$

$$\frac{N_t}{2.7 \times 10^{21}} = e^{-0.22}$$

$$N_t = 2.2 \times 10^{21} \text{ Ra nuclei}$$

Finally, from the number of nuclei remaining after 500 years and the rate constant, we can calculate the activity.

$$\text{activity} = kN$$

$$\textbf{activity} = (1.4 \times 10^{-11} \text{ /s})(2.2 \times 10^{21} \text{ nuclei}) = \textbf{3.1} \times \textbf{10}^{\textbf{10}} \textbf{ nuclear disintegrations/s}$$

13.114 **(a)** The rate law for the reaction is:

$$\text{rate} = k[\text{Hb}][\text{O}_2]$$

We are given the rate constant and the concentration of Hb and O_2, so we can substitute in these quantities to solve for rate.

$$\text{rate} = (2.1 \times 10^6 \text{ /M·s})(8.0 \times 10^{-6} \text{ M})(1.5 \times 10^{-6} \text{ M})$$

$$\textbf{rate} = \textbf{2.5} \times \textbf{10}^{\textbf{-5}} \textbf{ M/s}$$

(b) If HbO_2 is being formed at the rate of 2.5×10^{-5} M/s, then O_2 is being consumed at the same rate, $\mathbf{2.5 \times 10^{-5}}$ $\boldsymbol{M}$**/s.** Note the 1:1 mole ratio between O_2 and HbO_2.

(c) The rate of formation of HbO_2 increases, but the concentration of Hb remains the same. Assuming that temperature is constant, we can use the same rate constant as in part (a). We substitute rate, [Hb], and the rate constant into the rate law to solve for O_2 concentration.

$$\text{rate} = k[\text{Hb}][O_2]$$

$$1.4 \times 10^{-4} \ M/s = (2.1 \times 10^6 \ /M \cdot s)(8.0 \times 10^{-6} \ M)[O_2]$$

$$\mathbf{[O_2] = 8.3 \times 10^{-6} \ \boldsymbol{M}}$$

13.115 Initially, the rate increases with increasing pressure (concentration) of NH_3. The straight-line relationship in the first half of the plot shows that the rate of reaction is directly proportional to the concentration of ammonia. Rate $= k[NH_3]$. The more ammonia that is adsorbed on the tungsten surface, the faster the reaction. At a certain pressure (concentration), the rate is no longer dependent on the concentration of ammonia (horizontal portion of plot). The reaction is now zero-order in NH_3 concentration. At a certain concentration of NH_3, all the reactive sites on the metal surface are occupied by NH_3 molecules, and the rate becomes constant. Increasing the concentration further has no effect on the rate.

13.116 λ_1 (the absorbance of A) decreases with time. This would happen for all the mechanisms shown. Note that λ_2 (the absorbance of B) increases with time and then decreases. Therefore, B cannot be a product as shown in mechanisms (a) or (b). If B were a product its absorbance would increase with time and level off, but it would not decrease. Since the concentration of B increases and then after some time begins to decrease, it must mean that it is produced and then it reacts to produce product as in mechanisms (c) and (d). In mechanism (c), two products are C and D, so we would expect to see an increase in absorbance for two species. Since we see an increase in absorbance for only one species, then the mechanism that is consistent with the data must be **(d)**. λ_3 is the absorbance of C.

CHAPTER 14
CHEMICAL EQUILIBRIUM

14.13 $K_c = \dfrac{[B]}{[A]}$

(1) With $K_c = 10$, products are favored at equilibrium. Because the coefficients for both A and B are one, we expect the concentration of B to be 10 times that of A at equilibrium. Choice **(a)** is the best choice with 10 B molecules and 1 A molecule.

(2) With $K_c = 0.10$, reactants are favored at equilibrium. Because the coefficients for both A and B are one, we expect the concentration of A to be 10 times that of B at equilibrium. Choice **(d)** is the best choice with 10 A molecules and 1 B molecule.

You can calculate K_c in each case without knowing the volume of the container because the mole ratio between A and B is the same. Volume will cancel from the K_c expression. Only moles of each component are needed to calculate K_c.

14.14 Note that we are comparing similar reactions at equilibrium – two reactants producing one product, all with coefficients of one in the balanced equation.

(a) The reaction, $A + C \rightleftharpoons AC$ has the largest equilibrium constant. Of the three diagrams, there is the most product present at equilibrium.

(b) The reaction, $A + D \rightleftharpoons AD$ has the smallest equilibrium constant. Of the three diagrams, there is the least amount of product present at equilibrium.

14.15 When the equation for a reversible reaction is written in the opposite direction, the equilibrium constant becomes the reciprocal of the original equilibrium constant.

$$K' = \frac{1}{K} = \frac{1}{4.17 \times 10^{-34}} = 2.40 \times 10^{33}$$

14.16 The problem states that the system is at equilibrium, so we simply substitute the equilibrium concentrations into the equilibrium constant expression to calculate K_c.

Step 1: Calculate the concentrations of the components in units of mol/L. The molarities can be calculated by simply dividing the number of moles by the volume of the flask.

$$[H_2] = \frac{2.50 \text{ mol}}{12.0 \text{ L}} = 0.208 \; M$$

$$[S_2] = \frac{1.35 \times 10^{-5} \text{ mol}}{12.0 \text{ L}} = 1.13 \times 10^{-6} \; M$$

$$[H_2S] = \frac{8.70 \text{ mol}}{12.0 \text{ L}} = 0.725 \; M$$

Step 2: Once the molarities are known, K_c can be found by substituting the molarities into the equilibrium constant expression.

$$K_c = \frac{[H_2S]^2}{[H_2]^2[S_2]} = \frac{(0.725)^2}{(0.208)^2(1.13 \times 10^{-6})} = 1.08 \times 10^7$$

If you forget to convert moles to moles/liter, will you get a different answer? Under what circumstances will the two answers be the same?

14.17 Using Equation (14.5) of the text: $K_p = K_c(0.0821\,T)^{\Delta n}$

where, $\Delta n = 2 - 3 = -1$
and $T = (1273 + 273)$ K = 1546 K

$$K_p = (2.24 \times 10^{22})(0.0821 \times 1546)^{-1} = 1.76 \times 10^{20}$$

14.18 **Strategy:** The relationship between K_c and K_p is given by Equation (14.5) of the text. What is the change in the number of moles of gases from reactant to product? Recall that

$$\Delta n = \text{moles of gaseous products} - \text{moles of gaseous reactants}$$

What unit of temperature should we use?

Solution: The relationship between K_c and K_p is given by Equation (14.5) of the text.

$$K_p = K_c(0.0821\,T)^{\Delta n}$$

Rearrange the equation relating K_p and K_c, solving for K_c.

$$K_c = \frac{K_p}{(0.0821\,T)^{\Delta n}}$$

Because $T = 623$ K and $\Delta n = 3 - 2 = 1$, we have:

$$K_c = \frac{K_p}{(0.0821\,T)^{\Delta n}} = \frac{1.8 \times 10^{-5}}{(0.0821)(623\ \text{K})} = 3.5 \times 10^{-7}$$

14.19 We can write the equilibrium constant expression from the balanced equation and substitute in the pressures.

$$K_p = \frac{P_{NO}^2}{P_{N_2}P_{O_2}} = \frac{(0.050)^2}{(0.15)(0.33)} = 0.051$$

Do we need to know the temperature?

14.20 The equilibrium constant expressions are:

(a) $K_c = \dfrac{[NH_3]^2}{[N_2][H_2]^3}$

(b) $K_c = \dfrac{[NH_3]}{[N_2]^{\frac{1}{2}}[H_2]^{\frac{3}{2}}}$

Substituting the given equilibrium concentration gives:

(a) $K_c = \dfrac{(0.25)^2}{(0.11)(1.91)^3} = 0.082$

(b) $K_c = \dfrac{(0.25)}{(0.11)^{\frac{1}{2}}(1.91)^{\frac{3}{2}}} = 0.29$

Is there a relationship between the K_c values from parts (a) and (b)?

14.21   The equilibrium constant expression for the two forms of the equation are:

$$K_c = \frac{[I]^2}{[I_2]} \quad \text{and} \quad K_c' = \frac{[I_2]}{[I]^2}$$

The relationship between the two equilibrium constants is

$$K_c' = \frac{1}{K_c} = \frac{1}{3.8 \times 10^{-5}} = 2.6 \times 10^4$$

K_p can be found as shown below.

$$K_p = K_c'(0.0821\,T)^{\Delta n} = (2.6 \times 10^4)(0.0821 \times 1000)^{-1} = 3.2 \times 10^2$$

14.22   Because pure solids do not enter into an equilibrium constant expression, we can calculate K_p directly from the pressure that is due solely to $CO_2(g)$.

$$K_p = P_{CO_2} = 0.105$$

Now, we can convert K_p to K_c using the following equation.

$$K_p = K_c(0.0821\,T)^{\Delta n}$$

$$K_c = \frac{K_p}{(0.0821\,T)^{\Delta n}}$$

$$K_c = \frac{0.105}{(0.0821 \times 623)^{(1-0)}} = 2.05 \times 10^{-3}$$

14.23   We substitute the given pressures into the reaction quotient expression.

$$Q_p = \frac{P_{PCl_3}\,P_{Cl_2}}{P_{PCl_5}} = \frac{(0.223)(0.111)}{(0.177)} = 0.140$$

The calculated value of Q_p is less than K_p for this system. The system will change in a way to increase Q_p until it is equal to K_p. To achieve this, the pressures of PCl_3 and Cl_2 must *increase*, and the pressure of PCl_5 must *decrease*.

Could you actually determine the final pressure of each gas?

14.24 **Strategy:** Because they are constant quantities, the concentrations of solids and liquids do not appear in the equilibrium constant expressions for heterogeneous systems. The total pressure at equilibrium that is given is due to both NH_3 and CO_2. Note that for every 1 atm of CO_2 produced, 2 atm of NH_3 will be produced due to the stoichiometry of the balanced equation. Using this ratio, we can calculate the partial pressures of NH_3 and CO_2 at equilibrium.

Solution: The equilibrium constant expression for the reaction is

$$K_p = P_{NH_3}^2 P_{CO_2}$$

The total pressure in the flask (0.363 atm) is a sum of the partial pressures of NH_3 and CO_2.

$$P_T = P_{NH_3} + P_{CO_2} = 0.363 \text{ atm}$$

Let the partial pressure of $CO_2 = x$. From the stoichiometry of the balanced equation, you should find that $P_{NH_3} = 2P_{CO_2}$. Therefore, the partial pressure of $NH_3 = 2x$. Substituting into the equation for total pressure gives:

$$P_T = P_{NH_3} + P_{CO_2} = 2x + x = 3x$$

$$3x = 0.363 \text{ atm}$$

$$x = P_{CO_2} = 0.121 \text{ atm}$$

$$P_{NH_3} = 2x = 0.242 \text{ atm}$$

Substitute the equilibrium pressures into the equilibrium constant expression to solve for K_p.

$$K_p = P_{NH_3}^2 P_{CO_2} = (0.242)^2 (0.121) = \mathbf{7.09 \times 10^{-3}}$$

14.25 Of the original 1.05 moles of Br_2, 1.20% has dissociated. The amount of Br_2 dissociated in molar concentration is:

$$[Br_2] = 0.0120 \times \frac{1.05 \text{ mol}}{0.980 \text{ L}} = 0.0129 \ M$$

Setting up a table:

	$Br_2(g)$	$\rightleftharpoons$	$2Br(g)$
Initial (M):	$\dfrac{1.05 \text{ mol}}{0.980 \text{ L}} = 1.07 \ M$		0
Change (M):	-0.0129		$+2(0.0129)$
Equilibrium (M):	1.06		0.0258

$$K_c = \frac{[Br]^2}{[Br_2]} = \frac{(0.0258)^2}{1.06} = \mathbf{6.3 \times 10^{-4}}$$

14.26   If the CO pressure at equilibrium is 0.497 atm, the balanced equation requires the chlorine pressure to have the same value. The initial pressure of phosgene gas can be found from the ideal gas equation.

$$P = \frac{nRT}{V} = \frac{(3.00 \times 10^{-2}\ \text{mol})(0.0821\ \text{L} \cdot \text{atm/mol} \cdot \text{K})(800\ \text{K})}{(1.50\ \text{L})} = 1.31\ \text{atm}$$

Since there is a 1:1 mole ratio between phosgene and CO, the partial pressure of CO formed (0.497 atm) equals the partial pressure of phosgene reacted. The phosgene pressure at equilibrium is:

	CO(g)	+	Cl₂(g)	⇌	COCl₂(g)
Initial (atm):	0		0		1.31
Change (atm):	+0.497		+0.497		−0.497
Equilibrium (atm):	0.497		0.497		0.81

The value of K_p is then found by substitution.

$$K_p = \frac{P_{COCl_2}}{P_{CO}P_{Cl_2}} = \frac{0.81}{(0.497)^2} = 3.3$$

14.27   Let x be the initial pressure of NOBr. Using the balanced equation, we can write expressions for the partial pressures at equilibrium.

$$P_{NOBr} = (1 - 0.34)x = 0.66x$$

$$P_{NO} = 0.34x$$

$$P_{Br_2} = 0.17x$$

The sum of these is the total pressure.

$$0.66x + 0.34x + 0.17x = 1.17x = 0.25\ \text{atm}$$

$$x = 0.21\ \text{atm}$$

The equilibrium pressures are then

$$P_{NOBr} = 0.66(0.21) = 0.14\ \text{atm}$$

$$P_{NO} = 0.34(0.21) = 0.071\ \text{atm}$$

$$P_{Br_2} = 0.17(0.21) = 0.036\ \text{atm}$$

We find K_p by substitution.

$$K_p = \frac{(P_{NO})^2 P_{Br_2}}{(P_{NOBr})^2} = \frac{(0.071)^2(0.036)}{(0.14)^2} = 9.3 \times 10^{-3}$$

The relationship between K_p and K_c is given by

$$K_p = K_c(RT)^{\Delta n}$$

We find K_c (for this system $\Delta n = +1$)

$$K_c = \frac{K_p}{(RT)^{\Delta n}} = \frac{K_p}{RT} = \frac{9.3 \times 10^{-3}}{(0.0821 \times 298)^1} = 3.8 \times 10^{-4}$$

14.28   In this problem, you are asked to calculate K_c.

Step 1: Calculate the initial concentration of NOCl

$$[NOCl]_0 = \frac{2.50 \text{ mol}}{1.50 \text{ L}} = 1.67 \ M$$

Step 2: Let's represent the change in concentration of NOCl as $-2x$. Setting up a table:

	$2NOCl(g)$	$\rightleftharpoons$	$2NO(g)$	$+$	$Cl_2(g)$
Initial (M):	1.67		0		0
Change (M):	$-2x$		$+2x$		$+x$
Equilibrium (M):	$1.67 - 2x$		$2x$		x

If 28.0 percent of the NOCl has dissociated at equilibrium, the amount reacted is:

$$(0.280)(1.67 \ M) = 0.468 \ M$$

In the table above, we have represented the amount of NOCl that reacts as $2x$. Therefore,

$$2x = 0.468 \ M$$

$$x = 0.234 \ M$$

The equilibrium concentrations of NOCl, NO, and Cl_2 are:

$$[NOCl] = (1.67 - 2x)M = (1.67 - 0.468)M = 1.20 \ M$$
$$[NO] = 2x = 0.468 \ M$$
$$[Cl_2] = x = 0.234 \ M$$

Step 3: The equilibrium constant K_c can be calculated by substituting the above concentrations into the equilibrium constant expression.

$$K_c = \frac{[NO]^2[Cl_2]}{[NOCl]^2} = \frac{(0.468)^2(0.234)}{(1.20)^2} = 0.0356$$

14.29   The target equation is the sum of the first two.

$$H_2S \rightleftharpoons H^+ + HS$$
$$HS \rightleftharpoons H^+ + S^{2-}$$
$$\overline{\phantom{H_2S \rightleftharpoons 2H^+ + S^{2-}}}$$
$$H_2S \rightleftharpoons 2H^+ + S^{2-}$$

Since this is the case, the equilibrium constant for the combined reaction is the product of the constants for the component reactions (Section 14.2 of the text). The equilibrium constant is therefore:

$$K_c = K_c'K_c'' = 9.5 \times 10^{-27}$$

What happens in the special case when the two component reactions are the same? Can you generalize this relationship to adding more than two reactions? What happens if one takes the difference between two reactions?

14.30 $K = K'K''$

$K = (6.5 \times 10^{-2})(6.1 \times 10^{-5})$

$\boldsymbol{K = 4.0 \times 10^{-6}}$

14.31   Given:

$$K_p' = \frac{P_{CO}^2}{P_{CO_2}} = 1.3 \times 10^{14} \qquad\qquad K_p'' = \frac{P_{COCl_2}}{P_{CO}\,P_{Cl_2}} = 6.0 \times 10^{-3}$$

For the overall reaction:

$$\boldsymbol{K_p} = \frac{P_{COCl_2}^2}{P_{CO_2}\,P_{Cl_2}^2} = K_p'(K_p'')^2 = (1.3 \times 10^{14})(6.0 \times 10^{-3})^2 = \boldsymbol{4.7 \times 10^9}$$

14.32   To obtain $2SO_2$ as a reactant in the final equation, we must reverse the first equation and multiply by two. For the equilibrium, $2SO_2(g) \rightleftharpoons 2S(s) + 2O_2(g)$

$$K_c''' = \left(\frac{1}{K_c'}\right)^2 = \left(\frac{1}{4.2 \times 10^{52}}\right)^2 = 5.7 \times 10^{-106}$$

Now we can add the above equation to the second equation to obtain the final equation. Since we add the two equations, the equilibrium constant is the product of the equilibrium constants for the two reactions.

$2SO_2(g) \rightleftharpoons 2S(s) + 2O_2(g)$	$K_c''' = 5.7 \times 10^{-106}$
$2S(s) + 3O_2(g) \rightleftharpoons 2SO_3(g)$	$K_c'' = 9.8 \times 10^{128}$
$2SO_2(g) + O_2(g) \rightleftharpoons 2SO_3(g)$	$\boldsymbol{K_c = K_c''' \times K_c'' = 5.6 \times 10^{23}}$

14.35   **(a)**   Assuming the self-ionization of water occurs by a single elementary step mechanism, the equilibrium constant is just the ratio of the forward and reverse rate constants.

$$\boldsymbol{K} = \frac{k_f}{k_r} = \frac{k_1}{k_{-1}} = \frac{2.4 \times 10^{-5}}{1.3 \times 10^{11}} = \boldsymbol{1.8 \times 10^{-16}}$$

(b)   The product can be written as:

$$[H^+][OH^-] = K[H_2O]$$

What is $[H_2O]$? It is the concentration of pure water. One liter of water has a mass of 1000 g (density = 1.00 g/mL). The number of moles of H_2O is:

$$1000 \text{ g} \times \frac{1 \text{ mol}}{18.0 \text{ g}} = 55.5 \text{ mol}$$

The concentration of water is 55.5 mol/1.00 L or 55.5 M. The product is:

$$[H^+][OH^-] = (1.8 \times 10^{-16})(55.5) = 1.0 \times 10^{-14}$$

We assume the concentration of hydrogen ion and hydroxide ion are equal.

$$[H^+] = [OH^-] = (1.0 \times 10^{-14})^{1/2} = 1.0 \times 10^{-7} \, M$$

14.36   At equilibrium, the value of K_c is equal to the ratio of the forward rate constant to the rate constant for the reverse reaction.

$$K_c = \frac{k_f}{k_r} = \frac{k_f}{5.1 \times 10^{-2}} = 12.6$$

$$k_f = (12.6)(5.1 \times 10^{-2}) = 0.64$$

The forward reaction is third order, so the units of k_f must be:

$$\text{rate} = k_f[A]^2[B]$$

$$k_f = \frac{\text{rate}}{(\text{concentration})^3} = \frac{M/s}{M^3} = 1/M^2 \cdot s$$

$$k_f = \textbf{0.64} \, /M^2 \cdot \textbf{s}$$

14.39   Given:

$$K_p = \frac{P_{SO_3}^2}{P_{SO_2}^2 \, P_{O_2}} = 5.60 \times 10^4$$

Initially, the total pressure is (0.350 + 0.762) atm or 1.112 atm. As the reaction progresses from left to right toward equilibrium there will be a decrease in the number of moles of molecules present. (Note that 2 moles of SO_2 react with 1 mole of O_2 to produce 2 moles of SO_3, or, at constant pressure, three atmospheres of reactants forms two atmospheres of products.) Since pressure is directly proportional to the number of molecules present, at equilibrium the total pressure will be less than 1.112 atm.

14.40   **Strategy:** We are given the initial concentrations of the gases, so we can calculate the reaction quotient (Q_c). How does a comparison of Q_c with K_c enable us to determine if the system is at equilibrium or, if not, in which direction the net reaction will proceed to reach equilibrium?

Solution: Recall that for a system to be at equilibrium, $Q_c = K_c$. Substitute the given concentrations into the equation for the reaction quotient to calculate Q_c.

$$Q_c = \frac{[NH_3]_0^2}{[N_2]_0[H_2]_0^3} = \frac{[0.48]^2}{[0.60][0.76]^3} = 0.87$$

Comparing Q_c to K_c, we find that $Q_c < K_c$ (0.87 < 1.2). The ratio of initial concentrations of products to reactants is too small. To reach equilibrium, reactants must be converted to products. The system proceeds from left to right (consuming reactants, forming products) to reach equilibrium.

Therefore, **[NH$_3$] will increase** and **[N$_2$] and [H$_2$] will decrease** at equilibrium.

14.41   The balanced equation shows that one mole of carbon monoxide will combine with one mole of water to form hydrogen and carbon dioxide. Let x be the depletion in the concentration of either CO or H$_2$O at equilibrium (why can x serve to represent either quantity?). The equilibrium concentration of hydrogen must then also be equal to x. The changes are summarized as shown in the table.

	H$_2$	+	CO$_2$	$\rightleftharpoons$	H$_2$O	+	CO
Initial (*M*):	0		0		0.0300		0.0300
Change (*M*):	+x		+x		−x		−x
Equilibrium (*M*):	x		x		(0.0300 − x)		(0.0300 − x)

The equilibrium constant is:

$$K_c = \frac{[H_2O][CO]}{[H_2][CO_2]} = 0.534$$

Substituting,

$$\frac{(0.0300 - x)^2}{x^2} = 0.534$$

Taking the square root of both sides, we obtain:

$$\frac{(0.0300 - x)}{x} = \sqrt{0.534} = 0.731$$

$$x = 0.0173 \, M$$

The number of moles of H$_2$ formed is:

$$0.0173 \text{ mol/L} \times 10.0 \text{ L} = \textbf{0.173 mol H}_2$$

14.42   **Strategy:** The equilibrium constant K_p is given, and we start with pure NO$_2$. The partial pressure of O$_2$ at equilibrium is 0.25 atm. From the stoichiometry of the reaction, we can determine the partial pressure of NO at equilibrium. Knowing K_p and the partial pressures of both O$_2$ and NO, we can solve for the partial pressure of NO$_2$.

Solution: Since the reaction started with only pure NO$_2$, the equilibrium concentration of NO must be twice the equilibrium concentration of O$_2$, due to the 2:1 mole ratio of the balanced equation. Therefore, the equilibrium partial pressure of **NO** is (2 × 0.25 atm) = **0.50 atm**.

We can find the equilibrium NO$_2$ pressure by rearranging the equilibrium constant expression, then substituting in the known values.

$$K_p = \frac{P_{NO}^2 \, P_{O_2}}{P_{NO_2}^2}$$

$$P_{NO_2} = \sqrt{\frac{P_{NO}^2 \, P_{O_2}}{K_p}} = \sqrt{\frac{(0.50)^2 (0.25)}{158}} = \textbf{0.020 atm}$$

14.43   Notice that the balanced equation requires that for every two moles of HBr consumed, one mole of H_2 and one mole of Br_2 must be formed. Let $2x$ be the depletion in the concentration of HBr at equilibrium. The equilibrium concentrations of H_2 and Br_2 must therefore each be x. The changes are shown in the table.

	H_2	+	Br_2	$\rightleftharpoons$	2HBr
Initial (M):	0		0		0.267
Change (M):	$+x$		$+x$		$-2x$
Equilibrium (M):	x		x		$(0.267 - 2x)$

The equilibrium constant relationship is given by:

$$K_c = \frac{[HBr]^2}{[H_2][Br_2]}$$

Substitution of the equilibrium concentration expressions gives

$$K_c = \frac{(0.267 - 2x)^2}{x^2} = 2.18 \times 10^6$$

Taking the square root of both sides we obtain:

$$\frac{0.267 - 2x}{x} = 1.48 \times 10^3$$

$$x = 1.80 \times 10^{-4}$$

The equilibrium concentrations are:

$$[H_2] = [Br_2] = 1.80 \times 10^{-4}\ M$$

$$[HBr] = 0.267 - 2(1.80 \times 10^{-4}) = 0.267\ M$$

If the depletion in the concentration of HBr at equilibrium were defined as x, rather than $2x$, what would be the appropriate expressions for the equilibrium concentrations of H_2 and Br_2? Should the final answers be different in this case?

14.44   **Strategy:** We are given the initial amount of I_2 (in moles) in a vessel of known volume (in liters), so we can calculate its molar concentration. Because initially no I atoms are present, the system could not be at equilibrium. Therefore, some I_2 will dissociate to form I atoms until equilibrium is established.

Solution: We follow the procedure outlined in Section 14.4 of the text to calculate the equilibrium concentrations.

Step 1: The initial concentration of I_2 is 0.0456 mol/2.30 L = 0.0198 M. The stoichiometry of the problem shows 1 mole of I_2 dissociating to 2 moles of I atoms. Let x be the amount (in mol/L) of I_2 dissociated. It follows that the equilibrium concentration of I atoms must be $2x$. We summarize the changes in concentrations as follows:

	$I_2(g)$	$\rightleftharpoons$	$2I(g)$
Initial (M):	0.0198		0.000
Change (M):	$-x$		$+2x$
Equilibrium (M):	$(0.0198 - x)$		$2x$

Step 2: Write the equilibrium constant expression in terms of the equilibrium concentrations. Knowing the value of the equilibrium constant, solve for x.

$$K_c = \frac{[I]^2}{[I_2]} = \frac{(2x)^2}{(0.0198 - x)} = 3.80 \times 10^{-5}$$

$$4x^2 + (3.80 \times 10^{-5})x - (7.52 \times 10^{-7}) = 0$$

The above equation is a quadratic equation of the form $ax^2 + bx + c = 0$. The solution for a quadratic equation is

$$x = \frac{-b \pm \sqrt{b^2 - 4ac}}{2a}$$

Here, we have $a = 4$, $b = 3.80 \times 10^{-5}$, and $c = -7.52 \times 10^{-7}$. Substituting into the above equation,

$$x = \frac{(-3.80 \times 10^{-5}) \pm \sqrt{(3.80 \times 10^{-5})^2 - 4(4)(-7.52 \times 10^{-7})}}{2(4)}$$

$$x = \frac{(-3.80 \times 10^{-5}) \pm (3.47 \times 10^{-3})}{8}$$

$$x = 4.29 \times 10^{-4} \, M \quad \text{or} \quad x = -4.39 \times 10^{-4} \, M$$

The second solution is physically impossible because you cannot have a negative concentration. The first solution is the correct answer.

Step 3: Having solved for x, calculate the equilibrium concentrations of all species.

$$[I] = 2x = (2)(4.29 \times 10^{-4} \, M) = \mathbf{8.58 \times 10^{-4} \, M}$$

$$[I_2] = (0.0198 - x) = [0.0198 - (4.29 \times 10^{-4})] \, M = \mathbf{0.0194 \, M}$$

Tip: We could have simplified this problem by assuming that x was small compared to 0.0198. We could then assume that $0.0198 - x \approx 0.0198$. By making this assumption, we could have avoided solving a quadratic equation.

14.45 Since equilibrium pressures are desired, we calculate K_p.

$$K_p = K_c(0.0821 \, T)^{\Delta n} = (4.63 \times 10^{-3})(0.0821 \times 800)^1 = 0.304$$

	$COCl_2(g)$	$\rightleftharpoons$	$CO(g)$	$+$	$Cl_2(g)$
Initial (atm):	0.760		0.000		0.000
Change (atm):	$-x$		$+x$		$+x$
Equilibrium (atm):	$(0.760 - x)$		x		x

$$\frac{x^2}{(0.760 - x)} = 0.304$$

$$x^2 + 0.304x - 0.231 = 0$$

$$x = 0.352 \text{ atm}$$

At equilibrium:

$$P_{COCl_2} = (0.760 - 0.352)\text{atm} = \textbf{0.408 atm}$$

$$P_{CO} = \textbf{0.352 atm}$$

$$P_{Cl_2} = \textbf{0.352 atm}$$

14.46 **(a)** The equilibrium constant, K_c, can be found by simple substitution.

$$K_c = \frac{[H_2O][CO]}{[CO_2][H_2]} = \frac{(0.040)(0.050)}{(0.086)(0.045)} = \textbf{0.52}$$

(b) The magnitude of the reaction quotient Q_c for the system after the concentration of CO_2 becomes 0.50 mol/L, but before equilibrium is reestablished, is:

$$Q_c = \frac{(0.040)(0.050)}{(0.50)(0.045)} = 0.089$$

The value of Q_c is smaller than K_c; therefore, the system will shift to the right, increasing the concentrations of CO and H_2O and decreasing the concentrations of CO_2 and H_2. Let x be the depletion in the concentration of CO_2 at equilibrium. The stoichiometry of the balanced equation then requires that the decrease in the concentration of H_2 must also be x, and that the concentration increases of CO and H_2O be equal to x as well. The changes in the original concentrations are shown in the table.

	CO_2	+	H_2	$\rightleftharpoons$	CO	+	H_2O
Initial (M):	0.50		0.045		0.050		0.040
Change (M):	$-x$		$-x$		$+x$		$+x$
Equilibrium (M):	$(0.50 - x)$		$(0.045 - x)$		$(0.050 + x)$		$(0.040 + x)$

The equilibrium constant expression is:

$$K_c = \frac{[H_2O][CO]}{[CO_2][H_2]} = \frac{(0.040 + x)(0.050 + x)}{(0.50 - x)(0.045 - x)} = 0.52$$

$$0.52(x^2 - 0.545x + 0.0225) = x^2 + 0.090x + 0.0020$$

$$0.48x^2 + 0.373x - (9.7 \times 10^{-3}) = 0$$

The positive root of the equation is $x = 0.025$.

The equilibrium concentrations are:

$$[CO_2] = (0.50 - 0.025)\, M = \textbf{0.48}\, \textbf{\textit{M}}$$
$$[H_2] = (0.045 - 0.025)\, M = \textbf{0.020}\, \textbf{\textit{M}}$$
$$[CO] = (0.050 + 0.025)\, M = \textbf{0.075}\, \textbf{\textit{M}}$$
$$[H_2O] = (0.040 + 0.025)\, M = \textbf{0.065}\, \textbf{\textit{M}}$$

14.47 The equilibrium constant expression for the system is:

$$K_p = \frac{(P_{CO})^2}{P_{CO_2}}$$

The total pressure can be expressed as:

$$P_{total} = P_{CO_2} + P_{CO}$$

If we let the partial pressure of CO be x, then the partial pressure of CO_2 is:

$$P_{CO_2} = P_{total} - x = (4.50 - x)\,atm$$

Substitution gives the equation:

$$K_p = \frac{(P_{CO})^2}{P_{CO_2}} = \frac{x^2}{(4.50 - x)} = 1.52$$

This can be rearranged to the quadratic:

$$x^2 + 1.52x - 6.84 = 0$$

The solutions are $x = 1.96$ and $x = -3.48$; only the positive result has physical significance (why?). The equilibrium pressures are

$$P_{CO} = x = \mathbf{1.96\ atm}$$

$$P_{CO_2} = (4.50 - 1.96) = \mathbf{2.54\ atm}$$

14.48   The initial concentrations are $[H_2] = 0.80\ mol/5.0\ L = 0.16\ M$ and $[CO_2] = 0.80\ mol/5.0\ L = 0.16\ M$.

	$H_2(g)$	$+$	$CO_2(g)$	$\rightleftharpoons$	$H_2O(g)$	$+$	$CO(g)$
Initial (M):	0.16		0.16		0.00		0.00
Change (M):	$-x$		$-x$		$+x$		$+x$
Equilibrium (M):	$0.16 - x$		$0.16 - x$		x		x

$$K_c = \frac{[H_2O][CO]}{[H_2][CO_2]} = 4.2 = \frac{x^2}{(0.16 - x)^2}$$

Taking the square root of both sides, we obtain:

$$\frac{x}{0.16 - x} = 2.0$$

$$x = 0.11\ M$$

The equilibrium concentrations are:

$$[H_2] = [CO_2] = (0.16 - 0.11)\ M = \mathbf{0.05\ M}$$

$$[H_2O] = [CO] = \mathbf{0.11\ M}$$

14.53   **(a)**   Addition of more $Cl_2(g)$ (a reactant) would shift the position of equilibrium to the **right**.

(b)   Removal of $SO_2Cl_2(g)$ (a product) would shift the position of equilibrium to the **right**.

(c)   Removal of $SO_2(g)$ (a reactant) would shift the position of equilibrium to the **left**.

14.54   **(a)**   Removal of $CO_2(g)$ from the system would shift the position of equilibrium to the **right**.

        (b)   Addition of more solid Na_2CO_3 would have **no effect**. $[Na_2CO_3]$ does not appear in the equilibrium constant expression.

        (c)   Removal of some of the solid $NaHCO_3$ would have **no effect**. Same reason as (b).

14.55   **(a)**   This reaction is endothermic. (Why?) According to Section 14.5, an increase in temperature favors an endothermic reaction, so the equilibrium constant should become **larger**.

        (b)   This reaction is exothermic. Such reactions are favored by decreases in temperature. The magnitude of K_c should **decrease**.

        (c)   In this system heat is neither absorbed nor released. A change in temperature should have **no effect** on the magnitude of the equilibrium constant.

14.56   **Strategy:** A change in pressure can affect only the volume of a gas, but not that of a solid or liquid because solids and liquids are much less compressible. The stress applied is an increase in pressure. According to Le Châtelier's principle, the system will adjust to partially offset this stress. In other words, the system will adjust to decrease the pressure. This can be achieved by shifting to the side of the equation that has fewer moles of gas. Recall that pressure is directly proportional to moles of gas: $PV = nRT$ so $P \propto n$.

        Solution:
        (a)   Changes in pressure ordinarily do not affect the concentrations of reacting species in condensed phases because liquids and solids are virtually incompressible. Pressure change should have **no effect** on this system.

        (b)   Same situation as (a).

        (c)   Only the product is in the gas phase. An increase in pressure should favor the reaction that decreases the total number of moles of gas. The equilibrium should shift to the **left**, that is, the amount of B should decrease and that of A should increase.

        (d)   In this equation there are equal moles of gaseous reactants and products. A shift in either direction will have no effect on the total number of moles of gas present. There will be **no change** when the pressure is increased.

        (e)   A shift in the direction of the reverse reaction (**left**) will have the result of decreasing the total number of moles of gas present.

14.57   **(a)**   A pressure increase will favor the reaction (forward or reverse?) that decreases the total number of moles of gas. The equilibrium should shift to the **right**, i.e., more I_2 will be produced at the expense of I.

        (b)   If the concentration of I_2 is suddenly altered, the system is no longer at equilibrium. Evaluating the magnitude of the reaction quotient Q_c allows us to predict the direction of the resulting equilibrium shift. The reaction quotient for this system is:

$$Q_c = \frac{[I_2]_0}{[I]_0^2}$$

        Increasing the concentration of I_2 will increase Q_c. The equilibrium will be reestablished in such a way that Q_c is again equal to the equilibrium constant. More I will form. The system shifts to the **left** to establish equilibrium.

        (c)   The forward reaction is exothermic. A decease in temperature will shift the system to the **right** to reestablish equilibrium.

14.58   **Strategy:** (a) What does the sign of $\Delta H°$ indicate about the heat change (endothermic or exothermic) for the forward reaction? (b) The stress is the addition of Cl_2 gas. How will the system adjust to partially offset the stress? (c) The stress is the removal of PCl_3 gas. How will the system adjust to partially offset the stress? (d) The stress is an increase in pressure. The system will adjust to decrease the pressure. Remember, pressure is directly proportional to moles of gas. (e) What is the function of a catalyst? How does it affect a reacting system not at equilibrium? at equilibrium?

Solution:

(a)   The stress applied is the heat added to the system. Note that the reaction is endothermic ($\Delta H° > 0$). Endothermic reactions absorb heat from the surroundings; therefore, we can think of heat as a reactant.

$$\text{heat} + PCl_5(g) \rightleftharpoons PCl_3(g) + Cl_2(g)$$

The system will adjust to remove some of the added heat by undergoing a decomposition reaction (from **left to right**)

(b)   The stress is the addition of Cl_2 gas. The system will shift in the direction to remove some of the added Cl_2. The system shifts from **right to left** until equilibrium is reestablished.

(c)   The stress is the removal of PCl_3 gas. The system will shift to replace some of the PCl_3 that was removed. The system shifts from **left to right** until equilibrium is reestablished.

(d)   The stress applied is an increase in pressure. The system will adjust to remove the stress by decreasing the pressure. Recall that pressure is directly proportional to the number of moles of gas. In the balanced equation we see 1 mole of gas on the reactants side and 2 moles of gas on the products side. The pressure can be decreased by shifting to the side with the fewer moles of gas. The system will shift from **right to left** to reestablish equilibrium.

(e)   The function of a catalyst is to increase the rate of a reaction. If a catalyst is added to the reacting system not at equilibrium, the system will reach equilibrium faster than if left undisturbed. If a system is already at equilibrium, as in this case, the addition of a catalyst will not affect either the concentrations of reactant and product, or the equilibrium constant.

14.59   (a)   Increasing the temperature favors the endothermic reaction so that the concentrations of SO_2 and O_2 will increase while that of SO_3 will decrease.

(b)   Increasing the pressure favors the reaction that decreases the number of moles of gas. The concentration of SO_3 will increase.

(c)   Increasing the concentration of SO_2 will lead to an increase in the concentration of SO_3 and a decrease in the concentration of O_2.

(d)   A catalyst has no effect on the position of equilibrium.

(e)   Adding an inert gas at constant volume has no effect on the position of equilibrium.

14.60   There will be no change in the pressures. A catalyst has no effect on the position of the equilibrium.

14.61   (a)   If helium gas is added to the system without changing the pressure or the temperature, the volume of the container must necessarily be increased. This will decrease the partial pressures of all the reactants and products. A pressure decrease will favor the reaction that increases the number of moles of gas. The position of equilibrium will shift to the **left**.

(b)   If the volume remains unchanged, the partial pressures of all the reactants and products will remain the same. The reaction quotient Q_c will still equal the equilibrium constant, and there will be **no change** in the position of equilibrium.

14.62   For this system, $K_p = [CO_2]$.

This means that to remain at equilibrium, the pressure of carbon dioxide must stay at a fixed value as long as the temperature remains the same.

(a)   If the volume is increased, the pressure of CO_2 will drop (Boyle's law, pressure and volume are inversely proportional). Some $CaCO_3$ will break down to form more CO_2 and CaO. **(Shift right)**

(b)   Assuming that the amount of added solid CaO is not so large that the volume of the system is altered significantly, there should be **no change** at all. If a huge amount of CaO were added, this would have the effect of reducing the volume of the container. What would happen then?

(c)   Assuming that the amount of $CaCO_3$ removed doesn't alter the container volume significantly, there should be **no change**. Removing a huge amount of $CaCO_3$ will have the effect of increasing the container volume. The result in that case will be the same as in part (a).

(d)   The pressure of CO_2 will be greater and will exceed the value of K_p. Some CO_2 will combine with CaO to form more $CaCO_3$. **(Shift left)**

(e)   Carbon dioxide combines with aqueous NaOH according to the equation

$$CO_2(g) + NaOH(aq) \rightarrow NaHCO_3(aq)$$

This will have the effect of reducing the CO_2 pressure and causing more $CaCO_3$ to break down to CO_2 and CaO. **(Shift right)**

(f)   Carbon dioxide does not react with hydrochloric acid, but $CaCO_3$ does.

$$CaCO_3(s) + 2HCl(aq) \rightarrow CaCl_2(aq) + CO_2(g) + H_2O(l)$$

The CO_2 produced by the action of the acid will combine with CaO as discussed in (d) above. **(Shift left)**

(g)   This is a decomposition reaction. Decomposition reactions are endothermic. Increasing the temperature will favor this reaction and produce more CO_2 and CaO. **(Shift right)**

14.63   **(i)**   The temperature of the system is not given.
          (ii)   It is not stated whether the equilibrium constant is K_p or K_c (would they be different for this reaction?).
          (iii) A balanced equation is not given.
          (iv)   The phases of the reactants and products are not given.

14.64   **(a)**   Since the total pressure is 1.00 atm, the sum of the partial pressures of NO and Cl_2 is

1.00 atm − partial pressure of NOCl = 1.00 atm − 0.64 atm = 0.36 atm

The stoichiometry of the reaction requires that the partial pressure of NO be twice that of Cl_2. Hence, the partial pressure of NO is **0.24 atm** and the partial pressure of Cl_2 is **0.12 atm**.

(b)   The equilibrium constant K_p is found by substituting the partial pressures calculated in part (a) into the equilibrium constant expression.

$$K_p = \frac{P_{NO}^2 P_{Cl_2}}{P_{NOCl}^2} = \frac{(0.24)^2(0.12)}{(0.64)^2} = \mathbf{0.017}$$

14.65   **(a)** $K_p = \dfrac{P_{NO}^2}{P_{N_2} P_{O_2}} = \dfrac{P_{NO}^2}{(3.0)(0.012)} = 2.9 \times 10^{-11}$

$P_{NO} = 1.0 \times 10^{-6} \text{ atm}$

(b) $4.0 \times 10^{-31} = \dfrac{P_{NO}^2}{(0.78)(0.21)}$

$P_{NO} = 2.6 \times 10^{-16} \text{ atm}$

(c)   Since K_p increases with temperature, it is endothermic.

(d)   Lightening. The electrical energy promotes the endothermic reaction.

14.66   The equilibrium expression for this system is given by:

$$K_p = P_{CO_2} P_{H_2O}$$

(a)   In a closed vessel the decomposition will stop when the product of the partial pressures of CO_2 and H_2O equals K_p. Adding more sodium bicarbonate will have **no effect**.

(b)   In an open vessel, $CO_2(g)$ and $H_2O(g)$ will escape from the vessel, and the partial pressures of CO_2 and H_2O will never become large enough for their product to equal K_p. Therefore, equilibrium will never be established. Adding more sodium bicarbonate will result in the production of **more CO_2 and H_2O**.

14.67   The relevant relationships are:

$$K_c = \frac{[B]^2}{[A]} \quad \text{and} \quad K_p = \frac{P_B^2}{P_A}$$

$$K_p = K_c(0.0821\,T)^{\Delta n} = K_c(0.0821\,T) \qquad \Delta n = +1$$

We set up a table for the calculated values of K_c and K_p.

$T(°C)$	K_c	K_p
200	$\dfrac{(0.843)^2}{(0.0125)} = 56.9$	$56.9(0.0821 \times 473) = 2.21 \times 10^3$
300	$\dfrac{(0.764)^2}{(0.171)} = 3.41$	$3.41(0.0821 \times 573) = 1.60 \times 10^2$
400	$\dfrac{(0.724)^2}{(0.250)} = 2.10$	$2.10(0.0821 \times 673) = 116$

Since K_c (and K_p) decrease with temperature, the reaction is exothermic.

14.68 **(a)** The equation that relates K_p and K_c is:

$$K_p = K_c(0.0821\,T)^{\Delta n}$$

For this reaction, $\Delta n = 3 - 2 = 1$

$$K_c = \frac{K_p}{(0.0821T)} = \frac{2 \times 10^{-42}}{(0.0821 \times 298)} = \mathbf{8 \times 10^{-44}}$$

(b) Because of a very large activation energy, the reaction of hydrogen with oxygen is infinitely slow without a catalyst or an initiator. The action of a single spark on a mixture of these gases results in the explosive formation of water.

14.69 Using data from Appendix 3 we calculate the enthalpy change for the reaction.

$$\Delta H° = 2\Delta H_f°(NOCl) - 2\Delta H_f°(NO) - \Delta H_f°(Cl_2) = 2(51.7\text{ kJ/mol}) - 2(90.4\text{ kJ/mol}) - (0) = -77.4\text{ kJ/mol}$$

The enthalpy change is negative, so the reaction is exothermic. The formation of NOCl will be favored by low temperature.

A pressure increase favors the reaction forming fewer moles of gas. The formation of NOCl will be favored by high pressure.

14.70 **(a)** Calculate the value of K_p by substituting the equilibrium partial pressures into the equilibrium constant expression.

$$K_p = \frac{P_B}{P_A^2} = \frac{(0.60)}{(0.60)^2} = \mathbf{1.7}$$

(b) The total pressure is the sum of the partial pressures for the two gaseous components, A and B. We can write:

$$P_A + P_B = 1.5\text{ atm}$$

and

$$P_B = 1.5 - P_A$$

Substituting into the expression for K_p gives:

$$K_p = \frac{(1.5 - P_A)}{P_A^2} = 1.7$$

$$1.7P_A^2 + P_A - 1.5 = 0$$

Solving the quadratic equation, we obtain:

$$P_A = \mathbf{0.69\text{ atm}}$$

and by difference,

$$P_B = \mathbf{0.81\text{ atm}}$$

Check that substituting these equilibrium concentrations into the equilibrium constant expression gives the equilibrium constant calculated in part (a).

$$K_p = \frac{P_B}{P_A^2} = \frac{0.81}{(0.69)^2} = 1.7$$

14.71 **(a)** The balanced equation shows that equal amounts of ammonia and hydrogen sulfide are formed in this decomposition. The partial pressures of these gases must just be half the total pressure, i.e., 0.355 atm. The value of K_p is

$$K_p = P_{NH_3} P_{H_2S} = (0.355)^2 = \mathbf{0.126}$$

(b) We find the number of moles of ammonia (or hydrogen sulfide) and ammonium hydrogen sulfide.

$$n_{NH_3} = \frac{PV}{RT} = \frac{(0.355 \text{ atm})(4.000 \text{ L})}{(0.0821 \text{ L} \cdot \text{atm/K} \cdot \text{mol})(297 \text{ K})} = 0.0582 \text{ mol}$$

$$n_{NH_4HS} = 6.1589 \text{ g} \times \frac{1 \text{ mol}}{51.12 \text{ g}} = 0.1205 \text{ mol (before decomposition)}$$

From the balanced equation the percent decomposed is

$$\frac{0.0582 \text{ mol}}{0.1205 \text{ mol}} \times 100\% = \mathbf{48.3\%}$$

(c) If the temperature does not change, K_p has the same value. The total pressure will still be 0.709 atm at equilibrium. In other words the amounts of ammonia and hydrogen sulfide will be twice as great, and the amount of solid ammonium hydrogen sulfide will be:

$$[0.1205 - 2(0.0582)]\text{mol} = \mathbf{0.0041 \text{ mol } NH_4HS}$$

14.72 Total number of moles of gas is:

$$0.020 + 0.040 + 0.96 = 1.02 \text{ mol of gas}$$

You can calculate the partial pressure of each gaseous component from the mole fraction and the total pressure.

$$P_{NO} = X_{NO} P_T = \frac{0.040}{1.02} \times 0.20 = 0.0078 \text{ atm}$$

$$P_{O_2} = X_{O_2} P_T = \frac{0.020}{1.02} \times 0.20 = 0.0039 \text{ atm}$$

$$P_{NO_2} = X_{NO_2} P_T = \frac{0.96}{1.02} \times 0.20 = 0.19 \text{ atm}$$

Calculate K_p by substituting the partial pressures into the equilibrium constant expression.

$$K_p = \frac{P_{NO_2}^2}{P_{NO}^2 P_{O_2}} = \frac{(0.19)^2}{(0.0078)^2 (0.0039)} = \mathbf{1.5 \times 10^5}$$

14.73 Since the reactant is a solid, we can write:

$$K_p = (P_{NH_3})^2 P_{CO_2}$$

The total pressure is the sum of the ammonia and carbon dioxide pressures.

$$P_{total} = P_{NH_3} + P_{CO_2}$$

From the stoichiometry,

$$P_{NH_3} = 2P_{CO_2}$$

Therefore:

$$P_{total} = 2P_{CO_2} + P_{CO_2} = 3P_{CO_2} = 0.318 \text{ atm}$$

$$P_{CO_2} = 0.106 \text{ atm}$$

$$P_{NH_3} = 0.212 \text{ atm}$$

Substituting into the equilibrium expression:

$$\boldsymbol{K_p} = (0.212)^2(0.106) = \boldsymbol{4.76 \times 10^{-3}}$$

14.74   Set up a table that contains the initial concentrations, the change in concentrations, and the equilibrium concentration. Assume that the vessel has a volume of 1 L.

	H_2	+	Cl_2	$\rightleftharpoons$	$2HCl$
Initial (*M*):	0.47		0		3.59
Change (*M*):	+x		+x		−2x
Equilibrium (*M*):	(0.47 + x)		x		(3.59 − 2x)

Substitute the equilibrium concentrations into the equilibrium constant expression, then solve for *x*. Since $\Delta n = 0$, $K_c = K_p$.

$$K_c = \frac{[HCl]^2}{[H_2][Cl_2]} = \frac{(3.59 - 2x)^2}{(0.47 + x)x} = 193$$

Solving the quadratic equation,

$$x = 0.10$$

Having solved for *x*, calculate the equilibrium concentrations of all species.

$$[H_2] = 0.57 \, M \qquad [Cl_2] = 0.10 \, M \qquad [HCl] = 3.39 \, M$$

Since we assumed that the vessel had a volume of 1 L, the above molarities also correspond to the number of moles of each component.

From the mole fraction of each component and the total pressure, we can calculate the partial pressure of each component.

$$\text{Total number of moles} = 0.57 + 0.10 + 3.39 = 4.06 \text{ mol}$$

$$P_{H_2} = \frac{0.57}{4.06} \times 2.00 = \textbf{0.28 atm}$$

$$P_{Cl_2} = \frac{0.10}{4.06} \times 2.00 = \textbf{0.049 atm}$$

$$P_{HCl} = \frac{3.39}{4.06} \times 2.00 = \textbf{1.67 atm}$$

14.75   **(a)**   From the balanced equation

$$N_2O_4 \rightleftharpoons 2NO_2$$

Initial (mol):	1	0
Change (mol):	$-\alpha$	$+2\alpha$
Equilibrium (mol):	$(1 - \alpha)$	2α

The total moles in the system = (moles N_2O_4 + moles NO_2) = $[(1 - \alpha) + 2\alpha] = 1 + \alpha$. If the total pressure in the system is P, then:

$$P_{N_2O_4} = \frac{1 - \alpha}{1 + \alpha} P \quad \text{and} \quad P_{NO_2} = \frac{2\alpha}{1 + \alpha} P$$

$$K_p = \frac{P_{NO_2}^2}{P_{N_2O_4}} = \frac{\left(\frac{2\alpha}{1 + \alpha}\right)^2 P^2}{\left(\frac{1 - \alpha}{1 + \alpha}\right) P}$$

$$K_p = \frac{\left(\frac{4\alpha^2}{1 + \alpha}\right) P}{1 - \alpha} = \frac{4\alpha^2}{1 - \alpha^2} P$$

(b)   Rearranging the K_p expression:

$$4\alpha^2 P = K_p - \alpha^2 K_p$$

$$\alpha^2(4P + K_p) = K_p$$

$$\alpha^2 = \frac{K_p}{4P + K_p}$$

$$\alpha = \sqrt{\frac{K_p}{4P + K_p}}$$

K_p is a constant (at constant temperature). Thus, as P increases, α must decrease, indicating that the system shifts to the **left**. This is also what one would predict based on Le Châtelier's principle.

14.76   This is a difficult problem. Express the equilibrium number of moles in terms of the initial moles and the change in number of moles (x). Next, calculate the mole fraction of each component. Using the mole fraction, you should come up with a relationship between partial pressure and total pressure for each component. Substitute the partial pressures into the equilibrium constant expression to solve for the total pressure, P_T.

The reaction is:

	N_2	$+$	$3 H_2$	$\rightleftharpoons$	$2 NH_3$
Initial (mol):	1		3		0
Change (mol):	$-x$		$-3x$		$2x$
Equilibrium (mol):	$(1 - x)$		$(3 - 3x)$		$2x$

$$\text{Mole fraction of NH}_3 = \frac{\text{mol of NH}_3}{\text{total number of moles}}$$

$$X_{\text{NH}_3} = \frac{2x}{(1-x) + (3-3x) + 2x} = \frac{2x}{4-2x}$$

$$0.21 = \frac{2x}{4-2x}$$

$$x = 0.35 \text{ mol}$$

Substituting x into the following mole fraction equations, the mole fractions of N_2 and H_2 can be calculated.

$$X_{\text{N}_2} = \frac{1-x}{4-2x} = \frac{1-0.35}{4-2(0.35)} = 0.20$$

$$X_{\text{H}_2} = \frac{3-3x}{4-2x} = \frac{3-3(0.35)}{4-2(0.35)} = 0.59$$

The partial pressures of each component are equal to the mole fraction multiplied by the total pressure.

$$P_{\text{NH}_3} = 0.21 P_T \qquad P_{\text{N}_2} = 0.20 P_T \qquad P_{\text{H}_2} = 0.59 P_T$$

Substitute the partial pressures above (in terms of P_T) into the equilibrium constant expression, and solve for P_T.

$$K_p = \frac{P_{\text{NH}_3}^2}{P_{\text{H}_2}^3 P_{\text{N}_2}}$$

$$4.31 \times 10^{-4} = \frac{(0.21)^2 P_T^2}{(0.59 P_T)^3 (0.20 P_T)}$$

$$4.31 \times 10^{-4} = \frac{1.07}{P_T^2}$$

$$\boldsymbol{P_T = 5.0 \times 10^1 \text{ atm}}$$

14.77 For the balanced equation: $K_c = \dfrac{[\text{H}_2]^2 [\text{S}_2]}{[\text{H}_2\text{S}]^2}$

$$[\text{S}_2] = \frac{[\text{H}_2\text{S}]^2}{[\text{H}_2]^2} K_c = \left(\frac{4.84 \times 10^{-3}}{1.50 \times 10^{-3}} \right)^2 (2.25 \times 10^{-4}) = \boldsymbol{2.34 \times 10^{-3} \, M}$$

14.78 The initial molarity of SO_2Cl_2 is:

$$[\text{SO}_2\text{Cl}_2] = \frac{6.75 \text{ g SO}_2\text{Cl}_2 \times \dfrac{1 \text{ mol SO}_2\text{Cl}_2}{135.0 \text{ g SO}_2\text{Cl}_2}}{2.00 \text{ L}} = 0.0250 \, M$$

The concentration of SO_2 at equilibrium is:

$$[SO_2] = \frac{0.0345 \text{ mol}}{2.00 \text{ L}} = 0.01725 \ M$$

Since there is a 1:1 mole ratio between SO_2 and SO_2Cl_2, the concentration of SO_2 at equilibrium (0.01725 M) equals the concentration of SO_2Cl_2 reacted. The concentrations of SO_2Cl_2 and Cl_2 at equilibrium are:

	$SO_2Cl_2(g)$	$\rightleftharpoons$	$SO_2(g)$	+	$Cl_2(g)$
Initial (M):	0.0250		0		0
Change (M):	−0.01725		+0.01725		+0.01725
Equilibrium (M):	0.00775		0.01725		0.01725

Substitute the equilibrium concentrations into the equilibrium constant expression to calculate K_c.

$$K_c = \frac{[SO_2][Cl_2]}{[SO_2Cl_2]} = \frac{(0.0173)(0.0173)}{(0.00775)} = 3.86 \times 10^{-2}$$

14.79 For a 100% yield, 2.00 moles of SO_3 would be formed (why?). An 80% yield means 2.00 moles × (0.80) = 1.60 moles SO_3 is formed.

The amount of SO_2 remaining at equilibrium = (2.00 − 1.60)mol = 0.40 mol

The amount of O_2 reacted = $\frac{1}{2}$ × (amount of SO_2 reacted) = ($\frac{1}{2}$ × 1.60)mol = 0.80 mol

The amount of O_2 remaining at equilibrium = (2.00 − 0.80)mol = 1.20 mol

Total moles at equilibrium = moles SO_2 + moles O_2 + moles SO_3 = (0.40 + 1.20 + 1.60)mol = 3.20 moles

$$P_{SO_2} = \frac{0.40}{3.20} P_{total} = 0.125 \, P_{total}$$

$$P_{O_2} = \frac{1.20}{3.20} P_{total} = 0.375 \, P_{total}$$

$$P_{SO_3} = \frac{1.60}{3.20} P_{total} = 0.500 \, P_{total}$$

$$K_p = \frac{P_{SO_3}^{\ 2}}{P_{SO_2}^{\ 2} P_{O_2}}$$

$$0.13 = \frac{(0.500 \, P_{total})^2}{(0.125 \, P_{total})^2 (0.375 \, P_{total})}$$

$$P_{total} = 328 \text{ atm}$$

14.80 $I_2(g) \rightleftharpoons 2I(g)$

Assuming 1 mole of I_2 is present originally and α moles reacts, at equilibrium: $[I_2] = 1 - \alpha$, $[I] = 2\alpha$. The total number of moles present in the system $= (1 - \alpha) + 2\alpha = 1 + \alpha$. From Problem 14.75(a) in the text, we know that K_p is equal to:

$$K_p = \frac{4\alpha^2}{1 - \alpha^2} P \qquad (1)$$

If there were no dissociation, then the pressure would be:

$$P = \frac{nRT}{V} = \frac{\left(1.00 \text{ g} \times \dfrac{1 \text{ mol}}{253.8 \text{ g}}\right)\left(0.0821 \dfrac{\text{L} \cdot \text{atm}}{\text{mol} \cdot \text{K}}\right)(1473 \text{ K})}{0.500 \text{ L}} = 0.953 \text{ atm}$$

$$\frac{\text{observed pressure}}{\text{calculated pressure}} = \frac{1.51 \text{ atm}}{0.953 \text{ atm}} = \frac{1 + \alpha}{1}$$

$$\alpha = 0.584$$

Substituting in equation (1) above:

$$K_p = \frac{4\alpha^2}{1 - \alpha^2} P = \frac{(4)(0.584)^2}{1 - (0.584)^2} \times 1.51 = \mathbf{3.13}$$

14.81   Panting decreases the concentration of CO_2 because CO_2 is exhaled during respiration. This decreases the concentration of carbonate ions, shifting the equilibrium to the left. Less $CaCO_3$ is produced. Two possible solutions would be either to cool the chickens' environment or to feed them carbonated water.

14.82   According to the ideal gas law, pressure is directly proportional to the concentration of a gas in mol/L if the reaction is at constant volume and temperature. Therefore, pressure may be used as a concentration unit. The reaction is:

	N_2	+	$3H_2$	$\rightleftharpoons$	$2NH_3$
Initial (atm):	0.862		0.373		0
Change (atm):	$-x$		$-3x$		$+2x$
Equilibrium (atm):	$(0.862 - x)$		$(0.373 - 3x)$		$2x$

$$K_p = \frac{P_{NH_3}^2}{P_{H_2}^3 P_{N_2}}$$

$$4.31 \times 10^{-4} = \frac{(2x)^2}{(0.373 - 3x)^3(0.862 - x)}$$

At this point, we need to make two assumptions that $3x$ is very small compared to 0.373 and that x is very small compared to 0.862. Hence,

$$0.373 - 3x \approx 0.373$$

and

$$0.862 - x \approx 0.862$$

$$4.31 \times 10^{-4} \approx \frac{(2x)^2}{(0.373)^3 (0.862)}$$

Solving for x.

$$x = 2.20 \times 10^{-3} \text{ atm}$$

The equilibrium pressures are:

$$P_{N_2} = [0.862 - (2.20 \times 10^{-3})] \text{atm} = \textbf{0.860 atm}$$

$$P_{H_2} = [0.373 - (3)(2.20 \times 10^{-3})] \text{atm} = \textbf{0.366 atm}$$

$$P_{NH_3} = (2)(2.20 \times 10^{-3} \text{ atm}) = \textbf{4.40} \times \textbf{10}^{-3} \textbf{ atm}$$

Was the assumption valid that we made above? Typically, the assumption is considered valid if x is less than 5 percent of the number that we said it was very small compared to. Is this the case?

14.83 **(a)**   The sum of the mole fractions must equal one.

$$X_{CO} + X_{CO_2} = 1 \qquad \text{and} \qquad X_{CO_2} = 1 - X_{CO}$$

According to the hint, the average molar mass is the sum of the products of the mole fraction of each gas and its molar mass.

$$(X_{CO} \times 28.01 \text{ g}) + [(1 - X_{CO}) \times 44.01 \text{ g}] = 35 \text{ g}$$

Solving, $X_{CO} = \textbf{0.56}$ and $X_{CO_2} = \textbf{0.44}$

(b)   Solving for the pressures $P_{total} = P_{CO} + P_{CO_2} = 11 \text{ atm}$

$$P_{CO} = X_{CO} P_{total} = (0.56)(11 \text{ atm}) = 6.2 \text{ atm}$$

$$P_{CO_2} = X_{CO_2} P_{total} = (0.44)(11 \text{ atm}) = 4.8 \text{ atm}$$

$$K_p = \frac{P_{CO}^2}{P_{CO_2}} = \frac{(6.2)^2}{4.8} = \textbf{8.0}$$

14.84 **(a)**   The equation is:

	fructose	$\rightleftharpoons$	glucose
Initial (M):	0.244		0
Change (M):	−0.131		+0.131
Equilibrium (M):	0.113		0.131

Calculating the equilibrium constant,

$$K_c = \frac{[\text{glucose}]}{[\text{fructose}]} = \frac{0.131}{0.113} = \textbf{1.16}$$

(b)   **Percent converted** $= \dfrac{\text{amount of fructose converted}}{\text{original amount of fructose}} \times 100\%$

$$= \frac{0.131}{0.244} \times 100\% = \textbf{53.7\%}$$

14.85   If you started with radioactive iodine in the solid phase, then you should fine radioactive iodine in the vapor phase at equilibrium. Conversely, if you started with radioactive iodine in the vapor phase, you should find radioactive iodine in the solid phase. Both of these observations indicate a dynamic equilibrium between solid and vapor phase.

14.86   **(a)**   There is only one gas phase component, O_2. The equilibrium constant is simply

$$K_p = P_{O_2} = 0.49 \text{ atm}$$

(b)   From the ideal gas equation, we can calculate the moles of O_2 produced by the decomposition of CuO.

$$n_{O_2} = \frac{PV}{RT} = \frac{(0.49 \text{ atm})(2.0 \text{ L})}{(0.0821 \text{ L} \cdot \text{atm/K} \cdot \text{mol})(1297 \text{ K})} = 9.2 \times 10^{-3} \text{ mol } O_2$$

From the balanced equation,

$$(9.2 \times 10^{-3} \text{ mol } O_2) \times \frac{4 \text{ mol CuO}}{1 \text{ mol } O_2} = 3.7 \times 10^{-2} \text{ mol CuO decomposed}$$

$$\textbf{Fraction of CuO decomposed} = \frac{\text{amount of CuO lost}}{\text{original amount of CuO}}$$

$$= \frac{3.7 \times 10^{-2} \text{ mol}}{0.16 \text{ mol}} = \textbf{0.23}$$

(c)   If a 1.0 mol sample were used, the pressure of oxygen would still be the same (0.49 atm) and it would be due to the same quantity of O_2. Remember, a pure solid does not affect the equilibrium position. The moles of CuO lost would still be 3.7×10^{-2} mol. Thus the fraction decomposed would be:

$$\frac{0.037}{1.0} = \textbf{0.037}$$

(d)   If the number of moles of CuO were less than 3.7×10^{-2} mol, the equilibrium could not be established because the pressure of O_2 would be less than 0.49 atm. Therefore, the smallest number of moles of CuO needed to establish equilibrium must be slightly greater than 3.7×10^{-2} mol.

14.87   If there were 0.88 mole of CO_2 initially and at equilibrium there were 0.11 moles, then $(0.88 - 0.11)$ moles = 0.77 moles reacted.

	NO	+	CO_2	$\rightleftharpoons$	NO_2	+	CO
Initial (mol):	3.9		0.88		0		0
Change (mol):	−0.77		−0.77		+0.77		+0.77
Equilibrium (mol):	(3.9 − 0.77)		0.11		0.77		0.77

Solving for the equilibrium constant:

$$K_c = \frac{(0.77)(0.77)}{(3.9 - 0.77)(0.11)} = \textbf{1.7}$$

In the balanced equation there are equal number of moles of products and reactants; therefore, the volume of the container will not affect the calculation of K_c. We can solve for the equilibrium constant in terms of moles.

14.88   We first must find the initial concentrations of all the species in the system.

$$[H_2]_0 = \frac{0.714 \text{ mol}}{2.40 \text{ L}} = 0.298 \ M$$

$$[I_2]_0 = \frac{0.984 \text{ mol}}{2.40 \text{ L}} = 0.410 \ M$$

$$[HI]_0 = \frac{0.886 \text{ mol}}{2.40 \text{ L}} = 0.369 \ M$$

Calculate the reaction quotient by substituting the initial concentrations into the appropriate equation.

$$Q_c = \frac{[HI]_0^2}{[H_2]_0[I_2]_0} = \frac{(0.369)^2}{(0.298)(0.410)} = 1.11$$

We find that Q_c is less than K_c. The equilibrium will shift to the right, decreasing the concentrations of H_2 and I_2 and increasing the concentration of HI.

We set up the usual table. Let x be the decrease in concentration of H_2 and I_2.

	H_2	$+$	I_2	$\rightleftharpoons$	$2 HI$
Initial (M):	0.298		0.410		0.369
Change (M):	$-x$		$-x$		$+2x$
Equilibrium (M):	$(0.298 - x)$		$(0.410 - x)$		$(0.369 + 2x)$

The equilibrium constant expression is:

$$K_c = \frac{[HI]^2}{[H_2][I_2]} = \frac{(0.369 + 2x)^2}{(0.298 - x)(0.410 - x)} = 54.3$$

This becomes the quadratic equation

$$50.3x^2 - 39.9x + 6.48 = 0$$

The smaller root is $x = 0.228 \ M$. (The larger root is physically impossible.)

Having solved for x, calculate the equilibrium concentrations.

$$[H_2] = (0.298 - 0.228) \ M = \textbf{0.070} \ \textbf{\textit{M}}$$
$$[I_2] = (0.410 - 0.228) \ M = \textbf{0.182} \ \textbf{\textit{M}}$$
$$[HI] = [0.369 + 2(0.228)] \ M = \textbf{0.825} \ \textbf{\textit{M}}$$

14.89   Since we started with pure A, then any A that is lost forms equal amounts of B and C. Since the total pressure is P, the pressure of $B + C = P - 0.14 \ P = 0.86 \ P$. The pressure of $B = C = 0.43 \ P$.

$$K_p = \frac{P_B P_C}{P_A} = \frac{(0.43 \ P)(0.43 \ P)}{0.14 \ P} = \textbf{1.3} \ \textbf{\textit{P}}$$

14.90   The gas cannot be (a) because the color became lighter with heating. Heating (a) to 150°C would produce some HBr, which is colorless and would lighten rather than darken the gas.

The gas cannot be (b) because Br_2 doesn't dissociate into Br atoms at 150°C, so the color shouldn't change.

The gas must be (c). From 25°C to 150°C, heating causes N_2O_4 to dissociate into NO_2, thus darkening the color (NO_2 is a brown gas).

$$N_2O_4(g) \rightarrow 2NO_2(g)$$

Above 150°C, the NO_2 breaks up into colorless NO and O_2.

$$2NO_2(g) \rightarrow 2NO(g) + O_2(g)$$

An increase in pressure shifts the equilibrium back to the left, forming NO_2, thus darkening the gas again.

$$2NO(g) + O_2(g) \rightarrow 2NO_2(g)$$

14.91 Since the catalyst is exposed to the reacting system, it would catalyze the $2A \rightarrow B$ reaction. This shift would result in a decrease in the number of gas molecules, so the gas pressure decreases. The piston would be pushed down by the atmospheric pressure. When the cover is over the box, the catalyst is no longer able to favor the forward reaction. To reestablish equilibrium, the $B \rightarrow 2A$ step would dominate. This would increase the gas pressure so the piston rises and so on.

Conclusion: Such a catalyst would result in a perpetual motion machine (the piston would move up and down forever) which can be used to do work without input of energy or net consumption of chemicals. Such a machine cannot exist.

14.92 Given the following: $K_c = \dfrac{[NH_3]^2}{[N_2][H_2]^3} = 1.2$

(a) Temperature must have units of Kelvin.

$$K_p = K_c(0.0821\,T)^{\Delta n}$$
$$\boldsymbol{K_p = (1.2)(0.0821 \times 648)^{(2-4)} = 4.2 \times 10^{-4}}$$

(b) Recalling that,

$$K_{\text{forward}} = \frac{1}{K_{\text{reverse}}}$$

Therefore,

$$\boldsymbol{K_c' = \frac{1}{1.2} = 0.83}$$

(c) Since the equation

$$\tfrac{1}{2}N_2(g) + \tfrac{3}{2}H_2(g) \rightleftharpoons NH_3(g)$$

is equivalent to

$$\tfrac{1}{2}[N_2(g) + 3H_2(g) \rightleftharpoons 2NH_3(g)]$$

then, K_c' for the reaction:

$$\tfrac{1}{2}N_2(g) + \tfrac{3}{2}H_2(g) \rightleftharpoons NH_3(g)$$

equals $(K_c)^{\frac{1}{2}}$ for the reaction:

$$N_2(g) + 3H_2(g) \rightleftharpoons 2NH_3(g)$$

Thus,

$$K_c' = (K_c)^{\frac{1}{2}} = \sqrt{1.2} = \mathbf{1.1}$$

(d)   For K_p in part (b):

$$K_p = (0.83)(0.0821 \times 648)^{+2} = \mathbf{2.3 \times 10^3}$$

and for K_p in part (c):

$$K_p = (1.1)(0.0821 \times 648)^{-1} = \mathbf{0.021}$$

14.93   (a)   Color deepens   (b)   increases   (c)   decreases
        (d)   increases   (e)   unchanged

14.94   The vapor pressure of water is equivalent to saying the partial pressure of $H_2O(g)$.

$$K_p = P_{H_2O} = \mathbf{0.0231}$$

$$K_c = \frac{K_p}{(0.0821T)^{\Delta n}} = \frac{0.0231}{(0.0821 \times 293)^1} = \mathbf{9.60 \times 10^{-4}}$$

14.95   Potassium is more volatile than sodium. Therefore, its removal shifts the equilibrium from left to right.

14.96   We can calculate the average molar mass of the gaseous mixture from the density.

$$\mathcal{M} = \frac{dRT}{P}$$

Let $\overline{\mathcal{M}}$ be the average molar mass of NO_2 and N_2O_4. The above equation becomes:

$$\overline{\mathcal{M}} = \frac{dRT}{P} = \frac{(2.3 \text{ g/L})(0.0821 \text{ L} \cdot \text{atm/K} \cdot \text{mol})(347 \text{ K})}{1.3 \text{ atm}}$$

$$\overline{\mathcal{M}} = 50.4 \text{ g/mol}$$

The average molar mass is equal to the sum of the molar masses of each component times the respective mole fractions. Setting this up, we can calculate the mole fraction of each component.

$$\overline{\mathcal{M}} = X_{NO_2}\mathcal{M}_{NO_2} + X_{N_2O_4}\mathcal{M}_{N_2O_4} = 50.4 \text{ g/mol}$$

$$X_{NO_2}(46.01 \text{ g/mol}) + (1 - X_{NO_2})(92.01 \text{ g/mol}) = 50.4 \text{ g/mol}$$

$$X_{NO_2} = 0.905$$

We can now calculate the partial pressure of NO_2 from the mole fraction and the total pressure.

$$P_{NO_2} = X_{NO_2}P_T$$

$$P_{NO_2} = (0.905)(1.3 \text{ atm}) = \mathbf{1.2 \text{ atm}}$$

We can calculate the partial pressure of N_2O_4 by difference.

$$P_{N_2O_4} = P_T - P_{NO_2}$$

$$P_{N_2O_4} = (1.3 - 1.18)\,\text{atm} = \textbf{0.12 atm}$$

Finally, we can calculate K_p for the dissociation of N_2O_4. .

$$K_p = \frac{P_{NO_2}^2}{P_{N_2O_4}} = \frac{(1.2)^2}{0.12} = \textbf{12}$$

14.97   **(a)**   Since both reactions are endothermic ($\Delta H°$ is positive), according to Le Châtelier's principle the products would be favored at high temperatures. Indeed, the steam-reforming process is carried out at very high temperatures (between 800°C and 1000°C). It is interesting to note that in a plant that uses natural gas (methane) for both hydrogen generation and heating, about one-third of the gas is burned to maintain the high temperatures.

In each reaction there are more moles of products than reactants; therefore, we expect products to be favored at low pressures. In reality, the reactions are carried out at high pressures. The reason is that when the hydrogen gas produced is used captively (usually in the synthesis of ammonia), high pressure leads to higher yields of ammonia.

(b)   **(i)**   The relation between K_c and K_p is given by Equation (14.5) of the text:

$$K_p = K_c(0.0821\,T)^{\Delta n}$$

Since $\Delta n = 4 - 2 = 2$, we write:

$$K_p = (18)(0.0821 \times 1073)^2 = \textbf{1.4} \times \textbf{10}^{\textbf{5}}$$

(ii)   Let x be the amount of CH_4 and H_2O (in atm) reacted. We write:

	CH_4	+	H_2O	$\rightleftharpoons$	CO	+	$3H_2$
Initial (atm):	15		15		0		0
Change (atm):	$-x$		$-x$		$+x$		$+3x$
Equilibrium (atm):	$15-x$		$15-x$		x		$3x$

The equilibrium constant is given by:

$$K_p = \frac{P_{CO}P_{H_2}^3}{P_{CH_4}P_{H_2O}}$$

$$1.4 \times 10^5 = \frac{(x)(3x)^3}{(15-x)(15-x)} = \frac{27x^4}{(15-x)^2}$$

Taking the square root of both sides, we obtain:

$$3.7 \times 10^2 = \frac{5.2x^2}{15-x}$$

which can be expressed as

$$5.2x^2 + (3.7 \times 10^2 x) - (5.6 \times 10^3) = 0$$

Solving the quadratic equation, we obtain

$$x = 13 \text{ atm}$$

(The other solution for x is negative and is physically impossible.)

At equilibrium, the pressures are:

$$P_{CH_4} = (15 - 13) = \textbf{2 atm}$$

$$P_{H_2O} = (15 - 13) = \textbf{2 atm}$$

$$P_{CO} = \textbf{13 atm}$$

$$P_{H_2} = 3(13 \text{ atm}) = \textbf{39 atm}$$

14.98   **(a)**   shifts to right   **(b)**   shifts to right   **(c)**   no change   **(d)**   no change
            (e)   no change   **(f)**   shifts to left

14.99 $K_p = P_{NH_3} P_{HCl}$

$$P_{NH_3} = P_{HCl} = \frac{2.2}{2} = 1.1 \text{ atm}$$

$$\textbf{\textit{K}}_\textbf{p} = (1.1)(1.1) = \textbf{1.2}$$

14.100   The equilibrium is: $N_2O_4(g) \rightleftharpoons 2NO_2(g)$

$$K_p = \frac{(P_{NO_2})^2}{P_{N_2O_4}} = \frac{0.15^2}{0.20} = 0.113$$

Volume is doubled so pressure is halved. Let's calculate Q_p and compare it to K_p.

$$Q_p = \frac{\left(\dfrac{0.15}{2}\right)^2}{\left(\dfrac{0.20}{2}\right)} = 0.0563 < K_p$$

Equilibrium will shift to the right. Some N_2O_4 will react, and some NO_2 will be formed. Let x = amount of N_2O_4 reacted.

	$N_2O_4(g)$	$\rightleftharpoons$	$2NO_2(g)$
Initial (atm):	0.10		0.075
Change (atm):	$-x$		$+2x$
Equilibrium (atm):	$0.10 - x$		$0.075 + 2x$

Substitute into the K_p expression to solve for x.

$$K_p = 0.113 = \frac{(0.075 + 2x)^2}{0.10 - x}$$

$$4x^2 + 0.413x - 5.67 \times 10^{-3} = 0$$

$$x = 0.0123$$

At equilibrium:

$$P_{NO_2} = 0.075 + 2(0.0123) = 0.0996 \approx \textbf{0.10 atm}$$

$$P_{N_2O_4} = 0.10 - 0.0123 = \textbf{0.088 atm}$$

Check:

$$K_p = \frac{(0.10)^2}{0.088} = 0.114 \qquad \text{close enough to } 0.113$$

14.101 **(a)** React Ni with CO above 50°C. Pump away the Ni(CO)$_4$ vapor (shift equilibrium to right), leaving the solid impurities behind.

(b) Consider the reverse reaction:

$$Ni(CO)_4(g) \rightarrow Ni(s) + 4CO(g)$$

$$\Delta H° = 4\Delta H_f°(CO) - \Delta H_f°[Ni(CO)_4]$$

$$\Delta H° = (4)(-110.5 \text{ kJ/mol}) - (1)(-602.9 \text{ kJ/mol}) = 160.9 \text{ kJ/mol}$$

The decomposition is endothermic, which is favored at high temperatures. Heat Ni(CO)$_4$ above 200°C to convert it back to Ni.

14.102 **(a)** Molar mass of PCl$_5$ = 208.2 g/mol

$$P = \frac{nRT}{V} = \frac{\left(2.50 \text{ g} \times \dfrac{1 \text{ mol}}{208.2 \text{ g}}\right)\left(0.0821\dfrac{\text{L} \cdot \text{atm}}{\text{mol} \cdot \text{K}}\right)(523 \text{ K})}{0.500 \text{ L}} = \textbf{1.03 atm}$$

(b)

	PCl$_5$	$\rightleftharpoons$	PCl$_3$	+	Cl$_2$
Initial (atm)	1.03		0		0
Change (atm)	$-x$		$+x$		$+x$
Equilibrium (atm)	$1.03 - x$		x		x

$$K_p = 1.05 = \frac{x^2}{1.03 - x}$$

$$x^2 + 1.05x - 1.08 = 0$$

$$x = 0.639$$

At equilibrium:

$$P_{PCl_5} = 1.03 - 0.639 = \textbf{0.39 atm}$$

(c) $P_T = (1.03 - x) + x + x = 1.03 + 0.639 = \textbf{1.67 atm}$

(d) $\dfrac{0.639 \text{ atm}}{1.03 \text{ atm}} = \textbf{0.620}$

14.103 (a)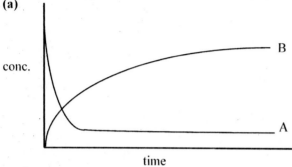

conc.

(b)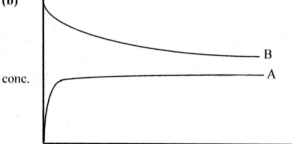

conc.

(c)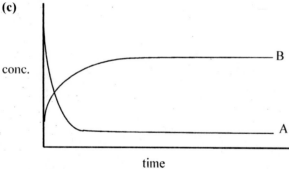

conc.

14.104 (a) $K_p = P_{Hg} = 0.0020 \text{ mmHg} = 2.6 \times 10^{-6} \text{ atm} = \textbf{2.6} \times \textbf{10}^{-6}$ (equil. constants are expressed without units)

$$K_c = \frac{K_p}{(0.0821T)^{\Delta n}} = \frac{2.6 \times 10^{-6}}{(0.0821 \times 299)^1} = \textbf{1.1} \times \textbf{10}^{-7}$$

(b) Volume of lab $= (6.1 \text{ m})(5.3 \text{ m})(3.1 \text{ m}) = 100 \text{ m}^3$

[Hg] $= K_c$

$$\text{Total mass of Hg vapor} \ = \ \frac{1.1 \times 10^{-7} \text{ mol}}{1 \text{ L}} \times \frac{200.6 \text{ g}}{1 \text{ mol}} \times \frac{1 \text{ L}}{1000 \text{ cm}^3} \times \left(\frac{1 \text{ cm}}{0.01 \text{ m}}\right)^3 \times 100 \text{ m}^3 \ = \ \textbf{2.2 g}$$

The concentration of mercury vapor in the room is:

$$\frac{2.2 \text{ g}}{100 \text{ m}^3} \ = \ 0.022 \text{ g/m}^3 \ = \ \textbf{22 mg/m}^3$$

Yes! This concentration exceeds the safety limit of 0.05 mg/m^3. Better clean up the spill!

14.105   Initially, at equilibrium: $[NO_2] = 0.0475\ M$ and $[N_2O_4] = 0.487\ M$. At the instant the volume is halved, the concentrations double.

$[NO_2] = 2(0.0475\ M) = 0.0950\ M$ and $[N_2O_4] = 2(0.487\ M) = 0.974\ M$. The system is no longer at equilibrium. The system will shift to the left to offset the increase in pressure when the volume is halved. When a new equilibrium position is established, we write:

$$N_2O_4 \quad \rightleftharpoons \quad 2\ NO_2$$
$$0.974\ M + x \qquad 0.0950\ M - 2x$$

$$K_c \ = \ \frac{[NO_2]^2}{[N_2O_4]} \ = \ \frac{(0.0950 - 2x)^2}{(0.982 + x)}$$

$$4x^2 - 0.3846x + 4.52 \times 10^{-3} \ = \ 0$$

Solving $x = 0.0824\ M$ (impossible)   and $x = 0.0137\ M$

At the new equilibrium,

$$[N_2O_4] \ = \ 0.974 + 0.0137 \ = \ \textbf{0.988\ M}$$
$$[NO_2] \ = \ 0.0950 - (2 \times 0.0137) \ = \ \textbf{0.0676\ M}$$

As we can see, the new equilibrium concentration of NO$_2$ is *greater* than the initial equilibrium concentration (0.0475 M). Therefore, the gases should look *darker*!

14.106   There is a temporary dynamic equilibrium between the melting ice cubes and the freezing of water between the ice cubes.

14.107   **(a)**   A catalyst speeds up the rates of the forward and reverse reactions to the same extent.

   (b)   A catalyst would not change the energies of the reactant and product.

   (c)   The first reaction is exothermic. Raising the temperature would favor the reverse reaction, increasing the amount of reactant and decreasing the amount of product at equilibrium. The equilibrium constant, K, would decrease. The second reaction is endothermic. Raising the temperature would favor the forward reaction, increasing the amount of product and decreasing the amount of reactant at equilibrium. The equilibrium constant, K, would increase.

   (d)   A catalyst lowers the activation energy for the forward and reverse reactions to the same extent. Adding a catalyst to a reaction mixture will simply cause the mixture to reach equilibrium sooner. The same equilibrium mixture could be obtained without the catalyst, but we might have to wait longer for equilibrium to be reached. If the same equilibrium position is reached, with or without a catalyst, then the equilibrium constant is the same.

14.108 First, let's calculate the initial concentration of ammonia.

$$[NH_3] = \frac{14.6 \text{ g} \times \dfrac{1 \text{ mol NH}_3}{17.03 \text{ g NH}_3}}{4.00 \text{ L}} = 0.214 \text{ } M$$

Let's set up a table to represent the equilibrium concentrations. We represent the amount of NH_3 that reacts as $2x$.

	$2NH_3(g)$	$\rightleftharpoons$	$N_2(g)$	$+$	$3H_2(g)$
Initial (M):	0.214		0		0
Change (M):	$-2x$		$+x$		$+3x$
Equilibrium (M):	$0.214 - 2x$		x		$3x$

Substitute into the equilibrium constant expression to solve for x.

$$K_c = \frac{[N_2][H_2]^3}{[NH_3]^2}$$

$$0.83 = \frac{(x)(3x)^3}{(0.214 - 2x)^2} = \frac{27x^4}{(0.214 - 2x)^2}$$

Taking the square root of both sides of the equation gives:

$$0.91 = \frac{5.20x^2}{0.214 - 2x}$$

Rearranging,

$$5.20x^2 + 1.82x - 0.195 = 0$$

Solving the quadratic equation gives the solutions:

$$x = 0.086 \text{ } M \text{ and } x = -0.44 \text{ } M$$

The positive root is the correct answer. The equilibrium concentrations are:

$$[NH_3] = 0.214 - 2(0.086) = \textbf{0.042 } \textbf{\textit{M}}$$
$$[N_2] = \textbf{0.086 } \textbf{\textit{M}}$$
$$[H_2] = 3(0.086) = \textbf{0.26 } \textbf{\textit{M}}$$

CHAPTER 15
ACIDS AND BASES

15.3 Table 15.2 of the text contains a list of important Brønsted acids and bases. **(a)** both (why?), **(b)** base, **(c)** acid, **(d)** base, **(e)** acid, **(f)** base, **(g)** base, **(h)** base, **(i)** acid, **(j)** acid.

15.4 Recall that the conjugate base of a Brønsted acid is the species that remains when *one* proton has been removed from the acid.

    (a) nitrite ion: NO_2^-

    (b) hydrogen sulfate ion (also called bisulfate ion): HSO_4^-

    (c) hydrogen sulfide ion (also called bisulfide ion): HS^-

    (d) cyanide ion: CN^-

    (e) formate ion: $HCOO^-$

15.5 In general the components of the conjugate acid–base pair are on opposite sides of the reaction arrow. The base always has one fewer proton than the acid.

    (a) The conjugate acid–base pairs are (1) HCN (acid) and CN^- (base) and (2) CH_3COO^- (base) and CH_3COOH (acid).

    (b) (1) HCO_3^- (acid) and CO_3^{2-} (base) and (2) HCO_3^- (base) and H_2CO_3 (acid).

    (c) (1) $H_2PO_4^-$ (acid) and HPO_4^{2-} (base) and (2) NH_3 (base) and NH_4^+ (acid).

    (d) (1) HClO (acid) and ClO^- (base) and (2) CH_3NH_2 (base) and $CH_3NH_3^+$ (acid).

    (e) (1) H_2O (acid) and OH^- (base) and (2) CO_3^{2-} (base) and HCO_3^- (acid).

15.6 The conjugate acid of any base is just the base with a proton added.

    (a) H_2S **(b)** H_2CO_3 **(c)** HCO_3^- **(d)** H_3PO_4 **(e)** $H_2PO_4^-$

    (f) HPO_4^{2-} **(g)** H_2SO_4 **(h)** HSO_4^- **(i)** HSO_3^-

15.7 **(a)** The Lewis structures are

    (b) H^+ and $C_2H_2O_4$ can act only as acids, $C_2HO_4^-$ can act as both an acid and a base, and $C_2O_4^{2-}$ can act only as a base.

15.8 The conjugate base of any acid is simply the acid minus one proton.

    (a) CH_2ClCOO^- **(b)** IO_4^- **(c)** $H_2PO_4^-$ **(d)** HPO_4^{2-} **(e)** PO_4^{3-}

    (f) HSO_4^- **(g)** SO_4^{2-} **(h)** IO_3^- **(i)** SO_3^{2-} **(j)** NH_3

    (k) HS^- **(l)** S^{2-} **(m)** OCl^-

15.15 $[H^+] = 1.4 \times 10^{-3} \, M$

$$[OH^-] = \frac{K_w}{[H^+]} = \frac{1.0 \times 10^{-14}}{1.4 \times 10^{-3}} = \mathbf{7.1 \times 10^{-12} \, M}$$

15.16 $[OH^-] = 0.62 \, M$

$$[H^+] = \frac{K_w}{[OH^-]} = \frac{1.0 \times 10^{-14}}{0.62} = \mathbf{1.6 \times 10^{-14} \, M}$$

15.17 **(a)** HCl is a strong acid, so the concentration of hydrogen ion is also 0.0010 M. (What is the concentration of chloride ion?) We use the definition of pH.

$$pH = -\log[H^+] = -\log(0.0010) = \mathbf{3.00}$$

(b) KOH is an ionic compound and completely dissociates into ions. We first find the concentration of hydrogen ion.

$$[H^+] = \frac{K_w}{[OH^-]} = \frac{1.0 \times 10^{-14}}{0.76} = 1.3 \times 10^{-14} \, M$$

The pH is then found from its defining equation

$$pH = -\log[H^+] = -\log[1.3 \times 10^{-14}] = \mathbf{13.89}$$

15.18 **(a)** Ba(OH)$_2$ is ionic and fully ionized in water. The concentration of the hydroxide ion is $5.6 \times 10^{-4} \, M$ (Why? What is the concentration of Ba^{2+}?) We find the hydrogen ion concentration.

$$[H^+] = \frac{K_w}{[OH^-]} = \frac{1.0 \times 10^{-14}}{5.6 \times 10^{-4}} = 1.8 \times 10^{-11} \, M$$

The pH is then: $\quad pH = -\log[H^+] = -\log(1.8 \times 10^{-11}) = \mathbf{10.74}$

(b) Nitric acid is a strong acid, so the concentration of hydrogen ion is also $5.2 \times 10^{-4} \, M$. The pH is:

$$pH = -\log[H^+] = -\log(5.2 \times 10^{-4}) = \mathbf{3.28}$$

15.19 Since pH $= -\log[H^+]$, we write $[H^+] = 10^{-pH}$

(a) $[H^+] = 10^{-2.42} = \mathbf{3.8 \times 10^{-3} \, M}$ **(c)** $[H^+] = 10^{-6.96} = \mathbf{1.1 \times 10^{-7} \, M}$

(b) $[H^+] = 10^{-11.21} = \mathbf{6.2 \times 10^{-12} \, M}$ **(d)** $[H^+] = 10^{-15.00} = \mathbf{1.0 \times 10^{-15} \, M}$

15.20 **Strategy:** Here we are given the pH of a solution and asked to calculate $[H^+]$. Because pH is defined as pH $= -\log[H^+]$, we can solve for $[H^+]$ by taking the antilog of the pH; that is, $[H^+] = 10^{-pH}$.

Solution: From Equation (15.4) of the text:

(a) pH $= -\log[H^+] = 5.20$

$\log[H^+] = -5.20$

To calculate $[H^+]$, we need to take the antilog of -5.20.

$$[H^+] = 10^{-5.20} = 6.3 \times 10^{-6} \, M$$

Check: Because the pH is between 5 and 6, we can expect $[H^+]$ to be between $1 \times 10^{-5} \, M$ and $1 \times 10^{-6} \, M$. Therefore, the answer is reasonable.

(b) $pH = -\log [H^+] = 16.00$

$\log[H^+] = -16.00$

$[H^+] = 10^{-16.00} = 1.0 \times 10^{-16} \, M$

(c)

Strategy: We are given the concentration of OH^- ions and asked to calculate $[H^+]$. The relationship between $[H^+]$ and $[OH^-]$ in water or an aqueous solution is given by the ion-product of water, K_W [Equation (15.3) of the text].

Solution: The ion product of water is applicable to all aqueous solutions. At 25°C,

$$K_W = 1.0 \times 10^{-14} = [H^+][OH^-]$$

Rearranging the equation to solve for $[H^+]$, we write

$$[H^+] = \frac{1.0 \times 10^{-14}}{[OH^-]} = \frac{1.0 \times 10^{-14}}{3.7 \times 10^{-9}} = 2.7 \times 10^{-6} \, M$$

Check: Since the $[OH^-] < 1 \times 10^{-7} \, M$ we expect the $[H^+]$ to be greater than $1 \times 10^{-7} \, M$.

15.21

pH	$[H^+]$	Solution is:
< 7	$> 1.0 \times 10^{-7} \, M$	acid
> 7	$< 1.0 \times 10^{-7} \, M$	basic
$= 7$	$= 1.0 \times 10^{-7} \, M$	neutral

15.22 **(a)** acidic   **(b)** neutral   **(c)** basic

15.23 The pH can be found using Equation (15.6) of the text.

$$pH = 14.00 - pOH = 14.00 - 9.40 = 4.60$$

The hydrogen ion concentration can be found as in Example 15.4 of the text.

$$4.60 = -\log[H^+]$$

Taking the antilog of both sides:

$$[H^+] = 2.5 \times 10^{-5} \, M$$

15.24 $5.50 \text{ mL} \times \dfrac{1 \text{ L}}{1000 \text{ mL}} \times \dfrac{0.360 \text{ mol}}{1 \text{ L}} = \mathbf{1.98 \times 10^{-3} \text{ mol KOH}}$

KOH is a strong base and therefore ionizes completely. The OH^- concentration equals the KOH concentration, because there is a 1:1 mole ratio between KOH and OH^-.

$$[OH^-] = 0.360 \ M$$

$$\textbf{pOH} = -\log[OH^-] = \mathbf{0.444}$$

15.25   We can calculate the OH^- concentration from the pOH.

$$pOH = 14.00 - pH = 14.00 - 10.00 = 4.00$$

$$[OH^-] = 10^{-pOH} = 1.0 \times 10^{-4} \ M$$

Since NaOH is a strong base, it ionizes completely. The OH^- concentration equals the initial concentration of NaOH.

$$[NaOH] = 1.0 \times 10^{-4} \text{ mol/L}$$

So, we need to prepare 546 mL of $1.0 \times 10^{-4} \ M$ NaOH.

This is a dimensional analysis problem. We need to perform the following unit conversions.

$$\text{mol/L} \rightarrow \text{mol NaOH} \rightarrow \text{grams NaOH}$$

$546 \text{ mL} = 0.546 \text{ L}$

$$\textbf{? g NaOH} = 546 \text{ mL} \times \dfrac{1.0 \times 10^{-4} \text{ mol NaOH}}{1000 \text{ mL soln}} \times \dfrac{40.00 \text{ g NaOH}}{1 \text{ mol NaOH}} = \mathbf{2.2 \times 10^{-3} \text{ g NaOH}}$$

15.26   Molarity of the HCl solution is: $\dfrac{18.4 \text{ g HCl} \times \dfrac{1 \text{ mol HCl}}{36.46 \text{ g HCl}}}{662 \times 10^{-3} \text{ L}} = 0.762 \ M$

$$\textbf{pH} = -\log(0.762) = \mathbf{0.118}$$

15.31   A strong acid, such as HCl, will be completely ionized, choice **(b)**.

A weak acid will only ionize to a lesser extent compared to a strong acid, choice **(c)**.

A very weak acid will remain almost exclusively as the acid molecule in solution. Choice **(d)** is the best choice.

15.32   **(1)**   The two steps in the ionization of a weak diprotic acid are:

$$H_2A(aq) + H_2O(l) \rightleftharpoons H_3O^+(aq) + HA^-(aq)$$

$$HA^-(aq) + H_2O(l) \rightleftharpoons H_3O^+(aq) + A^{2-}(aq)$$

The diagram that represents a weak diprotic acid is **(c)**. In this diagram, we only see the first step of the ionization, because HA^- is a much weaker acid than H_2A.

(2)   Both **(b)** and **(d)** are chemically implausible situations. Because HA^- is a much weaker acid than H_2A, you would not see a higher concentration of A^{2-} compared to HA^-.

15.33   **(a)**   strong acid,   **(b)**   weak acid,   **(c)**   strong acid (first stage of ionization),
            (d)   weak acid,   **(e)**   weak acid,   **(f)**   weak acid,
            (g)   strong acid,   **(h)**   weak acid,   **(i)**   weak acid.

15.34   **(a)** strong base   **(b)** weak base   **(c)** weak base   **(d)** weak base   **(e)** strong base

15.35   The maximum possible concentration of hydrogen ion in a 0.10 M solution of HA is 0.10 M. This is the case if HA is a strong acid. If HA is a weak acid, the hydrogen ion concentration is less than 0.10 M. The pH corresponding to 0.10 M $[H^+]$ is 1.00. (Why three digits?) For a smaller $[H^+]$ the pH is larger than 1.00 (why?).

(a)   false, the pH is greater than 1.00   **(b)**   false, they are equal   **(c)**   true   **(d)**   false

15.36   **(a)**   false, they are equal   **(b)**   true, find the value of log(1.00) on your calculator
            (c)   true   **(d)**   false, if the acid is strong, $[HA] = 0.00$ M

15.37   The direction should favor formation of $F^-(aq)$ and $H_2O(l)$. Hydroxide ion is a stronger base than fluoride ion, and hydrofluoric acid is a stronger acid than water.

15.38 Cl^- is the conjugate base of the strong acid, HCl. It is a negligibly weak base and has no affinity for protons. Therefore, the reaction will *not* proceed from left to right to any measurable extent.

Another way to think about this problem is to consider the possible products of the reaction.

$$CH_3COOH(aq) + Cl^-(aq) \rightarrow HCl(aq) + CH_3COO^-(aq)$$

The favored reaction is the one that proceeds from right to left. HCl is a strong acid and will ionize completely, donating all its protons to the base, CH_3COO^-.

15.43   We set up a table for the dissociation.

	$C_6H_5COOH(aq)$	$\rightleftharpoons$	$H^+(aq)$	$+$	$C_6H_5COO^-(aq)$
Initial (M):	0.10		0.00		0.00
Change (M):	$-x$		$+x$		$+x$
Equilibrium (M):	$(0.10 - x)$		x		x

$$K_a = \frac{[H^+][C_6H_5COO^-]}{[C_6H_5COOH]}$$

$$6.5 \times 10^{-5} = \frac{x^2}{(0.10 - x)}$$

$$x^2 + (6.5 \times 10^{-5})x - (6.5 \times 10^{-6}) = 0$$

Solving the quadratic equation:

$$x = 2.5 \times 10^{-3} \, M = [\text{H}^+]$$

$$\text{pH} = -\log(2.5 \times 10^{-3}) = \textbf{2.60}$$

This problem could be solved more easily if we could assume that $(0.10 - x) \approx 0.10$. If the assumption is mathematically valid, then it would not be necessary to solve a quadratic equation, as we did above. Re-solve the problem above, making the assumption. Was the assumption valid? What is our criterion for deciding?

15.44   Strategy: Recall that a weak acid only partially ionizes in water. We are given the initial quantity of a weak acid (CH_3COOH) and asked to calculate the concentrations of H^+, CH_3COO^-, and CH_3COOH at equilibrium. First, we need to calculate the initial concentration of CH_3COOH. In determining the H^+ concentration, we ignore the ionization of H_2O as a source of H^+, so the major source of H^+ ions is the acid. We follow the procedure outlined in Section 15.5 of the text.

Solution:
Step 1: Calculate the concentration of acetic acid before ionization.

$$0.0560 \text{ g acetic acid} \times \frac{1 \text{ mol acetic acid}}{60.05 \text{ g acetic acid}} = 9.33 \times 10^{-4} \text{ mol acetic acid}$$

$$\frac{9.33 \times 10^{-4} \text{ mol}}{0.0500 \text{ L soln}} = 0.0187 \, M \text{ acetic acid}$$

Step 2: We ignore water's contribution to $[\text{H}^+]$. We consider CH_3COOH as the only source of H^+ ions.

Step 3: Letting x be the equilibrium concentration of H^+ and CH_3COO^- ions in mol/L, we summarize:

	$CH_3COOH(aq)$	$\rightleftharpoons$	$\text{H}^+(aq)$	+	$CH_3COO^-(aq)$
Initial (M):	0.0187		0		0
Change (M):	$-x$		$+x$		$+x$
Equilibrium (M):	$0.0187 - x$		x		x

Step 3: Write the ionization constant expression in terms of the equilibrium concentrations. Knowing the value of the equilibrium constant (K_a), solve for x. You can look up the K_a value in Table 15.3 of the text.

$$K_a = \frac{[\text{H}^+][CH_3COO^-]}{[CH_3COOH]}$$

$$1.8 \times 10^{-5} = \frac{(x)(x)}{(0.0187 - x)}$$

At this point, we can make an assumption that x is very small compared to 0.0187. Hence,

$$0.0187 - x \approx 0.0187$$

$$1.8 \times 10^{-5} = \frac{(x)(x)}{0.0187}$$

$$x = \textbf{5.8} \times \textbf{10}^{-4} \, \textbf{\textit{M}} = \textbf{|H}^+\textbf{|} = \textbf{|CH}_3\textbf{COO}^-\textbf{|}$$

$$[CH_3COOH] = (0.0187 - 5.8 \times 10^{-4})M = \textbf{0.0181} \, \textbf{\textit{M}}$$

Check: Testing the validity of the assumption,

$$\frac{5.8 \times 10^{-4}}{0.0187} \times 100\% = 3.1\% < 5\%$$

The assumption is valid.

15.45 First we find the hydrogen ion concentration.

$$[H^+] = 10^{-pH} = 10^{-6.20} = 6.3 \times 10^{-7} \, M$$

If the concentration of $[H^+]$ is $6.3 \times 10^{-7} \, M$, that means that $6.3 \times 10^{-7} \, M$ of the weak acid, HA, ionized because of the 1:1 mole ratio between HA and H^+. Setting up a table:

	$HA(aq)$	$\rightleftharpoons$	$H^+(aq)$	+	$A^-(aq)$
Initial (*M*):	0.010		0		0
Change (*M*):	-6.3×10^{-7}		$+6.3 \times 10^{-7}$		$+6.3 \times 10^{-7}$
Equilibrium (*M*):	≈ 0.010		6.3×10^{-7}		6.3×10^{-7}

Substituting into the acid ionization constant expression:

$$K_a = \frac{[H^+][A^-]}{[HA]} = \frac{(6.3 \times 10^{-7})(6.3 \times 10^{-7})}{0.010} = 4.0 \times 10^{-11}$$

We have omitted the contribution to $[H^+]$ due to water.

15.46 A pH of 3.26 corresponds to a $[H^+]$ of $5.5 \times 10^{-4} \, M$. Let the original concentration of formic acid be *I*. If the concentration of $[H^+]$ is $5.5 \times 10^{-4} \, M$, that means that $5.5 \times 10^{-4} \, M$ of HCOOH ionized because of the 1:1 mole ratio between HCOOH and H^+.

	$HCOOH(aq)$	$\rightleftharpoons$	$H^+(aq)$	+	$HCOO^-(aq)$
Initial (*M*):	*I*		0		0
Change (*M*):	-5.5×10^{-4}		$+5.5 \times 10^{-4}$		$+5.5 \times 10^{-4}$
Equilibrium (*M*):	$I - (5.5 \times 10^{-4})$		5.5×10^{-4}		5.5×10^{-4}

Substitute K_a and the equilibrium concentrations into the ionization constant expression to solve for *I*.

$$K_a = \frac{[H^+][HCOO^-]}{[HCOOH]}$$

$$1.7 \times 10^{-4} = \frac{(5.5 \times 10^{-4})^2}{x - (5.5 \times 10^{-4})}$$

$$I = [HCOOH] = 2.3 \times 10^{-3} \, M$$

15.47 (a) Set up a table showing initial and equilibrium concentrations.

	$C_6H_5COOH(aq)$	$\rightleftharpoons$	$H^+(aq)$	+	$C_6H_5COO^-(aq)$
Initial (*M*):	0.20		0.00		0.00
Change (*M*):	$-x$		$+x$		$+x$
Equilibrium (*M*):	$(0.20 - x)$		x		x

Using the value of K_a from Table 15.3 of the text:

$$K_a = \frac{[H^+][C_6H_5COO^-]}{[C_6H_5COOH]}$$

$$6.5 \times 10^{-5} = \frac{x^2}{(0.20 - x)}$$

We assume that x is small so $(0.20 - x) \approx 0.20$

$$6.5 \times 10^{-5} = \frac{x^2}{0.20}$$

$$x = 3.6 \times 10^{-3} \, M = [H^+] = [C_6H_5COO^-]$$

Percent ionization $= \dfrac{3.6 \times 10^{-3} \, M}{0.20 \, M} \times 100\% = \mathbf{1.8\%}$

(b)   Set up a table as above.

	$C_6H_5COOH(aq)$	$\rightleftharpoons$	$H^+(aq)$	$+$	$C_6H_5COO^-(aq)$
Initial (M):	0.00020		0.00000		0.00000
Change (M):	$-x$		$+x$		$+x$
Equilibrium (M):	$(0.00020 - x)$		x		x

Using the value of K_a from Table 15.3 of the text:

$$K_a = \frac{[H^+][C_6H_5COO^-]}{[C_6H_5COOH]}$$

$$6.5 \times 10^{-5} = \frac{x^2}{(0.00020 - x)}$$

In this case we cannot apply the approximation that $(0.00020 - x) \approx 0.00020$ (see the discussion in Example 15.8 of the text). We obtain the quadratic equation:

$$x^2 + (6.5 \times 10^{-5})x - (1.3 \times 10^{-8}) = 0$$

The positive root of the equation is $x = 8.6 \times 10^{-5} \, M$. (Is this less than 5% of the original concentration, $0.00020 \, M$? That is, is the acid more than 5% ionized?) The percent ionization is then:

Percent ionization $= \dfrac{8.6 \times 10^{-5} \, M}{0.00020 \, M} \times 100\% = \mathbf{43\%}$

Note that the extent to which a weak acid ionizes depends on the initial concentration of the acid. The more dilute the solution, the greater the percent ionization (see Figure 15.4 of the text).

15.48   Percent ionization is defined as:

$$\text{percent ionization} = \frac{\text{ionized acid concentration at equilibrium}}{\text{initial concentration of acid}} \times 100\%$$

For a monoprotic acid, HA, the concentration of acid that undergoes ionization is equal to the concentration of H^+ ions or the concentration of A^- ions at equilibrium. Thus, we can write:

$$\text{percent ionization} = \frac{[H^+]}{[HA]_0} \times 100\%$$

(a) First, recognize that hydrofluoric acid is a weak acid. It is not one of the six strong acids, so it must be a weak acid.

Step 1: Express the equilibrium concentrations of all species in terms of initial concentrations and a single unknown x, that represents the change in concentration. Let $(-x)$ be the depletion in concentration (mol/L) of HF. From the stoichiometry of the reaction, it follows that the increase in concentration for both H^+ and F^- must be x. Complete a table that lists the initial concentrations, the change in concentrations, and the equilibrium concentrations.

	$HF(aq)$	$\rightleftharpoons$	$H^+(aq)$	$+$	$F^-(aq)$
Initial (M):	0.60		0		0
Change (M):	$-x$		$+x$		$+x$
Equilibrium (M):	$0.60 - x$		x		x

Step 2: Write the ionization constant expression in terms of the equilibrium concentrations. Knowing the value of the equilibrium constant (K_a), solve for x.

$$K_a = \frac{[H^+][F^-]}{[HF]}$$

You can look up the K_a value for hydrofluoric acid in Table 15.3 of your text.

$$7.1 \times 10^{-4} = \frac{(x)(x)}{(0.60 - x)}$$

At this point, we can make an assumption that x is very small compared to 0.60. Hence,

$$0.60 - x \approx 0.60$$

Oftentimes, assumptions such as these are valid if K is very small. A very small value of K means that a very small amount of reactants go to products. Hence, x is small. If we did not make this assumption, we would have to solve a quadratic equation.

$$7.1 \times 10^{-4} = \frac{(x)(x)}{0.60}$$

Solving for x.

$$x = 0.021 \ M = [H^+]$$

Step 3: Having solved for the $[H^+]$, calculate the percent ionization.

$$\textbf{percent ionization} = \frac{[H^+]}{[HF]_0} \times 100\%$$

$$= \frac{0.021 \ M}{0.60 \ M} \times 100\% = \textbf{3.5\%}$$

(b) – (c) are worked in a similar manner to part (a). However, as the initial concentration of HF becomes smaller, the assumption that x is very small compared to this concentration will no longer be valid. You must solve a quadratic equation.

(b) $K_a = \dfrac{[H^+][F^-]}{[HF]} = \dfrac{x^2}{(0.0046 - x)} = 7.1 \times 10^{-4}$

$x^2 + (7.1 \times 10^{-4})x - (3.3 \times 10^{-6}) = 0$

$x = 1.5 \times 10^{-3}\ M$

Percent ionization $= \dfrac{1.5 \times 10^{-3}\ M}{0.0046\ M} \times 100\% = \mathbf{33\%}$

(c) $K_a = \dfrac{[H^+][F^-]}{[HF]} = \dfrac{x^2}{(0.00028 - x)} = 7.1 \times 10^{-4}$

$x^2 + (7.1 \times 10^{-4})x - (2.0 \times 10^{-7}) = 0$

$x = 2.2 \times 10^{-4}\ M$

Percent ionization $= \dfrac{2.2 \times 10^{-4}\ M}{0.00028\ M} \times 100\% = \mathbf{79\%}$

As the solution becomes more dilute, the percent ionization increases.

15.49   Given 14% ionization, the concentrations must be:

$[H^+] = [A^-] = 0.14 \times 0.040\ M = 0.0056\ M$

$[HA] = (0.040 - 0.0056)\ M = 0.034\ M$

The value of K_a can be found by substitution.

$K_a = \dfrac{[H^+][A^-]}{[HA]} = \dfrac{(0.0056)^2}{0.034} = \mathbf{9.2 \times 10^{-4}}$

15.50   The equilibrium is:

$$C_9H_8O_4(aq) \rightleftharpoons H^+(aq) + C_9H_7O_4^-(aq)$$

	$C_9H_8O_4(aq)$	$H^+(aq)$	$C_9H_7O_4^-(aq)$
Initial (M):	0.20	0	0
Change (M):	$-x$	$+x$	$+x$
Equilibrium (M):	$0.20 - x$	x	x

(a) $K_a = \dfrac{[H^+][C_9H_7O_4^-]}{[C_9H_8O_4]}$

$3.0 \times 10^{-4} = \dfrac{x^2}{(0.20 - x)}$

Assuming $(0.20 - x) \approx 0.20$

$$x = [H^+] = 7.7 \times 10^{-3} \, M$$

Percent ionization $= \dfrac{x}{0.20} \times 100\% = \dfrac{7.7 \times 10^{-3} \, M}{0.20 \, M} \times 100\% = \mathbf{3.9\%}$

(b)   At pH 1.00 the concentration of hydrogen ion is $0.10 \, M$ ($[H^+] = 10^{-pH}$). The extra hydrogen ions will tend to suppress the ionization of the weak acid (LeChâtelier's principle, Section 14.5 of the text). The position of equilibrium is shifted in the direction of the un-ionized acid. Let's set up a table of concentrations with the initial concentration of H^+ equal to $0.10 \, M$.

	$C_9H_8O_4(aq)$	$\rightleftharpoons$	$H^+(aq)$	$+ \, C_9H_7O_4^-(aq)$
Initial (M):	0.20		0.10	0
Change (M):	$-x$		$+x$	$+x$
Equilibrium (M):	$0.20 - x$		$0.10 + x$	x

$$K_a = \frac{[H^+][C_9H_7O_4^-]}{[C_9H_8O_4]}$$

$$3.0 \times 10^{-4} = \frac{x(0.10 + x)}{(0.20 - x)}$$

Assuming $(0.20 - x) \approx 0.20$ and $(0.10 + x) \approx 0.10$

$$x = 6.0 \times 10^{-4} \, M$$

Percent ionization $= \dfrac{x}{0.20} \times 100\% = \dfrac{6.0 \times 10^{-4} \, M}{0.20 \, M} \times 100\% = \mathbf{0.30\%}$

The high acidity of the gastric juices appears to enhance the rate of absorption of unionized aspirin molecules through the stomach lining. In some cases this can irritate these tissues and cause bleeding.

15.53   **(a)**   We construct the usual table.

	$NH_3(aq)$	$+ \, H_2O(l)$	$\rightleftharpoons$	$NH_4^+(aq)$	$+ \, OH^-(aq)$
Initial (M):	0.10			0.00	0.00
Change (M):	$-x$			$+x$	$+x$
Equilibrium (M):	$(0.10 - x)$			x	x

$$K_b = \frac{[NH_4^+][OH^-]}{[NH_3]}$$

$$1.8 \times 10^{-5} = \frac{x^2}{(0.10 - x)}$$

Assuming $(0.10 - x) \approx 0.10$, we have:

$$1.8 \times 10^{-5} = \frac{x^2}{0.10}$$

$$x = 1.3 \times 10^{-3} M = [OH^-]$$

$$pOH = -\log(1.3 \times 10^{-3}) = 2.89$$

$$\mathbf{pH} = 14.00 - 2.89 = \mathbf{11.11}$$

By following the identical procedure, we can show: **(b) pH = 8.96**.

15.54   Strategy: Weak bases only partially ionize in water.

$$B(aq) + H_2O(l) \rightleftharpoons BH^+(aq) + OH^-(aq)$$

Note that the concentration of the weak base given refers to the initial concentration before ionization has started. The pH of the solution, on the other hand, refers to the situation at equilibrium. To calculate K_b, we need to know the concentrations of all three species, [B], [BH$^+$], and [OH$^-$] at equilibrium. We ignore the ionization of water as a source of OH$^-$ ions.

Solution: We proceed as follows.

Step 1: The major species in solution are B, OH$^-$, and the conjugate acid BH$^+$.

Step 2: First, we need to calculate the hydroxide ion concentration from the pH value. Calculate the pOH from the pH. Then, calculate the OH$^-$ concentration from the pOH.

$$pOH = 14.00 - pH = 14.00 - 10.66 = 3.34$$

$$pOH = -\log[OH^-]$$

$$-pOH = \log[OH^-]$$

Taking the antilog of both sides of the equation,

$$10^{-pOH} = [OH^-]$$

$$[OH^-] = 10^{-3.34} = 4.6 \times 10^{-4} M$$

Step 3: If the concentration of OH$^-$ is 4.6×10^{-4} M at equilibrium, that must mean that 4.6×10^{-4} M of the base ionized. We summarize the changes.

	B(aq)	+ H$_2$O(l) $\rightleftharpoons$	BH$^+$(aq)	+	OH$^-$(aq)
Initial (*M*):	0.30		0		0
Change (*M*):	-4.6×10^{-4}		$+4.6 \times 10^{-4}$		$+4.6 \times 10^{-4}$
Equilibrium (*M*):	$0.30 - (4.6 \times 10^{-4})$		4.6×10^{-4}		4.6×10^{-4}

Step 4: Substitute the equilibrium concentrations into the ionization constant expression to solve for K_b.

$$K_b = \frac{[BH^+][OH^-]}{[B]}$$

$$K_b = \frac{(4.6 \times 10^{-4})^2}{(0.30)} = \mathbf{7.1 \times 10^{-7}}$$

15.55   A pH of 11.22 corresponds to a $[H^+]$ of 6.03×10^{-12} M and a $[OH^-]$ of 1.66×10^{-3} M.

Setting up a table:

	$NH_3(aq)$ + $H_2O(l)$ $\rightleftharpoons$	$NH_4^+(aq)$ +	$OH^-(aq)$
Initial (M):	I	0.00	0.00
Change (M):	-1.66×10^{-3}	$+1.66 \times 10^{-3}$	$+1.66 \times 10^{-3}$
Equilibrium (M):	$I - (1.66 \times 10^{-3})$	1.66×10^{-3}	1.66×10^{-3}

$$K_b = \frac{[NH_4^+][OH^-]}{[NH_3]}$$

$$1.8 \times 10^{-5} = \frac{(1.66 \times 10^{-3})(1.66 \times 10^{-3})}{I - (1.66 \times 10^{-3})}$$

Assuming 1.66×10^{-3} is small relative to x, then

$$x = 0.15 \, M = [NH_3]$$

15.56   The reaction is:

	$NH_3(aq) + H_2O(l)$ $\rightleftharpoons$	$NH_4^+(aq) +$	$OH^-(aq)$
Initial (M):	0.080	0	0
Change (M):	$-x$	$+x$	$+x$
Equilibrium (M):	$0.080 - x$	x	x

At equilibrium we have:

$$K_a = \frac{[NH_4^+][OH^-]}{[NH_3]}$$

$$1.8 \times 10^{-5} = \frac{x^2}{(0.080 - x)} \approx \frac{x^2}{0.080}$$

$$x = 1.2 \times 10^{-3} \, M$$

$$\textbf{Percent NH}_3 \textbf{ present as NH}_4^+ = \frac{1.2 \times 10^{-3}}{0.080} \times 100\% = \textbf{1.5\%}$$

15.61   If $K_{a_1} \gg K_{a_2}$, we can assume that the equilibrium concentration of hydrogen ion results only from the first stage of ionization. In the second stage this always leads to an expression of the type:

$$\frac{(c + y)(y)}{(c - y)} = K_{a_2}$$

where c represents the equilibrium hydrogen ion concentration found in the first stage. If $c \gg K_{a_2}$, we can assume $(c \pm y) \approx c$, and consequently $y = K_{a_2}$.

Is this conclusion also true for the second stage ionization of a triprotic acid like H_3PO_4?

15.62   The pH of a 0.040 M HCl solution (strong acid) is: pH = $-\log(0.040)$ = **1.40**.

Strategy: Determining the pH of a diprotic acid in aqueous solution is more involved than for a monoprotic acid. The first stage of ionization for H_2SO_4 goes to completion. We follow the procedure for determining the pH of a strong acid for this stage. The conjugate base produced in the first ionization (HSO_4^-) is a weak acid. We follow the procedure for determining the pH of a weak acid for this stage.

Solution: We proceed according to the following steps.

Step 1: H_2SO_4 is a strong acid. The first ionization stage goes to completion. The ionization of H_2SO_4 is

$$H_2SO_4(aq) \rightarrow H^+(aq) + HSO_4^-(aq)$$

The concentrations of all the species (H_2SO_4, H^+, and HSO_4^-) before and after ionization can be represented as follows.

	$H_2SO_4(aq)$	$\rightarrow$	$H^+(aq)$	+	$HSO_4^-(aq)$
Initial (M):	0.040		0		0
Change (M):	−0.040		+0.040		+0.040
Final (M):	0		0.040		0.040

Step 2: Now, consider the second stage of ionization. HSO_4^- is a weak acid. Set up a table showing the concentrations for the second ionization stage. Let x be the change in concentration. Note that the initial concentration of H^+ is 0.040 M from the first ionization.

	$HSO_4^-(aq)$	$\rightleftharpoons$	$H^+(aq)$	+	$SO_4^{2-}(aq)$
Initial (M):	0.040		0.040		0
Change (M):	$-x$		$+x$		$+x$
Equilibrium (M):	$0.040 - x$		$0.040 + x$		x

Write the ionization constant expression for K_a. Then, solve for x. You can find the K_a value in Table 15.5 of the text.

$$K_a = \frac{[H^+][SO_4^{2-}]}{[HSO_4^-]}$$

$$1.3 \times 10^{-2} = \frac{(0.040 + x)(x)}{(0.040 - x)}$$

Since K_a is quite large, we cannot make the assumptions that

$$0.040 - x \approx 0.040 \quad \text{and} \quad 0.040 + x \approx 0.040$$

Therefore, we must solve a quadratic equation.

$$x^2 + 0.053x - (5.2 \times 10^{-4}) = 0$$

$$x = \frac{-0.053 \pm \sqrt{(0.053)^2 - 4(1)(-5.2 \times 10^{-4})}}{2(1)}$$

$$x = \frac{-0.053 \pm 0.070}{2}$$

$$x = 8.5 \times 10^{-3} \, M \quad \text{or} \quad x = -0.062 \, M$$

The second solution is physically impossible because you cannot have a negative concentration. The first solution is the correct answer.

Step 3: Having solved for x, we can calculate the H^+ concentration at equilibrium. We can then calculate the pH from the H^+ concentration.

$$[H^+] = 0.040\ M + x = [0.040 + (8.5 \times 10^{-3})]M = 0.049\ M$$

$$\mathbf{pH} = -\log(0.049) = \mathbf{1.31}$$

Without doing any calculations, could you have known that the pH of the sulfuric acid would be lower (more acidic) than that of the hydrochloric acid?

15.63   There is no H_2SO_4 in the solution because HSO_4^- has no tendency to accept a proton to produce H_2SO_4. (Why?) We are only concerned with the ionization:

	$HSO_4^-(aq)$ $\rightleftharpoons$	$H^+(aq)$ +	$SO_4^{2-}(aq)$
Initial (M):	0.20	0.00	0.00
Change (M):	$-x$	$+x$	$+x$
Equilibrium (M):	$(0.20 - x)$	$+x$	$+x$

$$K_a = \frac{[H^+][SO_4^{2-}]}{[HSO_4^-]}$$

$$1.3 \times 10^{-2} = \frac{(x)(x)}{(0.20 - x)}$$

Solving the quadratic equation:

$$x = [H^+] = [SO_4^{2-}] = \mathbf{0.045\ M}$$

$$[HSO_4^-] = (0.20 - 0.045)\ M = \mathbf{0.16\ M}$$

15.64   For the first stage of ionization:

	$H_2CO_3(aq)$ $\rightleftharpoons$	$H^+(aq)$ +	$HCO_3^-(aq)$
Initial (M):	0.025	0.00	0.00
Change (M):	$-x$	$+x$	$+x$
Equilibrium (M):	$(0.025 - x)$	x	x

$$K_{a_1} = \frac{[H^+][HCO_3^-]}{[H_2CO_3]}$$

$$4.2 \times 10^{-7} = \frac{x^2}{(0.025 - x)} \approx \frac{x^2}{0.025}$$

$$x = 1.0 \times 10^{-4}\ M$$

For the second ionization,

$$HCO_3^-(aq) \rightleftharpoons H^+(aq) + CO_3^{2-}(aq)$$

	$HCO_3^-(aq)$	$H^+(aq)$	$CO_3^{2-}(aq)$
Initial (M):	1.0×10^{-4}	1.0×10^{-4}	0.00
Change (M):	$-x$	$+x$	$+x$
Equilibrium (M):	$(1.0 \times 10^{-4}) - x$	$(1.0 \times 10^{-4}) + x$	x

$$K_{a_2} = \frac{[H^+][CO_3^{2-}]}{[HCO_3^-]}$$

$$4.8 \times 10^{-11} = \frac{[(1.0 \times 10^{-4}) + x](x)}{(1.0 \times 10^{-4}) - x} \approx \frac{(1.0 \times 10^{-4})(x)}{(1.0 \times 10^{-4})}$$

$$x = 4.8 \times 10^{-11} \, M$$

Since HCO_3^- is a very weak acid, there is little ionization at this stage. Therefore we have:

$$[H^+] = [HCO_3^-] = 1.0 \times 10^{-4} \, M \text{ and } [CO_3^{2-}] = x = 4.8 \times 10^{-11} \, M$$

15.67   The strength of the H–X bond is the dominant factor in determining the strengths of binary acids. As with the hydrogen halides (see Section 15.9 of the text), the H–X bond strength decreases going down the column in Group 6A. The compound with the weakest H–X bond will be the strongest binary acid: **$H_2Se > H_2S > H_2O$**.

15.68   All the listed pairs are oxoacids that contain different central atoms whose elements are in the same group of the periodic table and have the same oxidation number. In this situation the acid with the most electronegative central atom will be the strongest.

(a) $H_2SO_4 > H_2SeO_4$.

(b) $H_3PO_4 > H_3AsO_4$

15.69   The $CHCl_2COOH$ is a stronger acid than $CH_2ClCOOH$. Having two electronegative chlorine atoms compared to one, will draw more electron density toward itself, making the O–H bond more polar. The hydrogen atom in $CHCl_2COOH$ is more easily ionized compared to the hydrogen atom in $CH_2ClCOOH$.

15.70   The conjugate bases are $C_6H_5O^-$ from phenol and CH_3O^- from methanol. The $C_6H_5O^-$ is stabilized by resonance:

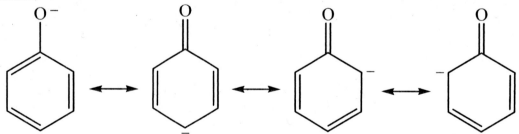

The CH_3O^- ion has no such resonance stabilization. A more stable conjugate base means an increase in the strength of the acid.

15.75 **(a)** The K^+ cation does not hydrolyze. The Br^- anion is the conjugate base of the strong acid HBr. Therefore, Br^- will not hydrolyze either, and the solution is neutral, **pH ≈ 7.**

(b) Al^{3+} is a small metal cation with a high charge, which hydrolyzes to produce H^+ ions. The NO_3^- anion does not hydrolyze. It is the conjugate base of the strong acid, HNO_3. The solution will be acidic, **pH < 7.**

(c) The Ba^{2+} cation does not hydrolyze. The Cl^- anion is the conjugate base of the strong acid HCl. Therefore, Cl^- will not hydrolyze either, and the solution is neutral, **pH ≈ 7.**

(d) Bi^{3+} is a small metal cation with a high charge, which hydrolyzes to produce H^+ ions. The NO_3^- anion does not hydrolyze. It is the conjugate base of the strong acid, HNO_3. The solution will be acidic, **pH < 7.**

15.76 **Strategy:** In deciding whether a salt will undergo hydrolysis, ask yourself the following questions: Is the cation a highly charged metal ion or an ammonium ion? Is the anion the conjugate base of a weak acid? If yes to either question, then hydrolysis will occur. In cases where both the cation and the anion react with water, the pH of the solution will depend on the relative magnitudes of K_a for the cation and K_b for the anion (see Table 15.7 of the text).

Solution: We first break up the salt into its cation and anion components and then examine the possible reaction of each ion with water.

(a) The Na^+ cation does not hydrolyze. The Br^- anion is the conjugate base of the strong acid HBr. Therefore, Br^- will not hydrolyze either, and the solution is **neutral**.

(b) The K^+ cation does not hydrolyze. The SO_3^{2-} anion is the conjugate base of the weak acid HSO_3^- and will hydrolyze to give HSO_3^- and OH^-. The solution will be **basic**.

(c) Both the NH_4^+ and NO_2^- ions will hydrolyze. NH_4^+ is the conjugate acid of the weak base NH_3, and NO_2^- is the conjugate base of the weak acid HNO_2. From Tables 15.3 and 15.4 of the text, we see that the K_a of NH_4^+ (5.6×10^{-10}) is greater than the K_b of NO_2^- (2.2×10^{-11}). Therefore, the solution will be **acidic**.

(d) Cr^{3+} is a small metal cation with a high charge, which hydrolyzes to produce H^+ ions. The NO_3^- anion does not hydrolyze. It is the conjugate base of the strong acid, HNO_3. The solution will be **acidic**.

15.77 There are two possibilities: (i) MX is the salt of a strong acid and a strong base so that neither the cation nor the anion react with water to alter the pH and (ii) MX is the salt of a weak acid and a weak base with K_a for the acid equal to K_b for the base. The hydrolysis of one would be exactly offset by the hydrolysis of the other.

15.78 There is an inverse relationship between acid strength and conjugate base strength. As acid strength decreases, the proton accepting power of the conjugate base increases. In general the weaker the acid, the stronger the conjugate base. All three of the potassium salts ionize completely to form the conjugate base of the respective acid. The greater the pH, the stronger the conjugate base, and therefore, the weaker the acid.

The order of increasing acid strength is **HZ < HY < HX**.

15.79   The salt, sodium acetate, completely dissociates upon dissolution, producing 0.36 M [Na^+] and 0.36 M [CH_3COO^-] ions. The [CH_3COO^-] ions will undergo hydrolysis because they are a weak base.

$$CH_3COO^-(aq) + H_2O(l) \rightleftharpoons CH_3COOH(aq) + OH^-(aq)$$

	CH_3COO^-		CH_3COOH	OH^-
Initial (M):	0.36		0.00	0.00
Change (M):	$-x$		$+x$	$+x$
Equilibrium (M):	$(0.36 - x)$		$+x$	$+x$

$$K_b = \frac{[CH_3COOH][OH^-]}{[CH_3COO^-]}$$

$$5.6 \times 10^{-10} = \frac{x^2}{(0.36 - x)}$$

Assuming $(0.36 - x) \approx 0.36$, then

$$x = [OH^-] = 1.4 \times 10^{-5}$$

$$pOH = -\log(1.4 \times 10^{-5}) = 4.85$$

$$\mathbf{pH} = 14.00 - 4.85 = \mathbf{9.15}$$

15.80   The salt ammonium chloride completely ionizes upon dissolution, producing 0.42 M [NH_4^+] and 0.42 M [Cl^-] ions. NH_4^+ will undergo hydrolysis because it is a weak acid (NH_4^+ is the conjugate acid of the weak base, NH_3).

Step 1:   Express the equilibrium concentrations of all species in terms of initial concentrations and a single unknown x, that represents the change in concentration. Let $(-x)$ be the depletion in concentration (mol/L) of NH_4^+. From the stoichiometry of the reaction, it follows that the increase in concentration for both H_3O^+ and NH_3 must be x. Complete a table that lists the initial concentrations, the change in concentrations, and the equilibrium concentrations.

$$NH_4^+(aq) + H_2O(l) \rightleftharpoons NH_3(aq) + H_3O^+(aq)$$

	NH_4^+		NH_3	H_3O^+
Initial (M):	0.42		0.00	0.00
Change (M):	$-x$		$+x$	$+x$
Equilibrium (M):	$(0.42 - x)$		x	x

Step 2:   You can calculate the K_a value for NH_4^+ from the K_b value of NH_3. The relationship is

$$K_a \times K_b = K_w$$

or

$$K_a = \frac{K_w}{K_b} = \frac{1.0 \times 10^{-14}}{1.8 \times 10^{-5}} = 5.6 \times 10^{-10}$$

Step 3:   Write the ionization constant expression in terms of the equilibrium concentrations. Knowing the value of the equilibrium constant (K_a), solve for x.

$$K_a = \frac{[NH_3][H_3O^+]}{[NH_4^+]}$$

$$5.6 \times 10^{-10} = \frac{x^2}{0.42 - x} \approx \frac{x^2}{0.42}$$

$$x = [H^+] = 1.5 \times 10^{-5} \, M$$

$$pH = -\log(1.5 \times 10^{-5}) = \textbf{4.82}$$

Since NH_4Cl is the salt of a weak base (aqueous ammonia) and a strong acid (HCl), we expect the solution to be slightly acidic, which is confirmed by the calculation.

15.81 $\quad HCO_3^- \rightleftharpoons H^+ + CO_3^{2-}$ $\qquad\qquad K_a = 4.8 \times 10^{-11}$

$HCO_3^- + H_2O \rightleftharpoons H_2CO_3 + OH^-$ $\qquad K_b = \dfrac{K_w}{K_a} = \dfrac{1.0 \times 10^{-14}}{4.2 \times 10^{-7}} = 2.4 \times 10^{-8}$

HCO_3^- has a greater tendency to hydrolyze than to ionize ($K_b > K_a$). The solution will be basic (pH > 7).

15.82 The acid and base reactions are:

acid: $\quad HPO_4^{2-}(aq) \rightleftharpoons H^+(aq) + PO_4^{3-}(aq)$

base: $\quad HPO_4^{2-}(aq) + H_2O(l) \rightleftharpoons H_2PO_4^-(aq) + OH^-(aq)$

K_a for HPO_4^{2-} is 4.8×10^{-13}. Note that HPO_4^{2-} is the conjugate base of $H_2PO_4^-$, so K_b is 1.6×10^{-7}. Comparing the two K's, we conclude that the monohydrogen phosphate ion is a much stronger proton acceptor (base) than a proton donor (acid). The solution will be **basic**.

15.85 Metal ions with high oxidation numbers are unstable. Consequently, these metals tend to form covalent bonds (rather than ionic bonds) with oxygen. Covalent metal oxides are acidic while ionic metal oxides are basic. The latter oxides contain the O^{2-} ion which reacts with water as follows:

$$O^{2-} + H_2O \rightarrow 2OH^-$$

15.86 The most basic oxides occur with metal ions having the lowest positive charges (or lowest oxidation numbers).

(a) $\quad Al_2O_3 < BaO < K_2O$ $\qquad$ **(b)** $\quad CrO_3 < Cr_2O_3 < CrO$

15.87 **(a)** $\quad 2HCl(aq) + Zn(OH)_2(s) \rightarrow ZnCl_2(aq) + 2H_2O(l)$

(b) $\quad 2OH^-(aq) + Zn(OH)_2(s) \rightarrow Zn(OH)_4^{2-}(aq)$

15.88 $Al(OH)_3$ is an amphoteric hydroxide. The reaction is:

$$Al(OH)_3(s) + OH^-(aq) \rightarrow Al(OH)_4^-(aq)$$

This is a Lewis acid-base reaction. Can you identify the acid and base?

15.91 **(a)** Lewis acid; see the reaction with water shown in Section 15.12 of the text.

(b) Lewis base; water combines with H^+ to form H_3O^+.

(c)   Lewis base.

(d)   Lewis acid; SO_2 reacts with water to form H_2SO_3. Compare to CO_2 above. Actually, SO_2 can also act as a Lewis base under some circumstances.

(e)   Lewis base; see the reaction with H^+ to form ammonium ion.

(f)   Lewis base; see the reaction with H^+ to form water.

(g)   Lewis acid; does H^+ have any electron pairs to donate?

(h)   Lewis acid; compare to the example of NH_3 reacting with BF_3.

15.92 $AlCl_3$ is a Lewis acid with an incomplete octet of electrons and Cl^- is the Lewis base donating a pair of electrons.

15.93   **(a)**   Both molecules have the same acceptor atom (boron) and both have exactly the same structure (trigonal planar). Fluorine is more electronegative than chlorine so we would predict based on electronegativity arguments that boron trifluoride would have a greater affinity for unshared electron pairs than boron trichloride.

(b)   Since it has the larger positive charge, iron(III) should be a stronger Lewis acid than iron(II).

15.94   By definition Brønsted acids are proton donors, therefore such compounds must contain at least one hydrogen atom. In Problem 15.91, Lewis acids that do not contain hydrogen, and therefore are not Brønsted acids, are CO_2, SO_2, and BCl_3. Can you name others?

15.95   The ionization of any acid is an endothermic process. The higher the temperature, the greater the K_a value. Formic acid will be a stronger acid at 40°C than at 25°C.

15.96   We first find the number of moles of CO_2 produced in the reaction:

$$0.350 \text{ g NaHCO}_3 \times \frac{1 \text{ mol NaHCO}_3}{84.01 \text{ g NaHCO}_3} \times \frac{1 \text{ mol CO}_2}{1 \text{ mol NaHCO}_3} = 4.17 \times 10^{-3} \text{ mol CO}_2$$

$$V_{CO_2} = \frac{n_{CO_2} RT}{P} = \frac{(4.17 \times 10^{-3} \text{ mol})(0.0821 \text{ L} \cdot \text{atm/K} \cdot \text{mol})(37.0 + 273)\text{K}}{(1.00 \text{ atm})} = \textbf{0.106 L}$$

15.97   Choice **(c)** because 0.70 M KOH has a higher pH than 0.60 M NaOH. Adding an equal volume of 0.60 M NaOH lowers the $[OH^-]$ to 0.65 M, hence lowering the pH.

15.98   If we assume that the unknown monoprotic acid is a strong acid that is 100% ionized, then the $[H^+]$ concentration will be 0.0642 M.

$$pH = -\log (0.0642) = 1.19$$

Since the actual pH of the solution is higher, the acid must be a weak acid.

15.99   **(a)**   For the forward reaction NH_4^+ and NH_3 are the conjugate acid and base pair, respectively. For the reverse reaction NH_3 and NH_2^- are the conjugate acid and base pair, respectively.

   (b) H^+ corresponds to NH_4^+; OH^- corresponds to NH_2^-. For the neutral solution, $[NH_4^+] = [NH_2^-]$.

15.100   The reaction of a weak acid with a strong base is driven to completion by the formation of water. Irrespective of whether the strong base is reacting with a strong monoprotic acid or a weak monoprotic acid, the same number of moles of acid is required to react with a constant number of moles of base. Therefore the volume of base required to react with the same concentration of acid solutions (either both weak, both strong, or one strong and one weak) will be the same.

15.101 $$K_a = \frac{[H^+][A^-]}{[HA]}$$

$[HA] \approx 0.1 \text{ M}$
$[A^-] \approx 0.1 \text{ M}$

Therefore,

$$K_a = [H^+] = \frac{K_w}{[OH^-]}$$

$$[OH^-] = \frac{K_w}{K_a}$$

15.102   High oxidation state leads to covalent compounds and low oxidation state leads to ionic compounds. Therefore, CrO is ionic and basic and CrO_3 is covalent and acidic.

15.103 $HCOOH \rightleftharpoons HCOO^- + H^+$ $K_a = 1.7 \times 10^{-4}$

   $H^+ + OH^- \rightleftharpoons H_2O$ $K_w' = \dfrac{1}{K_w} = \dfrac{1}{1.0 \times 10^{-14}} = 1.0 \times 10^{14}$

   ───────────────────────────────────────

   $HCOOH + OH^- \rightleftharpoons HCOO^- + H_2O$

   $$K = K_a K_w' = (1.7 \times 10^{-4})(1.0 \times 10^{14}) = 1.7 \times 10^{10}$$

15.104   We can write two equilibria that add up to the equilibrium in the problem.

$$CH_3COOH(aq) \rightleftharpoons H^+(aq) + CH_3COO^-(aq) \qquad K_a = \frac{[H^+][CH_3COO^-]}{[CH_3COOH]} = 1.8 \times 10^{-5}$$

$$H^+(aq) + NO_2^-(aq) \rightleftharpoons HNO_2(aq) \qquad K_a' = \frac{1}{K_a(HNO_2)} = \frac{1}{4.5 \times 10^{-4}} = 2.2 \times 10^3$$

$$K_a' = \frac{[HNO_2]}{[H^+][NO_2^-]}$$

$$CH_3COOH(aq) + NO_2^-(aq) \rightleftharpoons CH_3COO^-(aq) + HNO_2(aq) \qquad K = \frac{[CH_3COO^-][HNO_2]}{[CH_3COOH][NO_2^-]} = K_a \times K_a'$$

The equilibrium constant for this sum is the product of the equilibrium constants of the component reactions.

$$\boldsymbol{K} = K_a \times K_a' = (1.8 \times 10^{-5})(2.2 \times 10^3) = \boldsymbol{4.0 \times 10^{-2}}$$

15.105   (a)

H^-	$+$	H_2O	$\rightarrow$	OH^-	$+$	H_2
base₁		acid₂		base₂		acid₁

H^- $+$ H_2O $\rightarrow$ OH^- $+$ H_2
base$_1$ acid$_2$ base$_2$ acid$_1$

(b) H^- is the reducing agent and H_2O is the oxidizing agent.

15.106   In this specific case the K_a of ammonium ion is the same as the K_b of acetate ion $[K_a(NH_4^+) = 5.6 \times 10^{-10}$, $K_b(CH_3COO^-) = 5.6 \times 10^{-10}]$. The two are of exactly (to two significant figures) equal strength. The solution will have **pH 7.00**.

What would the pH be if the concentration were 0.1 *M* in ammonium acetate?   0.4 *M*?

15.107 $K_b = 8.91 \times 10^{-6}$

$$K_a = \frac{K_w}{K_b} = 1.1 \times 10^{-9}$$

$$pH = 7.40$$
$$[H^+] = 10^{-7.40} = 3.98 \times 10^{-8}$$

$$K_a = \frac{[H^+][\text{conjugate base}]}{[\text{acid}]}$$

Therefore,

$$\frac{[\text{conjugate base}]}{[\text{acid}]} = \frac{K_a}{[H^+]} = \frac{1.1 \times 10^{-9}}{3.98 \times 10^{-8}} = \boldsymbol{0.028}$$

15.108   The fact that fluorine attracts electrons in a molecule more strongly than hydrogen should cause NF_3 to be a poor electron pair donor and a poor base.   **NH_3 is the stronger base.**

15.109 Because the P–H bond is weaker, there is a greater tendency for PH_4^+ to ionize. Therefore, PH_3 is a weaker base than NH_3.

15.110 The autoionization for deuterium-substituted water is: $D_2O \rightleftharpoons D^+ + OD^-$

$$[D^+][OD^-] = 1.35 \times 10^{-15} \qquad (1)$$

(a) The definition of pD is: $\quad \textbf{pD} = -\log[D^+] = -\log\sqrt{1.35 \times 10^{-15}} = \textbf{7.43}$

(b) To be acidic, the **pD** must be **< 7.43**.

(c) Taking –log of both sides of equation (1) above:

$$-\log[D^+] + -\log[OD^-] = -\log(1.35 \times 10^{-15})$$

$$\textbf{pD} + \textbf{pOD} = \textbf{14.87}$$

15.111 **(a)** HNO_2 **(b)** HF **(c)** BF_3 **(d)** NH_3 **(e)** H_2SO_3
(f) HCO_3^- and CO_3^{2-}

The reactions for (f) are: $\quad HCO_3^-(aq) + H^+(aq) \rightarrow CO_2(g) + H_2O(l)$

$$CO_3^{2-}(aq) + 2H^+(aq) \rightarrow CO_2(g) + H_2O(l)$$

15.112 First we must calculate the molarity of the trifluoromethane sulfonic acid. (Molar mass = 150.1 g/mol)

$$\text{Molarity} = \frac{0.616 \text{ g} \times \dfrac{1 \text{ mol}}{150.1 \text{ g}}}{0.250 \text{ L}} = 0.0164 \ M$$

Since trifluoromethane sulfonic acid is a strong acid and is 100% ionized, the $[H^+]$ is 0.0165 *M*.

$$\textbf{pH} = -\log(0.0164) = \textbf{1.79}$$

15.113 **(a)** The Lewis structure of H_3O^+ is:

$$\left[\ H-\overset{\displaystyle ..}{\underset{\displaystyle |}{O}}-H \ \right]^+$$
$$\qquad\quad H$$

Note that this structure is very similar to the Lewis structure of NH_3. The geometry is **trigonal pyramidal**.

(b) H_4O^{2+} does *not* exist because the positively charged H_3O^+ has no affinity to accept the positive H^+ ion. If H_4O^{2+} existed, it would have a tetrahedral geometry.

15.114   The reactions are $HF \rightleftharpoons H^+ + F^-$ (1)

$$F^- + HF \rightleftharpoons HF_2^-$$ (2)

Note that for equation (2), the equilibrium constant is relatively large with a value of 5.2. This means that the equilibrium lies to the right. Applying Le Châtelier's principle, as HF ionizes in the first step, the F^- that is produced is partially removed in the second step. More HF must ionize to compensate for the removal of the F^-, at the same time producing more H^+.

15.115   The equations are: $Cl_2(g) + H_2O(l) \rightleftharpoons HCl(aq) + HClO(aq)$

$$HCl(aq) + AgNO_3(aq) \rightleftharpoons AgCl(s) + HNO_3(aq)$$

In the presence of OH^- ions, the first equation is shifted to the right:

$$H^+ \text{ (from HCl)} + OH^- \longrightarrow H_2O$$

Therefore, the concentration of HClO increases. (The 'bleaching action' is due to ClO^- ions.)

15.116   (a)   We must consider both the complete ionization of the strong acid, and the partial ionization of water.

$$HA \longrightarrow H^+ + A^-$$
$$H_2O \rightleftharpoons H^+ + OH^-$$

From the above two equations, the $[H^+]$ in solution is:

$$[H^+] = [A^-] + [OH^-]$$ (1)

We can also write:

$$[H^+][OH^-] = K_w$$

$$[OH^-] = \frac{K_w}{[H^+]}$$

Substituting into Equation (1):

$$[H^+] = [A^-] + \frac{K_w}{[H^+]}$$

$$[H^+]^2 = [A^-][H^+] + K_w$$

$$[H^+]^2 - [A^-][H^+] - K_w = 0$$

Solving a quadratic equation:

$$[H^+] = \frac{[A^-] \pm \sqrt{[A^-]^2 + 4K_w}}{2}$$

(b) For the strong acid, HCl, with a concentration of 1.0×10^{-7} M, the $[Cl^-]$ will also be 1.0×10^{-7} M.

$$[H^+] = \frac{[Cl^-] \pm \sqrt{[Cl^-]^2 + 4K_w}}{2} = \frac{1 \times 10^{-7} \pm \sqrt{(1 \times 10^{-7})^2 + 4(1 \times 10^{-14})}}{2}$$

$[H^+] = 1.6 \times 10^{-7}$ M (or -6.0×10^{-8} M, which is impossible)

pH $= -\log[1.6 \times 10^{-7}] = $ **6.80**

15.117 Given the equation: $HbH^+ + O_2 \rightleftharpoons HbO_2 + H^+$

(a) From the equilibrium equation, high oxygen concentration puts stress on the left side of the equilibrium and thus shifts the concentrations to the right to compensate. **HbO_2 is favored.**

(b) High acid, H^+ concentration, places stress on the right side of the equation forcing concentrations on the left side to increase, thus releasing oxygen and increasing the concentration of **HbH^+**.

(c) Removal of CO_2 decreases H^+ (in the form of carbonic acid), thus shifting the reaction to the **right**. More HbO_2 will form. Breathing into a paper bag increases the concentration of CO_2 (re-breathing the exhaled CO_2), thus causing more O_2 to be released as explained above.

15.118 The solution for the first step is standard:

$$H_3PO_4(aq) \rightleftharpoons H^+(aq) + H_2PO_4^-(aq)$$

	$H_3PO_4(aq)$	$H^+(aq)$	$H_2PO_4^-(aq)$
Initial (M):	0.100	0.000	0.000
Change (M):	$-x$	$+x$	$+x$
Equil. (M):	$(0.100 - x)$	x	x

$$K_{a_1} = \frac{[H^+][H_2PO_4^-]}{[H_3PO_4]}$$

$$7.5 \times 10^{-3} = \frac{x^2}{(0.100 - x)}$$

In this case we probably cannot say that $(0.100 - x) \approx 0.100$ due to the magnitude of K_a. We obtain the quadratic equation:

$$x^2 + (7.5 \times 10^{-3})x - (7.5 \times 10^{-4}) = 0$$

The positive root is $x = 0.0239$ M. We have:

$$[H^+] = [H_2PO_4^-] = 0.0239 \ M$$

$$[H_3PO_4] = (0.100 - 0.0239) \ M = 0.076 \ M$$

For the second ionization:

$$H_2PO_4^-(aq) \rightleftharpoons H^+(aq) + HPO_4^{2-}(aq)$$

	$H_2PO_4^-(aq)$	$H^+(aq)$	$HPO_4^{2-}(aq)$
Initial (M):	0.0239	0.0239	0.000
Change (M):	$-y$	$+y$	$+y$
Equil (M):	$(0.0239 - y)$	$(0.0239 + y)$	y

$$K_{a_2} = \frac{[\text{H}^+][\text{HPO}_4^{2-}]}{[\text{H}_2\text{PO}_4^-]}$$

$$6.2 \times 10^{-8} = \frac{(0.0239 + y)(y)}{(0.0239 - y)} \approx \frac{(0.0239)(y)}{(0.0239)}$$

$$y = 6.2 \times 10^{-8}\ M.$$

Thus,

$$[\text{H}^+] = [\text{H}_2\text{PO}_4^-] = 0.0239\ M$$

$$[\text{HPO}_4^{2-}] = y = 6.2 \times 10^{-8}\ M$$

We set up the problem for the third ionization in the same manner.

$$\text{HPO}_4^{2-}(aq) \quad \rightleftharpoons \quad \text{H}^+(aq) \ + \ \text{PO}_4^{3-}(aq)$$

	$\text{HPO}_4^{2-}(aq)$	$\text{H}^+(aq)$	$\text{PO}_4^{3-}(aq)$
Initial (M):	6.2×10^{-8}	0.0239	0
Change (M):	$-z$	$+z$	$+z$
Equil. (M):	$(6.2 \times 10^{-8}) - z$	$0.0239 + z$	z

$$K_{a_3} = \frac{[\text{H}^+][\text{PO}_4^{3-}]}{[\text{HPO}_4^{2-}]}$$

$$4.8 \times 10^{-13} = \frac{(0.0239 + z)(z)}{(6.2 \times 10^{-8}) - z} \approx \frac{(0.239)(z)}{(6.2 \times 10^{-8})}$$

$$z = 1.2 \times 10^{-18}\ M$$

The equilibrium concentrations are:

$$[\text{H}^+] = [\text{H}_2\text{PO}_4^-] = 0.0239\ M$$

$$[\text{H}_3\text{PO}_4] = 0.076\ M$$

$$[\text{HPO}_4^{2-}] = 6.2 \times 10^{-8}\ M$$

$$[\text{PO}_4^{3-}] = 1.2 \times 10^{-18}\ M$$

15.119 (a) Number of moles NaOH $= M \times$ vol (L) $= 0.0568\ M \times 0.0138\ \text{L} = 7.84 \times 10^{-4}$ mol

If the acid were all dimer, then:

$$\text{mol of dimer} = \frac{\text{mol NaOH}}{2} = \frac{7.84 \times 10^{-4}\ \text{mol}}{2} = 3.92 \times 10^{-4}\ \text{mol}$$

If the acetic acid were all dimer, the pressure that would be exerted would be:

$$P = \frac{nRT}{V} = \frac{(3.92 \times 10^{-4}\ \text{mol})(0.0821\ \text{L} \cdot \text{atm/K} \cdot \text{mol})(324\ \text{K})}{0.360\ \text{L}} = 0.0290\ \text{atm}$$

However, the actual pressure is 0.0342 atm. If α mol of dimer dissociates to monomers, then 2α monomer forms.

$$(CH_3COOH)_2 \;\rightleftharpoons\; 2CH_3COOH$$
$$1-\alpha \qquad\qquad\qquad 2\alpha$$

The total moles of acetic acid is:

$$\text{moles dimer} + \text{monomer} = (1-\alpha) + 2\alpha = 1 + \alpha$$

Using partial pressures:

$$P_{observed} = P(1+\alpha)$$
$$0.0342 \text{ atm} = (0.0290 \text{ atm})(1+\alpha)$$
$$\boldsymbol{\alpha = 0.179}$$

(b) The equilibrium constant is:

$$K_p = \frac{P_{CH_3COOH}^2}{P_{(CH_3COOH)_2}} = \frac{\left(\dfrac{2\alpha}{1+\alpha}\right)^2 \left(P_{observed}\right)^2}{\left(\dfrac{1-\alpha}{1+\alpha}\right)P_{observed}} = \frac{4\alpha^2 P_{observed}}{1-\alpha^2} = \boldsymbol{4.53 \times 10^{-3}}$$

15.120 $0.100\,M\,\text{Na}_2\text{CO}_3 \;\rightarrow\; 0.200\,M\,\text{Na}^+ + 0.100\,M\,\text{CO}_3^{2-}$

<u>First stage:</u>

	$CO_3^{2-}(aq) + H_2O(l)$	$\rightleftharpoons$	$HCO_3^-(aq)$	$+$	$OH^-(aq)$
Initial (M):	0.100		0		0
Change (M):	$-x$		$+x$		$+x$
Equilibrium (M):	$0.100 - x$		x		x

$$K_1 = \frac{K_w}{K_2} = \frac{1.0 \times 10^{-14}}{4.8 \times 10^{-11}} = 2.1 \times 10^{-4}$$

$$K_1 = \frac{[HCO_3^-][OH^-]}{[CO_3^{2-}]}$$

$$2.1 \times 10^{-4} = \frac{x^2}{0.100 - x} \approx \frac{x^2}{0.100}$$

$$x = 4.6 \times 10^{-3}\,M = [HCO_3^-] = [OH^-]$$

<u>Second stage:</u>

	$HCO_3^-(aq) + H_2O(l)$	$\rightleftharpoons$	$H_2CO_3(aq)$	$+$	$OH^-(aq)$
Initial (M):	4.6×10^{-3}		0		4.6×10^{-3}
Change (M):	$-y$		$+y$		$+y$
Equilibrium (M):	$(4.6 \times 10^{-3}) - y$		y		$(4.6 \times 10^{-3}) + y$

$$K_2 = \frac{[H_2CO_3][OH^-]}{[HCO_3^-]}$$

$$2.4 \times 10^{-8} = \frac{y[(4.6 \times 10^{-3}) + y]}{(4.6 \times 10^{-3}) - y} \approx \frac{(y)(4.6 \times 10^{-3})}{(4.6 \times 10^{-3})}$$

$$y = 2.4 \times 10^{-8} \ M$$

At equilibrium:

$$[Na^+] = 0.200 \ M$$

$$[HCO_3^-] = (4.6 \times 10^{-3}) \ M - (2.4 \times 10^{-8}) \ M \approx 4.6 \times 10^{-3} \ M$$

$$[H_2CO_3] = 2.4 \times 10^{-8} \ M$$

$$[OH^-] = (4.6 \times 10^{-3}) \ M + (2.4 \times 10^{-8}] \ M \approx 4.6 \times 10^{-3} \ M$$

$$[H^+] = \frac{1.0 \times 10^{-14}}{4.6 \times 10^{-3}} = 2.2 \times 10^{-12} \ M$$

15.121 $[CO_2] = kP = (2.28 \times 10^{-3} \text{ mol/L·atm})(3.20 \text{ atm}) = 7.30 \times 10^{-3} \ M$

$$CO_2(aq) \quad + \quad H_2O(l) \quad \rightleftharpoons \quad H^+(aq) \quad + \quad HCO_3^-(aq)$$
$$(7.30 \times 10^{-3} - x) \ M \qquad\qquad\qquad x \ M \qquad\qquad x \ M$$

$$K_a = \frac{[H^+][HCO_3^-]}{[CO_2]}$$

$$4.2 \times 10^{-7} = \frac{x^2}{(7.30 \times 10^{-3}) - x} \approx \frac{x^2}{7.30 \times 10^{-3}}$$

$$x = 5.5 \times 10^{-5} \ M = [H^+]$$

$$\textbf{pH} = \textbf{4.26}$$

15.122 When NaCN is treated with HCl, the following reaction occurs.

$$NaCN + HCl \rightarrow NaCl + HCN$$

HCN is a very weak acid, and only partially ionizes in solution.

$$HCN(aq) \rightleftharpoons H^+(aq) + CN^-(aq)$$

The main species in solution is HCN which has a tendency to escape into the gas phase.

$$HCN(aq) \rightleftharpoons HCN(g)$$

Since the HCN(g) that is produced is a highly poisonous compound, it would be dangerous to treat NaCN with acids without proper ventilation.

15.123 When the pH is 10.00, the pOH is 4.00 and the concentration of hydroxide ion is 1.0×10^{-4} *M*. The concentration of HCN must be the same. (Why?) If the concentration of NaCN is *x*, the table looks like:

$$CN^-(aq) + H_2O(l) \rightleftharpoons HCN(aq) + OH^-(aq)$$

Initial (M):	x	0	0
Change (M):	-1.0×10^{-4}	$+1.0 \times 10^{-4}$	$+1.0 \times 10^{-4}$
Equilibrium (M):	$(x - 1.0 \times 10^{-4})$	(1.0×10^{-4})	(1.0×10^{-4})

$$K_b = \frac{[HCN][OH^-]}{[CN^-]}$$

$$2.0 \times 10^{-5} = \frac{(1.0 \times 10^{-4})^2}{(x - 1.0 \times 10^{-4})}$$

$$x = 6.0 \times 10^{-4} \ M = [CN^-]_0$$

Amount of NaCN $= 250 \ mL \times \dfrac{6.0 \times 10^{-4} \ mol \ NaCN}{1000 \ mL} \times \dfrac{49.01 \ g \ NaCN}{1 \ mol \ NaCN} = \mathbf{7.4 \times 10^{-3} \ g \ NaCN}$

15.124 $pH = 2.53 = -\log[H^+]$

$[H^+] = 2.95 \times 10^{-3} \ M$

Since the concentration of H^+ at equilibrium is $2.95 \times 10^{-3} \ M$, that means that $2.95 \times 10^{-3} \ M$ HCOOH ionized. Let' represent the initial concentration of HCOOH as I. The equation representing the ionization of formic acid is:

$$HCOOH(aq) \rightleftharpoons H^+(aq) + HCOO^-(aq)$$

Initial (M):	I	0	0
Change (M):	-2.95×10^{-3}	$+2.95 \times 10^{-3}$	$+2.95 \times 10^{-3}$
Equilibrium (M):	$I - (2.95 \times 10^{-3})$	2.95×10^{-3}	2.95×10^{-3}

$$K_a = \frac{[H^+][HCOO^-]}{[HCOOH]}$$

$$1.7 \times 10^{-4} = \frac{(2.95 \times 10^{-3})^2}{I - (2.95 \times 10^{-3})}$$

$$I = 0.054 \ M$$

There are 0.054 moles of formic acid in 1000 mL of solution. The mass of formic acid in 100 mL is:

$$100 \ mL \times \frac{0.054 \ mol \ formic \ acid}{1000 \ mL \ soln} \times \frac{46.03 \ g \ formic \ acid}{1 \ mol \ formic \ acid} = \mathbf{0.25 \ g \ formic \ acid}$$

15.125 The equilibrium is established:

$$CH_3COOH(aq) \rightleftharpoons CH_3COO^-(aq) + H^+(aq)$$

Initial (M):	0.150	0	0.100
Change (M):	$-x$	$+x$	$+x$
Equilibrium (M):	$(0.150 - x)$	x	$(0.100 + x)$

$$K_a = \frac{[CH_3COO^-][H^+]}{[CH_3COOH]}$$

$$1.8 \times 10^{-5} = \frac{x(0.100 + x)}{0.150 - x} \approx \frac{0.100x}{0.150}$$

$$x = 2.7 \times 10^{-5} \, M$$

$2.7 \times 10^{-5} \, M$ is the $[H^+]$ contributed by CH_3COOH. HCl is a strong acid that completely ionizes. It contributes a $[H^+]$ of 0.100 M to the solution.

$$[H^+]_{total} = [0.100 + (2.7 \times 10^{-5})] \, M \approx 0.100 \, M$$

pH = 1.000

The pH is totally determined by the HCl and is independent of the CH_3COOH.

15.126 The balanced equation is: $Mg + 2HCl \rightarrow MgCl_2 + H_2$

$$\text{mol of Mg} = 1.87 \text{ g Mg} \times \frac{1 \text{ mol Mg}}{24.31 \text{ g Mg}} = 0.0769 \text{ mol}$$

From the balanced equation:

$$\text{mol of HCl required for reaction} = 2 \times \text{mol Mg} = (2)(0.0769 \text{ mol}) = 0.154 \text{ mol HCl}$$

The concentration of HCl:

pH = −0.544, thus $[H^+]$ = 3.50 M

initial mol HCl = $M \times$ Vol (L) = (3.50 M)(0.0800 L) = 0.280 mol HCl

Moles of HCl left after reaction:

initial mol HCl − mol HCl reacted = 0.280 mol − 0.154 mol = 0.126 mol HCl

Molarity of HCl left after reaction:

M = mol/L = 0.126 mol/0.080 L = 1.58 M

pH = −log(1.58) = −0.20

15.127 **(a)** The pH of the solution of HA would be lower. (Why?)

(b) The electrical conductance of the HA solution would be greater. (Why?)

(c) The rate of hydrogen evolution from the HA solution would be greater. Presumably, the rate of the reaction between the metal and hydrogen ion would depend on the hydrogen ion concentration (i.e., this would be part of the rate law). The hydrogen ion concentration will be greater in the HA solution.

15.128 The important equation is the hydrolysis of NO_2^-: $NO_2^- + H_2O \rightleftharpoons HNO_2 + OH^-$

(a) Addition of HCl will result in the reaction of the H^+ from the HCl with the OH^- that was present in the solution. The OH^- will effectively be removed and the equilibrium will **shift to the right** to compensate (more hydrolysis).

(b) Addition of NaOH is effectively addition of more OH^- which places stress on the right hand side of the equilibrium. The equilibrium will **shift to the left** (less hydrolysis) to compensate for the addition of OH^-.

(c) Addition of NaCl will have **no effect**.

(d) Recall that the percent ionization of a weak acid increases with dilution (see Figure 15.4 of the text). The same is true for weak bases. Thus dilution will cause more hydrolysis, shifting the equilibrium to the **right**.

15.129 Like carbon dioxide, sulfur dioxide behaves as a Lewis acid by accepting a pair of electrons from the Lewis base water. The Lewis acid-base adduct rearranges to form sulfurous acid in a manner exactly analogous to the rearrangement of the carbon dioxide-water adduct to form carbonic acid that is presented on page 665 of the textbook.

15.130 In Chapter 11, we found that salts with their formal electrostatic intermolecular attractions had low vapor pressures and thus high boiling points. Ammonia and its derivatives (amines) are molecules with dipole-dipole attractions; as long as the nitrogen has one direct N–H bond, the molecule will have hydrogen bonding. Even so, these molecules will have much higher vapor pressures than ionic species. Thus, if we could convert the neutral ammonia-type molecules into salts, their vapor pressures, and thus associated odors, would decrease. Lemon juice contains acids which can react with neutral ammonia-type (amine) molecules to form ammonium salts.

$$NH_3 + H^+ \rightarrow NH_4^+$$

$$RNH_2 + H^+ \rightarrow RNH_3^+$$

15.131 pH = 10.64

$$[H^+] = 2.3 \times 10^{-11}\ M$$

$$[OH^-] = 4.3 \times 10^{-4}\ M$$

$$
\begin{array}{ccccccc}
CH_3NH_2(aq) & + & H_2O(l) & \rightleftharpoons & CH_3NH_3^+(aq) & + & OH^-(aq) \\
(x - 4.3 \times 10^{-4})\ M & & & & 4.3 \times 10^{-4}\ M & & 4.3 \times 10^{-4}\ M
\end{array}
$$

$$K_b = \frac{[CH_3NH_3^+][OH^-]}{[CH_3NH_2]}$$

$$4.4 \times 10^{-4} = \frac{(4.3 \times 10^{-4})(4.3 \times 10^{-4})}{x - (4.3 \times 10^{-4})}$$

$$4.4 \times 10^{-4}x - 1.9 \times 10^{-7} = 1.8 \times 10^{-7}$$

$$x = 8.4 \times 10^{-4}\ M$$

The molar mass of CH_3NH_2 is 31.06 g/mol.

The mass of CH_3NH_2 in 100.0 mL is:

$$100.0 \text{ mL} \times \frac{8.4 \times 10^{-4} \text{ mol } CH_3NH_2}{1000 \text{ mL}} \times \frac{31.06 \text{ g } CH_3NH_2}{1 \text{ mol } CH_3NH_2} = 2.6 \times 10^{-3} \text{ g } CH_3NH_2$$

15.132
$$HCOOH \rightleftharpoons H^+ + HCOO^-$$

Initial (M):	0.400	0	0
Change (M):	$-x$	$+x$	$+x$
Equilibrium (M):	$0.400 - x$	x	x

Total concentration of particles in solution: $(0.400 - x) + x + x = 0.400 + x$

Assuming the molarity of the solution is equal to the molality, we can write:

$$\Delta T_f = K_f m$$

$$0.758 = (1.86)(0.400 + x)$$

$$0.408 = 0.400 + x$$

$$x = 0.00800 = [H^+] = [HCOO^-]$$

$$K_a = \frac{[H^+][HCOO^-]}{[HCOOH]} = \frac{(0.00800)(0.00800)}{0.400 - 0.00800} = 1.6 \times 10^{-4}$$

15.133   **(a)** $NH_2^- + H_2O \rightarrow NH_3 + OH^-$

$$N^{3-} + 3H_2O \rightarrow NH_3 + 3OH^-$$

(b) N^{3-} is the stronger base since each ion produces 3 OH^- ions.

15.134 $SO_2(g) + H_2O(l) \rightleftharpoons H^+(aq) + HSO_3^-(aq)$

Recall that 0.12 ppm SO_2 would mean 0.12 parts SO_2 per 1 million (10^6) parts of air by volume. The number of particles of SO_2 per volume will be directly related to the pressure.

$$P_{SO_2} = \frac{0.12 \text{ parts } SO_2}{10^6 \text{ parts air}} \text{ atm} = 1.2 \times 10^{-7} \text{ atm}$$

We can now calculate the $[H^+]$ from the equilibrium constant expression.

$$K = \frac{[H^+][HSO_3^-]}{P_{SO_2}}$$

$$1.3 \times 10^{-2} = \frac{x^2}{1.2 \times 10^{-7}}$$

$$x^2 = (1.3 \times 10^{-2})(1.2 \times 10^{-7})$$

$$x = 3.9 \times 10^{-5}\ M = [\text{H}^+]$$

$$\textbf{pH} = -\log(3.9 \times 10^{-5}) = \textbf{4.40}$$

15.135 $\dfrac{[\text{H}^+][\text{ClO}^-]}{[\text{HClO}]} = 3.0 \times 10^{-8}$

A pH of 7.8 corresponds to $[\text{H}^+] = 1.6 \times 10^{-8}\ M$

Substitute $[\text{H}^+]$ into the equation above to solve for the $\dfrac{[\text{ClO}^-]}{[\text{HClO}]}$ ratio.

$$\frac{[\text{ClO}^-]}{[\text{HClO}]} = \frac{3.0 \times 10^{-8}}{1.6 \times 10^{-8}} = 1.9$$

This indicates that to obtain a pH of 7.8, the $[\text{ClO}^-]$ must be 1.9 times greater than the $[\text{HClO}]$. We can write:

$$\textbf{\%ClO}^- = \frac{\text{part ClO}^-}{\text{part ClO}^- + \text{part HClO}} \times 100\% = \frac{1.9}{1.9 + 1.0} \times 100\% = \textbf{66\%}$$

By difference, **%HClO = 34%**

15.136 In inhaling the smelling salt, some of the powder dissolves in the basic solution. The ammonium ions react with the base as follows:

$$\text{NH}_4^+(aq) + \text{OH}^-(aq) \rightarrow \text{NH}_3(aq) + \text{H}_2\text{O}$$

It is the pungent odor of ammonia that prevents a person from fainting.

15.137 (a) The overall equation is

$$\text{Fe}_2\text{O}_3(s) + 6\text{HCl}(aq) \longrightarrow 2\text{FeCl}_3(aq) + 3\text{H}_2\text{O}(l)$$

and the net ionic equation is

$$\text{Fe}_2\text{O}_3(s) + 6\text{H}^+(aq) \longrightarrow 2\text{Fe}^{3+}(aq) + 3\text{H}_2\text{O}(l)$$

Since HCl donates the H^+ ion, it is the Brønsted acid. Each Fe_2O_3 unit accepts six H^+ ions; therefore, it is the Brønsted base.

(b) The first stage is

$$\text{CaCO}_3(s) + \text{HCl}(aq) \longrightarrow \text{Ca}^{2+}(aq) + \text{HCO}_3^-(aq) + \text{Cl}^-(aq)$$

and the second stage is

$$\text{HCl}(aq) + \text{HCO}_3^-(aq) \longrightarrow \text{CO}_2(g) + \text{Cl}^-(aq) + \text{H}_2\text{O}(l)$$

The overall equation is

$$\text{CaCO}_3(s) + 2\text{HCl}(aq) \longrightarrow \text{CaCl}_2(aq) + \text{H}_2\text{O}(l) + \text{CO}_2(g)$$

The CaCl_2 formed is soluble in water.

(c)   We need to find the concentration of the HCl solution in order to determine its pH. Let's assume a volume of 1.000 L = 1000 mL. The mass of 1000 mL of solution is:

$$1000 \text{ mL} \times \frac{1.073 \text{ g}}{1 \text{ mL}} = 1073 \text{ g}$$

The number of moles of HCl in a 15 percent solution is:

$$\frac{15\% \text{ HCl}}{100\% \text{ soln}} \times 1073 \text{ g soln} = (1.6 \times 10^2 \text{ g HCl}) \times \frac{1 \text{ mol HCl}}{36.46 \text{ g HCl}} = 4.4 \text{ mol HCl}$$

Thus, there are 4.4 moles of HCl in one liter of solution, and the concentration is 4.4 M. The pH of the solution is

pH = −log(4.4) = −0.64

This is a highly acidic solution (note that the pH is negative), which is needed to dissolve large quantities of rocks in the oil recovery process.

15.138 **(c)** does not represent a Lewis acid-base reaction. In this reaction, the F–F single bond is broken and single bonds are formed between P and each F atom. For a Lewis acid-base reaction, the Lewis acid is an electron-pair acceptor and the Lewis base is an electron-pair donor.

15.139 **(a)**   **False**. A Lewis acid such as CO_2 is not a Brønsted acid. It does not have a hydrogen ion to donate.

(b)   **False**. Consider the weak acid, NH_4^+. The conjugate base of this acid is NH_3, which is neutral.

(c)   **False**. The percent ionization of a base decreases with increasing concentration of base in solution.

(d)   **False**. A solution of barium fluoride is basic. The fluoride ion, F^-, is the conjugate base of a weak acid. It will hydrolyze to produce OH^- ions.

15.140 From the given pH's, we can calculate the $[H^+]$ in each solution.

Solution (1): $[H^+] = 10^{-pH} = 10^{-4.12} = 7.6 \times 10^{-5} M$

Solution (2): $[H^+] = 10^{-5.76} = 1.7 \times 10^{-6} M$

Solution (3): $[H^+] = 10^{-5.34} = 4.6 \times 10^{-6} M$

We are adding solutions (1) and (2) to make solution (3). The volume of solution (2) is 0.528 L. We are going to add a given volume of solution (1) to solution (2). Let's call this volume x. The moles of H^+ in solutions (1) and (2) will equal the moles of H^+ in solution (3).

mol H^+ soln (1) + mol H^+ soln (2) = mol H^+ soln (3)

Recall that mol = $M \times$ L. We have:

$$(7.6 \times 10^{-5} \text{ mol/L})(x \text{ L}) + (1.7 \times 10^{-6} \text{ mol/L})(0.528 \text{ L}) = (4.6 \times 10^{-6} \text{ mol/L})(0.528 + x)\text{L}$$

$$(7.6 \times 10^{-5})x + (9.0 \times 10^{-7}) = (2.4 \times 10^{-6}) + (4.6 \times 10^{-6})x$$

$$(7.1 \times 10^{-5})x = 1.5 \times 10^{-6}$$

$$x = 0.021 \text{ L} = \textbf{21 mL}$$

CHAPTER 16
ACID-BASE EQUILIBRIA AND SOLUBILITY EQUILIBRIA

16.5 **(a)** This is a weak acid problem. Setting up the standard equilibrium table:

$$CH_3COOH(aq) \rightleftharpoons H^+(aq) + CH_3COO^-(aq)$$

	$CH_3COOH(aq)$	$H^+(aq)$	$CH_3COO^-(aq)$
Initial (*M*):	0.40	0.00	0.00
Change (*M*):	$-x$	$+x$	$+x$
Equilibrium (*M*):	$(0.40 - x)$	x	x

$$K_a = \frac{[H^+][CH_3COO^-]}{[CH_3COOH]}$$

$$1.8 \times 10^{-5} = \frac{x^2}{(0.40 - x)} \approx \frac{x^2}{0.40}$$

$$x = [H^+] = 2.7 \times 10^3 \, M$$

$$\mathbf{pH = 2.57}$$

(b) In addition to the acetate ion formed from the ionization of acetic acid, we also have acetate ion formed from the sodium acetate dissolving.

$$CH_3COONa(aq) \rightarrow CH_3COO^-(aq) + Na^+(aq)$$

Dissolving 0.20 *M* sodium acetate initially produces 0.20 *M* CH_3COO^- and 0.20 *M* Na^+. The sodium ions are not involved in any further equilibrium (why?), but the acetate ions must be added to the equilibrium in part (a).

$$CH_3COOH(aq) \rightleftharpoons H^+(aq) + CH_3COO^-(aq)$$

	$CH_3COOH(aq)$	$H^+(aq)$	$CH_3COO^-(aq)$
Initial (*M*):	0.40	0.00	0.20
Change (*M*):	$-x$	$+x$	$+x$
Equilibrium (*M*):	$(0.40 - x)$	x	$(0.20 + x)$

$$K_a = \frac{[H^+][CH_3COO^-]}{[CH_3COOH]}$$

$$1.8 \times 10^{-5} = \frac{(x)(0.20 + x)}{(0.40 - x)} \approx \frac{x(0.20)}{0.40}$$

$$x = [H^+] = 3.6 \times 10^{-5} \, M$$

$$\mathbf{pH = 4.44}$$

Could you have predicted whether the pH should have increased or decreased after the addition of the sodium acetate to the pure 0.40 *M* acetic acid in part (a)?

An alternate way to work part (b) of this problem is to use the Henderson-Hasselbalch equation.

$$pH = pK_a + \log\frac{[\text{conjugate base}]}{[\text{acid}]}$$

$$pH = -\log(1.8 \times 10^{-5}) + \log\frac{0.20\ M}{0.40\ M} = 4.74 - 0.30 = \textbf{4.44}$$

16.6   **(a)**   This is a weak base calculation.

$$NH_3(aq) + H_2O(l) \rightleftharpoons NH_4^+(aq) + OH^-(aq)$$

Initial (M):	0.20	0	0
Change (M):	$-x$	$+x$	$+x$
Equilibrium (M):	$0.20 - x$	x	x

$$K_b = \frac{[NH_4^+][OH^-]}{[NH_3]}$$

$$1.8 \times 10^{-5} = \frac{(x)(x)}{0.20 - x} \approx \frac{x^2}{0.20}$$

$$x = 1.9 \times 10^{-3}\ M = [OH^-]$$

$$pOH = 2.72$$

$$\textbf{pH} = \textbf{11.28}$$

(b)   The initial concentration of NH_4^+ is 0.30 M from the salt NH_4Cl. We set up a table as in part (a).

$$NH_3(aq) + H_2O(l) \rightleftharpoons NH_4^+(aq) + OH^-(aq)$$

Initial (M):	0.20	0.30	0
Change (M):	$-x$	$+x$	$+x$
Equilibrium (M):	$0.20 - x$	$0.30 + x$	x

$$K_b = \frac{[NH_4^+][OH^-]}{[NH_3]}$$

$$1.8 \times 10^{-5} = \frac{(x)(0.30 + x)}{0.20 - x} \approx \frac{x(0.30)}{0.20}$$

$$x = 1.2 \times 10^{-5}\ M = [OH^-]$$

$$pOH = 4.92$$

$$\textbf{pH} = \textbf{9.08}$$

Alternatively, we could use the Henderson-Hasselbalch equation to solve this problem. Table 15.4 gives the value of K_a for the ammonium ion. Substituting into the Henderson-Hasselbalch equation gives:

$$pH = pK_a + \log\frac{[\text{conjugate base}]}{\text{acid}} = -\log(5.6 \times 10^{-10}) + \log\frac{(0.20)}{(0.30)}$$

$$\textbf{pH} = 9.25 - 0.18 = \textbf{9.07}$$

Is there any difference in the Henderson-Hasselbalch equation in the cases of a weak acid and its conjugate base and a weak base and its conjugate acid?

16.9 (a) HCl (hydrochloric acid) is a strong acid. A buffer is a solution containing both a weak acid and a weak base. Therefore, this is *not* a buffer system.

(b) H_2SO_4 (sulfuric acid) is a strong acid. A buffer is a solution containing both a weak acid and a weak base. Therefore, this is *not* a buffer system.

(c) This solution contains both a weak acid, $H_2PO_4^-$ and its conjugate base, HPO_4^{2-}. Therefore, this is a buffer system.

(d) HNO_2 (nitrous acid) is a weak acid, and its conjugate base, NO_2^- (nitrite ion, the anion of the salt KNO_2), is a weak base. Therefore, this is a buffer system.

16.10 **Strategy:** What constitutes a buffer system? Which of the preceding solutions contains a weak acid and its salt (containing the weak conjugate base)? Which of the preceding solutions contains a weak base and its salt (containing the weak conjugate acid)? Why is the conjugate base of a strong acid not able to neutralize an added acid?

Solution: The criteria for a buffer system are that we must have a weak acid and its salt (containing the weak conjugate base) or a weak base and its salt (containing the weak conjugate acid).

(a) HCN is a weak acid, and its conjugate base, CN^-, is a weak base. Therefore, this is a buffer system.

(b) HSO_4^- is a weak acid, and its conjugate base, SO_4^{2-} is a weak base (see Table 15.5 of the text). Therefore, this is a buffer system.

(c) NH_3 (ammonia) is a weak base, and its conjugate acid, NH_4^+ is a weak acid. Therefore, this is a buffer system.

(d) Because HI is a strong acid, its conjugate base, I^-, is an extremely weak base. This means that the I^- ion will not combine with a H^+ ion in solution to form HI. Thus, this system cannot act as a buffer system.

16.11 $NH_4^+(aq) \rightleftharpoons NH_3(aq) + H^+(aq)$

$K_a = 5.6 \times 10^{-10}$

$pK_a = 9.25$

$pH = pK_a + \log\dfrac{[NH_3]}{[NH_4^+]} = 9.25 + \log\dfrac{0.15\ M}{0.35\ M} = \textbf{8.88}$

16.12 **Strategy:** The pH of a buffer system can be calculated in a similar manner to a weak acid equilibrium problem. The difference is that a common-ion is present in solution. The K_a of CH_3COOH is 1.8×10^{-5} (see Table 15.3 of the text).

Solution:

(a) We summarize the concentrations of the species at equilibrium as follows:

	$CH_3COOH(aq)$	$\rightleftharpoons$	$H^+(aq)$	+	$CH_3COO^-(aq)$
Initial (*M*):	2.0		0		2.0
Change (*M*):	$-x$		$+x$		$+x$
Equilibrium (*M*):	$2.0 - x$		x		$2.0 + x$

$$K_a = \frac{[H^+][CH_3COO^-]}{[CH_3COOH]}$$

$$K_a = \frac{[H^+](2.0 + x)}{(2.0 - x)} \approx \frac{[H^+](2.0)}{2.0}$$

$$K_a = [H^+]$$

Taking the $-\log$ of both sides,

$$pK_a = pH$$

Thus, for a buffer system in which the [weak acid] = [weak base],

$$pH = pK_a$$

$$\mathbf{pH} = -\log(1.8 \times 10^{-5}) = \mathbf{4.74}$$

(b)   Similar to part (a),

$$\mathbf{pH} = \mathbf{p}K_a = \mathbf{4.74}$$

Buffer (a) will be a more effective buffer because the concentrations of acid and base components are ten times higher than those in (b). Thus, buffer (a) can neutralize 10 times more added acid or base compared to buffer (b).

16.13 $H_2CO_3(aq) \rightleftharpoons HCO_3^-(aq) + H^+(aq)$

$$K_{a_1} = 4.2 \times 10^{-7}$$

$$pK_{a_1} = 6.38$$

$$pH = pK_a + \log\frac{[HCO_3^-]}{[H_2CO_3]}$$

$$8.00 = 6.38 + \log\frac{[HCO_3^-]}{[H_2CO_3]}$$

$$\log\frac{[HCO_3^-]}{[H_2CO_3]} = 1.62$$

$$\frac{[HCO_3^-]}{[H_2CO_3]} = 41.7$$

$$\frac{[H_2CO_3]}{[HCO_3^-]} = \mathbf{0.024}$$

16.14   *Step 1:* Write the equilibrium that occurs between $H_2PO_4^-$ and HPO_4^{2-}. Set up a table relating the initial concentrations, the change in concentration to reach equilibrium, and the equilibrium concentrations.

$$H_2PO_4^-(aq) \rightleftharpoons H^+(aq) + HPO_4^{2-}(aq)$$

	$H_2PO_4^-(aq)$	$H^+(aq)$	$HPO_4^{2-}(aq)$
Initial (M):	0.15	0	0.10
Change (M):	$-x$	$+x$	$+x$
Equilibrium (M):	$0.15 - x$	x	$0.10 + x$

Step 2: Write the ionization constant expression in terms of the equilibrium concentrations. Knowing the value of the equilibrium constant (K_a), solve for x.

$$K_a = \frac{[H^+][HPO_4^{2-}]}{[H_2PO_4^-]}$$

You can look up the K_a value for dihydrogen phosphate in Table 15.5 of your text.

$$6.2 \times 10^{-8} = \frac{(x)(0.10 + x)}{(0.15 - x)}$$

$$6.2 \times 10^{-8} \approx \frac{(x)(0.10)}{(0.15)}$$

$$x = [H^+] = 9.3 \times 10^{-8} \, M$$

Step 3: Having solved for the $[H^+]$, calculate the pH of the solution.

$$\mathbf{pH} = -\log[H^+] = -\log(9.3 \times 10^{-8}) = \mathbf{7.03}$$

16.15 Using the Henderson–Hasselbalch equation:

$$pH = pK_a + \log\frac{[CH_3COO^-]}{[CH_3COOH]}$$

$$4.50 = 4.74 + \log\frac{[CH_3COO^-]}{[CH_3COOH]}$$

Thus,

$$\frac{[CH_3COO^-]}{[CH_3COOH]} = \mathbf{0.58}$$

16.16 We can use the Henderson-Hasselbalch equation to calculate the ratio $[HCO_3^-]/[H_2CO_3]$. The Henderson-Hasselbalch equation is:

$$pH = pK_a + \log\frac{[\text{conjugate base}]}{[\text{acid}]}$$

For the buffer system of interest, HCO_3^- is the conjugate base of the acid, H_2CO_3. We can write:

$$pH = 7.40 = -\log(4.2 \times 10^{-7}) + \log\frac{[HCO_3^-]}{[H_2CO_3]}$$

$$7.40 = 6.38 + \log\frac{[HCO_3^-]}{[H_2CO_3]}$$

The [conjugate base]/[acid] ratio is:

$$\log \frac{[HCO_3^-]}{[H_2CO_3]} = 7.40 - 6.38 = 1.02$$

$$\frac{[HCO_3^-]}{[H_2CO_3]} = 10^{1.02} = \mathbf{1.0 \times 10^1}$$

The buffer should be more effective against an added acid because ten times more base is present compared to acid. Note that a pH of 7.40 is only a two significant figure number (Why?); the final result should only have two significant figures.

16.17   For the first part we use K_a for ammonium ion. (Why?) The Henderson–Hasselbalch equation is

$$\mathbf{pH} = -\log(5.6 \times 10^{-10}) + \log \frac{(0.20 \ M)}{(0.20 \ M)} = \mathbf{9.25}$$

For the second part, the acid–base reaction is

$$NH_3(g) + H^+(aq) \rightarrow NH_4^+(aq)$$

We find the number of moles of HCl added

$$10.0 \ mL \times \frac{0.10 \ mol \ HCl}{1000 \ mL \ soln} = 0.0010 \ mol \ HCl$$

The number of moles of NH_3 and NH_4^+ originally present are

$$65.0 \ mL \times \frac{0.20 \ mol}{1000 \ mL \ soln} = 0.013 \ mol$$

Using the acid-base reaction, we find the number of moles of NH_3 and NH_4^+ after addition of the HCl.

	$NH_3(aq)$	+ $H^+(aq)$	$\rightarrow$ $NH_4^+(aq)$
Initial (mol):	0.013	0.0010	0.013
Change (mol):	−0.0010	−0.0010	+0.0010
Final (mol):	0.012	0	0.014

We find the new pH:

$$\mathbf{pH} = 9.25 + \log \frac{(0.012)}{(0.014)} = \mathbf{9.18}$$

16.18   As calculated in Problem 16.12, the pH of this buffer system is equal to pK_a.

$$pH = pK_a = -\log(1.8 \times 10^{-5}) = 4.74$$

(a)   The added NaOH will react completely with the acid component of the buffer, CH_3COOH. NaOH ionizes completely; therefore, 0.080 mol of OH^- are added to the buffer.

Step 1: The neutralization reaction is:

$$CH_3COOH(aq) \; + \; OH^-(aq) \longrightarrow CH_3COO^-(aq) \; + \; H_2O(l)$$

	CH₃COOH	OH⁻	CH₃COO⁻
Initial (mol):	1.00	0.080	1.00
Change (mol):	−0.080	−0.080	+0.080
Final (mol):	0.92	0	1.08

Step 2: Now, the acetic acid equilibrium is reestablished. Since the volume of the solution is 1.00 L, we can convert directly from moles to molar concentration.

$$CH_3COOH(aq) \; \rightleftharpoons \; H^+(aq) \; + \; CH_3COO^-(aq)$$

	CH₃COOH	H⁺	CH₃COO⁻
Initial (*M*):	0.92	0	1.08
Change (*M*):	−x	+x	+x
Equilibrium (*M*):	0.92 − x	x	1.08 + x

Write the K_a expression, then solve for x.

$$K_a = \frac{[H^+][CH_3COO^-]}{[CH_3COOH]}$$

$$1.8 \times 10^{-5} = \frac{(x)(1.08 + x)}{(0.92 - x)} \approx \frac{x(1.08)}{0.92}$$

$$x = [H^+] = 1.5 \times 10^{-5} \, M$$

Step 3: Having solved for the $[H^+]$, calculate the pH of the solution.

$$\mathbf{pH = -\log[H^+] = -\log(1.5 \times 10^{-5}) = 4.82}$$

The pH of the buffer increased from 4.74 to 4.82 upon addition of 0.080 mol of strong base.

(b)   The added acid will react completely with the base component of the buffer, CH_3COO^-. HCl ionizes completely; therefore, 0.12 mol of H^+ ion are added to the buffer

Step 1: The neutralization reaction is:

$$CH_3COO^-(aq) + H^+(aq) \longrightarrow CH_3COOH(aq)$$

	CH₃COO⁻	H⁺	CH₃COOH
Initial (mol):	1.00	0.12	1.00
Change (mol):	−0.12	−0.12	+0.12
Final (mol):	0.88	0	1.12

Step 2: Now, the acetic acid equilibrium is reestablished. Since the volume of the solution is 1.00 L, we can convert directly from moles to molar concentration.

$$CH_3COOH(aq) \; \rightleftharpoons \; H^+(aq) \; + \; CH_3COO^-(aq)$$

	CH₃COOH	H⁺	CH₃COO⁻
Initial (*M*):	1.12	0	0.88
Change (*M*):	−x	+x	+x
Equilibrium (*M*):	1.12 − x	x	0.88 + x

Write the K_a expression, then solve for x.

$$K_a = \frac{[H^+][CH_3COO^-]}{[CH_3COOH]}$$

$$1.8 \times 10^{-5} = \frac{(x)(0.88 + x)}{(1.12 - x)} \approx \frac{x(0.88)}{1.12}$$

$$x = [H^+] = 2.3 \times 10^{-5} \, M$$

Step 3: Having solved for the $[H^+]$, calculate the pH of the solution.

$$\textbf{pH} = -\log[H^+] = -\log(2.3 \times 10^{-5}) = \textbf{4.64}$$

The pH of the buffer decreased from 4.74 to 4.64 upon addition of 0.12 mol of strong acid.

16.19   We write

$$K_{a_1} = 1.1 \times 10^{-3} \qquad pK_{a_1} = 2.96$$

$$K_{a_2} = 2.5 \times 10^{-6} \qquad pK_{a_2} = 5.60$$

In order for the buffer solution to behave effectively, the pK_a of the acid component must be close to the desired pH. Therefore, the proper buffer system is **Na₂A/NaHA**.

16.20   **Strategy:** For a buffer to function effectively, the concentration of the acid component must be roughly equal to the conjugate base component. According to Equation (16.4) of the text, when the desired pH is close to the pK_a of the acid, that is, when pH $\approx$ pK_a,

$$\log \frac{[\text{conjugate base}]}{[\text{acid}]} \approx 0$$

or

$$\frac{[\text{conjugate base}]}{[\text{acid}]} \approx 1$$

Solution: To prepare a solution of a desired pH, we should choose a weak acid with a pK_a value close to the desired pH. Calculating the pK_a for each acid:

For HA, $pK_a = -\log(2.7 \times 10^{-3}) = 2.57$

For HB, $pK_a = -\log(4.4 \times 10^{-6}) = 5.36$

For HC, $pK_a = -\log(2.6 \times 10^{-9}) = 8.59$

The buffer solution with a pK_a closest to the desired pH is HC. Thus, **HC** is the best choice to prepare a buffer solution with pH $= 8.60$.

16.23   Since the acid is monoprotic, the number of moles of KOH is equal to the number of moles of acid.

$$\text{Moles acid} = 16.4 \text{ mL} \times \frac{0.08133 \text{ mol}}{1000 \text{ mL}} = 0.00133 \text{ mol}$$

$$\textbf{Molar mass} = \frac{0.2688 \text{ g}}{0.00133 \text{ mol}} = \textbf{202 g/mol}$$

16.24   We want to calculate the molar mass of the diprotic acid. The mass of the acid is given in the problem, so we need to find moles of acid in order to calculate its molar mass.

$$\underset{\text{want to calculate}}{\text{molar mass of } H_2A} = \frac{\overset{\text{given}}{g\ H_2A}}{\underset{\text{need to find}}{\text{mol } H_2A}}$$

The neutralization reaction is:

$$2KOH(aq) + H_2A(aq) \longrightarrow K_2A(aq) + 2H_2O(l)$$

From the volume and molarity of the base needed to neutralize the acid, we can calculate the number of moles of H_2A reacted.

$$11.1 \text{ mL KOH} \times \frac{1.00 \text{ mol KOH}}{1000 \text{ mL}} \times \frac{1 \text{ mol } H_2A}{2 \text{ mol KOH}} = 5.55 \times 10^{-3} \text{ mol } H_2A$$

We know that 0.500 g of the diprotic acid were reacted (1/10 of the 250 mL was tested). Divide the number of grams by the number of moles to calculate the molar mass.

$$\mathcal{M}(H_2A) = \frac{0.500 \text{ g } H_2A}{5.55 \times 10^{-3} \text{ mol } H_2A} = \textbf{90.1 g/mol}$$

16.25   The neutralization reaction is:

$$H_2SO_4(aq) + 2NaOH(aq) \rightarrow Na_2SO_4(aq) + 2H_2O(l)$$

Since one mole of sulfuric acid combines with two moles of sodium hydroxide, we write:

$$\text{mol NaOH} = 12.5 \text{ mL } H_2SO_4 \times \frac{0.500 \text{ mol } H_2SO_4}{1000 \text{ mL soln}} \times \frac{2 \text{ mol NaOH}}{1 \text{ mol } H_2SO_4} = 0.0125 \text{ mol NaOH}$$

$$\textbf{concentration of NaOH} = \frac{0.0125 \text{ mol NaOH}}{50.0 \times 10^{-3} \text{ L soln}} = \textbf{0.25 } \textbf{\textit{M}}$$

16.26   We want to calculate the molarity of the $Ba(OH)_2$ solution. The volume of the solution is given (19.3 mL), so we need to find the moles of $Ba(OH)_2$ to calculate the molarity.

$$M \text{ of } Ba(OH)_2 = \frac{\overset{\text{need to find}}{\text{mol } Ba(OH)_2}}{\underset{\text{given}}{\text{L of } Ba(OH)_2 \text{ soln}}}$$

want to calculate

The neutralization reaction is:

$$2HCOOH + Ba(OH)_2 \rightarrow (HCOO)_2Ba + 2H_2O$$

From the volume and molarity of HCOOH needed to neutralize $Ba(OH)_2$, we can determine the moles of $Ba(OH)_2$ reacted.

$$20.4 \text{ mL HCOOH} \times \frac{0.883 \text{ mol HCOOH}}{1000 \text{ mL}} \times \frac{1 \text{ mol } Ba(OH)_2}{2 \text{ mol HCOOH}} = 9.01 \times 10^{-3} \text{ mol } Ba(OH)_2$$

The molarity of the $Ba(OH)_2$ solution is:

$$\frac{9.01 \times 10^{-3} \text{ mol } Ba(OH)_2}{19.3 \times 10^{-3} \text{ L}} = \textbf{0.467 } \textbf{\textit{M}}$$

16.27   **(a)**   Since the acid is monoprotic, the moles of acid equals the moles of base added.

$$HA(aq) + NaOH(aq) \longrightarrow NaA(aq) + H_2O(l)$$

$$\text{Moles acid} = 18.4 \text{ mL} \times \frac{0.0633 \text{ mol}}{1000 \text{ mL soln}} = 0.00116 \text{ mol}$$

We know the mass of the unknown acid in grams and the number of moles of the unknown acid.

$$\textbf{Molar mass} = \frac{0.1276 \text{ g}}{0.00116 \text{ mol}} = \textbf{1.10} \times \textbf{10}^2 \textbf{ g/mol}$$

(b)   The number of moles of NaOH in 10.0 mL of solution is

$$10.0 \text{ mL} \times \frac{0.0633 \text{ mol}}{1000 \text{ mL soln}} = 6.33 \times 10^{-4} \text{ mol}$$

The neutralization reaction is:

	$HA(aq)$	$+$	$NaOH(aq)$	$\longrightarrow$	$NaA(aq)$	$+$	$H_2O(l)$
Initial (mol):	0.00116		6.33×10^{-4}		0		
Change (mol):	-6.33×10^{-4}		-6.33×10^{-4}		$+6.33 \times 10^{-4}$		
Final (mol):	5.3×10^{-4}		0		6.33×10^{-4}		

Now, the weak acid equilibrium will be reestablished. The total volume of solution is 35.0 mL.

$$[HA] = \frac{5.3 \times 10^{-4}\ \text{mol}}{0.035\ \text{L}} = 0.015\ M$$

$$[A^-] = \frac{6.33 \times 10^{-4}\ \text{mol}}{0.035\ \text{L}} = 0.0181\ M$$

We can calculate the $[H^+]$ from the pH.

$$[H^+] = 10^{-pH} = 10^{-5.87} = 1.35 \times 10^{-6}\ M$$

	HA(aq)	$\rightleftharpoons$	H$^+$(aq)	+	A$^-$(aq)
Initial (M):	0.015		0		0.0181
Change (M):	-1.35×10^{-6}		$+1.35 \times 10^{-6}$		$+1.35 \times 10^{-6}$
Equilibrium (M):	0.015		1.35×10^{-6}		0.0181

Substitute the equilibrium concentrations into the equilibrium constant expression to solve for K_a.

$$K_a = \frac{[H^+][A^-]}{[HA]} = \frac{(1.35 \times 10^{-6})(0.0181)}{0.015} = 1.6 \times 10^{-6}$$

16.28 The resulting solution is not a buffer system. There is excess NaOH and the neutralization is well past the equivalence point.

$$\text{Moles NaOH} = 0.500\ \text{L} \times \frac{0.167\ \text{mol}}{1\ \text{L}} = 0.0835\ \text{mol}$$

$$\text{Moles CH}_3\text{COOH} = 0.500\ \text{L} \times \frac{0.100\ \text{mol}}{1\ \text{L}} = 0.0500\ \text{mol}$$

	CH$_3$COOH(aq) + NaOH(aq) $\rightarrow$ CH$_3$COONa(aq) + H$_2$O(l)		
Initial (mol):	0.0500	0.0835	0
Change (mol):	-0.0500	-0.0500	$+0.0500$
Final (mol):	0	0.0335	0.0500

The volume of the resulting solution is 1.00 L (500 mL + 500 mL = 1000 mL).

$$[OH^-] = \frac{0.0335\ \text{mol}}{1.00\ \text{L}} = 0.0335\ M$$

$$[Na^+] = \frac{(0.0335 + 0.0500)\,\text{mol}}{1.00\ \text{L}} = 0.0835\ M$$

$$[H^+] = \frac{K_w}{[OH^-]} = \frac{1.0 \times 10^{-14}}{0.0335} = 3.0 \times 10^{-13}\ M$$

$$[CH_3COO^-] = \frac{0.0500\ \text{mol}}{1.00\ \text{L}} = 0.0500\ M$$

$$CH_3COO^-(aq) + H_2O(l) \rightleftharpoons CH_3COOH(aq) + OH^-(aq)$$

Initial (M):	0.0500	0	0.0335
Change (M):	$-x$	$+x$	$+x$
Equilibrium (M):	$0.0500 - x$	x	$0.0335 + x$

$$K_b = \frac{[CH_3COOH][OH^-]}{[CH_3COO^-]}$$

$$5.6 \times 10^{-10} = \frac{(x)(0.0335 + x)}{(0.0500 - x)} \approx \frac{(x)(0.0335)}{(0.0500)}$$

$$x = [CH_3COOH] = 8.4 \times 10^{-10}\ M$$

16.29 $HCl(aq) + CH_3NH_2(aq) \rightleftharpoons CH_3NH_3^+(aq) + Cl^-(aq)$

Since the concentrations of acid and base are equal, equal volumes of each solution will need to be added to reach the equivalence point. Therefore, the solution volume is doubled at the equivalence point, and the concentration of the conjugate acid from the salt, $CH_3NH_3^+$, is:

$$\frac{0.20\ M}{2} = 0.10\ M$$

The conjugate acid undergoes hydrolysis.

$$CH_3NH_3^+(aq) + H_2O(l) \rightleftharpoons H_3O^+(aq) + CH_3NH_2(aq)$$

Initial (M):	0.10	0	0
Change (M):	$-x$	$+x$	$+x$
Equilibrium (M):	$0.10 - x$	x	x

$$K_a = \frac{[H_3O^+][CH_3NH_2]}{[CH_3NH_3^+]}$$

$$2.3 \times 10^{-11} = \frac{x^2}{0.10 - x}$$

Assuming that, $0.10 - x \approx 0.10$

$$x = [H_3O^+] = 1.5 \times 10^{-6}\ M$$

$$\textbf{pH} = \textbf{5.82}$$

16.30   Let's assume we react 1 L of HCOOH with 1 L of NaOH.

$$HCOOH(aq) + NaOH(aq) \rightarrow HCOONa(aq) + H_2O(l)$$

Initial (mol):	0.10	0.10	0
Change (mol):	-0.10	-0.10	$+0.10$
Final (mol):	0	0	0.10

The solution volume has doubled (1 L + 1 L = 2 L). The concentration of HCOONa is:

$$M \text{ (HCOONa)} = \frac{0.10 \text{ mol}}{2 \text{ L}} = 0.050 \ M$$

$HCOO^-(aq)$ is a weak base. The hydrolysis is:

$$HCOO^-(aq) + H_2O(l) \rightleftharpoons HCOOH(aq) + OH^-(aq)$$

	$HCOO^-$	$HCOOH$	OH^-
Initial (M):	0.050	0	0
Change (M):	$-x$	$+x$	$+x$
Equilibrium (M):	$0.050 - x$	x	x

$$K_b = \frac{[HCOOH][OH^-]}{[HCOO^-]}$$

$$5.9 \times 10^{-11} = \frac{x^2}{0.050 - x} \approx \frac{x^2}{0.050}$$

$$x = 1.7 \times 10^{-6} \ M = [OH^-]$$

$$pOH = 5.77$$

$$\mathbf{pH = 8.23}$$

16.31   The reaction between CH_3COOH and KOH is:

$$CH_3COOH(aq) + KOH(aq) \rightarrow CH_3COOK(aq) + H_2O(l)$$

We see that 1 mole $CH_3COOH \simeq 1$ mol KOH. Therefore, at every stage of titration, we can calculate the number of moles of acid reacting with base, and the pH of the solution is determined by the excess acid or base left over. At the equivalence point, however, the neutralization is complete, and the pH of the solution will depend on the extent of the hydrolysis of the salt formed, which is CH_3COOK.

(a)   No KOH has been added. This is a weak acid calculation.

$$CH_3COOH(aq) + H_2O(l) \rightleftharpoons H_3O^+(aq) + CH_3COO^-(aq)$$

	CH_3COOH	H_3O^+	CH_3COO^-
Initial (M):	0.100	0	0
Change (M):	$-x$	$+x$	$+x$
Equilibrium (M):	$0.100 - x$	x	x

$$K_a = \frac{[H_3O^+][CH_3COO^-]}{[CH_3COOH]}$$

$$1.8 \times 10^{-5} = \frac{(x)(x)}{0.100 - x} \approx \frac{x^2}{0.100}$$

$$x = 1.34 \times 10^{-3} \ M = [H_3O^+]$$

$$\mathbf{pH = 2.87}$$

(b)   The number of moles of CH_3COOH originally present in 25.0 mL of solution is:

$$25.0 \text{ mL} \times \frac{0.100 \text{ mol } CH_3COOH}{1000 \text{ mL } CH_3COOH \text{ soln}} = 2.50 \times 10^{-3} \text{ mol}$$

The number of moles of KOH in 5.0 mL is:

$$5.0 \text{ mL} \times \frac{0.200 \text{ mol KOH}}{1000 \text{ mL KOH soln}} = 1.00 \times 10^{-3} \text{ mol}$$

We work with moles at this point because when two solutions are mixed, the solution volume increases. As the solution volume increases, molarity will change, but the number of moles will remain the same. The changes in number of moles are summarized.

	$CH_3COOH(aq)$	+	$KOH(aq)$	$\rightarrow$	$CH_3COOK(aq)$ + $H_2O(l)$
Initial (mol):	2.50×10^{-3}		1.00×10^{-3}		0
Change (mol):	-1.00×10^{-3}		-1.00×10^{-3}		$+1.00 \times 10^{-3}$
Final (mol):	1.50×10^{-3}		0		1.00×10^{-3}

At this stage, we have a buffer system made up of CH_3COOH and CH_3COO^- (from the salt, CH_3COOK). We use the Henderson-Hasselbalch equation to calculate the pH.

$$pH = pK_a + \log \frac{\text{[conjugate base]}}{\text{[acid]}}$$

$$pH = -\log(1.8 \times 10^{-5}) + \log \left(\frac{1.00 \times 10^{-3}}{1.50 \times 10^{-3}} \right)$$

pH = 4.56

(c)   This part is solved similarly to part (b).

The number of moles of KOH in 10.0 mL is:

$$10.0 \text{ mL} \times \frac{0.200 \text{ mol KOH}}{1000 \text{ mL KOH soln}} = 2.00 \times 10^{-3} \text{ mol}$$

The changes in number of moles are summarized.

	$CH_3COOH(aq)$	+	$KOH(aq)$	$\rightarrow$	$CH_3COOK(aq)$ + $H_2O(l)$
Initial (mol):	2.50×10^{-3}		2.00×10^{-3}		0
Change (mol):	-2.00×10^{-3}		-2.00×10^{-3}		$+2.00 \times 10^{-3}$
Final (mol):	0.50×10^{-3}		0		2.00×10^{-3}

At this stage, we have a buffer system made up of CH_3COOH and CH_3COO^- (from the salt, CH_3COOK). We use the Henderson-Hasselbalch equation to calculate the pH.

$$pH = pK_a + \log \frac{\text{[conjugate base]}}{\text{[acid]}}$$

$$pH = -\log(1.8 \times 10^{-5}) + \log \left(\frac{2.00 \times 10^{-3}}{0.50 \times 10^{-3}} \right)$$

pH = 5.34

(d) We have reached the equivalence point of the titration. 2.50×10^{-3} mole of CH_3COOH reacts with 2.50×10^{-3} mole KOH to produce 2.50×10^{-3} mole of CH_3COOK. The only major species present in solution at the equivalence point is the salt, CH_3COOK, which contains the conjugate base, CH_3COO^-. Let's calculate the molarity of CH_3COO^-. The volume of the solution is: (25.0 mL + 12.5 mL = 37.5 mL = 0.0375 L).

$$M(CH_3COO^-) = \frac{2.50 \times 10^{-3} \text{ mol}}{0.0375 \text{ L}} = 0.0667 \ M$$

We set up the hydrolysis of CH_3COO^-, which is a weak base.

$$CH_3COO^-(aq) + H_2O(l) \rightleftharpoons CH_3COOH(aq) + OH^-(aq)$$

	$CH_3COO^-(aq) + H_2O(l) \rightleftharpoons$	$CH_3COOH(aq)$	$OH^-(aq)$
Initial (*M*):	0.0667	0	0
Change (*M*):	$-x$	$+x$	$+x$
Equilibrium (*M*):	$0.0667 - x$	x	x

$$K_b = \frac{[CH_3COOH][OH^-]}{[CH_3COO^-]}$$

$$5.6 \times 10^{-10} = \frac{(x)(x)}{0.0667 - x} \approx \frac{x^2}{0.0667}$$

$$x = 6.1 \times 10^{-6} \ M = [OH^-]$$

$$pOH = 5.21$$

$$\mathbf{pH = 8.79}$$

(e) We have passed the equivalence point of the titration. The excess strong base, KOH, will determine the pH at this point. The moles of KOH in 15.0 mL are:

$$15.0 \text{ mL} \times \frac{0.200 \text{ mol KOH}}{1000 \text{ mL KOH soln}} = 3.00 \times 10^{-3} \text{ mol}$$

The changes in number of moles are summarized.

	$CH_3COOH(aq)$	+	$KOH(aq)$	$\rightarrow$	$CH_3COOK(aq) + H_2O(l)$
Initial (mol):	2.50×10^{-3}		3.00×10^{-3}		0
Change (mol):	-2.50×10^{-3}		-2.50×10^{-3}		$+2.50 \times 10^{-3}$
Final (mol):	0		0.50×10^{-3}		2.50×10^{-3}

Let's calculate the molarity of the KOH in solution. The volume of the solution is now 40.0 mL = 0.0400 L.

$$M(KOH) = \frac{0.50 \times 10^{-3} \text{ mol}}{0.0400 \text{ L}} = 0.0125 \ M$$

KOH is a strong base. The pOH is:

$$pOH = -\log(0.0125) = 1.90$$

$$\mathbf{pH = 12.10}$$

16.32   The reaction between NH_3 and HCl is:

$$NH_3(aq) + HCl(aq) \rightarrow NH_4Cl(aq)$$

We see that 1 mole $NH_3 \simeq 1$ mol HCl. Therefore, at every stage of titration, we can calculate the number of moles of base reacting with acid, and the pH of the solution is determined by the excess base or acid left over. At the equivalence point, however, the neutralization is complete, and the pH of the solution will depend on the extent of the hydrolysis of the salt formed, which is NH_4Cl.

(a)   No HCl has been added. This is a weak base calculation.

$$NH_3(aq) + H_2O(l) \rightleftharpoons NH_4^+(aq) + OH^-(aq)$$

Initial (M):	0.300	0	0
Change (M):	$-x$	$+x$	$+x$
Equilibrium (M):	$0.300 - x$	x	x

$$K_b = \frac{[NH_4^+][OH^-]}{[NH_3]}$$

$$1.8 \times 10^{-5} = \frac{(x)(x)}{0.300 - x} \approx \frac{x^2}{0.300}$$

$$x = 2.3 \times 10^{-3} \, M = [OH^-]$$

$$pOH = 2.64$$

$$\textbf{pH = 11.36}$$

(b)   The number of moles of NH_3 originally present in 10.0 mL of solution is:

$$10.0 \text{ mL} \times \frac{0.300 \text{ mol } NH_3}{1000 \text{ mL } NH_3 \text{ soln}} = 3.00 \times 10^{-3} \text{ mol}$$

The number of moles of HCl in 10.0 mL is:

$$10.0 \text{ mL} \times \frac{0.100 \text{ mol } HCl}{1000 \text{ mL } HCl \text{ soln}} = 1.00 \times 10^{-3} \text{ mol}$$

We work with moles at this point because when two solutions are mixed, the solution volume increases. As the solution volume increases, molarity will change, but the number of moles will remain the same. The changes in number of moles are summarized.

	$NH_3(aq)$	+	$HCl(aq)$	$\rightarrow$	$NH_4Cl(aq)$
Initial (mol):	3.00×10^{-3}		1.00×10^{-3}		0
Change (mol):	-1.00×10^{-3}		-1.00×10^{-3}		$+1.00 \times 10^{-3}$
Final (mol):	2.00×10^{-3}		0		1.00×10^{-3}

At this stage, we have a buffer system made up of NH_3 and NH_4^+ (from the salt, NH_4Cl). We use the Henderson-Hasselbalch equation to calculate the pH.

$$pH = pK_a + \log \frac{[\text{conjugate base}]}{[\text{acid}]}$$

$$pH = -\log(5.6 \times 10^{-10}) + \log\left(\frac{2.00 \times 10^{-3}}{1.00 \times 10^{-3}}\right)$$

pH = 9.55

(c) This part is solved similarly to part (b).

The number of moles of HCl in 20.0 mL is:

$$20.0 \text{ mL} \times \frac{0.100 \text{ mol HCl}}{1000 \text{ mL HCl soln}} = 2.00 \times 10^{-3} \text{ mol}$$

The changes in number of moles are summarized.

	$NH_3(aq)$	+	$HCl(aq)$	$\rightarrow$	$NH_4Cl(aq)$
Initial (mol):	3.00×10^{-3}		2.00×10^{-3}		0
Change (mol):	-2.00×10^{-3}		-2.00×10^{-3}		$+2.00 \times 10^{-3}$
Final (mol):	1.00×10^{-3}		0		2.00×10^{-3}

At this stage, we have a buffer system made up of NH_3 and NH_4^+ (from the salt, NH_4Cl). We use the Henderson-Hasselbalch equation to calculate the pH.

$$pH = pK_a + \log\frac{[\text{conjugate base}]}{[\text{acid}]}$$

$$pH = -\log(5.6 \times 10^{-10}) + \log\left(\frac{1.00 \times 10^{-3}}{2.00 \times 10^{-3}}\right)$$

pH = 8.95

(d) We have reached the equivalence point of the titration. 3.00×10^{-3} mole of NH_3 reacts with 3.00×10^{-3} mole HCl to produce 3.00×10^{-3} mole of NH_4Cl. The only major species present in solution at the equivalence point is the salt, NH_4Cl, which contains the conjugate acid, NH_4^+. Let's calculate the molarity of NH_4^+. The volume of the solution is: (10.0 mL + 30.0 mL = 40.0 mL = 0.0400 L).

$$M(NH_4^+) = \frac{3.00 \times 10^{-3} \text{ mol}}{0.0400 \text{ L}} = 0.0750 \ M$$

We set up the hydrolysis of NH_4^+, which is a weak acid.

	$NH_4^+(aq) + H_2O(l)$	$\rightleftharpoons$	$H_3O^+(aq)$	+	$NH_3(aq)$
Initial (*M*):	0.0750		0		0
Change (*M*):	$-x$		$+x$		$+x$
Equilibrium (*M*):	$0.0750 - x$		x		x

$$K_a = \frac{[H_3O^+][NH_3]}{[NH_4^+]}$$

$$5.6 \times 10^{-10} = \frac{(x)(x)}{0.0750 - x} \approx \frac{x^2}{0.0750}$$

$$x = 6.5 \times 10^{-6}\,M = [\text{H}_3\text{O}^+]$$

pH = 5.19

(e)   We have passed the equivalence point of the titration. The excess strong acid, HCl, will determine the pH at this point. The moles of HCl in 40.0 mL are:

$$40.0\ \text{mL} \times \frac{0.100\ \text{mol HCl}}{1000\ \text{mL HCl soln}} = 4.00 \times 10^{-3}\ \text{mol}$$

The changes in number of moles are summarized.

	$\text{NH}_3(aq)$	+	$\text{HCl}(aq)$	$\rightarrow$	$\text{NH}_4\text{Cl}(aq)$
Initial (mol):	3.00×10^{-3}		4.00×10^{-3}		0
Change (mol):	-3.00×10^{-3}		-3.00×10^{-3}		$+3.00 \times 10^{-3}$
Final (mol):	0		1.00×10^{-3}		3.00×10^{-3}

Let's calculate the molarity of the HCl in solution. The volume of the solution is now 50.0 mL = 0.0500 L.

$$M\ (\text{HCl}) = \frac{1.00 \times 10^{-3}\ \text{mol}}{0.0500\ \text{L}} = 0.0200\ M$$

HCl is a strong acid. The pH is:

pH = −log(0.0200) = 1.70

16.35   **(a)**   HCOOH is a weak acid and NaOH is a strong base. Suitable indicators are cresol red and phenolphthalein.

(b)   HCl is a strong acid and KOH is a strong base. Suitable indicators are all those listed with the exceptions of thymol blue, bromophenol blue, and methyl orange.

(c) HNO_3 is a strong acid and CH_3NH_2 is a weak base. Suitable indicators are bromophenol blue, methyl orange, methyl red, and chlorophenol blue.

16.36 CO_2 in the air dissolves in the solution:

$$\text{CO}_2 + \text{H}_2\text{O} \rightleftharpoons \text{H}_2\text{CO}_3$$

The carbonic acid neutralizes the NaOH.

16.37   The weak acid equilibrium is

$$\text{HIn}(aq) \rightleftharpoons \text{H}^+(aq) + \text{In}^-(aq)$$

We can write a K_a expression for this equilibrium.

$$K_a = \frac{[\text{H}^+][\text{In}^-]}{[\text{HIn}]}$$

Rearranging,

$$\frac{[\text{HIn}]}{[\text{In}^-]} = \frac{[\text{H}^+]}{K_a}$$

From the pH, we can calculate the H^+ concentration.

$$[H^+] = 10^{-pH} = 10^{-4} = 1.0 \times 10^{-4}\ M$$

$$\frac{[HIn]}{[In^-]} = \frac{[H^+]}{K_a} = \frac{1.0 \times 10^{-4}}{1.0 \times 10^{-6}} = 100$$

Since the concentration of HIn is 100 times greater than the concentration of In^-, the color of the solution will be that of HIn, the nonionized formed. The color of the solution will be **red**.

16.38   According to Section 16.5 of the text, when $[HIn] \approx [In^-]$ the indicator color is a mixture of the colors of HIn and In^-. In other words, the indicator color changes at this point. When $[HIn] \approx [In^-]$ we can write:

$$\frac{[In^-]}{[HIn]} = \frac{K_a}{[H^+]} = 1$$

$$[H^+] = K_a = 2.0 \times 10^{-6}$$

$$\mathbf{pH = 5.70}$$

16.45   **(a)**   The solubility equilibrium is given by the equation

$$AgI(s) \rightleftharpoons Ag^+(aq) + I^-(aq)$$

The expression for K_{sp} is given by

$$K_{sp} = [Ag^+][I^-]$$

The value of K_{sp} can be found in Table 16.2 of the text. If the equilibrium concentration of silver ion is the value given, the concentration of iodide ion must be

$$[I^-] = \frac{K_{sp}}{[Ag^+]} = \frac{8.3 \times 10^{-17}}{9.1 \times 10^{-9}} = 9.1 \times 10^{-9}\ M$$

(b)   The value of K_{sp} for aluminum hydroxide can be found in Table 16.2 of the text. The equilibrium expressions are:

$$Al(OH)_3(s) \rightleftharpoons Al^{3+}(aq) + 3OH^-(aq)$$

$$K_{sp} = [Al^{3+}][OH^-]^3$$

Using the given value of the hydroxide ion concentration, the equilibrium concentration of aluminum ion is:

$$[Al^{3+}] = \frac{K_{sp}}{[OH^-]^3} = \frac{1.8 \times 10^{-33}}{(2.9 \times 10^{-9})^3} = 7.4 \times 10^{-8}\ M$$

What is the pH of this solution? Will the aluminum concentration change if the pH is altered?

16.46   **Strategy:** In each part, we can calculate the number of moles of compound dissolved in one liter of solution (the molar solubility). Then, from the molar solubility, s, we can determine K_{sp}.

Solution:

(a) $\dfrac{7.3 \times 10^{-2}\ \text{g SrF}_2}{1\ \text{L soln}} \times \dfrac{1\ \text{mol SrF}_2}{125.6\ \text{g SrF}_2} = 5.8 \times 10^{-4}\ \text{mol/L} = s$

Consider the dissociation of SrF_2 in water. Let s be the molar solubility of SrF_2.

$$SrF_2(s) \rightleftharpoons Sr^{2+}(aq) + 2F^-(aq)$$

Initial (M):		0	0
Change (M):	$-s$	$+s$	$+2s$
Equilibrium (M):		s	$2s$

$$K_{sp} = [Sr^{2+}][F^-]^2 = (s)(2s)^2 = 4s^3$$

The molar solubility (s) was calculated above. Substitute into the equilibrium constant expression to solve for K_{sp}.

$$K_{sp} = [Sr^{2+}][F^-]^2 = 4s^3 = 4(5.8 \times 10^{-4})^3 = \mathbf{7.8 \times 10^{-10}}$$

(b) $\dfrac{6.7 \times 10^{-3}\ \text{g Ag}_3\text{PO}_4}{1\ \text{L soln}} \times \dfrac{1\ \text{mol Ag}_3\text{PO}_4}{418.7\ \text{g Ag}_3\text{PO}_4} = 1.6 \times 10^{-5}\ \text{mol/L} = s$

(b) is solved in a similar manner to (a)

The equilibrium equation is:

$$Ag_3PO_4(s) \rightleftharpoons 3Ag^+(aq) + PO_4^{3-}(aq)$$

Initial (M):		0	0
Change (M):	$-s$	$+3s$	$+s$
Equilibrium (M):		$3s$	s

$$K_{sp} = [Ag^+]^3[PO_4^{3-}] = (3s)^3(s) = 27s^4 = 27(1.6 \times 10^{-5})^4 = \mathbf{1.8 \times 10^{-18}}$$

16.47 For $MnCO_3$ dissolving, we write

$$MnCO_3(s) \rightleftharpoons Mn^{2+}(aq) + CO_3^{2-}(aq)$$

For every mole of $MnCO_3$ that dissolves, one mole of Mn^{2+} will be produced and one mole of CO_3^{2-} will be produced. If the molar solubility of $MnCO_3$ is s mol/L, then the concentrations of Mn^{2+} and CO_3^{2-} are:

$$[Mn^{2+}] = [CO_3^{2-}] = s = 4.2 \times 10^{-6}\ M$$

$$K_{sp} = [Mn^{2+}][CO_3^{2-}] = s^2 = (4.2 \times 10^{-6})^2 = \mathbf{1.8 \times 10^{-11}}$$

16.48 First, we can convert the solubility of MX in g/L to mol/L.

$$\dfrac{4.63 \times 10^{-3}\ \text{g MX}}{1\ \text{L soln}} \times \dfrac{1\ \text{mol MX}}{346\ \text{g MX}} = 1.34 \times 10^{-5}\ \text{mol/L} = s\ \text{(molar solubility)}$$

The equilibrium reaction is:

$$MX(s) \rightleftharpoons M^{n+}(aq) + X^{n-}(aq)$$

Initial (M):		0	0
Change (M):	$-s$	$+s$	$+s$
Equilibrium (M):		s	s

$$K_{sp} = [M^{n+}][X^{n-}] = s^2 = (1.34 \times 10^{-5})^2 = \mathbf{1.80 \times 10^{-10}}$$

16.49 The charges of the M and X ions are +3 and −2, respectively (are other values possible?). We first calculate the number of moles of M_2X_3 that dissolve in 1.0 L of water

$$\text{Moles } M_2X_3 = (3.6 \times 10^{-17} \text{ g}) \times \frac{1 \text{ mol}}{288 \text{ g}} = 1.3 \times 10^{-19} \text{ mol}$$

The molar solubility, s, of the compound is therefore 1.3×10^{-19} M. At equilibrium the concentration of M^{3+} must be $2s$ and that of X^{2-} must be $3s$. (See Table 16.3 of the text.)

$$K_{sp} = [M^{3+}]^2[X^{2-}]^3 = [2s]^2[3s]^3 = 108s^5$$

Since these are equilibrium concentrations, the value of K_{sp} can be found by simple substitution

$$K_{sp} = 108s^5 = 108(1.3 \times 10^{-19})^5 = \mathbf{4.0 \times 10^{-93}}$$

16.50 **Strategy:** We can look up the K_{sp} value of CaF_2 in Table 16.2 of the text. Then, setting up the dissociation equilibrium of CaF_2 in water, we can solve for the molar solubility, s.

Solution: Consider the dissociation of CaF_2 in water.

$$CaF_2(s) \rightleftharpoons Ca^{2+}(aq) + 2F^-(aq)$$

Initial (M):		0	0
Change (M):	$-s$	$+s$	$+2s$
Equilibrium (M):		s	$2s$

Recall, that the concentration of a pure solid does not enter into an equilibrium constant expression. Therefore, the concentration of CaF_2 is not important.

Substitute the value of K_{sp} and the concentrations of Ca^{2+} and F^- in terms of s into the solubility product expression to solve for s, the molar solubility.

$$K_{sp} = [Ca^{2+}][F^-]^2$$
$$4.0 \times 10^{-11} = (s)(2s)^2$$
$$4.0 \times 10^{-11} = 4s^3$$
$$s = \text{molar solubility} = \mathbf{2.2 \times 10^{-4} \text{ mol/L}}$$

The molar solubility indicates that 2.2×10^{-4} mol of CaF_2 will dissolve in 1 L of an aqueous solution.

16.51   Let s be the molar solubility of $Zn(OH)_2$. The equilibrium concentrations of the ions are then

$$[Zn^{2+}] = s \text{ and } [OH^-] = 2s$$

$$K_{sp} = [Zn^{2+}][OH^-]^2 = (s)(2s)^2 = 4s^3 = 1.8 \times 10^{-14}$$

$$s = \left(\frac{1.8 \times 10^{-14}}{4}\right)^{\frac{1}{3}} = 1.7 \times 10^{-5}$$

$$[OH^-] = 2s = 3.4 \times 10^{-5} \, M \text{ and pOH} = 4.47$$

$$\mathbf{pH} = 14.00 - 4.47 = \mathbf{9.53}$$

If the K_{sp} of $Zn(OH)_2$ were smaller by many more powers of ten, would $2s$ still be the hydroxide ion concentration in the solution?

16.52   First we can calculate the OH^- concentration from the pH.

$$pOH = 14.00 - pH$$

$$pOH = 14.00 - 9.68 = 4.32$$

$$[OH^-] = 10^{-pOH} = 10^{-4.32} = 4.8 \times 10^{-5} \, M$$

The equilibrium equation is:

$$MOH(s) \rightleftharpoons M^+(aq) + OH^-(aq)$$

From the balanced equation we know that $[M^+] = [OH^-]$

$$\mathbf{\mathit{K}_{sp}} = [M^+][OH^-] = (4.8 \times 10^{-5})^2 = \mathbf{2.3 \times 10^{-9}}$$

16.53   According to the solubility rules, the only precipitate that might form is $BaCO_3$.

$$Ba^{2+}(aq) + CO_3^{2-}(aq) \rightarrow BaCO_3(s)$$

The number of moles of Ba^{2+} present in the original 20.0 mL of $Ba(NO_3)_2$ solution is

$$20.0 \text{ mL} \times \frac{0.10 \text{ mol } Ba^{2+}}{1000 \text{ mL soln}} = 2.0 \times 10^{-3} \text{ mol } Ba^{2+}$$

The total volume after combining the two solutions is 70.0 mL. The concentration of Ba^{2+} in 70 mL is

$$[Ba^{2+}] = \frac{2.0 \times 10^{-3} \text{ mol } Ba^{2+}}{70.0 \times 10^{-3} \text{ L}} = 2.9 \times 10^{-2} \, M$$

The number of moles of CO_3^{2-} present in the original 50.0 mL Na_2CO_3 solution is

$$50.0 \text{ mL} \times \frac{0.10 \text{ mol } CO_3^{2-}}{1000 \text{ mL soln}} = 5.0 \times 10^{-3} \text{ mol } CO_3^{2-}$$

The concentration of CO_3^{2-} in the 70.0 mL of combined solution is

$$[CO_3^{2-}] = \frac{5.0 \times 10^{-3} \text{ mol } CO_3^{2-}}{70.0 \times 10^{-3} \text{ L}} = 7.1 \times 10^{-2} \, M$$

Now we must compare Q and K_{sp}. From Table 16.2 of the text, the K_{sp} for $BaCO_3$ is 8.1×10^{-9}. As for Q,

$$Q = [Ba^{2+}]_0[CO_3^{2-}]_0 = (2.9 \times 10^{-2})(7.1 \times 10^{-2}) = 2.1 \times 10^{-3}$$

Since $(2.1 \times 10^{-3}) > (8.1 \times 10^{-9})$, then $Q > K_{sp}$. Therefore, $BaCO_3$ will precipitate.

16.54 The net ionic equation is:

$$Sr^{2+}(aq) + 2F^-(aq) \longrightarrow SrF_2(s)$$

Let's find the limiting reagent in the precipitation reaction.

$$\text{Moles } F^- = 75 \text{ mL} \times \frac{0.060 \text{ mol}}{1000 \text{ mL soln}} = 0.0045 \text{ mol}$$

$$\text{Moles } Sr^{2+} = 25 \text{ mL} \times \frac{0.15 \text{ mol}}{1000 \text{ mL soln}} = 0.0038 \text{ mol}$$

From the stoichiometry of the balanced equation, twice as many moles of F^- are required to react with Sr^{2+}. This would require 0.0076 mol of F^-, but we only have 0.0045 mol. Thus, F^- is the limiting reagent.

Let's assume that the above reaction goes to completion. Then, we will consider the equilibrium that is established when SrF_2 partially dissociates into ions.

	$Sr^{2+}(aq)$	+ 2 $F^-(aq)$	$\longrightarrow$ $SrF_2(s)$
Initial (mol):	0.0038	0.0045	0
Change (mol):	−0.00225	−0.0045	+0.00225
Final (mol):	0.00155	0	0.00225

Now, let's establish the equilibrium reaction. The total volume of the solution is 100 mL = 0.100 L. Divide the above moles by 0.100 L to convert to molar concentration.

	$SrF_2(s)$	$\rightleftharpoons$ $Sr^{2+}(aq)$	+ 2$F^-(aq)$
Initial (M):	0.0225	0.0155	0
Change (M):	−s	+s	+2s
Equilibrium (M):	0.0225 − s	0.0155 + s	2s

Write the solubility product expression, then solve for s.

$$K_{sp} = [Sr^{2+}][F^-]^2$$

$$2.0 \times 10^{-10} = (0.0155 + s)(2s)^2 \approx (0.0155)(2s)^2$$

$$s = 5.7 \times 10^{-5} \, M$$

$$[F^-] = 2s = \mathbf{1.1 \times 10^{-4} \, M}$$

$$[Sr^{2+}] = 0.0155 + s = \mathbf{0.016 \, M}$$

Both sodium ions and nitrate ions are spectator ions and therefore do not enter into the precipitation reaction.

$$[NO_3^-] = \frac{2(0.0038)\,mol}{0.10\ L} = 0.076\ M$$

$$[Na^+] = \frac{0.0045\ mol}{0.10\ L} = 0.045\ M$$

16.55 (a) The solubility product expressions for both substances have exactly the same mathematical form and are therefore directly comparable. The substance having the smaller K_{sp} (**AgI**) will precipitate first. (Why?)

(b) When CuI just begins to precipitate the solubility product expression will just equal K_{sp} (saturated solution). The concentration of Cu^+ at this point is 0.010 M (given in the problem), so the concentration of iodide ion must be:

$$K_{sp} = [Cu^+][I^-] = (0.010)[I^-] = 5.1 \times 10^{-12}$$

$$[I^-] = \frac{5.1 \times 10^{-12}}{0.010} = 5.1 \times 10^{-10}\ M$$

Using this value of $[I^-]$, we find the silver ion concentration

$$[Ag^+] = \frac{K_{sp}}{[I^-]} = \frac{8.3 \times 10^{-17}}{5.1 \times 10^{-10}} = 1.6 \times 10^{-7}\ M$$

(c) The percent of silver ion remaining in solution is:

$$\%\ Ag^+(aq) = \frac{1.6 \times 10^{-7}\ M}{0.010\ M} \times 100\% = 0.0016\%\ \text{or}\ 1.6 \times 10^{-3}\%$$

Is this an effective way to separate silver from copper?

16.56 For Fe(OH)$_3$, $K_{sp} = 1.1 \times 10^{-36}$. When $[Fe^{3+}] = 0.010\ M$, the $[OH^-]$ value is:

$$K_{sp} = [Fe^{3+}][OH^-]^3$$

or

$$[OH^-] = \left(\frac{K_{sp}}{[Fe^{3+}]}\right)^{\frac{1}{3}}$$

$$[OH^-] = \left(\frac{1.1 \times 10^{-36}}{0.010}\right)^{\frac{1}{3}} = 4.8 \times 10^{-12}\ M$$

This $[OH^-]$ corresponds to a pH of 2.68. In other words, Fe(OH)$_3$ will begin to precipitate from this solution at pH of 2.68.

For Zn(OH)$_2$, $K_{sp} = 1.8 \times 10^{-14}$. When $[Zn^{2+}] = 0.010\ M$, the $[OH^-]$ value is:

$$[OH^-] = \left(\frac{K_{sp}}{[Zn^{2+}]}\right)^{\frac{1}{2}}$$

$$[OH^-] = \left(\frac{1.8 \times 10^{-14}}{0.010}\right)^{\frac{1}{2}} = 1.3 \times 10^{-6} \, M$$

This corresponds to a pH of 8.11. In other words $Zn(OH)_2$ will begin to precipitate from the solution at pH = 8.11. These results show that $Fe(OH)_3$ will precipitate when the pH just exceeds 2.68 and that $Zn(OH)_2$ will precipitate when the pH just exceeds 8.11. Therefore, to selectively remove iron as $Fe(OH)_3$, the pH must be *greater than* **2.68** but *less than* **8.11**.

16.59   First let s be the molar solubility of $CaCO_3$ in this solution.

$$CaCO_3(s) \rightleftharpoons Ca^{2+}(aq) + CO_3^{2-}(aq)$$

Initial (*M*):		0.050	0
Change (*M*):	$-s$	$+s$	$+s$
Equilibrium (*M*):		$(0.050 + s)$	s

$$K_{sp} = [Ca^{2+}][CO_3^{2-}] = (0.050 + s)s = 8.7 \times 10^{-9}$$

We can assume $0.050 + s \approx 0.050$, then

$$s = \frac{8.7 \times 10^{-9}}{0.050} = 1.7 \times 10^{-7} \, M$$

The mass of $CaCO_3$ can then be found.

$$(3.0 \times 10^2 \, \text{mL}) \times \frac{1.7 \times 10^{-7} \, \text{mol}}{1000 \, \text{mL soln}} \times \frac{100.1 \, \text{g CaCO}_3}{1 \, \text{mol}} = \textbf{5.1} \times \textbf{10}^{-6} \, \textbf{g CaCO}_3$$

16.60   **Strategy:** In parts (b) and (c), this is a common-ion problem. In part (b), the common ion is Br^-, which is supplied by both $PbBr_2$ and KBr. Remember that the presence of a common ion will affect only the solubility of $PbBr_2$, but not the K_{sp} value because it is an equilibrium constant. In part (c), the common ion is Pb^{2+}, which is supplied by both $PbBr_2$ and $Pb(NO_3)_2$.

Solution:

(a)   Set up a table to find the equilibrium concentrations in pure water.

$$PbBr_2(s) \rightleftharpoons Pb^{2+}(aq) + 2Br^-(aq)$$

Initial (*M*)		0	0
Change (*M*)	$-s$	$+s$	$+2s$
Equilibrium (*M*)		s	$2s$

$$K_{sp} = [Pb^{2+}][Br^-]^2$$
$$8.9 \times 10^{-6} = (s)(2s)^2$$
$$s = \text{molar solubility} = \textbf{0.013} \, \textbf{M}$$

(b)   Set up a table to find the equilibrium concentrations in 0.20 *M* KBr. KBr is a soluble salt that ionizes completely giving an initial concentration of $Br^- = 0.20 \, M$.

$$PbBr_2(s) \rightleftharpoons Pb^{2+}(aq) + 2Br^-(aq)$$

Initial (M)		0	0.20
Change (M)	$-s$	$+s$	$+2s$
Equilibrium (M)		s	$0.20 + 2s$

$$K_{sp} = [Pb^{2+}][Br^-]^2$$

$$8.9 \times 10^{-6} = (s)(0.20 + 2s)^2$$

$$8.9 \times 10^{-6} \approx (s)(0.20)^2$$

$$s = \text{molar solubility} = \mathbf{2.2 \times 10^{-4}}\ \boldsymbol{M}$$

Thus, the molar solubility of $PbBr_2$ is reduced from 0.013 M to 2.2×10^{-4} M as a result of the common ion (Br^-) effect.

(c)   Set up a table to find the equilibrium concentrations in 0.20 M $Pb(NO_3)_2$. $Pb(NO_3)_2$ is a soluble salt that dissociates completely giving an initial concentration of $[Pb^{2+}] = 0.20\ M$.

$$PbBr_2(s) \rightleftharpoons Pb^{2+}(aq) + 2Br^-(aq)$$

Initial (M):	0.20	0	
Change (M):	$-s$	$+s$	$+2s$
Equilibrium (M):		$0.20 + s$	$2s$

$$K_{sp} = [Pb^{2+}][Br^-]^2$$

$$8.9 \times 10^{-6} = (0.20 + s)(2s)^2$$

$$8.9 \times 10^{-6} \approx (0.20)(2s)^2$$

$$s = \text{molar solubility} = \mathbf{3.3 \times 10^{-3}}\ \boldsymbol{M}$$

Thus, the molar solubility of $PbBr_2$ is reduced from 0.013 M to 3.3×10^{-3} M as a result of the common ion (Pb^{2+}) effect.

Check: You should also be able to predict the decrease in solubility due to a common-ion using Le Châtelier's principle. Adding Br^- or Pb^{2+} ions shifts the system to the left, thus decreasing the solubility of $PbBr_2$.

16.61   We first calculate the concentration of chloride ion in the solution.

$$[Cl^-] = \frac{10.0\ \text{g } CaCl_2}{1\ \text{L soln}} \times \frac{1\ \text{mol } CaCl_2}{111.0\ \text{g } CaCl_2} \times \frac{2\ \text{mol } Cl^-}{1\ \text{mol } CaCl_2} = 0.180\ M$$

$$AgCl(s) \rightleftharpoons Ag^+(aq) + Cl^-(aq)$$

Initial (M):		0.000	0.180
Change (M):	$-s$	$+s$	$+s$
Equilibrium (M):		s	$(0.180 + s)$

If we assume that $(0.180 + s) \approx 0.180$, then

$$K_{sp} = [Ag^+][Cl^-] = 1.6 \times 10^{-10}$$

$$[Ag^+] = \frac{K_{sp}}{[Cl^-]} = \frac{1.6 \times 10^{-10}}{0.180} = 8.9 \times 10^{-10} \, M = s$$

The molar solubility of AgCl is $8.9 \times 10^{-10} \, M$.

16.62 **(a)** The equilibrium reaction is:

$$BaSO_4(s) \rightleftharpoons Ba^{2+}(aq) + SO_4^{2-}(aq)$$

Initial (*M*):		0	0
Change (*M*):	−s	+s	+s
Equilibrium (*M*):		s	s

$$K_{sp} = [Ba^{2+}][SO_4^{2-}]$$
$$1.1 \times 10^{-10} = s^2$$
$$s = 1.0 \times 10^{-5} \, M$$

The molar solubility of $BaSO_4$ in pure water is 1.0×10^{-5} mol/L.

(b) The initial concentration of SO_4^{2-} is 1.0 *M*.

$$BaSO_4(s) \rightleftharpoons Ba^{2+}(aq) + SO_4^{2-}(aq)$$

Initial (*M*):		0	1.0
Change (*M*):	−s	+s	+s
Equilibrium (*M*):		s	1.0 + s

$$K_{sp} = [Ba^{2+}][SO_4^{2-}]$$
$$1.1 \times 10^{-10} = (s)(1.0 + s) \approx (s)(1.0)$$
$$s = 1.1 \times 10^{-10} \, M$$

Due to the common ion effect, the molar solubility of $BaSO_4$ decreases to 1.1×10^{-10} mol/L in $1.0 \, M \, SO_4^{2-}(aq)$ compared to 1.0×10^{-5} mol/L in pure water.

16.63 When the anion of a salt is a base, the salt will be more soluble in acidic solution because the hydrogen ion decreases the concentration of the anion (Le Chatelier's principle):

$$B^-(aq) + H^+(aq) \rightleftharpoons HB(aq)$$

(a) $BaSO_4$ will be slightly more soluble because SO_4^{2-} is a base (although a weak one).

(b) The solubility of $PbCl_2$ in acid is unchanged over the solubility in pure water because HCl is a strong acid, and therefore Cl^- is a negligibly weak base.

(c) $Fe(OH)_3$ will be more soluble in acid because OH^- is a base.

(d) $CaCO_3$ will be more soluble in acidic solution because the CO_3^{2-} ions react with H^+ ions to form CO_2 and H_2O. The CO_2 escapes from the solution, shifting the equilibrium. Although it is not important in this case, the carbonate ion is also a base.

16.64 **(b)** $SO_4^{2-}(aq)$ is a weak base

(c) $OH^-(aq)$ is a strong base

(d) $C_2O_4^{2-}(aq)$ is a weak base

(e) $PO_4^{3-}(aq)$ is a weak base.

The solubilities of the above will increase in acidic solution. Only (a), which contains an extremely weak base (I^- is the conjugate base of the strong acid HI) is unaffected by the acid solution.

16.65   In water:

$$Mg(OH)_2 \rightleftharpoons Mg^{2+} + 2OH^-$$
$$\quad\quad\quad\quad\quad\quad s \quad\quad\quad 2s$$

$$K_{sp} = 4s^3 = 1.2 \times 10^{-11}$$

$$s = 1.4 \times 10^{-4}\ M$$

In a buffer at pH $= 9.0$

$$[H^+] = 1.0 \times 10^{-9}$$

$$[OH^-] = 1.0 \times 10^{-5}$$

$$1.2 \times 10^{-11} = (s)(1.0 \times 10^{-5})^2$$

$$s = 0.12\ M$$

16.66   From Table 16.2, the value of K_{sp} for iron(II) is 1.6×10^{-14}.

(a)   At pH = 8.00, pOH = 14.00 − 8.00 = 6.00, and $[OH^-] = 1.0 \times 10^{-6}\ M$

$$[Fe^{2+}] = \frac{K_{sp}}{[OH^-]^2} = \frac{1.6 \times 10^{-14}}{(1.0 \times 10^{-6})^2} = 0.016\ M$$

The *molar solubility* of iron(II) hydroxide at pH = 8.00 is **0.016 *M***

(b)   At pH = 10.00, pOH = 14.00 − 10.00 = 4.00, and $[OH^-] = 1.0 \times 10^{-4}\ M$

$$[Fe^{2+}] = \frac{K_{sp}}{[OH^-]^2} = \frac{1.6 \times 10^{-14}}{(1.0 \times 10^{-4})^2} = 1.6 \times 10^{-6}\ M$$

The *molar solubility* of iron(II) hydroxide at pH = 10.00 is **1.6×10^{-6} *M***.

16.67   The solubility product expression for magnesium hydroxide is

$$K_{sp} = [Mg^{2+}][OH^-]^2 = 1.2 \times 10^{-11}$$

We find the hydroxide ion concentration when $[Mg^{2+}]$ is $1.0 \times 10^{-10}\ M$

$$[OH^-] = \left(\frac{1.2 \times 10^{-11}}{1.0 \times 10^{-10}} \right)^{\frac{1}{2}} = 0.35\ M$$

Therefore the concentration of OH^- must be slightly greater than 0.35 M.

16.68 We first determine the effect of the added ammonia. Let's calculate the concentration of NH_3. This is a dilution problem.

$$M_iV_i = M_fV_f$$
$$(0.60\ M)(2.00\ \text{mL}) = M_f(1002\ \text{mL})$$
$$M_f = 0.0012\ M\ NH_3$$

Ammonia is a weak base ($K_b = 1.8 \times 10^{-5}$).

	$NH_3 + H_2O$	$\rightleftharpoons$	NH_4^+	$+$	OH^-
Initial (M):	0.0012		0		0
Change (M):	$-x$		$+x$		$+x$
Equil. (M):	$0.0012 - x$		x		x

$$K_b = \frac{[NH_4^+][OH^-]}{[NH_3]}$$

$$1.8 \times 10^{-5} = \frac{x^2}{(0.0012 - x)}$$

Solving the resulting quadratic equation gives $x = 0.00014$, or $[OH^-] = 0.00014\ M$

This is a solution of iron(II) sulfate, which contains Fe^{2+} ions. These Fe^{2+} ions could combine with OH^- to precipitate $Fe(OH)_2$. Therefore, we must use K_{sp} for iron(II) hydroxide. We compute the value of Q_c for this solution.

$$Fe(OH)_2(s) \rightleftharpoons Fe^{2+}(aq) + 2OH^-(aq)$$

$$Q = [Fe^{2+}]_0[OH^-]_0^2 = (1.0 \times 10^{-3})(0.00014)^2 = 2.0 \times 10^{-11}$$

Q is larger than K_{sp} $[Fe(OH)_2] = 1.6 \times 10^{-14}$. The concentrations of the ions in solution are greater than the equilibrium concentrations; the solution is saturated. The system will shift left to reestablish equilibrium; therefore, **a precipitate of $Fe(OH)_2$ will form.**

16.71 First find the molarity of the copper(II) ion

$$\text{Moles } CuSO_4 = 2.50\ \text{g} \times \frac{1\ \text{mol}}{159.6\ \text{g}} = 0.0157\ \text{mol}$$

$$[Cu^{2+}] = \frac{0.0157\ \text{mol}}{0.90\ \text{L}} = 0.0174\ M$$

As in Example 16.15 of the text, the position of equilibrium will be far to the right. We assume essentially all the copper ion is complexed with NH_3. The NH_3 consumed is $4 \times 0.0174\ M = 0.0696\ M$. The uncombined NH_3 remaining is $(0.30 - 0.0696)\ M$, or 0.23 M. The equilibrium concentrations of $Cu(NH_3)_4^{2+}$ and NH_3 are therefore **0.0174 M** and **0.23 M**, respectively. We find $[Cu^{2+}]$ from the formation constant expression.

$$K_f = \frac{[Cu(NH_3)_4^{2+}]}{[Cu^{2+}][NH_3]^4} = 5.0 \times 10^{13} = \frac{0.0174}{[Cu^{2+}](0.23)^4}$$

$$[Cu^{2+}] = 1.2 \times 10^{-13}\ M$$

16.72   **Strategy:** The addition of $Cd(NO_3)_2$ to the NaCN solution results in complex ion formation. In solution, Cd^{2+} ions will complex with CN^- ions. The concentration of Cd^{2+} will be determined by the following equilibrium

$$Cd^{2+}(aq) + 4CN^-(aq) \rightleftharpoons Cd(CN)_4^{2-}$$

From Table 16.4 of the text, we see that the formation constant (K_f) for this reaction is very large ($K_f = 7.1 \times 10^{16}$). Because K_f is so large, the reaction lies mostly to the right. At equilibrium, the concentration of Cd^{2+} will be very small. As a good approximation, we can assume that essentially all the dissolved Cd^{2+} ions end up as $Cd(CN)_4^{2-}$ ions. What is the initial concentration of Cd^{2+} ions? A very small amount of Cd^{2+} will be present at equilibrium. Set up the K_f expression for the above equilibrium to solve for $[Cd^{2+}]$.

Solution: Calculate the initial concentration of Cd^{2+} ions.

$$[Cd^{2+}]_0 = \frac{0.50\ g \times \dfrac{1\ mol\ Cd(NO_3)_2}{236.42\ g\ Cd(NO_3)_2} \times \dfrac{1\ mol\ Cd^{2+}}{1\ mol\ Cd(NO_3)_2}}{0.50\ L} = 4.2 \times 10^{-3}\ M$$

If we assume that the above equilibrium goes to completion, we can write

	$Cd^{2+}(aq)$	$+$	$4CN^-(aq)$	$\longrightarrow$	$Cd(CN)_4^{2-}(aq)$
Initial (M):	4.2×10^{-3}		0.50		0
Change (M):	-4.2×10^{-3}		$-4(4.2 \times 10^{-3})$		$+4.2 \times 10^{-3}$
Final (M):	0		0.48		4.2×10^{-3}

To find the concentration of free Cd^{2+} at equilibrium, use the formation constant expression.

$$K_f = \frac{[Cd(CN)_4^{2-}]}{[Cd^{2+}][CN^-]^4}$$

Rearranging,

$$[Cd^{2+}] = \frac{[Cd(CN)_4^{2-}]}{K_f[CN^-]^4}$$

Substitute the equilibrium concentrations calculated above into the formation constant expression to calculate the equilibrium concentration of Cd^{2+}.

$$[Cd^{2+}] = \frac{[Cd(CN)_4^{2-}]}{K_f[CN^-]^4} = \frac{4.2 \times 10^{-3}}{(7.1 \times 10^{16})(0.48)^4} = 1.1 \times 10^{-18}\ M$$

$$[CN^-] = 0.48\ M + 4(1.1 \times 10^{-18}\ M) = 0.48\ M$$

$$[Cd(CN)_4^{2-}] = (4.2 \times 10^{-3}\ M) - (1.1 \times 10^{-18}) = 4.2 \times 10^{-3}\ M$$

Check: Substitute the equilibrium concentrations calculated into the formation constant expression to calculate K_f. Also, the small value of $[Cd^{2+}]$ at equilibrium, compared to its initial concentration of 4.2×10^{-3} M, certainly justifies our approximation that almost all the Cd^{2+} ions react.

16.73   The reaction

$$Al(OH)_3(s) + OH^-(aq) \;\rightleftharpoons\; Al(OH)_4^-(aq)$$

is the sum of the two known reactions

$$Al(OH)_3(s) \;\rightleftharpoons\; Al^{3+}(aq) + 3OH^-(aq) \qquad\qquad K_{sp} = 1.8 \times 10^{-33}$$

$$Al^{3+}(aq) + 4OH^-(aq) \;\rightleftharpoons\; Al(OH)_4^-(aq) \qquad\qquad K_f = 2.0 \times 10^{33}$$

The equilibrium constant is

$$K = K_{sp}K_f = (1.8 \times 10^{-33})(2.0 \times 10^{33}) = 3.6 = \frac{[Al(OH)_4^-]}{[OH^-]}$$

When pH = 14.00, $[OH^-] = 1.0$ M, therefore

$$[Al(OH)_4^-] = K[OH^-] = (3.6)(1\ M) = 3.6\ M$$

This represents the maximum possible concentration of the complex ion at pH 14.00. Since this is much larger than the initial 0.010 M, the complex ion will be the predominant species.

16.74   Silver iodide is only slightly soluble. It dissociates to form a small amount of Ag^+ and I^- ions. The Ag^+ ions then complex with NH_3 in solution to form the complex ion $Ag(NH_3)_2^+$. The balanced equations are:

$$AgI(s) \;\rightleftharpoons\; Ag^+(aq) + I^-(aq) \qquad\qquad K_{sp} = [Ag^+][I^-] = 8.3 \times 10^{-17}$$

$$Ag^+(aq) + 2NH_3(aq) \;\rightleftharpoons\; Ag(NH_3)_2^+(aq) \qquad\qquad K_f = \frac{[Ag(NH_3)_2^+]}{[Ag^+][NH_3]^2} = 1.5 \times 10^7$$

Overall: $AgI(s) + 2NH_3(aq) \;\rightleftharpoons\; Ag(NH_3)_2^+(aq) + I^-(aq)$ $\qquad K = K_{sp} \times K_f = 1.2 \times 10^{-9}$

If s is the molar solubility of AgI then,

	AgI(s)	+	2NH_3(aq)	$\rightleftharpoons$	Ag(NH_3)_2^+(aq)	+	I^-(aq)
Initial (M):			1.0		0.0		0.0
Change (M):	−s		−2s		+s		+s
Equilibrium (M):			(1.0 − 2s)		s		s

Because K_f is large, we can assume all of the silver ions exist as $Ag(NH_3)_2^+$. Thus,

$$[Ag(NH_3)_2^+] = [I^-] = s$$

We can write the equilibrium constant expression for the above reaction, then solve for s.

$$K = 1.2 \times 10^{-9} = \frac{(s)(s)}{(1.0 - 2s)^2} \approx \frac{(s)(s)}{(1.0)^2}$$

$$s = 3.5 \times 10^{-5}\ M$$

At equilibrium, 3.5×10^{-5} moles of AgI dissolves in 1 L of 1.0 M NH$_3$ solution.

16.75   The balanced equations are:

$$Ag^+(aq) + 2NH_3(aq) \rightleftharpoons Ag(NH_3)_2^+(aq)$$
$$Zn^{2+}(aq) + 4NH_3(aq) \rightleftharpoons Zn(NH_3)_4^{2+}(aq)$$

Zinc hydroxide forms a complex ion with excess OH$^-$ and silver hydroxide does not; therefore, zinc hydroxide is soluble in 6 M NaOH.

16.76   **(a)**   The equations are as follows:

$$CuI_2(s) \rightleftharpoons Cu^{2+}(aq) + 2I^-(aq)$$

$$\mathbf{Cu^{2+}(aq) + 4NH_3(aq) \rightleftharpoons [Cu(NH_3)_4]^{2+}(aq)}$$

The ammonia combines with the Cu^{2+} ions formed in the first step to form the complex ion [Cu(NH$_3$)$_4$]$^{2+}$, effectively removing the Cu^{2+} ions, causing the first equilibrium to shift to the right (resulting in more CuI$_2$ dissolving).

(b)   Similar to part (a):

$$AgBr(s) \rightleftharpoons Ag^+(aq) + Br^-(aq)$$

$$\mathbf{Ag^+(aq) + 2CN^-(aq) \rightleftharpoons [Ag(CN)_2]^-(aq)}$$

(c)   Similar to parts (a) and (b).

$$HgCl_2(s) \rightleftharpoons Hg^{2+}(aq) + 2Cl^-(aq)$$

$$\mathbf{Hg^{2+}(aq) + 4Cl^-(aq) \rightleftharpoons [HgCl_4]^{2-}(aq)}$$

16.79   Silver chloride will dissolve in aqueous ammonia because of the formation of a complex ion. Lead chloride will not dissolve; it doesn't form an ammonia complex.

16.80   Since some PbCl$_2$ precipitates, the solution is saturated. From Table 16.2, the value of K_{sp} for lead(II) chloride is 2.4×10^{-4}. The equilibrium is:

$$PbCl_2(aq) \rightleftharpoons Pb^{2+}(aq) + 2Cl^-(aq)$$

We can write the solubility product expression for the equilibrium.

$$K_{sp} = [Pb^{2+}][Cl^-]^2$$

K_{sp} and $[Cl^-]$ are known. Solving for the Pb^{2+} concentration,

$$[Pb^{2+}] = \frac{K_{sp}}{[Cl^-]^2} = \frac{2.4 \times 10^{-4}}{(0.15)^2} = 0.011 \ M$$

16.81   Ammonium chloride is the salt of a weak base (ammonia). It will react with strong aqueous hydroxide to form ammonia (Le Châtelier's principle).

$$NH_4Cl(s) + OH^-(aq) \rightarrow NH_3(g) + H_2O(l) + Cl^-(aq)$$

The human nose is an excellent ammonia detector. Nothing happens between KCl and strong aqueous NaOH.

16.82   Chloride ion will precipitate Ag^+ but not Cu^{2+}. So, dissolve some solid in H_2O and add HCl. If a precipitate forms, the salt was $AgNO_3$. A flame test will also work. Cu^{2+} gives a green flame test.

16.83   According to the Henderson-Hasselbalch equation:

$$pH = pK_a + \log \frac{[\text{conjugate base}]}{[\text{acid}]}$$

If: $\dfrac{[\text{conjugate base}]}{[\text{acid}]} = 10$, then:

$$pH = pK_a + 1$$

If: $\dfrac{[\text{conjugate base}]}{[\text{acid}]} = 0.1$, then:

$$pH = pK_a - 1$$

Therefore, the range of the ratio is:

$$0.1 < \frac{[\text{conjugate base}]}{[\text{acid}]} < 10$$

16.84   We can use the Henderson-Hasselbalch equation to solve for the pH when the indicator is 90% acid / 10% conjugate base and when the indicator is 10% acid / 90% conjugate base.

$$pH = pK_a + \log \frac{[\text{conjugate base}]}{[\text{acid}]}$$

Solving for the pH with 90% of the indicator in the HIn form:

$$pH = 3.46 + \log \frac{[10]}{[90]} = 3.46 - 0.95 = 2.51$$

Next, solving for the pH with 90% of the indicator in the In^- form:

$$pH = 3.46 + \log \frac{[90]}{[10]} = 3.46 + 0.95 = 4.41$$

Thus the pH range varies from **2.51 to 4.41** as the [HIn] varies from 90% to 10%.

16.85   Referring to Figure 16.5, at the half-equivalence point, [weak acid] = [conjugate base]. Using the Henderson-Hasselbalch equation:

$$pH = pK_a + \log \frac{[\text{conjugate base}]}{[\text{acid}]}$$

so,

$$pH = pK_a$$

16.86   First, calculate the pH of the 2.00 M weak acid (HNO_2) solution before any NaOH is added.

	$HNO_2(aq)$	$\rightleftharpoons$	$H^+(aq)$	$+$	$NO_2^-(aq)$
Initial (M):	2.00		0		0
Change (M):	$-x$		$+x$		$+x$
Equilibrium (M):	$2.00 - x$		x		x

$$K_a = \frac{[H^+][NO_2^-]}{[HNO_2]}$$

$$4.5 \times 10^{-4} = \frac{x^2}{2.00 - x} \approx \frac{x^2}{2.00}$$

$$x = [H^+] = 0.030 \ M$$

$$pH = -\log(0.030) = 1.52$$

Since the pH after the addition is 1.5 pH units greater, the new pH = 1.52 + 1.50 = 3.02.

From this new pH, we can calculate the $[H^+]$ in solution.

$$[H^+] = 10^{-pH} = 10^{-3.02} = 9.55 \times 10^{-4} \ M$$

When the NaOH is added, we dilute our original 2.00 M HNO_2 solution to:

$$M_i V_i = M_f V_f$$
$$(2.00 \ M)(400 \ \text{mL}) = M_f(600 \ \text{mL})$$
$$M_f = 1.33 \ M$$

Since we have not reached the equivalence point, we have a buffer solution. The reaction between HNO_2 and NaOH is:

$$HNO_2(aq) + NaOH(aq) \longrightarrow NaNO_2(aq) + H_2O(l)$$

Since the mole ratio between HNO_2 and NaOH is 1:1, the decrease in $[HNO_2]$ is the same as the decrease in [NaOH].

We can calculate the decrease in $[HNO_2]$ by setting up the weak acid equilibrium. From the pH of the solution, we know that the $[H^+]$ at equilibrium is $9.55 \times 10^{-4} \ M$.

	$HNO_2(aq)$	$\rightleftharpoons$	$H^+(aq)$	$+$	$NO_2^-(aq)$
Initial (M):	1.33		0		0
Change (M):	$-x$				$+x$
Equilibrium (M):	$1.33 - x$		9.55×10^{-4}		x

We can calculate x from the equilibrium constant expression.

$$K_a = \frac{[H^+][NO_2^-]}{[HNO_2]}$$

$$4.5 \times 10^{-4} = \frac{(9.55 \times 10^{-4})(x)}{1.33 - x}$$

$$x = 0.426 \ M$$

Thus, x is the decrease in $[HNO_2]$ which equals the concentration of added OH^-. However, this is the concentration of NaOH after it has been diluted to 600 mL. We need to correct for the dilution from 200 mL to 600 mL to calculate the concentration of the original NaOH solution.

$$M_i V_i = M_f V_f$$
$$M_i(200 \ \text{mL}) = (0.426 \ M)(600 \ \text{mL})$$
$$[\textbf{NaOH}] = M_i = \textbf{1.28} \ \boldsymbol{M}$$

16.87 The K_a of butyric acid is obtained by taking the antilog of 4.7 ($10^{-4.7}$) which is 2×10^{-5}. The value of K_b is:

$$K_b = \frac{K_w}{K_a} = \frac{1.0 \times 10^{-14}}{2 \times 10^{-5}} = \textbf{5} \times \textbf{10}^{-\textbf{10}}$$

16.88 The resulting solution is not a buffer system. There is excess NaOH and the neutralization is well past the equivalence point.

$$\text{Moles NaOH} = 0.500 \ \text{L} \times \frac{0.167 \ \text{mol}}{1 \ \text{L}} = 0.0835 \ \text{mol}$$

$$\text{Moles HCOOH} = 0.500 \ \text{L} \times \frac{0.100 \ \text{mol}}{1 \ \text{L}} = 0.0500 \ \text{mol}$$

	HCOOH(aq) +	NaOH(aq) $\rightarrow$	HCOONa(aq) + H$_2$O(l)
Initial (mol):	0.0500	0.0835	0
Change (mol):	−0.0500	−0.0500	+0.0500
Final (mol):	0	0.0335	0.0500

The volume of the resulting solution is 1.00 L (500 mL + 500 mL = 1000 mL).

$$[\textbf{OH}^-] = \frac{0.0335 \ \text{mol}}{1.00 \ \text{L}} = \textbf{0.0335} \ \boldsymbol{M}$$

$$[\textbf{Na}^+] = \frac{(0.0335 + 0.0500) \ \text{mol}}{1.00 \ \text{L}} = \textbf{0.0835} \ \boldsymbol{M}$$

$$[\textbf{H}^+] = \frac{K_w}{[OH^-]} = \frac{1.0 \times 10^{-14}}{0.0335} = \textbf{3.0} \times \textbf{10}^{-\textbf{13}} \ \boldsymbol{M}$$

$$[\textbf{HCOO}^-] = \frac{0.0500 \ \text{mol}}{1.00 \ \text{L}} = \textbf{0.0500} \ \boldsymbol{M}$$

$$HCOO^-(aq) + H_2O(l) \rightleftharpoons HCOOH(aq) + OH^-(aq)$$

Initial (M):	0.0500	0	0.0335
Change (M):	$-x$	$+x$	$+x$
Equilibrium (M):	$0.0500 - x$	x	$0.0335 + x$

$$K_b = \frac{[HCOOH][OH^-]}{[HCOO^-]}$$

$$5.9 \times 10^{-11} = \frac{(x)(0.0335 + x)}{(0.0500 - x)} \approx \frac{(x)(0.0335)}{(0.0500)}$$

$$x = [HCOOH] = 8.8 \times 10^{-11}\ M$$

16.89   Most likely the increase in solubility is due to complex ion formation:

$$Cd(OH)_2(s) + 2OH^- \rightleftharpoons Cd(OH)_4^{2-}(aq)$$

This is a Lewis acid-base reaction.

16.90   The number of moles of $Ba(OH)_2$ present in the original 50.0 mL of solution is:

$$50.0\ \text{mL} \times \frac{1.00\ \text{mol Ba(OH)}_2}{1000\ \text{mL soln}} = 0.0500\ \text{mol Ba(OH)}_2$$

The number of moles of H_2SO_4 present in the original 86.4 mL of solution, assuming complete dissociation, is:

$$86.4\ \text{mL} \times \frac{0.494\ \text{mol H}_2\text{SO}_4}{1000\ \text{mL soln}} = 0.0427\ \text{mol H}_2\text{SO}_4$$

The reaction is:

	$Ba(OH)_2(aq)$	$+\ H_2SO_4(aq)$	$\rightarrow$	$BaSO_4(s)$	$+\ 2H_2O(l)$
Initial (mol):	0.0500	0.0427		0	
Change (mol):	-0.0427	-0.0427		$+0.0427$	
Final (mol):	0.0073	0		0.0427	

Thus the mass of $BaSO_4$ formed is:

$$0.0427\ \text{mol BaSO}_4 \times \frac{233.4\ \text{g BaSO}_4}{1\ \text{mol BaSO}_4} = \textbf{9.97 g BaSO}_4$$

The pH can be calculated from the excess OH^- in solution. First, calculate the molar concentration of OH^-. The total volume of solution is 136.4 mL = 0.1364 L.

$$[OH^-] = \frac{0.0073\ \text{mol Ba(OH)}_2 \times \dfrac{2\ \text{mol OH}^-}{1\ \text{mol Ba(OH)}_2}}{0.1364\ \text{L}} = 0.11\ M$$

$$pOH = -\log(0.11) = 0.96$$

$$\textbf{pH} = 14.00 - pOH = 14.00 - 0.96 = \textbf{13.04}$$

16.91 A solubility equilibrium is an equilibrium between a solid (reactant) and its components (products: ions, neutral molecules, etc.) in solution. Only **(d)** represents a solubility equilibrium.

Consider part (b). Can you write the equilibrium constant for this reaction in terms of K_{sp} for calcium phosphate?

16.92 First, we calculate the molar solubility of $CaCO_3$.

$$CaCO_3(s) \rightleftharpoons Ca^{2+}(aq) + CO_3^{2-}(aq)$$

Initial (M):		0	0
Change (M):	$-s$	$+s$	$+s$
Equil. (M):		s	s

$$K_{sp} = [Ca^{2+}][CO_3^{2-}] = s^2 = 8.7 \times 10^{-9}$$
$$s = 9.3 \times 10^{-5} M = 9.3 \times 10^{-5} \text{ mol/L}$$

The moles of $CaCO_3$ in the kettle are:

$$116 \text{ g} \times \frac{1 \text{ mol } CaCO_3}{100.1 \text{ g } CaCO_3} = 1.16 \text{ mol } CaCO_3$$

The volume of distilled water needed to dissolve 1.16 moles of $CaCO_3$ is:

$$1.16 \text{ mol } CaCO_3 \times \frac{1 \text{ L}}{9.3 \times 10^{-5} \text{ mol } CaCO_3} = 1.2 \times 10^4 \text{ L}$$

The number of times the kettle would have to be filled is:

$$(1.2 \times 10^4 \text{ L}) \times \frac{1 \text{ filling}}{2.0 \text{ L}} = \textbf{6.0} \times \textbf{10}^\textbf{3} \textbf{ fillings}$$

Note that the very important assumption is made that each time the kettle is filled, the calcium carbonate is allowed to reach equilibrium before the kettle is emptied.

16.93 Since equal volumes of the two solutions were used, the initial molar concentrations will be halved.

$$[Ag^+] = \frac{0.12 \, M}{2} = 0.060 \, M$$

$$[Cl^-] = \frac{2(0.14 \, M)}{2} = 0.14 \, M$$

Let's assume that the Ag^+ ions and Cl^- ions react completely to form $AgCl(s)$. Then, we will reestablish the equilibrium between $AgCl$, Ag^+, and Cl^-.

	$Ag^+(aq)$	$+ \ Cl^-(aq)$	$\longrightarrow \ AgCl(s)$
Initial (M):	0.060	0.14	0
Change (M):	-0.060	-0.060	$+0.060$
Final (M):	0	0.080	0.060

Now, setting up the equilibrium,

$$AgCl(s) \rightleftharpoons Ag^+(aq) + Cl^-(aq)$$

	AgCl(s)	Ag⁺(aq)	Cl⁻(aq)
Initial (*M*):	0.060	0	0.080
Change (*M*):	−s	+s	+s
Equilibrium (*M*):	0.060 − s	s	0.080 + s

Set up the K_{sp} expression to solve for s.

$$K_{sp} = [Ag^+][Cl^-]$$

$$1.6 \times 10^{-10} = (s)(0.080 + s)$$

$$s = 2.0 \times 10^{-9} \, M$$

$$[Ag^+] = s = \mathbf{2.0 \times 10^{-9} \, M}$$
$$[Cl^-] = 0.080 \, M + s = \mathbf{0.080 \, M}$$
$$[Zn^{2+}] = \frac{0.14 \, M}{2} = \mathbf{0.070 \, M}$$
$$[NO_3^-] = \frac{0.12 \, M}{2} = \mathbf{0.060 \, M}$$

16.94   First we find the molar solubility and then convert moles to grams. The solubility equilibrium for silver carbonate is:

$$Ag_2CO_3(s) \rightleftharpoons 2Ag^+(aq) + CO_3^{2-}(aq)$$

	Ag₂CO₃(s)	2Ag⁺(aq)	CO₃²⁻(aq)
Initial (*M*):		0	0
Change (*M*):	−s	+2s	+s
Equilibrium (*M*):		2s	s

$$K_{sp} = [Ag^+]^2[CO_3^{2-}] = (2s)^2(s) = 4s^3 = 8.1 \times 10^{-12}$$

$$s = \left(\frac{8.1 \times 10^{-12}}{4}\right)^{\frac{1}{3}} = 1.3 \times 10^{-4} \, M$$

Converting from mol/L to g/L:

$$\frac{1.3 \times 10^{-4} \, \text{mol}}{1 \, \text{L soln}} \times \frac{275.8 \, \text{g}}{1 \, \text{mol}} = \mathbf{0.036 \, g/L}$$

16.95   For Fe(OH)₃, $K_{sp} = 1.1 \times 10^{-36}$. When $[Fe^{3+}] = 0.010 \, M$, the $[OH^-]$ value is

$$K_{sp} = [Fe^{3+}][OH^-]^3$$

or

$$[OH^-] = \left(\frac{K_{sp}}{[Fe^{3+}]}\right)^{\frac{1}{3}}$$

$$[OH^-] = \left(\frac{1.1 \times 10^{-36}}{0.010}\right)^{\frac{1}{3}} = 4.8 \times 10^{-12} \, M$$

This [OH⁻] corresponds to a pH of 2.68. In other words, $Fe(OH)_3$ will begin to precipitate from this solution at pH of 2.68.

For $Zn(OH)_2$, $K_{sp} = 1.8 \times 10^{-14}$. When $[Zn^{2+}] = 0.010$ M, the [OH⁻] value is

$$[OH^-] = \left(\frac{K_{sp}}{[Zn^{2+}]} \right)^{\frac{1}{2}}$$

$$[OH^-] = \left(\frac{1.8 \times 10^{-14}}{0.010} \right)^{\frac{1}{2}} = 1.3 \times 10^{-6} \ M$$

This corresponds to a pH of 8.11. In other words $Zn(OH)_2$ will begin to precipitate from the solution at pH = 8.11. These results show that $Fe(OH)_3$ will precipitate when the pH just exceeds 2.68 and that $Zn(OH)_2$ will precipitate when the pH just exceeds 8.11. Therefore, to selectively remove iron as $Fe(OH)_3$, the pH must be greater than 2.68 but less than 8.11.

16.96 **(a)** To 2.50×10^{-3} mole HCl (that is, 0.0250 L of 0.100 M solution) is added 1.00×10^{-3} mole CH_3NH_2 (that is, 0.0100 L of 0.100 M solution).

	$HCl(aq)$	+	$CH_3NH_2(aq)$	→	$CH_3NH_3Cl(aq)$
Initial (mol):	2.50×10^{-3}		1.00×10^{-3}		0
Change (mol):	-1.00×10^{-3}		-1.00×10^{-3}		$+1.00 \times 10^{-3}$
Equilibrium (mol):	1.50×10^{-3}		0		1.00×10^{-3}

After the acid-base reaction, we have 1.50×10^{-3} mol of HCl remaining. Since HCl is a strong acid, the [H⁺] will come from the HCl. The total solution volume is 35.0 mL = 0.0350 L.

$$[H^+] = \frac{1.50 \times 10^{-3} \text{ mol}}{0.0350 \text{ L}} = 0.0429 \ M$$

pH = 1.37

(b) When a total of 25.0 mL of CH_3NH_2 is added, we reach the equivalence point. That is, 2.50×10^{-3} mol HCl reacts with 2.50×10^{-3} mol CH_3NH_2 to form 2.50×10^{-3} mol CH_3NH_3Cl. Since there is a total of 50.0 mL of solution, the concentration of $CH_3NH_3^+$ is:

$$[CH_3NH_3^+] = \frac{2.50 \times 10^{-3} \text{ mol}}{0.0500 \text{ L}} = 5.00 \times 10^{-2} \ M$$

This is a problem involving the hydrolysis of the weak acid $CH_3NH_3^+$.

	$CH_3NH_3^+(aq)$	⇌	$H^+(aq)$	+	$CH_3NH_2(aq)$
Initial (M):	5.00×10^{-2}		0		0
Change (M):	$-x$		$+x$		$+x$
Equilibrium (M):	$(5.00 \times 10^{-2}) - x$		x		x

$$K_a = \frac{[CH_3NH_2][H^+]}{[CH_3NH_3^+]}$$

$$2.3 \times 10^{-11} = \frac{x^2}{(5.00 \times 10^{-2}) - x} \approx \frac{x^2}{5.00 \times 10^{-2}}$$

$$1.15 \times 10^{-12} = x^2$$

$$x = 1.07 \times 10^{-6} \, M = [H^+]$$

pH = 5.97

(c)   35.0 mL of 0.100 M CH$_3$NH$_2$ (3.50×10^{-3} mol) is added to the 25 mL of 0.100 M HCl (2.50×10^{-3} mol).

	HCl(aq)	+	CH$_3$NH$_2$(aq)	$\rightarrow$	CH$_3$NH$_3$Cl(aq)
Initial (mol):	2.50×10^{-3}		3.50×10^{-3}		0
Change (mol):	-2.50×10^{-3}		-2.50×10^{-3}		$+2.50 \times 10^{-3}$
Equilibrium (mol):	0		1.00×10^{-3}		2.50×10^{-3}

This is a buffer solution. Using the Henderson-Hasselbalch equation:

$$pH = pK_a + \log\frac{[\text{conjugate base}]}{[\text{acid}]}$$

$$pH = -\log(2.3 \times 10^{-11}) + \log\frac{(1.00 \times 10^{-3})}{(2.50 \times 10^{-3})} = \mathbf{10.24}$$

16.97   The equilibrium reaction is:

	Pb(IO$_3$)$_2$(aq)	$\rightleftharpoons$	Pb^{2+}(aq)	+	2IO$_3^-$(aq)
Initial (M):			0		0.10
Change (M):	-2.4×10^{-11}		$+2.4 \times 10^{-11}$		$+2(2.4 \times 10^{-11})$
Equilibrium (M):			2.4×10^{-11}		≈ 0.10

Substitute the equilibrium concentrations into the solubility product expression to calculate K_{sp}.

$$K_{sp} = [Pb^{2+}][IO_3^-]^2$$

$$\mathbf{K_{sp}} = (2.4 \times 10^{-11})(0.10)^2 = \mathbf{2.4 \times 10^{-13}}$$

16.98   The precipitate is HgI$_2$.

$$Hg^{2+}(aq) + 2I^-(aq) \longrightarrow HgI_2(s)$$

With further addition of I$^-$, a soluble complex ion is formed and the precipitate redissolves.

$$HgI_2(s) + 2I^-(aq) \longrightarrow HgI_4^{2-}(aq)$$

16.99 $BaSO_4(s) \rightleftharpoons Ba^{2+}(aq) + SO_4^{2-}(aq)$

$$K_{sp} = [Ba^{2+}][SO_4^{2-}] = 1.1 \times 10^{-10}$$

$$[Ba^{2+}] = 1.0 \times 10^{-5} \, M$$

In 5.0 L, the number of moles of Ba^{2+} is

$$(5.0 \, L)(1.0 \times 10^{-5} \, mol/L) = 5.0 \times 10^{-5} \, mol \, Ba^{2+} = 5.0 \times 10^{-5} \, mol \, BaSO_4$$

The number of grams of $BaSO_4$ dissolved is

$$(5.0 \times 10^{-5} \, mol \, BaSO_4) \times \frac{233.4 \, g \, BaSO_4}{1 \, mol \, BaSO_4} = 0.012 \, g \, BaSO_4$$

In practice, even less $BaSO_4$ will dissolve because the $BaSO_4$ is not in contact with the entire volume of blood. $Ba(NO_3)_2$ is too soluble to be used for this purpose.

16.100   We can use the Henderson-Hasselbalch equation to solve for the pH when the indicator is 95% acid / 5% conjugate base and when the indicator is 5% acid / 95% conjugate base.

$$pH = pK_a + \log\frac{[conjugate \, base]}{[acid]}$$

Solving for the pH with 95% of the indicator in the HIn form:

$$pH = 9.10 + \log\frac{[5]}{[95]} = 9.10 - 1.28 = 7.82$$

Next, solving for the pH with 95% of the indicator in the In^- form:

$$pH = 9.10 + \log\frac{[95]}{[5]} = 9.10 + 1.28 = 10.38$$

Thus the pH range varies from **7.82 to 10.38** as the [HIn] varies from 95% to 5%.

16.101   **(a)**   The solubility product expressions for both substances have exactly the same mathematical form and are therefore directly comparable. The substance having the smaller K_{sp} (**AgBr**) will precipitate first. (Why?)

(b)   When CuBr just begins to precipitate the solubility product expression will just equal K_{sp} (saturated solution). The concentration of Cu^+ at this point is 0.010 M (given in the problem), so the concentration of bromide ion must be:

$$K_{sp} = [Cu^+][Br^-] = (0.010)[Br^-] = 4.2 \times 10^{-8}$$

$$[Br^-] = \frac{4.2 \times 10^{-8}}{0.010} = 4.2 \times 10^{-6} \, M$$

Using this value of $[Br^-]$, we find the silver ion concentration

$$[Ag^+] = \frac{K_{sp}}{[Br^-]} = \frac{7.7 \times 10^{-13}}{4.2 \times 10^{-6}} = 1.8 \times 10^{-7}\ M$$

(c)   The percent of silver ion remaining in solution is:

$$\%\ Ag^+(aq) = \frac{1.8 \times 10^{-7}\ M}{0.010\ M} \times 100\% = 0.0018\%\ \text{or}\ 1.8 \times 10^{-3}\%$$

Is this an effective way to separate silver from copper?

16.102 (a)   We abbreviate the name of cacodylic acid to CacH. We set up the usual table.

$$CacH(aq) \rightleftharpoons Cac^-(aq) + H^+(aq)$$

	CacH(aq)	Cac⁻(aq)	H⁺(aq)
Initial (*M*):	0.10	0	0
Change (*M*):	$-x$	$+x$	$+x$
Equilibrium (*M*):	$0.10 - x$	x	x

$$K_a = \frac{[H^+][Cac^-]}{[CacH]}$$

$$6.4 \times 10^{-7} = \frac{x^2}{0.10 - x} \approx \frac{x^2}{0.10}$$

$$x = 2.5 \times 10^{-4}\ M = [H^+]$$

$$pH = -\log(2.5 \times 10^{-4}) = 3.60$$

(b)   We set up a table for the hydrolysis of the anion:

$$Cac^-(aq) + H_2O(l) \rightleftharpoons CacH(aq) + OH^-(aq)$$

	Cac⁻(aq)	H₂O(l)	CacH(aq)	OH⁻(aq)
Initial (*M*):	0.15		0	0
Change (*M*):	$-x$		$+x$	$+x$
Equilibrium (*M*):	$0.15 - x$		x	x

The ionization constant, K_b, for Cac⁻ is:

$$K_b = \frac{K_w}{K_a} = \frac{1.0 \times 10^{-14}}{6.4 \times 10^{-7}} = 1.6 \times 10^{-8}$$

$$K_b = \frac{[CacH][OH^-]}{[Cac^-]}$$

$$1.6 \times 10^{-8} = \frac{x^2}{0.15 - x} \approx \frac{x^2}{0.15}$$

$$x = 4.9 \times 10^{-5}\ M$$

$$pOH = -\log(4.9 \times 10^{-5}) = 4.31$$

$$pH = 14.00 - 4.31 = 9.69$$

(c) Number of moles of CacH from (a) is:

$$50.0 \text{ mL CacH} \times \frac{0.10 \text{ mol CacH}}{1000 \text{ mL}} = 5.0 \times 10^{-3} \text{ mol CacH}$$

Number of moles of Cac$^-$ from (b) is:

$$25.0 \text{ mL CacNa} \times \frac{0.15 \text{ mol CacNa}}{1000 \text{ mL}} = 3.8 \times 10^{-3} \text{ mol CacNa}$$

At this point we have a buffer solution.

$$\text{pH} = \text{pK}_a + \log\frac{[\text{Cac}^-]}{[\text{CacH}]} = -\log(6.4 \times 10^{-7}) + \log\frac{3.8 \times 10^{-3}}{5.0 \times 10^{-3}} = \mathbf{6.07}$$

16.103 The initial number of moles of Ag$^+$ is

$$\text{mol Ag}^+ = 50.0 \text{ mL} \times \frac{0.010 \text{ mol Ag}^+}{1000 \text{ mL soln}} = 5.0 \times 10^{-4} \text{ mol Ag}^+$$

We can use the counts of radioactivity as being proportional to concentration. Thus, we can use the ratio to determine the quantity of Ag$^+$ still in solution. However, since our original 50 mL of solution has been diluted to 500 mL, the counts per mL will be reduced by ten. Our diluted solution would then produce 7402.5 counts per minute if no removal of Ag$^+$ had occurred.

The number of moles of Ag$^+$ that correspond to 44.4 counts are:

$$44.4 \text{ counts} \times \frac{5.0 \times 10^{-4} \text{ mol Ag}^+}{7402.5 \text{ counts}} = 3.0 \times 10^{-6} \text{ mol Ag}^+$$

$$\text{Original mol of IO}_3^- = 100 \text{ mL} \times \frac{0.030 \text{ mol IO}_3^-}{1000 \text{ mL soln}} = 3.0 \times 10^{-3} \text{ mol}$$

The quantity of IO$_3^-$ remaining after reaction with Ag$^+$:

(original moles − moles reacted with Ag$^+$) = $(3.0 \times 10^{-3} \text{ mol}) - [(5.0 \times 10^{-4} \text{ mol}) - (3.0 \times 10^{-6} \text{ mol})]$

$$= 2.5 \times 10^{-3} \text{ mol IO}_3^-$$

The total final volume is 500 mL or 0.50 L.

$$[\text{Ag}^+] = \frac{3.0 \times 10^{-6} \text{ mol Ag}^+}{0.50 \text{ L}} = 6.0 \times 10^{-6} \text{ } M$$

$$[\text{IO}_3^-] = \frac{2.5 \times 10^{-3} \text{ mol IO}_3^-}{0.50 \text{ L}} = 5.0 \times 10^{-3} \text{ } M$$

$$\text{AgIO}_3(s) \rightleftharpoons \text{Ag}^+(aq) + \text{IO}_3^-(aq)$$

$$\mathbf{\mathit{K}_{sp}} = [\text{Ag}^+][\text{IO}_3^-] = (6.0 \times 10^{-6})(5.0 \times 10^{-3}) = \mathbf{3.0 \times 10^{-8}}$$

16.104 **(a)** $MCO_3 + 2HCl \rightarrow MCl_2 + H_2O + CO_2$

$HCl + NaOH \rightarrow NaCl + H_2O$

(b) We are given the mass of the metal carbonate, so we need to find moles of the metal carbonate to calculate its molar mass. We can find moles of MCO_3 from the moles of HCl reacted.

Moles of HCl reacted with MCO_3 = Total moles of HCl – Moles of excess HCl

$$\text{Total moles of HCl} = 20.00 \text{ mL} \times \frac{0.0800 \text{ mol}}{1000 \text{ mL soln}} = 1.60 \times 10^{-3} \text{ mol HCl}$$

$$\text{Moles of excess HCl} = 5.64 \text{ mL} \times \frac{0.1000 \text{ mol}}{1000 \text{ mL soln}} = 5.64 \times 10^{-4} \text{ mol HCl}$$

Moles of HCl reacted with MCO_3 = $(1.60 \times 10^{-3} \text{ mol}) - (5.64 \times 10^{-4} \text{ mol}) = 1.04 \times 10^{-3} \text{ mol HCl}$

$$\text{Moles of } MCO_3 \text{ reacted} = (1.04 \times 10^{-3} \text{ mol HCl}) \times \frac{1 \text{ mol } MCO_3}{2 \text{ mol HCl}} = 5.20 \times 10^{-4} \text{ mol } MCO_3$$

$$\textbf{Molar mass of } MCO_3 = \frac{0.1022 \text{ g}}{5.20 \times 10^{-4} \text{ mol}} = \textbf{197 g/mol}$$

Molar mass of CO_3 = 60.01 g

Molar mass of M = 197 g/mol – 60.01 g/mol = 137 g/mol

The metal, M, is **Ba**!

16.105 **(a)** $H^+ + OH^- \rightarrow H_2O$ $\qquad K = 1.0 \times 10^{14}$

(b) $H^+ + NH_3 \rightarrow NH_4^+$

$$K = \frac{1}{K_a} = \frac{1}{5.6 \times 10^{-10}} = 1.8 \times 10^9$$

(c) $CH_3COOH + OH^- \rightarrow CH_3COO^- + H_2O$

Broken into 2 equations:

$CH_3COOH \rightarrow CH_3COO^- + H^+$ $\qquad K_a$

$H^+ + OH^- \rightarrow H_2O$ $\qquad\qquad 1/K_w$

$$K = \frac{K_a}{K_w} = \frac{1.8 \times 10^{-5}}{1.0 \times 10^{-14}} = 1.8 \times 10^9$$

(d) $CH_3COOH + NH_3 \rightarrow CH_3COONH_4$

Broken into 2 equations:

$$CH_3COOH \rightarrow CH_3COO^- + H^+ \qquad K_a$$

$$NH_3 + H^+ \rightarrow NH_4^+ \qquad \frac{1}{K_a'}$$

$$K = \frac{K_a}{K_a'} = \frac{1.8 \times 10^{-5}}{5.6 \times 10^{-10}} = 3.2 \times 10^4$$

16.106 The number of moles of NaOH reacted is:

$$15.9 \text{ mL NaOH} \times \frac{0.500 \text{ mol NaOH}}{1000 \text{ mL soln}} = 7.95 \times 10^{-3} \text{ mol NaOH}$$

Since two moles of NaOH combine with one mole of oxalic acid, the number of moles of oxalic acid reacted is 3.98×10^{-3} mol. This is the number of moles of oxalic acid hydrate in 25.0 mL of solution. In 250 mL, the number of moles present is 3.98×10^{-2} mol. Thus the molar mass is:

$$\frac{5.00 \text{ g}}{3.98 \times 10^{-2} \text{ mol}} = 126 \text{ g/mol}$$

From the molecular formula we can write:

$$2(1.008)\text{g} + 2(12.01)\text{g} + 4(16.00)\text{g} + x(18.02)\text{g} = 126 \text{ g}$$

Solving for x:

$$x = 2$$

16.107 (a) Mix 500 mL of 0.40 M CH$_3$COOH with 500 mL of 0.40 M CH$_3$COONa. Since the final volume is 1.00 L, then the concentrations of the two solutions that were mixed must be one-half of their initial concentrations.

(b) Mix 500 mL of 0.80 M CH$_3$COOH with 500 mL of 0.40 M NaOH. (Note: half of the acid reacts with all of the base to make a solution identical to that in part (a) above.)

$$CH_3COOH + NaOH \rightarrow CH_3COONa + H_2O$$

(c) Mix 500 mL of 0.80 M CH$_3$COONa with 500 mL of 0.40 M HCl. (Note: half of the salt reacts with all of the acid to make a solution identical to that in part (a) above.)

$$CH_3COO^- + H^+ \rightarrow CH_3COOH$$

16.108 (a)

$$pH = pK_a + \log\frac{\text{[conjugate base]}}{\text{[acid]}}$$

$$8.00 = 9.10 + \log\frac{\text{[ionized]}}{\text{[un-ionized]}}$$

$$\frac{\text{[un-ionized]}}{\text{[ionized]}} = 12.6 \qquad\qquad (1)$$

(b) First, let's calculate the total concentration of the indicator. 2 drops of the indicator are added and each drop is 0.050 mL.

$$2 \text{ drops} \times \frac{0.050 \text{ mL phenolphthalein}}{1 \text{ drop}} = 0.10 \text{ mL phenolphthalein}$$

This 0.10 mL of phenolphthalein of concentration 0.060 M is diluted to 50.0 mL.

$$M_i V_i = M_f V_f$$
$$(0.060 \ M)(0.10 \text{ mL}) = M_f(50.0 \text{ mL})$$
$$M_f = 1.2 \times 10^{-4} \ M$$

Using equation (1) above and letting y = [ionized], then [un-ionized] = $(1.2 \times 10^{-4}) - y$.

$$\frac{(1.2 \times 10^{-4}) - y}{y} = 12.6$$

$$y = \mathbf{8.8 \times 10^{-6} \ M}$$

16.109 The sulfur-containing air-pollutants (like H_2S) reacts with Pb^{2+} to form PbS, which gives paintings a darkened look.

16.110 **(a)** Add sulfate. Na_2SO_4 is soluble, $BaSO_4$ is not.

(b) Add sulfide. K_2S is soluble, PbS is not

(c) Add iodide. ZnI_2 is soluble, HgI_2 is not.

16.111 Strontium sulfate is the more soluble of the two compounds. Therefore, we can assume that all of the SO_4^{2-} ions come from $SrSO_4$.

$$SrSO_4(s) \rightleftharpoons Sr^{2+}(aq) + SO_4^{2-}(aq)$$

$$K_{sp} = [Sr^{2+}][SO_4^{2-}] = s^2 = 3.8 \times 10^{-7}$$

$$s = [Sr^{2+}] = [SO_4^{2-}] = \sqrt{3.8 \times 10^{-7}} = \mathbf{6.2 \times 10^{-4} \ M}$$

For $BaSO_4$:

$$\mathbf{[Ba^{2+}]} = \frac{K_{sp}}{[SO_4^{2-}]} = \frac{1.1 \times 10^{-10}}{6.2 \times 10^{-4}} = \mathbf{1.8 \times 10^{-7} \ M}$$

16.112 The amphoteric oxides cannot be used to prepare buffer solutions because they are insoluble in water.

16.113 $CaSO_4 \rightleftharpoons Ca^{2+} + SO_4^{2-}$

$$s^2 = 2.4 \times 10^{-5}$$

$$s = 4.9 \times 10^{-3} \ M$$

$$\text{Solubility} = \frac{4.9 \times 10^{-3} \text{ mol}}{1 \text{ L}} \times \frac{136.2 \text{ g}}{1 \text{ mol}} = \mathbf{0.67 \text{ g/L}}$$

$$Ag_2SO_4 \;\rightleftharpoons\; 2Ag^+ + SO_4^{2-}$$
$$\qquad\qquad 2s \qquad s$$

$$1.4 \times 10^{-5} = 4s^3$$

$$s = 0.015 \; M$$

Solubility $= \dfrac{0.015 \text{ mol}}{1 \text{ L}} \times \dfrac{311.1 \text{ g}}{1 \text{ mol}} = \textbf{4.7 g/L}$

Note: Ag_2SO_4 has a larger solubility.

16.114 The ionized polyphenols have a dark color. In the presence of citric acid from lemon juice, the anions are converted to the lighter-colored acids.

16.115 $H_2PO_4^- \;\rightleftharpoons\; H^+ + HPO_4^{2-}$

$$K_a = 6.2 \times 10^{-8}$$

$$pK_a = 7.20$$

$$7.50 = 7.20 + \log \frac{[HPO_4^{2-}]}{[H_2PO_4^-]}$$

$$\frac{[HPO_4^{2-}]}{[H_2PO_4^-]} = 2.0$$

We need to add enough NaOH so that

$$[HPO_4^{2-}] = 2[H_2PO_4^-]$$

Initially there was

$$0.200 \text{ L} \times 0.10 \text{ mol/L} = 0.020 \text{ mol NaH}_2PO_4 \text{ present.}$$

For $[HPO_4^{2-}] = 2[H_2PO_4^-]$, we must add enough NaOH to react with 2/3 of the $H_2PO_4^-$. After reaction with NaOH, we have:

$$\frac{0.020}{3} \text{ mol } H_2PO_4^- = 0.0067 \text{ mol } H_2PO_4^-$$

$$\text{mol } HPO_4^{2-} = 2 \times 0.0067 \text{ mol} = 0.013 \text{ mol } HPO_4^{2-}$$

The moles of NaOH reacted is equal to the moles of HPO_4^{2-} produced because the mole ratio between OH^- and HPO_4^{2-} is 1:1.

$$OH^- + H_2PO_4^- \rightarrow HPO_4^{2-} + H_2O$$

$$V_{NaOH} = \frac{\text{mol}_{NaOH}}{M_{NaOH}} = \frac{0.013 \text{ mol}}{1.0 \text{ mol/L}} = 0.013 \text{ L} = \textbf{13 mL}$$

16.116   Assuming the density of water to be 1.00 g/mL, 0.05 g Pb^{2+} per 10^6 g water is equivalent to 5×10^{-5} g Pb^{2+}/L

$$\frac{0.05 \text{ g } Pb^{2+}}{1 \times 10^6 \text{ g } H_2O} \times \frac{1 \text{ g } H_2O}{1 \text{ mL } H_2O} \times \frac{1000 \text{ mL } H_2O}{1 \text{ L } H_2O} = 5 \times 10^{-5} \text{ g } Pb^{2+}/L$$

$$PbSO_4 \rightleftharpoons Pb^{2+} + SO_4^{2-}$$

Initial (*M*):		0	0
Change (*M*):	$-s$	$+s$	$+s$
Equilibrium (*M*):		s	s

$$K_{sp} = [Pb^{2+}][SO_4^{2-}]$$

$$1.6 \times 10^{-8} = s^2$$

$$s = 1.3 \times 10^{-4} M$$

The solubility of $PbSO_4$ in g/L is:

$$\frac{1.3 \times 10^{-4} \text{ mol}}{1 \text{ L}} \times \frac{303.3 \text{ g}}{1 \text{ mol}} = 4.0 \times 10^{-2} \text{ g/L}$$

Yes. The $[Pb^{2+}]$ exceeds the safety limit of 5×10^{-5} g Pb^{2+}/L.

16.117   **(a)**   The acidic hydrogen is from the carboxyl group.

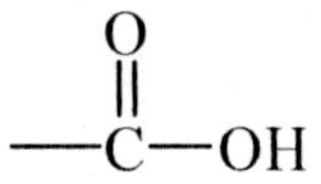

$$\begin{array}{c} O \\ \parallel \\ -C-OH \end{array}$$

(b)   At pH 6.50, Equation (16.4) of the text can be written as:

$$6.50 = 2.76 + \log\frac{[P^-]}{[HP]}$$

$$\frac{[P^-]}{[HP]} = 5.5 \times 10^3$$

Thus, nearly all of the penicillin G will be in the ionized form. The ionized form is more soluble in water because it bears a net charge; penicillin G is largely nonpolar and therefore much less soluble in water. (Both penicillin G and its salt are effective antibiotics.)

(c)   First, the dissolved NaP salt completely dissociates in water as follows:

$$NaP \xrightarrow{H_2O} Na^+ + P^-$$
$$\qquad\qquad 0.12 \ M \quad 0.12 \ M$$

We need to concentrate only on the hydrolysis of the P^- ion.

Step 1: Let x be the equilibrium concentrations of HP and OH$^-$ due to the hydrolysis of P$^-$ ions. We summarize the changes:

$$P^-(aq) + H_2O(l) \rightleftharpoons HP(aq) + OH^-(aq)$$

	P$^-(aq)$ + H$_2$O(l) $\rightleftharpoons$ HP(aq) +		OH$^-(aq)$
Initial (*M*):	0.12	0	0
Change (*M*):	$-x$	$+x$	$+x$
Equilibrium (*M*):	$0.12 - x$	x	x

Step 2: $K_b = \dfrac{K_w}{K_a} = \dfrac{1.00 \times 10^{-14}}{1.64 \times 10^{-3}} = 6.10 \times 10^{-12}$

$$K_b = \frac{[HP][OH^-]}{[P^-]}$$

$$6.10 \times 10^{-12} = \frac{x^2}{0.12 - x}$$

Assuming that $0.12 - x \approx 0.12$, we write:

$$6.10 \times 10^{-12} = \frac{x^2}{0.12}$$

$$x = 8.6 \times 10^{-7}\ M$$

Step 3: At equilibrium:

$$[OH^-] = 8.6 \times 10^{-7}\ M$$

$$pOH = -\log(8.6 \times 10^{-7}) = 6.07$$

$$\textbf{pH} = 14.00 - 6.07 = \textbf{7.93}$$

Because HP is a relatively strong acid, P$^-$ is a weak base. Consequently, only a small fraction of P$^-$ undergoes hydrolysis and the solution is slightly basic.

16.118 **(c)** has the highest [H$^+$]

$$F^- + SbF_5 \rightarrow SbF_6^-$$

Removal of F$^-$ promotes further ionization of HF.

16.119

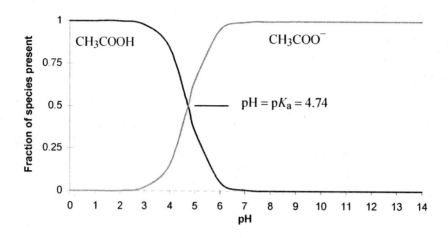

16.120 (a)   This is a common ion (CO_3^{2-}) problem.

The dissociation of Na_2CO_3 is:

$$Na_2CO_3(s) \xrightarrow{H_2O} 2Na^+(aq) \quad + \quad CO_3^{2-}(aq)$$
$$2(0.050 \ M) \qquad 0.050 \ M$$

Let s be the molar solubility of $CaCO_3$ in Na_2CO_3 solution. We summarize the changes as:

	$CaCO_3(s)$ $\rightleftharpoons$	$Ca^{2+}(aq)$ +	$CO_3^{2-}(aq)$
Initial (M):		0.00	0.050
Change (M):		$+s$	$+s$
Equil. (M):		$+s$	$0.050 + s$

$$K_{sp} = [Ca^{2+}][CO_3^{2-}]$$

$$8.7 \times 10^{-9} = s(0.050 + s)$$

Since s is small, we can assume that $0.050 + s \approx 0.050$

$$8.7 \times 10^{-9} = 0.050s$$

$$s = 1.7 \times 10^{-7} \ M$$

Thus, the addition of washing soda to permanent hard water removes most of the Ca^{2+} ions as a result of the common ion effect.

(b) Mg^{2+} is not removed by this procedure, because $MgCO_3$ is fairly soluble ($K_{sp} = 4.0 \times 10^{-5}$).

(c)   The K_{sp} for $Ca(OH)_2$ is 8.0×10^{-6}.

	$Ca(OH)_2 \rightleftharpoons$	$Ca^{2+} +$	$2OH^-$
At equil.:		s	$2s$

$$K_{sp} = 8.0 \times 10^{-6} = [Ca^{2+}][OH^-]^2$$

$$4s^3 = 8.0 \times 10^{-6}$$

$$s = 0.0126 \ M$$

$$[OH^-] = 2s = 0.0252 \ M$$

$$pOH = -\log(0.0252) = 1.60$$

$$\textbf{pH = 12.40}$$

(d)   The $[OH^-]$ calculated above is $0.0252 \ M$. At this rather high concentration of OH^-, most of the Mg^{2+} will be removed as $Mg(OH)_2$. The small amount of Mg^{2+} remaining in solution is due to the following equilibrium:

$$Mg(OH)_2(s) \rightleftharpoons Mg^{2+}(aq) + 2OH^-(aq)$$

$$K_{sp} = [Mg^{2+}][OH^-]^2$$

$$1.2 \times 10^{-11} = [Mg^{2+}](0.0252)^2$$

$$\mathbf{[Mg^{2+}] = 1.9 \times 10^{-8} \ M}$$

(e)   Remove Ca^{2+} first because it is present in larger amounts.

16.121 $pH = pK_a + \log\dfrac{[In^-]}{[HIn]}$

For acid color:

$$pH = pK_a + \log\frac{1}{10}$$

$$pH = pK_a - \log 10$$

$$pH = pK_a - 1$$

For base color:

$$pH = pK_a + \log\frac{10}{1}$$

$$pH = pK_a + 1$$

Combining these two equations:

$$\textbf{pH = p}\boldsymbol{K}_{\textbf{a}} \pm \textbf{1}$$

16.122 $pH = pK_a + \log \dfrac{[\text{conjugate base}]}{[\text{acid}]}$

At pH = 1.0,

-COOH $1.0 = 2.3 + \log \dfrac{[-COO^-]}{[-COOH]}$

$\dfrac{[-COOH]}{[-COO^-]} = 20$

$-NH_3^+$ $1.0 = 9.6 + \log \dfrac{[-NH_2]}{[-NH_3^+]}$

$\dfrac{[-NH_3^+]}{[-NH_2]} = 4 \times 10^8$

Therefore the **predominant species** is: $^+NH_3 - CH_2 - COOH$

At pH = 7.0,

-COOH $7.0 = 2.3 + \log \dfrac{[-COO^-]}{[-COOH]}$

$\dfrac{[-COO^-]}{[-COOH]} = 5 \times 10^4$

$-NH_3^+$ $7.0 = 9.6 + \log \dfrac{[-NH_2]}{[-NH_3^+]}$

$\dfrac{[-NH_3^+]}{[-NH_2]} = 4 \times 10^2$

Predominant species: $^+NH_3 - CH_2 - COO^-$

At pH = 12.0,

-COOH $12.0 = 2.3 + \log \dfrac{[-COO^-]}{[-COOH]}$

$\dfrac{[-COO^-]}{[-COOH]} = 5 \times 10^9$

$-NH_3^+$ $12.0 = 9.6 + \log \dfrac{[-NH_2]}{[-NH_3^+]}$

$\dfrac{[-NH_2]}{[-NH_3^+]} = 2.5 \times 10^2$

Predominant species: $NH_2 - CH_2 - COO^-$

16.123 **(a)** The pK_b value can be determined at the half-equivalence point of the titration (half the volume of added acid needed to reach the equivalence point). At this point in the titration $pH = pK_a$, where K_a refers to the acid ionization constant of the conjugate acid of the weak base. The Henderson-Hasselbalch equation reduces to $pH = pK_a$ when [acid] = [conjugate base]. Once the pK_a value is determined, the pK_b value can be calculated as follows:

$$pK_a + pK_b = 14.00$$

(b) Let B represent the base, and BH^+ represents its conjugate acid.

$$B(aq) + H_2O(l) \rightleftharpoons BH^+(aq) + OH^-(aq)$$

$$K_b = \frac{[BH^+][OH^-]}{[B]}$$

$$[OH^-] = \frac{K_b[B]}{[BH^+]}$$

Taking the negative logarithm of both sides of the equation gives:

$$-\log[OH^-] = -\log K_b - \log\frac{[B]}{[BH^+]}$$

$$pOH = pK_b + \log\frac{[BH^+]}{[B]}$$

The titration curve would look very much like Figure 16.5 of the text, except the y-axis would be pOH and the x-axis would be volume of strong acid added. The pK_b value can be determined at the half-equivalence point of the titration (half the volume of added acid needed to reach the equivalence point). At this point in the titration, the concentrations of the buffer components, [B] and $[BH^+]$, are equal, and hence $pOH = pK_b$.

CHAPTER 17
CHEMISTRY IN THE ATMOSPHERE

17.5 For ideal gases, mole fraction is the same as volume fraction. From Table 17.1 of the text, CO_2 is 0.033% of the composition of dry air, by volume. The value 0.033% means 0.033 volumes (or moles, in this case) out of 100 or

$$X_{CO_2} = \frac{0.033}{100} = 3.3 \times 10^{-4}$$

To change to parts per million (ppm), we multiply the mole fraction by one million.

$$(3.3 \times 10^{-4})(1 \times 10^6) = \textbf{330 ppm}$$

17.6 Using the information in Table 17.1 and Problem 17.5, 0.033 percent of the volume (and therefore the pressure) of dry air is due to CO_2. The partial pressure of CO_2 is:

$$P_{CO_2} = X_{CO_2} P_T = (3.3 \times 10^{-4})(754 \text{ mmHg}) \times \frac{1 \text{ atm}}{760 \text{ mmHg}} = \textbf{3.3} \times \textbf{10}^{-4} \textbf{ atm}$$

17.7 In the stratosphere, the air temperature rises with altitude. This warming effect is the result of exothermic reactions triggered by UV radiation from the sun. For further discussion, see Section 17.3 of the text.

17.8 From Problem 5.98, the total mass of air is 5.25×10^{18} kg. Table 17.1 lists the composition of air by volume. Under the same conditions of P and T, $V \propto n$ (Avogadro's law).

$$\text{Total moles of gases} = (5.25 \times 10^{21} \text{ g}) \times \frac{1 \text{ mol}}{29.0 \text{ g}} = 1.81 \times 10^{20} \text{ mol}$$

Mass of N_2 (78.03%):

$$(0.7803)(1.81 \times 10^{20} \text{ mol}) \times \frac{28.02 \text{ g}}{1 \text{ mol}} = 3.96 \times 10^{21} \text{ g} = \textbf{3.96} \times \textbf{10}^{18} \textbf{ kg}$$

Mass of O_2 (20.99%):

$$(0.2099)(1.81 \times 10^{20} \text{ mol}) \times \frac{32.00 \text{ g}}{1 \text{ mol}} = 1.22 \times 10^{21} \text{ g} = \textbf{1.22} \times \textbf{10}^{18} \textbf{ kg}$$

Mass of CO_2 (0.033%):

$$(3.3 \times 10^{-4})(1.81 \times 10^{20} \text{ mol}) \times \frac{44.01 \text{ g}}{1 \text{ mol}} = 2.63 \times 10^{18} \text{ g} = \textbf{2.63} \times \textbf{10}^{15} \textbf{ kg}$$

17.11 The energy of one photon is:

$$\frac{460 \times 10^3 \text{ J}}{1 \text{ mol}} \times \frac{1 \text{ mol}}{6.022 \times 10^{23} \text{ photons}} = 7.64 \times 10^{-19} \text{ J/photon}$$

The wavelength can now be calculated.

$$\lambda = \frac{hc}{E} = \frac{(6.63 \times 10^{-34} \text{ J} \cdot \text{s})(3.00 \times 10^8 \text{ m/s})}{7.64 \times 10^{-19} \text{ J}} = 2.60 \times 10^{-7} \text{ m} = \textbf{260 nm}$$

17.12   **Strategy:** We are given the wavelength of the emitted photon and asked to calculate its energy. Equation (7.2) of the text relates the energy and frequency of an electromagnetic wave.

$$E = h\nu$$

First, we calculate the frequency from the wavelength, then we can calculate the energy difference between the two levels.

Solution: Calculate the frequency from the wavelength.

$$\nu = \frac{c}{\lambda} = \frac{3.00 \times 10^8 \text{ m/s}}{558 \times 10^{-9} \text{ m}} = 5.38 \times 10^{14} \text{ /s}$$

Now, we can calculate the energy difference from the frequency.

$$\Delta E = h\nu = (6.63 \times 10^{-34} \text{ J} \cdot \text{s})(5.38 \times 10^{14} \text{ /s})$$
$$\Delta E = \textbf{3.57} \times \textbf{10}^{-19} \textbf{ J}$$

17.21   The formula for the volume is $4\pi r^2 h$, where $r = 6.371 \times 10^6$ m and $h = 3.0 \times 10^{-3}$ m (or 3.0 mm).

$$V = 4\pi (6.371 \times 10^6 \text{ m})^2 (3.0 \times 10^{-3} \text{ m}) = 1.5 \times 10^{12} \text{ m}^3 \times \frac{1000 \text{ L}}{1 \text{ m}^3} = 1.5 \times 10^{15} \text{ L}$$

Recall that at STP, one mole of gas occupies 22.41 L.

$$\text{moles O}_3 = (1.5 \times 10^{15} \text{ L}) \times \frac{1 \text{ mol}}{22.41 \text{ L}} = 6.7 \times 10^{13} \text{ mol O}_3$$

$$\textbf{molecules O}_3 = (6.7 \times 10^{13} \text{ mol O}_3) \times \frac{6.022 \times 10^{23} \text{ molecules}}{1 \text{ mol}} = \textbf{4.0} \times \textbf{10}^{37} \textbf{ molecules}$$

$$\textbf{mass O}_3 \textbf{ (kg)} = (6.7 \times 10^{13} \text{ mol O}_3) \times \frac{48.00 \text{ g O}_3}{1 \text{ mol O}_3} \times \frac{1 \text{ kg}}{1000 \text{ g}} = \textbf{3.2} \times \textbf{10}^{12} \textbf{ kg O}_3$$

17.22   The quantity of ozone lost is:

$$(0.06)(3.2 \times 10^{12} \text{ kg}) = 1.9 \times 10^{11} \text{ kg of O}_3$$

Assuming no further deterioration, the kilograms of O_3 that would have to be manufactured on a daily basis are:

$$\frac{1.9 \times 10^{11} \text{ kg O}_3}{100 \text{ yr}} \times \frac{1 \text{ yr}}{365 \text{ days}} = \textbf{5.2} \times \textbf{10}^6 \textbf{ kg/day}$$

The standard enthalpy of formation (from Appendix 3 of the text) for ozone:

$$\tfrac{3}{2} O_2 \rightarrow O_3 \qquad \Delta H_f^{\circ} = 142.2 \text{ kJ/mol}$$

The *total* energy required is:

$$(1.9 \times 10^{14} \text{ g of O}_3) \times \frac{1 \text{ mol O}_3}{48.00 \text{ g O}_3} \times \frac{142.2 \text{ kJ}}{1 \text{ mol O}_3} = \textbf{5.6} \times \textbf{10}^{\textbf{14}} \textbf{ kJ}$$

17.23   The formula for Freon-11 is $CFCl_3$ and for Freon-12 is CF_2Cl_2. The equations are:

$$CCl_4 + HF \rightarrow CFCl_3 + HCl$$

$$CFCl_3 + HF \rightarrow CF_2Cl_2 + HCl$$

A catalyst is necessary for both reactions.

17.24   The energy of the photons of UV radiation in the troposphere is insufficient (that is, the wavelength is too long and the frequency is too small) to break the bonds in CFCs.

17.25 $\lambda = 250$ nm

$$\nu = \frac{3.00 \times 10^8 \text{ m/s}}{250 \times 10^{-9} \text{ m}} = 1.20 \times 10^{15} \text{ /s}$$

$$E = h\nu = (6.63 \times 10^{-34} \text{ J·s})(1.20 \times 10^{15} \text{ /s}) = 7.96 \times 10^{-19} \text{ J}$$

Converting to units of kJ/mol:

$$\frac{7.96 \times 10^{-19} \text{ J}}{1 \text{ photon}} \times \frac{6.022 \times 10^{23} \text{ photons}}{1 \text{ mol}} \times \frac{1 \text{ kJ}}{1000 \text{ J}} = \textbf{479 kJ/mol}$$

Solar radiation preferentially breaks the C–Cl bond. There is not enough energy to break the C–F bond.

17.26   First, we need to calculate the energy needed to break one bond.

$$\frac{276 \times 10^3 \text{ J}}{1 \text{ mol}} \times \frac{1 \text{ mol}}{6.022 \times 10^{23} \text{ molecules}} = 4.58 \times 10^{-19} \text{ J/molecule}$$

The longest wavelength required to break this bond is:

$$\lambda = \frac{hc}{E} = \frac{(3.00 \times 10^8 \text{ m/s})(6.63 \times 10^{-34} \text{ J·s})}{4.58 \times 10^{-19} \text{ J}} = 4.34 \times 10^{-7} \text{ m} = \textbf{434 nm}$$

434 nm is in the visible region of the electromagnetic spectrum; therefore, CF_3Br will be decomposed in **both** the troposphere and stratosphere.

17.27   The Lewis structures for chlorine nitrate and chlorine monoxide are:

17.28 The Lewis structure of HCFC–123 is:

$$\begin{array}{ccc} F & H & \\ | & | & \\ F-C-C-Cl \\ | & | & \\ F & Cl & \end{array}$$

The Lewis structure for CF_3CFH_2 is:

$$\begin{array}{ccc} F & H & \\ | & | & \\ F-C-C-H \\ | & | & \\ F & F & \end{array}$$

Lone pairs on the outer atoms have been omitted.

17.39 The equation is: $2ZnS + 3O_2 \rightarrow 2ZnO + 2SO_2$

$$(4.0 \times 10^4 \text{ ton ZnS}) \times \frac{1 \text{ ton} \cdot \text{mol ZnS}}{97.46 \text{ ton ZnS}} \times \frac{1 \text{ ton} \cdot \text{mol SO}_2}{1 \text{ ton} \cdot \text{mol ZnS}} \times \frac{64.07 \text{ ton SO}_2}{1 \text{ ton} \cdot \text{mol SO}_2} = \textbf{2.6} \times \textbf{10}^\textbf{4} \textbf{ tons SO}_\textbf{2}$$

17.40 **Strategy:** Looking at the balanced equation, how do we compare the amounts of CaO and CO_2? We can compare them based on the mole ratio from the balanced equation.

Solution: Because the balanced equation is given in the problem, the mole ratio between CaO and CO_2 is known: 1 mole CaO $\simeq$ 1 mole CO_2. If we convert grams of CaO to moles of CaO, we can use this mole ratio to convert to moles of CO_2. Once moles of CO_2 are known, we can convert to grams CO_2.

$$\text{mass CO}_2 = (1.7 \times 10^{13} \text{ g CaO}) \times \frac{1 \text{ mol CaO}}{56.08 \text{ g CaO}} \times \frac{1 \text{ mol CO}_2}{1 \text{ mol CaO}} \times \frac{44.01 \text{ g}}{1 \text{ mol CO}_2}$$

$$= 1.3 \times 10^{13} \text{ g CO}_2 = \textbf{1.3} \times \textbf{10}^\textbf{10} \textbf{ kg CO}_\textbf{2}$$

17.41 Total amount of heat absorbed is:

$$(1.8 \times 10^{20} \text{ mol}) \times \frac{29.1 \text{ J}}{\text{K} \cdot \text{mol}} \times 3 \text{ K} = 1.6 \times 10^{22} \text{ J} = \textbf{1.6} \times \textbf{10}^\textbf{19} \textbf{ kJ}$$

The heat of fusion of ice in units of J/kg is:

$$\frac{6.01 \times 10^3 \text{ J}}{1 \text{ mol}} \times \frac{1 \text{ mol}}{18.02 \text{ g}} \times \frac{1000 \text{ g}}{1 \text{ kg}} = 3.3 \times 10^5 \text{ J/kg}$$

The amount of ice melted by the temperature rise:

$$(1.6 \times 10^{22} \text{ J}) \times \frac{1 \text{ kg}}{3.3 \times 10^5 \text{ J}} = \textbf{4.8} \times \textbf{10}^\textbf{16} \textbf{ kg}$$

17.42 Ethane and propane are greenhouse gases. They would contribute to global warming.

17.49 $(3.1 \times 10^{10} \text{ g}) \times \dfrac{2.4}{100} \times \dfrac{1 \text{ mol S}}{32.07 \text{ g S}} \times \dfrac{1 \text{ mol SO}_2}{1 \text{ mol S}} = 2.3 \times 10^7 \text{ mol SO}_2$

$$V = \frac{nRT}{P} = \frac{(2.3 \times 10^7 \text{ mol})(0.0821 \text{ L}\cdot\text{atm/mol}\cdot\text{K})(273 \text{ K})}{1 \text{ atm}} = \mathbf{5.2 \times 10^8 \text{ L}}$$

17.50   Recall that ppm means the number of parts of substance per 1,000,000 parts. We can calculate the partial pressure of SO_2 in the troposphere.

$$P_{SO_2} = \frac{0.16 \text{ molecules of SO}_2}{10^6 \text{ parts of air}} \times 1 \text{ atm} = 1.6 \times 10^{-7} \text{ atm}$$

Next, we need to set up the equilibrium constant expression to calculate the concentration of H^+ in the rainwater. From the concentration of H^+, we can calculate the pH.

$$SO_2 \quad + \quad H_2O \; \rightleftharpoons \; H^+ \; + \; HSO_3^-$$

Equilibrium: 1.6×10^{-7} atm $\qquad\qquad x \qquad x$

$$K = \frac{[H^+][HSO_3^-]}{P_{SO_2}} = 1.3 \times 10^{-2}$$

$$1.3 \times 10^{-2} = \frac{x^2}{1.6 \times 10^{-7}}$$

$$x^2 = 2.1 \times 10^{-9}$$

$$x = 4.6 \times 10^{-5} \, M = [H^+]$$

$$\mathbf{pH} = -\log(4.6 \times 10^{-5}) = \mathbf{4.34}$$

17.57   **(a)**   Since this is an elementary reaction, the rate law is:

$$\text{Rate} = k[NO]^2[O_2]$$

(b)   Since $[O_2]$ is very large compared to $[NO]$, then the reaction is a pseudo second-order reaction and the rate law can be simplified to:

$$\text{Rate} = k'[NO]^2$$

where $k' = k[O_2]$

(c)   Since for a second-order reaction

$$t_{\frac{1}{2}} = \frac{1}{k[A]_0}$$

then,

$$\frac{\left(t_{\frac{1}{2}}\right)_1}{\left(t_{\frac{1}{2}}\right)_2} = \frac{[(A_0)_2]}{[(A_0)_1]}$$

$$\frac{6.4 \times 10^3 \text{ min}}{\left(t_{\frac{1}{2}}\right)_2} = \frac{10 \text{ ppm}}{2 \text{ ppm}}$$

Solving, the new half life is:

$$\left(t_{\frac{1}{2}}\right)_2 = 1.3 \times 10^3 \text{ min}$$

You could also solve for k using the half-life and concentration (2 ppm). Then substitute k and the new concentration (10 ppm) into the half-life equation to solve for the new half-life. Try it.

17.58   **Strategy:** This problem gives the volume, temperature, and pressure of PAN. Is the gas undergoing a change in any of its properties? What equation should we use to solve for moles of PAN? Once we have determined moles of PAN, we can convert to molarity and use the first-order rate law to solve for rate.

Solution: Because no changes in gas properties occur, we can use the ideal gas equation to calculate the moles of PAN. 0.55 ppm by volume means:

$$\frac{V_{PAN}}{V_T} = \frac{0.55 \text{ L}}{1 \times 10^6 \text{ L}}$$

Rearranging Equation (5.8) of the text, at STP, the number of moles of PAN in 1.0 L of air is:

$$n = \frac{PV}{RT} = \frac{(1 \text{ atm})\left(\dfrac{0.55 \text{ L}}{1 \times 10^6 \text{ L}} \times 1.0 \text{ L}\right)}{(0.0821 \text{ L} \cdot \text{atm/K} \cdot \text{mol})(273 \text{ K})} = 2.5 \times 10^{-8} \text{ mol}$$

Since the decomposition follows first-order kinetics, we can write:

$$\text{rate} = k[\text{PAN}]$$

$$\textbf{rate} = (4.9 \times 10^{-4} \text{ /s})\left(\frac{2.5 \times 10^{-8} \text{ mol}}{1.0 \text{ L}}\right) = \textbf{1.2} \times \textbf{10}^{-11} \textbf{ M/s}$$

17.59   The volume a gas occupies is directly proportional to the number of moles of gas. Therefore, 0.42 ppm by volume can also be expressed as a mole fraction.

$$X_{O_3} = \frac{n_{O_3}}{n_{\text{total}}} = \frac{0.42}{1 \times 10^6} = 4.2 \times 10^{-7}$$

The partial pressure of ozone can be calculated from the mole fraction and the total pressure.

$$\textbf{P}_{\textbf{O}_3} = X_{O_3} P_T = (4.2 \times 10^{-7})(748 \text{ mmHg}) = (3.14 \times 10^{-4} \text{ mmHg}) \times \frac{1 \text{ atm}}{760 \text{ mmHg}} = \textbf{4.1} \times \textbf{10}^{-7} \textbf{ atm}$$

Substitute into the ideal gas equation to calculate moles of ozone.

$$n_{O_3} = \frac{P_{O_3} V}{RT} = \frac{(4.1 \times 10^{-7} \text{ atm})(1 \text{ L})}{(0.0821 \text{ L} \cdot \text{atm/mol} \cdot \text{K})(293 \text{ K})} = 1.7 \times 10^{-8} \text{ mol}$$

Number of O_3 molecules:

$$(1.7 \times 10^{-8} \text{ mol } O_3) \times \frac{6.022 \times 10^{23} \text{ } O_3 \text{ molecules}}{1 \text{ mol } O_3} = \mathbf{1.0 \times 10^{16} \text{ } O_3 \text{ molecules}}$$

17.60   The Gobi desert lacks the primary pollutants (nitric oxide, carbon monoxide, hydrocarbons) to have photochemical smog. The primary pollutants are present both in New York City and in Boston. However, the sunlight that is required for the conversion of the primary pollutants to the secondary pollutants associated with smog is more likely in a July afternoon than one in January. Therefore, answer **(b)** is correct.

17.65   The room volume is:

$$17.6 \text{ m} \times 8.80 \text{ m} \times 2.64 \text{ m} = 4.09 \times 10^2 \text{ m}^3$$

Since $1 \text{ m}^3 = 1 \times 10^3$ L, then the volume of the container is 4.09×10^5 L. The quantity, 8.00×10^2 ppm is:

$$\frac{8.00 \times 10^2}{1 \times 10^6} = 8.00 \times 10^{-4} = \text{mole fraction of CO}$$

The pressure of the CO(atm) is:

$$P_{CO} = X_{CO}P_T = (8.00 \times 10^{-4})(756 \text{ mmHg}) \times \frac{1 \text{ atm}}{760 \text{ mmHg}} = 7.96 \times 10^{-4} \text{ atm}$$

The moles of CO is:

$$n = \frac{PV}{RT} = \frac{(7.96 \times 10^{-4} \text{ atm})(4.09 \times 10^5 \text{ L})}{(0.0821 \text{ L} \cdot \text{atm/K} \cdot \text{mol})(293 \text{ K})} = 13.5 \text{ mol}$$

The mass of CO in the room is:

$$\mathbf{mass} = 13.5 \text{ mol} \times \frac{28.01 \text{ g CO}}{1 \text{ mol CO}} = \mathbf{378 \text{ g CO}}$$

17.66   **Strategy:** After writing a balanced equation, how do we compare the amounts of $CaCO_3$ and CO_2? We can compare them based on the mole ratio from the balanced equation. Once we have moles of CO_2, we can then calculate moles of air using the ideal gas equation. From the moles of CO_2 and the moles of air, we can calculate the percentage of CO_2 in the air.

Solution: First, we need to write a balanced equation.

$$CO_2 + Ca(OH)_2 \rightarrow CaCO_3 + H_2O$$

The mole ratio between $CaCO_3$ and CO_2 is: 1 mole $CaCO_3 \simeq$ 1 mole CO_2. If we convert grams of $CaCO_3$ to moles of $CaCO_3$, we can use this mole ratio to convert to moles of CO_2. Once moles of CO_2 are known, we can convert to grams CO_2.

Moles of CO_2 reacted:

$$0.026 \text{ g CaCO}_3 \times \frac{1 \text{ mol CaCO}_3}{100.1 \text{ g CaCO}_3} \times \frac{1 \text{ mol CO}_2}{1 \text{ mol CaCO}_3} = 2.6 \times 10^{-4} \text{ mol CO}_2$$

The total number of moles of air can be calculated using the ideal gas equation.

$$n = \frac{PV}{RT} = \frac{\left(747 \text{ mmHg} \times \dfrac{1 \text{ atm}}{760 \text{ mmHg}}\right)(5.0 \text{ L})}{(0.0821 \text{ L} \cdot \text{atm/mol} \cdot \text{K})(291 \text{ K})} = 0.21 \text{ mol air}$$

The percentage by volume of CO_2 in air is:

$$\frac{V_{CO_2}}{V_{air}} \times 100\% = \frac{n_{CO_2}}{n_{air}} \times 100\% = \frac{2.6 \times 10^{-4} \text{ mol}}{0.21 \text{ mol}} \times 100\% = \textbf{0.12\%}$$

17.67 The chapter sections where these gases are discussed are:

O_3: Section 17.7 SO_2: Section 17.6 NO_2: Sections 17.5, 17.7

Rn: Section 17.8   PAN: Section 17.7   CO: Sections 17.5, 17.7, 17.8

17.68 An increase in temperature has shifted the system to the right; the equilibrium constant has increased with an increase in temperature. If we think of heat as a reactant (endothermic)

$$\text{heat} + N_2 + O_2 \rightleftharpoons 2 NO$$

based on Le Châtelier's principle, adding heat would indeed shift the system to the right. Therefore, the reaction is **endothermic**.

17.69 **(a)** From the balanced equation:

$$K_c = \frac{[O_2][HbCO]}{[CO][HbO_2]}$$

(b) Using the information provided:

$$212 = \frac{[O_2][HbCO]}{[CO][HbO_2]} = \frac{[8.6 \times 10^{-3}][HbCO]}{[1.9 \times 10^{-6}][HbO_2]}$$

Solving, the ratio of $HbCO$ to HbO_2 is:

$$\frac{[\textbf{HbCO}]}{[\textbf{HbO}_2]} = \frac{(212)(1.9 \times 10^{-6})}{(8.6 \times 10^{-3})} = \textbf{0.047}$$

17.70 The concentration of O_2 could be monitored. Formation of CO_2 must deplete O_2.

17.71 **(a)**

$$N_2O + O \rightleftharpoons 2NO$$

$$\underline{2NO + 2O_3 \rightleftharpoons 2O_2 + 2NO_2}$$

$$\text{Overall: } N_2O + O + 2O_3 \rightleftharpoons 2O_2 + 2NO_2$$

(b) Compounds with a permanent dipole moment such as N_2O are more effective greenhouse gases than nonpolar species such as CO_2 (Section 17.5 of the text).

(c)   The moles of adipic acid are:

$$(2.2 \times 10^9 \text{ kg adipic acid}) \times \frac{1000 \text{ g}}{1 \text{ kg}} \times \frac{1 \text{ mol adipic acid}}{146.1 \text{ g adipic acid}} = 1.5 \times 10^{10} \text{ mol adipic acid}$$

The number of moles of adipic acid is given as being equivalent to the moles of N_2O produced, and from the overall balanced equation, one mole of N_2O will react with two moles of O_3. Thus,

$$1.5 \times 10^{10} \text{ mol adipic acid} \rightarrow 1.5 \times 10^{10} \text{ mol } N_2O \text{ which reacts with } \textbf{3.0} \times \textbf{10}^{\textbf{10}} \textbf{ mol } \textbf{O}_3.$$

17.72   In Problem 17.6, we determined the partial pressure of CO_2 in dry air to be 3.3×10^{-4} atm. Using Henry's law, we can calculate the concentration of CO_2 in water.

$$c = kP$$

$$[CO_2] = (0.032 \text{ mol/L·atm})(3.3 \times 10^{-4} \text{ atm}) = 1.06 \times 10^{-5} \text{ mol/L}$$

We assume that all of the dissolved CO_2 is converted to H_2CO_3, thus giving us 1.06×10^{-5} mol/L of H_2CO_3. H_2CO_3 is a weak acid. Setup the equilibrium of this acid in water and solve for $[H^+]$.

The equilibrium expression is:

	H_2CO_3	$\rightleftharpoons$	H^+	+	HCO_3^-
Initial (M):	1.06×10^{-5}		0		0
Change (M):	$-x$		$+x$		$+x$
Equilibrium (M):	$(1.06 \times 10^{-5}) - x$		x		x

$$K \text{ (from Table 15.5)} = 4.2 \times 10^{-7} = \frac{[H^+][HCO_3^-]}{[H_2CO_3]} = \frac{x^2}{(1.06 \times 10^{-5}) - x}$$

Solving the quadratic equation:

$$x = 1.9 \times 10^{-6} M = [H^+]$$

$$\textbf{pH} = -\log(1.9 \times 10^{-6}) = \textbf{5.72}$$

17.73   First we calculate the number of ^{222}Rn atoms.

$$\text{Volume of basement} = (14 \text{ m} \times 10 \text{ m} \times 3.0 \text{ m}) = 4.2 \times 10^2 \text{ m}^3 = 4.2 \times 10^5 \text{ L}$$

$$n_{\text{air}} = \frac{PV}{RT} = \frac{(1.0 \text{ atm})(4.2 \times 10^5 \text{ L})}{(0.0821 \text{ L·atm/mol·K})(273 \text{ K})} = 1.9 \times 10^4 \text{ mol air}$$

$$n_{\text{Rn}} = \frac{P_{\text{Rn}}}{P_{\text{air}}} \times (1.9 \times 10^4) = \frac{1.2 \times 10^{-6} \text{ mmHg}}{760 \text{ mmHg}} \times (1.9 \times 10^4 \text{ mol}) = 3.0 \times 10^{-5} \text{ mol Rn}$$

Number of ^{222}Rn atoms at the beginning:

$$(3.0 \times 10^{-5} \text{ mol Rn}) \times \frac{6.022 \times 10^{23} \text{ Rn atoms}}{1 \text{ mol Rn}} = \textbf{1.8} \times \textbf{10}^{\textbf{19}} \textbf{ Rn atoms}$$

$$k = \frac{0.693}{3.8 \text{ d}} = 0.182 \text{ d}^{-1}$$

From Equation (13.3) of the text:

$$\ln \frac{[A]_t}{[A]_0} = -kt$$

$$\ln \frac{x}{1.8 \times 10^{19}} = -(0.182 \text{ d}^{-1})(31 \text{ d})$$

$$x = 6.4 \times 10^{16} \text{ Rn atoms}$$

17.74   **Strategy:** From ΔH_f° values given in Appendix 3 of the text, we can calculate ΔH° for the reaction

$$NO_2 \rightarrow NO + O$$

Then, we can calculate ΔE° from ΔH°. The ΔE° calculated will have units of kJ/mol. If we can convert this energy to units of J/molecule, we can calculate the wavelength required to decompose NO_2.

Solution: We use the ΔH_f° values in Appendix 3 and Equation (6.18) of the text.

$$\Delta H_{rxn}^\circ = \sum n \Delta H_f^\circ (\text{products}) - \sum m \Delta H_f^\circ (\text{reactants})$$

Consider reaction (1):

$$\Delta H^\circ = \Delta H_f^\circ(NO) + \Delta H_f^\circ(O) - \Delta H_f^\circ(NO_2)$$

$$\Delta H^\circ = (1)(90.4 \text{ kJ/mol}) + (1)(249.4 \text{ kJ/mol}) - (1)(33.85 \text{ kJ/mol})$$

$$\Delta H^\circ = 306 \text{ kJ/mol}$$

From Equation (6.10) of the text, $\Delta E^\circ = \Delta H^\circ - RT\Delta n$

$$\Delta E^\circ = (306 \times 10^3 \text{ J/mol}) - (8.314 \text{ J/mol·K})(298 \text{ K})(1)$$

$$\Delta E^\circ = 304 \times 10^3 \text{ J/mol}$$

This is the energy needed to dissociate 1 mole of NO_2. We need the energy required to dissociate *one molecule* of NO_2.

$$\frac{304 \times 10^3 \text{ J}}{1 \text{ mol } NO_2} \times \frac{1 \text{ mol } NO_2}{6.022 \times 10^{23} \text{ molecules } NO_2} = 5.05 \times 10^{-19} \text{ J/molecule}$$

The longest wavelength that can dissociate NO_2 is:

$$\lambda = \frac{hc}{E} = \frac{(6.63 \times 10^{-34} \text{ J·s})(3.00 \times 10^8 \text{ m/s})}{5.05 \times 10^{-19} \text{ J}} = 3.94 \times 10^{-7} \text{ m} = 394 \text{ nm}$$

17.75   **(a)**   Its small concentration is the result of its high reactivity.

(b)   OH has a great tendency to abstract an H atom from another compound because of the large energy of the O–H bond (see Table 9.4 of the text).

(c) $NO_2 + OH \rightarrow HNO_3$

(d) $OH + SO_2 \rightarrow HSO_3$

$HSO_3 + O_2 + H_2O \rightarrow H_2SO_4 + HO_2$

17.76   This reaction has a high activation energy.

17.77   The blackened bucket has a large deposit of elemental carbon. When heated over the burner, it forms poisonous carbon monoxide.

$$C + CO_2 \rightarrow 2CO$$

A smaller amount of CO is also formed as follows:

$$2C + O_2 \rightarrow 2CO$$

17.78   The size of tree rings can be related to CO_2 content, where the number of rings indicates the age of the tree. The amount of CO_2 in ice can be directly measured from portions of polar ice in different layers obtained by drilling. The "age" of CO_2 can be determined by radiocarbon dating and other methods.

17.79   The use of the aerosol can liberate CFC's that destroy the ozone layer.

17.80 $Cl_2 + O_2 \rightarrow 2ClO$

$\Delta H^\circ = \Sigma BE(\text{reactants}) - \Sigma BE(\text{products})$

$\Delta H^\circ = (1)(242.7 \text{ kJ/mol}) + (1)(498.7 \text{ kJ/mol}) - (2)(206 \text{ kJ/mol})$

$\Delta H^\circ = 329 \text{ kJ/mol}$

$\Delta H^\circ = 2\Delta H_f^\circ(ClO) - 2\Delta H_f^\circ(Cl_2) - 2\Delta H_f^\circ(O_2)$

$329 \text{ kJ/mol} = 2\Delta H_f^\circ(ClO) - 0 - 0$

$\Delta H_f^\circ(ClO) = \dfrac{329 \text{ kJ/mol}}{2} = \textbf{165 kJ/mol}$

CHAPTER 18
ENTROPY, FREE ENERGY,
AND EQUILIBRIUM

18.6 The probability (P) of finding all the molecules in the same flask becomes progressively smaller as the number of molecules increases. An equation that relates the probability to the number of molecules is given in the text.

$$P = \left(\frac{1}{2}\right)^N$$

where,

N is the total number of molecules present.

Using the above equation, we find:

(a) $P = 0.02$ (b) $P = 9 \times 10^{-19}$ (c) $P = 2 \times 10^{-181}$

18.9 (a) This is easy. The liquid form of any substance always has greater entropy (more microstates).

(b) This is hard. At first glance there may seem to be no apparent difference between the two substances that might affect the entropy (molecular formulas identical). However, the first has the –O–H structural feature which allows it to participate in hydrogen bonding with other molecules. This allows a more ordered arrangement of molecules in the liquid state. The standard entropy of CH_3OCH_3 is larger.

(c) This is also difficult. Both are monatomic species. However, the Xe atom has a greater molar mass than Ar. Xenon has the higher standard entropy.

(d) Same argument as part (c). Carbon dioxide gas has the higher standard entropy (see Appendix 3).

(e) O_3 has a greater molar mass than O_2 and thus has the higher standard entropy.

(f) Using the same argument as part (c), one mole of N_2O_4 has a larger standard entropy than one mole of NO_2. Compare values in Appendix 3.

Use the data in Appendix 3 to compare the standard entropy of one mole of N_2O_4 with that of two moles of NO_2. In this situation the number of atoms is the same for both. Which is higher and why?

18.10 In order of increasing entropy per mole at 25°C:

(c) < (d) < (e) < (a) < (b)

(c) Na(s): ordered, crystalline material.
(d) NaCl(s): ordered crystalline material, but with more particles per mole than Na(s).
(e) H_2: a diatomic gas, hence of higher entropy than a solid.
(a) Ne(g): a monatomic gas of higher molar mass than H_2.
(b) $SO_2(g)$: a polyatomic gas of higher molar mass than Ne.

18.11 Using Equation (18.7) of the text to calculate ΔS_{rxn}°

(a) $\Delta S_{rxn}^{\circ} = S^{\circ}(SO_2) - [S^{\circ}(O_2) + S^{\circ}(S)]$

$$\Delta S^\circ_{rxn} = (1)(248.5 \text{ J/K} \cdot \text{mol}) - (1)(205.0 \text{ J/K} \cdot \text{mol}) - (1)(31.88 \text{ J/K} \cdot \text{mol}) = \textbf{11.6 J/K} \cdot \textbf{mol}$$

(b) $\Delta S^\circ_{rxn} = S^\circ(\text{MgO}) + S^\circ(\text{CO}_2) - S^\circ(\text{MgCO}_3)$

$$\Delta S^\circ_{rxn} = (1)(26.78 \text{ J/K} \cdot \text{mol}) + (1)(213.6 \text{ J/K} \cdot \text{mol}) - (1)(65.69 \text{ J/K} \cdot \text{mol}) = \textbf{174.7 J/K} \cdot \textbf{mol}$$

18.12   **Strategy:** To calculate the standard entropy change of a reaction, we look up the standard entropies of reactants and products in Appendix 3 of the text and apply Equation (18.7). As in the calculation of enthalpy of reaction, the stoichiometric coefficients have no units, so ΔS°_{rxn} is expressed in units of J/K·mol.

Solution: The standard entropy change for a reaction can be calculated using the following equation.

$$\Delta S^\circ_{rxn} = \Sigma n S^\circ(\text{products}) - \Sigma m S^\circ(\text{reactants})$$

(a) $\Delta S^\circ_{rxn} = S^\circ(\text{Cu}) + S^\circ(\text{H}_2\text{O}) - [S^\circ(\text{H}_2) + S^\circ(\text{CuO})]$

$$= (1)(33.3 \text{ J/K·mol}) + (1)(188.7 \text{ J/K·mol}) - [(1)(131.0 \text{ J/K·mol}) + (1)(43.5 \text{ J/K·mol})]$$

$$= \textbf{47.5 J/K·mol}$$

(b) $\Delta S^\circ_{rxn} = S^\circ(\text{Al}_2\text{O}_3) + 3S^\circ(\text{Zn}) - [2S^\circ(\text{Al}) + 3S^\circ(\text{ZnO})]$

$$= (1)(50.99 \text{ J/K·mol}) + (3)(41.6 \text{ J/K·mol}) - [(2)(28.3 \text{ J/K·mol}) + (3)(43.9 \text{ J/K·mol})]$$

$$= \textbf{-12.5 J/K·mol}$$

(c) $\Delta S^\circ_{rxn} = S^\circ(\text{CO}_2) + 2S^\circ(\text{H}_2\text{O}) - [S^\circ(\text{CH}_4) + 2S^\circ(\text{O}_2)]$

$$= (1)(213.6 \text{ J/K·mol}) + (2)(69.9 \text{ J/K·mol}) - [(1)(186.2 \text{ J/K·mol}) + (2)(205.0 \text{ J/K·mol})]$$

$$= \textbf{-242.8 J/K·mol}$$

Why was the entropy value for water different in parts (a) and (c)?

18.13   All parts of this problem rest on two principles. First, the entropy of a solid is always less than the entropy of a liquid, and the entropy of a liquid is always much smaller than the entropy of a gas. Second, in comparing systems in the same phase, the one with the most complex particles has the higher entropy.

(a)   Positive entropy change (increase). One of the products is in the gas phase (more microstates).

(b)   Negative entropy change (decrease). Liquids have lower entropies than gases.

(c)   Positive. Same as (a).

(d)   Positive. There are two gas-phase species on the product side and only one on the reactant side.

18.14   **(a)** $\Delta S < 0$; gas reacting with a liquid to form a solid (decrease in number of moles of gas, hence a decrease in microstates).

(b) $\Delta S > 0$; solid decomposing to give a liquid and a gas (an increase in microstates).

(c) $\Delta S > 0$; increase in number of moles of gas (an increase in microstates).

(d) $\Delta S < 0$; gas reacting with a solid to form a solid (decrease in number of moles of gas, hence a decrease in microstates).

18.17 Using Equation (18.12) of the text to solve for the change in standard free energy,

(a) $\Delta G° = 2\Delta G_f°(NO) - \Delta G_f°(N_2) - \Delta G_f°(O_2) = (2)(86.7 \text{ kJ/mol}) - 0 - 0 = $ **173.4 kJ/mol**

(b) $\Delta G° = \Delta G_f°[H_2O(g)] - \Delta G_f°[H_2O(l)] = (1)(-228.6 \text{ kJ/mol}) - (1)(-237.2 \text{ kJ/mol}) = $ **8.6 kJ/mol**

(c) $\Delta G° = 4\Delta G_f°(CO_2) + 2\Delta G_f°(H_2O) - 2\Delta G_f°(C_2H_2) - 5\Delta G_f°(O_2)$

$= (4)(-394.4 \text{ kJ/mol}) + (2)(-237.2 \text{ kJ/mol}) - (2)(209.2 \text{ kJ/mol}) - (5)(0) = $ **−2470 kJ/mol**

18.18 **Strategy:** To calculate the standard free-energy change of a reaction, we look up the standard free energies of formation of reactants and products in Appendix 3 of the text and apply Equation (18.12). Note that all the stoichiometric coefficients have no units so $\Delta G_{rxn}°$ is expressed in units of kJ/mol. The standard free energy of formation of any element in its stable allotropic form at 1 atm and 25°C is zero.

Solution: The standard free energy change for a reaction can be calculated using the following equation.

$$\Delta G_{rxn}° = \Sigma n \Delta G_f°(\text{products}) - \Sigma m \Delta G_f°(\text{reactants})$$

(a) $\Delta G_{rxn}° = 2\Delta G_f°(MgO) - [2\Delta G_f°(Mg) + \Delta G_f°(O_2)]$

$\Delta G_{rxn}° = (2)(-569.6 \text{ kJ/mol}) - [(2)(0) + (1)(0)] = $ **−1139 kJ/mol**

(b) $\Delta G_{rxn}° = 2\Delta G_f°(SO_3) - [2\Delta G_f°(SO_2) + \Delta G_f°(O_2)]$

$\Delta G_{rxn}° = (2)(-370.4 \text{ kJ/mol}) - [(2)(-300.4 \text{ kJ/mol}) + (1)(0)] = $ **−140.0 kJ/mol**

(c) $\Delta G_{rxn}° = 4\Delta G_f°[CO_2(g)] + 6\Delta G_f°[H_2O(l)] - \{2\Delta G_f°[C_2H_6(g)] + 7\Delta G_f°[O_2(g)]\}$

$\Delta G_{rxn}° = (4)(-394.4 \text{ kJ/mol}) + (6)(-237.2 \text{ kJ/mol}) - [(2)(-32.89 \text{ kJ/mol}) + (7)(0)] = $ **−2935.0 kJ/mol**

18.19 Reaction A: First apply Equation (18.10) of the text to compute the free energy change at 25°C (298 K)

$$\Delta G = \Delta H - T\Delta S = 10,500 \text{ J/mol} - (298 \text{ K})(30 \text{ J/K·mol}) \doteq 1560 \text{ J/mol}$$

The +1560 J/mol shows the reaction is not spontaneous at 298 K. The ΔG will change sign (i.e., the reaction will become spontaneous) above the temperature at which $\Delta G = 0$.

$$T = \frac{\Delta H}{\Delta S} = \frac{10500 \text{ J/mol}}{30 \text{ J/K·mol}} = \textbf{350 K}$$

Reaction B: Calculate ΔG.

$$\Delta G = \Delta H - T\Delta S = 1800 \text{ J/mol} - (298 \text{ K})(-113 \text{ J/K·mol}) = \textbf{35,500 J/mol}$$

The free energy change is positive, which shows that the reaction is not spontaneous at 298 K. Since both terms are positive, there is no temperature at which their sum is negative. The reaction is not spontaneous at any temperature.

18.20   Reaction A: Calculate ΔG from ΔH and ΔS.

$$\Delta G = \Delta H - T\Delta S = -126,000 \text{ J/mol} - (298 \text{ K})(84 \text{ J/K} \cdot \text{mol}) = -151,000 \text{ J/mol}$$

The free energy change is negative so the reaction is spontaneous at 298 K. Since ΔH is negative and ΔS is positive, **the reaction is spontaneous at all temperatures**.

Reaction B: Calculate ΔG.

$$\Delta G = \Delta H - T\Delta S = -11,700 \text{ J/mol} - (298 \text{ K})(-105 \text{ J/K} \cdot \text{mol}) = +19,600 \text{ J}$$

The free energy change is positive at 298 K which means the reaction is not spontaneous at that temperature. The positive sign of ΔG results from the large negative value of ΔS. At lower temperatures, the $-T\Delta S$ term will be smaller thus allowing the free energy change to be negative.

ΔG will equal zero when $\Delta H = T\Delta S$.

Rearranging,

$$T = \frac{\Delta H}{\Delta S} = \frac{-11700 \text{ J/mol}}{-105 \text{ J/K} \cdot \text{mol}} = \textbf{111 K}$$

At temperatures **below 111 K**, ΔG will be negative and the reaction will be spontaneous.

18.23   Find the value of K by solving Equation (18.14) of the text.

$$K_p = e^{\frac{-\Delta G^\circ}{RT}} = e^{\frac{-2.60 \times 10^3 \text{ J/mol}}{(8.314 \text{ J/K} \cdot \text{mol})(298 \text{ K})}} = e^{-1.05} = \textbf{0.35}$$

18.24   Strategy: According to Equation (18.14) of the text, the equilibrium constant for the reaction is related to the standard free energy change; that is, $\Delta G^\circ = -RT \ln K$. Since we are given the equilibrium constant in the problem, we can solve for ΔG°. What temperature unit should be used?

Solution: The equilibrium constant is related to the standard free energy change by the following equation.

$$\Delta G^\circ = -RT \ln K$$

Substitute K_w, R, and T into the above equation to calculate the standard free energy change, ΔG°. The temperature at which $K_w = 1.0 \times 10^{-14}$ is 25°C = 298 K.

$$\Delta G^\circ = -RT \ln K_w$$

$$\Delta G^\circ = -(8.314 \text{ J/mol} \cdot \text{K})(298 \text{ K}) \ln(1.0 \times 10^{-14}) = \textbf{8.0} \times \textbf{10}^4 \text{ J/mol} = \textbf{8.0} \times \textbf{10}^1 \text{ kJ/mol}$$

18.25 $K_{sp} = [Fe^{2+}][OH^-]^2 = 1.6 \times 10^{-14}$

$$\Delta G^\circ = -RT \ln K_{sp} = -(8.314 \text{ J/K} \cdot \text{mol})(298 \text{ K}) \ln(1.6 \times 10^{-14}) = \textbf{7.9} \times \textbf{10}^4 \text{ J/mol} = \textbf{79 kJ/mol}$$

18.26   Use standard free energies of formation from Appendix 3 to find the standard free energy difference.

$$\Delta G^\circ_{rxn} = 2\Delta G^\circ_f[H_2(g)] + \Delta G^\circ_f[O_2(g)] - 2\Delta G^\circ_f[H_2O(g)]$$

$$\Delta G^\circ_{rxn} = (2)(0) + (1)(0) - (2)(-228.6 \text{ kJ/mol})$$

$$\Delta G^\circ_{rxn} = \textbf{457.2 kJ/mol} = \textbf{4.572} \times \textbf{10}^5 \text{ J/mol}$$

We can calculate K_p using the following equation.

$$\Delta G° = -RT \ln K_p$$

$$4.572 \times 10^5 \text{ J/mol} = -(8.314 \text{ J/mol·K})(298 \text{ K}) \ln K_p$$

$$-185 = \ln K_p$$

Taking the antiln of both sides,

$$e^{-185} = K_p$$

$$K_p = 4.5 \times 10^{-81}$$

18.27 **(a)** We first find the standard free energy change of the reaction.

$$\Delta G°_{rxn} = \Delta G°_f[PCl_3(g)] + \Delta G°_f[Cl_2(g)] - \Delta G°_f[PCl_5(g)]$$

$$= (1)(-286 \text{ kJ/mol}) + (1)(0) - (1)(-325 \text{ kJ/mol}) = \textbf{39 kJ/mol}$$

We can calculate K_p using Equation (18.14) of the text.

$$K_p = e^{\frac{-\Delta G°}{RT}} = e^{\frac{-39 \times 10^3 \text{ J/mol}}{(8.314 \text{ J/K·mol})(298 \text{ K})}} = e^{-16} = \textbf{1} \times \textbf{10}^{-7}$$

(b) We are finding the free energy difference between the reactants and the products at their nonequilibrium values. The result tells us the direction of and the potential for further chemical change. We use the given nonequilibrium pressures to compute Q_p.

$$Q_p = \frac{P_{PCl_3} P_{Cl_2}}{P_{PCl_5}} = \frac{(0.27)(0.40)}{0.0029} = 37$$

The value of ΔG (notice that this is not the standard free energy difference) can be found using Equation (18.13) of the text and the result from part (a).

$$\Delta G = \Delta G° + RT \ln Q = (39 \times 10^3 \text{ J/mol}) + (8.314 \text{ J/K·mol})(298 \text{ K}) \ln(37) = \textbf{48 kJ/mol}$$

Which way is the direction of spontaneous change for this system? What would be the value of ΔG if the given data were equilibrium pressures? What would be the value of Q_p in that case?

18.28 **(a)** The equilibrium constant is related to the standard free energy change by the following equation.

$$\Delta G° = -RT \ln K$$

Substitute K_p, R, and T into the above equation to the standard free energy change, $\Delta G°$.

$$\Delta G° = -RT \ln K_p$$

$$\Delta G° = -(8.314 \text{ J/mol·K})(2000 \text{ K}) \ln(4.40) = \textbf{-2.46} \times \textbf{10}^4 \text{ J/mol} = \textbf{-24.6 kJ/mol}$$

(b)
Strategy: From the information given we see that neither the reactants nor products are at their standard state of 1 atm. We use Equation (18.13) of the text to calculate the free-energy change under non-standard-state conditions. Note that the partial pressures are expressed as dimensionless quantities in the reaction quotient Q_p.

Solution: Under non-standard-state conditions, ΔG is related to the reaction quotient Q by the following equation.

$$\Delta G = \Delta G° + RT \ln Q_p$$

We are using Q_p in the equation because this is a gas-phase reaction.

Step 1: $\Delta G°$ was calculated in part (a). We must calculate Q_p.

$$Q_p = \frac{P_{H_2O} \cdot P_{CO}}{P_{H_2} \cdot P_{CO_2}} = \frac{(0.66)(1.20)}{(0.25)(0.78)} = 4.1$$

Step 2: Substitute $\Delta G° = -2.46 \times 10^4$ J/mol and Q_p into the following equation to calculate ΔG.

$$\Delta G = \Delta G° + RT \ln Q_p$$

$$\Delta G = -2.46 \times 10^4 \text{ J/mol} + (8.314 \text{ J/mol·K})(2000 \text{ K}) \ln(4.1)$$

$$\Delta G = (-2.46 \times 10^4 \text{ J/mol}) + (2.35 \times 10^4 \text{ J/mol})$$

$$\mathbf{\Delta G = -1.10 \times 10^3 \text{ J/mol} = -1.10 \text{ kJ/mol}}$$

18.29   The expression of K_p is: $K_p = P_{CO_2}$

Thus you can predict the equilibrium pressure directly from the value of the equilibrium constant. The only task at hand is computing the values of K_p using Equations (18.10) and (18.14) of the text.

(a)   At 25°C, $\Delta G° = \Delta H° - T\Delta S° = (177.8 \times 10^3 \text{ J/mol}) - (298 \text{ K})(160.5 \text{ J/K·mol}) = 130.0 \times 10^3$ J/mol

$$P_{CO_2} = K_p = e^{\frac{-\Delta G°}{RT}} = e^{\frac{-130.0 \times 10^3 \text{ J/mol}}{(8.314 \text{ J/K·mol})(298 \text{ K})}} = e^{-52.47} = \mathbf{1.6 \times 10^{-23} \text{ atm}}$$

(b)   At 800°C, $\Delta G° = \Delta H° - T\Delta S° = (177.8 \times 10^3 \text{ J/mol}) - (1073 \text{ K})(160.5 \text{ J/K·mol}) = 5.58 \times 10^3$ J/mol

$$P_{CO_2} = K_p = e^{\frac{-\Delta G°}{RT}} = e^{\frac{-5.58 \times 10^3 \text{ J/mol}}{(8.314 \text{ J/K·mol})(1073 \text{ K})}} = e^{-0.625} = \mathbf{0.535 \text{ atm}}$$

What assumptions are made in the second calculation?

18.30   We use the given K_p to find the standard free energy change.

$$\Delta G° = -RT \ln K$$

$$\Delta G° = -(8.314 \text{ J/K·mol})(298 \text{ K}) \ln(5.62 \times 10^{35}) = 2.04 \times 10^5 \text{ J/mol} = -204 \text{ kJ/mol}$$

The standard free energy of formation of one mole of $COCl_2$ can now be found using the standard free energy of reaction calculated above and the standard free energies of formation of $CO(g)$ and $Cl_2(g)$.

$$\Delta G°_{rxn} = \Sigma n \Delta G°_f(\text{products}) - \Sigma m \Delta G°_f(\text{reactants})$$

$$\Delta G°_{rxn} = \Delta G°_f[COCl_2(g)] - \{\Delta G°_f[CO(g)] + \Delta G°_f[Cl_2(g)]\}$$

$$-204 \text{ kJ/mol} = (1)\Delta G°_f[COCl_2(g)] - [(1)(-137.3 \text{ kJ/mol}) + (1)(0)]$$

$$\mathbf{\Delta G°_f[COCl_2(g)] = -341 \text{ kJ/mol}}$$

18.31   The equilibrium constant expression is: $K_p = P_{H_2O}$

We are actually finding the equilibrium vapor pressure of water (compare to Problem 18.29). We use Equation (18.14) of the text.

$$P_{H_2O} = K_p = e^{\frac{-\Delta G^\circ}{RT}} = e^{\frac{-8.6 \times 10^3 \text{ J/mol}}{(8.314 \text{ J/K}\cdot\text{mol})(298 \text{ K})}} = e^{-3.47} = \mathbf{3.1 \times 10^{-2} \text{ atm}}$$

The positive value of ΔG° implies that reactants are favored at equilibrium at 25°C. Is that what you would expect?

18.32   The standard free energy change is given by:

$$\Delta G^\circ_{rxn} = \Delta G^\circ_f(\text{graphite}) - \Delta G^\circ_f(\text{diamond})$$

You can look up the standard free energy of formation values in Appendix 3 of the text.

$$\Delta G^\circ_{rxn} = (1)(0) - (1)(2.87 \text{ kJ/mol}) = \mathbf{-2.87 \text{ kJ/mol}}$$

Thus, the formation of graphite from diamond is **favored** under standard-state conditions at 25°C. However, the rate of the diamond to graphite conversion is very slow (due to a high activation energy) so that it will take millions of years before the process is complete.

18.35 $C_6H_{12}O_6 + 6O_2 \rightarrow 6CO_2 + 6H_2O$ $\Delta G^\circ = -2880 \text{ kJ/mol}$

$ADP + H_3PO_4 \rightarrow ATP + H_2O$ $\Delta G^\circ = +31 \text{ kJ/mol}$

Maximum number of ATP molecules synthesized:

$$2880 \text{ kJ/mol} \times \frac{1 \text{ ATP molecule}}{31 \text{ kJ/mol}} = \mathbf{93 \text{ ATP molecules}}$$

18.36   The equation for the coupled reaction is:

$$\textbf{glucose} + \textbf{ATP} \rightarrow \textbf{glucose 6–phosphate} + \textbf{ADP}$$

$$\Delta G^\circ = 13.4 \text{ kJ/mol} - 31 \text{ kJ/mol} = -18 \text{ kJ/mol}$$

As an estimate:

$$\ln K = \frac{-\Delta G^\circ}{RT}$$

$$\ln K = \frac{-(-18 \times 10^3 \text{ J/mol})}{(8.314 \text{ J/K}\cdot\text{mol})(298 \text{ K})} = 7.3$$

$$K = \mathbf{1 \times 10^3}$$

18.37   When Humpty broke into pieces, he became more disordered (spontaneously). The king was unable to reconstruct Humpty.

18.38   In each part of this problem we can use the following equation to calculate ΔG.

$$\Delta G = \Delta G^\circ + RT \ln Q$$

or,

$$\Delta G = \Delta G^\circ + RT \ln [H^+][OH^-]$$

(a)   In this case, the given concentrations are equilibrium concentrations at 25°C. Since the reaction is at equilibrium, $\Delta G = 0$. This is advantageous, because it allows us to calculate ΔG°. Also recall that at equilibrium, $Q = K$. We can write:

$$\Delta G^\circ = -RT \ln K_w$$

$$\Delta G^\circ = -(8.314 \text{ J/K·mol})(298 \text{ K}) \ln(1.0 \times 10^{-14}) = 8.0 \times 10^4 \text{ J/mol}$$

(b) $\Delta G = \Delta G^\circ + RT \ln Q = \Delta G^\circ + RT \ln [H^+][OH^-]$

$$\Delta G = (8.0 \times 10^4 \text{ J/mol}) + (8.314 \text{ J/K·mol})(298 \text{ K}) \ln [(1.0 \times 10^{-3})(1.0 \times 10^{-4})] = \mathbf{4.0 \times 10^4 \text{ J/mol}}$$

(c) $\Delta G = \Delta G^\circ + RT \ln Q = \Delta G^\circ + RT \ln [H^+][OH^-]$

$$\Delta G = (8.0 \times 10^4 \text{ J/mol}) + (8.314 \text{ J/K·mol})(298 \text{ K}) \ln [(1.0 \times 10^{-12})(2.0 \times 10^{-8})] = \mathbf{-3.2 \times 10^4 \text{ J/mol}}$$

(d) $\Delta G = \Delta G^\circ + RT \ln Q = \Delta G^\circ + RT \ln [H^+][OH^-]$

$$\Delta G = (8.0 \times 10^4 \text{ J/mol}) + (8.314 \text{ J/K·mol})(298 \text{ K}) \ln [(3.5)(4.8 \times 10^{-4})] = \mathbf{6.4 \times 10^4 \text{ J/mol}}$$

18.39   Only E and H are associated with the first law alone.

18.40   One possible explanation is simply that no reaction is possible, namely that there is an unfavorable free energy difference between products and reactants ($\Delta G > 0$).

A second possibility is that the potential for spontaneous change is there ($\Delta G < 0$), but that the reaction is extremely slow (very large activation energy).

A remote third choice is that the student accidentally prepared a mixture in which the components were already at their equilibrium concentrations.

Which of the above situations would be altered by the addition of a catalyst?

18.41   **(a)**   An ice cube melting in a glass of water at 20°C. The value of ΔG for this process is negative so it must be spontaneous.

(b)   A "perpetual motion" machine. In one version, a model has a flywheel which, once started up, drives a generator which drives a motor which keeps the flywheel running at a constant speed and also lifts a weight.

(c)   A perfect air conditioner; it extracts heat energy from the room and warms the outside air without using any energy to do so. (Note: this process does not violate the first law of thermodynamics.)

(d)   Same example as (a).

(e)   A closed flask at 25°C containing $NO_2(g)$ and $N_2O_4(g)$ at equilibrium.

18.42   For a solid to liquid phase transition (melting) the entropy always increases ($\Delta S > 0$) and the reaction is always endothermic ($\Delta H > 0$).

(a)   Melting is always spontaneous above the melting point, so $\Delta G < 0$.

(b)   At the melting point (−77.7°C), solid and liquid are in equilibrium, so $\Delta G = 0$.

(c)   Melting is not spontaneous below the melting point, so $\Delta G > 0$.

18.43   For a reaction to be spontaneous, ΔG must be negative. If ΔS is negative, as it is in this case, then the reaction must be exothermic (why?). When water freezes, it gives off heat (exothermic). Consequently, the entropy of the surroundings increases and $\Delta S_{universe} > 0$.

18.44   If the process is *spontaneous* as well as *endothermic*, the signs of ΔG and ΔH must be negative and positive, respectively. Since $\Delta G = \Delta H - T\Delta S$, the sign of **$\Delta S$ must be positive ($\Delta S > 0$)** for ΔG to be negative.

18.45   The equation is: $BaCO_3(s) \rightleftharpoons BaO(s) + CO_2(g)$

$$\Delta G° = \Delta G_f°(BaO) + \Delta G_f°(CO_2) - \Delta G_f°(BaCO_3)$$

$$\Delta G° = (1)(-528.4 \text{ kJ/mol}) + (1)(-394.4 \text{ kJ/mol}) - (1)(-1138.9 \text{ kJ/mol}) = 216.1 \text{ kJ/mol}$$

$$\Delta G° = -RT\ln K_p$$

$$\ln K_p = \frac{-2.16 \times 10^5 \text{ J/mol}}{(8.314 \text{ J/K} \cdot \text{mol})(298 \text{ K})} = -87.2$$

$$K_p = P_{CO_2} = e^{-87.2} = 1 \times 10^{-38} \text{ atm}$$

18.46   **(a)**   Using the relationship:

$$\frac{\Delta H_{vap}}{T_{b.p.}} = \Delta S_{vap} \approx 90 \text{ J/K} \cdot \text{mol}$$

benzene $\Delta S_{vap} = 87.8 \text{ J/K·mol}$

hexane $\Delta S_{vap} = 90.1 \text{ J/K·mol}$

mercury $\Delta S_{vap} = 93.7 \text{ J/K·mol}$

toluene $\Delta S_{vap} = 91.8 \text{ J/K·mol}$

Most liquids have ΔS_{vap} approximately equal to a constant value because the order of the molecules in the liquid state is similar. The order of most gases is totally random; thus, ΔS for liquid → vapor should be similar for most liquids.

(b)   Using the data in Table 11.6 of the text, we find:

ethanol $\Delta S_{vap} = 111.9 \text{ J/K·mol}$

water $\Delta S_{vap} = 109.4 \text{ J/K·mol}$

Both water and ethanol have a larger ΔS_{vap} because the liquid molecules are more ordered due to hydrogen bonding (there are fewer microstates in these liquids).

18.47   Evidence shows that HF, which is strongly hydrogen-bonded in the liquid phase, is still considerably hydrogen-bonded in the vapor state such that its ΔS_{vap} is smaller than most other substances.

18.48   **(a)** $2CO + 2NO \rightarrow 2CO_2 + N_2$

(b)   The oxidizing agent is NO; the reducing agent is CO.

(c) $\Delta G° = 2\Delta G_f°(CO_2) + \Delta G_f°(N_2) - 2\Delta G_f°(CO) - 2\Delta G_f°(NO)$

$\Delta G° = (2)(-394.4 \text{ kJ/mol}) + (0) - (2)(-137.3 \text{ kJ/mol}) - (2)(86.7 \text{ kJ/mol}) = -687.6 \text{ kJ/mol}$

$\Delta G° = -RT\ln K_p$

$\ln K_p = \dfrac{6.876 \times 10^5 \text{ J/mol}}{(8.314 \text{ J/K} \cdot \text{mol})(298 \text{ K})} = 278$

$K_p = 1 \times 10^{121}$

(d) $Q_p = \dfrac{P_{N_2} P_{CO_2}^2}{P_{CO}^2 P_{NO}^2} = \dfrac{(0.80)(0.030)^2}{(5.0 \times 10^{-5})^2 (5.0 \times 10^{-7})^2} = 1.2 \times 10^{18}$

Since $Q_p \ll K_p$, the reaction will proceed from **left to right**.

(e) $\Delta H° = 2\Delta H_f°(CO_2) + \Delta H_f°(N_2) - 2\Delta H_f°(CO) - 2\Delta H_f°(NO)$

$\Delta H° = (2)(-393.5 \text{ kJ/mol}) + (0) - (2)(-110.5 \text{ kJ/mol}) - (2)(90.4 \text{ kJ/mol}) = -746.8 \text{ kJ/mol}$

Since $\Delta H°$ is negative, raising the temperature will decrease K_p, thereby increasing the amount of reactants and decreasing the amount of products. **No**, the formation of N_2 and CO_2 is not favored by raising the temperature.

18.49   **(a)**   At two different temperatures T_1 and T_2,

$$\Delta G_1° = \Delta H° - T_1\Delta S° = -RT\ln K_1 \qquad (1)$$

$$\Delta G_2° = \Delta H° - T_2\Delta S° = -RT\ln K_2 \qquad (2)$$

Rearranging Equations (1) and (2),

$$\ln K_1 = \dfrac{-\Delta H°}{RT_1} + \dfrac{\Delta S°}{R} \qquad (3)$$

$$\ln K_2 = \dfrac{-\Delta H°}{RT_2} + \dfrac{\Delta S°}{R} \qquad (4)$$

Subtracting equation (3) from equation (4) gives,

$$\ln K_2 - \ln K_1 = \left(\dfrac{-\Delta H°}{RT_2} + \dfrac{\Delta S°}{R}\right) - \left(\dfrac{-\Delta H°}{RT_1} + \dfrac{\Delta S°}{R}\right)$$

$$\ln \dfrac{K_2}{K_1} = \dfrac{\Delta H°}{R}\left(\dfrac{1}{T_1} - \dfrac{1}{T_2}\right)$$

$$\ln\frac{K_2}{K_1} = \frac{\Delta H^\circ}{R}\left(\frac{T_2 - T_1}{T_1 T_2}\right)$$

(b)   Using the equation that we just derived, we can calculate the equilibrium constant at 65°C.

$K_1 = 4.63 \times 10^{-3}$ $T_1 = 298$ K

$K_2 = ?$ $T_2 = 338$ K

$$\ln\frac{K_2}{4.63 \times 10^{-3}} = \frac{58.0 \times 10^3 \text{ J/mol}}{8.314 \text{ J/K} \cdot \text{mol}}\left(\frac{338 \text{ K} - 298 \text{ K}}{(338 \text{ K})(298 \text{ K})}\right)$$

$$\ln\frac{K_2}{4.63 \times 10^{-3}} = 2.77$$

Taking the antiln of both sides of the equation,

$$\frac{K_2}{4.63 \times 10^{-3}} = e^{2.77}$$

$$K_2 = 0.074$$

$K_2 > K_1$, as we would predict for a positive ΔH°. Recall that an increase in temperature will shift the equilibrium towards the endothermic reaction; that is, the decomposition of N_2O_4.

18.50   The equilibrium reaction is:

$$AgCl(s) \rightleftharpoons Ag^+(aq) + Cl^-(aq)$$

$$K_{sp} = [Ag^+][Cl^-] = 1.6 \times 10^{-10}$$

We can calculate the standard enthalpy of reaction from the standard enthalpies of formation in Appendix 3 of the text.

$$\Delta H^\circ = \Delta H_f^\circ(Ag^+) + \Delta H_f^\circ(Cl^-) - \Delta H_f^\circ(AgCl)$$

$$\Delta H^\circ = (1)(105.9 \text{ kJ/mol}) + (1)(-167.2 \text{ kJ/mol}) - (1)(-127.0 \text{ kJ/mol}) = 65.7 \text{ kJ/mol}$$

From Problem 18.49(a):

$$\ln\frac{K_2}{K_1} = \frac{\Delta H^\circ}{R}\left(\frac{T_2 - T_1}{T_1 T_2}\right)$$

$K_1 = 1.6 \times 10^{-10}$ $T_1 = 298$ K

$K_2 = ?$ $T_2 = 333$ K

$$\ln\frac{K_2}{1.6 \times 10^{-10}} = \frac{6.57 \times 10^4 \text{ J}}{8.314 \text{ J/K} \cdot \text{mol}}\left(\frac{333 \text{ K} - 298 \text{ K}}{(333 \text{ K})(298 \text{ K})}\right)$$

$$\ln\frac{K_2}{1.6 \times 10^{-10}} = 2.79$$

$$\frac{K_2}{1.6 \times 10^{-10}} = e^{2.79}$$

$$K_2 = 2.6 \times 10^{-9}$$

The increase in K indicates that the solubility increases with temperature.

18.51   At absolute zero. A substance can never have a negative entropy.

18.52   Assuming that both $\Delta H°$ and $\Delta S°$ are temperature independent, we can calculate both $\Delta H°$ and $\Delta S°$.

$$\Delta H° = \Delta H_f^\circ(CO) + \Delta H_f^\circ(H_2) - [\Delta H_f^\circ(H_2O) + \Delta H_f^\circ(C)]$$

$$\Delta H° = (1)(-110.5 \text{ kJ/mol}) + (1)(0)] - [(1)(-241.8 \text{ kJ/mol}) + (1)(0)]$$

$$\Delta H° = 131.3 \text{ kJ/mol}$$

$$\Delta S° = S°(CO) + S°(H_2) - [S°(H_2O) + S°(C)]$$

$$\Delta S° = [(1)(197.9 \text{ J/K·mol}) + (1)(131.0 \text{ J/K·mol})] - [(1)(188.7 \text{ J/K·mol}) + (1)(5.69 \text{ J/K·mol})]$$

$$\Delta S° = 134.5 \text{ J/K·mol}$$

It is obvious from the given conditions that the reaction must take place at a fairly high temperature (in order to have red–hot coke). Setting $\Delta G° = 0$

$$0 = \Delta H° - T\Delta S°$$

$$T = \frac{\Delta H°}{\Delta S°} = \frac{131.3 \text{ kJ/mol} \times \dfrac{1000 \text{ J}}{1 \text{ kJ}}}{134.5 \text{ J/K·mol}} = \textbf{976 K} = \textbf{703°C}$$

The temperature must be greater than 703°C for the reaction to be spontaneous.

18.53   **(a)**   We know that HCl is a strong acid and HF is a weak acid. Thus, the equilibrium constant will be less than 1 ($K < 1$).

(b)   The number of particles on each side of the equation is the same, so $\Delta S° \approx 0$. Therefore $\Delta H°$ will dominate.

(c)   HCl is a weaker bond than HF (see Table 9.4 of the text), therefore $\Delta H° > 0$.

18.54   For a reaction to be spontaneous at constant temperature and pressure, $\Delta G < 0$. The process of crystallization proceeds with more order (less disorder), so $\Delta S < 0$. We also know that

$$\Delta G = \Delta H - T\Delta S$$

Since ΔG must be negative, and since the entropy term will be positive ($-T\Delta S$, where ΔS is negative), then ΔH must be negative ($\Delta H < 0$). The reaction will be exothermic.

18.55 For the reaction: $CaCO_3(s) \rightleftharpoons CaO(s) + CO_2(g)$ $K_p = P_{CO_2}$

Using the equation from Problem 18.49:

$$\ln\frac{K_2}{K_1} = \frac{\Delta H^\circ}{R}\left(\frac{1}{T_1} - \frac{1}{T_2}\right) = \frac{\Delta H^\circ}{R}\left(\frac{T_2 - T_1}{T_1 T_2}\right)$$

Substituting,

$$\ln\frac{1829}{22.6} = \frac{\Delta H^\circ}{8.314 \text{ J/K}\cdot\text{mol}}\left(\frac{1223 \text{ K} - 973 \text{ K}}{(973 \text{ K})(1223 \text{ K})}\right)$$

Solving,

$$\Delta H^\circ = 1.74 \times 10^5 \text{ J/mol} = 174 \text{ kJ/mol}$$

18.56 For the reaction to be spontaneous, ΔG must be negative.

$$\Delta G = \Delta H - T\Delta S$$

Given that $\Delta H = 19$ kJ/mol = 19,000 J/mol, then

$$\Delta G = 19,000 \text{ J/mol} - (273 \text{ K} + 72 \text{ K})(\Delta S)$$

Solving the equation with the value of $\Delta G = 0$

$$0 = 19,000 \text{ J/mol} - (273 \text{ K} + 72 \text{ K})(\Delta S)$$

$$\Delta S = 55 \text{ J/K}\cdot\text{mol}$$

This value of ΔS which we solved for is the value needed to produce a ΔG value of zero. The *minimum* value of ΔS that will produce a spontaneous reaction will be any value of entropy *greater than* 55 J/K·mol.

18.57 (a) $\Delta S > 0$ (b) $\Delta S < 0$ (c) $\Delta S > 0$ (d) $\Delta S > 0$

18.58 The second law states that the entropy of the universe must increase in a spontaneous process. But the entropy of the universe is the sum of two terms: the entropy of the system plus the entropy of the surroundings. One of the entropies can decrease, but not both. In this case, the decrease in system entropy is offset by an increase in the entropy of the surroundings. The reaction in question is exothermic, and the heat released raises the temperature (and the entropy) of the surroundings.

Could this process be spontaneous if the reaction were endothermic?

18.59 At the temperature of the normal boiling point the free energy difference between the liquid and gaseous forms of mercury (or any other substances) is zero, i.e. the two phases are in equilibrium. We can therefore use Equation (18.10) of the text to find this temperature. For the equilibrium,

$$Hg(l) \rightleftharpoons Hg(g)$$

$$\Delta G = \Delta H - T\Delta S = 0$$

$$\Delta H = \Delta H_f^\circ[Hg(g)] - \Delta H_f^\circ[Hg(l)] = 60,780 \text{ J/mol} - 0 = 60780 \text{ J/mol}$$

$$\Delta S = S^\circ[Hg(g)] - S^\circ[Hg(l)] = 174.7 \text{ J/K}\cdot\text{mol} - 77.4 \text{ J/K}\cdot\text{mol} = 97.3 \text{ J/K}\cdot\text{mol}$$

$$T_{bp} = \frac{\Delta H}{\Delta S} = \frac{60780 \text{ J/mol}}{97.3 \text{ J/K} \cdot \text{mol}} = 625 \text{ K} = \textbf{352°C}$$

What assumptions are made? Notice that the given enthalpies and entropies are at standard conditions, namely 25°C and 1.00 atm pressure. In performing this calculation we have tacitly assumed that these quantities don't depend upon temperature. The actual normal boiling point of mercury is 356.58°C. Is the assumption of the temperature independence of these quantities reasonable?

18.60   **Strategy:** At the boiling point, liquid and gas phase ethanol are at equilibrium, so $\Delta G = 0$. From equation (18.10) of the text, we have $\Delta G = 0 = \Delta H - T\Delta S$ or $\Delta S = \Delta H/T$. To calculate the entropy change for the liquid ethanol $\rightarrow$ gas ethanol transition, we write $\Delta S_{vap} = \Delta H_{vap}/T$. What temperature unit should we use?

Solution: The entropy change due to the phase transition (the vaporization of ethanol), can be calculated using the following equation. Recall that the temperature must be in units of Kelvin (78.3°C = 351 K).

$$\Delta S_{vap} = \frac{\Delta H_{vap}}{T_{b.p.}}$$

$$\Delta S_{vap} = \frac{39.3 \text{ kJ/mol}}{351 \text{ K}} = 0.112 \text{ kJ/mol} \cdot \text{K} = 112 \text{ J/mol} \cdot \text{K}$$

The problem asks for the change in entropy for the vaporization of 0.50 moles of ethanol. The ΔS calculated above is for 1 mole of ethanol.

ΔS for 0.50 mol = (112 J/mol·K)(0.50 mol) = **56 J/K**

18.61   There is no connection between the spontaneity of a reaction predicted by ΔG and the rate at which the reaction occurs. A negative free energy change tells us that a reaction has the potential to happen, but gives no indication of the rate.

Does the fact that a reaction occurs at a measurable rate mean that the free energy difference ΔG is negative?

18.62   For the given reaction we can calculate the standard free energy change from the standard free energies of formation (see Appendix 3 of the text). Then, we can calculate the equilibrium constant, K_p, from the standard free energy change.

$$\Delta G° = \Delta G_f°[\text{Ni(CO)}_4] - [4\Delta G_f°(\text{CO}) + \Delta G_f°(\text{Ni})]$$

$$\Delta G° = (1)(-587.4 \text{ kJ/mol}) - [(4)(-137.3 \text{ kJ/mol}) + (1)(0)] = -38.2 \text{ kJ/mol} = -3.82 \times 10^4 \text{ J/mol}$$

Substitute $\Delta G°$, R, and T (in K) into the following equation to solve for K_p.

$$\Delta G° = -RT\ln K_p$$

$$\ln K_p = \frac{-\Delta G°}{RT} = \frac{-(-3.82 \times 10^4 \text{ J/mol})}{(8.314 \text{ J/K} \cdot \text{mol})(353 \text{ K})}$$

$$K_p = \textbf{4.5} \times \textbf{10}^5$$

18.63   (a) $\Delta G° = 2\Delta G_f°(\text{HBr}) - \Delta G_f°(\text{H}_2) - \Delta G_f°(\text{Br}_2) = (2)(-53.2 \text{ kJ/mol}) - (1)(0) - (1)(0)$

$$\Delta G° = -106.4 \text{ kJ/mol}$$

$$\ln K_p = \frac{-\Delta G°}{RT} = \frac{106.4 \times 10^3 \text{ J/mol}}{(8.314 \text{ J/K·mol})(298 \text{ K})} = 42.9$$

$$K_p = 4 \times 10^{18}$$

(b) $\Delta G° = \Delta G_f°(\text{HBr}) - \frac{1}{2}\Delta G_f°(\text{H}_2) - \frac{1}{2}\Delta G_f°(\text{Br}_2) = (1)(-53.2 \text{ kJ/mol}) - (\frac{1}{2})(0) - (\frac{1}{2})(0)$

$$\Delta G° = -53.2 \text{ kJ/mol}$$

$$\ln K_p = \frac{-\Delta G°}{RT} = \frac{53.2 \times 10^3 \text{ J/mol}}{(8.314 \text{ J/K·mol})(298 \text{ K})} = 21.5$$

$$K_p = 2 \times 10^9$$

The K_p in (a) is the square of the K_p in (b). Both $\Delta G°$ and K_p depend on the number of moles of reactants and products specified in the balanced equation.

18.64   The equilibrium constant is related to the standard free energy change by the following equation:

$$\Delta G° = -RT\ln K_p$$

$$2.12 \times 10^5 \text{ J/mol} = -(8.314 \text{ J/mol·K})(298 \text{ K})\ln K_p$$

$$-85.6 = \ln K_p$$

$$K_p = 6.7 \times 10^{-38}$$

We can write the equilibrium constant expression for the reaction.

$$K_p = \sqrt{P_{\text{O}_2}}$$

$$P_{\text{O}_2} = (K_p)^2$$

$$\boldsymbol{P_{\text{O}_2} = (6.7 \times 10^{-38})^2 = 4.5 \times 10^{-75} \text{ atm}}$$

This pressure is far too small to measure.

18.65   Talking involves various biological processes (to provide the necessary energy) that lead to a increase in the entropy of the universe. Since the overall process (talking) is spontaneous, the entropy of the universe must increase.

18.66   Both (a) and (b) apply to a reaction with a negative $\Delta G°$ value. Statement (c) is not always true. An endothermic reaction that has a positive $\Delta S°$ (increase in entropy) will have a negative $\Delta G°$ value at high temperatures.

18.67   **(a)**   If $\Delta G°$ for the reaction is 173.4 kJ/mol,

then, $\Delta G_f° = \dfrac{173.4 \text{ kJ/mol}}{2} = \textbf{86.7 kJ/mol}$

(b) $\Delta G° = -RT\ln K_p$

$$173.4 \times 10^3 \text{ J/mol} = -(8.314 \text{ J/K·mol})(298 \text{ K})\ln K_p$$

$$K_p = 4 \times 10^{-31}$$

(c) $\Delta H°$ for the reaction is $2 \times \Delta H_f^\circ (NO) = (2)(86.7 \text{ kJ/mol}) = 173.4 \text{ kJ/mol}$

Using the equation in Problem 18.49:

$$\ln \frac{K_2}{4 \times 10^{-31}} = \frac{173.4 \times 10^3 \text{ J/mol}}{8.314 \text{ J/mol} \cdot \text{K}} \left(\frac{1373 \text{ K} - 298 \text{ K}}{(1373 \text{ K})(298 \text{ K})} \right)$$

$$K_2 = 3 \times 10^{-7}$$

(d)   Lightning promotes the formation of NO (from N_2 and O_2 in the air) which eventually leads to the formation of nitrate ion (NO_3^-), an essential nutrient for plants.

18.68   We write the two equations as follows. The standard free energy change for the overall reaction will be the sum of the two steps.

$$\begin{array}{ll}
CuO(s) \rightleftharpoons Cu(s) + \frac{1}{2}O_2(g) & \Delta G° = 127.2 \text{ kJ/mol} \\[6pt]
C(\text{graphite}) + \frac{1}{2}O_2(g) \rightleftharpoons CO(g) & \Delta G° = -137.3 \text{ kJ/mol} \\
\hline
\textbf{CuO + C(graphite)} \rightleftharpoons \textbf{Cu(s) + CO(g)} & \Delta G° = -10.1 \text{ kJ/mol}
\end{array}$$

We can now calculate the equilibrium constant from the standard free energy change, $\Delta G°$.

$$\ln K = \frac{-\Delta G°}{RT} = \frac{-(-10.1 \times 10^3 \text{ J/mol})}{(8.314 \text{ J/K} \cdot \text{mol})(673 \text{ K})}$$

$$\ln K = 1.81$$

$$K = 6.1$$

18.69   Using the equation in the Chemistry in Action entitled "The Efficiency of Heat Engines" in Chapter 18:

$$\text{Efficiency} = \frac{T_2 - T_1}{T_2} = \frac{2473 \text{ K} - 1033 \text{ K}}{2473 \text{ K}} = 0.5823$$

The work done by moving the car:

$$mgh = (1200 \text{ kg})(9.81 \text{ m/s}^2) \times h = \text{heat generated by the engine.}$$

The heat generated by the gas:

$$1.0 \text{ gal} \times \frac{3.1 \text{ kg}}{1 \text{ gal}} \times \frac{1000 \text{ g}}{1 \text{ kg}} \times \frac{1 \text{ mol}}{114.2 \text{ g}} \times \frac{5510 \times 10^3 \text{ J}}{1 \text{ mol}} = 1.5 \times 10^8 \text{ J}$$

The maximum use of the energy generated by the gas is:

$$(\text{energy})(\text{efficiency}) = (1.5 \times 10^8 \text{ J})(0.5823) = 8.7 \times 10^7 \text{ J}$$

Setting the (useable) energy generated by the gas equal to the work done moving the car:

$$8.7 \times 10^7 \text{ J} = (1200 \text{ kg})(9.81 \text{ m/s}^2) \times h$$

$$h = 7.4 \times 10^3 \text{ m}$$

18.70   1.   Crystal structure has disorder.

2.   There is impurity present in the crystal.

18.71   **(a)**   The first law states that energy can neither be created nor destroyed. We cannot obtain energy out of nowhere.

(b)   If we calculate the efficiency of such an engine, we find that $T_h = T_c$, so the efficiency is zero! See Chemistry in Action on p. 777 of the text.

18.72   **(a)** $\Delta G° = \Delta G_f°(H_2) + \Delta G_f°(Fe^{2+}) - \Delta G_f°(Fe) - 2\Delta G_f°(H^+)]$

$\Delta G° = (1)(0) + (1)(-84.9 \text{ kJ/mol}) - (1)(0) - (2)(0)$

$\Delta G° = -84.9 \text{ kJ/mol}$

$\Delta G° = -RT\ln K$

$-84.9 \times 10^3 \text{ J/mol} = -(8.314 \text{ J/mol·K})(298 \text{ K})\ln K$

$\boldsymbol{K = 7.6 \times 10^{14}}$

(b) $\Delta G° = \Delta G_f°(H_2) + \Delta G_f°(Cu^{2+}) - \Delta G_f°(Cu) - 2\Delta G_f°(H^+)]$

$\Delta G° = 64.98 \text{ kJ/mol}$

$\Delta G° = -RT\ln K$

$64.98 \times 10^3 \text{ J/mol} = -(8.314 \text{ J/mol·K})(298 \text{ K})\ln K$

$\boldsymbol{K = 4.1 \times 10^{-12}}$

The activity series is correct. The very large value of K for reaction (a) indicates that *products* are highly favored; whereas, the very small value of K for reaction (b) indicates that *reactants* are highly favored.

18.73 $2NO + O_2 \underset{k_r}{\overset{k_f}{\rightleftharpoons}} 2NO_2$

$\Delta G° = (2)(51.8 \text{ kJ/mol}) - (2)(86.7 \text{ kJ/mol}) - 0 = -69.8 \text{ kJ/mol}$

$\Delta G° = -RT\ln K$

$-69.8 \times 10^3 \text{ J/mol} = -(8.314 \text{ J/mol·K})(298 \text{ K})\ln K$

$K = 1.7 \times 10^{12} \ M^{-1}$

$K = \dfrac{k_f}{k_r}$

$1.7 \times 10^{12} \ M^{-1} = \dfrac{7.1 \times 10^9 \ M^{-2}\text{s}^{-1}}{k_r}$

$\boldsymbol{k_r = 4.2 \times 10^{-3} \ M^{-1}\text{s}^{-1}}$

18.74   **(a)**   It is a "reverse" disproportionation redox reaction.

(b) $\Delta G° = (2)(-228.6 \text{ kJ/mol}) - (2)(-33.0 \text{ kJ/mol}) - (1)(-300.4 \text{ kJ/mol})$

$\Delta G° = -90.8 \text{ kJ/mol}$

$-90.8 \times 10^3 \text{ J/mol} = -(8.314 \text{ J/mol·K})(298 \text{ K}) \ln K$

$K = 8.2 \times 10^{15}$

Because of the large value of K, this method is efficient for removing SO_2.

(c) $\Delta H° = (2)(-241.8 \text{ kJ/mol}) + (3)(0) - (2)(-20.15 \text{ kJ/mol}) - (1)(-296.1 \text{ kJ/mol})$

$\Delta H° = -147.2 \text{ kJ/mol}$

$\Delta S° = (2)(188.7 \text{ J/K·mol}) + (3)(31.88 \text{ J/K·mol}) - (2)(205.64 \text{ J/K·mol}) - (1)(248.5 \text{ J/K·mol})$

$\Delta S° = -186.7 \text{ J/K·mol}$

$\Delta G° = \Delta H° - T\Delta S°$

Due to the negative entropy change, $\Delta S°$, the free energy change, $\Delta G°$, will become positive at higher temperatures. Therefore, the reaction will be **less effective** at high temperatures.

18.75   **(1)**   Measure K and use $\Delta G° = -RT \ln K$

(2)   Measure $\Delta H°$ and $\Delta S°$ and use $\Delta G° = \Delta H° - T\Delta S°$

18.76 $2O_3 \rightleftharpoons 3O_2$

$\Delta G° = 3\Delta G_f°(O_2) - 2\Delta G_f°(O_3) = 0 - (2)(163.4 \text{ kJ/mol})$

$\Delta G° = -326.8 \text{ kJ/mol}$

$-326.8 \times 10^3 \text{ J/mol} = -(8.314 \text{ J/mol·K})(243 \text{ K}) \ln K_p$

$K_p = 1.8 \times 10^{70}$

Due to the large magnitude of K, you would expect this reaction to be spontaneous in the forward direction. However, this reaction has a **large activation energy**, so the rate of reaction is extremely slow.

18.77   First convert to moles of ice.

$$74.6 \text{ g } H_2O(s) \times \frac{1 \text{ mol } H_2O(s)}{18.02 \text{ g } H_2O(s)} = 4.14 \text{ mol } H_2O(s)$$

For a phase transition:

$$\Delta S_{sys} = \frac{\Delta H_{sys}}{T}$$

$$\Delta S_{sys} = \frac{(4.14)(6060 \text{ J/mol})}{273 \text{ K}} = \textbf{91.1 J/K·mol}$$

$$\Delta S_{surr} = \frac{-\Delta H_{sys}}{T}$$

$$\Delta S_{surr} = \frac{-(4.14)(6060 \text{ J/mol})}{273 \text{ K}} = -91.1 \text{ J/K} \cdot \text{mol}$$

$$\Delta S_{univ} = \Delta S_{sys} + \Delta S_{surr} = 0$$

This is an equilibrium process. There is no net change.

18.78 Heating the ore alone is not a feasible process. Looking at the coupled process:

$Cu_2S \rightarrow 2Cu + S$	$\Delta G° = 86.1$ kJ/mol
$S + O_2 \rightarrow SO_2$	$\Delta G° = -300.4$ kJ/mol
$Cu_2S + O_2 \rightarrow 2Cu + SO_2$	$\Delta G° = -214.3$ **kJ/mol**

Since $\Delta G°$ is a large negative quantity, the coupled reaction is feasible for extracting sulfur.

18.79 Since we are dealing with the same ion (K^+), Equation (18.13) of the text can be written as:

$$\Delta G = \Delta G° + RT \ln Q$$

$$\Delta G = 0 + (8.314 \text{ J/mol} \cdot \text{K})(310 \text{ K}) \ln \left(\frac{400 \text{ m}M}{15 \text{ m}M} \right)$$

$$\Delta G = 8.5 \times 10^3 \text{ J/mol} = 8.5 \text{ kJ/mol}$$

18.80 First, we need to calculate $\Delta H°$ and $\Delta S°$ for the reaction in order to calculate $\Delta G°$.

$$\Delta H° = -41.2 \text{ kJ/mol} \qquad\qquad \Delta S° = -42.0 \text{ J/K} \cdot \text{mol}$$

Next, we calculate $\Delta G°$ at 300°C or 573 K, assuming that $\Delta H°$ and $\Delta S°$ are temperature independent.

$$\Delta G° = \Delta H° - T\Delta S°$$

$$\Delta G° = -41.2 \times 10^3 \text{ J/mol} - (573 \text{ K})(-42.0 \text{ J/K} \cdot \text{mol})$$

$$\Delta G° = -1.71 \times 10^4 \text{ J/mol}$$

Having solved for $\Delta G°$, we can calculate K_p.

$$\Delta G° = -RT \ln K_p$$

$$-1.71 \times 10^4 \text{ J/mol} = -(8.314 \text{ J/K} \cdot \text{mol})(573 \text{ K}) \ln K_p$$

$$\ln K_p = 3.59$$

$$K_p = 36$$

Due to the negative entropy change calculated above, we expect that $\Delta G°$ will become positive at some temperature higher than 300°C. We need to find the temperature at which $\Delta G°$ becomes zero. This is the temperature at which reactants and products are equally favored ($K_p = 1$).

$$\Delta G° = \Delta H° - T\Delta S°$$

$$0 = \Delta H° - T\Delta S°$$

$$T = \frac{\Delta H^\circ}{\Delta S^\circ} = \frac{-41.2 \times 10^3 \text{ J/mol}}{-42.0 \text{ J/K} \cdot \text{mol}}$$

$$T = 981 \text{ K} = 708°C$$

This calculation shows that at 708°C, $\Delta G^\circ = 0$ and the equilibrium constant $K_p = 1$. Above 708°C, ΔG° is positive and K_p will be smaller than 1, meaning that reactants will be favored over products. Note that the temperature 708°C is only an estimate, as we have assumed that both ΔH° and ΔS° are independent of temperature.

Using a more efficient catalyst will *not* increase K_p at a given temperature, because the catalyst will speed up both the forward and reverse reactions. The value of K_p will stay the same.

18.81   **(a)** ΔG° for CH_3COOH:

$$\Delta G^\circ = -(8.314 \text{ J/mol} \cdot \text{K})(298 \text{ K})\ln(1.8 \times 10^{-5})$$

$$\Delta G^\circ = 2.7 \times 10^4 \text{ J/mol} = \textbf{27 kJ/mol}$$

ΔG° for $CH_2ClCOOH$:

$$\Delta G^\circ = -(8.314 \text{ J/mol} \cdot \text{K})(298 \text{ K})\ln(1.4 \times 10^{-3})$$

$$\Delta G^\circ = 1.6 \times 10^4 \text{ J/mol} = \textbf{16 kJ/mol}$$

(b)   The $T\Delta S^\circ$ is the dominant term.

(c)   The breaking of the O–H bond in ionization of the acid and the forming of the O–H bond in H_3O^+.

(d)   The CH_3COO^- ion is smaller than CH_2ClCOO^- and can participate in hydration to a greater extent, leading to a more ordered solution.

18.82   butane $\rightarrow$ isobutane

$$\Delta G^\circ = \Delta G_f^\circ(\text{isobutane}) - \Delta G_f^\circ(\text{butane})$$

$$\Delta G^\circ = (1)(-18.0 \text{ kJ/mol}) - (1)(-15.9 \text{ kJ/mol})$$

$$\Delta G^\circ = -2.1 \text{ kJ/mol}$$

For a mixture at equilibrium at 25°C:

$$\Delta G^\circ = -RT\ln K_p$$

$$-2.1 \times 10^3 \text{ J/mol} = -(8.314 \text{ J/mol} \cdot \text{K})(298 \text{ K})\ln K_p$$

$$K_p = 2.3$$

$$K_p = \frac{P_{\text{isobutane}}}{P_{\text{butane}}} \; \alpha \; \frac{\text{mol isobutane}}{\text{mol butane}}$$

$$2.3 = \frac{\text{mol isobutane}}{\text{mol butane}}$$

This shows that there are 2.3 times as many moles of isobutane as moles of butane. Or, we can say for every one mole of butane, there are 2.3 moles of isobutane.

$$\textbf{mol \% isobutane} = \frac{2.3 \text{ mol}}{2.3 \text{ mol} + 1.0 \text{ mol}} \times 100\% = \textbf{70\%}$$

By difference, the mole % of butane is **30%**.

Yes, this result supports the notion that straight-chain hydrocarbons like butane are less stable than branched-chain hydrocarbons like isobutane.

18.83   Heat is absorbed by the rubber band, so ΔH is positive. Since the contraction occurs spontaneously, ΔG is negative. For the reaction to be spontaneous, ΔS must be positive meaning that the rubber becomes more disordered upon heating. This is consistent with what we know about the structure of rubber; The rubber molecules become more disordered upon contraction (See the Figure in the Chemistry in Action Essay on p. 787 of the text).

18.84   We can calculate K_p from $\Delta G°$.

$$\Delta G° = (1)(-394.4 \text{ kJ/mol}) + (0) - (1)(-137.3 \text{ kJ/mol}) - (1)(-255.2 \text{ kJ/mol})$$

$$\Delta G° = -1.9 \text{ kJ/mol}$$

$$-1.9 \times 10^3 \text{ J/mol} = -(8.314 \text{ J/mol·K})(1173 \text{ K}) \ln K_p$$

$$K_p = 1.2$$

Now, from K_p, we can calculate the mole fractions of CO and CO_2.

$$K_p = \frac{P_{CO_2}}{P_{CO}} = 1.2 \qquad P_{CO_2} = 1.2 P_{CO}$$

$$X_{CO} = \frac{P_{CO}}{P_{CO} + P_{CO_2}} = \frac{P_{CO}}{P_{CO} + 1.2 P_{CO}} = \frac{1}{2.2} = \textbf{0.45}$$

$$X_{CO_2} = 1 - 0.45 = \textbf{0.55}$$

We assumed that $\Delta G°$ calculated from $\Delta G_f°$ values was temperature independent. The $\Delta G_f°$ values in Appendix 3 of the text are measured at 25°C, but the temperature of the reaction is 900°C.

18.85 $\Delta G° = -RT \ln K$

and,

$\Delta G = \Delta G° + RT \ln Q$

Substituting,

$\Delta G = -RT \ln K + RT \ln Q$
$\Delta G = RT(\ln Q - \ln K)$
$\Delta G = RT \ln \left(\dfrac{Q}{K} \right)$

If $Q > K$, $\Delta G > 0$, and the net reaction will proceed from right to left (see Figure 14.4 of the text).

If $Q < K$, $\Delta G < 0$, and the net reaction will proceed from left to right.

If $Q = K$, $\Delta G = 0$. The system is at equilibrium.

18.86   For a phase transition, $\Delta G = 0$. We write:

$$\Delta G = \Delta H - T\Delta S$$

$$0 = \Delta H - T\Delta S$$

$$\Delta S_{sub} = \frac{\Delta H_{sub}}{T}$$

Substituting ΔH and the temperature, $(-78° + 273°)K = 195$ K, gives

$$\Delta S_{sub} = \frac{\Delta H_{sub}}{T} = \frac{25.2 \times 10^3 \text{ J}}{195 \text{ K}} = 129 \text{ J/K} \cdot \text{mol}$$

This value of ΔS_{sub} is for the sublimation of 1 mole of CO_2. We convert to the ΔS value for the sublimation of 84.8 g of CO_2.

$$84.8 \text{ g CO}_2 \times \frac{1 \text{ mol CO}_2}{44.01 \text{ g CO}_2} \times \frac{129 \text{ J}}{\text{K} \cdot \text{mol}} = \textbf{249 J/K}$$

18.87   The second law of thermodynamics states that the entropy of the universe increases in a spontaneous process and remains unchanged in an equilibrium process. Therefore, the entropy of the universe is increasing with time, and thus entropy could be used to determine the forward direction of time.

18.88   First, let's convert the age of the universe from units of years to units of seconds.

$$(13 \times 10^9 \text{ yr}) \times \frac{365 \text{ days}}{1 \text{ yr}} \times \frac{24 \text{ h}}{1 \text{ day}} \times \frac{3600 \text{ s}}{1 \text{ h}} = 4.1 \times 10^{17} \text{ s}$$

The probability of finding all 100 molecules in the same flask is 8×10^{-31}. Multiplying by the number of seconds gives:

$$(8 \times 10^{-31})(4.1 \times 10^{17} \text{ s}) = \textbf{3} \times \textbf{10}^{-13} \textbf{ s}$$

18.89   Equation (18.10) represents the standard free-energy change for a reaction, and not for a particular compound like CO_2. The correct form is:

$$\Delta G° = \Delta H° - T\Delta S°$$

For a given reaction, $\Delta G°$ and $\Delta H°$ would need to be calculated from standard formation values (graphite, oxygen, and carbon dioxide) first, before plugging into the equation. Also, $\Delta S°$ would need to be calculated from standard entropy values.

$$C(graphite) + O_2(g) \rightarrow CO_2(g)$$

18.90   We can calculate ΔS_{sys} from standard entropy values in Appendix 3 of the text.   We can calculate ΔS_{surr} from the ΔH_{sys} value given in the problem.   Finally, we can calculate ΔS_{univ} from the ΔS_{sys} and ΔS_{surr} values.

$$\Delta S_{sys} = (2)(69.9 \text{ J/K·mol}) - [(2)(131.0 \text{ J/K·mol}) + (1)(205.0 \text{ J/K·mol})] = \mathbf{-327 \text{ J/K·mol}}$$

$$\Delta S_{surr} = \frac{-\Delta H_{sys}}{T} = \frac{-(-571.6 \times 10^3 \text{ J/mol})}{298 \text{ K}} = \mathbf{1918 \text{ J/K · mol}}$$

$$\Delta S_{univ} = \Delta S_{sys} + \Delta S_{surr} = (-327 + 1918) \text{ J/K·mol} = \mathbf{1591 \text{ J/K·mol}}$$

18.91 $\Delta H°$ is endothermic. Heat must be added to denature the protein. Denaturation leads to more disorder (an increase in microstates). The magnitude of $\Delta S°$ is fairly large (1600 J/K·mol). Proteins are large molecules and therefore denaturation would lead to a large increase in microstates. The temperature at which the process favors the denatured state can be calculated by setting $\Delta G°$ equal to zero.

$$\Delta G° = \Delta H° - T\Delta S°$$

$$0 = \Delta H° - T\Delta S°$$

$$T = \frac{\Delta H°}{\Delta S°} = \frac{512 \text{ kJ/mol}}{1.60 \text{ kJ/K · mol}} = \mathbf{320 \text{ K} = 47°C}$$

18.92 q, and w are *not* state functions. Recall that state functions represent properties that are determined by the state of the system, regardless of how that condition is achieved. Heat and work are not state functions because they are not properties of the system. They manifest themselves only during a process (during a change). Thus their values depend on the path of the process and vary accordingly.

18.93   **(d)** will not lead to an increase in entropy of the system. The gas is returned to its original state. The entropy of the system does not change.

18.94   Since the adsorption is spontaneous, ΔG must be negative ($\Delta G < 0$). When hydrogen bonds to the surface of the catalyst, the system becomes more ordered ($\Delta S < 0$). Since there is a decrease in entropy, the adsorption must be exothermic for the process to be spontaneous ($\Delta H < 0$).

CHAPTER 19
ELECTROCHEMISTRY

19.1   We follow the steps are described in detail in Section 19.1 of the text.

(a)   The problem is given in ionic form, so combining Steps 1 and 2, the half-reactions are:

         oxidation: $Fe^{2+} \rightarrow Fe^{3+}$

         reduction: $H_2O_2 \rightarrow H_2O$

Step 3: We balance each half-reaction for number and type of atoms and charges.

The *oxidation half-reaction* is already balanced for Fe atoms. There are three net positive charges on the right and two net positive charges on the left, we add one electrons to the right side to balance the charge.

$$Fe^{2+} \rightarrow Fe^{3+} + e^-$$

Reduction half-reaction: we add one H_2O to the right-hand side of the equation to balance the O atoms.

$$H_2O_2 \rightarrow 2H_2O$$

To balance the H atoms, we add $2H^+$ to the left-hand side.

$$H_2O_2 + 2H^+ \rightarrow 2H_2O$$

There are two net positive charges on the left, so we add two electrons to the same side to balance the charge.

$$H_2O_2 + 2H^+ + 2e^- \rightarrow 2H_2O$$

Step 4: We now add the oxidation and reduction half-reactions to give the overall reaction. In order to equalize the number of electrons, we need to multiply the oxidation half-reaction by 2.

$$\begin{array}{l} 2(Fe^{2+} \rightarrow Fe^{3+} + e^-) \\ \underline{H_2O_2 + 2H^+ + 2e^- \rightarrow 2H_2O} \\ 2Fe^{2+} + H_2O_2 + 2H^+ + 2e^- \rightarrow 2Fe^{3+} + 2H_2O + 2e^- \end{array}$$

The electrons on both sides cancel, and we are left with the balanced net ionic equation in acidic medium.

$$\mathbf{2Fe^{2+} + H_2O_2 + 2H^+ \rightarrow 2Fe^{3+} + 2H_2O}$$

(b)   The problem is given in ionic form, so combining Steps 1 and 2, the half-reactions are:

         oxidation: $Cu \rightarrow Cu^{2+}$

         reduction: $HNO_3 \rightarrow NO$

Step 3: We balance each half-reaction for number and type of atoms and charges.

The *oxidation half-reaction* is already balanced for Cu atoms. There are two net positive charges on the right, so we add two electrons to the right side to balance the charge.

$$Cu \rightarrow Cu^{2+} + 2e^-$$

Reduction half-reaction: we add two H_2O to the right-hand side of the equation to balance the O atoms.

$$HNO_3 \rightarrow NO + 2H_2O$$

To balance the H atoms, we add $3H^+$ to the left-hand side.

$$3H^+ + HNO_3 \rightarrow NO + 2H_2O$$

There are three net positive charges on the left, so we add three electrons to the same side to balance the charge.

$$3H^+ + HNO_3 + 3e^- \rightarrow NO + 2H_2O$$

Step 4: We now add the oxidation and reduction half-reactions to give the overall reaction. In order to equalize the number of electrons, we need to multiply the oxidation half-reaction by 3 and the reduction half-reaction by 2.

$$\frac{\begin{array}{l} 3(Cu \rightarrow Cu^{2+} + 2e^-) \\ 2(3H^+ + HNO_3 + 3e^- \rightarrow NO + 2H_2O) \end{array}}{3Cu + 6H^+ + 2HNO_3 + 6e^- \rightarrow 3Cu^{2+} + 2NO + 4H_2O + 6e^-}$$

The electrons on both sides cancel, and we are left with the balanced net ionic equation in acidic medium.

$$3Cu + 6H^+ + 2HNO_3 \rightarrow 3Cu^{2+} + 2NO + 4H_2O$$

(c) $\quad 3CN^- + 2MnO_4^- + H_2O \rightarrow 3CNO^- + 2MnO_2 + 2OH^-$

(d) $\quad 3Br_2 + 6OH^- \rightarrow BrO_3^- + 5Br^- + 3H_2O$

(e) $\quad$ Half-reactions balanced for S and I:

$$\text{oxidation:} \quad 2S_2O_3^{2-} \rightarrow S_4O_6^{2-}$$
$$\text{reduction:} \quad I_2 \rightarrow 2I^-$$

Both half-reactions are already balanced for O, so we balance charge with electrons

$$2S_2O_3^{2-} \rightarrow S_4O_6^{2-} + 2e^-$$
$$I_2 + 2e^- \rightarrow 2I^-$$

The electron count is the same on both sides. We add the equations, canceling electrons, to obtain the balanced equation.

$$2S_2O_3^{2-} + I_2 \rightarrow S_4O_6^{2-} + 2I^-$$

19.2 **Strategy:** We follow the procedure for balancing redox reactions presented in Section 19.1 of the text.

Solution:
(a)
Step 1: The unbalanced equation is given in the problem.

$$Mn^{2+} + H_2O_2 \longrightarrow MnO_2 + H_2O$$

Step 2: The two half-reactions are:

$$Mn^{2+} \xrightarrow{\text{oxidation}} MnO_2$$

$$H_2O_2 \xrightarrow{\text{reduction}} H_2O$$

Step 3: We balance each half-reaction for number and type of atoms and charges.

The *oxidation half-reaction* is already balanced for Mn atoms. To balance the O atoms, we add two water molecules on the left side.

$$Mn^{2+} + 2H_2O \longrightarrow MnO_2$$

To balance the H atoms, we add 4 H^+ to the right-hand side.

$$Mn^{2+} + 2H_2O \longrightarrow MnO_2 + 4H^+$$

There are four net positive charges on the right and two net positive charge on the left, we add two electrons to the right side to balance the charge.

$$Mn^{2+} + 2H_2O \longrightarrow MnO_2 + 4H^+ + 2e^-$$

Reduction half-reaction: we add one H_2O to the right-hand side of the equation to balance the O atoms.

$$H_2O_2 \longrightarrow 2H_2O$$

To balance the H atoms, we add $2H^+$ to the left-hand side.

$$H_2O_2 + 2H^+ \longrightarrow 2H_2O$$

There are two net positive charges on the left, so we add two electrons to the same side to balance the charge.

$$H_2O_2 + 2H^+ + 2e^- \longrightarrow 2H_2O$$

Step 4: We now add the oxidation and reduction half-reactions to give the overall reaction. Note that the number of electrons gained and lost is equal.

$$Mn^{2+} + 2H_2O \longrightarrow MnO_2 + 4H^+ + 2e^-$$
$$H_2O_2 + 2H^+ + 2e^- \longrightarrow 2H_2O$$
$$\overline{Mn^{2+} + H_2O_2 + 2e^- \longrightarrow MnO_2 + 2H^+ + 2e^-}$$

The electrons on both sides cancel, and we are left with the balanced net ionic equation in acidic medium.

$$Mn^{2+} + H_2O_2 \longrightarrow MnO_2 + 2H^+$$

Because the problem asks to balance the equation in basic medium, we add one OH^- to both sides for each H^+ and combine pairs of H^+ and OH^- on the same side of the arrow to form H_2O.

$$Mn^{2+} + H_2O_2 + 2OH^- \longrightarrow MnO_2 + 2H^+ + 2OH^-$$

Combining the H^+ and OH^- to form water we obtain:

$$\mathbf{Mn^{2+} + H_2O_2 + 2OH^- \longrightarrow MnO_2 + 2H_2O}$$

Step 5: Check to see that the equation is balanced by verifying that the equation has the same types and numbers of atoms and the same charges on both sides of the equation.

(b)   This problem can be solved by the same methods used in part (a).

$$2Bi(OH)_3 + 3SnO_2^{2-} \longrightarrow 2Bi + 3H_2O + 3SnO_3^{2-}$$

(c)
Step 1: The unbalanced equation is given in the problem.

$$Cr_2O_7^{2-} + C_2O_4^{2-} \longrightarrow Cr^{3+} + CO_2$$

Step 2: The two half-reactions are:

$$C_2O_4^{2-} \xrightarrow{\text{oxidation}} CO_2$$
$$Cr_2O_7^{2-} \xrightarrow{\text{reduction}} Cr^{3+}$$

Step 3: We balance each half-reaction for number and type of atoms and charges.

In the *oxidation half-reaction*, we first need to balance the C atoms.

$$C_2O_4^{2-} \longrightarrow 2CO_2$$

The O atoms are already balanced. There are two net negative charges on the left, so we add two electrons to the right to balance the charge.

$$C_2O_4^{2-} \longrightarrow 2CO_2 + 2e^-$$

In the *reduction half-reaction*, we first need to balance the Cr atoms.

$$Cr_2O_7^{2-} \longrightarrow 2Cr^{3+}$$

We add seven H_2O molecules on the right to balance the O atoms.

$$Cr_2O_7^{2-} \longrightarrow 2Cr^{3+} + 7H_2O$$

To balance the H atoms, we add $14H^+$ to the left-hand side.

$$Cr_2O_7^{2-} + 14H^+ \longrightarrow 2Cr^{3+} + 7H_2O$$

There are twelve net positive charges on the left and six net positive charges on the right. We add six electrons on the left to balance the charge.

$$Cr_2O_7^{2-} + 14H^+ + 6e^- \longrightarrow 2Cr^{3+} + 7H_2O$$

Step 4: We now add the oxidation and reduction half-reactions to give the overall reaction. In order to equalize the number of electrons, we need to multiply the oxidation half-reaction by 3.

$$3(C_2O_4^{2-} \longrightarrow 2CO_2 + 2e^-)$$
$$\underline{Cr_2O_7^{2-} + 14H^+ + 6e^- \longrightarrow 2Cr^{3+} + 7H_2O}$$
$$3C_2O_4^{2-} + Cr_2O_7^{2-} + 14H^+ + 6e^- \longrightarrow 6CO_2 + 2Cr^{3+} + 7H_2O + 6e^-$$

The electrons on both sides cancel, and we are left with the balanced net ionic equation in acidic medium.

$$3C_2O_4^{2-} + Cr_2O_7^{2-} + 14H^+ \longrightarrow 6CO_2 + 2Cr^{3+} + 7H_2O$$

Step 5: Check to see that the equation is balanced by verifying that the equation has the same types and numbers of atoms and the same charges on both sides of the equation.

(d)   This problem can be solved by the same methods used in part (c).

$$2Cl^- + 2ClO_3^- + 4H^+ \longrightarrow Cl_2 + 2ClO_2 + 2H_2O$$

19.11

Half-reaction	$E°(V)$
$Mg^{2+}(aq) + 2e^- \rightarrow Mg(s)$	-2.37
$Cu^{2+}(aq) + 2e^- \rightarrow Cu(s)$	$+0.34$

The overall equation is: $Mg(s) + Cu^{2+}(aq) \rightarrow Mg^{2+}(aq) + Cu(s)$

$$E° = 0.34 \text{ V} - (-2.37 \text{ V}) = \textbf{2.71 V}$$

19.12   **Strategy:** At first, it may not be clear how to assign the electrodes in the galvanic cell. From Table 19.1 of the text, we write the standard reduction potentials of Al and Ag and apply the diagonal rule to determine which is the anode and which is the cathode.

Solution: The standard reduction potentials are:

$$Ag^+(1.0 \ M) + e^- \rightarrow Ag(s) \qquad E° = 0.80 \text{ V}$$
$$Al^{3+}(1.0 \ M) + 3e^- \rightarrow Al(s) \qquad E° = -1.66 \text{ V}$$

Applying the diagonal rule, we see that Ag^+ will oxidize Al.

Anode (oxidation):	$Al(s) \rightarrow Al^{3+}(1.0 \ M) + 3e^-$
Cathode (reduction):	$3Ag^+(1.0 \ M) + 3e^- \rightarrow 3Ag(s)$
Overall:	$\textbf{Al}(s) + \textbf{3Ag}^+(\textbf{1.0 } M) \rightarrow \textbf{Al}^{3+}(\textbf{1.0 } M) + \textbf{3Ag}(s)$

Note that in order to balance the overall equation, we multiplied the reduction of Ag^+ by 3. We can do so because, as an intensive property, $E°$ is not affected by this procedure. We find the emf of the cell using Equation (19.1) and Table 19.1 of the text.

$$E°_{cell} = E°_{cathode} - E°_{anode} = E°_{Ag^+/Ag} - E°_{Al^{3+}/Al}$$

$$E°_{cell} = 0.80 \text{ V} - (-1.66 \text{ V}) = \textbf{+2.46 V}$$

Check: The positive value of $E°$ shows that the forward reaction is favored.

19.13   The appropriate half-reactions from Table 19.1 are

$$I_2(s) + 2e^- \rightarrow 2I^-(aq) \qquad\qquad E°_{anode} = 0.53 \text{ V}$$
$$Fe^{3+}(aq) + e^- \rightarrow Fe^{2+}(aq) \qquad\qquad E°_{cathode} = 0.77 \text{ V}$$

Thus iron(III) should oxidize iodide ion to iodine. This makes the iodide ion/iodine half-reaction the anode. The standard emf can be found using Equation (19.1).

$$E^\circ_{cell} = E^\circ_{cathode} - E^\circ_{anode} = 0.77 \text{ V} - 0.53 \text{ V} = 0.24 \text{ V}$$

(The emf was not required in this problem, but the fact that it is positive confirms that the reaction should favor products at equilibrium.)

19.14 The half–reaction for oxidation is:

$$2H_2O(l) \xrightarrow{\text{oxidation (anode)}} O_2(g) + 4H^+(aq) + 4e^- \qquad E^\circ_{anode} = +1.23 \text{ V}$$

The species that can oxidize water to molecular oxygen must have an E°_{red} more positive than +1.23 V. From Table 19.1 of the text we see that only **$Cl_2(g)$** and **$MnO_4^-(aq)$** in acid solution can oxidize water to oxygen.

19.15 The overall reaction is:

$$5NO_3^-(aq) + 3Mn^{2+}(aq) + 2H_2O(l) \rightarrow 5NO(g) + 3MnO_4^-(aq) + 4H^+(aq)$$

$$E^\circ_{cell} = E^\circ_{cathode} - E^\circ_{anode} = 0.96 \text{ V} - 1.51 \text{ V} = -0.55 \text{ V}$$

The negative emf indicates that reactants are favored at equilibrium. NO_3^- will not oxidize Mn^{2+} to MnO_4^- under standard-state conditions.

19.16 **Strategy:** E°_{cell} is *positive* for a spontaneous reaction. In each case, we can calculate the standard cell emf from the potentials for the two half-reactions.

$$E^\circ_{cell} = E^\circ_{cathode} - E^\circ_{anode}$$

Solution:

(a) $E^\circ = -0.40 \text{ V} - (-2.87 \text{ V}) = \textbf{2.47 V}$. The reaction is spontaneous.

(b) $E^\circ = -0.14 \text{ V} - 1.07 \text{ V} = \textbf{-1.21 V}$. The reaction is not spontaneous.

(c) $E^\circ = -0.25 \text{ V} - 0.80 \text{ V} = \textbf{-1.05 V}$. The reaction is not spontaneous.

(d) $E^\circ = 0.77 \text{ V} - 0.15 \text{ V} = \textbf{0.62 V}$. The reaction is spontaneous.

19.17 From Table 19.1 of the text, we compare the standard reduction potentials for the half-reactions. The more positive the potential, the better the substance as an oxidizing agent.

(a) Au^{3+} (b) Ag^+ (c) Cd^{2+} (d) O_2 in acidic solution.

19.18 **Strategy:** The greater the tendency for the substance to be oxidized, the stronger its tendency to act as a reducing agent. The species that has a stronger tendency to be oxidized will have a smaller reduction potential.

Solution: In each pair, look for the one with the smaller reduction potential. This indicates a greater tendency for the substance to be oxidized.

(a) Li   (b) H_2 (c) Fe^{2+} (d) Br^-

19.21   We find the standard reduction potentials in Table 19.1 of the text.

$$E_{cell}^{\circ} = E_{cathode}^{\circ} - E_{anode}^{\circ} = -0.76 \text{ V} - (-2.37 \text{ V}) = 1.61 \text{ V}$$

$$E_{cell}^{\circ} = \frac{0.0257 \text{ V}}{n} \ln K$$

$$\ln K = \frac{nE_{cell}^{\circ}}{0.0257 \text{ V}}$$

$$K = e^{\frac{nE_{cell}^{\circ}}{0.0257 \text{ V}}}$$

$$K = e^{\frac{(2)(1.61 \text{ V})}{0.0257 \text{ V}}}$$

$$K = 3 \times 10^{54}$$

19.22   **Strategy:** The relationship between the equilibrium constant, K, and the standard emf is given by Equation (19.5) of the text: $E_{cell}^{\circ} = (0.0257 \text{ V} / n) \ln K$. Thus, knowing n (the moles of electrons transferred) and the equilibrium constant, we can determine E_{cell}°.

Solution: The equation that relates K and the standard cell emf is:

$$E_{cell}^{\circ} = \frac{0.0257 \text{ V}}{n} \ln K$$

We see in the reaction that Mg goes to Mg^{2+} and Zn^{2+} goes to Zn. Therefore, two moles of electrons are transferred during the redox reaction. Substitute the equilibrium constant and the moles of e^- transferred ($n = 2$) into the above equation to calculate E°.

$$E^{\circ} = \frac{(0.0257 \text{ V}) \ln K}{n} = \frac{(0.0257 \text{ V}) \ln(2.69 \times 10^{12})}{2} = \textbf{0.368 V}$$

19.23   In each case we use standard reduction potentials from Table 19.1 together with Equation (19.5) of the text.

(a) $E_{cell}^{\circ} = E_{cathode}^{\circ} - E_{anode}^{\circ} = 1.07 \text{ V} - 0.53 \text{ V} = 0.54 \text{ V}$

$$\ln K = \frac{nE_{cell}^{\circ}}{0.0257 \text{ V}}$$

$$K = e^{\frac{nE_{cell}^{\circ}}{0.0257 \text{ V}}}$$

$$K = e^{\frac{(2)(0.54 \text{ V})}{0.0257 \text{ V}}} = 2 \times 10^{18}$$

(b) $E_{cell}^{\circ} = E_{cathode}^{\circ} - E_{anode}^{\circ} = 1.61 \text{ V} - 1.36 \text{ V} = 0.25 \text{ V}$

$$K = e^{\frac{(2)(0.25 \text{ V})}{0.0257 \text{ V}}} = 3 \times 10^{8}$$

(c) $E^{\circ}_{cell} = E^{\circ}_{cathode} - E^{\circ}_{anode} = 1.51\ V - 0.77\ V = 0.74\ V$

$$K = e^{\frac{(5)(0.74\ V)}{0.0257\ V}} = 3 \times 10^{62}$$

19.24 **(a)** We break the equation into two half–reactions:

$$Mg(s) \xrightarrow{\text{oxidation (anode)}} Mg^{2+}(aq) + 2e^{-} \qquad E^{\circ}_{anode} = -2.37\ V$$

$$Pb^{2+}(aq) + 2e^{-} \xrightarrow{\text{reduction (cathode)}} Pb(s) \qquad E^{\circ}_{cathode} = -0.13\ V$$

The standard emf is given by

$$E^{\circ}_{cell} = E^{\circ}_{cathode} - E^{\circ}_{anode} = -0.13\ V - (-2.37\ V) = 2.24\ V$$

We can calculate ΔG° from the standard emf.

$$\Delta G^{\circ} = -nFE^{\circ}_{cell}$$

$$\Delta G^{\circ} = -(2)(96500\ J/V \cdot mol)(2.24\ V) = -432\ kJ/mol$$

Next, we can calculate K using Equation (19.5) of the text.

$$E^{\circ}_{cell} = \frac{0.0257\ V}{n} \ln K$$

or

$$\ln K = \frac{nE^{\circ}_{cell}}{0.0257\ V}$$

and

$$K = e^{\frac{nE^{\circ}}{0.0257}}$$

$$K = e^{\frac{(2)(2.24)}{0.0257}} = 5 \times 10^{75}$$

Tip: You could also calculate K_c from the standard free energy change, ΔG°, using the equation: $\Delta G^{\circ} = -RT \ln K_c$.

(b) We break the equation into two half–reactions:

$$Br_2(l) + 2e^{-} \xrightarrow{\text{reduction (cathode)}} 2Br^{-}(aq) \qquad E^{\circ}_{cathode} = 1.07\ V$$

$$2I^{-}(aq) \xrightarrow{\text{oxidation (anode)}} I_2(s) + 2e^{-} \qquad E^{\circ}_{anode} = 0.53\ V$$

The standard emf is

$$E^{\circ}_{cell} = E^{\circ}_{cathode} - E^{\circ}_{anode} = 1.07\ V - 0.53\ V = 0.54\ V$$

We can calculate ΔG° from the standard emf.

$$\Delta G^{\circ} = -nFE^{\circ}_{cell}$$

$$\Delta G^{\circ} = -(2)(96500\ J/V \cdot mol)(0.54\ V) = -104\ kJ/mol$$

Next, we can calculate K using Equation (19.5) of the text.

$$K = e^{\frac{nE^\circ}{0.0257}}$$

$$K = e^{\frac{(2)(0.54)}{0.0257}} = 2 \times 10^{18}$$

(c) This is worked in an analogous manner to parts (a) and (b).

$$E^\circ_{cell} = E^\circ_{cathode} - E^\circ_{anode} = 1.23 \text{ V} - 0.77 \text{ V} = 0.46 \text{ V}$$

$$\Delta G^\circ = -nFE^\circ_{cell}$$

$$\Delta G^\circ = -(4)(96500 \text{ J/V·mol})(0.46 \text{ V}) = \mathbf{-178 \text{ kJ/mol}}$$

$$K = e^{\frac{nE^\circ}{0.0257}}$$

$$K = e^{\frac{(4)(0.46)}{0.0257}} = \mathbf{1 \times 10^{31}}$$

(d) This is worked in an analogous manner to parts (a), (b), and (c).

$$E^\circ_{cell} = E^\circ_{cathode} - E^\circ_{anode} = 0.53 \text{ V} - (-1.66 \text{ V}) = 2.19 \text{ V}$$

$$\Delta G^\circ = -nFE^\circ_{cell}$$

$$\Delta G^\circ = -(6)(96500 \text{ J/V·mol})(2.19 \text{ V}) = \mathbf{-1.27 \times 10^3 \text{ kJ/mol}}$$

$$K = e^{\frac{nE^\circ}{0.0257}}$$

$$K = e^{\frac{(6)(2.19)}{0.0257}} = \mathbf{8 \times 10^{211}}$$

19.25   The half-reactions are: $Fe^{3+}(aq) + e^- \rightarrow Fe^{2+}(aq)$ $E^\circ_{anode} = 0.77 \text{ V}$

$Ce^{4+}(aq) + e^- \rightarrow Ce^{3+}(aq)$ $E^\circ_{cathode} = 1.61 \text{ V}$

Thus, Ce^{4+} will oxidize Fe^{2+} to Fe^{3+}; this makes the Fe^{2+}/Fe^{3+} half-reaction the anode. The standard cell emf is found using Equation (19.1) of the text.

$$E^\circ_{cell} = E^\circ_{cathode} - E^\circ_{anode} = 1.61 \text{ V} - 0.77 \text{ V} = 0.84 \text{ V}$$

The values of ΔG° and K_c are found using Equations (19.3) and (19.5) of the text.

$$\Delta G^\circ = -nFE^\circ_{cell} = -(1)(96500 \text{ J/V·mol})(0.84 \text{ V}) = \mathbf{-81 \text{ kJ/mol}}$$

$$\ln K = \frac{nE^\circ_{cell}}{0.0257 \text{ V}}$$

$$K_c = e^{\frac{nE^\circ_{cell}}{0.0257 \text{ V}}} = e^{\frac{(1)(0.84 \text{ V})}{0.0257 \text{ V}}} = 2 \times 10^{14}$$

19.26   **Strategy:** The relationship between the standard free energy change and the standard emf of the cell is given by Equation (19.3) of the text: $\Delta G^\circ = -nFE^\circ_{cell}$. The relationship between the equilibrium constant, K, and the standard emf is given by Equation (19.5) of the text: $E^\circ_{cell} = (0.0257 \text{ V}/n)\ln K$. Thus, if we can determine E°_{cell}, we can calculate ΔG° and K. We can determine the E°_{cell} of a hypothetical galvanic cell made up of two couples (Cu^{2+}/Cu^+ and Cu^+/Cu) from the standard reduction potentials in Table 19.1 of the text.

Solution: The half-cell reactions are:

Anode (oxidation):	$Cu^+(1.0 \text{ } M) \rightarrow Cu^{2+}(1.0 \text{ } M) + e^-$
Cathode (reduction):	$Cu^+(1.0 \text{ } M) + e^- \rightarrow Cu(s)$
Overall:	$2Cu^+(1.0 \text{ } M) \rightarrow Cu^{2+}(1.0 \text{ } M) + Cu(s)$

$$E^\circ_{cell} = E^\circ_{cathode} - E^\circ_{anode} = E^\circ_{Cu^+/Cu} - E^\circ_{Cu^{2+}/Cu^+}$$

$$E^\circ_{cell} = 0.52 \text{ V} - 0.15 \text{ V} = \textbf{0.37 V}$$

Now, we use Equation (19.3) of the text. The overall reaction shows that $n = 1$.

$$\Delta G^\circ = -nFE^\circ_{cell}$$

$$\Delta G^\circ = -(1)(96500 \text{ J/V·mol})(0.37 \text{ V}) = \textbf{−36 kJ/mol}$$

Next, we can calculate K using Equation (19.5) of the text.

$$E^\circ_{cell} = \frac{0.0257 \text{ V}}{n}\ln K$$

or

$$\ln K = \frac{nE^\circ_{cell}}{0.0257 \text{ V}}$$

and

$$K = e^{\frac{nE^\circ}{0.0257}}$$

$$K = e^{\frac{(1)(0.37)}{0.0257}} = e^{14.4} = \textbf{2} \times \textbf{10}^\textbf{6}$$

Check: The negative value of ΔG° and the large positive value of K, both indicate that the reaction favors products at equilibrium. The result is consistent with the fact that E° for the galvanic cell is positive.

19.29   If this were a standard cell, the concentrations would all be 1.00 M, and the voltage would just be the standard emf calculated from Table 19.1 of the text. Since cell emf's depend on the concentrations of the reactants and products, we must use the Nernst equation [Equation (19.8) of the text] to find the emf of a nonstandard cell.

$$E = E° - \frac{0.0257\ V}{n} \ln Q$$

$$E = 1.10\ V - \frac{0.0257\ V}{2} \ln \frac{[Zn^{2+}]}{[Cu^{2+}]}$$

$$E = 1.10\ V - \frac{0.0257\ V}{2} \ln \frac{0.25}{0.15}$$

$$E = \mathbf{1.09\ V}$$

How did we find the value of 1.10 V for $E°$?

19.30 **Strategy:** The standard emf ($E°$) can be calculated using the standard reduction potentials in Table 19.1 of the text. Because the reactions are not run under standard-state conditions (concentrations are not 1 M), we need Nernst's equation [Equation (19.8) of the text] to calculate the emf (E) of a hypothetical galvanic cell. Remember that solids do not appear in the reaction quotient (Q) term in the Nernst equation. We can calculate ΔG from E using Equation (19.2) of the text: $\Delta G = -nFE_{cell}$.

Solution:

(a) The half-cell reactions are:

Anode (oxidation): $Mg(s) \rightarrow Mg^{2+}(1.0\ M) + 2e^-$
Cathode (reduction): $Sn^{2+}(1.0\ M) + 2e^- \rightarrow Sn(s)$
_____
Overall: $Mg(s) + Sn^{2+}(1.0\ M) \rightarrow Mg^{2+}(1.0\ M) + Sn(s)$

$$E°_{cell} = E°_{cathode} - E°_{anode} = E°_{Sn^{2+}/Sn} - E°_{Mg^{2+}/Mg}$$

$$E°_{cell} = -0.14\ V - (-2.37\ V) = \mathbf{2.23\ V}$$

From Equation (19.8) of the text, we write:

$$E = E° - \frac{0.0257\ V}{n} \ln Q$$

$$E = E° - \frac{0.0257\ V}{n} \ln \frac{[Mg^{2+}]}{[Sn^{2+}]}$$

$$E = 2.23\ V - \frac{0.0257\ V}{2} \ln \frac{0.045}{0.035} = \mathbf{2.23\ V}$$

We can now find the free energy change at the given concentrations using Equation (19.2) of the text. Note that in this reaction, $n = 2$.

$$\Delta G = -nFE_{cell}$$

$$\Delta G = -(2)(96500\ J/V \cdot mol)(2.23\ V) = \mathbf{-430\ kJ/mol}$$

(b) The half-cell reactions are:

Anode (oxidation): $3[Zn(s) \rightarrow Zn^{2+}(1.0\ M) + 2e^-]$
Cathode (reduction): $2[Cr^{3+}(1.0\ M) + 3e^- \rightarrow Cr(s)]$
_____
Overall: $3Zn(s) + 2Cr^{3+}(1.0\ M) \rightarrow 3Zn^{2+}(1.0\ M) + 2Cr(s)$

$$E^{\circ}_{cell} = E^{\circ}_{cathode} - E^{\circ}_{anode} = E^{\circ}_{Cr^{3+}/Cr} - E^{\circ}_{Zn^{2+}/Zn}$$

$$E^{\circ}_{cell} = -0.74 \text{ V} - (-0.76 \text{ V}) = \textbf{0.02 V}$$

From Equation (19.8) of the text, we write:

$$E = E^{\circ} - \frac{0.0257 \text{ V}}{n} \ln Q$$

$$E = E^{\circ} - \frac{0.0257 \text{ V}}{n} \ln \frac{[Zn^{2+}]^3}{[Cr^{3+}]^2}$$

$$E = 0.02 \text{ V} - \frac{0.0257 \text{ V}}{6} \ln \frac{(0.0085)^3}{(0.010)^2} = \textbf{0.04 V}$$

We can now find the free energy change at the given concentrations using Equation (19.2) of the text. Note that in this reaction, $n = 6$.

$$\Delta G = -nFE_{cell}$$

$$\Delta G = -(6)(96500 \text{ J/V·mol})(0.04 \text{ V}) = \textbf{-23 kJ/mol}$$

19.31 The overall reaction is: $Zn(s) + 2H^+(aq) \rightarrow Zn^{2+}(aq) + H_2(g)$

$$E^{\circ}_{cell} = E^{\circ}_{cathode} - E^{\circ}_{anode} = 0.00 \text{ V} - (-0.76 \text{ V}) = \textbf{0.76 V}$$

$$E = E^{\circ} - \frac{0.0257 \text{ V}}{n} \ln \frac{[Zn^{2+}]P_{H_2}}{[H^+]^2}$$

$$E = 0.76 \text{ V} - \frac{0.0257 \text{ V}}{2} \ln \frac{(0.45)(2.0)}{(1.8)^2} = \textbf{0.78 V}$$

19.32 Let's write the two half-reactions to calculate the standard cell emf. (Oxidation occurs at the Pb electrode.)

$$Pb(s) \xrightarrow{\text{oxidation (anode)}} Pb^{2+}(aq) + 2e^- \qquad E^{\circ}_{anode} = -0.13 \text{ V}$$

$$2H^+(aq) + 2e^- \xrightarrow{\text{reduction (cathode)}} H_2(g) \qquad E^{\circ}_{cathode} = 0.00 \text{ V}$$

$$\overline{2H^+(aq) + Pb(s) \longrightarrow H_2(g) + Pb^{2+}(aq)}$$

$$E^{\circ}_{cell} = E^{\circ}_{cathode} - E^{\circ}_{anode} = 0.00 \text{ V} - (-0.13 \text{ V}) = 0.13 \text{ V}$$

Using the Nernst equation, we can calculate the cell emf, E.

$$E = E^{\circ} - \frac{0.0257 \text{ V}}{n} \ln \frac{[Pb^{2+}]P_{H_2}}{[H^+]^2}$$

$$E = 0.13 \text{ V} - \frac{0.0257 \text{ V}}{2} \ln \frac{(0.10)(1.0)}{(0.050)^2} = \textbf{0.083 V}$$

19.33   As written, the reaction is not spontaneous under standard state conditions; the cell emf is negative.

$$E^{\circ}_{cell} = E^{\circ}_{cathode} - E^{\circ}_{anode} = -0.76 \text{ V} - 0.34 \text{ V} = -1.10 \text{ V}$$

The reaction will become spontaneous when the concentrations of zinc(II) and copper(II) ions are such as to make the emf positive. The turning point is when the emf is zero. We solve the Nernst equation for the $[Cu^{2+}]/[Zn^{2+}]$ ratio at this point.

$$E_{cell} = E^{\circ} - \frac{0.0257 \text{ V}}{n} \ln Q$$

$$0 = -1.10 \text{ V} - \frac{0.0257 \text{ V}}{2} \ln \frac{[Cu^{2+}]}{[Zn^{2+}]}$$

$$\ln \frac{[Cu^{2+}]}{[Zn^{2+}]} = -85.6$$

$$\frac{[Cu^{2+}]}{[Zn^{2+}]} = e^{-85.6} = \mathbf{6.7 \times 10^{-38}}$$

In other words for the reaction to be spontaneous, the $[Cu^{2+}]/[Zn^{2+}]$ ratio must be less than 6.7×10^{-38}. Is the reduction of zinc(II) by copper metal a practical use of copper?

19.34   All concentration cells have the same standard emf: *zero* volts.

$$Mg^{2+}(aq) + 2e^- \xrightarrow{\text{reduction (cathode)}} Mg(s) \qquad E^{\circ}_{cathode} = -2.37 \text{ V}$$

$$Mg(s) \xrightarrow{\text{oxidation (anode)}} Mg^{2+}(aq) + 2e^- \qquad E^{\circ}_{anode} = -2.37 \text{ V}$$

$$E^{\circ}_{cell} = E^{\circ}_{cathode} - E^{\circ}_{anode} = -2.37 \text{ V} - (-2.37 \text{ V}) = 0.00 \text{ V}$$

We use the Nernst equation to compute the emf. There are two moles of electrons transferred from the reducing agent to the oxidizing agent in this reaction, so $n = 2$.

$$E = E^{\circ} - \frac{0.0257 \text{ V}}{n} \ln Q$$

$$E = E^{\circ} - \frac{0.0257 \text{ V}}{n} \ln \frac{[Mg^{2+}]_{ox}}{[Mg^{2+}]_{red}}$$

$$E = 0 \text{ V} - \frac{0.0257 \text{ V}}{2} \ln \frac{0.24}{0.53} = \mathbf{0.010 \text{ V}}$$

What is the direction of spontaneous change in all concentration cells?

19.37   **(a)**   The total charge passing through the circuit is

$$3.0 \text{ h} \times \frac{8.5 \text{ C}}{1 \text{ s}} \times \frac{3600 \text{ s}}{1 \text{ h}} = 9.2 \times 10^4 \text{ C}$$

From the anode half-reaction we can find the amount of hydrogen.

$$(9.2 \times 10^4 \text{ C}) \times \frac{2 \text{ mol H}_2}{4 \text{ mol e}^-} \times \frac{1 \text{ mol e}^-}{96500 \text{ C}} = 0.48 \text{ mol H}_2$$

The volume can be computed using the ideal gas equation

$$V = \frac{nRT}{P} = \frac{(0.48 \text{ mol})(0.0821 \text{ L} \cdot \text{atm/K} \cdot \text{mol})(298 \text{ K})}{155 \text{ atm}} = \textbf{0.076 L}$$

(b)   The charge passing through the circuit in one minute is

$$\frac{8.5 \text{ C}}{1 \text{ s}} \times \frac{60 \text{ s}}{1 \text{ min}} = 510 \text{ C/min}$$

We can find the amount of oxygen from the cathode half-reaction and the ideal gas equation.

$$\frac{510 \text{ C}}{1 \text{ min}} \times \frac{1 \text{ mol e}^-}{96500 \text{ C}} \times \frac{1 \text{ mol O}_2}{4 \text{ mol e}^-} = 1.3 \times 10^{-3} \text{ mol O}_2/\text{min}$$

$$V = \frac{nRT}{P} = \left(\frac{1.3 \times 10^{-3} \text{ mol O}_2}{1 \text{ min}} \right) \left(\frac{(0.0821 \text{ L} \cdot \text{atm/K} \cdot \text{mol})(298 \text{ K})}{1 \text{ atm}} \right) = 0.032 \text{ L O}_2/\text{min}$$

$$\frac{0.032 \text{ L O}_2}{1 \text{ min}} \times \frac{1.0 \text{ L air}}{0.20 \text{ L O}_2} = \textbf{0.16 L of air/min}$$

19.38   We can calculate the standard free energy change, $\Delta G°$, from the standard free energies of formation, $\Delta G_f°$ using Equation (18.12) of the text. Then, we can calculate the standard cell emf, $E_{cell}°$, from $\Delta G°$.

The overall reaction is:

$$C_3H_8(g) + 5O_2(g) \longrightarrow 3CO_2(g) + 4H_2O(l)$$

$$\Delta G_{rxn}° = 3\Delta G_f°[CO_2(g)] + 4\Delta G_f°[H_2O(l)] - \{\Delta G_f°[C_3H_8(g)] + 5\Delta G_f°[O_2(g)]\}$$

$$\Delta G_{rxn}° = (3)(-394.4 \text{ kJ/mol}) + (4)(-237.2 \text{ kJ/mol}) - [(1)(-23.5 \text{ kJ/mol}) + (5)(0)] = -2108.5 \text{ kJ/mol}$$

We can now calculate the standard emf using the following equation:

$$\Delta G° = -nFE_{cell}°$$

or

$$E_{cell}° = \frac{-\Delta G°}{nF}$$

Check the half-reactions on p. 820 of the text to determine that 20 moles of electrons are transferred during this redox reaction.

$$E_{cell}° = \frac{-(-2108.5 \times 10^3 \text{ J/mol})}{(20)(96500 \text{ J/V} \cdot \text{mol})} = \textbf{1.09 V}$$

Does this suggest that, in theory, it should be possible to construct a galvanic cell (battery) based on any conceivable spontaneous reaction?

19.45 $\text{Mass Mg} = 1.00 \ F \times \dfrac{1 \text{ mol Mg}}{2 \text{ mol e}^-} \times \dfrac{24.31 \text{ g Mg}}{1 \text{ mol Mg}} = \mathbf{12.2 \text{ g Mg}}$

19.46   **(a)**   The only ions present in molten $BaCl_2$ are Ba^{2+} and Cl^-. The electrode reactions are:

anode: $2Cl^-(aq) \longrightarrow Cl_2(g) + 2e^-$

cathode: $Ba^{2+}(aq) + 2e^- \longrightarrow Ba(s)$

This cathode half-reaction tells us that 2 moles of e^- are required to produce 1 mole of $Ba(s)$.

(b)
Strategy: According to Figure 19.20 of the text, we can carry out the following conversion steps to calculate the quantity of Ba in grams.

$$\text{current} \times \text{time} \rightarrow \text{coulombs} \rightarrow \text{mol } e^- \rightarrow \text{mol Ba} \rightarrow \text{g Ba}$$

This is a large number of steps, so let's break it down into two parts. First, we calculate the coulombs of electricity that pass through the cell. Then, we will continue on to calculate grams of Ba.

Solution: First, we calculate the coulombs of electricity that pass through the cell.

$$0.50 \text{ A} \times \frac{1 \text{ C}}{1 \text{ A}\cdot\text{s}} \times \frac{60 \text{ s}}{1 \text{ min}} \times 30 \text{ min} = 9.0 \times 10^2 \text{ C}$$

We see that for every mole of Ba formed at the cathode, 2 moles of electrons are needed. The grams of Ba produced at the cathode are:

$$\textbf{? g Ba} = (9.0 \times 10^2 \text{ C}) \times \frac{1 \text{ mol } e^-}{96,500 \text{ C}} \times \frac{1 \text{ mol Ba}}{2 \text{ mol } e^-} \times \frac{137.3 \text{ g Ba}}{1 \text{ mol Ba}} = \mathbf{0.64 \text{ g Ba}}$$

19.47   The half-reactions are: $Na^+ + e^- \rightarrow Na$
$Al^{3+} + 3e^- \rightarrow Al$

Since 1 g is the same idea as 1 ton as long as we are comparing two quantities, we can write:

$$1 \text{ g Na} \times \frac{1 \text{ mol}}{22.99 \text{ g Na}} \times 1 \ e^- = 0.043 \text{ mol } e^-$$

$$1 \text{ g Al} \times \frac{1 \text{ mol}}{26.98 \text{ g Al}} \times 3 \ e^- = 0.11 \text{ mol } e^-$$

It is cheaper to prepare 1 ton of sodium by electrolysis.

19.48   The cost for producing various metals is determined by the moles of electrons needed to produce a given amount of metal. For each reduction, let's first calculate the number of tons of metal produced per 1 mole of electrons (1 ton = 9.072×10^5 g). The reductions are:

$$Mg^{2+} + 2e^- \longrightarrow Mg \qquad \frac{1 \text{ mol Mg}}{2 \text{ mol } e^-} \times \frac{24.31 \text{ g Mg}}{1 \text{ mol Mg}} \times \frac{1 \text{ ton}}{9.072 \times 10^5 \text{ g}} = 1.340 \times 10^{-5} \text{ ton Mg/mol } e^-$$

$$Al^{3+} + 3e^- \longrightarrow Al \qquad \frac{1 \text{ mol Al}}{3 \text{ mol } e^-} \times \frac{26.98 \text{ g Al}}{1 \text{ mol Al}} \times \frac{1 \text{ ton}}{9.072 \times 10^5 \text{ g}} = 9.913 \times 10^{-6} \text{ ton Al/mol } e^-$$

$$Na^+ + e^- \longrightarrow Na \qquad \frac{1 \text{ mol Na}}{1 \text{ mol } e^-} \times \frac{22.99 \text{ g Na}}{1 \text{ mol Na}} \times \frac{1 \text{ ton}}{9.072 \times 10^5 \text{ g}} = 2.534 \times 10^{-5} \text{ ton Na/mol } e^-$$

$$Ca^{2+} + 2e^- \longrightarrow Ca \qquad \frac{1 \text{ mol Ca}}{2 \text{ mol } e^-} \times \frac{40.08 \text{ g Ca}}{1 \text{ mol Ca}} \times \frac{1 \text{ ton}}{9.072 \times 10^5 \text{ g}} = 2.209 \times 10^{-5} \text{ ton Ca/mol } e^-$$

Now that we know the tons of each metal produced per mole of electrons, we can convert from $155/ton Mg to the cost to produce the given amount of each metal.

(a) For aluminum :

$$\frac{\$155}{1 \text{ ton Mg}} \times \frac{1.340 \times 10^{-5} \text{ ton Mg}}{1 \text{ mol } e^-} \times \frac{1 \text{ mol } e^-}{9.913 \times 10^{-6} \text{ ton Al}} \times 10.0 \text{ tons Al} = \mathbf{\$2.10 \times 10^3}$$

(b) For sodium:

$$\frac{\$155}{1 \text{ ton Mg}} \times \frac{1.340 \times 10^{-5} \text{ ton Mg}}{1 \text{ mol } e^-} \times \frac{1 \text{ mol } e^-}{2.534 \times 10^{-5} \text{ ton Na}} \times 30.0 \text{ tons Na} = \mathbf{\$2.46 \times 10^3}$$

(c) For calcium:

$$\frac{\$155}{1 \text{ ton Mg}} \times \frac{1.340 \times 10^{-5} \text{ ton Mg}}{1 \text{ mol } e^-} \times \frac{1 \text{ mol } e^-}{2.209 \times 10^{-5} \text{ ton Ca}} \times 50.0 \text{ tons Ca} = \mathbf{\$4.70 \times 10^3}$$

19.49   Find the amount of oxygen using the ideal gas equation

$$n = \frac{PV}{RT} = \frac{\left(755 \text{ mmHg} \times \dfrac{1 \text{ atm}}{760 \text{ mmHg}}\right)(0.076 \text{ L})}{(0.0821 \text{ L} \cdot \text{atm/K} \cdot \text{mol})(298 \text{ K})} = 3.1 \times 10^{-3} \text{ mol O}_2$$

Since the half-reaction shows that one mole of oxygen requires four faradays of electric charge, we write

$$(3.1 \times 10^{-3} \text{ mol O}_2) \times \frac{4 \text{ } F}{1 \text{ mol O}_2} = \mathbf{0.012 \text{ } F}$$

19.50   **(a)**   The half–reaction is:

$$2H_2O(l) \longrightarrow O_2(g) + 4H^+(aq) + 4e^-$$

First, we can calculate the number of moles of oxygen produced using the ideal gas equation.

$$n_{O_2} = \frac{PV}{RT}$$

$$n_{O_2} = \frac{(1.0 \text{ atm})(0.84 \text{ L})}{(0.0821 \text{ L} \cdot \text{atm/mol} \cdot \text{K})(298 \text{ K})} = 0.034 \text{ mol O}_2$$

Since 4 moles of electrons are needed for every 1 mole of oxygen, we will need 4 F of electrical charge to produce 1 mole of oxygen.

$$? \ F \ = \ 0.034 \ \text{mol O}_2 \times \frac{4 \ F}{1 \ \text{mol O}_2} \ = \ \textbf{0.14} \ \textbf{\textit{F}}$$

(b)   The half–reaction is:

$$2\text{Cl}^-(aq) \ \longrightarrow \ \text{Cl}_2(g) + 2e^-$$

The number of moles of chlorine produced is:

$$n_{\text{Cl}_2} \ = \ \frac{PV}{RT}$$

$$n_{\text{Cl}_2} \ = \ \frac{\left(750 \ \text{mmHg} \times \dfrac{1 \ \text{atm}}{760 \ \text{mmHg}}\right)(1.50 \ \text{L})}{(0.0821 \ \text{L} \cdot \text{atm/mol} \cdot \text{K})(298 \ \text{K})} \ = \ 0.0605 \ \text{mol Cl}_2$$

Since 2 moles of electrons are needed for every 1 mole of chlorine gas, we will need 2 F of electrical charge to produce 1 mole of chlorine gas.

$$? \ F \ = \ 0.0605 \ \text{mol Cl}_2 \times \frac{2 \ F}{1 \ \text{mol Cl}_2} \ = \ \textbf{0.121} \ \textbf{\textit{F}}$$

(c)   The half–reaction is:

$$\text{Sn}^{2+}(aq) + 2e^- \ \longrightarrow \ \text{Sn}(s)$$

The number of moles of Sn(s) produced is

$$? \ \text{mol Sn} \ = \ 6.0 \ \text{g Sn} \times \frac{1 \ \text{mol Sn}}{118.7 \ \text{g Sn}} \ = \ 0.051 \ \text{mol Sn}$$

Since 2 moles of electrons are needed for every 1 mole of Sn, we will need 2 F of electrical charge to reduce 1 mole of Sn^{2+} ions to Sn metal.

$$? \ F \ = \ 0.051 \ \text{mol Sn} \times \frac{2 \ F}{1 \ \text{mol Sn}} \ = \ \textbf{0.10} \ \textbf{\textit{F}}$$

19.51   The half-reactions are: $\text{Cu}^{2+}(aq) + 2e^- \rightarrow \text{Cu}(s)$

$$2\text{Br}^-(aq) \rightarrow \text{Br}_2(l) + 2e^-$$

The mass of copper produced is:

$$4.50 \ \text{A} \times 1 \ \text{h} \times \frac{3600 \ \text{s}}{1 \ \text{h}} \times \frac{1 \ \text{C}}{1 \ \text{A} \cdot \text{s}} \times \frac{1 \ \text{mol e}^-}{96500 \ \text{C}} \times \frac{1 \ \text{mol Cu}}{2 \ \text{mol e}^-} \times \frac{63.55 \ \text{g Cu}}{1 \ \text{mol Cu}} \ = \ \textbf{5.33 g Cu}$$

The mass of bromine produced is:

$$4.50 \ \text{A} \times 1 \ \text{h} \times \frac{3600 \ \text{s}}{1 \ \text{h}} \times \frac{1 \ \text{C}}{1 \ \text{A} \cdot \text{s}} \times \frac{1 \ \text{mol e}^-}{96500 \ \text{C}} \times \frac{1 \ \text{mol Br}_2}{2 \ \text{mol e}^-} \times \frac{159.8 \ \text{g Br}_2}{1 \ \text{mol Br}_2} \ = \ \textbf{13.4 g Br}_2$$

19.52 **(a)** The half–reaction is:

$$Ag^+(aq) + e^- \longrightarrow Ag(s)$$

(b) Since this reaction is taking place in an aqueous solution, the probable oxidation is the oxidation of water. (Neither Ag^+ nor NO_3^- can be further oxidized.)

$$2H_2O(l) \longrightarrow O_2(g) + 4H^+(aq) + 4e^-$$

(c) The half-reaction tells us that 1 mole of electrons is needed to reduce 1 mol of Ag^+ to Ag metal. We can set up the following strategy to calculate the quantity of electricity (in C) needed to deposit 0.67 g of Ag.

$$\text{grams Ag} \rightarrow \text{mol Ag} \rightarrow \text{mol } e^- \rightarrow \text{coulombs}$$

$$0.67 \text{ g Ag} \times \frac{1 \text{ mol Ag}}{107.9 \text{ g Ag}} \times \frac{1 \text{ mol } e^-}{1 \text{ mol Ag}} \times \frac{96500 \text{ C}}{1 \text{ mol } e^-} = \mathbf{6.0 \times 10^2 \text{ C}}$$

19.53 The half-reaction is: $Co^{2+} + 2e^- \rightarrow Co$

$$2.35 \text{ g Co} \times \frac{1 \text{ mol Co}}{58.93 \text{ g Co}} \times \frac{2 \text{ mol } e^-}{1 \text{ mol Co}} \times \frac{96500 \text{ C}}{1 \text{ mol } e^-} = \mathbf{7.70 \times 10^3 \text{ C}}$$

19.54 **(a)** First find the amount of charge needed to produce 2.00 g of silver according to the half–reaction:

$$Ag^+(aq) + e^- \longrightarrow Ag(s)$$

$$2.00 \text{ g Ag} \times \frac{1 \text{ mol Ag}}{107.9 \text{ g Ag}} \times \frac{1 \text{ mol } e^-}{1 \text{ mol Ag}} \times \frac{96500 \text{ C}}{1 \text{ mol } e^-} = 1.79 \times 10^3 \text{ C}$$

The half–reaction for the reduction of copper(II) is:

$$Cu^{2+}(aq) + 2e^- \longrightarrow Cu(s)$$

From the amount of charge calculated above, we can calculate the mass of copper deposited in the second cell.

$$(1.79 \times 10^3 \text{ C}) \times \frac{1 \text{ mol } e^-}{96500 \text{ C}} \times \frac{1 \text{ mol Cu}}{2 \text{ mol } e^-} \times \frac{63.55 \text{ g Cu}}{1 \text{ mol Cu}} = \mathbf{0.589 \text{ g Cu}}$$

(b) We can calculate the current flowing through the cells using the following strategy.

$$\text{Coulombs} \rightarrow \text{Coulombs/hour} \rightarrow \text{Coulombs/second}$$

Recall that $1 \text{ C} = 1 \text{ A·s}$

The current flowing through the cells is:

$$(1.79 \times 10^3 \text{ A·s}) \times \frac{1 \text{ h}}{3600 \text{ s}} \times \frac{1}{3.75 \text{ h}} = \mathbf{0.133 \text{ A}}$$

19.55   The half-reaction for the oxidation of chloride ion is:

$$2Cl^-(aq) \rightarrow Cl_2(g) + 2e^-$$

First, let's calculate the moles of e^- flowing through the cell in one hour.

$$1500 \text{ A} \times \frac{1 \text{ C}}{1 \text{ A} \cdot \text{s}} \times \frac{3600 \text{ s}}{1 \text{ h}} \times \frac{1 \text{ mol } e^-}{96500 \text{ C}} = 55.96 \text{ mol } e^-$$

Next, let's calculate the hourly production rate of chlorine gas (in kg). Note that the anode efficiency is 93.0%.

$$55.96 \text{ mol } e^- \times \frac{1 \text{ mol Cl}_2}{2 \text{ mol } e^-} \times \frac{0.07090 \text{ kg Cl}_2}{1 \text{ mol Cl}_2} \times \frac{93.0\%}{100\%} = \textbf{1.84 kg Cl}_2\textbf{/h}$$

19.56   *Step 1:* Balance the half–reaction.

$$Cr_2O_7^{2-}(aq) + 14H^+(aq) + 12e^- \longrightarrow 2Cr(s) + 7H_2O(l)$$

Step 2: Calculate the quantity of chromium metal by calculating the volume and converting this to mass using the given density.

$$\text{Volume Cr } = \text{ thickness} \times \text{surface area}$$

$$\text{Volume Cr } = (1.0 \times 10^{-2} \text{ mm}) \times \frac{1 \text{ m}}{1000 \text{ mm}} \times 0.25 \text{ m}^2 = 2.5 \times 10^{-6} \text{ m}^3$$

Converting to cm^3,

$$(2.5 \times 10^{-6} \text{ m}^3) \times \left(\frac{1 \text{ cm}}{0.01 \text{ m}} \right)^3 = 2.5 \text{ cm}^3$$

Next, calculate the mass of Cr.

$$\text{Mass } = \text{ density} \times \text{volume}$$

$$\text{Mass Cr } = 2.5 \text{ cm}^3 \times \frac{7.19 \text{ g}}{1 \text{ cm}^3} = 18 \text{ g Cr}$$

Step 3: Find the number of moles of electrons required to electrodeposit 18 g of Cr from solution. The half-reaction is:

$$Cr_2O_7^{2-}(aq) + 14H^+(aq) + 12e^- \longrightarrow 2Cr(s) + 7H_2O(l)$$

Six moles of electrons are required to reduce 1 mol of Cr metal. But, we are electrodepositing less than 1 mole of Cr (*s*). We need to complete the following conversions:

$$\text{g Cr } \rightarrow \text{ mol Cr } \rightarrow \text{ mol } e^-$$

$$? \text{ faradays } = 18 \text{ g Cr} \times \frac{1 \text{ mol Cr}}{52.00 \text{ g Cr}} \times \frac{6 \text{ mol } e^-}{1 \text{ mol Cr}} = 2.1 \text{ mol } e^-$$

Step 4: Determine how long it will take for 2.1 moles of electrons to flow through the cell when the current is 25.0 C/s. We need to complete the following conversions:

$$\text{mol } e^- \rightarrow \text{ coulombs } \rightarrow \text{ seconds } \rightarrow \text{ hours}$$

$$? \text{ h} = 2.1 \text{ mol e}^- \times \frac{96,500 \text{ C}}{1 \text{ mol e}^-} \times \frac{1 \text{ s}}{25.0 \text{ C}} \times \frac{1 \text{ h}}{3600 \text{ s}} = \mathbf{2.3 \text{ h}}$$

Would any time be saved by connecting several bumpers together in a series?

19.57 The quantity of charge passing through the solution is:

$$0.750 \text{ A} \times \frac{1 \text{ C}}{1 \text{ A} \cdot \text{s}} \times \frac{60 \text{ s}}{1 \text{ min}} \times \frac{1 \text{ mol e}^-}{96500 \text{ C}} \times 25.0 \text{ min} = 1.17 \times 10^{-2} \text{ mol e}^-$$

Since the charge of the copper ion is +2, the number of moles of copper formed must be:

$$(1.17 \times 10^{-2} \text{ mol e}^-) \times \frac{1 \text{ mol Cu}}{2 \text{ mol e}^-} = 5.85 \times 10^{-3} \text{ mol Cu}$$

The units of molar mass are grams per mole. The molar mass of copper is:

$$\frac{0.369 \text{ g}}{5.85 \times 10^{-3} \text{ mol}} = \mathbf{63.1 \text{ g/mol}}$$

19.58 Based on the half-reaction, we know that one faraday will produce half a mole of copper.

$$Cu^{2+}(aq) + 2e^- \longrightarrow Cu(s)$$

First, let's calculate the charge (in C) needed to deposit 0.300 g of Cu.

$$(3.00 \text{ A})(304 \text{ s}) \times \frac{1 \text{ C}}{1 \text{ A} \cdot \text{s}} = 912 \text{ C}$$

We know that one faraday will produce half a mole of copper, but we don't have a half a mole of copper. We have:

$$0.300 \text{ g Cu} \times \frac{1 \text{ mol Cu}}{63.55 \text{ g Cu}} = 4.72 \times 10^{-3} \text{ mol}$$

We calculated the number of coulombs (912 C) needed to produce 4.72×10^{-3} mol of Cu. How many coulombs will it take to produce 0.500 moles of Cu? This will be Faraday's constant.

$$\frac{912 \text{ C}}{4.72 \times 10^{-3} \text{ mol Cu}} \times 0.500 \text{ mol Cu} = \mathbf{9.66 \times 10^4 \text{ C}} = 1 \text{ } F$$

19.59 The number of faradays supplied is:

$$1.44 \text{ g Ag} \times \frac{1 \text{ mol Ag}}{107.9 \text{ g Ag}} \times \frac{1 \text{ mol e}^-}{1 \text{ mol Ag}} = 0.0133 \text{ mol e}^-$$

Since we need three faradays to reduce one mole of X^{3+}, the molar mass of X must be:

$$\frac{0.120 \text{ g X}}{0.0133 \text{ mol e}^-} \times \frac{3 \text{ mol e}^-}{1 \text{ mol X}} = \mathbf{27.1 \text{ g/mol}}$$

19.60   First we can calculate the number of moles of hydrogen produced using the ideal gas equation.

$$n_{H_2} = \frac{PV}{RT}$$

$$n_{H_2} = \frac{\left(782 \text{ mmHg} \times \frac{1 \text{ atm}}{760 \text{ mmHg}}\right)(0.845 \text{ L})}{(0.0821 \text{ L} \cdot \text{atm/K} \cdot \text{mol})(298 \text{ K})} = 0.0355 \text{ mol}$$

The number of faradays passed through the solution is:

$$0.0355 \text{ mol } H_2 \times \frac{2 \text{ } F}{1 \text{ mol } H_2} = \textbf{0.0710 } \textit{F}$$

19.61   **(a)**   The half-reactions are:

$$H_2(g) \rightarrow 2H^+(aq) + 2e^-$$
$$\underline{Ni^{2+}(aq) + 2e^- \rightarrow Ni(s)}$$

The complete balanced equation is: $Ni^{2+}(aq) + H_2(g) \rightarrow Ni(s) + 2H^+(aq)$

Ni(s) is below and to the right of $H^+(aq)$ in Table 19.1 of the text (see the half-reactions at −0.25 and 0.00 V). Therefore, the spontaneous reaction is the reverse of the above reaction, that is:

$$Ni(s) + 2H^+(aq) \rightarrow Ni^{2+}(aq) + H_2(g)$$

(b)   The half-reactions are:

$$MnO_4^-(aq) + 8H^+(aq) + 5e^- \rightarrow Mn^{2+}(aq) + 4H_2O$$
$$\underline{2Cl^-(aq) \rightarrow Cl_2(g) + 2e^-}$$

The complete balanced equation is:

$$2MnO_4^-(aq) + 16H^+(aq) + 10Cl^-(aq) \rightarrow 2Mn^{2+}(aq) + 8H_2O + 5Cl_2(g)$$

In Table 19.1 of the text, $Cl^-(aq)$ is below and to the right of $MnO_4^-(aq)$; therefore the spontaneous reaction is as written.

(c)   The half-reactions are:

$$Cr(s) \rightarrow Cr^{3+}(aq) + 3e^-$$
$$\underline{Zn^{2+}(aq) + 2e^- \rightarrow Zn(s)}$$

The complete balanced equation is: $2Cr(s) + 3Zn^{2+}(aq) \rightarrow 2Cr^{3+}(aq) + 3Zn(s)$

In Table 19.1 of the text, Zn(s) is below and to the right of $Cr^{3+}(aq)$; therefore the spontaneous reaction is the reverse of the reaction as written.

19.62   The balanced equation is:

$$\textbf{Cr}_2\textbf{O}_7{}^{2-} + \textbf{6 Fe}^{2+} + \textbf{14H}^+ \longrightarrow \textbf{2Cr}^{3+} + \textbf{6Fe}^{3+} + \textbf{7H}_2\textbf{O}$$

The remainder of this problem is a solution stoichiometry problem.

The number of moles of potassium dichromate in 26.0 mL of the solution is:

$$26.0 \text{ mL} \times \frac{0.0250 \text{ mol}}{1000 \text{ mL soln}} = 6.50 \times 10^{-4} \text{ mol } K_2Cr_2O_7$$

From the balanced equation it can be seen that 1 mole of dichromate is stoichiometrically equivalent to 6 moles of iron(II). The number of moles of iron(II) oxidized is therefore

$$(6.50 \times 10^{-4} \text{ mol } Cr_2O_7^{2-}) \times \frac{6 \text{ mol } Fe^{2+}}{1 \text{ mol } Cr_2O_7^{2-}} = 3.90 \times 10^{-3} \text{ mol } Fe^{2+}$$

Finally, the molar concentration of Fe^{2+} is:

$$\frac{3.90 \times 10^{-3} \text{ mol}}{25.0 \times 10^{-3} \text{ L}} = 0.156 \text{ mol/L} = \mathbf{0.156 \ \mathit{M} \ Fe^{2+}}$$

19.63   The balanced equation is:

$$5SO_2(g) + 2MnO_4^-(aq) + 2H_2O(l) \rightarrow 5SO_4^{2-}(aq) + 2Mn^{2+}(aq) + 4H^+(aq)$$

The mass of SO_2 in the water sample is given by

$$7.37 \text{ mL} \times \frac{0.00800 \text{ mol } KMnO_4}{1000 \text{ mL soln}} \times \frac{5 \text{ mol } SO_2}{2 \text{ mol } KMnO_4} \times \frac{64.07 \text{ g } SO_2}{1 \text{ mol } SO_2} = \mathbf{9.44 \times 10^{-3} \text{ g } SO_2}$$

19.64   The balanced equation is:

$$MnO_4^- + 5Fe^{2+} + 8H^+ \longrightarrow Mn^{2+} + 5Fe^{3+} + 4H_2O$$

First, let's calculate the number of moles of potassium permanganate in 23.30 mL of solution.

$$23.30 \text{ mL} \times \frac{0.0194 \text{ mol}}{1000 \text{ mL soln}} = 4.52 \times 10^{-4} \text{ mol } KMnO_4$$

From the balanced equation it can be seen that 1 mole of permanganate is stoichiometrically equivalent to 5 moles of iron(II). The number of moles of iron(II) oxidized is therefore

$$(4.52 \times 10^{-4} \text{ mol } MnO_4^-) \times \frac{5 \text{ mol } Fe^{2+}}{1 \text{ mol } MnO_4^-} = 2.26 \times 10^{-3} \text{ mol } Fe^{2+}$$

The mass of Fe^{2+} oxidized is:

$$\text{mass } Fe^{2+} = (2.26 \times 10^{-3} \text{ mol } Fe^{2+}) \times \frac{55.85 \text{ g } Fe^{2+}}{1 \text{ mol } Fe^{2+}} = 0.126 \text{ g } Fe^{2+}$$

Finally, the mass percent of iron in the ore can be calculated.

$$\text{mass \% Fe} = \frac{\text{mass of iron}}{\text{total mass of sample}} \times 100\%$$

$$\text{\%Fe} = \frac{0.126 \text{ g}}{0.2792 \text{ g}} \times 100\% = \mathbf{45.1\%}$$

19.65   **(a)**   The balanced equation is:

$$2MnO_4^- + 5H_2O_2 + 6H^+ \rightarrow 5O_2 + 2Mn^{2+} + 8H_2O$$

(b)   The number of moles of potassium permanganate in 36.44 mL of the solution is

$$36.44 \text{ mL} \times \frac{0.01652 \text{ mol}}{1000 \text{ mL soln}} = 6.020 \times 10^{-4} \text{ mol of } KMnO_4$$

From the balanced equation it can be seen that in this particular reaction 2 moles of permanganate is stoichiometrically equivalent to 5 moles of hydrogen peroxide. The number of moles of H_2O_2 oxidized is therefore

$$(6.020 \times 10^{-4} \text{ mol } MnO_4^-) \times \frac{5 \text{ mol } H_2O_2}{2 \text{ mol } MnO_4^-} = 1.505 \times 10^{-3} \text{ mol } H_2O_2$$

The molar concentration of H_2O_2 is:

$$[H_2O_2] = \frac{1.505 \times 10^{-3} \text{ mol}}{25.0 \times 10^{-3} \text{ L}} = 0.0602 \text{ mol/L} = \textbf{0.0602 } \textit{\textbf{M}}$$

19.66   **(a)**   The half–reactions are:

(i) $MnO_4^-(aq) + 8H^+(aq) + 5e^- \longrightarrow Mn^{2+}(aq) + 4H_2O(l)$

(ii) $C_2O_4^{2-}(aq) \longrightarrow 2CO_2(g) + 2e^-$

We combine the half-reactions to cancel electrons, that is, [2 × equation (i)] + [5 × equation (ii)]

$$2MnO_4^-(aq) + 16H^+(aq) + 5C_2O_4^{2-}(aq) \longrightarrow 2Mn^{2+}(aq) + 10CO_2(g) + 8H_2O(l)$$

(b)   We can calculate the moles of $KMnO_4$ from the molarity and volume of solution.

$$24.0 \text{ mL } KMnO_4 \times \frac{0.0100 \text{ mol } KMnO_4}{1000 \text{ mL soln}} = 2.40 \times 10^{-4} \text{ mol } KMnO_4$$

We can calculate the mass of oxalic acid from the stoichiometry of the balanced equation. The mole ratio between oxalate ion and permanganate ion is 5:2.

$$(2.40 \times 10^{-4} \text{ mol } KMnO_4) \times \frac{5 \text{ mol } H_2C_2O_4}{2 \text{ mol } KMnO_4} \times \frac{90.04 \text{ g } H_2C_2O_4}{1 \text{ mol } H_2C_2O_4} = 0.0540 \text{ g } H_2C_2O_4$$

Finally, the percent by mass of oxalic acid in the sample is:

$$\textbf{\% oxalic acid} = \frac{0.0540 \text{ g}}{1.00 \text{ g}} \times 100\% = \textbf{5.40\%}$$

19.67

$\underline{E}$	$\underline{\Delta G}$	$\underline{\textit{Cell Reaction}}$
> 0	< 0	spontaneous
< 0	> 0	nonspontaneous
$= 0$	$= 0$	at equilibrium

19.68   The balanced equation is:

$$2MnO_4^- + 5C_2O_4^{2-} + 16H^+ \longrightarrow 2Mn^{2+} + 10CO_2 + 8H_2O$$

Therefore, 2 mol MnO_4^- reacts with 5 mol $C_2O_4^{2-}$

$$\text{Moles of } MnO_4^- \text{ reacted } = 24.2 \text{ mL} \times \frac{9.56 \times 10^{-4} \text{ mol } MnO_4^-}{1000 \text{ mL soln}} = 2.31 \times 10^{-5} \text{ mol } MnO_4^-$$

Recognize that the mole ratio of Ca^{2+} to $C_2O_4^{2-}$ is 1:1 in CaC_2O_4. The mass of Ca^{2+} in 10.0 mL is:

$$(2.31 \times 10^{-5} \text{ mol } MnO_4^-) \times \frac{5 \text{ mol } Ca^{2+}}{2 \text{ mol } MnO_4^-} \times \frac{40.08 \text{ g } Ca^{2+}}{1 \text{ mol } Ca^{2+}} = 2.31 \times 10^{-3} \text{ g } Ca^{2+}$$

Finally, converting to mg/mL, we have:

$$\frac{2.31 \times 10^{-3} \text{ g } Ca^{2+}}{10.0 \text{ mL}} \times \frac{1000 \text{ mg}}{1 \text{ g}} = \textbf{0.231 mg } Ca^{2+}\textbf{/mL blood}$$

19.69   The solubility equilibrium of AgBr is: $AgBr(s) \rightleftharpoons Ag^+(aq) + Br^-(aq)$

By reversing the second given half-reaction and adding it to the first, we obtain:

$Ag(s) \rightarrow Ag^+(aq) + e^-$	$E_{anode}^{\circ} = 0.80 \text{ V}$
$AgBr(s) + e^- \rightarrow Ag(s) + Br^-(aq)$	$E_{cathode}^{\circ} = 0.07 \text{ V}$
$AgBr(s) \rightleftharpoons Ag^+(aq) + Br^-(aq)$	$E_{cell}^{\circ} = E_{cathode}^{\circ} - E_{anode}^{\circ} = 0.07 \text{ V} - 0.80 \text{ V} = -0.73 \text{ V}$

At equilibrium, we have:

$$E = E^{\circ} - \frac{0.0257 \text{ V}}{n} \ln[Ag^+][Br^-]$$

$$0 = -0.73 \text{ V} - \frac{0.0257 \text{ V}}{1} \ln K_{sp}$$

$$\ln K_{sp} = -28.4$$

$$K_{sp} = \textbf{5} \times \textbf{10}^{-13}$$

(Note that this value differs from that given in Table 16.2 of the text, since the data quoted here were obtained from a student's lab report.)

19.70   **(a)**   The half–reactions are:

$2H^+(aq) + 2e^- \longrightarrow H_2(g)$	$E_{anode}^{\circ} = 0.00 \text{ V}$
$Ag^+(aq) + e^- \longrightarrow Ag(s)$	$E_{cathode}^{\circ} = 0.80 \text{ V}$

$$E_{cell}^{\circ} = E_{cathode}^{\circ} - E_{anode}^{\circ} = 0.80 \text{ V} - 0.00 \text{ V} = \textbf{0.80 V}$$

(b)   The spontaneous cell reaction under standard-state conditions is:

$$\textbf{2Ag}^+\textbf{(aq)} + \textbf{H}_2\textbf{(g)} \longrightarrow \textbf{2Ag(s)} + \textbf{2H}^+\textbf{(aq)}$$

(c) Using the Nernst equation we can calculate the cell potential under nonstandard-state conditions.

$$E = E° - \frac{0.0257 \text{ V}}{n} \ln \frac{[H^+]^2}{[Ag^+]^2 P_{H_2}}$$

(i) The potential is:

$$E = 0.80 \text{ V} - \frac{0.0257 \text{ V}}{2} \ln \frac{(1.0 \times 10^{-2})^2}{(1.0)^2(1.0)} = \textbf{0.92 V}$$

(ii) The potential is:

$$E = 0.80 \text{ V} - \frac{0.0257 \text{ V}}{2} \ln \frac{(1.0 \times 10^{-5})^2}{(1.0)^2(1.0)} = \textbf{1.10 V}$$

(d) From the results in part (c), we deduce that this cell is a pH meter; its potential is a sensitive function of the hydrogen ion concentration. Each 1 unit increase in pH causes a voltage increase of 0.060 V.

19.71 **(a)** If this were a standard cell, the concentrations would all be 1.00 M, and the voltage would just be the standard emf calculated from Table 19.1 of the text. Since cell emf's depend on the concentrations of the reactants and products, we must use the Nernst equation [Equation (19.8) of the text] to find the emf of a nonstandard cell.

$$E = E° - \frac{0.0257 \text{ V}}{n} \ln Q$$

$$E = 3.17 \text{ V} - \frac{0.0257 \text{ V}}{2} \ln \frac{[Mg^{2+}]}{[Ag^+]^2}$$

$$E = 3.17 \text{ V} - \frac{0.0257 \text{ V}}{2} \ln \frac{0.10}{[0.10]^2}$$

$$E = \textbf{3.14 V}$$

(b) First we calculate the concentration of silver ion remaining in solution after the deposition of 1.20 g of silver metal

Ag originally in solution: $\dfrac{0.100 \text{ mol Ag}^+}{1 \text{ L}} \times 0.346 \text{ L} = 3.46 \times 10^{-2} \text{ mol Ag}^+$

Ag deposited: $1.20 \text{ g Ag} \times \dfrac{1 \text{ mol}}{107.9 \text{ g}} = 1.11 \times 10^{-2} \text{ mol Ag}$

Ag remaining in solution: $(3.46 \times 10^{-2} \text{ mol Ag}) - (1.11 \times 10^{-2} \text{ mol Ag}) = 2.35 \times 10^{-2} \text{ mol Ag}$

$$[Ag^+] = \frac{2.35 \times 10^{-2} \text{ mol}}{0.346 \text{ L}} = 6.79 \times 10^{-2} M$$

The overall reaction is: $Mg(s) + 2Ag^+(aq) \rightarrow Mg^{2+}(aq) + 2Ag(s)$

We use the balanced equation to find the amount of magnesium metal suffering oxidation and dissolving.

$$(1.11 \times 10^{-2} \text{ mol Ag}) \times \frac{1 \text{ mol Mg}}{2 \text{ mol Ag}} = 5.55 \times 10^{-3} \text{ mol Mg}$$

The amount of magnesium originally in solution was

$$0.288 \text{ L} \times \frac{0.100 \text{ mol}}{1 \text{ L}} = 2.88 \times 10^{-2} \text{ mol}$$

The new magnesium ion concentration is:

$$\frac{[(5.55 \times 10^{-3}) + (2.88 \times 10^{-2})]\text{mol}}{0.288 \text{ L}} = 0.119 \text{ } M$$

The new cell emf is:

$$E = E° - \frac{0.0257 \text{ V}}{n} \ln Q$$

$$E = 3.17 \text{ V} - \frac{0.0257 \text{ V}}{2} \ln \frac{0.119}{(6.79 \times 10^{-2})^2} = \textbf{3.13 V}$$

19.72   The overvoltage of oxygen is not large enough to prevent its formation at the anode. Applying the diagonal rule, we see that water is oxidized before fluoride ion.

$$F_2(g) + 2e^- \longrightarrow 2F^-(aq) \qquad\qquad E° = 2.87 \text{ V}$$

$$O_2(g) + 4H^+(aq) + 4e^- \longrightarrow 2H_2O(l) \qquad\qquad E° = 1.23 \text{ V}$$

The very positive standard reduction potential indicates that F^- has essentially no tendency to undergo oxidation. The oxidation potential of chloride ion is much smaller (−1.36 V), and hence $Cl_2(g)$ can be prepared by electrolyzing a solution of NaCl.

This fact was one of the major obstacles preventing the discovery of fluorine for many years. HF was usually chosen as the substance for electrolysis, but two problems interfered with the experiment. First, any water in the HF was oxidized before the fluoride ion. Second, pure HF without any water in it is a nonconductor of electricity (HF is a weak acid!). The problem was finally solved by dissolving KF in liquid HF to give a conducting solution.

19.73   The cell voltage is given by:

$$E = E° - \frac{0.0257 \text{ V}}{2} \ln \frac{[Cu^{2+}]_{\text{dilute}}}{[Cu^{2+}]_{\text{concentrated}}}$$

$$E = 0 - \frac{0.0257 \text{ V}}{2} \ln \frac{0.080}{1.2} = \textbf{0.035 V}$$

19.74   We can calculate the amount of charge that 4.0 g of MnO_2 can produce.

$$4.0 \text{ g MnO}_2 \times \frac{1 \text{ mol}}{86.94 \text{ g}} \times \frac{2 \text{ mol e}^-}{2 \text{ mol MnO}_2} \times \frac{96500 \text{ C}}{1 \text{ mol e}^-} = 4.44 \times 10^3 \text{ C}$$

Since a current of one ampere represents a flow of one coulomb per second, we can find the time it takes for this amount of charge to pass.

$$0.0050 \text{ A} = 0.0050 \text{ C/s}$$

$$(4.44 \times 10^3 \text{ C}) \times \frac{1 \text{ s}}{0.0050 \text{ C}} \times \frac{1 \text{ h}}{3600 \text{ s}} = \mathbf{2.5 \times 10^2 \text{ h}}$$

19.75   The two electrode processes are:

anode: $2H_2O(l) \rightarrow O_2(g) + 4H^+(aq) + 4e^-$

cathode: $4H_2O(l) + 4e^- \rightarrow 2H_2(g) + 4OH^-(aq)$

The amount of hydrogen formed is twice the amount of oxygen. Notice that the solution at the anode will become acidic and that the solution at the cathode will become basic (test with litmus paper). What are the relative amounts of H^+ and OH^- formed in this process? Would the solutions surrounding the two electrodes neutralize each other exactly? If not, would the resulting solution be acidic or basic?

19.76   Since this is a concentration cell, the standard emf is zero. (Why?) Using the Nernst equation, we can write equations to calculate the cell voltage for the two cells.

$$(1) \quad E_{cell} = -\frac{RT}{nF} \ln Q = -\frac{RT}{2F} \ln \frac{[Hg_2^{2+}] \text{soln A}}{[Hg_2^{2+}] \text{soln B}}$$

$$(2) \quad E_{cell} = -\frac{RT}{nF} \ln Q = -\frac{RT}{1F} \ln \frac{[Hg^+] \text{soln A}}{[Hg^+] \text{soln B}}$$

In the first case, two electrons are transferred per mercury ion ($n = 2$), while in the second only one is transferred ($n = 1$). Note that the concentration ratio will be 1:10 in both cases. The voltages calculated at 18°C are:

$$(1) \quad E_{cell} = \frac{-(8.314 \text{ J/K} \cdot \text{mol})(291 \text{ K})}{2(96500 \text{ J} \cdot \text{V}^{-1}\text{mol}^{-1})} \ln 10^{-1} = 0.0289 \text{ V}$$

$$(2) \quad E_{cell} = \frac{-(8.314 \text{ J/K} \cdot \text{mol})(291 \text{ K})}{1(96500 \text{ J} \cdot \text{V}^{-1}\text{mol}^{-1})} \ln 10^{-1} = 0.0577 \text{ V}$$

Since the calculated cell potential for cell (1) agrees with the measured cell emf, we conclude that the mercury(I) ion exists as $\mathbf{Hg_2^{2+}}$ in solution.

19.77   According to the following standard reduction potentials:

$$O_2(g) + 4H^+(aq) + 4e^- \rightarrow 2H_2O \qquad E° = 1.23 \text{ V}$$

$$I_2(s) + 2e^- \rightarrow 2I^-(aq) \qquad E° = 0.53 \text{ V}$$

we see that it is easier to oxidize the iodide ion than water (because O_2 is a stronger oxidizing agent than I_2). Therefore, the anode reaction is:

$$2I^-(aq) \rightarrow I_2(s) + 2e^-$$

The solution surrounding the anode will become brown because of the formation of the triiodide ion:

$$I^- + I_2(s) \rightarrow I_3^-(aq)$$

The cathode reaction will be the same as in the NaCl electrolysis. (Why?) Since OH⁻ is a product, the solution around the cathode will become basic which will cause the phenolphthalein indicator to turn red.

19.78 We begin by treating this like an ordinary stoichiometry problem (see Chapter 3).

Step 1: Calculate the number of moles of Mg and Ag^+.

The number of moles of magnesium is:

$$1.56 \text{ g Mg} \times \frac{1 \text{ mol Mg}}{24.31 \text{ g Mg}} = 0.0642 \text{ mol Mg}$$

The number of moles of silver ion in the solution is:

$$\frac{0.100 \text{ mol Ag}^+}{1 \text{ L}} \times 0.1000 \text{ L} = 0.0100 \text{ mol Ag}^+$$

Step 2: Calculate the mass of Mg remaining by determining how much Mg reacts with Ag^+.

The balanced equation for the reaction is:

$$2Ag^+(aq) + Mg(s) \longrightarrow 2Ag(s) + Mg^{2+}(aq)$$

Since you need twice as much Ag^+ compared to Mg for complete reaction, Ag^+ is the limiting reagent. The amount of Mg consumed is:

$$0.0100 \text{ mol Ag}^+ \times \frac{1 \text{ mol Mg}}{2 \text{ mol Ag}^+} = 0.00500 \text{ mol Mg}$$

The amount of magnesium remaining is:

$$(0.0642 - 0.00500) \text{ mol Mg} \times \frac{24.31 \text{ g Mg}}{1 \text{ mol Mg}} = \textbf{1.44 g Mg}$$

Step 3: Assuming complete reaction, calculate the concentration of Mg^{2+} ions produced.

Since the mole ratio between Mg and Mg^{2+} is 1:1, the mol of Mg^{2+} formed will equal the mol of Mg reacted. The concentration of Mg^{2+} is:

$$[Mg^{2+}]_0 = \frac{0.00500 \text{ mol}}{0.100 \text{ L}} = 0.0500 \ M$$

Step 4: We can calculate the equilibrium constant for the reaction from the standard cell emf.

$$E^\circ_{cell} = E^\circ_{cathode} - E^\circ_{anode} = 0.80 \text{ V} - (-2.37 \text{ V}) = 3.17 \text{ V}$$

We can then compute the equilibrium constant.

$$K = e^{\frac{nE^\circ_{cell}}{0.0257}}$$

$$K = e^{\frac{(2)(3.17)}{0.0257}} = 1 \times 10^{107}$$

Step 5: To find equilibrium concentrations of Mg^{2+} and Ag^+, we have to solve an equilibrium problem.

Let x be the small amount of Mg^{2+} that reacts to achieve equilibrium. The concentration of Ag^+ will be $2x$ at equilibrium. Assume that essentially all Ag^+ has been reduced so that the initial concentration of Ag^+ is zero.

$$2Ag^+(aq) \;+\; Mg(s) \;\rightleftharpoons\; 2Ag(s) \;+\; Mg^{2+}(aq)$$

Initial (M):	0.0000	0.0500
Change (M):	+2x	−x
Equilibrium (M):	2x	(0.0500 − x)

$$K = \frac{[Mg^{2+}]}{[Ag^+]^2}$$

$$1 \times 10^{107} = \frac{(0.0500 - x)}{(2x)^2}$$

We can assume $0.0500 - x \approx 0.0500$.

$$1 \times 10^{107} \approx \frac{0.0500}{(2x)^2}$$

$$(2x)^2 = \frac{0.0500}{1 \times 10^{107}} = 0.0500 \times 10^{-107}$$

$$(2x)^2 = 5.00 \times 10^{-109} = 50.0 \times 10^{-110}$$

$$2x = 7 \times 10^{-55} \, M$$

$$[Ag^+] = 2x = 7 \times 10^{-55} \, M$$

$$[Mg^{2+}] = 0.0500 - x = 0.0500 \, M$$

19.79   Weigh the zinc and copper electrodes before operating the cell and re-weigh afterwards. The anode (Zn) should lose mass and the cathode (Cu) should gain mass.

19.80   **(a)**   Since this is an acidic solution, the gas must be hydrogen gas from the reduction of hydrogen ion. The two electrode reactions and the overall cell reaction are:

anode: $Cu(s) \longrightarrow Cu^{2+}(aq) + 2e^-$

cathode: $2H^+(aq) + 2e^- \longrightarrow H_2(g)$

$$Cu(s) + 2H^+(aq) \longrightarrow Cu^{2+}(aq) + H_2(g)$$

Since 0.584 g of copper was consumed, the amount of hydrogen gas produced is:

$$0.584 \text{ g Cu} \times \frac{1 \text{ mol Cu}}{63.55 \text{ g Cu}} \times \frac{1 \text{ mol H}_2}{1 \text{ mol Cu}} = 9.20 \times 10^{-3} \text{ mol H}_2$$

At STP, 1 mole of an ideal gas occupies a volume of 22.41 L. Thus, the volume of H_2 at STP is:

$$V_{H_2} = (9.20 \times 10^{-3} \text{ mol H}_2) \times \frac{22.41 \text{ L}}{1 \text{ mol}} = \textbf{0.206 L}$$

(b) From the current and the time, we can calculate the amount of charge:

$$1.18 \text{ A} \times \frac{1 \text{ C}}{1 \text{ A} \cdot \text{s}} \times (1.52 \times 10^3 \text{ s}) = 1.79 \times 10^3 \text{ C}$$

Since we know the charge of an electron, we can compute the number of electrons.

$$(1.79 \times 10^3 \text{ C}) \times \frac{1 \text{ e}^-}{1.6022 \times 10^{-19} \text{ C}} = 1.12 \times 10^{22} \text{ e}^-$$

Using the amount of copper consumed in the reaction and the fact that 2 mol of e⁻ are produced for every 1 mole of copper consumed, we can calculate Avogadro's number.

$$\frac{1.12 \times 10^{22} \text{ e}^-}{9.20 \times 10^{-3} \text{ mol Cu}} \times \frac{1 \text{ mol Cu}}{2 \text{ mol e}^-} = 6.09 \times 10^{23} \text{ /mol e}^-$$

In practice, Avogadro's number can be determined by electrochemical experiments like this. The charge of the electron can be found independently by Millikan's experiment.

19.81 The reaction is: $Al^{3+} + 3e^- \rightarrow Al$

First, let's calculate the number of coulombs of electricity that must pass through the cell to deposit 60.2 g of Al.

$$60.2 \text{ g Al} \times \frac{1 \text{ mol Al}}{26.98 \text{ g Al}} \times \frac{3 \text{ mol e}^-}{1 \text{ mol Al}} \times \frac{96500 \text{ C}}{1 \text{ mol e}^-} = 6.46 \times 10^5 \text{ C}$$

The time (in min) needed to pass this much charge is:

$$t_{\text{min}} = (6.46 \times 10^5 \text{ C}) \times \frac{1 \text{ A} \cdot \text{s}}{1 \text{ C}} \times \frac{1}{0.352 \text{ A}} \times \frac{1 \text{ min}}{60 \text{ s}} = 3.06 \times 10^4 \text{ min}$$

19.82 **(a)** We can calculate $\Delta G°$ from standard free energies of formation.

$$\Delta G° = 2\Delta G_f°(N_2) + 6\Delta G_f°(H_2O) - [4\Delta G_f°(NH_3) + 3\Delta G_f°(O_2)]$$

$$\Delta G = 0 + (6)(-237.2 \text{ kJ/mol}) - [(4)(-16.6 \text{ kJ/mol}) + 0]$$

$$\Delta G = -1356.8 \text{ kJ/mol}$$

(b) The half-reactions are:

$$4NH_3(g) \longrightarrow 2N_2(g) + 12H^+(aq) + 12e^-$$
$$3O_2(g) + 12H^+(aq) + 12e^- \longrightarrow 6H_2O(l)$$

The overall reaction is a 12-electron process. We can calculate the standard cell emf from the standard free energy change, $\Delta G°$.

$$\Delta G° = -nFE_{\text{cell}}°$$

$$E_{\text{cell}}° = \frac{-\Delta G°}{nF} = \frac{-\left(\dfrac{-1356.8 \text{ kJ}}{1 \text{ mol}} \times \dfrac{1000 \text{ J}}{1 \text{ kJ}}\right)}{(12)(96500 \text{ J/V} \cdot \text{mol})} = 1.17 \text{ V}$$

19.83 **(a)** The reaction is:

$$Zn + Cu^{2+} \rightarrow Zn^{2+} + Cu$$

Using the Nernst equation:

$$E = E° - \frac{0.0257}{2} \ln \frac{0.20}{0.20} = 1.10 \text{ V}$$

If NH_3 is added to the $CuSO_4$ solution:

$$Cu^{2+} + 4NH_3 \rightarrow Cu(NH_3)_4^{2+}$$

The concentration of copper ions $[Cu^{2+}]$ decreases, so the ln term becomes greater than 1 and E decreases. If NH_3 is added to the $ZnSO_4$ solution:

$$Zn^{2+} + 4NH_3 \rightarrow Zn(NH_3)_4^{2+}$$

The concentration of zinc ions $[Zn^{2+}]$ decreases, so the ln term becomes less than 1 and E increases.

(b) After addition of 25.0 mL of 3.0 M NH_3,

$$Cu^{2+} + 4NH_3 \rightarrow Cu(NH_3)_4^{2+}$$

Assume that all Cu^{2+} becomes $Cu(NH_3)_4^{2+}$:

$$[Cu(NH_3)_4^{2+}] = 0.10 \ M$$

$$[NH_3] = \frac{3.0 \ M}{2} - 0.40 \ M = 1.10 \ M$$

$$E = E° - \frac{0.0257}{2} \ln \frac{[Zn^{2+}]}{[Cu^{2+}]}$$

$$0.68 \text{ V} = 1.10 \text{ V} - \frac{0.0257}{2} \ln \frac{0.20}{[Cu^{2+}]}$$

$$[Cu^{2+}] = 1.3 \times 10^{-15} \ M$$

$$K_f = \frac{[Cu(NH_3)_4^{2+}]}{[Cu^{2+}][NH_3]^4} = \frac{0.10}{(1.3 \times 10^{-15})(1.1)^4} = 5.3 \times 10^{13}$$

Note: this value differs somewhat from that listed in Table 16.4 of the text.

19.84 The reduction of Ag^+ to Ag metal is:

$$Ag^+(aq) + e^- \longrightarrow Ag$$

We can calculate both the moles of Ag deposited and the moles of Au deposited.

$$? \text{ mol Ag} = 2.64 \text{ g Ag} \times \frac{1 \text{ mol Ag}}{107.9 \text{ g Ag}} = 2.45 \times 10^{-2} \text{ mol Ag}$$

$$? \text{ mol Au } = 1.61 \text{ g Au} \times \frac{1 \text{ mol Au}}{197.0 \text{ g Au}} = 8.17 \times 10^{-3} \text{ mol Au}$$

We do not know the oxidation state of Au ions, so we will represent the ions as Au^{n+}. If we divide the mol of Ag by the mol of Au, we can determine the ratio of Ag^+ reduced compared to Au^{n+} reduced.

$$\frac{2.45 \times 10^{-2} \text{ mol Ag}}{8.17 \times 10^{-3} \text{ mol Au}} = 3$$

That is, the same number of electrons that reduced the Ag^+ ions to Ag reduced only one-third the number of moles of the Au^{n+} ions to Au. Thus, each Au^{n+} required three electrons per ion for every one electron for Ag^+. The oxidation state for the gold ion is +3; the ion is Au^{3+}.

$$Au^{3+}(aq) + 3e^- \longrightarrow Au$$

19.85   Heating the garage will melt the snow on the car which is contaminated with salt. The aqueous salt will hasten corrosion.

19.86   We reverse the first half–reaction and add it to the second to come up with the overall balanced equation

$$Hg_2^{2+} \longrightarrow 2Hg^{2+} + 2e^- \qquad\qquad E^\circ_{\text{anode}} = +0.92 \text{ V}$$

$$\underline{Hg_2^{2+} + 2e^- \longrightarrow 2Hg \qquad\qquad E^\circ_{\text{cathode}} = +0.85 \text{ V}}$$

$$2Hg_2^{2+} \longrightarrow 2Hg^{2+} + 2Hg \qquad\qquad E^\circ_{\text{cell}} = 0.85 \text{ V} - 0.92 \text{ V} = -0.07 \text{ V}$$

Since the standard cell potential is an intensive property,

$$Hg_2^{2+}(aq) \longrightarrow Hg^{2+}(aq) + Hg(l) \qquad\qquad E^\circ_{\text{cell}} = -0.07 \text{ V}$$

We calculate ΔG° from E°.

$$\Delta G^\circ = -nFE^\circ = -(1)(96500 \text{ J/V·mol})(-0.07 \text{ V}) = \textbf{6.8 kJ/mol}$$

The corresponding equilibrium constant is:

$$K = \frac{[Hg^{2+}]}{[Hg_2^{2+}]}$$

We calculate K from ΔG°.

$$\Delta G^\circ = -RT\ln K$$

$$\ln K = \frac{-6.8 \times 10^3 \text{ J/mol}}{(8.314 \text{ J/K·mol})(298 \text{ K})}$$

$$\textbf{\textit{K} = 0.064}$$

19.87   (a)   Anode $2F^- \rightarrow F_2(g) + 2e^-$

Cathode $2H^+ + 2e^- \rightarrow H_2(g)$

Overall: $2H^+ + 2F^- \rightarrow H_2(g) + F_2(g)$

(b)   KF increases the electrical conductivity (what type of electrolyte is HF(l))? The K^+ is not reduced.

(c)   Calculating the moles of F_2

$$502 \text{ A} \times 15 \text{ h} \times \frac{3600 \text{ s}}{1 \text{ h}} \times \frac{1 \text{ C}}{1 \text{ A} \cdot \text{s}} \times \frac{1 \text{ mol e}^-}{96500 \text{ C}} \times \frac{1 \text{ mol F}_2}{2 \text{ mol e}^-} = 140 \text{ mol F}_2$$

Using the ideal gas law:

$$V = \frac{nRT}{P} = \frac{(140 \text{ mol})(0.0821 \text{ L} \cdot \text{atm/K} \cdot \text{mol})(297 \text{ K})}{1.2 \text{ atm}} = \textbf{2.8} \times \textbf{10}^3 \textbf{ L}$$

19.88   The reactions for the electrolysis of NaCl(aq) are:

Anode:	$2Cl^-(aq) \longrightarrow Cl_2(g) + 2e^-$
Cathode:	$2H_2O(l) + 2e^- \longrightarrow H_2(g) + 2OH^-(aq)$
Overall:	$2H_2O(l) + 2Cl^-(aq) \longrightarrow H_2(g) + Cl_2(g) + 2OH^-(aq)$

From the pH of the solution, we can calculate the OH^- concentration. From the [OH^-], we can calculate the moles of OH^- produced. Then, from the moles of OH^- we can calculate the average current used.

$$pH = 12.24$$
$$pOH = 14.00 - 12.24 = 1.76$$
$$[OH^-] = 1.74 \times 10^{-2} \, M$$

The moles of OH^- produced are:

$$\frac{1.74 \times 10^{-2} \text{ mol}}{1 \text{ L}} \times 0.300 \text{ L} = 5.22 \times 10^{-3} \text{ mol OH}^-$$

From the balanced equation, it takes 1 mole of e^- to produce 1 mole of OH^- ions.

$$(5.22 \times 10^{-3} \text{ mol OH}^-) \times \frac{1 \text{ mol e}^-}{1 \text{ mol OH}^-} \times \frac{96500 \text{ C}}{1 \text{ mol e}^-} = 504 \text{ C}$$

Recall that 1 C = 1 A·s

$$504 \text{ C} \times \frac{1 \text{ A} \cdot \text{s}}{1 \text{ C}} \times \frac{1 \text{ min}}{60 \text{ s}} \times \frac{1}{6.00 \text{ min}} = \textbf{1.4 A}$$

19.89   **(a)**   Anode: $Cu(s) \rightarrow Cu^{2+}(aq) + 2e^-$
Cathode: $Cu^{2+}(aq) + 2e^- \rightarrow Cu(s)$

The overall reaction is: $Cu(s) \rightarrow Cu(s)$ Cu is transferred from the anode to cathode.

(b)   Consulting Table 19.1 of the text, the Zn will be oxidized, but Zn^{2+} will not be reduced at the cathode. Ag will not be oxidized at the anode.

(c)   The moles of Cu: $1000 \text{ g Cu} \times \dfrac{1 \text{ mol Cu}}{63.55 \text{ g Cu}} = 15.7 \text{ mol Cu}$

The coulombs required: $15.7 \text{ mol Cu} \times \dfrac{2 \text{ mol e}^-}{1 \text{ mol Cu}} \times \dfrac{96500 \text{ C}}{1 \text{ mol e}^-} = 3.03 \times 10^6 \text{ C}$

The time required: $? \text{ s} = \dfrac{3.03 \times 10^6 \text{ C}}{18.9 \text{ A}} = 1.60 \times 10^5 \text{ s}$

$(1.60 \times 10^5 \text{ s}) \times \dfrac{1 \text{ hr}}{3600 \text{ s}} = \textbf{44.4 hr}$

19.90   The reaction is:

$$Pt^{n+} + ne^- \longrightarrow Pt$$

Thus, we can calculate the charge of the platinum ions by realizing that n mol of e⁻ are required per mol of Pt formed.

The moles of Pt formed are:

$$9.09 \text{ g Pt} \times \dfrac{1 \text{ mol Pt}}{195.1 \text{ g Pt}} = 0.0466 \text{ mol Pt}$$

Next, calculate the charge passed in C.

$$C = 2.00 \text{ h} \times \dfrac{3600 \text{ s}}{1 \text{ h}} \times \dfrac{2.50 \text{ C}}{1 \text{ s}} = 1.80 \times 10^4 \text{ C}$$

Convert to moles of electrons.

$$? \text{ mol e}^- = (1.80 \times 10^4 \text{ C}) \times \dfrac{1 \text{ mol e}^-}{96500 \text{ C}} = 0.187 \text{ mol e}^-$$

We now know the number of moles of electrons (0.187 mol e⁻) needed to produce 0.0466 mol of Pt metal. We can calculate the number of moles of electrons needed to produce 1 mole of Pt metal.

$$\dfrac{0.187 \text{ mol e}^-}{0.0466 \text{ mol Pt}} = 4.01 \text{ mol e}^-/\text{mol Pt}$$

Since we need 4 moles of electrons to reduce 1 mole of Pt ions, the charge on the Pt ions must be **+4.**

19.91   Using the standard reduction potentials found in Table 19.1

$$Cd^{2+}(aq + 2e^- \rightarrow Cd(s) \qquad E^\circ = -0.40 \text{ V}$$
$$Mg^{2+}(aq) + 2e^- \rightarrow Mg(s) \qquad E^\circ = -2.37 \text{ V}$$

Thus Cd^{2+} will oxidize Mg so that the magnesium half-reaction occurs at the anode.

$$Mg(s) + Cd^{2+}(aq) \rightarrow Mg^{2+}(aq) + Cd(s)$$

$$E^\circ_{cell} = E^\circ_{cathode} - E^\circ_{anode} = -0.40 \text{ V} - (-2.37 \text{ V}) = 1.97 \text{ V}$$

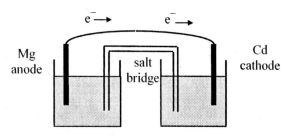

19.92   The half–reaction for the oxidation of water to oxygen is:

$$2H_2O(l) \xrightarrow{\text{oxidation (anode)}} O_2(g) + 4H^+(aq) + 4e^-$$

Knowing that one mole of any gas at STP occupies a volume of 22.41 L, we find the number of moles of oxygen.

$$4.26 \text{ L } O_2 \times \frac{1 \text{ mol}}{22.41 \text{ L}} = 0.190 \text{ mol } O_2$$

Since four electrons are required to form one oxygen molecule, the number of electrons must be:

$$0.190 \text{ mol } O_2 \times \frac{4 \text{ mol } e^-}{1 \text{ mol } O_2} \times \frac{6.022 \times 10^{23} \text{ e}^-}{1 \text{ mol}} = 4.58 \times 10^{23} \text{ e}^-$$

The amount of charge passing through the solution is:

$$6.00 \text{ A} \times \frac{1 \text{ C}}{1 \text{ A}\cdot\text{s}} \times \frac{3600 \text{ s}}{1 \text{ h}} \times 3.40 \text{ h} = 7.34 \times 10^4 \text{ C}$$

We find the electron charge by dividing the amount of charge by the number of electrons.

$$\frac{7.34 \times 10^4 \text{ C}}{4.58 \times 10^{23} \text{ e}^-} = 1.60 \times 10^{-19} \text{ C/e}^-$$

In actual fact, this sort of calculation can be used to find Avogadro's number, not the electron charge. The latter can be measured independently, and one can use this charge together with electrolytic data like the above to calculate the number of objects in one mole. See also Problem 19.80.

19.93   **(a)** $Au(s) + 3HNO_3(aq) + 4HCl(aq) \rightarrow HAuCl_4(aq) + 3H_2O(l) + 3NO_2(g)$

(b)   To increase the acidity and to form the stable complex ion, $AuCl_4^-$.

19.94   Cells of higher voltage require very reactive oxidizing and reducing agents, which are difficult to handle. (From Table 19.1 of the text, we see that 5.92 V is the theoretical limit of a cell made up of Li^+/Li and F_2/F^- electrodes under standard-state conditions.) Batteries made up of several cells in series are easier to use.

19.95   The overall cell reaction is:

$$2Ag^+(aq) + H_2(g) \rightarrow 2Ag(s) + 2H^+(aq)$$

We write the Nernst equation for this system.

$$E = E° - \frac{0.0257 \text{ V}}{n} \ln Q$$

$$E = 0.080 \text{ V} - \frac{0.0257 \text{ V}}{2} \ln \frac{[H^+]^2}{[Ag^+]^2 P_{H_2}}$$

The measured voltage is 0.589 V, and we can find the silver ion concentration as follows:

$$0.589 \text{ V} = 0.80 \text{ V} - \frac{0.0257 \text{ V}}{2} \ln \frac{1}{[Ag^+]^2}$$

$$\ln \frac{1}{[Ag^+]^2} = 16.42$$

$$\frac{1}{[Ag^+]^2} = 1.4 \times 10^7$$

$$[Ag^+] = 2.7 \times 10^{-4} \text{ } M$$

Knowing the silver ion concentration, we can calculate the oxalate ion concentration and the solubility product constant.

$$[C_2O_4^{2-}] = \frac{1}{2}[Ag^+] = 1.35 \times 10^{-4} \text{ } M$$

$$K_{sp} = [Ag^+]^2[C_2O_4^{2-}] = (2.7 \times 10^{-4})^2(1.35 \times 10^{-4}) = \mathbf{9.8 \times 10^{-12}}$$

19.96   The half-reactions are:

$Zn(s) + 4OH^-(aq) \rightarrow Zn(OH)_4^{2-}(aq) + 2e^-$	$E°_{anode} = -1.36$ V
$Zn^{2+}(aq) + 2e^- \rightarrow Zn(s)$	$E°_{cathode} = -0.76$ V

$$Zn^{2+}(aq) + 4OH^-(aq) \rightarrow Zn(OH)_4^{2-}(aq) \qquad E°_{cell} = -0.76 \text{ V} - (-1.36 \text{ V}) = 0.60 \text{ V}$$

$$E°_{cell} = \frac{0.0257 \text{ V}}{n} \ln K_f$$

$$K_f = e^{\frac{nE°}{0.0257}} = e^{\frac{(2)(0.60)}{0.0257}} = \mathbf{2 \times 10^{20}}$$

19.97   The half reactions are:

$H_2O_2(aq) \rightarrow O_2(g) + 2H^+(aq) + 2e^-$	$E°_{anode} = 0.68$ V
$H_2O_2(aq) + 2H^+(aq) + 2e^- \rightarrow 2H_2O(l)$	$E°_{cathode} = 1.77$ V

$$2H_2O_2(aq) \rightarrow 2H_2O(l) + O_2(g) \qquad E°_{cell} = E°_{cathode} - E°_{anode} = 1.77 \text{ V} - 0.68 \text{ V} = 1.09 \text{ V}$$

Thus, products are favored at equilibrium. H_2O_2 is not stable (it disproportionates).

19.98   **(a)**   Since electrons flow from X to SHE, $E°$ for X must be negative. Thus $E°$ for Y must be positive.

(b)
$$Y^{2+} + 2e^- \rightarrow Y \qquad\qquad E°_{cathode} = 0.34\ V$$
$$\underline{X \rightarrow X^{2+} + 2e^- \qquad\qquad E°_{anode} = -0.25\ V}$$
$$X + Y^{2+} \rightarrow X^{2+} + Y \qquad E°_{cell} = 0.34\ V - (-0.25\ V) = \mathbf{0.59\ V}$$

19.99   **(a)**   The half reactions are:

$$Sn^{4+}(aq) + 2e^- \rightarrow Sn^{2+}(aq) \qquad\qquad E°_{cathode} = 0.13\ V$$
$$\underline{2Tl(s) \rightarrow Tl^+(aq) + e^- \qquad\qquad E°_{anode} = -0.34\ V}$$
$$Sn^{4+}(aq) + 2Tl(s) \rightarrow Sn^{2+}(aq) + 2Tl^+(aq) \qquad E°_{cell} = E°_{cathode} - E°_{anode} = 0.13\ V - (-0.34\ V) = 0.47\ V$$

(b)
$$E°_{cell} = \frac{RT}{nF} \ln K$$

$$0.47\ V = \frac{(8.314)(298)}{(2)(96500)} \ln K$$

$$K = \mathbf{8 \times 10^{15}}$$

(c)
$$E = E° - \frac{0.0257}{2} \ln \frac{(1.0)(10.0)^2}{(1.0)} = (0.47 - 0.0592)\,V = \mathbf{0.41\ V}$$

19.100   (a)   Gold does not tarnish in air because the reduction potential for oxygen is insufficient to result in the oxidation of gold.

$$O_2 + 4H^+ + 4e^- \rightarrow 2H_2O \qquad\qquad E°_{cathode} = 1.23\ V$$

That is, $E°_{cell} = E°_{cathode} - E°_{anode} < 0$, for either oxidation by O_2 to Au^+ or Au^{3+}.

$$E°_{cell} = 1.23\ V - 1.50\ V\ < 0$$

or

$$E°_{cell} = 1.23\ V - 1.69\ V\ < 0$$

(b)
$$3(Au^+ + e^- \rightarrow Au) \qquad\qquad E°_{cathode} = 1.69\ V$$
$$\underline{Au \rightarrow Au^{3+} + 3e^- \qquad\qquad E°_{anode} = 1.50\ V}$$
$$3Au^+ \rightarrow 2Au + Au^{3+} \qquad E°_{cell} = 1.69\ V - 1.50\ V = 0.19\ V$$

Calculating ΔG,

$$\Delta G° = -nFE° = -(3)(96,500\ J/V \cdot mol)(0.19\ V) = -55.0\ kJ/mol$$

For spontaneous electrochemical equations, $\Delta G°$ must be negative. Thus, **the disproportionation occurs spontaneously.**

(c)   Since the most stable oxidation state for gold is Au^{3+}, the predicted reaction is:

$$2Au + 3F_2 \rightarrow 2AuF_3$$

19.101 It is mercury ion in solution that is extremely hazardous. Since mercury metal does not react with hydrochloric acid (the acid in gastric juice), it does not dissolve and passes through the human body unchanged. Nitric acid (not part of human gastric juices) dissolves mercury metal (see Problem 19.113); if nitric acid were secreted by the stomach, ingestion of mercury metal would be fatal.

19.102 The balanced equation is: $5Fe^{2+} + MnO_4^- + 8H^+ \longrightarrow Mn^{2+} + 5Fe^{3+} + 4 H_2O$

Calculate the amount of iron(II) in the original solution using the mole ratio from the balanced equation.

$$23.0 \text{ mL} \times \frac{0.0200 \text{ mol KMnO}_4}{1000 \text{ mL soln}} \times \frac{5 \text{ mol Fe}^{2+}}{1 \text{ mol KMnO}_4} = 0.00230 \text{ mol Fe}^{2+}$$

The concentration of iron(II) must be:

$$[Fe^{2+}] = \frac{0.00230 \text{ mol}}{0.0250 \text{ L}} = \textbf{0.0920 } \textbf{\textit{M}}$$

The total iron concentration can be found by simple proportion because the same sample volume (25.0 mL) and the same KMnO$_4$ solution were used.

$$[Fe]_{total} = \frac{40.0 \text{ mL KMnO}_4}{23.0 \text{ mL KMnO}_4} \times 0.0920 \text{ } M = 0.160 \text{ } M$$

$$[Fe^{3+}] = [Fe]_{total} - [Fe^{2+}] = \textbf{0.0680 } \textbf{\textit{M}}$$

Why are the two titrations with permanganate necessary in this problem?

19.103 Viewed externally, the anode looks negative because of the flow of the electrons (from $Zn \rightarrow Zn^{2+} + 2e^-$) toward the cathode. In solution, anions move toward the anode because they are attracted by the Zn^{2+} ions surrounding the anode.

19.104 **(a)** $1A \cdot h = 1A \times 3600s = 3600 \text{ C}$

(b) Anode: $Pb + SO_4^{2-} \rightarrow PbSO_4 + 2e^-$

Two moles of electrons are produced by 1 mole of Pb. Recall that the charge of 1 mol e$^-$ is 96,500 C. We can set up the following conversions to calculate the capacity of the battery.

mol Pb $\rightarrow$ mol e$^-$ $\rightarrow$ coulombs $\rightarrow$ ampere hour

$$406 \text{ g Pb} \times \frac{1 \text{ mol Pb}}{207.2 \text{ g Pb}} \times \frac{2 \text{ mol e}^-}{1 \text{ mol Pb}} \times \frac{96500 \text{ C}}{1 \text{ mol e}^-} = (3.74 \times 10^5 \text{ C}) \times \frac{1 \text{ h}}{3600 \text{ s}} = \textbf{104 A} \cdot \textbf{h}$$

This ampere·hour cannot be fully realized because the concentration of H$_2$SO$_4$ keeps decreasing.

(c) $E^\circ_{cell} = 1.70 \text{ V} - (-0.31 \text{ V}) = \textbf{2.01 V}$ (From Table 19.1 of the text)

$\Delta G^\circ = -nFE^\circ$

$\Delta G^\circ = -(2)(96500 \text{ J/V·mol})(2.01 \text{ V}) = \textbf{--3.88} \times \textbf{10}^5 \textbf{ J/mol}$

Spontaneous as expected.

19.105 **(a)**   The overall reaction is: $Pb + PbO_2 + H_2SO_4 \rightarrow 2PbSO_4 + 2H_2O$

Initial mass of H_2SO_4: $724 \text{ mL} \times \dfrac{1.29 \text{ g}}{1 \text{ mL}} \times 0.380 = 355 \text{ g}$

Final mass of H_2SO_4: $724 \text{ mL} \times \dfrac{1.19 \text{ g}}{1 \text{ mL}} \times 0.260 = 224 \text{ g}$

Mass of H_2SO_4 reacted $= 355 \text{ g} - 224 \text{ g} = 131 \text{ g}$

Moles of H_2SO_4 reacted $= 131 \text{ g} \times \dfrac{1 \text{ mol}}{98.09 \text{ g}} = 1.34 \text{ mol}$

$Q = 1.34 \text{ mol } H_2SO_4 \times \dfrac{2 \text{ mol e}^-}{2 \text{ mol } H_2SO_4} \times \dfrac{96500 \text{ C}}{1 \text{ mol e}^-} = \mathbf{1.29 \times 10^5 \text{ C}}$

(b) $t = \dfrac{Q}{I} = \dfrac{1.29 \times 10^5 \text{ C}}{22.4 \text{ A}} = 5.76 \times 10^3 \text{ s} = \mathbf{1.60 \text{ hr}}$

19.106 **(a)** unchanged   **(b)** unchanged   **(c)** squared   **(d)** doubled   **(e)** doubled

19.107 **(a)**

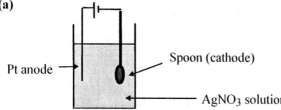

Pt anode —

Spoon (cathode)

$AgNO_3$ solution

(b) $t = \dfrac{Q}{I} = \dfrac{\left(0.884 \text{ g} \times \dfrac{1 \text{ mol}}{107.9 \text{ g}}\right)\left(\dfrac{96500 \text{ C}}{1 \text{ mol}}\right)}{18.5 \times 10^{-3} \text{ A}} = 4.27 \times 10^4 \text{ s} = \mathbf{11.9 \text{ hr}}$

19.108 $F_2 (g) + 2H^+ (aq) + 2e^- \rightarrow 2HF (g)$

$E = E° - \dfrac{RT}{2F} \ln \dfrac{P_{HF}^2}{P_{F_2}[H^+]^2}$

With increasing $[H^+]$, E will be larger. F_2 will become a **stronger oxidizing agent**.

19.109 <u>**Advantages:**</u>

(a) No start-up problems,   **(b)** much quieter,   **(c)** no pollution (smog),

(d) more energy efficient in the sense that when the car is not moving (for example at a traffic light), no electricity is consumed.

Disadvantages:

(a) Driving range is more limited than automobiles,

(b) total mass of batteries is appreciable,

(c) production of electricity needed to charge the batteries leads to pollution.

19.110 $\quad Pb \rightarrow Pb^{2+} + 2e^-$ $\qquad\qquad E^{\circ}_{anode} = -0.13$ V

$\qquad\quad \dfrac{2H^+ + 2e^- \rightarrow H_2}{Pb + 2H^+ \rightarrow Pb^{2+} + H_2} \qquad \dfrac{E^{\circ}_{cathode} = 0.00 \text{ V}}{E^{\circ}_{cell} = 0.00 \text{ V} - (-0.13 \text{ V}) = 0.13 \text{ V}}$

$$pH = 1.60$$

$$[H^+] = 10^{-1.60} = 0.025 \ M$$

$$E = E^{\circ} - \frac{RT}{nF} \ln \frac{[Pb^{2+}]P_{H_2}}{[H^+]^2}$$

$$0 = 0.13 - \frac{0.0257 \text{ V}}{2} \ln \frac{(0.035)P_{H_2}}{0.025^2}$$

$$\frac{0.26}{0.0257} = \ln \frac{(0.035)P_{H_2}}{0.025^2}$$

$$P_{H_2} = 4.4 \times 10^2 \text{ atm}$$

19.111 **(a)** At the anode (Mg): $\qquad Mg \rightarrow Mg^{2+} + 2e^-$

$\qquad\quad$ Also: $\qquad\qquad\qquad Mg + 2HCl \rightarrow MgCl_2 + H_2$

$\qquad\quad$ At the cathode (Cu): $\quad 2H^+ + 2e^- \rightarrow H_2$

(b) The solution does not turn blue.

(c) After all the HCl has been neutralized, the white precipitate is:

$$Mg^{2+} + 2OH^- \rightarrow Mg(OH)_2(s)$$

19.112 **(a)** The half-reactions are:

$\qquad$ Anode: $\quad Zn \rightarrow Zn^{2+} + 2e^-$

$\qquad\dfrac{\text{Cathode:} \quad \frac{1}{2}O_2 + 2e^- \rightarrow O^{2-}}{\text{Overall:} \quad Zn + \frac{1}{2}O_2 \rightarrow ZnO}$

To calculate the standard emf, we first need to calculate ΔG° for the reaction. From Appendix 3 of the text we write:

$$\Delta G^{\circ} = \Delta G^{\circ}_f(ZnO) - [\Delta G^{\circ}_f(Zn) + \tfrac{1}{2}\Delta G^{\circ}_f(O_2)]$$

$$\Delta G^{\circ} = -318.2 \text{ kJ/mol} - [0 + 0]$$

$$\Delta G^{\circ} = -318.2 \text{ kJ/mol}$$

$$\Delta G^\circ = -nFE^\circ$$
$$-318.2 \times 10^3 \text{ J/mol} = -(2)(96,500 \text{ J/V·mol})E^\circ$$
$$E^\circ = \mathbf{1.65 \text{ V}}$$

(b)   We use the following equation:

$$E = E^\circ - \frac{RT}{nF}\ln Q$$

$$E = 1.65 \text{ V} - \frac{0.0257 \text{ V}}{2}\ln\frac{1}{P_{O_2}}$$

$$E = 1.65 \text{ V} - \frac{0.0257 \text{ V}}{2}\ln\frac{1}{0.21}$$

$$E = 1.65 \text{ V} - 0.020 \text{ V}$$

$$E = \mathbf{1.63 \text{ V}}$$

(c)   Since the free energy change represents the maximum work that can be extracted from the overall reaction, the maximum amount of energy that can be obtained from this reaction is the free energy change. To calculate the energy density, we multiply the free energy change by the number of moles of Zn present in 1 kg of Zn.

$$\textbf{energy density} = \frac{318.2 \text{ kJ}}{1 \text{ mol Zn}} \times \frac{1 \text{ mol Zn}}{65.39 \text{ g Zn}} \times \frac{1000 \text{ g Zn}}{1 \text{ kg Zn}} = \mathbf{4.87 \times 10^3 \text{ kJ/kg Zn}}$$

(d)   One ampere is 1 C/s. The charge drawn every second is given by nF.

$$\text{charge} = nF$$

$$2.1 \times 10^5 \text{ C} = n(96,500 \text{ C/mol e}^-)$$

$$n = 2.2 \text{ mol e}^-$$

From the overall balanced reaction, we see that 4 moles of electrons will reduce 1 mole of O_2; therefore, the number of moles of O_2 reduced by 2.2 moles of electrons is:

$$\text{mol } O_2 = 2.2 \text{ mol e}^- \times \frac{1 \text{ mol } O_2}{4 \text{ mol e}^-} = 0.55 \text{ mol } O_2$$

The volume of oxygen at 1.0 atm partial pressure can be obtained by using the ideal gas equation.

$$V_{O_2} = \frac{nRT}{P} = \frac{(0.55 \text{ mol})(0.0821 \text{ L·atm/mol·K})(298 \text{ K})}{(1.0 \text{ atm})} = 13 \text{ L}$$

Since air is 21 percent oxygen by volume, the volume of air required every second is:

$$V_{\text{air}} = 13 \text{ L } O_2 \times \frac{100\% \text{ air}}{21\% \text{ } O_2} = \mathbf{62 \text{ L of air}}$$

19.113 **(a)** HCl: First, we write the half-reactions.

Oxidation: $2Hg(l) \longrightarrow Hg_2^{2+}(1\ M) + 2e^-$

Reduction: $2H^+(1\ M) + 2e^- \longrightarrow H_2(1\ atm)$

Overall: $2Hg(l) + 2H^+(1\ M) \longrightarrow Hg_2^{2+}(1\ M) + H_2(1\ atm)$

The standard emf, $E°$, is given by

$$E° = E°_{cathode} - E°_{anode}$$

$$E° = 0 - 0.85\ V$$

$$E° = -0.85V$$

(We omit the subscript "cell" because this reaction is not carried out in an electrochemical cell.) Since $E°$ is negative, we conclude that mercury is not oxidized by hydrochloric acid under standard-state conditions.

(b) HNO$_3$: The reactions are:

Oxidation: $3[2Hg(l) \longrightarrow Hg_2^{2+}(1\ M) + 2e^-]$

Reduction: $2[NO_3^-(1\ M) + 4H^+(1\ M) + 3e^- \longrightarrow NO(1\ atm) + 2H_2O(l)]$

Overall: $6Hg(l) + 2NO_3^-(1\ M) + 8H^+(1\ M) \longrightarrow 3Hg_2^{2+}(1\ M) + 2NO(1\ atm) + 4H_2O(l)$

Thus,

$$E° = E°_{cathode} - E°_{anode}$$

$$E° = 0.96V - 0.85V$$

$$E° = 0.11V$$

Since $E°$ is positive, products are favored at equilibrium under standard-state conditions.

The test tube on the left contains HNO$_3$ and Hg. HNO$_3$ can oxidize Hg and the product NO reacts with oxygen to form the brown gas NO$_2$.

19.114 We can calculate $\Delta G°_{rxn}$ using the following equation.

$$\Delta G°_{rxn} = \Sigma n \Delta G°_f(products) - \Sigma m \Delta G°_f(reactants)$$

$$\Delta G°_{rxn} = 0 + 0 - [(1)(-293.8\ kJ/mol) + 0] = 293.8\ kJ/mol$$

Next, we can calculate $E°$ using the equation

$$\Delta G° = -nFE°$$

We use a more accurate value for Faraday's constant.

$$293.8 \times 10^3\ J/mol = -(1)(96485.3\ J/V{\cdot}mol)E°$$

$$E° = -3.05\ V$$

19.115   (a)   First, we calculate the coulombs of electricity that pass through the cell.

$$0.22 \text{ A} \times \frac{1 \text{ C}}{1 \text{ A} \cdot \text{s}} \times \frac{3600 \text{ s}}{1 \text{ h}} \times 31.6 \text{ h} = 2.5 \times 10^4 \text{ C}$$

We see that for every mole of Cu formed at the cathode, 2 moles of electrons are needed. The grams of Cu produced at the cathode are:

$$\textbf{? g Cu} = (2.5 \times 10^4 \text{ C}) \times \frac{1 \text{ mol e}^-}{96,500 \text{ C}} \times \frac{1 \text{ mol Cu}}{2 \text{ mol e}^-} \times \frac{63.55 \text{ g Cu}}{1 \text{ mol Cu}} = \textbf{8.2 g Cu}$$

(b)   8.2 g of Cu is 0.13 mole of Cu. The moles of Cu^{2+} in the original solution are:

$$0.218 \text{ L} \times \frac{1 \text{ mol Cu}^{2+}}{1 \text{ L soln}} = 0.218 \text{ mol Cu}^{2+}$$

The mole ratio between Cu^{2+} and Cu is 1:1, so the moles of Cu^{2+} remaining in solution are:

$$\text{moles Cu}^{2+} \text{ remaining} = 0.218 \text{ mol} - 0.13 \text{ mol} = 0.088 \text{ mol Cu}^{2+}$$

The concentration of Cu^{2+} remaining is:

$$[\text{Cu}^{2+}] = \frac{0.088 \text{ mol Cu}^{2+}}{0.218 \text{ L soln}} = \textbf{0.40 } \textbf{\textit{M}}$$

19.116   First, we need to calculate E°_{cell}, then we can calculate K from the cell potential.

$H_2(g) \rightarrow 2H^+(aq) + 2e^-$	$E^\circ_{\text{anode}} = 0.00 \text{ V}$
$2H_2O(l) + 2e^- \rightarrow H_2(g) + 2OH^-$	$E^\circ_{\text{cathode}} = -0.83 \text{ V}$
$2H_2O(l) \rightarrow 2H^+(aq) + 2OH^-(aq)$	$E^\circ_{\text{cell}} = -0.83 \text{ V} - 0.00 \text{ V} = -0.83 \text{ V}$

We want to calculate K for the reaction: $H_2O(l) \rightarrow H^+(aq) + OH^-(aq)$. The cell potential for this reaction will be the same as the above reaction, but the moles of electrons transferred, n, will equal one.

$$E^\circ_{\text{cell}} = \frac{0.0257 \text{ V}}{n} \ln K_w$$

$$\ln K_w = \frac{n E^\circ_{\text{cell}}}{0.0257 \text{ V}}$$

$$K_w = e^{\frac{n E^\circ}{0.0257}}$$

$$K_w = e^{\frac{(1)(-0.83)}{0.0257}} = e^{-32} = 1 \times 10^{-14}$$

CHAPTER 20
METALLURGY AND THE CHEMISTRY OF METALS

20.11 For the given reaction we can calculate the standard free energy change from the standard free energies of formation. Then, we can calculate the equilibrium constant, K_p, from the standard free energy change.

$$\Delta G° = \Delta G_f°[Ni(CO)_4] - [4\Delta G_f°(CO) + \Delta G_f°(Ni)]$$

$$\Delta G° = (1)(-587.4 \text{ kJ/mol}) - [(4)(-137.3 \text{ kJ/mol}) + (1)(0)] = -38.2 \text{ kJ/mol} = -3.82 \times 10^4 \text{ J/mol}$$

Substitute $\Delta G°$, R, and T (in K) into the following equation to solve for K_p.

$$\Delta G° = -RT\ln K_p$$

$$\ln K_p = \frac{-\Delta G°}{RT} = \frac{-(-3.82 \times 10^4 \text{ J/mol})}{(8.314 \text{ J/K} \cdot \text{mol})(353 \text{ K})}$$

$$K_p = 4.5 \times 10^5$$

20.12 The cathode reaction is: $Cu^{2+}(aq) + 2e^- \longrightarrow Cu(s)$

First, let's calculate the number of moles of electrons needed to reduce 5.0 kg of Cu.

$$5.00 \text{ kg Cu} \times \frac{1000 \text{ g}}{1 \text{ kg}} \times \frac{1 \text{ mol Cu}}{63.55 \text{ g Cu}} \times \frac{2 \text{ mol e}^-}{1 \text{ mol Cu}} = 1.57 \times 10^2 \text{ mol e}^-$$

Next, let's determine how long it will take for 1.57×10^2 moles of electrons to flow through the cell when the current is 37.8 C/s.

$$(1.57 \times 10^2 \text{ mol e}^-) \times \frac{96,500 \text{ C}}{1 \text{ mol e}^-} \times \frac{1 \text{ s}}{37.8 \text{ C}} \times \frac{1 \text{ h}}{3600 \text{ s}} = \mathbf{111 \text{ h}}$$

20.13 Table 19.1 of the text shows that Pb, Fe, Co, Zn are more easily oxidized (stronger reducing agents) than copper. The Ag, Au, and Pt are harder to oxidize and will not dissolve.

Would you throw away the sludge if you were in charge of the copper refining plant? Why is it still profitable to manufacture copper even though the market price is very low?

20.14 The sulfide ore is first roasted in air:

$$2ZnS(s) + 3O_2(g) \longrightarrow 2ZnO(s) + 2SO_2(g)$$

The zinc oxide is then mixed with coke and limestone in a blast furnace where the following reductions occur:

$$ZnO(s) + C(s) \longrightarrow Zn(g) + CO(g)$$
$$ZnO(s) + CO(g) \longrightarrow Zn(g) + CO_2(g)$$

The zinc vapor formed distills from the furnace into an appropriate receiver.

20.15   The trick in this process centers on the fact that $TiCl_4$ is a liquid with a boiling point ($136.4°C$), a little higher than that of water. The tetrachloride can be formed by treating the oxide (rutile) with chlorine gas at high temperature. The balanced equation is:

$$TiO_2(s) + 2Cl_2(g) \rightarrow TiCl_4(l) + O_2(g)$$

The liquid tetrachloride can be isolated and purified by simple distillation. Purified $TiCl_4$ is then reduced with magnesium (a stronger reducing agent that Ti) at high temperature.

$$TiCl_4(g) + 2Mg(l) \rightarrow Ti(s) + 2MgCl_2(l)$$

The other product, $MgCl_2$, can be separated easily from titanium metal by dissolving in water.

20.16   **(a)**   We first find the mass of ore containing 2.0×10^8 kg of copper.

$$(2.0 \times 10^8 \text{ kg Cu}) \times \frac{100\% \text{ ore}}{0.80\% \text{ Cu}} = 2.5 \times 10^{10} \text{ kg ore}$$

We can then compute the volume from the density of the ore.

$$(2.5 \times 10^{10} \text{ kg}) \times \frac{1000 \text{ g}}{1 \text{ kg}} \times \frac{1 \text{ cm}^3}{2.8 \text{ g}} = \mathbf{8.9 \times 10^{12} \text{ cm}^3}$$

(b)   From the formula of chalcopyrite it is clear that two moles of sulfur dioxide will be formed per mole of copper. The mass of sulfur dioxide formed will be:

$$(2.0 \times 10^8 \text{ kg Cu}) \times \frac{1 \text{ mol Cu}}{0.06355 \text{ kg Cu}} \times \frac{2 \text{ mol SO}_2}{1 \text{ mol Cu}} \times \frac{0.06407 \text{ kg SO}_2}{1 \text{ mol SO}_2} = \mathbf{4.0 \times 10^8 \text{ kg SO}_2}$$

20.17   Very electropositive metals (i.e., very strong reducing agents) can only be isolated from their compounds by electrolysis. No chemical reducing agent is strong enough. In the given list $CaCl_2$, NaCl, and Al_2O_3 would require electrolysis.

20.18   Iron can be produced by reduction with coke in a blast furnace; whereas, aluminum is usually produced electrolytically, which is a much more expensive process.

20.27   All of these reactions are discussed in Section 20.5 of the text.

(a) $2K(s) + 2H_2O(l) \rightarrow 2KOH(aq) + H_2(g)$

(c) $2Na(s) + O_2(g) \rightarrow Na_2O_2(s)$

(b) $NaH(s) + H_2O(l) \rightarrow NaOH(aq) + H_2(g)$

(d) $K(s) + O_2(g) \rightarrow KO_2(s)$

20.28   **(a)** $2Na(s) + 2H_2O(l) \longrightarrow 2NaOH(aq) + H_2(g)$

(b) $2NaOH(aq) + CO_2(g) \longrightarrow Na_2CO_3(aq) + H_2O(l)$

(c) $Na_2CO_3(s) + 2HCl(aq) \longrightarrow 2NaCl(aq) + CO_2(g) + H_2O(l)$

(d) $NaHCO_3(aq) + HCl(aq) \longrightarrow NaCl(aq) + CO_2(g) + H_2O(l)$

(e) $2NaHCO_3(s) \longrightarrow Na_2CO_3(s) + CO_2(g) + H_2O(g)$

(f) $Na_2CO_3(s) \longrightarrow$ no reaction. Unlike $CaCO_3(s)$, $Na_2CO_3(s)$ is not decomposed by moderate heating.

20.29 $NaH + H_2O \rightarrow NaOH + H_2$

20.30   The balanced equation is: $Na_2CO_3(s) + 2HCl(aq) \longrightarrow 2NaCl(aq) + CO_2(g) + H_2O(l)$

$$mol\ CO_2\ produced\ =\ 25.0\ g\ Na_2CO_3 \times \frac{1\ mol\ Na_2CO_3}{106.0\ g\ Na_2CO_3} \times \frac{1\ mol\ CO_2}{1\ mol\ Na_2CO_3} = 0.236\ mol\ CO_2$$

$$V_{CO_2} = \frac{nRT}{P} = \frac{(0.236\ mol)(0.0821\ L \cdot atm/K \cdot mol)(283\ K)}{\left(746\ mmHg \times \dfrac{1\ atm}{760\ mmHg}\right)} = \textbf{5.59 L}$$

20.33   **(a)** $\Delta H° = \Delta H_f°(MgO) + \Delta H_f°(CO_2) - \Delta H_f°(MgCO_3)$

$\Delta H° = (1)(-601.8\ kJ/mol) + (1)(-393.5\ kJ/mol) - (1)(-1112.9\ kJ/mol) = \textbf{117.6 kJ/mol}$

(b) $\Delta H° = \Delta H_f°(CaO) + \Delta H_f°(CO_2) - \Delta H_f°(CaCO_3)$

$\Delta H° = (1)(-635.5\ kJ/mol) + (1)(-393.5\ kJ/mol) - (1)(-1206.9\ kJ/mol) = \textbf{177.8 kJ/mol}$

$\Delta H°$ is less for $MgCO_3$; therefore, it is more easily decomposed by heat.

20.34   First magnesium is treated with concentrated nitric acid (redox reaction) to obtain magnesium nitrate.

$$3Mg(s) + 8HNO_3(aq) \longrightarrow 3Mg(NO_3)_2(aq) + 4H_2O(l) + 2NO(g)$$

The magnesium nitrate is recovered from solution by evaporation, dried, and heated in air to obtain magnesium oxide:

$$2Mg(NO_3)_2(s) \longrightarrow 2MgO(s) + 4NO_2(g) + O_2(g)$$

20.35   As described in Section 20.6 of the text, magnesium metal will combine with chlorine.

$$Mg(s) + Cl_2(g) \rightarrow MgCl_2(s)$$

Magnesium will also react with HCl.

$$Mg(s) + 2HCl(aq) \rightarrow MgCl_2(aq) + H_2(g)$$

Neither of the above methods are really practical because magnesium metal is expensive to produce (electrolysis of magnesium chloride!). Can you suggest a method starting with a magnesium compound like $MgCO_3$?

20.36   The electron configuration of magnesium is $[Ne]3s^2$. The $3s$ electrons are outside the neon core (shielded), so they have relatively low ionization energies. Removing the third electron means separating an electron from the neon (closed shell) core, which requires a great deal more energy.

20.37   The water solubilities of the sulfates increase in the order **Ra < Ba < Sr < Ca < Mg**. The trend in this series is clearly in the sense of smaller ionic radius favoring greater solubility. Probably the smaller ion size results in much greater hydration energy (Section 6.7 of the text). Which sulfate in this series should have the largest lattice energy (Section 9.3 of the text)? Which is the more important factor in determining solubility in this series: hydration energy or lattice energy?

According to the *Handbook of Chemistry and Physics*, $BeSO_4$ reacts with water to form "$BeSO_4 \cdot 4H_2O$". In this sense it is not strictly comparable with the other sulfates of the Group 2A metals. However, this compound is really comprised of a sulfate ion and a $Be(H_2O)_4^{2+}$ complex ion. The latter is just a very strongly hydrated Be^{2+} ion. The solubility of the "$BeSO_4 \cdot 4H_2O$" is higher than any of other Group 2A sulfates, so it really does fit at the high solubility end of the series.

20.38   Even though helium and the Group 2A metals have ns^2 outer electron configurations, helium has a closed shell noble gas configuration and the Group 2A metals do not. The electrons in He are much closer to and more strongly attracted by the nucleus. Hence, the electrons in He are not easily removed. Helium is inert.

20.39   The formation of calcium oxide is:

$$2Ca(s) + O_2(g) \rightarrow 2CaO(s)$$

The conversion of calcium oxide to calcium hydroxide is:

$$CaO(s) + H_2O(l) \rightarrow Ca(OH)_2(s)$$

The reaction of calcium hydroxide with carbon dioxide is:

$$Ca(OH)_2(s) + CO_2(g) \rightarrow CaCO_3(s) + H_2O(l)$$

If calcium metal were exposed to extremely humid air, do you think that the oxide would still form?

20.40   **(a)**   quicklime: $CaO(s)$ **(b)**   slaked lime: $Ca(OH)_2(s)$

(c)   limewater: an aqueous suspension of $Ca(OH)_2$

20.43   According to Table 19.1 of the text, the following metals can reduce aluminum ion to aluminum:

$$Be, Mg, Na, Ca, Sr, Ba, K, Li$$

In 2001, the cheapest of these metals (magnesium) costs about \$12.00 per lb. The current cost of aluminum is \$0.70 per lb. Is the Hall process an improvement?

20.44   The reduction reaction is: $Al^{3+}(aq) + 3e^- \rightarrow Al(s)$

First, we can calculate the amount of charge needed to deposit 664 g of Al.

$$664 \text{ g Al} \times \frac{1 \text{ mol Al}}{26.98 \text{ g Al}} \times \frac{3 \text{ mol } e^-}{1 \text{ mol Al}} \times \frac{96,500 \text{ C}}{1 \text{ mol } e^-} = 7.12 \times 10^6 \text{ C}$$

Since a current of one ampere represents a flow of one coulomb per second, we can find the time it takes to pass this amount of charge.

$$32.6 \text{ A} = 32.6 \text{ C/s}$$

$$(7.12 \times 10^6 \text{ C}) \times \frac{1 \text{ s}}{32.6 \text{ C}} \times \frac{1 \text{ h}}{3600 \text{ s}} = \textbf{60.7 h}$$

20.45   The two complex ions can be classified as AB$_4$ and AB$_6$ structures (no unshared electron pairs on Al and 4 or 6 attached atoms, respectively). Their VSEPR geometries are **tetrahedral** and **octahedral**.

The accepted explanation for the nonexistence of AlCl$_6{}^{3-}$ is that the chloride ion is too big to form an octahedral cluster around a very small Al^{3+} ion. What is your guess for the formulas of complex ions formed between Al^{3+} and bromide or iodide ions? What about Ga^{3+} and chloride ion?

20.46   **(a)**   The relationship between cell voltage and free energy difference is:

$$\Delta G = -nFE$$

In the given reaction $n = 6$. We write:

$$E = \frac{-\Delta G}{nF} = \frac{-594 \times 10^3 \text{ J/mol}}{(6)(96500 \text{ J/V} \cdot \text{mol})} = -1.03 \text{ V}$$

The balanced equation shows *two* moles of aluminum. Is this the voltage required to produce *one* mole of aluminum? If we divide everything in the equation by two, we obtain:

$$\tfrac{1}{2}\text{Al}_2\text{O}_3(s) + \tfrac{3}{2}\text{C}(s) \rightarrow \text{Al}(l) + \tfrac{3}{2}\text{CO}(g)$$

For the new equation $n = 3$ and ΔG is $\left(\dfrac{1}{2}\right)$(594 kJ/mol) = 297 kJ/mol. We write:

$$E = \frac{-\Delta G}{nF} = \frac{-297 \times 10^3 \text{ J/mol}}{(3)(96500 \text{ J/V} \cdot \text{mol})} = -1.03 \text{ V}$$

The minimum voltage that must be applied is **1.03 V** (a negative sign in the answers above means that 1.03 V is required to produce the Al). The voltage required to produce one mole or one thousand moles of aluminum is the same; the amount of *current* will be different in each case.

(b)   First we convert 1.00 kg (1000 g) of Al to moles.

$$(1.00 \times 10^3 \text{ g Al}) \times \frac{1 \text{ mol Al}}{26.98 \text{ g Al}} = 37.1 \text{ mol Al}$$

The reaction in part (a) shows us that three moles of electrons are required to produce one mole of aluminum. The voltage is three times the minimum calculated above (namely, −3.09 V or −3.09 J/C). We can find the electrical energy by using the same equation with the other voltage.

$$\Delta G = -nFE = -(37.1)\left(\frac{3 \text{ mol e}^-}{1 \text{ mol Al}} \times \frac{96500 \text{ C}}{1 \text{ mol e}^-}\right)\left(\frac{-3.09 \text{ J}}{1 \text{ C}}\right) = 3.32 \times 10^7 \text{ J/mol} = 3.32 \times 10^4 \text{ kJ/mol}$$

This equation can be used because electrical work can be calculated by multiplying the voltage by the amount of charge transported through the circuit (joules = volts × coulombs). The nF term in Equation (19.2) of the text used above represents the amount of charge.

What is the significance of the positive sign of the free energy change? Would the manufacturing of aluminum be a different process if the free energy difference were negative?

20.47   The half-reaction for the oxidation of Al to AlO_2^- in basic solution is:

$$Al(s) + 4OH^-(aq) \rightarrow AlO_2^-(aq) + 2H_2O(l) + 3e^-$$

(a)   The nitrate-ammonia half-reaction is:

$$NO_3^-(aq) + 6H_2O(l) + 8e^- \rightarrow NH_3(aq) + 9OH^-(aq)$$

Combining the equations:

$$8Al(s) + 5OH^-(aq) + 3NO_3^-(aq) + 2H_2O(l) \rightarrow 9AlO_2^-(aq) + 3NH_3(aq)$$

(b)   The water-hydrogen half-reaction is:

$$H_2O(l) + e^- \rightarrow OH^-(aq) + \tfrac{1}{2}H_2(g)$$

Combining the equations:

$$Al(s) + OH^-(aq) + H_2O(l) \rightarrow AlO_2^-(aq) + \tfrac{3}{2}H_2(g)$$

(c)   The SnO_3^{2-}–Sn half-reaction is:

$$SnO_3^{2-}(aq) + 3H_2O(l) + 4e^- \rightarrow Sn(s) + 6OH^-(aq)$$

Combining the equations:

$$4Al(s) + 3SnO_3^{2-}(aq) + H_2O(l) \rightarrow 4AlO_2^-(aq) + 3Sn(s) + 2OH^-(aq)$$

20.48 $4Al(NO_3)_3(s) \longrightarrow 2Al_2O_3(s) + 12NO_2(g) + 3O_2(g)$

20.49   Some of aluminum's useful properties are: low density (light weight), high tensile strength, high electrical conductivity, high thermal conductivity, inert protective oxide surface coating.

20.50   The "bridge" bonds in Al_2Cl_6 break at high temperature: $Al_2Cl_6(g) \rightleftharpoons 2AlCl_3(g)$.

This increases the number of molecules in the gas phase and causes the pressure to be higher than expected for pure Al_2Cl_6.

If you know the equilibrium constants for the above reaction at higher temperatures, could you calculate the expected pressure of the $AlCl_3$–Al_2Cl_6 mixture?

20.51   **(a)** $2Al(s) + 3Cl_2(g) \rightarrow Al_2Cl_6(s)$

(b) $4Al(s) + 3O_2(g) \rightarrow 2Al_2O_3(s)$

(c) $2Al(s) + 3H_2SO_4(aq) \rightarrow Al_2(SO_4)_3(aq) + 3H_2(g)$

(d) $Al_2(SO_4)_3(aq) + (NH_4)_2SO_4(aq) \rightarrow 2NH_4(Al(SO_4)_2 \cdot 12H_2O(s)$, followed by the evaporation of the solution.

20.52   In Al_2Cl_6, each aluminum atom is surrounded by 4 bonding pairs of electrons (AB_4–type molecule), and therefore each aluminum atom is sp^3 **hybridized**. VSEPR analysis shows $AlCl_3$ to be an AB_3–type molecule (no lone pairs on the central atom). The geometry should be trigonal planar, and the aluminum atom should therefore be sp^2 **hybridized**.

20.53   Both CaO and MgO are basic oxides; they react with the acidic oxides formed in the furnace as follows:

$$6CaO(s) + P_4O_{10}(s) \rightarrow 2Ca_3(PO_4)_2(s) \qquad\qquad MgO(s) + SO_2(g) \rightarrow MgSO_3(s)$$

20.54   The formulas of the metal oxide and sulfide are MO and MS (why?). The balanced equation must therefore be:

$$2MS(s) + 3O_2(g) \rightarrow 2MO(s) + 2SO_2(g)$$

The number of moles of MO and MS are equal. We let x be the molar mass of metal. The number of moles of metal oxide is:

$$0.972 \text{ g} \times \frac{1 \text{ mol}}{(x + 16.00)\text{ g}}$$

The number of moles of metal sulfide is:

$$1.164 \text{ g} \times \frac{1 \text{ mol}}{(x + 32.07)\text{ g}}$$

The moles of metal oxide equal the moles of metal sulfide.

$$\frac{0.972}{(x + 16.00)} = \frac{1.164}{(x + 32.07)}$$

We solve for x.

$$0.972(x + 32.07) = 1.164(x + 16.00)$$

$$x = \textbf{65.4 g/mol}$$

20.55   Metals conduct electricity. If electrons were localized in pairs, they could not move through the metal.

20.56   Copper(II) ion is more easily reduced than either water or hydrogen ion (How can you tell? See Section 19.3 of the text.) Copper metal is more easily oxidized than water. Water should not be affected by the copper purification process.

20.57   The balanced equation for the permanganate/iron(II) reaction is:

$$5Fe^{2+}(aq) + MnO_4^-(aq) + 8H^+(aq) \rightarrow 5Fe^{3+}(aq) + Mn^{2+}(aq) + 4H_2O(l)$$

Thus one mole of permanganate is stoichiometrically equivalent to five moles of iron(II). The original amount of iron(II) is (Note: one mole of iron(II) is equivalent to one mole of iron(III)).

$$50.0 \text{ mL} \times \frac{0.0800 \text{ mol Fe}^{2+}}{1000 \text{ mL soln}} = 4.00 \times 10^{-3} \text{ mol Fe}^{2+}$$

The excess amount of iron(II) is determined by using the balanced equation:

$$Cr_2O_7^{2-}(aq) + 14H^+(aq) + 6Fe^{2+}(aq) \rightarrow 2Cr^{3+}(aq) + 7H_2O(l) + 6Fe^{3+}(aq)$$

Thus one mole of dichromate is equivalent to six moles of iron(II). The excess iron(II) is:

$$22.4 \text{ mL} \times \frac{0.0100 \text{ mol Cr}_2\text{O}_7^{2-}}{1000 \text{ mL soln}} \times \frac{6 \text{ mol Fe}^{2+}}{1 \text{ mol Cr}_2\text{O}_7^{2-}} = 1.34 \times 10^{-3} \text{ mol Fe}^{2+}$$

The amount of iron(II) consumed is $(4.00 \times 10^{-3} \text{ mol}) - (1.34 \times 10^{-3} \text{ mol}) = 2.66 \times 10^{-3} \text{ mol}$

The mass of manganese is:

$$(2.66 \times 10^{-3} \text{ mol Fe}^{2+}) \times \frac{1 \text{ mol MnO}_4^-}{5 \text{ mol Fe}^{2+}} \times \frac{1 \text{ mol Mn}}{1 \text{ mol MnO}_4^-} \times \frac{54.94 \text{ g Mn}}{1 \text{ mol Mn}} = 0.0292 \text{ g Mn}$$

The percent by mass of Mn is:

$$\frac{0.0292 \text{ g}}{0.450 \text{ g}} \times 100\% = \textbf{6.49\%}$$

20.58   Using Equation 18.12 from the text:

(a) $\Delta G_{\text{rxn}}^{\circ} = 4\Delta G_{\text{f}}^{\circ}(\text{Fe}) + 3\Delta G_{\text{f}}^{\circ}(\text{O}_2) - 2\Delta G_{\text{f}}^{\circ}(\text{Fe}_2\text{O}_3)$

$\boldsymbol{\Delta G_{\text{rxn}}^{\circ}} = (4)(0) + (3)(0) - (2)(-741.0 \text{ kJ/mol}) = \textbf{1482 kJ/mol}$

(b) $\Delta G_{\text{rxn}}^{\circ} = 4\Delta G_{\text{f}}^{\circ}(\text{Al}) + 3\Delta G_{\text{f}}^{\circ}(\text{O}_2) - 2\Delta G_{\text{f}}^{\circ}(\text{Al}_2\text{O}_3)$

$\boldsymbol{\Delta G_{\text{rxn}}^{\circ}} = (4)(0) + (3)(0) - (2)(-1576.4 \text{ kJ/mol}) = \textbf{3152.8 kJ/mol}$

20.59   Amphoterism means the ability to act both as an acid and as a base. Al(OH)_3 is amphoteric, as shown below:

<u>As an acid:</u>   <u>As a base:</u>

$\text{Al(OH)}_3(s) + \text{NaOH}(aq) \rightarrow \text{NaAl(OH)}_4(aq)$ $\text{Al(OH}_3)(s) + 3\text{HCl}(aq) \rightarrow \text{AlCl}_3(aq) + 3\text{H}_2\text{O}(l)$

20.60   At high temperature, magnesium metal reacts with nitrogen gas to form magnesium nitride.

$$3\text{Mg}(s) + \text{N}_2(g) \longrightarrow \text{Mg}_3\text{N}_2(s)$$

Can you think of any gas other than a noble gas that could provide an inert atmosphere for processes involving magnesium at high temperature?

20.61   The sodium atom has an ns^1 valence shell electron configuration. The molecular orbital diagram for the Na_2 molecule is exactly analogous to that of the Li_2 molecule that is discussed in Section 10.7 of the text. The ns^1 valence electron from each sodium will occupy a bonding σ_{3s} molecular orbital, giving Na_2 two more electrons in bonding molecular orbitals than in antibonding molecular orbitals. The bond order is *one*.

The alkaline earth metals all possess an ns^2 valence shell electron configuration. As in the case of Be_2 that is discussed in Section 10.7 of the text, all dimers of the alkaline earth metals would have equal numbers of electrons in bonding and antibonding molecular orbitals: that is, $(\sigma_{ns})^2(\sigma_{ns}^{\star})^2$. The bond order is *zero*, and such dimers would not be expected to exist.

20.62 **(a)** In water the aluminum(III) ion causes an increase in the concentration of hydrogen ion (lower pH). This results from the effect of the small diameter and high charge (3+) of the aluminum ion on surrounding water molecules. The aluminum ion draws electrons in the O–H bonds to itself, thus allowing easy formation of H^+ ions.

(b) $Al(OH)_3$ is an amphoteric hydroxide. It will dissolve in strong base with the formation of a complex ion.

$$Al(OH)_3(s) + OH^-(aq) \longrightarrow Al(OH)_4^-(aq)$$

The concentration of OH^- in aqueous ammonia is too low for this reaction to occur.

20.63 The reactions are:

(a) $Al_2(CO_3)_3(s) \rightarrow Al_2O_3(s) + 3CO_2(g)$

(b) $AlCl_3(s) + 3K(s) \rightarrow Al(s) + 3KCl(s)$

(c) $Ca(OH)_2(aq) + Na_2CO_3(aq) \rightarrow CaCO_3(s) + 2NaOH(aq)$

20.64 Calcium oxide is a base. The reaction is a neutralization.

$$CaO(s) + 2HCl(aq) \longrightarrow CaCl_2(aq) + H_2O(l)$$

20.65 $MgO + CO \rightarrow Mg + CO_2$

$$\Delta G° = \Delta G_f°(CO_2) - \Delta G_f°(MgO) - \Delta G_f°(CO)$$

$$= (1)(-394.4 \text{ kJ/mol}) - (1)(-569.6 \text{ kJ/mol}) - (1)(-137.3 \text{ kJ/mol}) = +312.5 \text{ kJ/mol}$$

$$\Delta G° = -RT\ln K_p$$

$$312.5 \times 10^3 \text{ J/mol} = -(8.314 \text{ J/mol·K})(298 \text{ K})\ln K_p$$

$$K_p = 1.7 \times 10^{-55}$$

K_p is much too small, even at high temperatures. No product will be formed.

20.66 Metals have closely spaced energy levels and (referring to Figure 20.10 of the text) a very small energy gap between filled and empty levels. Consequently, many electronic transitions can take place with absorption and subsequent emission continually occurring. Some of these transitions fall in the visible region of the spectrum and give rise to the flickering appearance.

20.67 Essentially the same as the other alkali metals, except that it is more reactive. Fr reacts with water.

$$2Fr + 2H_2O \rightarrow 2FrOH + H_2$$

Like K, Rb, and Cs, Fr also forms superoxide (FrO_2), in addition to oxide and peroxide.

20.68   NaF is used in toothpaste to fight tooth decay.

Li_2CO_3 is used to treat mental illness.

$Mg(OH)_2$ is an antacid.

$CaCO_3$ is an antacid.

$BaSO_4$ is used to enhance X ray images of the digestive system.

$Al(OH)_2NaCO_3$ is an antacid.

20.69

Scheme I

$2Mg + O_2 \rightarrow 2MgO$	$A = MgO$
$MgO + 2HCl \rightarrow MgCl_2 + H_2O$	$B = MgCl_2(aq)$
$MgCl_2 + Na_2CO_3 \rightarrow MgCO_3 + 2NaCl$	$C = MgCO_3$
$MgCO_3 \rightarrow MgO + CO_2$	$D = MgO, E = CO_2$
$CO_2 + Ca(OH)_2 \rightarrow CaCO_3 + H_2O$	$F = CaCO_3$

Scheme II

$Mg + H_2SO_4 \rightarrow MgSO_4 + H_2$	$G = MgSO_4(aq)$
$MgSO_4 + 2NaOH \rightarrow Mg(OH)_2 + Na_2SO_4$	$H = Mg(OH)_2$
$Mg(OH)_2 + 2HNO_3 \rightarrow Mg(NO_3)_2 + 2H_2O$	$I = Mg(NO_3)_2$
$2Mg(NO_3)_2 \rightarrow 2MgO + 4NO_2 + O_2$	NO_2 is the brown gas.

20.70   Both Li and Mg form oxides (Li_2O and MgO). Other Group 1A metals (Na, K, etc.) also form peroxides and superoxides. In Group 1A, only Li forms nitride (Li_3N), like Mg (Mg_3N_2).

Li resembles Mg in that its carbonate, fluoride, and phosphate have low solubilities.

20.71 N_2, because Li reacts with nitrogen to form lithium nitride.

20.72   You might know that Ag, Cu, Au, and Pt are found as free elements in nature, which leaves **Zn** by process of elimination. You could also look at Table 19.1 of the text to find the metal that is easily oxidized. Looking at the table, the standard oxidation potential of Zn is +0.76 V. The positive value indicates that Zn is easily oxidized to Zn^{2+} and will not exist as a free element in nature.

CHAPTER 21
NONMETALLIC ELEMENTS AND THEIR COMPOUNDS

21.11 Element number 17 is the halogen, chlorine. Since it is a nonmetal, chlorine will form the molecular compound HCl. Element 20 is the alkaline earth metal calcium which will form an ionic hydride, CaH_2. A water solution of HCl is called hydrochloric acid. Calcium hydride will react according to the equation (see Section 21.2 of the text).

$$CaH_2(s) + 2H_2O(l) \rightarrow Ca(OH)_2(aq) + 2H_2(g)$$

21.12 **(a)** Hydrogen reacts with alkali metals to form ionic hydrides:

$$2Na(l) + H_2(g) \rightarrow 2NaH(s)$$

The oxidation number of hydrogen drops from 0 to −1 in this reaction.

(b) Hydrogen reacts with oxygen (combustion) to form water:

$$2H_2(g) + O_2(g) \rightarrow 2H_2O(l)$$

The oxidation number of hydrogen increases from 0 to +1 in this reaction.

21.13 NaH: Ionic compound. It reacts with water as follows:

$$NaH(s) + H_2O(l) \rightarrow NaOH(aq) + H_2(g)$$

CaH_2: Ionic compound: It reacts with water as follows:

$$CaH_2(g) + 2H_2O(l) \rightarrow Ca(OH)_2(s) + 2H_2(g)$$

CH_4: Covalent compound. Unreactive. It burns in air or oxygen:

$$CH_4(g) + 2O_2(g) \rightarrow CO_2(g) + 2H_2O(l)$$

NH_3: Covalent compound. It is a weak base in water:

$$NH_3(aq) + H_2O(l) \rightleftharpoons NH_4^+(aq) + OH^-(aq)$$

H_2O: Covalent compound. It forms strong intermolecular hydrogen bonds. It is a good solvent for both ionic compounds and substances capable of forming hydrogen bonds.

HCl: Covalent compound (polar). It acts as a strong acid in water:

$$HCl(g) + H_2O(l) \rightarrow H_3O^+(aq) + Cl^-(aq)$$

21.14 Hydrogen forms an interstitial hydride with palladium, which behaves almost like a solution of hydrogen atoms in the metal. At elevated temperatures hydrogen atoms can pass through solid palladium; other substances cannot.

21.15   The equation is: $CaH_2(s) + 2H_2O(l) \rightarrow Ca(OH)_2(aq) + 2H_2(g)$

First, let's calculate the moles of H_2 using the ideal gas law.

$$\text{mol } H_2 = \frac{\left(746 \text{ mmHg} \times \dfrac{1 \text{ atm}}{760 \text{ mmHg}}\right)(26.4 \text{ L})}{(0.0821 \text{ L} \cdot \text{atm/mol} \cdot \text{K})(293 \text{ K})} = 1.08 \text{ mol } H_2$$

Now, we can calculate the mass of CaH_2 using the correct mole ratio from the balanced equation.

$$\textbf{Mass CaH}_2 = 1.08 \text{ mol } H_2 \times \frac{1 \text{ mol CaH}_2}{2 \text{ mol } H_2} \times \frac{42.10 \text{ g}}{1 \text{ mol CaH}_2} = \textbf{22.7 g CaH}_2$$

21.16   The number of moles of deuterium gas is:

$$n = \frac{PV}{RT} = \frac{(0.90 \text{ atm})(2.0 \text{ L})}{(0.0821 \text{ L} \cdot \text{atm/K} \cdot \text{mol})(298 \text{ K})} = 0.074 \text{ mol}$$

If the abundance of deuterium is 0.015 percent, the number of moles of water must be:

$$0.074 \text{ mol } D_2 \times \frac{100\% \text{ H}_2O}{0.015\% \text{ D}_2} = 4.9 \times 10^2 \text{ mol } H_2O$$

At a recovery of 80 percent the amount of water needed is:

$$(0.80)(4.9 \times 10^2 \text{ mol } H_2O) \times \frac{0.018 \text{ kg H}_2O}{1.0 \text{ mol H}_2O} = \textbf{11 kg H}_2\textbf{O}$$

21.17   According to Table 19.1 of the text, H_2 can reduce Cu^{2+}, but not Na^+. (How can you tell?) The reaction is:

$$CuO(s) + H_2(g) \rightarrow Cu(s) + H_2O(l)$$

21.18   **(a)** $H_2 + Cl_2 \rightarrow 2HCl$

(b) $3H_2 + N_2 \rightarrow 2NH_3$

(c) $2Li + H_2 \rightarrow 2LiH$

$LiH + H_2O \rightarrow LiOH + H_2$

21.25   The reaction can be represented:

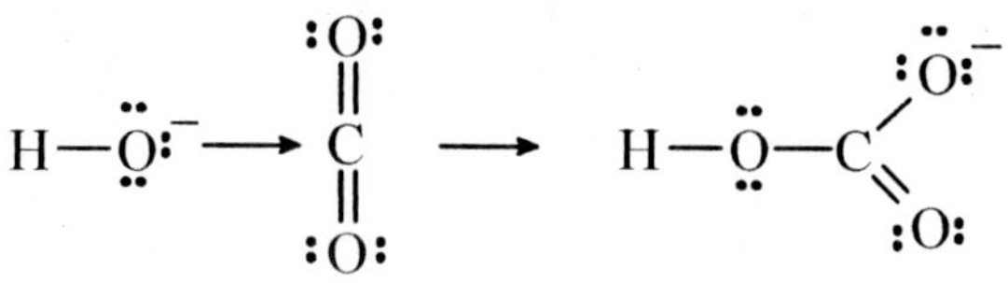

The lone pair on the hydroxide oxygen becomes a new carbon-oxygen bond. The octet rule requires that one of the electron pairs in the double bond be changed to a lone pair. What is the other resonance form of the product ion?

21.26   The Lewis structure is:

$$\left[:C\!\equiv\!C:\right]^{2-}$$

21.27   The reactions are:

  (a) $Be_2C(s) + 4H_2O(l) \rightarrow 2Be(OH)_2(aq) + CH_4(g)$

  (b) $CaC_2(s) + 2H_2O(l) \rightarrow Ca(OH)_2(aq) + C_2H_2(g)$

21.28   **(a)**   The reaction is: $2NaHCO_3(s) \rightarrow Na_2CO_3(s) + H_2O(g) + CO_2(g)$

       Is this an endo- or an exothermic process?

  (b)   The hint is generous. The reaction is:

$$Ca(OH)_2(aq) + CO_2(g) \rightarrow CaCO_3(s) + H_2O(l)$$

       The visual proof is the formation of a white precipitate of $CaCO_3$. Why would a water solution of
       NaOH be unsuitable to qualitatively test for carbon dioxide?

21.29   Magnesium and calcium carbonates are insoluble; the bicarbonates are soluble. Formation of a precipitate
       after addition of $MgCl_2$ solution would show the presence of Na_2CO_3.

       Assuming similar concentrations, which of the two sodium salt solutions would have a higher pH?

21.30   Heat causes bicarbonates to decompose according to the reaction:

$$2HCO_3^- \rightarrow CO_3^{2-} + H_2O + CO_2$$

       Generation of carbonate ion causes precipitation of the insoluble $MgCO_3$.

       Do you think there is much chance of finding natural mineral deposits of calcium or magnesium
       bicarbonates?

21.31   Bicarbonate ion can react with either H^+ or OH^-.

$$HCO_3^-(aq) + H^+(aq) \rightarrow H_2CO_3(aq)$$
$$HCO_3^-(aq) + OH^-(aq) \rightarrow CO_3^{2-}(aq) + H_2O(l)$$

       Since ammonia is a base, carbonate ion is formed which causes precipitation of $CaCO_3$.

21.32   The wet sodium hydroxide is first converted to sodium carbonate:

$$2NaOH(aq) + CO_2(g) \rightarrow Na_2CO_3(aq) + H_2O(l)$$

       and then to sodium hydrogen carbonate: $Na_2CO_3(aq) + H_2O(l) + CO_2(g) \rightarrow 2NaHCO_3(aq)$

       Eventually, the sodium hydrogen carbonate precipitates (the water solvent evaporates since $NaHCO_3$ is not
       hygroscopic). Thus, most of the white solid is $NaHCO_3$ plus some Na_2CO_3.

21.33   Table 19.1 of the text shows that magnesium metal has the potential to be an extremely powerful reducing agent. It appears inert at room temperature, but at high temperatures it can react with almost any source of oxygen atoms (including water!) to form MgO. In this case carbon dioxide is reduced to carbon.

$$2Mg(s) + CO_2(g) \rightarrow 2MgO(s) + C(s)$$

How does one extinguish a magnesium fire?

21.34   Carbon monoxide and molecular nitrogen are isoelectronic. Both have 14 electrons. What other diatomic molecules discussed in these problems are isoelectronic with CO?

21.39   The preparations are:

(a) $2NH_3(g) + 3CuO(s) \rightarrow N_2(g) + 3Cu(s) + 3H_2O(g)$

(b) $(NH_4)_2Cr_2O_7(s) \rightarrow N_2(g) + Cr_2O_3(s) + 4H_2O(g)$

21.40   **(a)** $2NaNO_3(s) \rightarrow 2NaNO_2(s) + O_2(g)$

(b) $NaNO_3(s) + C(s) \rightarrow NaNO_2(s) + CO(g)$

21.41   The balanced equation is: $NH_2^-(aq) + H_2O(l) \rightarrow NH_3(aq) + OH^-(aq)$

In this system the acid is H_2O (proton donor) and the base is NH_2^- (proton acceptor). What are the conjugate acid and the conjugate base?

21.42   The balanced equation is: $2NH_3(g) + CO_2(g) \rightarrow (NH_2)_2CO(s) + H_2O(l)$

If pressure increases, the position of equilibrium will shift in the direction with the smallest number of molecules in the gas phase, that is, to the right. Therefore, the reaction is best run at high pressure.

Write the expression for Q_p for this reaction. Does increasing pressure cause Q_p to increase or decrease? Is this consistent with the above prediction?

21.43   Lightening can cause N_2 and O_2 to react: $N_2(g) + O_2(g) \rightarrow 2NO(g)$

The NO formed naturally in this manner eventually suffers oxidation to nitric acid (see the Ostwald process in Section 21.4 of the text) which precipitates as rain. The nitrate ion is a natural source of nitrogen for growing plants.

21.44   The density of a gas depends on temperature, pressure, and the molar mass of the substance. When two gases are at the same pressure and temperature, the ratio of their densities should be the same as the ratio of their molar masses. The molar mass of ammonium chloride is 53.5 g/mol, and the ratio of this to the molar mass of molecular hydrogen (2.02 g/mol) is 26.8. The experimental value of 14.5 is roughly half this amount. Such results usually indicate breakup or dissociation into smaller molecules in the gas phase (note the temperature). The measured molar mass is the average of all the molecules in equilibrium.

$$NH_4Cl(g) \rightleftharpoons NH_3(g) + HCl(g)$$

Knowing that ammonium chloride is a stable substance at 298 K, is the above reaction exo- or endothermic?

21.45   The oxidation number of nitrogen in nitrous acid is +3. Since this value is between the extremes of a +5 and −3 for nitrogen, nitrous acid can be either oxidized or reduced. Nitrous acid can oxidize HI to I_2 (in other words HI acts as a reducing agent).

$$2HI(g) + 2HNO_2(aq) \rightarrow I_2(s) + 2NO(g) + 2H_2O(l)$$

A strong oxidizing agent can oxidize nitrous acid to nitric acid (oxidation number of nitrogen +5).

$$2Ce^{4+}(aq) + HNO_2(aq) + H_2O(l) \rightarrow 2Ce^{3+}(aq) + HNO_3(aq) + 2H^+(aq)$$

21.46   The highest oxidation state possible for a Group 5A element is +5. This is the oxidation state of nitrogen in nitric acid (HNO_3).

21.47   **(a)** $NH_4NO_3(s) \rightarrow N_2O(g) + 2H_2O(l)$ **(b)** $2KNO_3(s) \rightarrow 2KNO_2(s) + O_2(g)$

(c) $Pb(NO_3)_2(s) \rightarrow PbO(s) + 2NO_2(g) + O_2(g)$

21.48   Nitric acid is a strong oxidizing agent in addition to being a strong acid (see Table 19.1 of the text, $E_{red}^{\circ} = +0.96V$). The primary action of a good reducing agent like zinc is reduction of nitrate ion to ammonium ion.

$$4Zn(s) + NO_3^{-}(aq) + 10H^+(aq) \rightarrow 4Zn^{2+}(aq) + NH_4^{+}(aq) + 3H_2O(l)$$

21.49   The balanced equation is: $2KNO_3(s) + C(s) \rightarrow 2KNO_2(s) + CO_2(g)$

The maximum amount of potassium nitrite (theoretical yield) is:

$$57.0 \text{ g KNO}_3 \times \frac{1 \text{ mol KNO}_3}{101.1 \text{ g KNO}_3} \times \frac{1 \text{ mol KNO}_2}{1 \text{ mol KNO}_3} \times \frac{85.11 \text{ g KNO}_2}{1 \text{ mol KNO}_2} = \textbf{48.0 g KNO}_2$$

21.50   One of the best Lewis structures for nitrous oxide is:

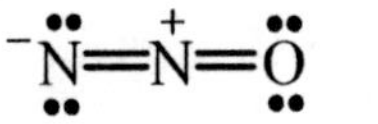

There are no lone pairs on the central nitrogen, making this an AB_2 VSEPR case. All such molecules are linear. Other resonance forms are:

Are all the resonance forms consistent with a linear geometry?

21.51   **(a)** $\Delta G_{rxn}^{\circ} = 2\Delta G_f^{\circ}(NO) - [0 + 0]$

173.4 kJ/mol $= 2\Delta G_f^{\circ}(NO)$

$\Delta G_f^{\circ}(NO) = \textbf{86.7 kJ/mol}$

(b)   From Equation (18.14) of the text:

$\Delta G^{\circ} = -RT\ln K_p$

173.4×10^3 J/mol $= -(8.314 \text{ J/K·mol})(298 \text{ K})\ln K_p$

$K_p = \textbf{4} \times \textbf{10}^{-31}$

(c)   Using Equation (14.5) of the text $[K_p = K_c(0.0821\ T)^{\Delta n}]$, $\Delta n = 0$, then $K_p = K_c = 4 \times 10^{-31}$

21.52 $\Delta H° = 4\Delta H_f°[NO(g)] + 6\Delta H_f°[H_2O(l)] - \{4\Delta H_f°[NH_3(g)] + 5\Delta H_f°[O_2(g)]\}$

$\Delta H° = (4)(90.4\ \text{kJ/mol}) + (6)(-285.8\ \text{kJ/mol}) - [(4)(-46.3\ \text{kJ/mol}) + (5)(0)] = \textbf{-1168 kJ/mol}$

21.53   The atomic radius of P (128 pm) is considerably larger than that of N (92 pm); consequently, the $3p$ orbital on a P atom cannot overlap effectively with a $3p$ orbital on a neighboring P atom to form a pi bond. Simply stated, the phosphorus is too large to allow effective overlap of the $3p$ orbitals to form π bonds.

21.54 $\Delta T_b = K_b m = 0.409°C$

$$\text{molality} = \frac{0.409°C}{2.34°C/m} = 0.175\ m$$

The number of grams of white phosphorus in 1 kg of solvent is:

$$\frac{1.645\ \text{g phosphorus}}{75.5\ \text{g CS}_2} \times \frac{1000\ \text{g}}{1\ \text{kg}} = 22.8\ \text{g phosphorus/kg CS}_2$$

The molar mass of white phosphorus is:

$$\frac{22.8\ \text{g phosphorus/kg CS}_2}{0.175\ \text{mol phosphorus/kg CS}_2} = \textbf{125 g/mol}$$

Let the molecular formula of white phosphorus be P_n so that:

$$n \times 30.97\ \text{g/mol} = 125\ \text{g/mol}$$

$$n = 4$$

The molecular formula of white phosphorus is $\textbf{P}_\textbf{4}$.

21.55   You won't find a reaction that starts with elemental phosphorus and ends with phosphoric acid. However there is more than one reaction having phosphoric acid as a product. One possibility is the reaction of P_4O_{10} with water.

$$P_4O_{10}(s) + 6H_2O(l) \rightarrow 4H_3PO_4(aq)$$

Can P_4O_{10} be formed from elemental phosphorus? Study of Section 21.4 of the text shows that P_4 combines with oxygen to form P_4O_{10}.

$$P_4(s) + 5O_2(g) \rightarrow P_4O_{10}(s)$$

The synthesis of phosphoric acid is the result of these two steps in sequence.

Can you come up with an alternative synthesis starting with elemental phosphorus and chlorine gas?

21.56   The balanced equation is:

$$\textbf{P}_\textbf{4}\textbf{O}_{\textbf{10}}\textbf{(s) + 4HNO}_\textbf{3}\textbf{(aq)} \rightarrow \textbf{2N}_\textbf{2}\textbf{O}_\textbf{5}\textbf{(g) + 4HPO}_\textbf{3}\textbf{(l)}$$

The theoretical yield of N_2O_5 is :

$$79.4\ \text{g P}_4O_{10} \times \frac{1\ \text{mol P}_4O_{10}}{283.9\ \text{g P}_4O_{10}} \times \frac{2\ \text{mol N}_2O_5}{1\ \text{mol P}_4O_{10}} \times \frac{108.0\ \text{g N}_2O_5}{1\ \text{mol N}_2O_5} = \textbf{60.4 g N}_\textbf{2}\textbf{O}_\textbf{5}$$

21.57 **(a)** Because the P–H bond is weaker, there is a greater tendency for PH_4^+ to ionize compared to NH_4^+. Therefore, PH_3 is a weaker base than NH_3.

(b) The electronegativity of nitrogen is greater than that of phosphorus. The N–H bond is much more polar than the P–H bond and can participate in hydrogen bonding. This increases intermolecular attractions and results in a higher boiling point.

(c) Elements in the second period never expand their octets. A common explanation is the absence of $2d$ atomic orbitals.

(d) The triple bond between two nitrogen atoms is one of the strongest atomic linkages known. The bonds in P_4 are highly strained because of the acute P–P–P angles and are more easily broken.

21.58 PH_4^+ is similar to NH_4^+. The hybridization of phosphorus in PH_4^+ is sp^3.

21.65 The molecular orbital energy level diagrams are:

Which of the three has the strongest bonding?

21.66 $\Delta G° = \Delta G_f°(NO_2) + \Delta G_f°(O_2) - [\Delta G_f°(NO) + \Delta G_f°(O_3)]$

$\Delta G° = (1)(51.8 \text{ kJ/mol}) + (0) - [(1)(86.7 \text{ kJ/mol}) + (1)(163.4 \text{ kJ/mol})] = -198.3 \text{ kJ/mol}$

$$\Delta G° = -RT \ln K_p$$

$$\ln K_p = \frac{-\Delta G°}{RT} = \frac{198.3 \times 10^3 \text{ J/mol}}{(8.314 \text{ J/K} \cdot \text{mol})(298 \text{ K})}$$

$$K_p = 6 \times 10^{34}$$

Since there is no change in the number of moles of gases, K_c is *equal* to K_p.

21.67 **(a)** As stated in the problem, the decomposition of hydrogen peroxide is accelerated by light. Storing solutions of the substance in dark-colored bottles helps to prevent this form of decomposition.

(b) The STP volume of oxygen gas formed is:

$$15.0 \text{ g soln} \times \frac{7.50\% \text{ H}_2\text{O}_2}{100\% \text{ soln}} \times \frac{1 \text{ mol H}_2\text{O}_2}{34.02 \text{ g H}_2\text{O}_2} \times \frac{1 \text{ mol O}_2}{2 \text{ mol H}_2\text{O}_2} \times \frac{22.41 \text{ L O}_2}{1 \text{ mol O}_2} = \textbf{0.371 L O}_2$$

21.68   Following the rules given in Section 4.4 of the text, we assign hydrogen an oxidation number of +1 and **fluorine** an oxidation number of −1. Since HFO is a neutral molecule, the oxidation number of **oxygen** is **zero**. Can you think of other compounds in which oxygen has this oxidation number?

21.69   Analogous to phosphorus in Problem 21.53, the $3p$ orbital overlap is poor for the formation of π bonds because of the relatively large size of sulfur compared to oxygen.

21.70   First, let's calculate the moles of sulfur in 48 million tons of sulfuric acid.

$$(48 \times 10^6 \text{ tons H}_2\text{SO}_4) \times \frac{2000 \text{ lb}}{1 \text{ ton}} \times \frac{453.6 \text{ g}}{1 \text{ lb}} \times \frac{1 \text{ mol H}_2\text{SO}_4}{98.09 \text{ g H}_2\text{SO}_4} \times \frac{1 \text{ mol S}}{1 \text{ mol H}_2\text{SO}_4} = \mathbf{4.4 \times 10^{11} \text{ mol S}}$$

Converting to grams of sulfur:

$$(4.4 \times 10^{11} \text{ mol S}) \times \frac{32.07 \text{ g S}}{1 \text{ mol S}} = \mathbf{1.4 \times 10^{13} \text{ g S}}$$

21.71   Each reaction uses $H_2SO_4(l)$ as a reagent.

(a) $HCOOH(l) \rightleftharpoons CO(g) + H_2O(l)$ (c) $2HNO_3(l) \rightleftharpoons N_2O_5(g) + H_2O(l)$

(b) $4H_3PO_4(l) \rightleftharpoons P_4O_{10}(s) + 6H_2O(l)$ (d) $2HClO_3(l) \rightleftharpoons Cl_2O_5(l) + H_2O(l)$

21.72   There are actually several steps involved in removing sulfur dioxide from industrial emissions with calcium carbonate. First calcium carbonate is heated to form carbon dioxide and calcium oxide.

$$CaCO_3(s) \rightleftharpoons CaO(s) + CO_2(g)$$

The CaO combines with sulfur dioxide to form calcium sulfite.

$$CaO(s) + SO_2(g) \rightarrow CaSO_3(s)$$

Alternatively, calcium sulfate forms if enough oxygen is present.

$$2CaSO_3(s) + O_2(g) \rightarrow 2CaSO_4(s)$$

The amount of calcium carbonate (limestone) needed in this problem is:

$$50.6 \text{ g SO}_2 \times \frac{1 \text{ mol SO}_2}{64.07 \text{ g SO}_2} \times \frac{1 \text{ mol CaCO}_3}{1 \text{ mol SO}_2} \times \frac{100.1 \text{ g CaCO}_3}{1 \text{ mol CaCO}_3} = \mathbf{79.1 \text{ g CaCO}_3}$$

The calcium oxide–sulfur dioxide reaction is an example of a Lewis acid-base reaction (see Section 15.12 of the text) between oxide ion and sulfur dioxide. Can you draw Lewis structures showing this process? Which substance is the Lewis acid and which is the Lewis base?

21.73   To form OF_6 there would have to be six bonds (twelve electrons) around the oxygen atom. This would violate the octet rule.

21.74 The usual explanation for the fact that no chemist has yet succeeded in making SCl_6, SBr_6 or SI_6 is based on the idea of excessive crowding of the six chlorine, bromine, or iodine atoms around the sulfur. Others suggest that sulfur in the +6 oxidation state would oxidize chlorine, bromine, or iodine in the −1 oxidation state to the free elements. In any case, none of these substances has been made as of the date of this writing.

It is of interest to point out that thirty years ago all textbooks confidently stated that compounds like ClF_5 could not be prepared.

Note that PCl_6^- is a known species. How different are the sizes of S and P?

21.75 The melting and boiling points of hydrogen sulfide are −85.5°C and −60.7°C, respectively. Those of water are, of course, 0°C and 100°C. Although hydrogen sulfide has more electrons and hence greater dispersion forces, water has the much higher melting and boiling points because of strong intermolecular hydrogen bonding. Water is a suitable solvent for many compounds, both ionic and molecular. Water has both acidic and basic properties. Hydrosulfuric acid, $H_2S(aq)$, is a weak diprotic acid. In its pure liquid form, hydrogen sulfide can dissolve a limited range of weakly polar substances. It is not amphoteric.

21.76 First we convert gallons of water to grams of water.

$$(2.0 \times 10^2 \text{ gal}) \times \frac{3.785 \text{ L}}{1 \text{ gal}} \times \frac{1000 \text{ mL}}{1 \text{ L}} \times \frac{1.00 \text{ g H}_2\text{O}}{1 \text{ mL}} = 7.6 \times 10^5 \text{ g H}_2\text{O}$$

An H_2S concentration of 22 ppm indicates that in 1 million grams of water, there will be 22 g of H_2S. First, let's calculate the number of moles of H_2S in 7.6×10^5 g of H_2O:

$$(7.6 \times 10^5 \text{ g H}_2\text{O}) \times \frac{22 \text{ g H}_2\text{S}}{1.0 \times 10^6 \text{ g H}_2\text{O}} \times \frac{1 \text{ mol H}_2\text{S}}{34.09 \text{ g H}_2\text{S}} = 0.49 \text{ mol H}_2\text{S}$$

The mass of chlorine required to react with 0.49 mol of H_2S is:

$$0.49 \text{ mol H}_2\text{S} \times \frac{1 \text{ mol Cl}_2}{1 \text{ mol H}_2\text{S}} \times \frac{70.90 \text{ g Cl}_2}{1 \text{ mol Cl}_2} = \textbf{35 g Cl}_2$$

21.77 Copper reacts with hot concentrated sulfuric acid to yield copper(II) sulfate, water, and sulfur dioxide. Concentrated sulfuric acid also reacts with carbon to produce carbon dioxide, water, and sulfur dioxide.

Can you write balanced equations for these processes?

21.78 A check of Table 19.1 of the text shows that sodium ion cannot be reduced by any of the substances mentioned in this problem; it is a "spectator ion". We focus on the substances that are actually undergoing oxidation or reduction and write half-reactions for each.

$$2I^-(aq) \rightarrow I_2(s)$$

$$H_2SO_4(aq) \rightarrow H_2S(g)$$

Balancing the oxygen, hydrogen, and charge gives:

$$2I^-(aq) \rightarrow I_2(s) + 2e^-$$

$$H_2SO_4(aq) + 8H^+(aq) + 8e^- \rightarrow H_2S(g) + 4H_2O(l)$$

Multiplying the iodine half-reaction by four and combining gives the balanced redox equation.

$$H_2SO_4(aq) + 8I^-(aq) + 8H^+(aq) \rightarrow H_2S(g) + 4I_2(s) + 4H_2O(l)$$

The hydrogen ions come from extra sulfuric acid. We add one sodium ion for each iodide ion to obtain the final equation.

$$\textbf{9H}_2\textbf{SO}_4\textbf{(aq) + 8NaI(aq)} \rightarrow \textbf{H}_2\textbf{S(g) + 4I}_2\textbf{(s) + 4H}_2\textbf{O(l) + 8NaHSO}_4\textbf{(aq)}$$

21.81   A number of methods are available for the preparation of metal chlorides.

(a) $Na(s) + Cl_2(g) \rightarrow 2NaCl(s)$

(b) $2HCl(aq) + Mg(s) \rightarrow MgCl_2(aq) + H_2(g)$

(c) $HCl(aq) + NaOH(aq) \rightarrow NaCl(aq) + H_2O(l)$

(d) $CaCO_3(s) + 2HCl(aq) \rightarrow CaCl_2(aq) + CO_2(g) + H_2O(l)$

(e) $AgNO_3(aq) + NaCl(aq) \rightarrow AgCl(s) + NaNO_3(aq)$

21.82   Sulfuric acid is added to solid sodium chloride, not aqueous sodium chloride. Hydrogen chloride is a gas at room temperature and can escape from the reacting mixture.

$$\textbf{H}_2\textbf{SO}_4\textbf{(l) + NaCl(s)} \rightarrow \textbf{HCl(g) + NaHSO}_4\textbf{(s)}$$

The reaction is driven to the right by the continuous loss of HCl(g) (Le Châtelier's principle).

What happens when sulfuric acid is added to a water solution of NaCl? Could you tell the difference between this solution and the one formed by adding hydrochloric acid to aqueous sodium sulfate?

21.83   We use X_2 to represent molecular chlorine, bromine, and iodine.

(a)   These halogens combine directly with hydrogen gas as follows:

$$X_2(g) + H_2(g) \rightleftharpoons 2HX(g)$$

The reaction is most energetic for chlorine and least for iodine.

(b)   The silver salts, AgX(s), are all insoluble in water. Any of them can be prepared by precipitation using $AgNO_3(aq)$ and NaX(aq) [see problem 21.81(e)].

(c)   Chlorine, bromine, and iodine are all good oxidizing agents. Their oxidizing power decreases from chlorine to iodine. For example, they oxidize many nonmetallic elements to the corresponding halides:

$$P_4 + 6X_2 \rightarrow 4PX_3$$

(d)   They react with NaOH(aq) to form a halide and a hypohalite:

$$X_2 + 2NaOH(aq) \rightarrow NaX(aq) + NaOX(aq) + H_2O(l)$$

(e)   The differences between fluorine and the other halogens are discussed in Section 21.6 of the text.

21.84   The reaction is: $2Br^-(aq) + Cl_2(g) \rightarrow 2Cl^-(aq) + Br_2(l)$

The number of moles of chlorine needed is:

$$167 \text{ g Br}^- \times \frac{1 \text{ mol Br}^-}{79.90 \text{ g Br}^-} \times \frac{1 \text{ mol Cl}_2}{2 \text{ mol Br}^-} = 1.05 \text{ mol Cl}_2(g)$$

Use the ideal gas equation to calculate the volume of Cl_2 needed.

$$V_{Cl_2} = \frac{nRT}{P} = \frac{(1.05 \text{ mol})(0.0821 \text{ L}\cdot\text{atm/K}\cdot\text{mol})(293 \text{ K})}{(1 \text{ atm})} = \textbf{25.3 L}$$

21.85   The structures are:

$$H-F\cdots H-F \qquad\qquad \left[\, F\cdots H\cdots F \,\right]^-$$

The HF_2^- ion has the strongest known hydrogen bond. More complex hydrogen bonded HF clusters are also known.

21.86   As with iodide salts, a redox reaction occurs between sulfuric acid and sodium bromide.

$$2H_2SO_4(aq) + 2NaBr\,(aq) \rightarrow SO_2(g) + Br_2(l) + 2H_2O(l) + Na_2SO_4(aq)$$

21.87   Fluoride, chloride, bromide, and iodide form complex ions (Sections 16.10 and 22.3 of the text) with many transition metals. With chloride the green $CuCl_4^{2-}$ forms:

$$Cu^{2+}(aq) + 4Cl^-(aq) \rightarrow CuCl_4^{2-}(aq)$$

With fluoride, copper(II) forms an insoluble green salt, CuF_2. Copper(II) cannot oxidize fluoride or chloride.

21.88   The balanced equation is:

$$Cl_2(g) + 2Br^-(aq) \rightarrow 2Cl^-(aq) + Br_2(g)$$

The number of moles of bromine is the same as the number of moles of chlorine, so this problem is essentially a gas law exercise in which P and T are changed for some given amount of gas.

$$V_2 = \frac{P_1 V_1}{T_1} \times \frac{T_2}{P_2} = \frac{(760 \text{ mmHg})(2.00 \text{ L})}{288 \text{ K}} \times \frac{373 \text{ K}}{700 \text{ mmHg}} = \textbf{2.81 L}$$

21.89   **(a)** I_3^- AB_2E_3 **linear**

   (b) $SiCl_4$ AB_4 **tetrahedral**

   (c) PF_5 AB_5 **trigonal bipyramidal**

   (d) SF_4 AB_4E **distorted tetrahedron**

21.90   The balanced equation is:

$$I_2O_5(s) + 5CO(g) \rightarrow I_2(s) + 5CO_2(g)$$

The oxidation number of iodine changes from +5 to 0 and the oxidation number of carbon changes from +2 to +4. **Iodine** is **reduced**; **carbon** is **oxidized**.

21.91   The balanced equations are:

(a) $2H_3PO_3(aq) \rightarrow H_3PO_4(aq) + PH_3(g) + O_2(g)$

(b) $Li_4C(s) + 4HCl(aq) \rightarrow 4LiCl(aq) + CH_4(g)$

(c) $2HI(g) + 2HNO_2(aq) \rightarrow I_2(s) + 2NO(g) + 2H_2O(l)$

(d) $H_2S(g) + 2Cl_2(g) \rightarrow 2HCl(g) + SCl_2(l)$

21.92   (a) $SiCl_4$ (b) F^- (c)   F   (d) CO_2

21.93   (a)   Physical properties: the substances have different molar masses; a molar mass or gas density measurement will distinguish between them. Nitrous oxide is polar, while molecular oxygen is not; a measurement of dipole moment will distinguish between the two. Can you think of other differences? How about magnetic properties?

(b)   Chemical properties: Oxygen reacts with NO to form the red-brown gas, NO_2; nitrous oxide doesn't react with NO. Can you think of other differences? What happens when you breathe nitrous oxide?

21.94   There is no change in oxidation number; it is zero for both compounds.

21.95   The acetylide ion has the electron configuration:

$$(\sigma_{1s})^2(\sigma_{1s}^\star)^2(\sigma_{2s})^2(\sigma_{2s}^\star)^2(\pi_{2p_y})^2(\pi_{2p_z})^2(\sigma_{2p_x})^2$$

21.96   (a) $2Na + 2D_2O \rightarrow 2NaOD + D_2$ (d) $CaC_2 + 2D_2O \rightarrow C_2D_2 + Ca(OD)_2$

(b) $2D_2O \xrightarrow{\text{electrolysis}} 2D_2 + O_2$ (e) $Be_2C + 4D_2O \rightarrow 2Be(OD)_2 + CD_4$

$D_2 + Cl_2 \rightarrow 2DCl$

(c) $Mg_3N_2 + 6D_2O \rightarrow 3Mg(OD)_2 + 2ND_3$ (f) $SO_3 + D_2O \rightarrow D_2SO_4$

21.97   The Lewis structures are shown below. In PCl_4^+, the phosphorus atom is sp^3 hybridized. In PCl_6^-, the phosphorus atom is sp^3d^2.

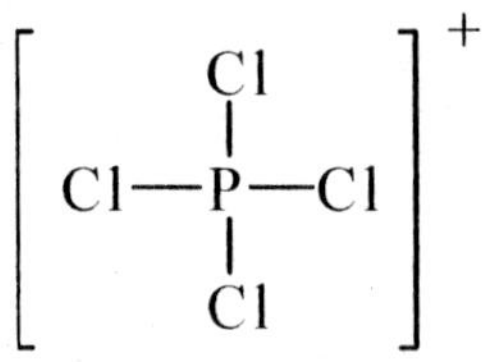

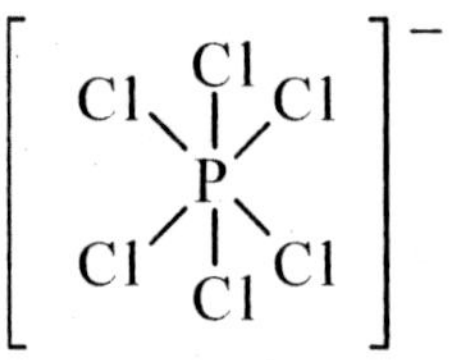

21.98   (a)   At elevated pressures, water boils above 100°C.

(b)   Water is sent down the outermost pipe so that it is able to melt a larger area of sulfur.

(c)   Sulfur deposits are structurally weak. There will be a danger of the sulfur mine collapsing.

21.99   It is a solid, insoluble in water, has metalloid properties. Chemically it resembles F, Cl, Br, and I, but is less reactive.

21.100 The oxidation is probably initiated by breaking a C–H bond (the rate-determining step). The C–D bond breaks at a slower rate than the C–H bond; therefore, replacing H by D decreases the rate of oxidation.

21.101 Light bulbs are frosted with hydrofluoric acid. HF(*aq*) is highly reactive towards silica and silicates.

$$6HF(aq) + SiO_2(s) \rightarrow H_2SiF_6(aq) + 2H_2O(l)$$

This reaction etches the glass, giving it a frosted appearance.

CHAPTER 22
TRANSITION METAL CHEMISTRY AND COORDINATION COMPOUNDS

22.11 **(a)** En is the abbreviation for **ethylenediamine** ($H_2NCH_2CH_2NH_2$).

(b) The oxidation number of Co is **+3**. (Why?)

(c) The coordination number of Co is **six**. (Why isn't this the same as the number of ligands?)

(d) **Ethylenediamine (en)** is a bidentate ligand. Could cyanide ion be a bidentate ligand? Ask your instructor.

22.12 **(a)** The oxidation number of Cr is **+3**.

(b) The coordination number of Cr is **6**.

(c) **Oxalate ion** ($C_2O_4^{2-}$) is a bidentate ligand.

22.13 **(a)** The net charge of the complex ion is the sum of the charges of the ligands and the central metal ion. In this case the complex ion has a −3 charge. (Potassium is always **+1**. Why?) Since the six cyanides are −1 each, the Fe must be **+3**.

(b) The complex ion has a −3 charge. Each oxalate ion has a −2 charge (Table 22.3 of the text). Therefore, the Cr must be **+3**.

(c) Since cyanide ion has a −1 charge, Ni must have a **+2** charge to make the complex ion carry a −2 net charge.

22.14 **Strategy:** The oxidation number of the metal atom is equal to its charge. First we look for known charges in the species. Recall that alkali metals are +1 and alkaline earth metals are +2. Also determine if the ligand is a charged or neutral species. From the known charges, we can deduce the net charge of the metal and hence its oxidation number.

Solution:
(a) Since **sodium** is always +1 and the oxygens are −2, **Mo** must have an oxidation number of **+6**.

(b) **Magnesium** is +2 and oxygen −2; therefore **W** is **+6**.

(c) CO ligands are neutral species, so the iron atom bears no net charge. The oxidation number of **Fe** is **0**.

22.15 **(a)** tetraamminedichlorocobalt(III)   **(c)** dibromobis(ethylenediamine)cobalt(III)

(b) triamminetrichlorochromium(III)   **(d)** hexaamminecobalt(III) chloride

22.16 **Strategy:** We follow the procedure for naming coordination compounds outlined in Section 22.3 of the text and refer to Tables 22.4 and 22.5 of the text for names of ligands and anions containing metal atoms.

Solution:
(a) Ethylenediamine is a neutral ligand, and each chloride has a −1 charge. Therefore, cobalt has a oxidation number of +3. The correct name for the ion is ***cis*−dichlorobis(ethylenediammine)cobalt(III)**. The prefix *bis* means two; we use this instead of *di* because *di* already appears in the name ethylenediamine.

(b)   There are four chlorides each with a −1 charge; therefore, Pt has a +4 charge. The correct name for the compound is **pentaamminechloroplatinum(IV) chloride**.

(c)   There are three chlorides each with a −1 charge; therefore, Co has a +3 charge. The correct name for the compound is **pentaamminechlorocobalt(III) chloride**.

22.17   The formulas are:

(a) $[Zn(OH)_4]^{2-}$ **(b)** $[CrCl(H_2O)_5]Cl_2$ **(c)** $[CuBr_4]^{2-}$ **(d)** $[Fe(EDTA)]^{2-}$

In (b), why two chloride ions at the end of the formula? In (d), does the "(II)" following ferrate refer to the −2 charge of the complex ion or the +2 charge of the iron atom?

22.18   **Strategy:** We follow the procedure in Section 22.3 of the text and refer to Tables 22.4 and 22.5 of the text for names of ligands and anions containing metal atoms.

Solution:

(a)   There are two ethylenediamine ligands and two chloride ligands. The correct formula is $[Cr(en)_2Cl_2]^+$.

(b)   There are five carbonyl (CO) ligands. The correct formula is $Fe(CO)_5$.

(c)   There are four cyanide ligands each with a −1 charge. Therefore, the complex ion has a −2 charge, and two K^+ ions are needed to balance the −2 charge of the anion. The correct formula is $K_2[Cu(CN)_4]$.

(d)   There are four NH_3 ligands and two H_2O ligands. Two chloride ions are needed to balance the +2 charge of the complex ion. The correct formula is $[Co(NH_3)_4(H_2O)Cl]Cl_2$.

22.23   The isomers are:

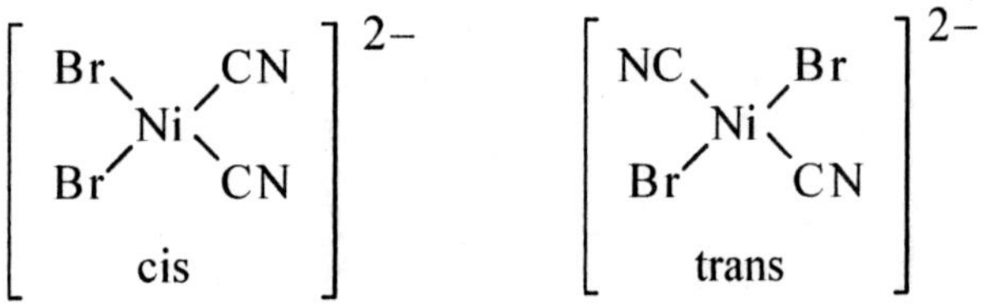

22.24   **(a)**   In general for any MA_2B_4 octahedral molecule, only **two** geometric isomers are possible. The only real distinction is whether the two A−ligands are *cis* or *trans*. In Figure 22.11 of the text, (a) and (c) are the same compound (Cl atoms *cis* in both), and (b) and (d) are identical (Cl atoms *trans* in both).

(b)   A model or a careful drawing is very helpful to understand the MA_3B_3 octahedral structure. There are only **two** possible geometric isomers. The first has all A's (and all B's) *cis*; this is called the facial isomer. The second has two A's (and two B's) at opposite ends of the molecule (*trans*). Try to make or draw other possibilities. What happens?

22.25   **(a)**   All six ligands are identical in this octahedral complex. There are no geometric or optical isomers.

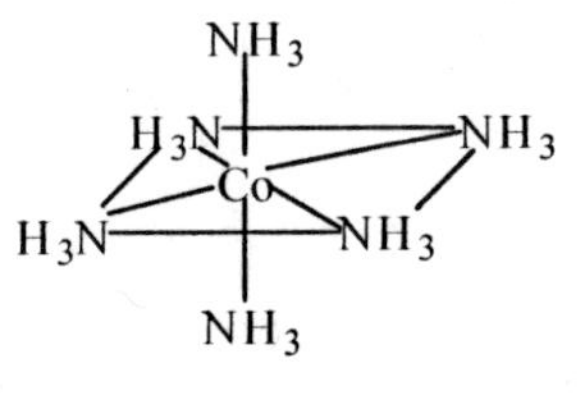

(b)   Again there are no geometric or optical isomers. To have *cis* and *trans* isomers there would have to be two chlorine ligands.

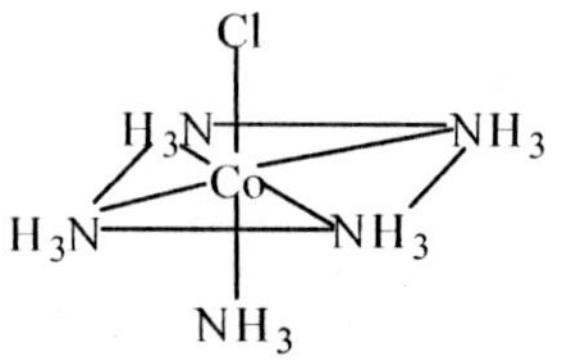

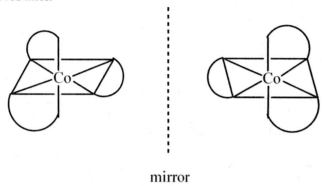

(c)   There are two optical isomers. They are like Figure 22.13 of the text with the two chlorine atoms replaced by one more bidentate ligand. The three bidentate oxalate ligands are represented by the curved lines.

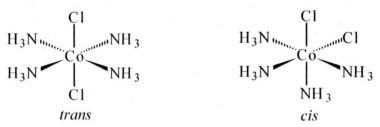

mirror

22.26   (a)   There are *cis* and *trans* geometric isomers (See Problem 22.24). No optical isomers.

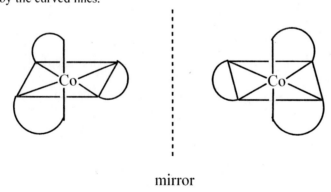

(b)   There are two optical isomers. See Figure 22.7 of the text. The three bidentate en ligands are represented by the curved lines.

mirror

22.33

$$— \quad d_{x^2-y^2}$$

$$\underline{\uparrow\downarrow} \quad d_{xy} \qquad\qquad \underline{\uparrow\downarrow} \quad \underline{\uparrow} \quad \underline{\uparrow}$$
$$\qquad\qquad\qquad\qquad d_{xy} \quad d_{xz} \quad d_{yz}$$

$$\underline{\uparrow\downarrow} \quad d_{z^2} \qquad\qquad \underline{\uparrow\downarrow} \quad \underline{\uparrow\downarrow}$$
$$\qquad\qquad\qquad\qquad d_{x^2-y^2} \quad d_{z^2}$$

$$d_{xz} \quad \underline{\uparrow\downarrow} \quad \underline{\uparrow\downarrow} \quad d_{yz}$$

$$[Ni(CN)_4]^{2-} \qquad\qquad\qquad [NiCl_4]^{2-}$$

22.34 When a substance appears to be yellow, it is absorbing light from the blue-violet, high energy end of the visible spectrum. Often this absorption is just the tail of a strong absorption in the ultraviolet. Substances that appear green or blue to the eye are absorbing light from the lower energy red or orange part of the spectrum.

Cyanide ion is a very strong field ligand. It causes a larger crystal field splitting than water, resulting in the absorption of higher energy (shorter wavelength) radiation when a d electron is excited to a higher energy d orbital.

22.35 **(a)** Each cyanide ligand has a -1 charge, so the oxidation state of Cr is $+2$. The electron configuration of Cr^{2+} is $[Ar]3d^4$. Cyanide is a strong field ligand (see Problem 22.34). The four $3d$ electrons should occupy the three lower orbitals as shown; there should be two unpaired electrons.

$$\overline{\quad} \quad \overline{\quad}$$
$$d_{z^2} \quad d_{x^2-y^2}$$

$$\underline{\uparrow\downarrow} \quad \underline{\uparrow} \quad \underline{\uparrow}$$
$$d_{xy} \quad d_{xz} \quad d_{yz}$$

$$[Cr(CN)_6]^{4-}$$

(b) Water is a weak field ligand. The four $3d$ electrons should occupy the five orbitals as shown. There should be four unpaired electrons.

$$\underline{\uparrow} \quad \overline{\quad}$$
$$d_{z^2} \quad d_{x^2-y^2}$$

$$\underline{\uparrow} \quad \underline{\uparrow} \quad \underline{\uparrow}$$
$$d_{xy} \quad d_{xz} \quad d_{yz}$$

$$[Cr(H_2O)_6]^{2+}$$

22.36 **(a)** Wavelengths of 470 nm fall between blue and blue-green, corresponding to an observed color in the **orange** part of the spectrum.

(b)   We convert wavelength to photon energy using the Planck relationship.

$$\Delta E = \frac{hc}{\lambda} = \frac{(6.63 \times 10^{-34} \text{ J} \cdot \text{s})(3.00 \times 10^{8} \text{ m/s})}{470 \times 10^{-9} \text{ m}} = 4.23 \times 10^{-19} \text{ J}$$

$$\frac{4.23 \times 10^{-19} \text{ J}}{1 \text{ photon}} \times \frac{6.022 \times 10^{23} \text{ photons}}{1 \text{ mol}} \times \frac{1 \text{ kJ}}{1000 \text{ J}} = \textbf{255 kJ/mol}$$

22.37   Recall the wavelength and energy are inversely proportional. Thus, absorption of longer wavelength radiation corresponds to a lower energy transition. Lower energy corresponds to a smaller crystal field splitting.

(a) H_2O is a weaker field ligand than NH_3. Therefore, the crystal field splitting will be smaller for the aquo complex, and it will absorb at longer wavelengths.

(b)   Fluoride is a weaker field ligand than cyanide. The fluoro complex will absorb at longer wavelengths.

(c)   Chloride is a weaker field ligand than NH_3. The chloro complex will absorb at longer wavelengths.

22.38   *Step 1:* The equation for freezing-point depression is

$$\Delta T_f = K_f m$$

Solve this equation algebraically for molality (m), then substitute ΔT_f and K_f into the equation to calculate the molality.

$$m = \frac{\Delta T_f}{K_f} = \frac{0.56°C}{1.86°C/m} = 0.30 \ m$$

Step 2: Multiplying the molality by the mass of solvent (in kg) gives moles of unknown solute. Then, dividing the mass of solute (in g) by the moles of solute, gives the molar mass of the unknown solute.

$$? \text{ mol of unknown solute} = \frac{0.30 \text{ mol solute}}{1 \text{ kg water}} \times 0.0250 \text{ kg water} = 0.0075 \text{ mol solute}$$

$$\text{molar mass of unknown} = \frac{0.875 \text{ g}}{0.0075 \text{ mol}} = 117 \text{ g/mol}$$

The molar mass of $Co(NH_3)_4Cl_3$ is 233.4 g/mol, which is twice the computed molar mass. This implies dissociation into two ions in solution; hence, there are **two moles** of ions produced per one mole of $Co(NH_3)_4Cl_3$. The formula must be:

[Co(NH3)4Cl2]Cl

which contains the complex ion $[Co(NH_3)_4Cl_2]^+$ and a chloride ion, Cl^-. Refer to Problem 22.26 (a) for a diagram of the structure of the complex ion.

22.41   Rust stain removal involves forming a water soluble oxalate ion complex of iron like $[Fe(C_2O_4)_3]^{3-}$. The overall reaction is:

$$Fe_2O_3(s) + 6H_2C_2O_4(aq) \rightarrow 2Fe(C_2O_4)_3^{3-}(aq) + 3H_2O(l) + 6H^+(aq)$$

Does this reaction depend on pH?

22.42   Use a radioactive label such as $^{14}CN^-$ (in NaCN). Add NaCN to a solution of $K_3Fe(CN)_6$. Isolate some of the $K_3Fe(CN)_6$ and check its radioactivity. If the complex shows radioactivity, then it must mean that the CN^- ion has participated in the exchange reaction.

22.43   The green precipitate in CuF_2. When KCl is added, the bright green solution is due to the formation of $CuCl_4^{2-}$:

$$Cu^{2+}(aq) + 4Cl^-(aq) \rightleftharpoons CuCl_4^{2-}(aq)$$

22.44   The white precipitate is copper(II) cyanide.

$$Cu^{2+}(aq) + 2CN^-(aq) \rightarrow Cu(CN)_2(s)$$

This forms a soluble complex with excess cyanide.

$$Cu(CN)_2(s) + 2CN^-(aq) \rightarrow Cu(CN)_4^{2-}(aq)$$

Copper(II) sulfide is normally a very insoluble substance. In the presence of excess cyanide ion, the concentration of the copper(II) ion is so low that CuS precipitation cannot occur. In other words, the cyanide complex of copper has a very large formation constant.

22.45   The overall reaction is:

$$\underset{\text{green}}{CuCl_4^{2-}(aq)} + 6H_2O(l) \rightleftharpoons \underset{\text{blue}}{Cu(H_2O)_6^{2+}} + 4Cl^-(aq)$$

Addition of excess water (dilution) shifts the equilibrium to the right (Le Châtelier's principle).

22.46   The formation constant expression is:

$$K_f = \frac{[Fe(H_2O)_5NCS^{2+}]}{[Fe(H_2O)_6^{3+}][SCN^-]}$$

Notice that the original volumes of the Fe(III) and SCN^- solutions were both 1.0 mL and that the final volume is 10.0 mL. This represents a tenfold dilution, and the concentrations of Fe(III) and SCN^- become 0.020 M and 1.0×10^{-4} M, respectively. We make a table.

	$Fe(H_2O)_6^{3+}$ +	SCN^- $\rightleftharpoons$	$Fe(H_2O)_5NCS^{2+}$ +	H_2O
Initial (M):	0.020	1.0×10^{-4}	0	
Change (M):	-7.3×10^{-5}	-7.3×10^{-5}	$+7.3 \times 10^{-5}$	
Equilibrium (M):	0.020	2.7×10^{-5}	7.3×10^{-5}	

$$K_f = \frac{7.3 \times 10^{-5}}{(0.020)(2.7 \times 10^{-5})} = 1.4 \times 10^2$$

22.47   The third ionization energy increases rapidly from left to right. Thus metals tend to form M^{2+} ions rather than M^{3+} ions.

22.48 Mn^{3+} is $3d^4$ and Cr^{3+} is $3d^5$. Therefore, **Mn^{3+}** has a greater tendency to accept an electron and is a stronger oxidizing agent. The $3d^5$ electron configuration of Cr^{3+} is a stable configuration.

22.49   **(a)**   Since carbon is less electronegative than oxygen, it is more likely that carbon will share electrons with Fe forming a metal-to-ligand sigma bond. The sp orbital on carbon containing the lone pair overlaps with the empty sp^3d^2 orbital on Fe.

   (b)

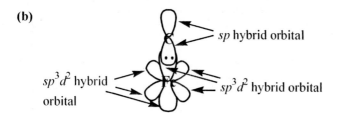

22.50   Ti is +3 and Fe is +3.

22.51   The three cobalt compounds would dissociate as follow:

$$[Co(NH_3)_6]Cl_3(aq) \rightarrow [Co(NH_3)_6]^{3+}(aq) + 3Cl^-(aq)$$

$$[Co(NH_3)_5Cl]Cl_2(aq) \rightarrow [Co(NH_3)_5Cl]^{2+}(aq) + 2Cl^-(aq)$$

$$[Co(NH_3)_4Cl_2]Cl(aq) \rightarrow [Co(NH_3)_4Cl_2]^+(aq) + Cl^-(aq)$$

In other words, the concentration of free ions in the three $1.00\ M$ solutions would be $4.00\ M$, $3.00\ M$, and $2.00\ M$, respectively. If you made up $1.00\ M$ solutions of $FeCl_3$, $MgCl_2$, and $NaCl$, these would serve as reference solutions in which the ion concentrations were $4.00\ M$, $3.00\ M$, and $2.00\ M$, respectively. A $1.00\ M$ solution of $[Co(NH_3)_5Cl]Cl_2$ would have an electrolytic conductivity close to that of the $MgCl_2$ solution, etc.

22.52   A 100.00 g sample of hemoglobin contains 0.34 g of iron. In moles this is:

$$0.34\ \text{g Fe} \times \frac{1\ \text{mol}}{55.85\ \text{g}} = 6.1 \times 10^{-3}\ \text{mol Fe}$$

The amount of hemoglobin that contains one mole of iron must be:

$$\frac{100.00\ \text{g hemoglobin}}{6.1 \times 10^{-3}\ \text{mol Fe}} = \textbf{1.6} \times \textbf{10}^4\ \textbf{g hemoglobin/mol Fe}$$

We compare this to the actual molar mass of hemoglobin:

$$\frac{6.5 \times 10^4\ \text{g hemoglobin}}{1\ \text{mol hemoglobin}} \times \frac{1\ \text{mol Fe}}{1.6 \times 10^4\ \text{g hemoglobin}} = 4\ \text{mol Fe/1 mol hemoglobin}$$

The discrepancy between our minimum value and the actual value can be explained by realizing that one hemoglobin molecule contains **four** iron atoms.

22.53   **(a)**   Zinc(II) has a completely filled $3d$ subshell giving the ion greater stability.

   (b)   Normally the colors of transition metal ions result from transitions within incompletely filled d subshells. The $3d$ subshell of zinc(II) ion is filled.

22.54   **(a)** $[Cr(H_2O)_6]Cl_3$,   **(b)** $[Cr(H_2O)_5Cl]Cl_2 \cdot H_2O$,   **(c)** $[Cr(H_2O)_4Cl_2]Cl \cdot 2H_2O$

The compounds can be identified by a conductance experiment. Compare the conductances of equal molar solutions of the three compounds with equal molar solutions of NaCl, $MgCl_2$, and $FeCl_3$. The solution that has similar conductance to the NaCl solution contains (c); the solution with the conductance similar to $MgCl_2$ contains (b); and the solution with conductance similar to $FeCl_3$ contains (a).

22.55   Reversing the first equation:

$$Ag(NH_3)_2^+(aq) \rightleftharpoons Ag^+(aq) + 2NH_3(aq) \qquad K_1 = \frac{1}{1.5 \times 10^7} = 6.7 \times 10^{-8}$$

$$Ag^+(aq) + 2CN^-(aq) \rightleftharpoons Ag(CN)_2^-(aq) \qquad K_2 = 1.0 \times 10^{21}$$

$$\overline{Ag(NH_3)_2^+(aq) + 2CN^-(aq) \rightleftharpoons Ag(CN)_2^-(aq) + 2NH_3(aq)}$$

$$K = K_1K_2 = (6.7 \times 10^{-8})(1.0 \times 10^{21}) = \mathbf{6.7 \times 10^{13}}$$

$$\Delta G° = -RT\ln K = -(8.314 \text{ J/mol·K})(298 \text{ K})\ln(6.7 \times 10^{13}) = \mathbf{-7.89 \times 10^4 \text{ J/mol}}$$

22.56 $Zn(s) \rightarrow Zn^{2+}(aq) + 2e^-$ $\qquad\qquad E°_{anode} = -0.76 \text{ V}$

$\qquad\quad 2[Cu^{2+}(aq) + e^- \rightarrow Cu^+(aq)]$ $\qquad E°_{cathode} = 0.15 \text{ V}$

$$\overline{Zn(s) + 2Cu^{2+}(aq) \rightarrow Zn^{2+}(aq) + 2Cu^+(aq) \quad E°_{cell} = E°_{cathode} - E°_{anode} = 0.15 \text{ V} - (-0.76 \text{ V}) = 0.91 \text{ V}}$$

$$\Delta G° = -nFE° = -(2)(96500 \text{ J/V·mol})(0.91 \text{ V}) = \mathbf{-1.8 \times 10^5 \text{ J/mol} = -1.8 \times 10^2 \text{ kJ/mol}}$$

$$\Delta G° = -RT\ln K$$

$$\ln K = \frac{-\Delta G°}{RT} = \frac{-(-1.8 \times 10^5 \text{ J/mol})}{(8.314 \text{ J/K} \cdot \text{mol})(298 \text{ K})}$$

$$\ln K = 72.7$$

$$K = e^{72.7} = \mathbf{4 \times 10^{31}}$$

22.57   The half-reactions are: $\qquad Pt^{2+}(aq) + 2e^- \rightarrow Pt(s) \qquad\qquad E°_{cathode} = 1.20 \text{ V}$

$$\qquad\qquad\qquad 2[Ag^+(aq) + e^- \rightarrow Ag(s)] \qquad E°_{anode} = 0.80 \text{ V}$$

$$\overline{\qquad\qquad 2Ag(s) + Pt^{2+}(aq) \rightarrow 2Ag^+(aq) + Pt(s)}$$

$$E°_{cell} = E°_{cathode} - E°_{anode} = 1.20 \text{ V} - 0.80 \text{ V} = \mathbf{0.40 \text{ V}}$$

Since the cell voltage is positive, products are favored at equilibrium.

At 25°C,

$$E°_{cell} = \frac{0.0257 \text{ V}}{n}\ln K$$

$$0.40 \text{ V} = \frac{0.0257 \text{ V}}{2}\ln K$$

$$\ln K = 31.1$$

$$\mathbf{\mathit{K}} = e^{31.1} = \mathbf{3 \times 10^{13}}$$

22.58   Iron is much more abundant than cobalt.

22.59   *Geometric isomers* are compounds with the same type and number of atoms and the same chemical bonds but different spatial arrangements; such isomers cannot be interconverted without breaking a chemical bond. *Optical isomers* are compounds that are nonsuperimposable mirror images.

22.60   Oxyhemoglobin absorbs higher energy light than deoxyhemoglobin. Oxyhemoglobin is diamagnetic (low spin), while deoxyhemoglobin is paramagnetic (high spin). These differences occur because oxygen (O_2) is a strong–field ligand. The crystal field splitting diagrams are:

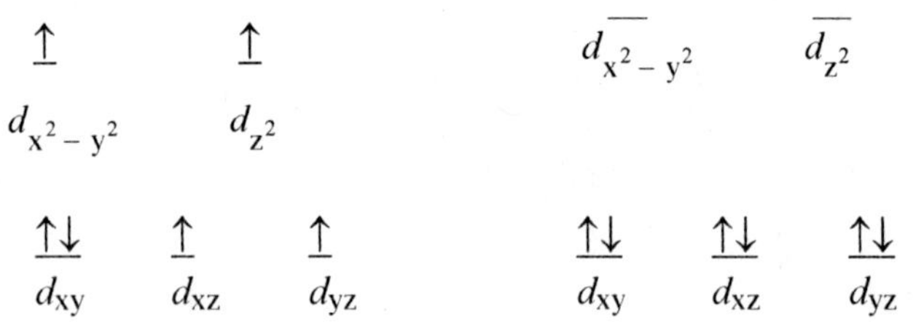

deoxyhemoglobin   oxyhemoglobin

22.61   The orbital splitting diagram below shows five unpaired electrons. The Pauli exclusion principle requires that an electron jumping to a higher energy $3d$ orbital would have to change its spin. A transition that involves a change in spin state is forbidden, and thus does not occur to any appreciable extent.

The colors of transition metal complexes are due to the emission of energy in the form of visible light when an excited d electron relaxes to the ground state. In Mn^{2+} complexes, the promotion of a d electron to an excited state is spin forbidden; therefore, there is little in the way of an emission spectrum and the complexes are practically colorless.

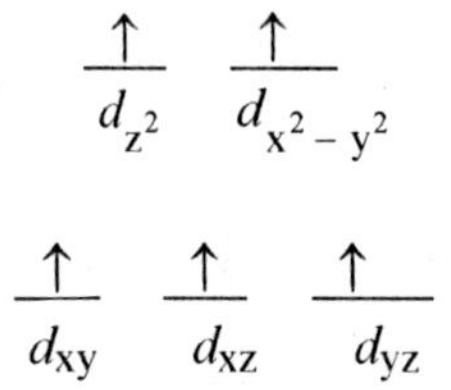

$$\mathbf{Mn(H_2O)_6^{2+}}$$

22.62   Complexes are expected to be colored when the highest occupied orbitals have between one and nine d electrons. Such complexes can therefore have $d \rightarrow d$ transitions (that are usually in the visible part of the electromagnetic radiation spectrum). The ions V^{5+}, Ca^{2+}, and Sc^{3+} have d^0 electron configurations and Cu^+, Zn^{2+}, and Pb^{2+} have d^{10} electron configurations: these complexes are colorless. The other complexes have outer electron configurations of d^1 to d^9 and are therefore colored.

22.63 Co^{2+} exists in solution as $CoCl_4^{2-}$ (blue) or $Co(H_2O)_6^{2+}$ (pink). The equilibrium is:

$$CoCl_4^{2-} + 6H_2O \rightleftharpoons Co(H_2O)_6^{2+} + 4Cl^- \qquad \Delta H° < 0$$
$$\text{blue} \qquad\qquad\qquad \text{pink}$$

Low temperature and low concentration of Cl^- ions favor the formation of $Co(H_2O)_6^{2+}$ ions. Adding HCl (more Cl^- ions) favors the formation of $CoCl_4^{2-}$. Adding $HgCl_2$ leads to:

$$HgCl_2 + 2Cl^- \rightarrow HgCl_4^{2-}$$

This reaction decreases $[Cl^-]$, so the pink color is restored.

22.64   Dipole moment measurement. Only the *cis* isomer has a dipole moment.

22.65 Fe^{2+} is $3d^6$; Fe^{3+} is $3d^5$.

Therefore, Fe^{3+} (like Mn^{2+}, see Problem 22.61) is nearly colorless, so it must be light yellow in color.

22.66   EDTA sequesters metal ions (like Ca^{2+} and Mg^{2+}) which are essential for growth and function, thereby depriving the bacteria to grow and multiply.

22.67   The Be complex exhibits optical isomerism. The Cu complex exhibits geometric isomerism.

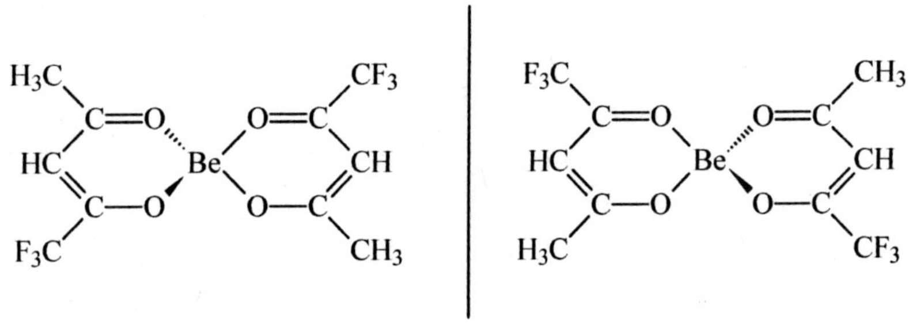

mirror

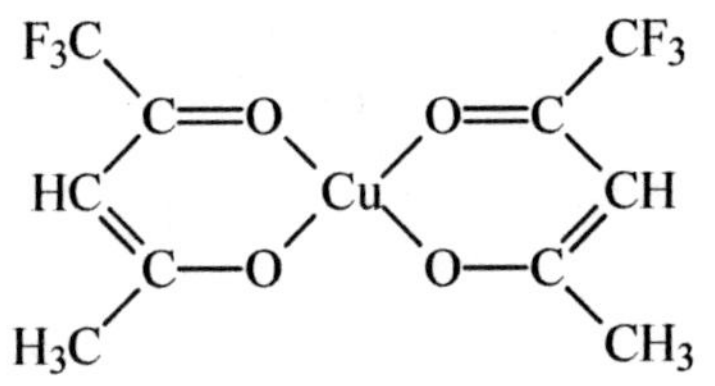

cis 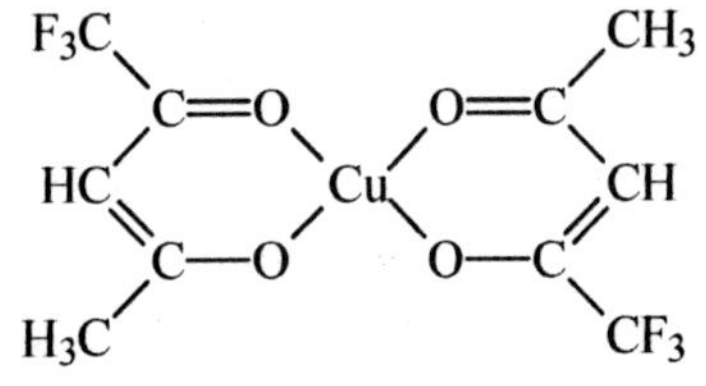   *trans*

22.68   The square planar complex shown in the problem has **3** geometric isomers. They are:

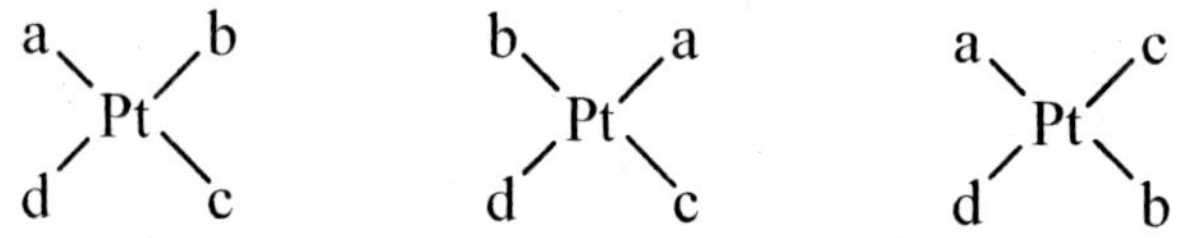

Note that in the first structure a is *trans* to c, in the second a is *trans* to d, and in the third a is *trans* to b. Make sure you realize that if we switch the positions of b and d in structure 1, we do not obtain another geometric isomer. A 180° rotation about the a–Pt–c axis gives structure 1.

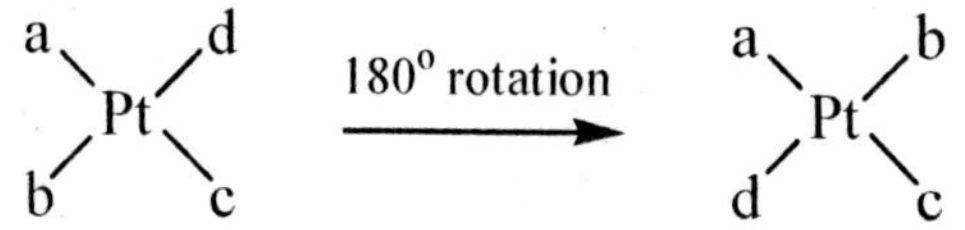

22.69   Isomer I must be the *cis* isomer.

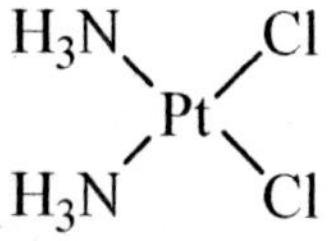

The chlorines must be *cis* to each other for one oxalate ion to complex with Pt.

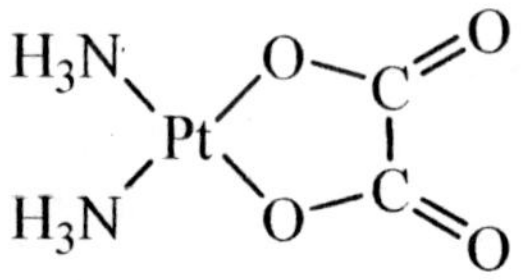

Isomer II must be the *trans* isomer.

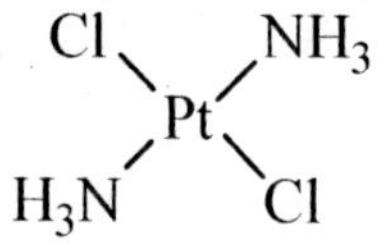

With the chlorines on opposite sides of the molecule, each Cl is replaced with a hydrogen oxalate ion.

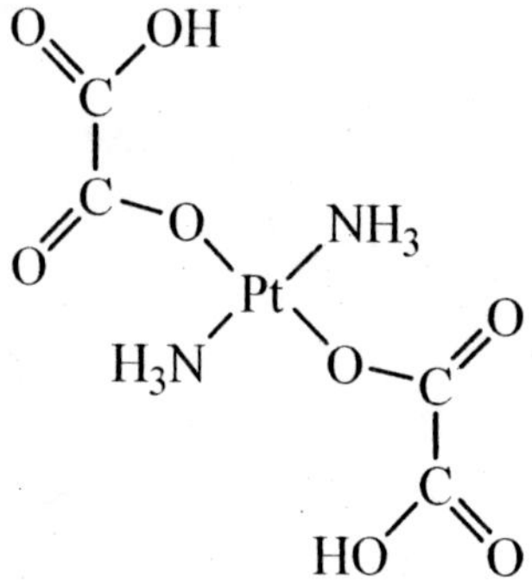

CHAPTER 23
NUCLEAR CHEMISTRY

23.5 **(a)** The atomic number sum and the mass number sum must remain the same on both sides of a nuclear equation. On the left side of this equation the atomic number sum is 13 (12 + 1) and the mass number sum is 27 (26 + 1). These sums must be the same on the right side. The atomic number of X is therefore 11 (13 − 2) and the mass number is 23 (27 − 4). X is sodium–23 ($^{23}_{11}Na$).

(b) X is $^{1}_{1}H$ or $^{1}_{1}p$ **(d)** X is $^{56}_{26}Fe$

(c) X is $^{1}_{0}n$ **(e)** X is $^{0}_{-1}\beta$

23.6 **Strategy:** In balancing nuclear equations, note that the sum of atomic numbers and that of mass numbers must match on both sides of the equation.

Solution:

(a) The sum of the mass numbers must be conserved. Thus, the unknown product will have a mass number of 0. The atomic number must be conserved. Thus, the nuclear charge of the unknown product must be −1. The particle is a β particle.

$$^{135}_{53}I \longrightarrow {}^{135}_{54}Xe + {}^{0}_{-1}\beta$$

(b) Balancing the mass numbers first, we find that the unknown product must have a mass of 40. Balancing the nuclear charges, we find that the atomic number of the unknown must be 20. Element number 20 is calcium (Ca).

$$^{40}_{19}K \longrightarrow {}^{0}_{-1}\beta + {}^{40}_{20}Ca$$

(c) Balancing the mass numbers, we find that the unknown product must have a mass of 4. Balancing the nuclear charges, we find that the nuclear charge of the unknown must be 2. The unknown particle is an alpha (α) particle.

$$^{59}_{27}Co + {}^{1}_{0}n \longrightarrow {}^{56}_{25}Mn + {}^{4}_{2}\alpha$$

(d) Balancing the mass numbers, we find that the unknown products must have a combined mass of 2. Balancing the nuclear charges, we find that the combined nuclear charge of the two unknown particles must be 0. The unknown particles are neutrons.

$$^{235}_{92}U + {}^{1}_{0}n \longrightarrow {}^{99}_{40}Sr + {}^{135}_{52}Te + 2{}^{1}_{0}n$$

23.13 We assume the nucleus to be spherical. The mass is:

$$235 \text{ amu} \times \frac{1 \text{ g}}{6.022 \times 10^{23} \text{ amu}} = 3.90 \times 10^{-22} \text{ g}$$

The volume is, $V = 4/3\pi r^3$.

$$V = \frac{4}{3}\pi \left((7.0 \times 10^{-3} \text{ pm}) \times \frac{1 \text{ cm}}{1 \times 10^{10} \text{ pm}} \right)^3 = 1.4 \times 10^{-36} \text{ cm}^3$$

The density is:

$$\frac{3.90 \times 10^{-22} \text{ g}}{1.4 \times 10^{-36} \text{ cm}^3} = 2.8 \times 10^{14} \text{ g/cm}^3$$

23.14   **Strategy:** The principal factor for determining the stability of a nucleus is the *neutron-to-proton ratio (n/p)*. For stable elements of low atomic number, the *n/p* ratio is close to 1. As the atomic number increases, the *n/p* ratios of stable nuclei become greater than 1. The following rules are useful in predicting nuclear stability.

1)   Nuclei that contain 2, 8, 20, 50, 82, or 126 protons or neutrons are generally more stable than nuclei that do not possess these numbers. These numbers are called *magic numbers*.

2)   Nuclei with even numbers of both protons and neutrons are generally more stable than those with odd numbers of these particles (see Table 23.2 of the text).

Solution:
(a)   **Lithium-9** should be less stable. The neutron-to-proton ratio is too high. For small atoms, the *n/p* ratio will be close to 1:1.

(b)   **Sodium-25** is less stable. Its neutron-to-proton ratio is probably too high.

(c)   **Scandium-48** is less stable because of odd numbers of protons and neutrons. We would not expect calcium-48 to be stable even though it has a magic number of protons. Its *n/p* ratio is too high.

23.15   Nickel, selenium, and cadmium have more stable isotopes. All three have even atomic numbers (see Table 23.2 of the text).

23.16   (a)   **Neon-17** should be radioactive. It falls below the belt of stability (low *n/p* ratio).

(b)   **Calcium-45** should be radioactive. It falls above the belt of stability (high *n/p* ratio).

(c)   All **technetium** isotopes are radioactive.

(d)   **Mercury-195** should be radioactive. Mercury-196 has an even number of both neutrons and protons.

(e)   All **curium** isotopes are unstable.

23.17   The mass change is:

$$\Delta m = \frac{\Delta E}{c^2} = \frac{-436400 \text{ J/mol}}{(3.00 \times 10^8 \text{ m/s})^2} = -4.85 \times 10^{-12} \text{ kg/mol H}_2$$

Is this mass measurable with ordinary laboratory analytical balances?

23.18   We can use the equation, $\Delta E = \Delta mc^2$, to solve the problem. Recall the following conversion factor:

$$1 \text{ J} = \frac{1 \text{ kg} \cdot \text{m}^2}{\text{s}^2}$$

The energy loss in one second is:

$$\Delta m = \frac{\Delta E}{c^2} = \frac{\dfrac{5 \times 10^{26} \text{ kg} \cdot \text{m}^2}{1 \text{ s}^2}}{\left(3.00 \times 10^8 \dfrac{\text{m}}{\text{s}}\right)^2} = 6 \times 10^9 \text{ kg}$$

Therefore the rate of mass loss is 6×10^9 **kg/s.**

23.19   We use the procedure shown is Example 23.2 of the text.

(a)   There are 4 neutrons and 3 protons in a Li–7 nucleus. The predicted mass is:

$$(3)(\text{mass of proton}) + (4)(\text{mass of neutron}) = (3)(1.007825 \text{ amu}) + (4)(1.008668 \text{ amu})$$

$$\text{predicted mass} = 7.058135 \text{ amu}$$

The mass defect, that is the difference between the predicted mass and the measured mass is:

$$\Delta m = 7.01600 \text{ amu} - 7.058135 \text{ amu} = -0.042135 \text{ amu}$$

The mass that is converted in energy, that is the energy released is:

$$\Delta E = \Delta mc^2 = \left(-0.042135 \text{ amu} \times \frac{1 \cdot \text{kg}}{6.022 \times 10^{26} \text{ amu}} \right) \left(3.00 \times 10^8 \text{ m/s} \right)^2 = -6.30 \times 10^{-12} \text{ J}$$

The nuclear binding energy is **6.30×10^{-12} J**. The binding energy per nucleon is:

$$\frac{6.30 \times 10^{-12} \text{ J}}{7 \text{ nucleons}} = 9.00 \times 10^{-13} \text{ J/nucleon}$$

Using the same procedure as in (a), using 1.007825 amu for $_1^1\text{H}$ and 1.008665 amu for $_0^1\text{n}$, we can show that:

(b)   For chlorine–35:   Nuclear binding energy = **4.92×10^{-11} J**

Nuclear binding energy per nucleon = **1.41×10^{-12} J/nucleon**

23.20   **Strategy:** To calculate the nuclear binding energy, we first determine the difference between the mass of the nucleus and the mass of all the protons and neutrons, which gives us the mass defect. Next, we apply Einstein's mass-energy relationship [$\Delta E = (\Delta m)c^2$].

Solution:

(a)   The binding energy is the energy required for the process

$$_2^4\text{He} \rightarrow 2\,_1^1\text{p} + 2\,_0^1\text{n}$$

There are 2 protons and 2 neutrons in the helium nucleus. The mass of 2 protons is

$$(2)(1.007825 \text{ amu}) = 2.015650 \text{ amu}$$

and the mass of 2 neutrons is

$$(2)(1.008665 \text{ amu}) = 2.017330 \text{ amu}$$

Therefore, the predicted mass of $_2^4\text{He}$ is 2.015650 + 2.017330 = 4.032980 amu, and the mass defect is

$$\Delta m = 4.032980 \text{ amu} - 4.0026 \text{ amu} = 0.0304 \text{ amu}$$

The energy change (ΔE) for the process is

$$\Delta E = (\Delta m)c^2$$

$$= (0.0304 \text{ amu})(3.00 \times 10^8 \text{ m/s})^2$$

$$= 2.74 \times 10^{15} \, \frac{\text{amu} \cdot \text{m}^2}{\text{s}^2}$$

Let's convert to more familiar energy units (J/He atom).

$$\frac{2.74 \times 10^{15} \, \text{amu} \cdot \text{m}^2}{1 \, \text{s}^2} \times \frac{1.00 \, \text{g}}{6.022 \times 10^{23} \, \text{amu}} \times \frac{1 \, \text{kg}}{1000 \, \text{g}} \times \frac{1 \, \text{J}}{\frac{1 \, \text{kg} \cdot \text{m}^2}{\text{s}^2}} = \mathbf{4.55 \times 10^{-12} \, J}$$

This is the nuclear binding energy. It's the energy required to break up one helium-4 nucleus into 2 protons and 2 neutrons.

When comparing the stability of any two nuclei we must account for the fact that they have different numbers of nucleons. For this reason, it is more meaningful to use the *nuclear binding energy per nucleon*, defined as

$$\text{nuclear binding energy per nucleon} \; = \; \frac{\text{nuclear binding energy}}{\text{number of nucleons}}$$

For the helium-4 nucleus,

$$\text{nuclear binding energy per nucleon} \; = \; \frac{4.55 \times 10^{-12} \, \text{J/He atom}}{4 \, \text{nucleons/He atom}} = \mathbf{1.14 \times 10^{-12} \, J/nucleon}$$

(b)   The binding energy is the energy required for the process

$$^{184}_{74}\text{W} \; \rightarrow \; 74 \, ^{1}_{1}\text{p} + 110 \, ^{1}_{0}\text{n}$$

There are 74 protons and 110 neutrons in the tungsten nucleus. The mass of 74 protons is

$$(74)(1.007825 \, \text{amu}) \; = \; 74.57905 \, \text{amu}$$

and the mass of 110 neutrons is

$$(110)(1.008665 \, \text{amu}) \; = \; 110.9532 \, \text{amu}$$

Therefore, the predicted mass of $^{184}_{74}\text{W}$ is $74.57905 + 110.9532 = 185.5323$ amu, and the mass defect is

$$\Delta m \; = \; 185.5323 \, \text{amu} - 183.9510 \, \text{amu} \; = \; 1.5813 \, \text{amu}$$

The energy change (ΔE) for the process is

$$\Delta E \; = \; (\Delta m)c^2$$

$$= \; (1.5813 \, \text{amu})(3.00 \times 10^8 \, \text{m/s})^2$$

$$= \; 1.42 \times 10^{17} \, \frac{\text{amu} \cdot \text{m}^2}{\text{s}^2}$$

Let's convert to more familiar energy units (J/W atom).

$$\frac{1.42 \times 10^{17} \, \text{amu} \cdot \text{m}^2}{1 \, \text{s}^2} \times \frac{1.00 \, \text{g}}{6.022 \times 10^{23} \, \text{amu}} \times \frac{1 \, \text{kg}}{1000 \, \text{g}} \times \frac{1 \, \text{J}}{\frac{1 \, \text{kg} \cdot \text{m}^2}{\text{s}^2}} = \mathbf{2.36 \times 10^{-10} \, J}$$

This is the nuclear binding energy. It's the energy required to break up one tungsten-184 nucleus into 74 protons and 110 neutrons.

When comparing the stability of any two nuclei we must account for the fact that they have different numbers of nucleons. For this reason, it is more meaningful to use the *nuclear binding energy per nucleon*, defined as

$$\text{nuclear binding energy per nucleon} = \frac{\text{nuclear binding energy}}{\text{number of nucleons}}$$

For the tungsten-184 nucleus,

$$\text{nuclear binding energy per nucleon} = \frac{2.36 \times 10^{-10} \text{ J/W atom}}{184 \text{ nucleons/W atom}} = \textbf{1.28} \times \textbf{10}^{-12} \textbf{ J/nucleon}$$

23.23 Alpha emission decreases the atomic number by two and the mass number by four. Beta emission increases the atomic number by one and has no effect on the mass number.

(a) $^{232}_{90}\text{Th} \xrightarrow{\alpha} {}^{228}_{88}\text{Ra} \xrightarrow{\beta} {}^{228}_{89}\text{Ac} \xrightarrow{\beta} {}^{228}_{90}\text{Th}$

(b) $^{235}_{92}\text{U} \xrightarrow{\alpha} {}^{231}_{90}\text{Th} \xrightarrow{\beta} {}^{231}_{91}\text{Pa} \xrightarrow{\alpha} {}^{227}_{89}\text{Ac}$

(c) $^{237}_{93}\text{Np} \xrightarrow{\alpha} {}^{233}_{91}\text{Pa} \xrightarrow{\beta} {}^{233}_{92}\text{U} \xrightarrow{\alpha} {}^{229}_{90}\text{Th}$

23.24 **Strategy:** According to Equation (13.3) of the text, the number of radioactive nuclei at time zero (N_0) and time t (N_t) is

$$\ln \frac{N_t}{N_0} = -\lambda t$$

and the corresponding half-life of the reaction is given by Equation (13.5) of the text:

$$t_{\frac{1}{2}} = \frac{0.693}{\lambda}$$

Using the information given in the problem and the first equation above, we can calculate the rate constant, λ. Then, the half-life can be calculated from the rate constant.

Solution: We can use the following equation to calculate the rate constant λ for each point.

$$\ln \frac{N_t}{N_0} = -\lambda t$$

From day 0 to day 1, we have

$$\ln \frac{389}{500} = -\lambda (1 \text{ d})$$

$$\lambda = 0.251 \text{ d}^{-1}$$

Following the same procedure for the other days,

t (d)	mass (g)	λ (d^{-1})
0	500	
1	389	0.251
2	303	0.250
3	236	0.250
4	184	0.250
5	143	0.250
6	112	0.249

The average value of λ is **0.250 d^{-1}**.

We use the average value of λ to calculate the half-life.

$$t_{\frac{1}{2}} = \frac{0.693}{\lambda} = \frac{0.693}{0.250 \text{ d}^{-1}} = \textbf{2.77 d}$$

23.25   The number of atoms decreases by half for each half-life. For ten half-lives we have:

$$(5.00 \times 10^{22} \text{ atoms}) \times \left(\frac{1}{2}\right)^{10} = \textbf{4.89} \times \textbf{10}^{19} \textbf{ atoms}$$

23.26   Since all radioactive decay processes have first–order rate laws, the decay rate is proportional to the amount of radioisotope at any time. The half-life is given by the following equation:

$$t_{\frac{1}{2}} = \frac{0.693}{\lambda} \qquad (1)$$

There is also an equation that relates the number of nuclei at time zero (N_0) and time t (N_t).

$$\ln \frac{N_t}{N_0} = -\lambda t$$

We can use this equation to solve for the rate constant, λ. Then, we can substitute λ into Equation (1) to calculate the half-life.

The time interval is:

$(2{:}15 \text{ p.m., } 12/17/92) - (1{:}00 \text{ p.m., } 12/3/92) = 14 \text{ d} + 1 \text{ hr} + 15 \text{ min} = 20{,}235 \text{ min}$

$$\ln \left(\frac{2.6 \times 10^4 \text{ dis/min}}{9.8 \times 10^5 \text{ dis/min}} \right) = -\lambda(20{,}235 \text{ min})$$

$$\lambda = 1.8 \times 10^{-4} \text{ min}^{-1}$$

Substitute λ into equation (1) to calculate the half-life.

$$t_{\frac{1}{2}} = \frac{0.693}{\lambda} = \frac{0.693}{1.8 \times 10^{-4} \text{ min}^{-1}} = \textbf{3.9} \times \textbf{10}^3 \textbf{ min or 2.7 d}$$

23.27 A truly first-order rate law implies that the mechanism is unimolecular; in other words the rate is determined only by the properties of the decaying atom or molecule and does not depend on collisions or interactions with other objects. This is why radioactive dating is reliable.

23.28 The equation for the overall process is:

$$^{232}_{90}\text{Th} \longrightarrow 6\,^4_2\text{He} + 4\,^0_{-1}\beta + \text{X}$$

The final product isotope must be $^{208}_{82}\text{Pb}$.

23.29 We start with the integrated first-order rate law, Equation (13.3) of the text:

$$\ln\frac{[A]_t}{[A]_0} = -\lambda t$$

We can calculate the rate constant, λ, from the half-life using Equation (13.6) of the text, and then substitute into Equation (13.3) to solve for the time.

$$t_{\frac{1}{2}} = \frac{0.693}{\lambda}$$

$$\lambda = \frac{0.693}{28.1 \text{ yr}} = 0.0247 \text{ yr}^{-1}$$

Substituting:

$$\ln\left(\frac{0.200}{1.00}\right) = -(0.0247 \text{ yr}^{-1})t$$

$$t = \textbf{65.2 yr}$$

23.30 Let's consider the decay of A first.

$$\lambda = \frac{0.693}{t_{\frac{1}{2}}} = \frac{0.693}{4.50 \text{ s}} = 0.154 \text{ s}^{-1}$$

Let's convert λ to units of day^{-1}.

$$0.154\frac{1}{s} \times \frac{3600 \text{ s}}{1 \text{ h}} \times \frac{24 \text{ h}}{1 \text{ d}} = 1.33 \times 10^4 \text{ d}^{-1}$$

Next, use the first-order rate equation to calculate the amount of A left after 30 days.

$$\ln\frac{[A]_t}{[A]_0} = -\lambda t$$

Let x be the amount of A left after 30 days.

$$\ln\frac{x}{100} = -(1.33 \times 10^4 \text{ d}^{-1})(30 \text{ d}) = -3.99 \times 10^5$$

$$\frac{x}{100} = e^{(-3.99 \times 10^5)}$$

$$x \approx 0$$

Thus, **no A remains**.

For B: As calculated above, all of A is converted to B in less than 30 days. In fact, essentially all of A is gone in less than 1 day! This means that at the beginning of the 30 day period, there is 1.00 mol of B present. The half life of B is 15 days, so that after two half-lives (30 days), there should be **0.25 mole of B** left.

For C: As in the case of A, the half-life of C is also very short. Therefore, at the end of the 30-day period, **no C is left**.

For D: D is not radioactive. 0.75 mol of B reacted in 30 days; therefore, due to a 1:1 mole ratio between B and D, there should be **0.75 mole of D** present after 30 days.

23.33   In the shorthand notation for nuclear reactions, the first symbol inside the parentheses is the "bombarding" particle (reactant) and the second symbol is the "ejected" particle (product).

(a) $^{15}_{7}N + ^{1}_{1}p \rightarrow ^{12}_{6}C + ^{4}_{2}\alpha$ X is $^{15}_{7}N$

(b) $^{27}_{13}Al + ^{2}_{1}d \rightarrow ^{25}_{12}Mg + ^{4}_{2}\alpha$ X is $^{25}_{12}Mg$

(c) $^{55}_{25}Mn + ^{1}_{0}n \rightarrow ^{56}_{25}Mn + \gamma$ X is $^{56}_{25}Mn$

23.34   (a) $^{80}_{34}Se + ^{2}_{1}H \longrightarrow ^{81}_{34}Se + ^{1}_{1}p$

(b) $^{9}_{4}Be + ^{2}_{1}H \longrightarrow ^{9}_{3}Li + 2^{1}_{1}p$

(c) $^{10}_{5}B + ^{1}_{0}n \longrightarrow ^{7}_{3}Li + ^{4}_{2}\alpha$

23.35   All you need is a high-intensity alpha particle emitter. Any heavy element like plutonium or curium will do. Place the bismuth-209 sample next to the alpha emitter and wait. The reaction is:

$$^{209}_{83}Bi + ^{4}_{2}\alpha \rightarrow ^{213}_{85}At \rightarrow ^{212}_{85}At + ^{1}_{0}n \rightarrow ^{211}_{85}At + ^{1}_{0}n$$

23.36   Upon bombardment with neutrons, mercury-198 is first converted to mercury-199, which then emits a proton. The reaction is:

$$^{198}_{80}Hg + ^{1}_{0}n \longrightarrow ^{199}_{80}Hg \longrightarrow ^{198}_{79}Au + ^{1}_{1}p$$

23.47   The easiest experiment would be to add a small amount of aqueous iodide containing some radioactive iodine to a saturated solution of lead(II) iodide. If the equilibrium is dynamic, radioactive iodine will eventually be detected in the solid lead(II) iodide.

Could this technique be used to investigate the forward and reverse rates of this reaction?

23.48   The fact that the radioisotope appears only in the I_2 shows that the IO_3^- is formed only from the IO_4^-. Does this result rule out the possibility that I_2 could be formed from IO_4^- as well? Can you suggest an experiment to answer the question?

23.49   On paper, this is a simple experiment. If one were to dope part of a crystal with a radioactive tracer, one could demonstrate diffusion in the solid state by detecting the tracer in a different part of the crystal at a later time. This actually happens with many substances. In fact, in some compounds one type of ion migrates easily while the other remains in fixed position!

23.50   Add iron-59 to the person's diet, and allow a few days for the iron–59 isotope to be incorporated into the person's body. Isolate red blood cells from a blood sample and monitor radioactivity from the hemoglobin molecules present in the red blood cells.

23.51   The design and operation of a Geiger counter are discussed in Figure 23.18 of the text.

23.52   Apparently there is a sort of Pauli exclusion principle for nucleons as well as for electrons. When neutrons pair with neutrons and when protons pair with protons, their spins cancel. Even–even nuclei are the only ones with no net spin.

23.53   **(a)**   The balanced equation is:

$$^{3}_{1}\text{H} \rightarrow {}^{3}_{2}\text{He} + {}^{0}_{-1}\beta$$

   (b)   The number of tritium (T) atoms in 1.00 kg of water is:

$$(1.00 \times 10^{3} \text{ g H}_2\text{O}) \times \frac{1 \text{ mol H}_2\text{O}}{18.02 \text{ g H}_2\text{O}} \times \frac{6.022 \times 10^{23} \text{ molecules H}_2\text{O}}{1 \text{ mol H}_2\text{O}} \times \frac{2 \text{ H atoms}}{1 \text{ H}_2\text{O}} \times \frac{1 \text{ T atom}}{1.0 \times 10^{17} \text{ H atoms}}$$

$$= 6.68 \times 10^{8} \text{ T atoms}$$

The number of disintegrations per minute will be:

$$\text{rate} = \lambda(\text{number of T atoms}) = \lambda N = \frac{0.693}{t_{\frac{1}{2}}} N$$

$$\text{rate} = \left(\frac{0.693}{12.5 \text{ yr}} \times \frac{1 \text{ yr}}{365 \text{ day}} \times \frac{1 \text{ day}}{24 \text{ h}} \times \frac{1 \text{ h}}{60 \text{ min}} \right) \left(6.68 \times 10^{8} \text{ T atoms} \right)$$

rate = 70.5 T atoms/min = 70.5 disintegrations/min

23.54   **(a)**   One millicurie represents 3.70×10^{7} disintegrations/s. The rate of decay of the isotope is given by the rate law: rate = λN, where N is the number of atoms in the sample. We find the value of λ in units of s^{-1}:

$$\lambda = \frac{0.693}{t_{\frac{1}{2}}} = \frac{0.693}{2.20 \times 10^{6} \text{ yr}} \times \frac{1 \text{ yr}}{365 \text{ d}} \times \frac{1 \text{ d}}{24 \text{ h}} \times \frac{1 \text{ h}}{3600 \text{ s}} = 9.99 \times 10^{-15} \text{ s}^{-1}$$

The number of atoms (N) in a 0.500 g sample of neptunium–237 is:

$$0.500 \text{ g} \times \frac{1 \text{ mol}}{237.0 \text{ g}} \times \frac{6.022 \times 10^{23} \text{ atoms}}{1 \text{ mol}} = 1.27 \times 10^{21} \text{ atoms}$$

rate of decay $= \lambda N$

$$= (9.99 \times 10^{-15} \text{ s}^{-1})(1.27 \times 10^{21} \text{ atoms}) = 1.27 \times 10^{7} \text{ atoms/s}$$

We can also say that:

$$\text{rate of decay} = 1.27 \times 10^7 \text{ disintegrations/s}$$

The activity in millicuries is:

$$(1.27 \times 10^7 \text{ disintegrations/s}) \times \frac{1 \text{ millicurie}}{3.70 \times 10^7 \text{ disintegrations/s}} = \textbf{0.343 millicuries}$$

(b)   The decay equation is:

$$^{237}_{93}\text{Np} \longrightarrow \, ^{4}_{2}\alpha + \, ^{233}_{91}\text{Pa}$$

23.55   **(a)** $^{235}_{92}\text{U} + \, ^{1}_{0}\text{n} \rightarrow \, ^{140}_{56}\text{Ba} + 3\,^{1}_{0}\text{n} + \, ^{93}_{36}\text{Kr}$ **(c)** $^{235}_{92}\text{U} + \, ^{1}_{0}\text{n} \rightarrow \, ^{87}_{35}\text{Br} + \, ^{146}_{57}\text{La} + 3\,^{1}_{0}\text{n}$

    (b) $^{235}_{92}\text{U} + \, ^{1}_{0}\text{n} \rightarrow \, ^{144}_{55}\text{Cs} + \, ^{90}_{37}\text{Rb} + 2\,^{1}_{0}\text{n}$ **(d)** $^{235}_{92}\text{U} + \, ^{1}_{0}\text{n} \rightarrow \, ^{160}_{62}\text{Sm} + \, ^{72}_{30}\text{Zn} + 4\,^{1}_{0}\text{n}$

23.56   We use the same procedure as in Problem 23.20.

	Isotope	Atomic Mass (amu)	Nuclear Binding Energy (J/nucleon)
(a)	^{10}B	10.0129	1.040×10^{-12}
(b)	^{11}B	11.00931	1.111×10^{-12}
(c)	^{14}N	14.00307	1.199×10^{-12}
(d)	^{56}Fe	55.9349	1.410×10^{-12}

23.57   The balanced nuclear equations are:

    (a) $^{3}_{1}\text{H} \rightarrow \, ^{3}_{2}\text{He} + \, ^{0}_{-1}\beta$ **(c)** $^{131}_{53}\text{I} \rightarrow \, ^{131}_{54}\text{Xe} + \, ^{0}_{-1}\beta$

    (b) $^{242}_{94}\text{Pu} \rightarrow \, ^{4}_{2}\alpha + \, ^{238}_{92}\text{U}$ **(d)** $^{251}_{98}\text{Cf} \rightarrow \, ^{247}_{96}\text{Cm} + \, ^{4}_{2}\alpha$

23.58   When an isotope is above the belt of stability, the neutron/proton ratio is too high. The only mechanism to correct this situation is beta emission; the process turns a neutron into a proton. Direct neutron emission does not occur.

$$^{18}_{7}\text{N} \longrightarrow \, ^{18}_{8}\text{O} + \, ^{0}_{-1}\beta$$

Oxygen–18 is a stable isotope.

23.59   Because both Ca and Sr belong to Group 2A, radioactive strontium that has been ingested into the human body becomes concentrated in bones (replacing Ca) and can damage blood cell production.

23.60   The age of the fossil can be determined by radioactively dating the age of the deposit that contains the fossil.

23.61   Normally the human body concentrates iodine in the thyroid gland. The purpose of the large doses of KI is to displace radioactive iodine from the thyroid and allow its excretion from the body.

23.62 **(a)** $^{209}_{83}\text{Bi} + ^{4}_{2}\alpha \longrightarrow ^{211}_{85}\text{At} + 2\,^{1}_{0}\text{n}$

(b) $^{209}_{83}\text{Bi}(\alpha, 2\text{n})^{211}_{85}\text{At}$

23.63 **(a)** The nuclear equation is: $^{14}_{7}\text{N} + ^{1}_{0}\text{n} \rightarrow ^{15}_{7}\text{N} + \gamma$

(b) X-ray analysis only detects shapes, particularly of metal objects. Bombs can be made in a variety of shapes and sizes and can be constructed of "plastic" explosives. Thermal neutron analysis is much more specific than X-ray analysis. However, articles that are high in nitrogen other than explosives (such as silk, wool, and polyurethane) will give "false positive" test results.

23.64 Because of the relative masses, the force of gravity on the sun is much greater than it is on Earth. Thus the nuclear particles on the sun are already held much closer together than the equivalent nuclear particles on the earth. Less energy (lower temperature) is required on the sun to force fusion collisions between the nuclear particles.

23.65 The neutron-to-proton ratio for tritium equals 2 and is thus outside the belt of stability. In a more elaborate analysis, it can be shown that the decay of tritium to ^{3}He is exothermic; thus, the total energy of the products is less than the reactant.

23.66 *Step 1:* The half-life of carbon-14 is 5730 years. From the half-life, we can calculate the rate constant, λ.

$$\lambda = \frac{0.693}{t_{\frac{1}{2}}} = \frac{0.693}{5730 \text{ yr}} = 1.21 \times 10^{-4} \text{ yr}^{-1}$$

Step 2: The age of the object can now be calculated using the following equation.

$$\ln \frac{N_t}{N_0} = -\lambda t$$

N = the number of radioactive nuclei. In the problem, we are given disintegrations per second per gram. The number of disintegrations is directly proportional to the number of radioactive nuclei. We can write,

$$\ln \frac{\text{decay rate of old sample}}{\text{decay rate of fresh sample}} = -\lambda t$$

$$\ln \frac{0.186 \text{ dps/g C}}{0.260 \text{ dps/g C}} = -(1.21 \times 10^{-4} \text{ yr}^{-1})t$$

$$t = 2.77 \times 10^3 \text{ yr}$$

23.67 $\ln \dfrac{N_t}{N_0} = -\lambda t$

$$\ln \frac{\text{mass of fresh sample}}{\text{mass of old sample}} = -(1.21 \times 10^{-4} \text{ yr}^{-1})(50000 \text{ yr})$$

$$\ln \frac{1.0 \text{ g}}{x \text{ g}} = -6.05$$

$$\frac{1.0}{x} = e^{-6.05}$$

$$x = 424$$

Percent of C-14 left $= \dfrac{1.0}{424} \times 100\% = \mathbf{0.24\%}$

23.68   **(a)**   The balanced equation is:

$$\,^{40}_{19}\text{K} \longrightarrow \,^{40}_{18}\text{Ar} + \,^{0}_{+1}\beta$$

(b)   First, calculate the rate constant λ.

$$\lambda = \frac{0.693}{t_{\frac{1}{2}}} = \frac{0.693}{1.2 \times 10^{9}\ \text{yr}} = 5.8 \times 10^{-10}\ \text{yr}^{-1}$$

Then, calculate the age of the rock by substituting λ into the following equation. ($N_t = 0.18 N_0$)

$$\ln \frac{N_t}{N_0} = -\lambda t$$

$$\ln \frac{0.18}{1.00} = -(5.8 \times 10^{-10}\ \text{yr}^{-1})t$$

$$t = \mathbf{3.0 \times 10^{9}\ yr}$$

23.69   All isotopes of radium are radioactive; therefore, radium is not naturally occurring and would not be found with barium. However, radium is a decay product of uranium–238, so it is found in uranium ores.

23.70   **(a)**   In the ^{90}Sr decay, the mass defect is:

$$\Delta m = (\text{mass } ^{90}\text{Y} + \text{mass } e^{-}) - \text{mass } ^{90}\text{Sr}$$

$$= [(89.907152\ \text{amu} + 5.4857 \times 10^{-4}\ \text{amu}) - 89.907738\ \text{amu}] = -3.743 \times 10^{-5}\ \text{amu}$$

$$= (-3.743 \times 10^{-5}\ \text{amu}) \times \frac{1\ \text{g}}{6.022 \times 10^{23}\ \text{amu}} = -6.216 \times 10^{-29}\ \text{g} = -6.216 \times 10^{-32}\ \text{kg}$$

The energy change is given by:

$$\Delta E = (\Delta m)c^{2}$$

$$= (-6.126 \times 10^{-32}\ \text{kg})(3.00 \times 10^{8}\ \text{m/s})^{2}$$

$$= -5.59 \times 10^{-15}\ \text{kg m}^{2}/\text{s}^{2} = -5.59 \times 10^{-15}\ \text{J}$$

Similarly, for the ^{90}Y decay, we have

$$\Delta m = (\text{mass } ^{90}\text{Zr} + \text{mass } e^{-}) - \text{mass } ^{90}\text{Y}$$

$$= [(89.904703\ \text{amu} + 5.4857 \times 10^{-4}\ \text{amu}) - 89.907152\ \text{amu}] = -1.900 \times 10^{-3}\ \text{amu}$$

$$= (-1.900 \times 10^{-3} \text{ amu}) \times \frac{1 \text{ g}}{6.022 \times 10^{23} \text{ amu}} = -3.156 \times 10^{-27} \text{ g} = -3.156 \times 10^{-30} \text{ kg}$$

and the energy change is:

$$\Delta E = (-3.156 \times 10^{-30} \text{ kg})(3.00 \times 10^{8} \text{ m/s})^2 = -2.84 \times 10^{-13} \text{ J}$$

The energy released in the above two decays is **5.59×10^{-15} J** and **2.84×10^{-13} J**. The total amount of energy released is:

$$(5.59 \times 10^{-15} \text{ J}) + (2.84 \times 10^{-13} \text{ J}) = 2.90 \times 10^{-13} \text{ J}.$$

(b) This calculation requires that we know the rate constant for the decay. From the half-life, we can calculate λ.

$$\lambda = \frac{0.693}{t_{\frac{1}{2}}} = \frac{0.693}{28.1 \text{ yr}} = 0.0247 \text{ yr}^{-1}$$

To calculate the number of moles of ^{90}Sr decaying in a year, we apply the following equation:

$$\ln \frac{N_t}{N_0} = -\lambda t$$

$$\ln \frac{x}{1.00} = -(0.0247 \text{ yr}^{-1})(1.00 \text{ yr})$$

where x is the number of moles of ^{90}Sr nuclei left over. Solving, we obtain:

$$x = 0.976 \text{ mol } ^{90}\text{Sr}$$

Thus the number of moles of nuclei which decay in a year is

$$(1.00 - 0.976) \text{ mol} = \mathbf{0.024 \text{ mol}}$$

This is a reasonable number since it takes 28.1 years for 0.5 mole of ^{90}Sr to decay.

(c) Since the half–life of ^{90}Y is much shorter than that of ^{90}Sr, we can safely assume that *all* the ^{90}Y formed from ^{90}Sr will be converted to ^{90}Zr. The energy changes calculated in part (a) refer to the decay of individual nuclei. In 0.024 mole, the number of nuclei that have decayed is:

$$0.024 \text{ mol} \times \frac{6.022 \times 10^{23} \text{ nuclei}}{1 \text{ mol}} = 1.4 \times 10^{22} \text{ nuclei}$$

Realize that there are two decay processes occurring, so we need to add the energy released for each process calculated in part (a). Thus, the heat released from 1 mole of ^{90}Sr waste in a year is given by:

$$\textbf{heat released} = (1.4 \times 10^{22} \text{ nuclei}) \times \frac{2.90 \times 10^{-13} \text{ J}}{1 \text{ nucleus}} = \mathbf{4.06 \times 10^{9} \text{ J}} = \mathbf{4.06 \times 10^{6} \text{ kJ}}$$

This amount is roughly equivalent to the heat generated by burning 50 tons of coal! Although the heat is released slowly during the course of a year, effective ways must be devised to prevent heat damage to the storage containers and subsequent leakage of radioactive material to the surroundings.

23.71   A radioactive isotope with a shorter half-life because more radiation would be emitted over a certain period of time.

23.72   First, let's calculate the number of disintegrations/s to which 7.4 mC corresponds.

$$7.4 \text{ mC} \times \frac{1 \text{ Ci}}{1000 \text{ mC}} \times \frac{3.7 \times 10^{10} \text{ disintegrations/s}}{1 \text{ Ci}} = 2.7 \times 10^8 \text{ disintegrations/s}$$

This is the rate of decay. We can now calculate the number of iodine-131 atoms to which this radioactivity corresponds. First, we calculate the half-life in seconds:

$$t_{\frac{1}{2}} = 8.1 \text{ d} \times \frac{24 \text{ h}}{1 \text{ d}} \times \frac{3600 \text{ s}}{1 \text{ h}} = 7.0 \times 10^5 \text{ s}$$

$$\lambda = \frac{0.693}{t_{\frac{1}{2}}} \qquad \text{Therefore, } \lambda = \frac{0.693}{7.0 \times 10^5 \text{ s}} = 9.9 \times 10^{-7} \text{ s}^{-1}$$

$$\text{rate} = \lambda N$$

$$2.7 \times 10^8 \text{ disintegrations/s} = (9.9 \times 10^{-7} \text{ s}^{-1})N$$

$$N = \textbf{2.7} \times \textbf{10}^{\textbf{14}} \textbf{ iodine-131 atoms}$$

23.73   The energy of irradiation is not sufficient to bring about nuclear transmutation.

23.74   One curie represents 3.70×10^{10} disintegrations/s. The rate of decay of the isotope is given by the rate law: rate $= \lambda N$, where N is the number of atoms in the sample and λ is the first-order rate constant. We find the value of λ in units of s^{-1}:

$$\lambda = \frac{0.693}{t_{\frac{1}{2}}} = \frac{0.693}{1.6 \times 10^3 \text{ yr}} = 4.3 \times 10^{-4} \text{ yr}^{-1}$$

$$\frac{4.3 \times 10^{-4}}{1 \text{ yr}} \times \frac{1 \text{ yr}}{365 \text{ d}} \times \frac{1 \text{ d}}{24 \text{ h}} \times \frac{1 \text{ h}}{3600 \text{ s}} = 1.4 \times 10^{-11} \text{ s}^{-1}$$

Now, we can calculate N, the number of Ra atoms in the sample.

$$\text{rate} = \lambda N$$

$$3.7 \times 10^{10} \text{ disintegrations/s} = (1.4 \times 10^{-11} \text{ s}^{-1})N$$

$$N = 2.6 \times 10^{21} \text{ Ra atoms}$$

By definition, 1 curie corresponds to exactly 3.7×10^{10} nuclear disintegrations per second which is the decay rate equivalent to that of *1 g of radium*. Thus, the mass of 2.6×10^{21} Ra atoms is 1 g.

$$\frac{2.6 \times 10^{21} \text{ Ra atoms}}{1.0 \text{ g Ra}} \times \frac{226.03 \text{ g Ra}}{1 \text{ mol Ra}} = \textbf{5.9} \times \textbf{10}^{\textbf{23}} \textbf{ atoms/mol} = N_A$$

23.75 $^{208}_{82}\text{Pb} + {}^{62}_{28}\text{Ni} \rightarrow {}^{270}_{110}\text{W}$

W and X are transition metals.

$^{209}_{83}\text{Bi} + {}^{64}_{28}\text{Ni} \rightarrow {}^{273}_{111}\text{X}$

$$^{208}_{82}\text{Pb} + {}^{66}_{30}\text{Zn} \rightarrow {}^{274}_{112}\text{Y} \qquad\qquad \text{Y resembles Zn, Cd, and Hg.}$$

$$^{244}_{94}\text{Pu} + {}^{48}_{20}\text{Ca} \rightarrow {}^{289}_{114}\text{Z} + 3\,{}^{1}_{0}\text{n} \qquad\qquad \text{Z is in the carbon family.}$$

23.76 All except gravitational have a nuclear origin.

23.77 There was radioactive material inside the box.

23.78 U–238, $t_{\frac{1}{2}} = 4.5 \times 10^{9}$ yr and Th–232, $t_{\frac{1}{2}} = 1.4 \times 10^{10}$ yr.

They are still present because of their long half lives.

23.79 **(a)** $^{238}_{92}\text{U} \rightarrow {}^{234}_{90}\text{Th} + {}^{4}_{2}\alpha$

$$\Delta m = 234.0436 + 4.0026 - 238.0508 = -0.0046 \text{ amu}$$

$$\Delta E = \Delta mc^2 = (-0.0046 \text{ amu})(3.00 \times 10^8 \text{ m/s})^2 = -4.14 \times 10^{14} \text{ amu}^2/\text{s}^2$$

$$\Delta E = \frac{-4.14 \times 10^{14} \text{ amu} \cdot \text{m}^2}{1 \text{ s}^2} \times \frac{1.00 \text{ kg}}{6.022 \times 10^{26} \text{ amu}} \times \frac{1 \text{ J}}{1 \text{ kg} \cdot \text{m}^2/\text{s}^2} = -6.87 \times 10^{-13} \text{ J}$$

(b) The smaller particle (α) will move away at a greater speed due to its lighter mass.

23.80 $E = \dfrac{hc}{\lambda}$

$$\lambda = \frac{hc}{E} = \frac{(3.00 \times 10^8 \text{ m/s})(6.63 \times 10^{-34} \text{ J} \cdot \text{s})}{2.4 \times 10^{-13} \text{ J}} = 8.3 \times 10^{-13} \text{ m} = 8.3 \times 10^{-4} \text{ nm}$$

This wavelength is clearly in the γ-ray region of the electromagnetic spectrum.

23.81 The α particles emitted by ^{241}Am ionize the air molecules between the plates. The voltage from the battery makes one plate positive and the other negative, so each plate attracts ions of opposite charge. This creates a current in the circuit attached to the plates. The presence of smoke particles between the plates reduces the current, because the ions that collide with smoke particles (or steam) are usually absorbed (and neutralized) by the particles. This drop in current triggers the alarm.

23.82 Only ^{3}H has a suitable half-life. The other half-lives are either too long or too short to accurately determine the time span of 6 years.

23.83 **(a)** The nuclear submarine can be submerged for a long period without refueling.

(b) Conventional diesel engines receive an input of oxygen. A nuclear reactor does not.

23.84 Obviously, a small scale chain reaction took place. Copper played the crucial role of reflecting neutrons from the splitting uranium-235 atoms back into the uranium sphere to trigger the chain reaction. Note that a sphere has the most appropriate geometry for such a chain reaction. In fact, during the implosion process prior to an atomic explosion, fragments of uranium-235 are pressed roughly into a sphere for the chain reaction to occur (see Section 23.5 of the text).

23.85 From the half-life, we can determine the rate constant, λ. Next, using the first-order integrated rate law, we can calculate the amount of copper remaining. Finally, from the initial amount of Cu and the amount remaining, we can calculate the amount of Zn produced.

$$t_{\frac{1}{2}} = \frac{0.693}{\lambda}$$

$$\lambda = \frac{0.693}{t_{\frac{1}{2}}} = \frac{0.693}{12.8 \text{ h}} = 0.0541 \text{ h}^{-1}$$

Next, plug the amount of copper, the time, and the rate constant into the first-order integrated rate law, to calculate the amount of copper remaining.

$$\ln \frac{N_t}{N_0} = -\lambda t$$

$$\ln \frac{\text{grams Cu remaining}}{84.0 \text{ g}} = -(0.0541 \text{ h}^{-1})(18.4 \text{ h})$$

$$\frac{\text{grams Cu remaining}}{84.0 \text{ g}} = e^{-(0.0541 \text{ h}^{-1})(18.4 \text{ h})}$$

$$\text{grams Cu remaining} = 31.0 \text{ g}$$

The quantity of Zn produced is:

g Zn $=$ initial g Cu $-$ g Cu remaining $=$ 84.0 g $-$ 31.0 g $=$ **53.0 g Zn**

23.86 In this problem, we are asked to calculate the molar mass of a radioactive isotope. Grams of sample are given in the problem, so if we can find moles of sample we can calculate the molar mass. The rate constant can be calculated from the half-life. Then, from the rate of decay and the rate constant, the number of radioactive nuclei can be calculated. The number of radioactive nuclei can be converted to moles.

First, we convert the half-life to units of minutes because the rate is given in dpm (disintegrations per minute). Then, we calculate the rate constant from the half-life.

$$(1.3 \times 10^9 \text{ yr}) \times \frac{365 \text{ days}}{1 \text{ yr}} \times \frac{24 \text{ h}}{1 \text{ day}} \times \frac{60 \text{ min}}{1 \text{ h}} = 6.8 \times 10^{14} \text{ min}$$

$$\lambda = \frac{0.693}{t_{\frac{1}{2}}} = \frac{0.693}{6.8 \times 10^{14} \text{ min}} = 1.0 \times 10^{-15} \text{ min}^{-1}$$

Next, we calculate the number of radioactive nuclei from the rate and the rate constant.

$$\text{rate} = \lambda N$$

$$2.9 \times 10^4 \text{ dpm} = (1.0 \times 10^{-15} \text{ min}^{-1}) N$$

$$N = 2.9 \times 10^{19} \text{ nuclei}$$

Convert to moles of nuclei, then determine the molar mass.

$$(2.9 \times 10^{19} \text{ nuclei}) \times \frac{1 \text{ mol}}{6.022 \times 10^{23} \text{ nuclei}} = 4.8 \times 10^{-5} \text{ mol}$$

$$\textbf{molar mass} = \frac{\text{g of substance}}{\text{mol of substance}} = \frac{0.0100 \text{ g}}{4.8 \times 10^{-5} \text{ mol}} = \textbf{2.1} \times \textbf{10}^2 \textbf{ g/mol}$$

CHAPTER 24
ORGANIC CHEMISTRY

24.11 The structures are as follows:

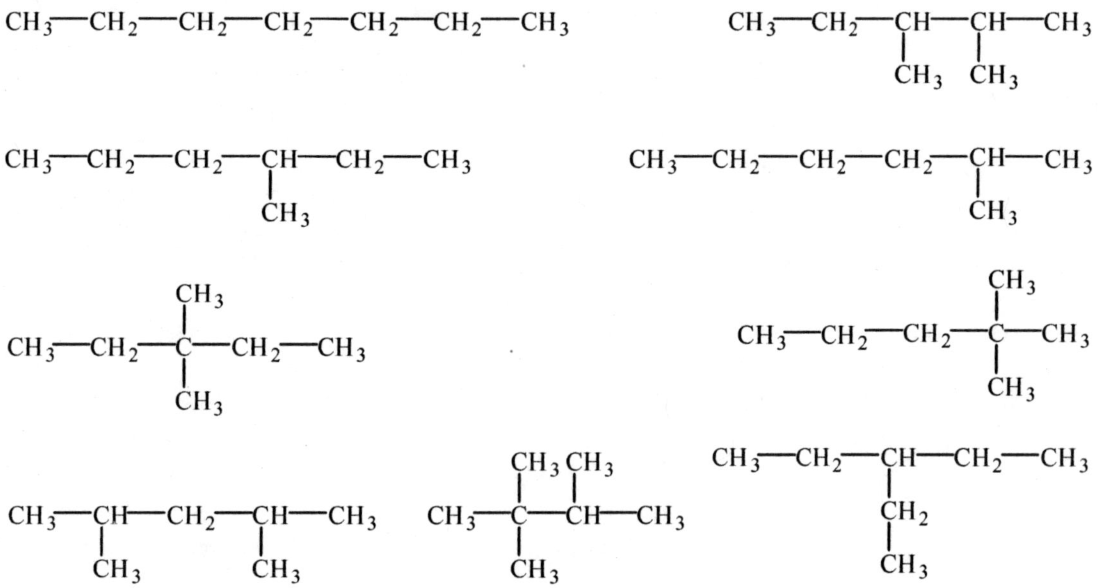

$$CH_3—CH_2—CH_2—CH_2—CH_2—CH_2—CH_3$$

$$CH_3—CH_2—CH—CH—CH_3$$
$$\qquad\qquad\quad | \quad\; |$$
$$\qquad\qquad\; CH_3 \; CH_3$$

$$CH_3—CH_2—CH_2—CH—CH_2—CH_3$$
$$\qquad\qquad\qquad\qquad |$$
$$\qquad\qquad\qquad\quad CH_3$$

$$CH_3—CH_2—CH_2—CH_2—CH—CH_3$$
$$\qquad\qquad\qquad\qquad\qquad\quad |$$
$$\qquad\qquad\qquad\qquad\qquad CH_3$$

$$CH_3$$
$$\quad |$$
$$CH_3—CH_2—C—CH_2—CH_3$$
$$\qquad\qquad\; |$$
$$\qquad\qquad CH_3$$

$$CH_3$$
$$\qquad\qquad\qquad\qquad\quad |$$
$$CH_3—CH_2—CH_2—C—CH_3$$
$$\qquad\qquad\qquad\qquad\quad |$$
$$\qquad\qquad\qquad\qquad\; CH_3$$

$$CH_3—CH—CH_2—CH—CH_3$$
$$\qquad\; | \qquad\qquad\; |$$
$$\quad CH_3 \qquad\;\; CH_3$$

$$CH_3 \; CH_3$$
$$\quad | \quad\; |$$
$$CH_3—C—CH—CH_3$$
$$\quad\; |$$
$$\quad CH_3$$

$$CH_3—CH_2—CH—CH_2—CH_3$$
$$\qquad\qquad\qquad |$$
$$\qquad\qquad\quad CH_2$$
$$\qquad\qquad\qquad |$$
$$\qquad\qquad\quad CH_3$$

24.12 **Strategy:** For small hydrocarbon molecules (eight or fewer carbons), it is relatively easy to determine the number of structural isomers by trial and error.

Solution: We are starting with *n*-pentane, so we do not need to worry about any branched chain structures. In the chlorination reaction, a Cl atom replaces one H atom. There are three different carbons on which the Cl atom can be placed. Hence, *three* structural isomers of chloropentane can be derived from *n*–pentane:

$$CH_3CH_2CH_2CH_2CH_2Cl \qquad\qquad CH_3CH_2CH_2CHClCH_3 \qquad\qquad CH_3CH_2CHClCH_2CH_3$$

24.13 The molecular formula shows the compound is either an alkene or a cycloalkane. (Why?) You can't tell which from the formula. The possible isomers are:

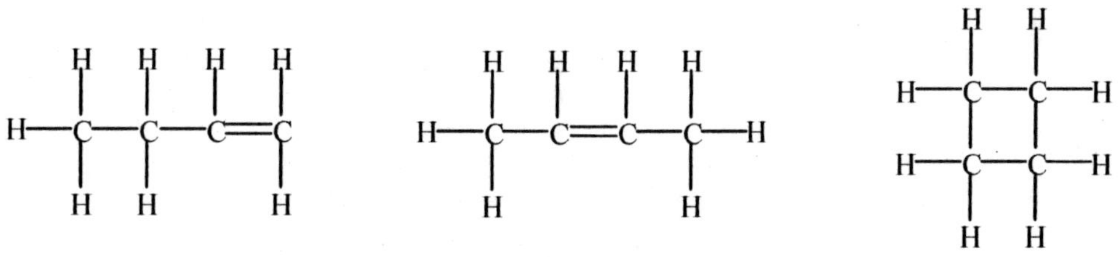

The structure in the middle (2–butene) can exist as *cis* or *trans* isomers. There are two more isomers. Can you find and draw them? Can you have an isomer with a double bond and a ring? What would the molecular formula be like in that case?

24.14   Both alkenes and cycloalkanes have the general formula C_nH_{2n}. Let's start with C_3H_6. It could be an alkene or a cycloalkane.

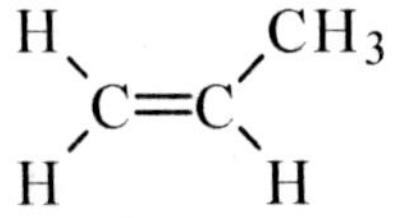

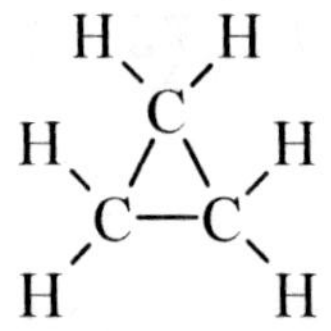

Now, let's replace one H with a Br atom to form C_3H_5Br. *Four* isomers are possible.

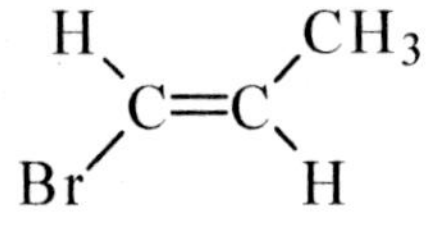

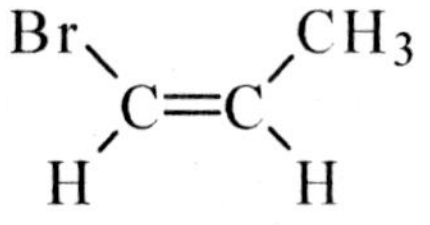

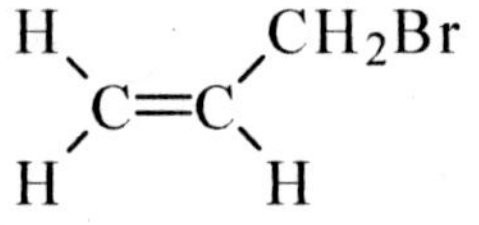

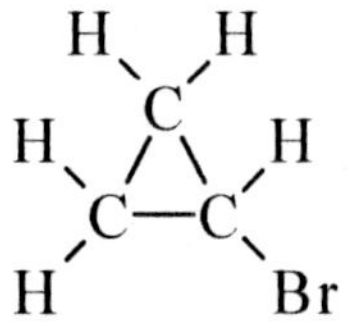

There is only one isomer for the cycloalkane. Note that all three carbons are equivalent in this structure.

24.15   The straight chain molecules have the highest boiling points and therefore the strongest intermolecular attractions. Theses chains can pack together more closely and efficiently than highly branched, cluster structures. This allows intermolecular forces to operate more effectively and cause stronger attractions.

24.16   **(a)**   This compound could be an **alkene** or a **cycloalkane**; both have the general formula, C_nH_{2n}.

(b)   This could be an **alkyne** with general formula, C_nH_{2n-2}. It could also be a hydrocarbon with two double bonds (a diene). It could be a cyclic hydrocarbon with one double bond (a cycloalkene).

(c)   This must be an **alkane**; the formula is of the C_nH_{2n+2} type.

(d)   This compound could be an **alkene** or a **cycloalkane**; both have the general formula, C_nH_{2n}.

(e)   This compound could be an **alkyne** with one triple bond, or it could be a cyclic alkene (unlikely because of ring strain).

24.17   The two isomers are:

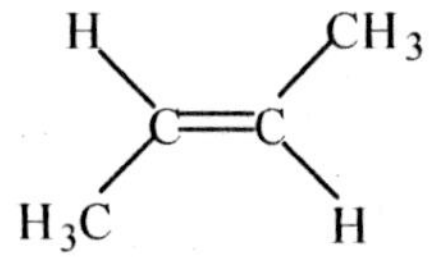

trans 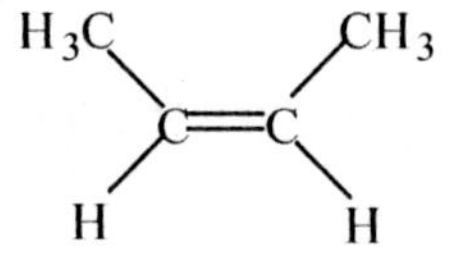   cis

A simplified method of presenting the structures is:

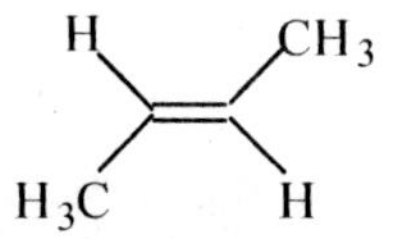

trans 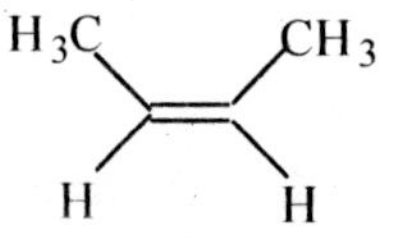   cis

The *cis* structure is more crowded and a little less stable. As a result, slightly more heat (energy) would be released when the alkene adds a molecule of hydrogen to form butane, C_4H_{10}. Note that butane is the product when either alkene is hydrogenated.

24.18   If cyclobutadiene were square or rectangular, the C–C–C angles must be 90°. If the molecule is diamond-shaped, two of the C–C–C angles must be less than 90°. Both of these situations result in a great deal of distortion and strain in the molecule. Cyclobutadiene is very unstable for these and other reasons.

24.19

cis-chlorofluoroethylene   trans-chlorofluoroethylene   1,1-chlorofluoroethylene

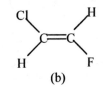

(a) and (b) are geometric isomers.

(c) is a structural isomer of both (a) and (b).

24.20   One compound is an alkane; the other is an alkene. Alkenes characteristically undergo addition reactions with hydrogen, with halogens (Cl_2, Br_2, I_2) and with hydrogen halides (HCl, HBr, HI). Alkanes do not react with these substances under ordinary conditions.

24.21   **(a)**   Ethylene is symmetrical; there is no preference in the addition.

$$CH_3-CH_2-OSO_3H$$

(b)   The positive part of the polar reagent adds to the carbon atom that already has the most hydrogen atoms.

$$\overset{\displaystyle OSO_3H}{\underset{}{\overset{|}{CH_3-CH-CH_3}}}$$

24.22   In this problem you are asked to calculate the standard enthalpy of reaction. This type of problem was covered in Chapter 6.

$$\Delta H^\circ_{rxn} = \Sigma n \Delta H^\circ_f (\text{products}) - \Sigma m \Delta H^\circ_f (\text{reactants})$$

$$\Delta H^\circ_{rxn} = \Delta H^\circ_f (C_6H_6) - 3\Delta H^\circ_f (C_2H_2)$$

You can look up ΔH°_f values in Appendix 3 of your textbook.

$$\Delta H^\circ_{rxn} = (1)(49.04 \text{ kJ/mol}) - (3)(226.6 \text{ kJ/mol}) = -630.8 \text{ kJ/mol}$$

24.23   **(a)** $CH_2=CH-CH_2-CH_3 + HBr \rightarrow CH_3-CHBr-CH_2-CH_3$

(b) $CH_3-CH=CH-CH_3 + HBr \rightarrow CH_3-CH_2-CHBr-CH_3$

(a) and (b) are the same.

24.24   In this problem you must distinguish between *cis* and *trans* isomers. Recall that *cis* means that two particular atoms (or groups of atoms) are adjacent to each other, and *trans* means that the atoms (or groups of atoms) are on opposite sides in the structural formula.

In (a), the Cl atoms are adjacent to each other. This is the *cis* isomer. In (b), the Cl atoms are on opposite sides of the structure. This is the *trans* isomer.

The names are: **(a)** *cis*-**1,2-dichlorocyclopropane**; and **(b)** *trans*-**1,2-dichlorocyclopropane**.

Are any other dichlorocyclopropane isomers possible?

24.25   **(a)** and **(c)**

24.26   **(a)**   This is a branched hydrocarbon. The name is based on the longest carbon chain. The name is **2–methylpentane**.

(b)   This is also a branched hydrocarbon. The longest chain includes the C_2H_5 group; the name is based on hexane, not pentane. This is an old trick. Carbon chains are flexible and don't have to lie in a straight line. The name is **2,3,4–trimethylhexane**. Why not 3,4,5–trimethylhexane?

(c)   How many carbons in the longest chain? It doesn't have to be straight! The name is **3–ethylhexane**.

(d)   An alkene with two double bonds is called a diene. The name is **3–methyl–1,4–pentadiene**.

(e)   The name is **2–pentyne**.

(f)   The name is **3–phenyl–1–pentene**.

24.27   **(a)**   This is a six-carbon chain with a methyl group on the third carbon.

$$CH_3-CH_2-\underset{\underset{\displaystyle CH_3}{|}}{CH}-CH_2-CH_2-CH_3$$

(b)   This is a six carbon ring with chlorine atoms on the 1,3, and 5 carbons.

Note: The carbon atoms in the ring have been omitted for simplicity.

(c)   This is a five carbon chain with methyl groups on the 2 and 3 carbons.

$$CH_3-\underset{\underset{\displaystyle CH_3}{|}}{CH}-\underset{\underset{\displaystyle CH_3}{|}}{CH}-CH_2-CH_3$$

(d)   This is a five carbon chain with a bromine atom on the second carbon and a phenyl group (a benzene molecule minus a hydrogen atom, C_6H_5) on the fourth carbon.

$$CH_3-CH-CH_2-CHBr-CH_3$$

(e)   This is an eight carbon chain with methyl groups on the 3, 4, and 5 carbons.

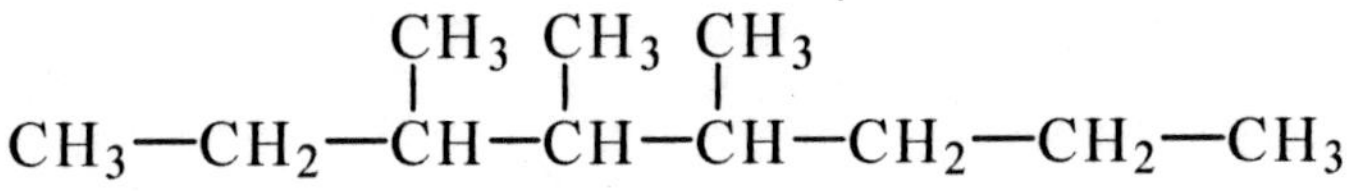

$$\underset{\overset{\displaystyle |}{\text{CH}_3}}{}\quad\underset{\overset{\displaystyle |}{\text{CH}_3}}{}\quad\underset{\overset{\displaystyle |}{\text{CH}_3}}{}$$

CH$_3$—CH$_2$—CH—CH—CH—CH$_2$—CH$_2$—CH$_3$

24.28   The hydrogen atoms have been omitted from the skeletal structure for simplicity.

(a)

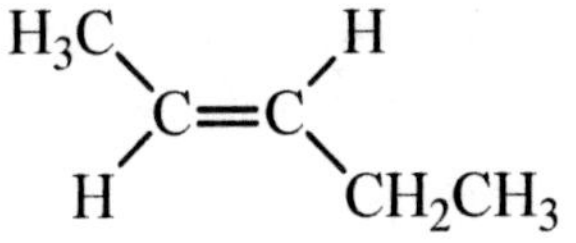

(b)

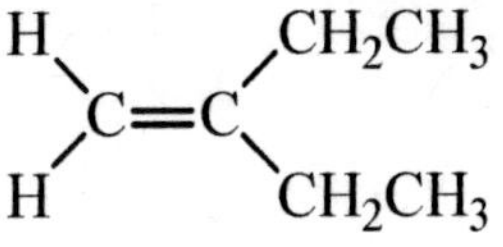

(c)

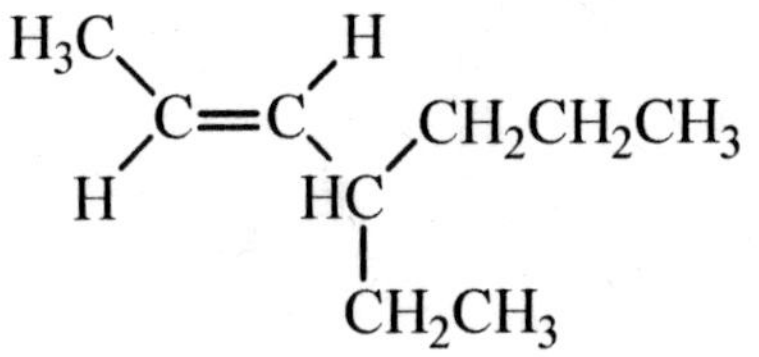

(d)

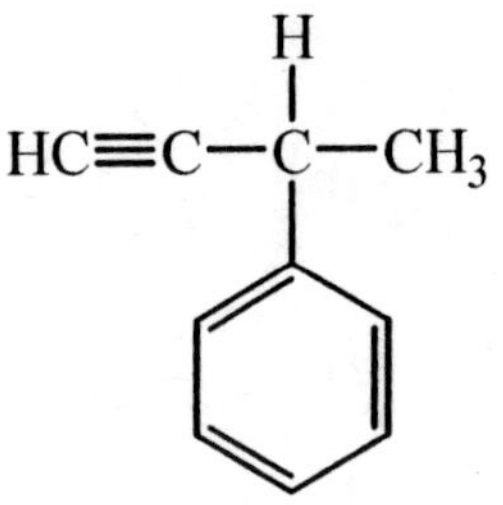

24.31

(a)

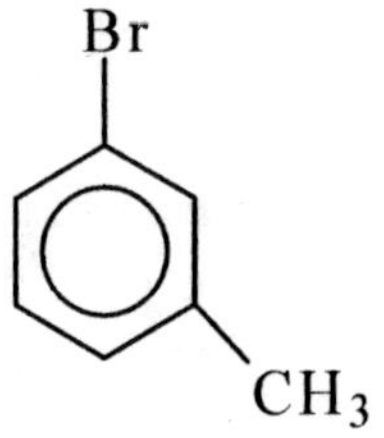

(b)

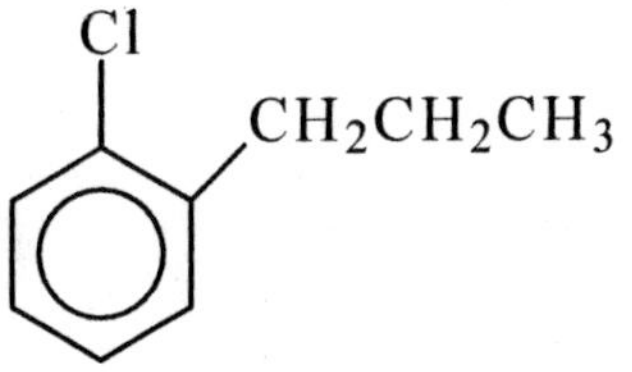

(c)

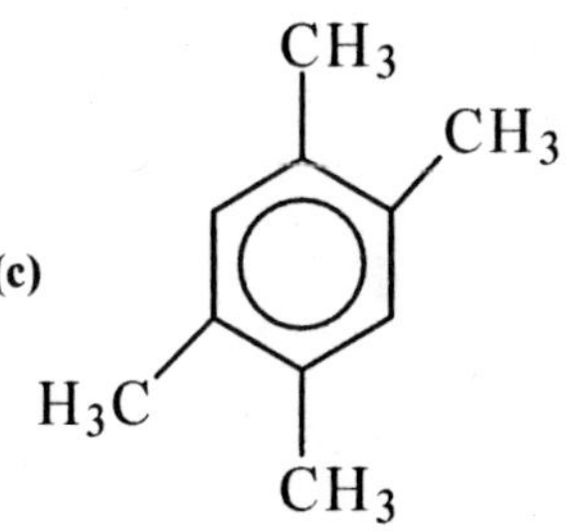

24.32   **Strategy:** We follow the IUPAC rules and use the information in Table 24.2 of the text. When a benzene ring has more than *two* substituents, you must specify the location of the substituents with numbers. Remember to number the ring so that you end up with the lowest numbering scheme as possible, giving preference to alphabetical order.

Solution:
(a)   Since a chloro group comes alphabetically before a methyl group, let's start by numbering the top carbon of the ring as 1. If we number clockwise, this places the second chloro group on carbon 3 and a methyl group on carbon 4.

This compound is **1,3–dichloro–4–methylbenzene**.

(b)   If we start numbering counterclockwise from the bottom carbon of the ring, the name is 2–ethyl–1,4–dinitrobenzene. Numbering clockwise from the top carbon gives 3–ethyl–1,4–dinitrobenzene.

Numbering as low as possible, the correct name is **2–ethyl–1,4–dinitrobenzene**.

(c)   Again, keeping the numbers as low as possible, the correct name for this compound is **1,2,4,5–tetramethylbenzene**. You should number clockwise from the top carbon of the ring.

24.35   **(a)**   There is only one isomer: CH_3OH

(b)   There are two structures with this molecular formula:

CH_3-CH_2-OH and CH_3-O-CH_3

(c)   The cyclic di-alcohol has geometric isomers.

$$CH_3CH_2-\overset{\overset{\displaystyle O}{\|}}{C}-OH \qquad CH_3-\overset{\overset{\displaystyle O}{\|}}{C}-O-CH_3 \qquad CH_3-\overset{\overset{\displaystyle O}{\|}}{C}-\overset{\overset{\displaystyle O}{\|}}{C}-CH_3 \qquad H-\overset{\overset{\displaystyle O}{\|}}{C}-CH_2CH_2-\overset{\overset{\displaystyle O}{\|}}{C}-H$$

(d)   There are two possible alcohols and one ether.

$$CH_3CH_2CH_2OH \qquad CH_3\underset{\underset{\displaystyle OH}{|}}{C}HCH_3 \qquad CH_3CH_2-O-CH_3$$

24.36   **Strategy:** Learning to recognize functional groups requires memorization of their structural formulas. Table 24.4 of the text shows a number of the important functional groups.

Solution:

(a) $H_3C-O-CH_2-CH_3$ contains a C–O–C group and is therefore an **ether**.

(b)   This molecule contains an RNH_2 group and is therefore an **amine**.

(c)   This molecule is an **aldehyde**. It contains a carbonyl group in which one of the atoms bonded to the carbonyl carbon is a hydrogen atom.

(d)   This molecule also contains a carbonyl group. However, in this case there are no hydrogen atoms bonded to the carbonyl carbon. This molecule is a **ketone**.

(e)   This molecule contains a carboxyl group. It is a **carboxylic acid**.

(f)   This molecule contains a hydroxyl group (–OH). It is an **alcohol**.

(g)   This molecule has both an RNH_2 group and a carboxyl group. It is therefore both an *amine* and a *carboxylic acid*, commonly called an **amino acid**.

24.37   Aldehydes can be oxidized easily to carboxylic acids. The oxidation reaction is:

$$CH_3-\overset{\overset{\displaystyle O}{\|}}{C}-H \xrightarrow{O_2} CH_3-\overset{\overset{\displaystyle O}{\|}}{C}-OH$$

Oxidation of a ketone requires that the carbon chain be broken:

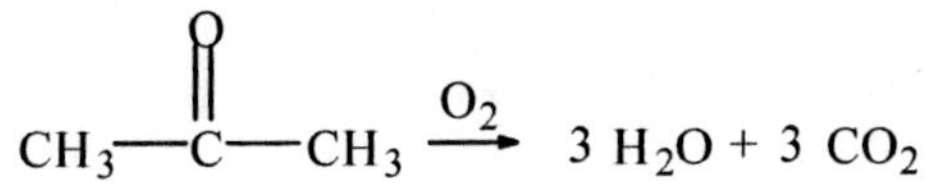

$$CH_3-\overset{\displaystyle O}{\overset{\|}{C}}-CH_3 \xrightarrow{\;O_2\;} 3\ H_2O + 3\ CO_2$$

24.38   Alcohols react with carboxylic acids to form esters. The reaction is:

$$HCOOH + CH_3OH \longrightarrow HCOOCH_3 + H_2O$$

The structure of the product is:

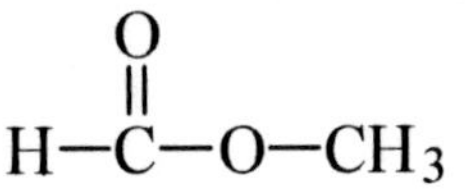

$$H-\overset{\displaystyle O}{\overset{\|}{C}}-O-CH_3 \qquad \text{(methyl formate)}$$

24.39   Alcohols can be oxidized to ketones under controlled conditions. The possible starting compounds are:

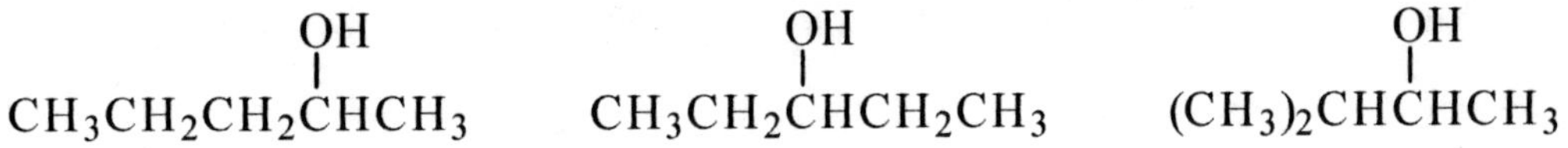

$$\overset{\displaystyle OH}{\overset{|}{CH_3CH_2CH_2CHCH_3}} \qquad \overset{\displaystyle OH}{\overset{|}{CH_3CH_2CHCH_2CH_3}} \qquad \overset{\displaystyle OH}{\overset{|}{(CH_3)_2CHCHCH_3}}$$

The corresponding products are:

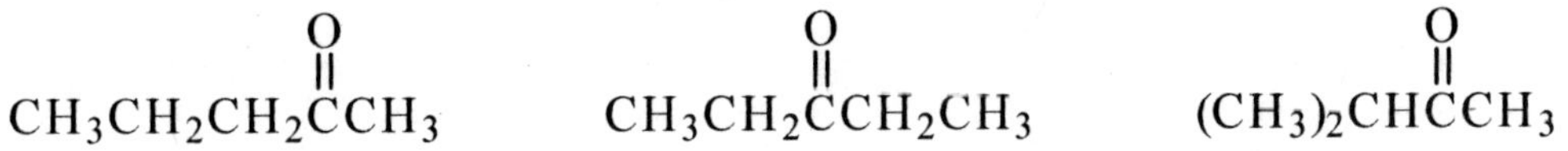

$$\overset{\displaystyle O}{\overset{\|}{CH_3CH_2CH_2CCH_3}} \qquad \overset{\displaystyle O}{\overset{\|}{CH_3CH_2CCH_2CH_3}} \qquad \overset{\displaystyle O}{\overset{\|}{(CH_3)_2CHCCH_3}}$$

Why isn't the alcohol $CH_3CH_2CH_2CH_2CH_2OH$ a possible starting compound?

24.40   The fact that the compound does not react with sodium metal eliminates the possibility that the substance is an alcohol. The only other possibility is the ether functional group. There are three ethers possible with this molecular formula:

$$CH_3-CH_2-O-CH_2-CH_3 \qquad CH_3-CH_2-CH_2-O-CH_3 \qquad (CH_3)_2CH-O-CH_3$$

Light–induced reaction with chlorine results in substitution of a chlorine atom for a hydrogen atom (the other product is HCl). For the first ether there are only two possible chloro derivatives:

$$ClCH_2-CH_2-O-CH_2-CH_3 \qquad\qquad CH_3-CHCl-O-CH_2-CH_3$$

For the second there are four possible chloro derivatives. Three are shown below. Can you draw the fourth?

$$CH_3-CHCl-CH_2-O-CH_3 \qquad CH_3-CH_2-CHCl-O-CH_3 \qquad CH_2Cl-CH_2-CH_2-O-CH_3$$

For the third there are three possible chloro derivatives:

$$\overset{\displaystyle CH_2Cl}{\overset{|}{CH_3-\!\!-\!\!CH-\!\!-O-\!\!-CH_3}} \qquad \overset{\displaystyle CH_3}{\overset{|}{CH_3-\!\!-\!\!CH-\!\!-O-\!\!-CH_2Cl}} \qquad \overset{\displaystyle Cl}{\overset{|}{(CH_3)_2-\!\!-\!\!CH-\!\!-O-\!\!-CH_3}}$$

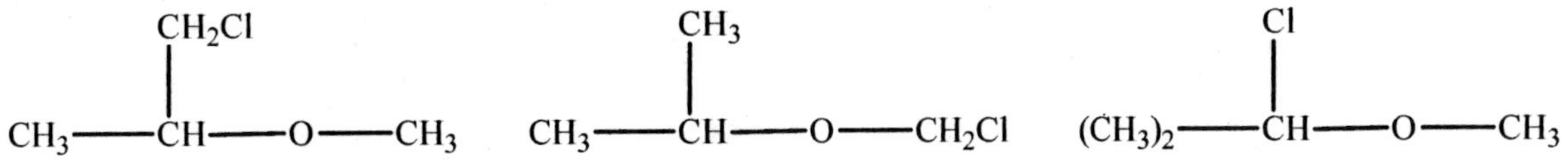

The **(CH₃)₂CH–O–CH₃** choice is the original compound.

24.41 **(a)** The product is similar to that in Problem 24.38.

$$CH_3CH_2O\!-\!\overset{\displaystyle O}{\overset{\|}{C}}\!-\!H$$

(b) Addition of hydrogen to an alkyne gives an alkene.

$$H\!-\!C\!\equiv\!C\!-\!CH_3 + H_2 \longrightarrow H_2C\!=\!CH\!-\!CH_3$$

The alkene can also add hydrogen to form an alkane.

$$H_2C\!=\!CH\!-\!CH_3 + H_2 \longrightarrow CH_3\!-\!CH_2\!-\!CH_3$$

(c) HBr will add to the alkene as shown (Note: the carbon atoms at the double bond have been omitted for simplicity).

$$\underset{H}{\overset{C_2H_5}{\diagdown}}\!=\!\underset{H}{\overset{H}{\diagup}} \quad + \quad HBr \quad \longrightarrow \quad C_2H_5\!-\!CHBr\!-\!CH_3$$

How do you know that the hydrogen adds to the CH_2 end of the alkene?

24.42 **(a)** ketone   **(b)** ester   **(c)** ether

24.43 The four isomers are:

24.44 This is a Hess's Law problem. See chapter 6.

If we rearrange the equations given and multiply times the necessary factors, we have:

$2CO_2(g) + 2H_2O(l) \longrightarrow C_2H_4(g) + 3O_2(g)$		$\Delta H° = 1411$ kJ/mol
$C_2H_2(g) + \frac{5}{2}O_2(g) \longrightarrow 2CO_2(g) + H_2O(l)$		$\Delta H° = -1299.5$ kJ/mol
$H_2(g) + \frac{1}{2}O_2(g) \longrightarrow H_2O(l)$		$\Delta H° = -285.8$ kJ/mol

$$C_2H_2(g) + H_2(g) \longrightarrow C_2H_4(g) \qquad\qquad \mathbf{\Delta H° = -174\ kJ/mol}$$

The heat of hydrogenation for acetylene is **−174 kJ/mol**.

24.45 **(a)** Cyclopropane because of the strained bond angles. (The C–C–C angle is 60° instead of 109.5°)

(b) Ethylene because of the C=C bond.

(c) Acetaldehyde (susceptible to oxidation).

24.46 To form a hydrogen bond *with water* a molecule must have at least one H–F, H–O, or H–N bond, *or* must contain an O, N, or F atom. The following can form hydrogen bonds with water:

(a) carboxylic acids   **(c)** ethers   **(d)** aldehydes   **(f)** amines

24.47 **(a)** The empirical formula is:

H: $3.2 \text{ g H} \times \dfrac{1 \text{ mol H}}{1.008 \text{ g H}} = 3.17 \text{ mol H}$

C: $37.5 \text{ g C} \times \dfrac{1 \text{ mol C}}{12.01 \text{ g C}} = 3.12 \text{ mol C}$

F: $59.3 \text{ g F} \times \dfrac{1 \text{ mol F}}{19.00 \text{ g F}} = 3.12 \text{ mol F}$

This gives the formula, $H_{3.17}C_{3.12}F_{3.12}$. Dividing by 3.12 gives the empirical formula, **HCF**.

(b) When temperature and amount of gas are constant, the product of pressure times volume is constant (Boyle's law).

$(2.00 \text{ atm})(0.322 \text{ L}) = 0.664 \text{ atm·L}$
$(1.50 \text{ atm})(0.409 \text{ L}) = 0.614 \text{ atm·L}$
$(1.00 \text{ atm})(0.564 \text{ L}) = 0.564 \text{ atm·L}$
$(0.50 \text{ atm})(1.028 \text{ L}) = 0.514 \text{ atm·L}$

The substance does not obey the ideal gas law.

(c) Since the gas does not obey the ideal gas equation exactly, the molar mass will only be approximate. Gases obey the ideal gas law best at lowest pressures. We use the 0.50 atm data.

$$n = \frac{PV}{RT} = \frac{(0.50 \text{ atm})(1.028 \text{ L})}{(0.0821 \text{ L·atm/K·mol})(363 \text{ K})} = 0.0172 \text{ mol}$$

$$\text{Molar mass} = \frac{1.00 \text{ g}}{0.0172 \text{ mol}} = 58.1 \text{ g/mol}$$

This is reasonably close to $C_2H_2F_2$ (64 g/mol).

(d) The $C_2H_2F_2$ formula is that of difluoroethylene. Three isomers are possible. The carbon atoms are omitted for simplicity (see Problem 24.17).

Only the third isomer has no dipole moment.

(e) The name is ***trans–difluoroethylene.***

24.48 **(a)** rubbing alcohol   **(b)** vinegar   **(c)** moth balls   **(d)** organic synthesis

      (e) organic synthesis   **(f)** antifreeze   **(g)** fuel (natural gas)   **(h)** synthetic polymers

24.49 In any stoichiometry problem, you must start with a balanced equation. The balanced equation for the combustion reaction is:

$$2C_8H_{18}(l) + 25O_2(g) \rightarrow 16CO_2(g) + 18H_2O(l)$$

To find the number of moles of octane in one liter, use density as a conversion factor to find grams of octane, then use the molar mass of octane to convert to moles of octane. The strategy is:

L octane $\rightarrow$ mL octane $\rightarrow$ g octane $\rightarrow$ mol octane

$$1.0 \text{ L} \times \frac{1000 \text{ mL}}{1 \text{ L}} \times \frac{0.70 \text{ g } C_8H_{18}}{1 \text{ mL } C_8H_{18}} \times \frac{1 \text{ mol } C_8H_{18}}{114.2 \text{ g } C_8H_{18}} = 6.13 \text{ mol } C_8H_{18}$$

Using the mole ratio from the balanced equation, the number of moles of oxygen used is:

$$6.13 \text{ mol } C_8H_{18} \times \frac{25 \text{ mol } O_2}{2 \text{ mol } C_8H_{18}} = 76.6 \text{ mol } O_2$$

From the ideal gas equation, we can calculate the volume of oxygen.

$$V_{O_2} = \frac{n_{O_2}RT}{P} = \frac{(76.6 \text{ mol})(293 \text{ K})}{1.00 \text{ atm}} \times \frac{0.0821 \text{ L} \cdot \text{atm}}{\text{mol} \cdot \text{K}} = 1.84 \times 10^3 \text{ L}$$

Air is only 22% O_2 by volume. Thus, 100 L of air will contain 22 L of O_2. Setting up the appropriate conversion factor, we find that the volume of air is:

$$? \textbf{ vol of air} = (1.84 \times 10^3 \text{ L } O_2) \times \frac{100 \text{ L air}}{22 \text{ L } O_2} = \textbf{8.4} \times \textbf{10}^3 \textbf{ L air}$$

24.50 **(a)** 2–butyne has **three** C–C sigma bonds.

      (b) Anthracene is:

There are **sixteen** C–C sigma bonds.

      (c)

There are **six** C–C sigma bonds.

24.51 **(a)** A benzene ring has six carbon-carbon bonds; hence, benzene has **six** C–C sigma bonds.

      (b) Cyclobutane has four carbon-carbon bonds; hence, cyclobutane has **four** sigma bonds.

(c)   Looking at the carbon skeleton of 2–methyl–3–ethylpentane, you should find **seven** C–C sigma bonds.

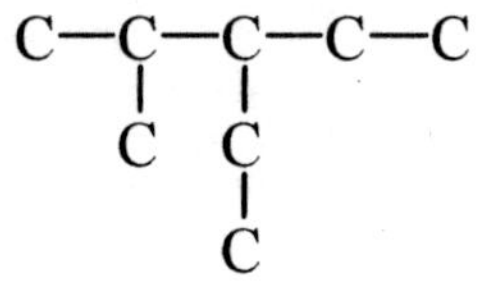

$$C-C-C-C-C$$
$$\qquad | \quad | $$
$$\qquad C \quad C$$
$$\qquad\qquad |$$
$$\qquad\qquad C$$

24.52   (a)   The easiest way to calculate the mg of C in CO_2 is by mass ratio. There are 12.01 g of C in 44.01 g CO_2 or 12.01 mg C in 44.01 mg CO_2.

$$? \,mg\, C \,=\, 57.94 \,mg\, CO_2 \times \frac{12.01 \,mg\, C}{44.01 \,mg\, CO_2} \,=\, \textbf{15.81 mg C}$$

Similarly,

$$? \,mg\, H \,=\, 11.85 \,mg\, H_2O \times \frac{2.016 \,mg\, H}{18.02 \,mg\, H_2O} \,=\, \textbf{1.326 mg H}$$

The mg of oxygen can be found by difference.

$$? \,mg\, O \,=\, 20.63 \,mg\, Y \,-\, 15.81 \,mg\, C \,-\, 1.326 \,mg\, H \,=\, \textbf{3.49 mg O}$$

(b)   *Step 1:* Calculate the number of moles of each element present in the sample. Use molar mass as a conversion factor.

$$? \,mol\, C \,=\, (15.81 \times 10^{-3} \,g\, C) \times \frac{1 \,mol\, C}{12.01 \,g\, C} \,=\, 1.316 \times 10^{-3} \,mol\, C$$

Similarly,

$$? \,mol\, H \,=\, (1.326 \times 10^{-3} \,g\, H) \times \frac{1 \,mol\, H}{1.008 \,g\, H} \,=\, 1.315 \times 10^{-3} \,mol\, H$$

$$? \,mol\, O \,=\, (3.49 \times 10^{-3} \,g\, O) \times \frac{1 \,mol\, O}{16.00 \,g\, O} \,=\, 2.18 \times 10^{-4} \,mol\, O$$

Thus, we arrive at the formula $C_{1.316 \times 10^{-3}} H_{1.315 \times 10^{-3}} O_{2.18 \times 10^{-4}}$, which gives the identity and the ratios of atoms present. However, chemical formulas are written with whole numbers.

Step 2: Try to convert to whole numbers by dividing all the subscripts by the smallest subscript.

C: $\dfrac{1.316 \times 10^{-3}}{2.18 \times 10^{-4}} = 6.04 \approx 6$ **H:** $\dfrac{1.315 \times 10^{-3}}{2.18 \times 10^{-4}} = 6.03 \approx 6$ **O:** $\dfrac{2.18 \times 10^{-4}}{2.18 \times 10^{-4}} = 1.00$

This gives us the empirical formula, **C_6H_6O**.

(c)   The presence of six carbons and a corresponding number of hydrogens suggests a benzene derivative. A plausible structure is shown below.

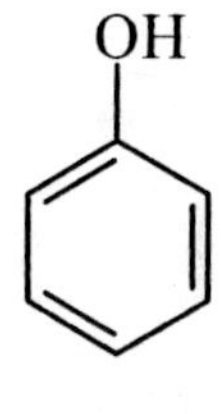

24.53   The structural isomers are:

1,2–dichlorobutane

$$CH_3 \!-\! CH_2 \!-\! \overset{*}{C}HCl \!-\! CH_2Cl$$

1,3–dichlorobutane

$$CH_3 \!-\! \overset{*}{C}HCl \!-\! CH_2 \!-\! CH_2Cl$$

2,3–dichlorobutane

$$CH_3 \!-\! \overset{*}{C}HCl \!-\! \overset{*}{C}HCl \!-\! CH_3$$

1,4–dichlorobutane

$$CH_2Cl \!-\! CH_2 \!-\! CH_2 \!-\! CH_2Cl$$

1,1–dichlorobutane

$$CH_3 \!-\! CH_2 \!-\! CH_2 \!-\! CHCl_2$$

2,2–dichlorobutane

$$CH_3 \!-\! CH_2 \!-\! CCl_2 \!-\! CH_3$$

1,3–dichloro–2–methylpropane

$$CH_2Cl \!-\! \underset{\underset{\textstyle CH_3}{|}}{CH} \!-\! CH_2Cl$$

1,2–dichloro–2–methylpropane

$$CH_3 \!-\! \underset{\underset{\textstyle CH_3}{|}}{CCl} \!-\! CH_2Cl$$

1,1–dichloro–2–methylpropane

$$CH_3 \!-\! \underset{\underset{\textstyle CH_3}{|}}{CH} \!-\! CHCl_2$$

The asterisk identifies the asymmetric carbon atom.

24.54   First, calculate the moles of each element.

C: $(9.708 \times 10^{-3} \text{ g CO}_2) \times \dfrac{1 \text{ mol CO}_2}{44.01 \text{ g CO}_2} \times \dfrac{1 \text{ mol C}}{1 \text{ mol CO}_2} = 2.206 \times 10^{-4} \text{ mol C}$

H: $(3.969 \times 10^{-3} \text{ g H}_2\text{O}) \times \dfrac{1 \text{ mol H}_2\text{O}}{18.02 \text{ g H}_2\text{O}} \times \dfrac{2 \text{ mol H}}{1 \text{ mol H}_2\text{O}} = 4.405 \times 10^{-4} \text{ mol H}$

The mass of oxygen is found by difference:

3.795 mg compound – (2.649 mg C + 0.445 mg H) = 0.701 mg O

O: $(0.701 \times 10^{-3} \text{ g O}) \times \dfrac{1 \text{ mol O}}{16.00 \text{ g O}} = 4.38 \times 10^{-5} \text{ mol O}$

This gives the formula is $C_{2.206 \times 10^{-4}} H_{4.405 \times 10^{-4}} O_{4.38 \times 10^{-5}}$. Dividing by the smallest number of moles gives the empirical formula, **C$_5$H$_{10}$O**.

We calculate moles using the ideal gas equation, and then calculate the molar mass.

$$n = \frac{PV}{RT} = \frac{(1.00 \text{ atm})(0.0898 \text{ L})}{(0.0821 \text{ L} \cdot \text{atm/K} \cdot \text{mol})(473 \text{ K})} = 0.00231 \text{ mol}$$

$$\textbf{molar mass} = \frac{\text{g of substance}}{\text{mol of substance}} = \frac{0.205 \text{ g}}{0.00231 \text{ mol}} = \textbf{88.7 g/mol}$$

The formula mass of $C_5H_{10}O$ is 86.13 g, so this is also the molecular formula. Three possible structures are:

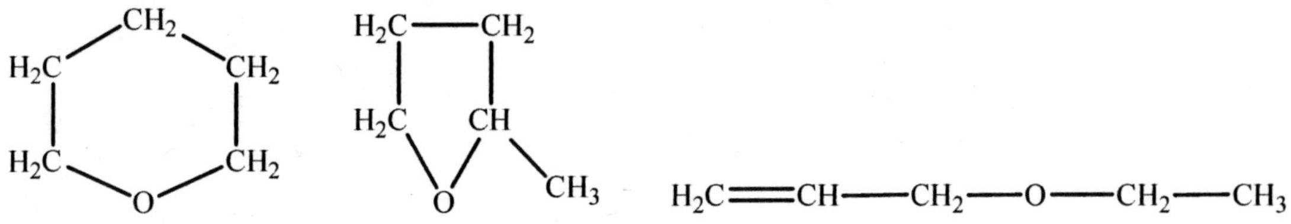

24.55   **(a)**   In comparing the compound in part (a) with the starting alkyne, it is clear that a molecule of HBr has been added to the triple bond. The reaction is:

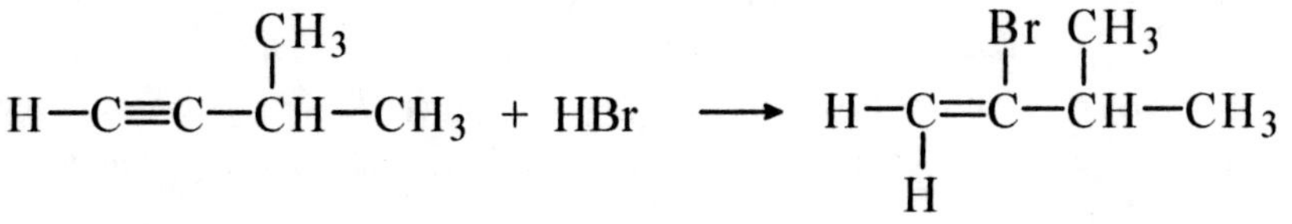

(b)   This compound can be made from the product formed in part (a) by addition of bromine to the double bond.

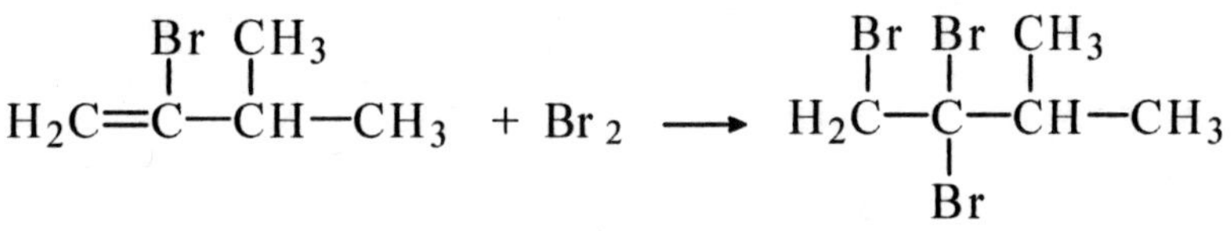

(c)   This compound can be made from the product of part (a) by addition of hydrogen to the double bond.

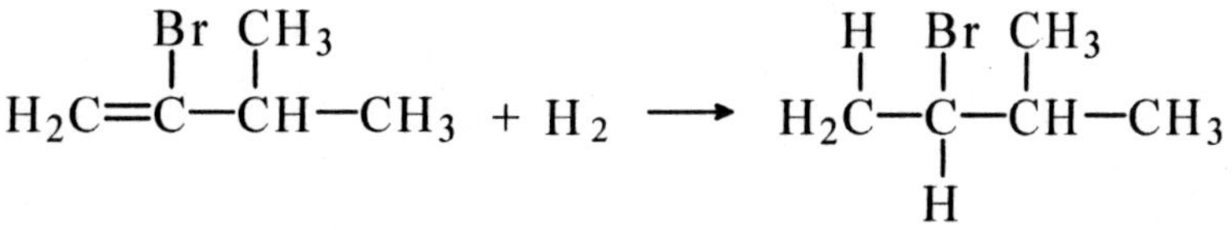

24.56   A carbon atom is asymmetric if it is bonded to four different atoms or groups. In the given structures the asymmetric carbons are marked with an asterisk (*).

(a) $CH_3-CH_2-\overset{*}{C}H-\overset{*}{C}H-\overset{O}{\overset{\|}{C}}-NH_2$ with CH_3 above the first starred CH and NH_2 below the second starred carbon

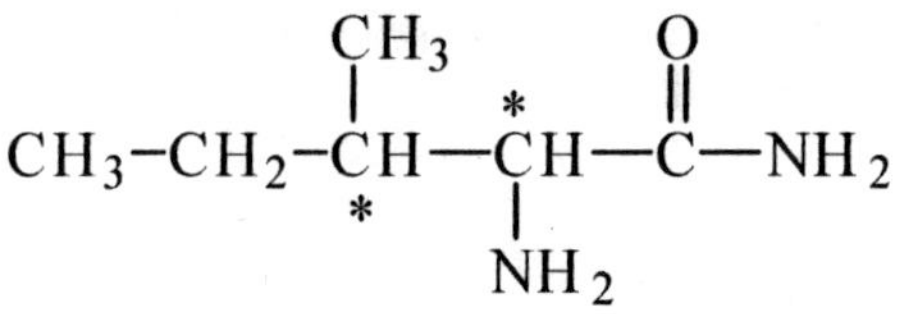

(b)

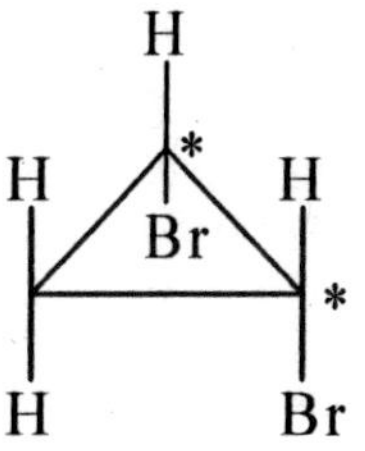

24.57   The isomers are:

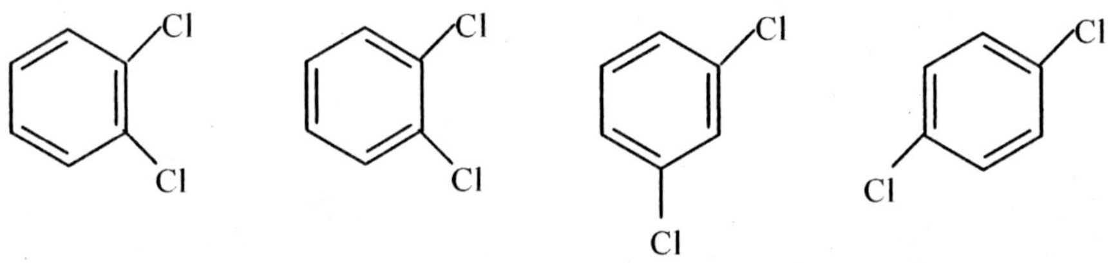

Did you have more isomers? Remember that benzene is a planar molecule; "turning over" a structure does not create a new isomer.

24.58   Acetone is a ketone with the formula, CH_3COCH_3. We must write the structure of an aldehyde that has the same number and types of atoms (C_3H_6O). Removing the aldehyde functional group (–CHO) from the formula leaves C_2H_5. This is the formula of an ethyl group. The aldehyde that is a structural isomer of acetone is:

$$CH_3CH_2\overset{\displaystyle O}{\overset{\|}{C}}{-}H$$

24.59   The structures are:

(a)

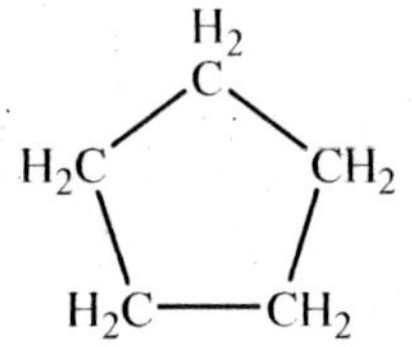

(b)

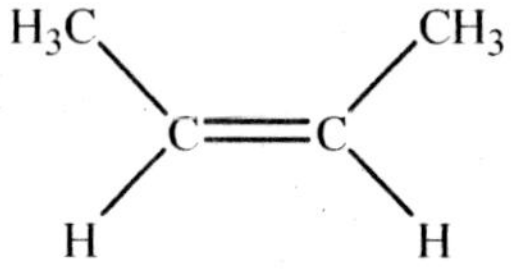

(c)

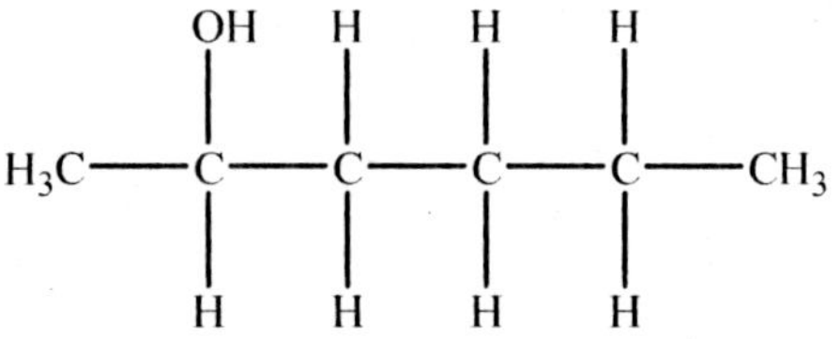

(d)

(e) $CH_3{-}C{\equiv}C{-}CH_3$

24.60   (a) alcohol   (b) ether   (c) aldehyde   (d) carboxylic acid   (e) amine

24.61   Ethanol has a melting point of $-117.3°C$, a boiling point of $+78.5°C$, and is miscible with water. Dimethyl ether has a melting point of $-138.5°C$, a boiling point of $-25°C$ (it is a gas at room temperature), and dissolves in water to the extent of 37 volumes of gas to one volume of water.

24.62   In Chapter 11, we found that salts with their electrostatic intermolecular attractions had low vapor pressures and thus high boiling points. Ammonia and its derivatives (amines) are molecules with dipole–dipole attractions. If the nitrogen has one direct N–H bond, the molecule will have hydrogen bonding. Even so, these molecules will have much weaker intermolecular attractions than ionic species and hence higher vapor pressures. Thus, if we could convert the neutral ammonia–type molecules into salts, their vapor pressures, and thus associated odors, would decrease. Lemon juice contains acids which can react with ammonia–type (amine) molecules to form ammonium salts.

$$NH_3 + H^+ \longrightarrow NH_4^+ \qquad\qquad RNH_2 + H^+ \longrightarrow RNH_3^+$$

24.63   Cyclohexane readily undergoes halogenation; for example, its reaction with bromine can be monitored by seeing the red color of bromine fading. Benzene does not react with halogens unless a catalyst is present.

24.64   Marsh gas (methane, CH_4); grain alcohol (ethanol, C_2H_5OH); wood alcohol (methanol, CH_3OH); rubbing alcohol (isopropyl alcohol, $(CH_3)_2CHOH$); antifreeze (ethylene glycol, CH_2OHCH_2OH); mothballs (naphthalene, $C_{10}H_8$); vinegar (acetic acid, CH_3COOH).

24.65   A mixture of *cis* and *trans* isomers would imply some sort of random addition mechanism in which one hydrogen atom at a time adds to the molecule.

The formation of pure *cis* or pure *trans* isomer indicates a more specific mechanism. For example, a pure *cis* product suggests simultaneous addition of both hydrogen atoms in the form of a hydrogen molecule to one side of the alkyne. In practice, the *cis* isomer is formed.

24.66   The asymmetric carbons are shown by asterisks:

(a)
$$H-\underset{\underset{H}{|}}{\overset{\overset{H}{|}}{C}}-\underset{\underset{Cl}{|}}{\overset{\overset{H}{|}}{\overset{*}{C}}}-\underset{\underset{H}{|}}{\overset{\overset{H}{|}}{C}}-Cl$$

(b)
$$CH_3-\underset{\underset{H}{|}}{\overset{\overset{OH}{|}}{\overset{*}{C}}}-\underset{\underset{H}{|}}{\overset{\overset{CH_3}{|}}{\overset{*}{C}}}-CH_2OH$$

(c)   All of the carbon atoms in the ring are asymmetric. Therefore there are **five** asymmetric carbon atoms.

24.67   **(a)**   Sulfuric acid ionizes as follows:

$$H_2SO_4(aq) \longrightarrow H^+(aq) + HSO_4^-(aq)$$

The cation (H^+) and anion (HSO_4^-) add to the double bond in propylene according to Markovnikov's rule:

$$CH_3-CH{=}CH_2 + H^+ + HSO_4^- \longrightarrow CH_3-\underset{\underset{H}{|}}{\overset{\overset{OSO_3H}{|}}{C}}-CH_3$$

Reaction of the intermediate with water yields isopropanol:

$$CH_3-\underset{\underset{H}{|}}{\overset{\overset{OSO_3H}{|}}{C}}-CH_3 + H_2O \longrightarrow CH_3-\underset{\underset{H}{|}}{\overset{\overset{OH}{|}}{C}}-CH_3 + H_2SO_4$$

Since sulfuric acid is regenerated, it plays the role of a catalyst.

(b)   The other structure containing the $-OH$ group is

$$CH_3-CH_2-CH_2-OH$$
propanol

(c)   From the structure of isopropanol shown above, we see that the molecule does not have an asymmetric carbon atom. Therefore, isopropanol is achiral.

(d)   Isopropanol is fairly volatile (b.p. = 82.5°C), and the –OH group allows it to form hydrogen bonds with water molecules. Thus, as it evaporates, it produces a cooling and soothing effect on the skin. It is also less toxic than methanol and less expensive than ethanol.

24.68   The red bromine vapor absorbs photons of blue light and dissociates to form bromine atoms.

$$Br_2 \rightarrow 2Br\bullet$$

The bromine atoms collide with methane molecules and abstract hydrogen atoms.

$$Br\bullet + CH_4 \rightarrow HBr + \bullet CH_3$$

The methyl radical then reacts with Br_2, giving the observed product and regenerating a bromine atom to start the process over again:

$$\bullet CH_3 + Br_2 \rightarrow CH_3Br + Br\bullet$$

$$Br\bullet + CH_4 \rightarrow HBr + \bullet CH_3 \qquad \text{and so on...}$$

24.69   **(a)**   Reaction between glycol and carboxylic acid (formation of an ester).

(b)

A fat or oil

(c)   Molecules having more C=C bonds are harder to pack tightly together. Consequently, the compound has a lower melting point.

(d) H_2 gas with either a heterogeneous or homogeneous catalyst would be used. See Section 13.6 of the text.

(e)   Number of moles of $Na_2S_2O_3$ reacted is:

$$20.6 \text{ mL} \times \frac{1 \text{ L}}{1000 \text{ mL}} \times \frac{0.142 \text{ mol } Na_2S_2O_3}{1 \text{ L}} = 2.93 \times 10^{-3} \text{ mol } Na_2S_2O_3$$

The mole ratio between I_2 and $Na_2S_2O_3$ is 1:2. The number of grams of I_2 left over is:

$$(2.93 \times 10^{-3} \text{ mol } Na_2S_2O_3) \times \frac{1 \text{ mol } I_2}{2 \text{ mol } Na_2S_2O_3} \times \frac{253.8 \text{ g } I_2}{1 \text{ mol } I_2} = 0.372 \text{ g } I_2$$

Number of grams of I_2 reacted is: $(43.8 - 0.372)\text{g} = 43.4 \text{ g } I_2$

The *iodine number* is the number of grams of iodine that react with 100 g of corn oil.

$$\textit{iodine number} = \frac{43.4 \text{ g } I_2}{35.3 \text{ g corn oil}} \times 100 \text{ g corn oil} = \mathbf{123}$$

CHAPTER 25
SYNTHETIC AND NATURAL ORGANIC POLYMERS

25.7 The reaction is initiated be a radical, R•

$$R• + CF_2{=}CF_2 \rightarrow R{-}CF_2{-}CF_2•$$

The product is also a radical, and the reaction continues.

$$R{-}CF_2{-}CF_2• + CF_2{=}CF_2 \rightarrow R{-}CF_2{-}CF_2{-}CF_2{-}CF_2• \quad \text{etc...}$$

25.8 The repeating structural unit of the polymer is:

Does each carbon atom still obey the octet rule?

25.9 The general reaction is a condensation to form an amide.

The polymer chain looks like:

Note that both reactants are disubstituted benzene derivatives with the substituents in the para or 1,4 positions.

25.10 Polystyrene is formed by an addition polymerization reaction with the monomer, styrene, which is a phenyl–substituted ethylene. The structures of styrene and polystyrene are shown in Table 25.1 of your text.

25.11 The structures are as shown.

(a) $CH_2{=}CF_2$ **(b)** $HO_2C{-}\bigcirc{-}CO_2H$ $H_2N{-}\bigcirc{-}NH_2$

25.12   The structures are shown.

(a) $H_2C{=}CH{-}CH{=}CH_2$

(b)

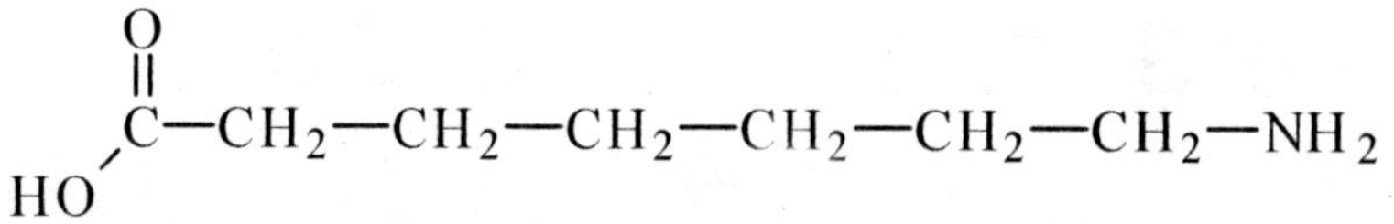

$$\underset{HO}{\overset{O}{\underset{}{\overset{\|}{C}}}}{-}CH_2{-}CH_2{-}CH_2{-}CH_2{-}CH_2{-}CH_2{-}NH_2$$

25.19   alanylglycine and glycylalanine are shown in Figure 25.8.

25.20   The main backbone of a polypeptide chain is made up of the α carbon atoms and the amide group repeating alternately along the chain.

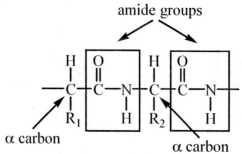

For each R group shown above, substitute the distinctive side groups of the two amino acids. Their are two possible dipeptides depending on how the two amino acids are connected, either glycine–lysine or lysine–glycine. The structures of the dipeptides are:

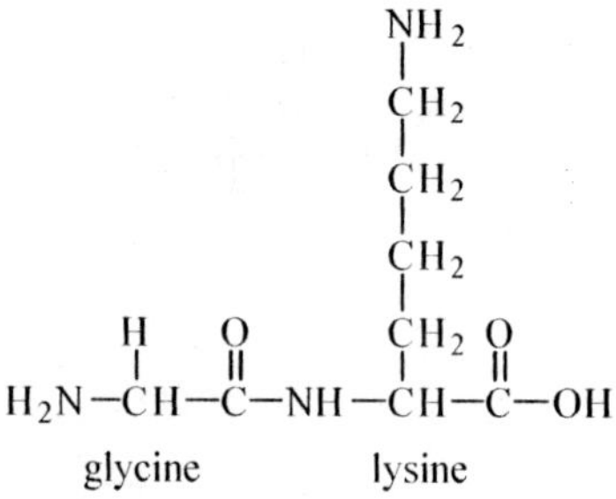

$$\begin{array}{c}
NH_2\\
|\\
CH_2\\
|\\
CH_2\\
|\\
CH_2\\
H\quad O\qquad CH_2\ O\\
|\quad \|\qquad\ |\quad \|\\
H_2N{-}CH{-}C{-}NH{-}CH{-}C{-}OH
\end{array}$$

glycine   lysine

and

$$\begin{array}{c}
NH_2\\
|\\
CH_2\\
|\\
CH_2\\
|\\
CH_2\\
CH_2\ O\qquad H\quad O\\
|\quad \|\qquad\ |\quad \|\\
H_2N{-}CH{-}C{-}NH{-}CH{-}C{-}OH
\end{array}$$

lysine   glycine

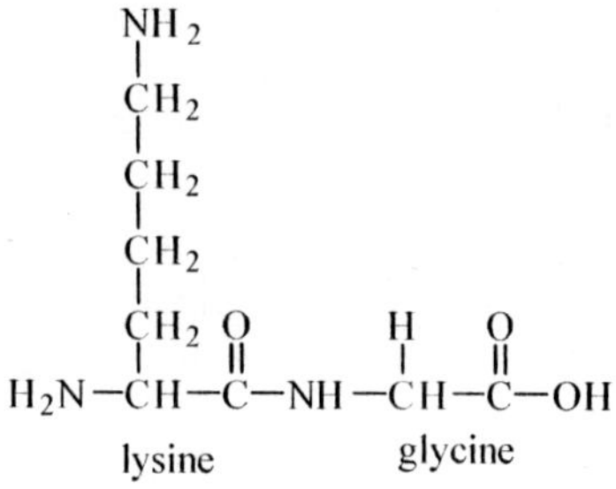

25.21   The structure of the polymer is:

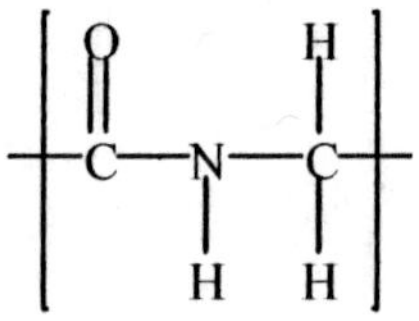

25.22   The rate increases in an expected manner from 10°C to 30°C and then drops rapidly. The probable reason for this is the loss of catalytic activity of the enzyme because of denaturation at high temperature.

25.27   There are two common structures for protein molecules, an α helix and a β–pleated sheet. The α–helical structure is stabilized by intramolecular hydrogen bonds between the NH and CO groups of the main chain, giving rise to an overall rodlike shape. The CO group of each amino acid is hydrogen-bonded to the NH group of the amino acid that is four residues away in the sequence. In this manner all the main-chain CO and NH groups take part in hydrogen bonding. The β–pleated structure is like a sheet rather than a rod. The polypeptide chain is almost fully extended, and each chain forms many intermolecular hydrogen bonds with adjacent chains. In general, then, the hydrogen bonding is responsible for the three dimensional geometry of the protein molecules.

In nucleic acids, the key to the double-helical structure is the formation of hydrogen bonds between bases in the two strands. Although hydrogen bonds can form between any two bases, called base pairs, the most favorable couplings are between adenine and thymine and between cytosine and guanine.

More information concerning the importance of hydrogen bonding in biological systems is in Sections 25.3 and 25.4 of the text.

25.28   Nucleic acids play an essential role in protein synthesis. Compared to proteins, which are made of up to 20 different amino acids, the composition of nucleic acids is considerably simpler. A DNA or RNA molecule contains only four types of building blocks: purines, pyrimidines, furanose sugars, and phosphate groups. Nucleic acids have simpler, uniform structures because they are primarily used for protein synthesis, whereas proteins have many uses.

25.29   When proteins are heated above body temperature they can lose some or all of their secondary and tertiary structure and become denatured. The denatured proteins no longer exhibit normal biological activity.

25.30   The sample that has the higher percentage of C–G base pairs has a higher melting point because C–G base pairs are held together by three hydrogen bonds. The A–T base pair interaction is relatively weaker because it has only two hydrogen bonds. Hydrogen bonds are represented by dashed lines in the structures below.

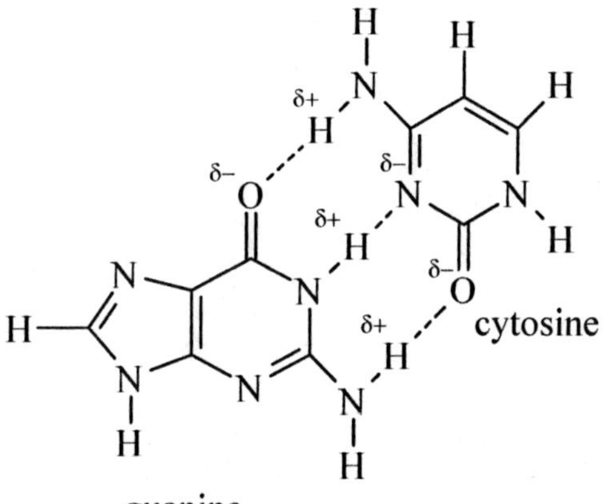

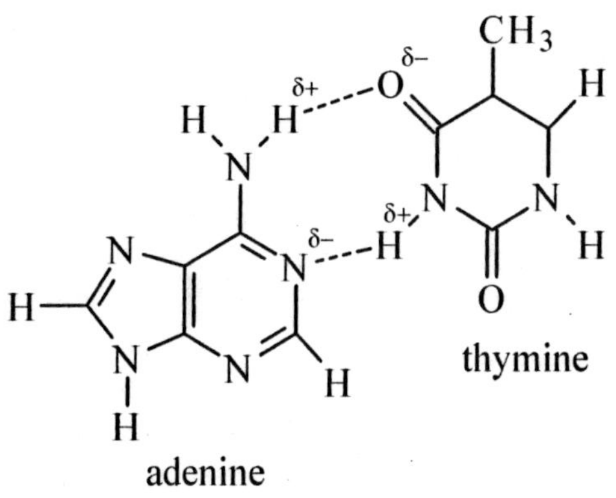

25.31   As is described in Section 25.3 of the text, acids *denature* enzymes. The citric acid in lemon juice denatures the enzyme that catalyzes the oxidation so as to inhibit the oxidation (browning).

25.32   Leg muscles are active having a high metabolism, which requires a high concentration of myoglobin. The high iron content from myoglobin makes the meat look dark after decomposition due to heating. The breast meat is "white" because of a low myoglobin content.

25.33   The cleavage reaction is:

$$-(CH_2)_4-\overset{O}{\overset{\|}{C}}-NH-(CH_2)_6-NH-\overset{O}{\overset{\|}{C}}- \xrightarrow{H^+} HOOC-(CH_2)_4-COOH + H_3\overset{+}{N}-(CH_2)_6-\overset{+}{N}H_3$$

25.34   Insects have blood that contains no hemoglobin. Thus, they rely on simple diffusion to supply oxygen. It is unlikely that a human-sized insect could obtain sufficient oxygen by diffusion alone to sustain its metabolic requirements.

25.35   The best way to attack this type of problem is with a systematic approach. Start with all the possible tripeptides with three lysines (one), then all possible tripeptides with two lysines and one alanine (three), one lysine and two alanines (three also ––Why the same number?), and finally three alanines (one).

      Lys–Lys–Lys

      Lys– Lys–Ala   Lys–Ala– Lys   Ala– Lys– Lys

      Lys–Ala–Ala   Ala–Lys–Ala   Ala–Ala–Lys

      Ala–Ala–Ala

Any other possibilities?

25.36   From the mass % Fe in hemoglobin, we can determine the mass of hemoglobin.

$$\% \text{ Fe} = \frac{\text{mass of Fe}}{\text{mass of compound (hemoglobin)}} \times 100\%$$

$$0.34\% = \frac{55.85 \text{ g}}{\text{mass of hemoglobin}} \times 100\%$$

minimum mass of hemoglobin $= 1.6 \times 10^4$ g

Hemoglobin must contain **four Fe atoms per molecule** for the actual molar mass to be four times the minimum value calculated.

25.37   The main interaction between water molecules and the amino acid residues is that of hydrogen bonding. In water the polar groups of the protein are on the exterior and the nonpolar groups are on the interior.

25.38   The type of intermolecular attractions that occur are mostly attractions between nonpolar groups. This type of intermolecular attraction is called a **dispersion force**.

25.39 **(a)**   deoxyribose and cytosine

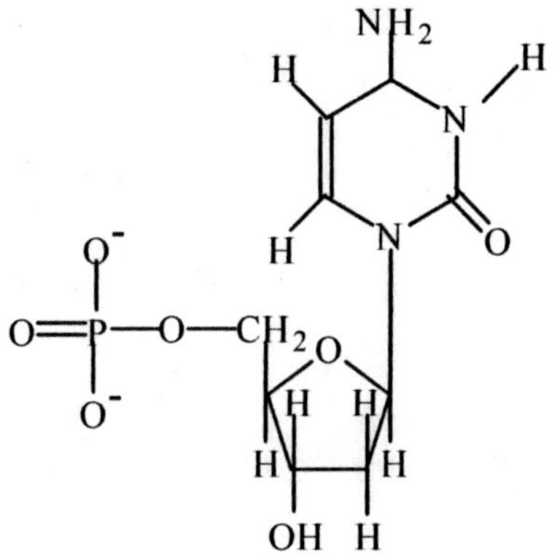

(b)   ribose and uracil

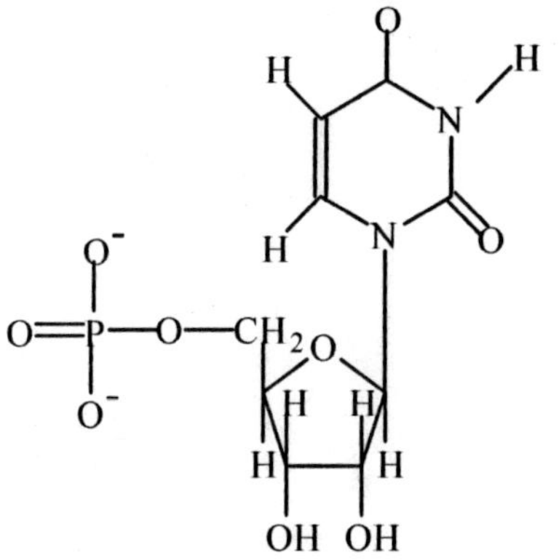

25.40 This is as much a puzzle as it is a chemistry problem. The puzzle involves breaking up a nine-link chain in various ways and trying to deduce the original chain sequence from the various pieces. Examine the pieces and look for patterns. Remember that depending on how the chain is cut, the same link (amino acid) can show up in more than one fragment.

Since there are only seven different amino acids represented in the fragments, at least one must appear more than once. The nonapeptide is:

Gly–Ala–Phe–Glu–His–Gly–Ala–Leu–Val

Do you see where all the pieces come from?

25.41 $pH = pK_a + \log\dfrac{[\text{conjugate base}]}{[\text{acid}]}$

At pH = 1,

–COOH $1 = 2.3 + \log\dfrac{[-COO^-]}{[-COOH]}$

$\dfrac{[-COOH]}{[-COO^-]} = 20$

$-NH_3^+$ $1 = 9.6 + \log\dfrac{[-NH_2]}{[-NH_3^+]}$

$$\dfrac{[-NH_3^+]}{[-NH_2]} = 4 \times 10^8$$

Therefore the **predominant species** is: $^+NH_3 - CH_2 - COOH$

At pH = 7,

$-COOH$ $7 = 2.3 + \log\dfrac{[-COO^-]}{[-COOH]}$

$$\dfrac{[-COO^-]}{[-COOH]} = 5 \times 10^4$$

$-NH_3^+$ $7 = 9.6 + \log\dfrac{[-NH_2]}{[-NH_3^+]}$

$$\dfrac{[-NH_3^+]}{[-NH_2]} = 4 \times 10^2$$

Predominant species: $^+NH_3 - CH_2 - COO^-$

At pH = 12,

$-COOH$ $12 = 2.3 + \log\dfrac{[-COO^-]}{[-COOH]}$

$$\dfrac{[-COO^-]}{[-COOH]} = 5 \times 10^9$$

$-NH_3^+$ $12 = 9.6 + \log\dfrac{[-NH_2]}{[-NH_3^+]}$

$$\dfrac{[-NH_2]}{[-NH_3^+]} = 2.5 \times 10^2$$

Predominant species: $NH_2 - CH_2 - COO^-$

25.42   No, the milk would *not* be fit to drink. Enzymes only act on one of two optical isomers of a compound.

25.43   **(a)**   The repeating unit in nylon 66 is

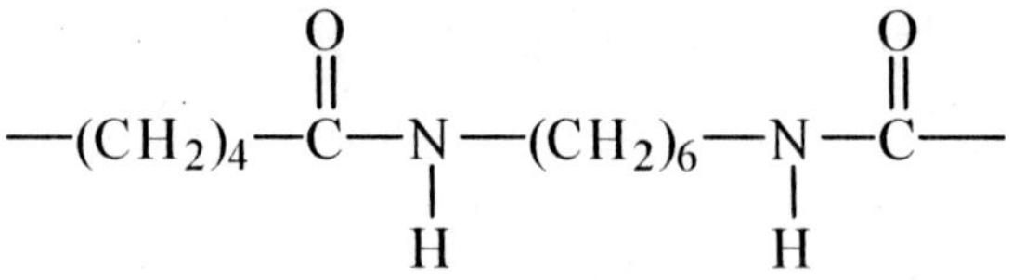

$$—(CH_2)_4 - \overset{\overset{\displaystyle O}{\|}}{C} - \underset{\underset{\displaystyle H}{|}}{N} - (CH_2)_6 - \underset{\underset{\displaystyle H}{|}}{N} - \overset{\overset{\displaystyle O}{\|}}{C} —$$

and the molar mass of the unit is 226.3 g/mol. Therefore, the number of repeating units (n) is

$$n = \frac{12000 \text{ g/mol}}{226.3 \text{ g/mol}} = 53$$

(b) The most obvious feature is the presence of the amide group in the repeating unit. Another important and related feature that makes the two types of polymers similar is the ability of the molecules to form intramolecular hydrogen bonds.

(c) We approach this question systematically. First, there are three tripeptides made up of only one type of amino acid:

Ala–Ala–Ala   Gly–Gly–Gly   Ser–Ser–Ser

Next, there are eighteen tripeptides made up of two types of amino acids.

Ala–Ala–Ser	Ser–Ser–Ala	Ala–Ala–Gly	Gly–Gly–Ala
Ala–Ser–Ala	Ser–Ala–Ser	Ala–Gly–Ala	Gly–Ala–Gly
Ser–Ala–Ala	Ala–Ser–Ser	Gly–Ala–Ala	Ala–Gly–Gly

Gly–Gly–Ser	Ser–Ser–Gly
Gly–Ser–Gly	Ser–Gly–Ser
Ser–Gly–Gly	Gly–Ser–Ser

Finally, there are six different tripeptides from three different amino acids.

Ala–Gly–Ser	Ser–Ala–Gly	Ala–Ser–Gly
Ser–Gly–Ala	Gly–Ala–Ser	Gly–Ser–Ala

Thus, there are a total of twenty-seven ways to synthesize a tripeptide from three amino acids. In silk, a basic six-residue unit repeats for long distances in the chain.

–Gly–Ser–Gly–Ala–Gly–Ala–

The ability of living organisms to reproduce the correct sequence is truly remarkable. It is also interesting to note that we can emulate the properties of silk with such a simple structure as nylon.

25.44 We assume $\Delta G = 0$, so that

$$\Delta G = \Delta H - T\Delta S$$

$$0 = \Delta H - T\Delta S$$

$$T = \frac{\Delta H}{\Delta S} = \frac{125 \times 10^3 \text{ J/mol}}{397 \text{ J/K} \cdot \text{mol}} = 315 \text{ K} = 42°C$$

25.45 In deoxyhemoglobin, it is believed that the Fe^{2+} ion has too large a radius to fit into the porphyrin ring (see Figure 25.15 of the text). When O_2 binds to Fe^{2+}, however, the ion shrinks somewhat so that it now fits into the plane of the ring. As the ion slips into the ring, it pulls the histidine residue toward the ring and thereby sets off a sequence of structural changes from one subunit to another. These structural changes occurring from one subunit to the next that cause deoxyhemoglobin crystals to shatter. Myoglobin is only made up of one of the four subunits and thus does not have the structural changes from subunit to subunit described above. Therefore, deoxymyoglobin crystals are unaffected by oxygen.

25.46